Superstring Construction

Current Physics – Sources and Comments

Coordinating editor

H. RUBINSTEIN

1. String theory in four dimensions, *edited by* M. Dine
2. Finite-size scaling, *edited by* J.L. Cardy
3. Regge theory of low-p_T hadronic interaetions, *edited by* L. Caneschi
4. Superstring construction, *edited by* A.N. Schellekens

NORTH-HOLLAND
AMSTERDAM · OXFORD · NEW YORK · TOKYO

Superstring Construction

Editor

Bert SCHELLEKENS

CERN
CH-1211 Geneva 23
Switzerland

1989
NORTH-HOLLAND
AMSTERDAM · OXFORD · NEW YORK · TOKYO

ISBN: 0 444 87492 5 (hardbound)
0 444 88016 x (paperback)

Published by:

North-Holland Physics
Elsevier Science Publishers B.V.
P.O. Box 103
1000 AC Amsterdam
The Netherlands

Sole distributors for the USA and Canada:

Elsevier Science Publishing Company, Inc.

655, Avenue of the Americas
New York, NY 10010
USA

Library of Congress Cataloging in Publication Data

Superstring construction/editor, A.N. Schellekens.
p. cm. – (Current physics; 4)
Includes bibliographical references.
ISBN 0-444-87492-5 (U.S.). – ISBN 0-444-88016-X (U.S.: pbk.)
1. Superstring theories. I. Schellekens, A.N. II. Series.
QC794.6.S85S86 1989
539.7′2 – dc20 89-16881
CIP

Printed in The Netherlands

Preface

String theory has made remarkable progress in the last four years. The most striking development was certainly the construction of many new string theories, some of which seem to have all the ingredients needed for phenomenological viability. Before 1984, just two consistent string theories with a stable ground state (i.e. no tachyons) were known, namely the type-IIA and type-IIB strings. These theories do not have gauge bosons in their spectrum, a fact that dimmed their phenomenological prospects. The phenomenologically more attractive type-I theories of open and closed strings were thought to be plagued by anomalies.

All this changed drastically in 1984, when Green and Schwarz discovered that the anomalies of type-I strings could be cancelled by a new mechanism that works only if the gauge group is SO(32). It soon became clear that there is a second solution to the conditions for cancellation of anomalies, namely the gauge group $E_8 \times E_8$. The fact that this gauge group, unlike SO(32), cannot be realized in type-I strings was a strong indication that an entire class of string theories had been overlooked.

This oversight was corrected quickly. Only a few months after the discovery of the anomaly cancellation the heterotic string, which can have both $E_8 \times E_8$ and SO(32) as its gauge group, was constructed.

The fact that the possibility of constructing the heterotic string was missed for so many years initiated a fresh look at the rules governing string construction. After the removal of some old prejudice and some misconceptions, it turned out that there were many ways of obtaining consistent string theories, especially below ten dimensions. These developments form the subject area of this reprint volume.

The original proposal suggested "compactification schemes" as the title, and this left me with a lot of freedom to choose the proper context. There are three common themes that came to mind: the general area of compactification of extra dimensions in field theory and string theory, string phenomenology, and string construction in four as well as ten dimensions. Since string "compactification" has completely overshadowed compactification in field theory, the first choice was not very attractive. The second choice, string phenomenology, is already covered by the first volume in this series, edited by M. Dine and entitled "String Theory in Four Dimensions". Therefore I decided to choose the third option, and to concentrate on papers about the art of string construction. Although the emphasis remains on four-dimensional strings, it is then natural to devote some attention to papers on ten-dimensional strings, in which many of the ideas were developed.

Even with such a sharply focused subject area, it was not easy to arrive at a selection that would fit within the page limits of this book. A first selection produced a total of 800 pages, and to cut this down to a more reasonable size some difficult choices had to be made. In general, I tried as much as possible to include those papers that contributed new ideas to string construction, rather than explorations of existing constructions. Furthermore papers on string-inspired field theory compactifications had to be dropped to leave enough room for real string constructions. Consequently this volume is devoted entirely to exactly solvable conformal field theories. This also reduced the overlap with Dine' volume, but there are a few papers in his volume that are simply too essential to omit.

Apart from two papers of historical importance, the time period covered in this volume is roughly from the end of 1984 to the beginning of 1987. This means that more recent developments, especially in the area of (rational) conformal field theory could not be covered. I have tried to correct that a little bit by providing a rather long introduction to conformal field theory applied to string "compactifications", which is intended both as a guide to the literature, and as the "glue" that keeps this book together.

I would like to thank all the people that have offered their suggestions for this selection, and I am grateful to H. Rubinstein for the proposal, and K. Jones for his advice and assistance.

A.N. Schellekens

Acknowledgements

The following article has been reprinted from Physical Review by kind permission of the publisher, The American Institute of Physics:
H. Kawai, D.C. Lewellen and S.-H.H. Tye, Classification of Closed Fermionic-String Models, Phys. Rev. D34 (1986) 3794–3805.

Contents

Preface v
Acknowledgements vii

1. Introduction 1
2. Conformal Field Theory for String Compactifications 3
   2.1. Consistency conditions 3
   2.2. Bosonic strings 5
      2.2.1. Virasoro representation theory 6
      2.2.2. Characters and modular invariance 7
      2.2.3. Extended Virasoro algebras 9
      2.2.4. Kac–Moody algebras 10
      2.2.5. Self-dual lattices 13
      2.2.6. Orbifolds 14
      2.2.7. Free fermions 14
      2.2.8. Tensor products 16
      2.2.9. The bosonic string spectrum 17
   2.3. Fermionic strings 19
      2.3.1. The bosonic string representation 19
      2.3.2. World-sheet supersymmetry 20
      2.3.3. Physical states and picture changing 22
      2.3.4. Heterotic string spectra 23
      2.3.5. Space–time supersymmetry 26
      2.3.6. Type-II spectra 28
      2.3.7. Anomalies 29
3. The Birth of Superstrings 31
   3.1. Introduction 31
   3.2. Papers
      Paper 1. F. Gliozzi, J. Scherk and D. Olive, Supersymmetry, supergravity theories and the dual spinor model, Nucl. Phys. B 122 (1977) 253. 33
      Paper 2. M.B. Green and J.H. Schwarz, Supersymmetrical string theories, Phys. Lett. B 109 (1982) 444. . . 71
      Paper 3. D.J. Gross, J.A. Harvey, E. Martinec and R. Rohm, Heterotic string theory (I). The free heterotic string, Nucl. Phys. B 256 (1985) 253. 76
4. Non-Supersymmetric Ten-Dimensional Strings 109
   4.1. Introduction 109
   4.2. Papers
      Paper 4. L.J. Dixon and J.A. Harvey, String theories in ten dimensions without spacetime supersymmetry, Nucl. Phys. B 274 (1986) 93. 111
      Paper 5. L. Alvarez-Gaumé, P. Ginsparg, G. Moore and C. Vafa, An $O(16) \times O(16)$ heterotic string, Phys. Lett. B 171 (1986) 155. 124
      Paper 6. N. Seiberg and E. Witten, Spin structures in string theory, Nucl. Phys. B 276 (1986) 272. 132
      Paper 7. H. Kawai, D.C. Lewellen and S.-H.H. Tye, Classification of closed fermionic-string models, Phys. Rev. D 34 (1986) 3794. 151
      Paper 8. W. Lerche, D. Lüst and A.N. Schellekens, Ten-dimensional heterotic strings from Niemeier lattices, Phys. Lett. B 181 (1986) 71; Erratum, Phys. Lett. B 184 (1987) 419. 162
5. Four-Dimensional Heterotic Strings 167
   5.1. Introduction 167
   5.2. Papers (I). Orbifolds
      Paper 9. L. Dixon, J.A. Harvey, C. Vafa and E. Witten, Strings on orbifolds, Nucl. Phys. B 261 (1985) 678. 172
      Paper 10. L. Dixon, J. Harvey, C. Vafa and E. Witten, Strings on orbifolds II, Nucl. Phys. B 274 (1986) 285. 181

Paper 11. L.E. Ibáñez, H.P. Nilles and F. Quevedo, Orbifolds and Wilson lines, Phys. Lett. B 187 (1987) 25. 211
Paper 12. K.S. Narain, M.H. Sarmadi and C. Vafa, Asymmetric orbifolds, Nucl. Phys. B 288 (1987) 551. 219
5.3. Papers (II). Self-dual lattices
Paper 13. K.S. Narain, New Heterotic string theories in uncompactified dimensions < 10, Phys. Lett. B 169 (1986) 41. 246
Paper 14. W. Lerche, D. Lüst and A.N. Schellekens, Chiral four-dimensional heterotic strings from self-dual lattices, Nucl. Phys. B 287 (1987) 477. 252
5.4. Papers (III). Free fermions
Paper 15. H. Kawai, D.C. Lewellen and S.-H.H. Tye, Construction of fermionic string models in four dimensions, Nucl. Phys. B 288 (1987) 1. 283
Paper 16. I. Antoniadis, C.P. Bachas and C. Kounnas, Four-dimensional superstrings, Nucl. Phys. B 289 (1987) 87. 359
5.5. Papers (IV). Discrete series tensor products
Paper 17. D. Gepner, Space-time supersymmetry in compactified string theory and superconformal models, Nucl. Phys. B 296 (1988) 757. 381
Paper 18. D. Gepner, Exactly solvable string compactifications on manifolds of SU (N) holonomy, Phys. Lett. B 199 (1987) 380. 403
6. Four-Dimensional Type-II Strings 413
6.1. Introduction 413
6.2. Papers
Paper 19. L. Castellani, R. d'Auria, F. Gliozzi and S. Sciuto, On the compactification of the closed supersymmetric string, Phys. Lett. B 168 (1986) 47. 415
Paper 20. R. Bluhm, L. Dolan and P. Goddard, A new method of incorporating symmetry into superstring theory, Nucl. Phys. B 289 (1987) 364. 421
Paper 21. H. Kawai, D.C. Lewellen and S.-H.H. Tye, Four-dimensional type II strings and their extensions: type III strings, Phys. Lett. B 191 (1987) 63. 442
Paper 22. W. Lerche, B.E.W. Nilsson and A.N. Schellekens, Covariant lattices, superconformal invariance and strings, Nucl. Phys. B 294 (1987) 136. 449
Paper 23. L.J. Dixon, V.S. Kaplunovksy and C. Vafa, On four-dimensional gauge theories from type II strings, Nucl. Phys. B 294 (1987) 43. 470
References 511

1. Introduction

Of course this introduction could begin with the usual elegant phrases about the beauty of having a possibly consistent theory of gravity with promising low-energy features. This would all be true, but it has been said many times before, and one can find many variations on this theme in the introductions of the papers in this volume. The formulation of these phrases has not required much change in the last few years. There has not been any reason to lower our expectations of string theory, but neither has there been any real progress towards fulfilling them.

The lack of substantial phenomenological progress can be attributed to two major problems. The first is that, although we have now a pretty good understanding of string perturbation theory, very little is known about non-perturbative physics. One of the reasons why this is a serious problem is that all known consistent string theories are supersymmetric, and hence contradict experiment. All indications are that supersymmetry is not broken perturbatively, which means that without a proper understanding of non-perturbative effects one is left to speculation. The second major problem is that string theory is not unique. If one could nail down the precise form of the theory at the Planck scale, one would know at least where to start extrapolating towards lower energies. Unfortunately, string theory has become a victim of its own success: the number of consistent four-dimensional string theories that can be constructed at present is enormous. The examples in this book are only the tip of the iceberg.

Despite this somewhat discouraging state of affairs, there is still a lot to be excited about. String theory continues to be an incredible rich field for new discoveries, although most of them are at first sight perhaps of more interest to mathematicians than to physicists. It would however not be the first time that the formal or mathematical breakthroughs are the ones that survive longest. Although the papers collected here had for the most part a phenomenological goal in mind, their lasting value may well lie in the new ideas and methods they introduced rather than in their concrete objectives. This is indeed precisely what has happened to the many papers that were written during the first string era. The techniques have survived, but the application to hadron physics does not receive nearly as much interest anymore.

For this reason the emphasis of this volume is on the "nuts and bolts" aspects of string construction rather than on phenomenological or philosophical questions (which are in any case already covered by the book edited by M. Dine in this series, which nicely complements the present volume). Subjects like conformal field theory, modular invariance, orbifold methods, self-dual lattices, Kac–Moody and super-Virasoro algebras and many others are certain to be discussed for many years to come. The attempts to construct new string theories, the results of which are collected in this book, gave a major impetus to many of these subjects. Studying these papers should be a worthwhile effort regardless of the ultimate fate of string phenomenology.

This volume is organized in the following way. The next chapter discusses the general idea of four-dimensional string construction from a perspective that developed mainly during the last few years, namely that of conformal field theory. The remaining four chapters contain the papers selected for this book, and are respectively about supersymmetric ten-dimensional strings, non-supersymmetric ten-dimensional strings, four-dimensional heterotic strings and four-dimensional type-II strings. Each of these chapters consists of a brief introduction, containing some historical remarks and, where necessary, a comparison of various approaches.

2. Conformal Field Theory for String Compactifications

According to our present understanding of string theory, all consistent theories are described by some 2-dimensional conformal field theory satisfying certain additional constraints. Conformal field theory (CFT) is the common language for all the papers in this volume. It provides the framework for comparing various approaches, but, more importantly, it can often be used to derive properties of an entire class of string theories in a construction-independent way (as is nicely illustrated by paper 23). This is a good place for putting some of the known facts together. The following is not intended as a pedagogical review of the subject, but rather as a user-friendly guide to the literature. For reviews of conformal field theory, and for details not explained here see, for example, refs. [1–7].

2.1. Consistency conditions

The papers in this volume (with the exception of the very first one) are all about oriented closed strings. Open strings are usually considered to be less appealing: their perturbation series is unpleasant, the Chan–Paton [8] method for adding gauge groups seems rather *ad hoc*, and open strings imply the existence of closed strings anyway, while the reverse is not true. Of course these are all matters of taste, and if interesting four-dimensional open strings could be constructed, they should be investigated. However, up to now the one-loop finiteness conditions (which selected the SO(32) theory in ten dimensions) do not seem to allow *any* four-dimensional solutions [9]. Should this somewhat surprising result hold under more general conditions than those considered in [9], then a whole class of string theories would have been ruled out by the simplest phenomenological constraint: the dimension of space–time. This should be enough motivation for a restriction to closed strings.

The conditions listed below are satisfied by all string theories in this book. We will not attempt to prove that all these requirements are necessary. No-go theorems have a bad reputation in physics. It is usually difficult to prove that the consequences of a violation of the rules are necessarily catastrophic. However, many years of work on string theory has led to the understanding that obeying the following rules of construction provides a simple way for avoiding all of the potential pitfalls that one faces in constructing string theories.

The first requirement is that the two-dimensional field theory describing a string theory in d dimensions must contain d free two-dimensional scalar fields X^{μ}, $\mu = 1, \ldots, d$, which must have both left- and right-moving modes.

The second requirement is that the theory should be invariant under reparametrizations of the coordinates on the world-sheet, in other words that it must have two-dimensional general coordinate invariance. The existence of this local symmetry implies the existence of a set of Faddeev–Popov ghosts b and c.

The third requirement is invariance under local rescalings of the world-sheet metric (Weyl invariance), which implies *conformal invariance*. Conformal transformations on the complex plane are generated on the fields in the theory by the energy–momentum tensor $T(z)$, which can be expanded in modes: $T(z)=\Sigma_n z^{-n-2}L_n$. These modes satisfy the *Virasoro algebra*

$$[L_n, L_m]=(n-m)+\frac{c}{12}n(n^2-1)\delta_{n,-m}. \tag{2.1}$$

There is a separate Virasoro algebra for left- and right-movers. In the following, quantities with a bar refer to left-movers, so that for example the left-moving Virasoro generators are denoted as $\bar{L}_n$.

In the quantized theory, conformal invariance may be broken by an anomaly. This anomaly receives contributions from all fields in the theory, and must be cancelled among them. The contribution to the conformal anomaly of some part of the theory is equal to the value of c (the central charge) in the Virasoro algebra for that part of the theory.

The fourth requirement applies only to string theories which have space–time fermions in their spectrum (we will refer to them as *fermionic strings* in the following; some authors (see for example paper 16) use this terminology for strings constructed using *world-sheet* fermions). Fermionic strings must, in addition to reparametrization invariance, have another two-dimensional gauge invariance, local supersymmetry. In principle one can have local (p, q) supersymmetry [10], where p is the number of supersymmetries of the left-moving modes and q the number of supersymmetries of the right-movers. The possible values of p and q are restricted by consistency conditions, but more seriously by phenomenogical requirements [11–14]: to be able to get four-dimensional strings one needs (provides one respects also the remaining rules) $p\leq 1$, $q\leq 1$. The existence of a supersymmetry in one or both chiral sectors of the world-sheet theory implies the existence of extra world-sheet fields. The scalars X^μ need fermionic partners ψ^μ, the Neveu–Schwarz–Ramond (NSR) fermions. Furthermore, the existence of an additional local symmetry implies the existence of additional ghosts, usually denoted as β and γ. There is one such pair of ghosts for every supersymmetry.

We will refer to the two-dimensional fields described so far as the *space–time* sector, as it determines the Lorentz representations of the space–time states. Thus there are two possible space–time sectors for each world-sheet chirality, namely an $N=0$ system (with fields X^μ, b, c), and an $N=1$ system with fields X^μ, b, c, ψ^μ, β, γ. In addition there is an *internal sector*, which is some conformal field theory whose conformal anomaly cancels that of the space–time sector (in units so that each scalar X contributes 1 to the conformal anomaly, each Majorana–Weyl fermion ψ contributes $\frac{1}{2}$, b and c together -26 and β and γ together 11). Denoting the internal sector as $\mathscr{C}_{c_L;c_R}$, where c_L and c_R are the conformal anomalies in the left- and right-moving sector respectively, we can have the following three classes of d-dimensional string theories:

Bosonic	$(N=0)_L(N=0)_R+\mathscr{C}_{26-d;26-d}$,	$d\leq 26$,
Heterotic	$(N=0)_L(N=1)_R+\mathscr{C}_{26-d;15-3d/2}$,	$d\leq 10$,
Type II	$(N=1)_L(N=1)_R+\mathscr{C}_{15-3d/2;15-3d/2}$,	$d\leq 10$.

This also specifies our convention for left- and right-movers in the heterotic string.

The internal conformal field theories have to satisfy some further conditions. First of all they should of course respect the ones already mentioned. Most of these are already contained in the requirement that they be conformal field theories. In addition that they should allow some realization of (p, q) supersymmetry (i.e. be superconformal field theories) whenever the space–time sector has such a supersymmetry.

The fifth requirement is unitarity. What we mean by this is that internal CFT's must be built out of unitary highest weight representations of the Virasoro algebra. This is a representation in a Hilbert space such that the Virasoro generators L_n satisfy $L_n^\dagger = L_{-n}$, and with an L_0 spectrum that is bounded from below. One consequence of unitarity is that c_L and c_R cannot be negative, which limits the space–time dimensions d to the values shown above. The only non-unitary CFT's that we allow to appear in string theory are the b, c and β, γ ghost systems, and the time-like components of X^μ and ψ^μ. Their non-unitarity is precisely needed to get a unitary space–time theory. Any additional violation of unitarity of the world-sheet theory would probably ruin the unitarity of the space–time theory.

The sixth condition is that the internal CFT should be *compact*, i.e. the spectrum of L_0 should be discrete. If this were not the case the d-dimensional mass spectrum would be continuous, which to a d-dimensional observer would indicate the presence of extra, uncompactified (though not necessarily flat) dimensions. There is nothing fundamentally wrong with this, but it is not what we would call a d-dimensional string theory.

The seventh and last condition is that of modular invariance of the partition function and all correlation functions on a Riemann surface of arbitrary genus. On all Riemann surfaces with at least one handle one can make coordinate transformations that cannot be connected continuously to the identity. Just as the local coordinate transformations, these global transformations must be "divided out" of the path integral, so that one only integrates over inequivalent surfaces. If the partition and correlation functions would not be invariant under modular transformations, one could not choose in an unambiguous way a region of integration that covers all inequivalent surfaces precisely once. Then the theory has a global anomaly, and does not make sense. One should not just have modular invariance for each genus separately, but the amplitudes at different genera should also have the correct relation with each other. In the limit where a Riemann surface falls apart into lower genus surfaces, the partition and correlation functions should factorize into the corresponding lower genus expressions. Finally, the contribution of the states in the spectrum to the one-loop partition function should be positive integers for space–time bosons and negative integers for space–time fermions, i.e. one should have a correct spin statistics relation.

These modular invariance conditions are certainly not all independent. For example, two-loop modular invariance is probably enough to guarantee multi-loop modular invariance; modular invariance of the partition function plus factorization probably guarantees modular invariance of the correlation functions; the spin statistics relation follows in special cases (see paper 6, [15–18]) and probably in general, from multi-loop modular invariance.

The art of string construction is to find solutions (conformal field theories $\mathscr{C}_{c_\mathrm{L};c_\mathrm{R}}$) to all these conditions. It would be nice if one could classify all the solutions, but we are still very far from doing that, despite the fact that a lot of progress has been made in the last few years.

2.2. Bosonic strings

Bosonic strings have no space–time fermions in their spectrum, and are therefore not of much interest if one has phenomenological applications in mind. However, they are the easiest ones to describe, and many of the results can be used in the discussion of heterotic and type-II strings. In particular the conditions for modular invariance for fermionic and bosonic strings are the same, in a sense which we will make more precise in section 2.3. This allows for a clean separation between problems associated with modular invariance (discussed here) and with world-sheet supersymmetry (discussed in section 2.3.2).

2.2.1. Virasoro representation theory

Consider first the internal CFT $\mathscr{C}_{26-d;26-d}$. The left-moving and right-moving Virasoro generators satisfy (2.1) with $c = 26 - d$. The Hilbert space is most conveniently described in terms of the eigenvalues of the operator L_0. Obviously L_n, $n < 0$, increases the L_0 eigenvalue when acting on a state. Since, by assumption, the L_0 spectrum is bounded from below, one can distinguish two kinds of states: the ground states $|\bar{h}, h\rangle$, which satisfy

$$L_0|\bar{h}, h\rangle = h|\bar{h}, h\rangle,$$

$$\bar{L}_0|\bar{h}, h\rangle = \bar{h}|\bar{h}, h\rangle,$$

$$L_n|\bar{h}, h\rangle = \bar{L}_n|\bar{h}, h\rangle = 0, \quad n > 0,$$

and all the other states that are not annihilated by all the positive L_n's. Hence all the latter states can be obtained from the ground states by the action of the operators L_{-n}, $n > 0$. These states are called the *descendants*.

The complete set of states obtained by acting on a given ground state with the operators L_{-n}, $n > 0$, form the *Verma module* V_h of that ground state, i.e.

$$V_h = \{L_{-n_1}L_{-n_2}, \ldots\}|\bar{h}, h\rangle, \quad n_i > 0.$$

A Verma module is a representation of the Virasoro algebra, and is completely determined by the values of c and h, in the same way as Lie algebra representations are completely determined by the highest weights (for this reason the ground states are sometimes called *highest weight states*). Since the right- and left-moving Virasoro algebras act independently, the complete set of states generated from $|\bar{h}, h\rangle$ is given by the direct product $V_{\bar{h}} \otimes V_h$ of the Verma modules for left- and right-movers.

The states in a Verma module do not necessarily all have positive norm. Those values of c and h for which negative norm states appear are of no interest in string theory, since the corresponding representations are not unitary. A necessary condition for the absence of negative norm states is $c \geq 0$, $h \geq 0$. A sufficient condition is $c > 1$, $h \geq 0$. In the region $0 < c < 1$ one obtains unitary representations only for special values of c and h [1,19–21], corresponding to the so-called *minimal conformal field theories*,

$$\begin{aligned} c &= 1 - \frac{6}{m(m+1)}, \quad m = 3, 4, 5, \ldots, \\ h &= \frac{((m+1)p - mq)^2 - 1}{4m(m+1)}, \quad p = 1, 2, \ldots, m-1; \quad q = 1, 2, \ldots, p. \end{aligned} \tag{2.2}$$

These values of c form the *discrete series*. Finally, for $c = 0$ the only unitary highest weight representation is the trivial one, $L_n|0\rangle = 0$ for *all n* [22].

A Verma module may furthermore contain *null vectors*, that is vectors $|\xi\rangle$ with zero norm. Such vectors satisfy $L_n|\xi\rangle = 0$, $n > 0$, and hence they are highest weight vectors of a submodule of V_h. The converse is also true: if V_h is not irreducible, it contains null vectors which are highest weight vectors of a submodule. One can always consistently set all null vectors equal to zero to obtain an irreducible representation of the Virasoro algebra that is free of null vectors.

Null vectors in unitary Virasoro representations appear only in the following cases:
- c, h as in (2.2),
- $c=1$, $h=\frac{1}{4}k^2$, $k\in\mathbb{Z}$.

Furthermore, the vector $L_{-1}|\bar{h}, 0\rangle$ is always a null vector. Therefore L_{-1} should always annihilate any state with $h=0$. The state $|0, 0\rangle$ is thus annihilated by all generators L_{-1}, L_0 and L_1 (and analogously for left-movers). This state is called the SL(2, C)-invariant vacuum because these operators generate a sub-algebra of the Virasoro algebra that is isomorphic to SL(2, C).

The ground states $|\bar{h}, h\rangle$ are in one-to-one correspondence with *primary fields* or *conformal fields* $\phi_{\bar{h},h}(\bar{z}, z)$, which, by definition, must transform in the following way under conformal transformations $z\to f(z)$:

$$\phi_{\bar{h},h}(\bar{z}, z)\to\phi'_{\bar{h},h}(\bar{z}, z)=\left(\frac{\partial\bar{f}}{\partial\bar{z}}\right)^{\bar{h}}\left(\frac{\partial f}{\partial z}\right)^{h}\phi_{\bar{h},h}(\bar{f}, f), \tag{2.3}$$

where $\bar{f}$ denotes $f(\bar{z})$. In general conformal fields are not simply a product of a left- and a right-moving field, although in many special cases studied in the literature they do factorize in that way. Infinitesimal conformal transformations $z\to z+\epsilon(z)$ are generated by the energy–momentum tensor, which has the following operator product with a primary field:

$$T(z)\phi_{\bar{h},h}(\bar{w}, w)=\frac{h}{(z-w)^2}\phi_{\bar{h},h}(\bar{w}, w)+\frac{1}{z-w}\partial_w\phi_{\bar{h},h}(\bar{w}, w).$$

There is an analogous expression for the left-moving component of the energy–momentum tensor.

The ground states are obtained by acting with the primary fields on the vacuum at $z=0$:

$$|\bar{h}, h\rangle=\lim_{z\to0}\phi_{\bar{h},h}(\bar{z}, z)|0, 0\rangle. \tag{2.4}$$

It is easy to verify that the requirements that L_n, $n>0$, should annihilate a ground state and that $\phi_{\bar{h},h}$ is a primary field are equivalent. The descendants correspond to combination of derivatives of the primary fields. Their operator product with the energy–momentum tensor contains poles of higher order in $(z-w)$.

For $c<1$ the number of unitary representation of the Visasoro algebra is finite, with h given by (2.2); for $c\geq1$ the number of unitary representations is infinite.

2.2.2. *Characters and modular invariance*

For each Verma module V_{h_l} (with null vectors removed) one can define a *character* χ_l

$$\chi_l(q)=\operatorname*{Tr}_{V_{h_l}}q^{L_0-c/24}.$$

From the expansion of $\chi_l(q)$ in powers of q one can read off the number of states with a certain L_0 eigenvalue.

In general only a subset of all the possible ground states can appear in a given string theory. One can choose a set of non-negative integers M_{kl} which give the multiplicity of the state $|h_k, h_l\rangle$ in the spectrum of the internal CFT. These integers are constrained by the requirement of modular invariance.

The first non-trivial modular invariance condition appears for one-loop diagrams without external lines. The one-loop vacuum amplitude for a bosonic string theory has the form

$$\int \frac{d^2\tau}{(\mathrm{Im}\ \tau)^2} \mathscr{P}^{\mathrm{st}}(\tau)\mathscr{P}^{\mathrm{int}}(\tau), \tag{2.5}$$

where the integral is over the inequivalent metrics on the torus. The integrand consists of two factors, representing the space–time and internal sectors. The one loop partition function for the internal sector can be expressed in terms of the characters of the Virasoro representations and their multiplicities,

$$\mathscr{P}^{\mathrm{int}}(\tau) = \sum M_{kl}\chi_k(\bar{q})\chi_l(q), \tag{2.6}$$

where $q = \mathrm{e}^{2\pi i\tau}$.

The integrand must be invariant under modular transformations, which for the torus have the simple form

$$\tau \to \frac{a\tau + b}{c\tau + d}, \quad a, b, c, d \in \mathbb{Z}, \quad ad - bc = 1.$$

This entire group of transformations is generated by two basic transformations:

$$T: \quad \tau \to \tau + 1,$$
$$S: \quad \tau \to -1/\tau.$$

Since they generate everything, it is sufficient to check invariance under just these two transformations. In bosonic string theories the internal sector has to be invariant by itself. The transformation of the characters under T is obvious:

$$\chi_l(\tau + 1) = \mathrm{e}^{2\pi i(h_l - c/24)}\chi_l(\tau). \tag{2.7}$$

This implies that if $M_{kl} \neq 0$, then $h_l - \bar{h}_k \in \mathbb{Z}$. The characters have a more complicated transformation under S:

$$\chi_l\left(-\frac{1}{\tau}\right) = \sum_k S_{lk}\chi_k(\tau), \tag{2.8}$$

where S is a unitary and symmetric matrix [23]. Invariance under this modular transformation leads thus to the matrix conditions $SMS^\dagger = M$, which is obviously satisfied if M is the identity (but in general this is by no means the only solution).

An interesting consequence of invariance under S and positivity of the matrix M is that the multiplicity of the vacuum (M_{00}) is non-zero:

$$M_{00} = \sum_{k,l} S_{0k} M_{kl} S_{l0}.$$

All terms in this sum have the same sign, because of the inequality [23]

$$\frac{S_{0k}}{S_{00}} = \lim_{q\to 0} \frac{S_{kl}\chi_l(q)}{S_{0l}\chi_l(q)} = \lim_{q\to 1} \frac{\chi_k(q)}{\chi_0(q)} > 0$$

(with implicit sums on l).

2.2.3. Extended Virasoro algebras

Knowledge of the Virasoro characters appearing in $\mathscr{P}^{\text{int}}$ completely specifies the one-loop vacuum amplitude of a compactified string theory, but is not sufficient to define a string theory completely. One also needs a prescription for determining the correlation functions on a surface of arbitrary genus. If all this information can be obtained at least in principle one speaks of an exactly solvable conformal field theory.

For $c<1$ the Virasoro algebra alone is enough to determine all correlation functions on the sphere [1] (see furthermore [24–27] for correlation functions on the torus and higher genus surfaces). This is due to the fact that one has a finite number of primary fields, but, more importantly, to the existence of null vectors which imply that the correlation functions satisfy certain differential equations. Unfortunately, the case of interest in string theory is $c \geq 1$ with $c \in \mathbb{Z}$. One can however consider special cases where the internal sector has more structure, allowing it to be an exactly solvable theory. This extra structure is, in all known cases, attributable to some set of *chiral operators*, which are primary fields with either $\bar{h}=0$ or $h=0$. A chiral operator $\phi_{0,h}$ must commute with $\bar{L}_{-1}$, and hence it follows that $\partial_{\bar{z}}\phi_{0,h}=0$, i.e. the operator depends only on z. Therefore we can concentrate on the right-moving sector and we will henceforth denote the operator as $\phi_h(z)$. We are not requiring that $h \in \mathbb{Z}$ or that the ground state $|0, h\rangle$ is part of the spectrum. The set of operators $\phi_h(z)$ generates an operator product algebra, which must be local (i.e. only integral powers of $(z-w)$ appear). The structure of this algebra depends on the value of h. The most well-known cases are $h=\frac{1}{2}$ (free fermions), $h=1$ (Kac–Moody algebras), $h=\frac{3}{2}$ (super-Virasoro algebras) $h=2$ (which includes the Virasoro algebra) and $h=3$ (the W algebra [28]).

The operator ϕ_h can be expanded in modes,

$$\phi_h(z) = \sum_r z^{-r-h}\phi_r; \qquad \phi_r = \oint \frac{\mathrm{d}z}{2\pi \mathrm{i}} z^{r+h-1}\phi_h(z). \tag{2.9}$$

The allowed values of r depend on the state on which ϕ acts, and can be determined by computing the operator product of ϕ with the primary field σ creating the state. If

$$\phi(z)\sigma(0) = z^{\alpha}\sigma'(0) + \dots, \tag{2.10}$$

then $r \in \mathbb{Z} - \alpha - h$, in order to make the mode-integral well-defined. Note that descendants of σ yield the same condition on the modes. Obviously, if $\phi(z)$ has a non-local operator product with any operator in the theory, it cannot be part of the theory itself. However, it may still play a role in determining the spectrum and the correlation functions. If the operator product is as in (2.10), the field ϕ picks up a phase $\mathrm{e}^{2\pi \mathrm{i}\alpha}$ when it moves along a contour around the position of σ. This behavior of ϕ is called the *monodromy* of ϕ with respect to the *twist field* σ (see [29–31]). In general, not every phase is allowed. Obviously the z-dependence of the higher-order terms on the right-hand side of (2.10) should differ by integral powers from the leading term;

otherwise it makes no sense to even discuss any possible rôle of ϕ in the theory. For the same reason, the right-hand side and the left-hand side of the operator product of the fields ϕ_h with each other must have the same monodromy. In other words, only twists that are an automorphism of the algebra are allowed.

The modes of ϕ have the following commutation relation with the Virasoro generators:

$$[L_n, \phi_r] = \{n(h-1) - r\}\phi_{r+n} .$$

Using this relation and the unitarity of the Virasoro representation one can show that ϕ is Hermitean, i.e. $\phi_r^\dagger = \phi_{-r}$ (more precisely, when there is more than one field with a given value of h one can always choose Hermitean linear combinations.)

The operators ϕ_r may form a new algebra that extends the Virasoro algebra. Such an *extended Virasoro algebra* may organize the combination of characters in the partition function in a more efficient way. Typically, one defines primary fields with respect to the extended algebra rather than just the Virasoro algebra. Such fields must satisfy the conditions for Virasoro primary fields, but in addition they must have a certain operator product with the operators $\phi_h(z)$, in such a way that ϕ_r, $r > 0$, annihilates the corresponding ground states. In operator language this means that the most singular term in the operator product of $\phi_h(z)$ with a primary field must behave like $(z-w)^{-\eta}$, $\eta \le h$. Descendants are defined by acting on these ground states with the operators L_{-n}, $n > 0$, and ϕ_{-r}, $r > 0$. In this way, some of the primary fields of the Virasoro algebra become descendants of the extended Virasoro algebra: the representations become larger, but there are fewer of them. Furthermore one typically gets additional null vectors, and hence more constraints on the correlation functions. For recent work on extended Virascro algebras see, for example, [32–35].

Instead of using Virasoro characters, one may express partition functions in terms of characters of an extended Virasoro algebra. It may happen that when the Virasoro characters are combined into characters of the extended algebra, the sum (2.6) reduces to a finite one. In that case one speaks of a *rational* conformal field theory (RCFT), a subject that has been studied extensively recently (see, e.g., [36–44]).

2.2.4. *Kac–Moody algebras*

The most familiar case is the extension of the Virasoro algebra by $h = 1$ chiral operators. Consider a set of such operators with local mutual operator products, and denote them as J^a, $a = 1, \ldots, D$. The operator product of two of such operators must have the form

$$J^a(z)J^b(w) = \frac{\frac{1}{2}k\delta^{ab}}{(z-w)^2} + \frac{1}{(z-w)} \mathrm{i} f_{abc} J^c + \text{finite terms.} \tag{2.11}$$

The form of the first term is a matter of normalization. It has been shown in [28] that in a unitary conformal field theory the coefficients f_{abc} are (completely anti-symmetric) structure constants of a Lie algebra. They are normalized here so that $f_{abc}f_{dbc} = C(A)\delta_{ab}$, where $C(A)$ is the quadratic Casimir eigenvalue of the adjoint representation (normalized to N for SU(N), $N+1$ for Sp($2N$), $N-2$ for SO(N) and 4, 9, 12, 18, 30 for, respectively, G_2, F_4, E_6, E_7, E_8). This fixes the normalization of J_a, and leaves an as yet undetermined central charge k.

The modes of $J^a(z)$, defined as in (2.9), satisfy

$$[J_n^a, J_m^b] = \mathrm{i} f_{abc} J_{m+n}^c + \tfrac{1}{2}km\delta^{ab}\delta_{m,-n} ,$$

so that the zero modes generate a Lie algebra $\mathscr{G}$. The complete algebra is called an *affine* algebra or *Kac–Moody* algebra (for reviews of this subject see [45,46]). One usually reserves this terminology for the non-Abelian part of the algebra $\mathscr{G}$; there may of course also be U(1) factors, for which f_{abc} is absent, so that k can be normalized to 1.

Primary fields of the Kac–Moody extended Virasoro algebra are defined as

$$T(z)\phi(w) = \frac{h}{(z-w)^2} + \frac{\partial_w\phi(w)}{z-w},$$

$$J^a(z)\phi(q) = \frac{1}{z-w} t^a\phi(w),$$

where t^a is a representation matrix of an irreducible representation R of the Lie algebra $\mathscr{G}$. The primary fields create ground states from the vacuum which are annihilated by all the positive modes of $J^a(z)$ as well as $T(z)$. One can build representations of the algebra by acting with the negative modes of $J^a(z)$ and $T(z)$. Using null vectors one can find a relation between the value of h and the representation R:

$$h(R) = \frac{C(R)}{C(A)+k},$$

where $C(R)$ is the quadratic Casimir eigenvalue of the representation R in the normalization given above.

Unitary representations are obtained only for special choices of the representation R, and are completely determined by this representation. The Lie algebra representation in its turn is completely determined by a highest weight vector λ. The condition for unitarity of the highest weight representation of the Kac–Moody algebra is

$$\lambda \cdot \psi \leq k, \tag{2.12}$$

where ψ is the highest root (i.e. the highest weight of the adjoint representation, which is always a long root). We normalize it so that $\psi \cdot \psi = 2$. If one chooses this normalization, unitarity restricts the values of k to positive integers (in the literature one finds very often the definition $m = 2k/\psi \cdot \psi$, where m is called the *level* of the representation. With our normalization, $m = k$).

In contrast to pure Virasoro algebras, the number of unitary highest weight representations is thus always finite for Kac–Moody (KM) algebras. This is especially helpful if the entire internal sector is spanned by products of KM algebras. To verify this one can compute the central charge of the Sugawara energy–momentum tensor of the Kac–Moody algebra, which is an explicit realization of $T(z)$ in terms of the currents:

$$T(z) = -\frac{1}{2[C(A)+k]} \sum_a J^a(z) J^a(z), \tag{2.13}$$

where the product is defined by point-splitting and subtraction of the singular terms. It is straightforward to compute the central charge:

$$c(\mathscr{G}, k) = \frac{k \dim \mathscr{G}}{C(A)+k}.$$

In the previous expressions we assumed that $\mathscr{G}$ is simple. If it consists of more than one factor one simply has to sum the central charge over all factors. If the result equals $26-d$ the entire internal sector can be written in terms of KM representations.

One can define characters of these representations in exactly the same way as for Virasoro representations, except that the trace includes now also all states created by modes of $J^a(z)$. One can in fact define a more refined kind of character, which not only contains information about the multiplicity of states at every level (where "level" refers to the eigenvalue of L_0), but also about the representations of $\mathscr{G}$. This Weyl–Kac character is defined as [47]

$$\chi_R(\boldsymbol{\theta}\,|\,q) = \mathrm{Tr}\; q^{L_0 - c/24}\, \mathrm{e}^{2\pi \mathrm{i}\boldsymbol{\theta}\cdot \boldsymbol{J}_0}, \tag{2.14}$$

where $\boldsymbol{J}_0$ denotes a maximal set of commuting generators (Cartan sub-algebra) of $\mathscr{G}$. The coefficient of $q^{h-c/24}$ in the expansion of this character is equal to the Chern character of the representation at that level.

If there are U(1) factors, they can each be represented by free bosons X, so that $J(z)=\partial_z X$. There are now an infinite number of primary fields that yield unitary representations: $\phi_p(z)=\mathrm{e}^{\mathrm{i}pX(z)}$, with $h=\frac{1}{2}p^2$, and charge p (where "charge" refers to the eigenvalue of J_0 on ϕ_p). Even though the number of primary fields is infinite, it is still advantageous to describe the spectrum in terms of $T(z)$ and $J(z)$ rather than $T(z)$ alone. One gets in that way the usual description of the bosonic string in terms of the modes α_n of the U(1) currents $J(z)$. These modes act independently without creating null vectors.

The characters of this extended algebra are

$$\chi_p(\nu\,|\,q) = \mathrm{e}^{2\pi \mathrm{i} p\nu} q^{1/2p^2} \eta(q)^{-1}, \tag{2.15}$$

where, analogous to (2.14), the first factor is the character of the U(1) representation (charge p), which is the same for all states built on the same ground state.

If the internal sector can be expressed completely in terms of KM representations, the infinite sum in (2.6) becomes a finite sum over KM characters of the form

$$\mathscr{P}^{\mathrm{int}}(\tau) = \sum_{kl} M'_{kl}\bar{\chi}_k(0,\bar{q})\chi_l(0,q).$$

Note that there is *a priori* no reason why the left- and right-moving KM algebras should be isomorphic. Hence the characters χ and $\bar{\chi}$ may be different; the bar does not denote complex conjugation.

Modular invariants

The modular transformations of KM characters (with first argument equal to zero) have the same general form as (2.7) and (2.8). The condition for invariance under S reads $M'=\bar{S}M'S$, where S is in general different from $\bar{S}$. The same formulas hold if U(1) factors are present, except that one gets an infinite sum.

Relatively little is known about the modular invariant combinations of these characters. Most work on this subject has been concentrated on invariants for identical left- and right-moving algebras, which is unfortunately by no means the most general case. The simplest way to construct a modular invariant is then to chose $M'_{kl}=\delta_{kl}$. This is called the *diagonal invariant*. An explicit realization of this kind of conformal field theory is provided by the Wess–Zumino–Witten models, corresponding to strings compactified on group manifolds [46,48].

Only for the SU(2) KM algebras are all positive invariants known for all values of k. The solutions admit a so-called A–D–E *classification* in terms of the Dynkin diagrams of the simply-laced algebras (i.e. A_n, D_n and E_6, E_7, E_8) [49], but this is only to be understood as a labelling of the solutions; the simply-laced algebras themselves do not play any known rôle.

2.2.5. Self-dual lattices

Another well-understood and important case is that of modular invariant combinations of level-1 simply-laced KM algebras. The currents of these algebras can be expressed in terms of free bosons by means of the Frenkel–Kac construction: the Cartan sub-algebra generators are represented by n currents $\partial_z X^I$, $I=1,\ldots,n$ (where n is the rank of $\mathscr{G}$), and the remaining generators by $e^{i\boldsymbol{r}\cdot X(z)}$, where $\boldsymbol{r}$ is one of the roots (normalized to have length $\sqrt{2}$). It is convenient to consider the ensemble of all free bosons that appear due to other simply-laced, level-1 KM factors or U(1)'s. Assume for simplicity that the entire internal sector can be described in this way, so that there are $26-d$ left- and right-moving bosons $X_R(z)$ and $X_L(\bar{z})$. The primary fields with respect to $\partial_z X_R$, $\partial_{\bar{z}} X_L$ and the Virasoro algebra are the exponentials $e^{i\boldsymbol{p}_R\cdot X_r}$ $e^{i\boldsymbol{p}_L\cdot X_L}$. Closure of the operator algebra requires that the set of vectors $(\boldsymbol{p}_L, \boldsymbol{p}_R)$ must close under addition. Compactness of the internal CFT implies that they must be discrete in $\mathbb{R}^{2N}$ $(N=26-d)$, so that they form a lattice. Invariance under the modular transformation T implies that $\boldsymbol{p}_L^2-\boldsymbol{p}_R^2=0 \bmod 2$ for all lattice points, which means that the lattice must be even and integral with respect to a Lorentzian metric $(+)^N(-)^N$. Finally, invariance under S implies that it must be self-dual with respect to the same metric. Conformal field theories constructed with even self-dual Lorentzian lattices provide the simplest known examples of bosonic strings in dimensions less than 26. It is straightforward to compute correlation functions or higher genus partition functions in such theories, and multi-loop modular invariance can be checked explicitly (see, e.g., [50–54]).

This description of modular invariant theories in terms of even self-dual Lorentzian lattices (ESDL) is due to Narain (paper 13). A special case was discovered by Englert and Neveu [55], who observed that a modular invariant theory with identical left- and right-moving simply-laced KM algebras $\mathscr{G}$ can be constructed by pairing the conjugacy classes of the algebras. The resulting lattice is of the form $\{(\boldsymbol{w}, \boldsymbol{w}'), \boldsymbol{w}, \boldsymbol{w}' \in \Lambda_W, \boldsymbol{w}-\boldsymbol{w}' \in \Lambda_R\}$, where Λ_W and Λ_R are, respectively, the root and weight lattices of $\mathscr{G}$. This solution can also be described as the diagonal invariant of $\mathscr{G}$, since for level-1 simply-laced algebras the conjugacy classes are in one-to-one correspondence with the highest weight representations of the Kac–Moody algebra.

Since the consistency conditions are all expressed in terms of a Lorentzian metric, SO(N, N) Lorentz transformations map different ESDL's into each other. An SO(N) × SO(N) subgroup consisting of separate rotations on left- or right-movers does not alter the spectrum of $(\bar{h}, h)=(\frac{1}{2}\boldsymbol{w}_R^2, \frac{1}{2}\boldsymbol{w}_L^2)$ values of the internal CFT. However all other Lorentz rotations *do* change the spectrum. Hence the physically inequivalent lattices are parameterized by the coset SO(N, N)/SO(N) × SO(N). Generic points on this coset manifold have as (0, 1) or (1, 0) operators only the currents $\partial_z X_R^I$ and $\partial_{\bar{z}} X_L^I$. In special points one may have vectors $(\boldsymbol{r}, 0)$ or $(0, \boldsymbol{r})$ with $\boldsymbol{r}^2=2$, which implies the existence of additional (1, 0) and (0, 1) fields in the theory. Those fields are Frenkel–Kac generators of a level-1 KM algebra. (See [56] for an explicit parametrization of such a lattice).

Bosonic strings compactified on even self-dual lattices $\Gamma_{N;N}$ can be described alternatively by "ordinary" torus compactification on a torus defined by an N-dimensional lattice Λ_N, in the presence of an anti-symmetric tensor B_{IJ} background field (this was shown in [57] for the slightly more complicated case of heterotic strings).

As was mentioned before, as long as one writes the partition function in terms of U(1) characters, it will be an infinite sum of combinations of such characters. Hence a generic lattice does not correspond to an RCFT. On the other hand, if the entire internal sector is built out of KM algebras, it is always an RCFT. This is certainly not the only possibility for making RCFT's by means of lattices. In general, an even self-dual Lorentzian lattice corresponds to an RCFT if and only if a basis exists so that all $2N$ U(1) charges are quantized. Lattice theories are solvable CFT's whether they are rational or not. This shows that rationality may be helpful, but is not necessary for solvability.

2.2.6. Orbifolds

The class of conformal field theories described by free bosons can be generalized further, from lattices to *orbifolds*. (Historically the name was associated with singular manifolds used in heterotic string compactification, but it has acquired a broader meaning.) Self-dual lattices are the most general CFT's in which all $2N$ U(1) bosons are present in the spectrum. Orbifolds are characterized by the presence of twist fields, with respect to which some of the currents ∂X have non-trivial monodromy. This prevents these currents from being part of the spectrum, but not from governing the properties of the theory.

To obtain modular invariant theories of this type one may choose a self-dual lattice which has some automorphism (which can be a discrete rotation (*twist*) or a translation (*shift*), or a combination of such operations). A new theory is then made by "modding out" by the automorphism, that is to remove all states from the spectrum that are not invariant under it. Whenever such an operation is performed in string theory, modular invariance requires that new states are added to replace the ones that were removed. These states (the *twisted sectors*) are built on ground states created by twist operators. The addition of twisted sectors is needed to guarantee invariance under the S transformation. Invariance under T leads to an additional constraint on the allowed twist, known as *level matching*. This is discussed in detail in paper 10. Multi-loop modular invariance is discussed in [58], but has not yet been proved in full generality.

In principle one can use automorphisms that act differently on the left- and the right-movers of a lattice $\Gamma_{N;N}$. The result is called an *asymmetric* defined orbifold. *Symmetric* orbifolds are constructed by modding out automorphisms of the torus by Λ_N (see section 2.2.5) which in terms of $\Gamma_{N;N}$ act symmetrically. Symmetric orbifolds are thus a special case of asymmetric ones. The application of these ideas to heterotic strings is discussed in papers 9–12.

If $c=1$ the requirement that the twist should be an automorphism of the algebra generated by ∂X allows only one twist, $X \to -X$. Although from the point of view of the U(1) algebra one could make such a twist only for left- and right-movers, one can only satisfy level-matching by applying it simultaneously for X_L and X_R. Such a twist is automatically an automorphism of any ESDL $\Gamma_{1;1}$, and the resulting orbifold theories form a new one-parameter family of theories, in addition to lattices. It turns out that these theories (lattices plus orbifolds) describe almost the entire space of solutions at $c=1$ [59,60]; the only other known solutions are three isolated points, obtained by modding the SU(2), level-1 modular invariant theory by certain discrete subgroups of SU(2) [61]. This may exhaust the list of $c=1$ CFT's. The $c=1$ theories can be used to construct bosonic strings in 25 dimensions. This is the only dimension where (perhaps) a complete classification of all internal sectors is presently known.

2.2.7. Free fermions

One may think of free fermions as an extension of the Virasoro algebra by operators $\psi(z)$ with $h=\frac{1}{2}$. The fact that h is non-integer implies that the fermions cannot appear as chiral operators in the spectrum of a string theory, but this does not really matter; they can still determine the structure of a theory, analogous to the operators ∂X in orbifold theories. On general grounds one can conclude that the operator product of a set of fermions ψ^a can only have the form

$$\psi^a(z)\psi^b(z) = \frac{\delta^{ab}}{z-w}, \tag{2.16}$$

where of course we have chosen an orthonormal basis. This implies that the modes b_r^a satisfy $\{b_r^a, b_s^b\} = \delta^{ab}\delta_{r,-s}$. The monodromy of the fermions is restricted by the condition that (2.16) must be respected. This implies that the set of fermions may transform by an orthogonal transformation, $\psi^a \to \Sigma_b R_{ab}\psi^b(z)$, around some twist field σ. For a given σ the matrix R can be diagonalized by means of a unitary transformation. The fact that R is orthogonal implies that the eigenvalues come in complex conjugate pairs, plus in general some eigenvalues ± 1.

If one has just one fermion, the only non-trivial automorphism is multiplication by -1, i.e. fermions can be periodic or anti-periodic around $z = 0$. Anti-periodic fermions are said to belong to the Ramond sector, periodic ones to the Neveu–Schwarz sector (note that because of the conformal factor, all boundary conditions of half-integer spin fields are reversed when going from the cylinder to the complex plane). Primary fields for the Virasoro algebra, combined with one free fermion ψ can have at most a $(z-w)^{-1/2}$ singularity with ψ. There are just two possibilities for the value of h, namely $h = 0$ (for the vacuum) and $h = \frac{1}{16}$. (ψ is a Virasoro primary field, but it is not primary with respect to itself; extended symmetry generators never are.) The $h = \frac{1}{16}$ primary field has a square root branch cut in its operator product with ψ. Acting on the vacuum it creates thus a ground state for the Ramond sector.

In order to construct modular invariants one has to define, in addition to the normal characters for these two ground states, also characters with an additional factor $(-)^F$ in the trace, where F is the two-dimensional fermion number. To such characters states created with an odd number of fermionic oscillators contribute with a $-$ sign. Altogether we get thus four characters, which correspond to the four spin structures (the periodicities along the two independent non-contractible loops) on the torus (in fact the Ramond character with $(-)^F$ factor vanishes, but one can give a meaning to its contribution by considering multi-loop diagrams).

A single fermion has $c = \frac{1}{2}$. To get larger values of c one can tensor many such systems with each other, for left-movers as well as right-movers. For simplicity one may still allow only periodic and anti-periodic boundary conditions, although for more than one fermion this is certainly not the most general solution. For $2N$ left- and right-moving fermions ($c = N$) one obtains 4^{4N} possible combinations of boundary conditions. The partition function is a linear combination of the corresponding 4^{4N} characters, and the coefficients must be chosen so as to get a modular invariant sum. This is known as *spin structure summation*.

String consistency imposes further constraints. It turns out that there are two equivalent choices that can be made: one may impose the condition of modular invariance at higher loops plus factorization, or one can require a "physically sensible projection" (which means that the partition function can be interpreted in terms of space–time bosons, each contributing $+1$ to the partition function). Clearly both these conditions are necessary for a consistent string theory. After some initial confusion, it appears now that these two conditions are in fact equivalent (see paper 6 and [15–18]). There are two groups which have formulated solutions to this problem: Kawai et al. (paper 15) chose to impose the second condition, Antoniadis et al. (paper 16) the first one. (Both papers are of course about heterotic strings rather than bosonic strings, but the issue of their equivalence can be settled completely by considering bosonic strings.)

One can generalize all the above to boundary conditions defined by allowing arbitrary SO($2N$) rotations in the monodromy, for left- and right-movers independently. For a given ground state such a matrix can always be diagonalized (on a complex basis), so that one gets pairs of fermions with conjugate monodromy phases, plus, in general, some periodic and anti-periodic ones. In the special case that the monodromy matrices for all ground states commute with each other (so that they can be simultaneously diagonalized) the problem of finding modular invariant partition functions has been solved (see papers 15, 16 and [62]).

The simplest subclass of fermionically constructed theories is obtained if all fermion boundary conditions are simultaneously diagonalizable on a basis of pairs of complex conjugate fermions ψ^+ and ψ^- whose monodromy eigenvalues are complex conjugate (note that because the monodromy matrix is orthogonal ψ^+ and ψ^- must have conjugate eigenvalues, except when eigenvalues ± 1 occur). In that case fermion pair can be replaced by a single boson X, by means of the well-know expression

$$\psi^+ = e^{iX}, \qquad \psi^- = e^{-iX},$$

and the fermion boundary conditions correspond to a certain value of the momentum eigenvalue of each boson. Hence we are back to the situation already discussed in the previous subsection: a fermionic modular invariant of this kind corresponds to bosons with momenta on a self-dual lattice (since this is the only way to make modular invariants with free bosons).

Conversely any $c_L = c_R = N$ system constructed out of free bosons with momenta on a self-dual lattice can be described by free fermions with some set of boundary conditions (the number of terms in the spin structure summation is however in general infinite) or by fermions with additional Thirring interactions (see [63] and paper 15). The lattice description has the advantage that all primary fields that occur have an explicit expression in terms of free bosons, whereas in fermionic language one does not have a simple expression for the spin fields or their generalization (a nice example is discussed in [64]).

2.2.8. *Tensor products*

A simple way to obtain solvable conformal field theories for $c > 1$ is to tensor several copies of elements of the discrete series (or other solvable CFT's). The Virasoro generators of the tensor product are simply the sum of the Virasoro generators of the components, and hence the central charge is the sum of the central charges of the components. One obtains automatically a class of extended Virasoro algebras, since the "orthogonal" linear combinations of the Virasoro generators are chiral operators in the theory. For example, if one considers the tensor product of two commuting Virasoro algebras with generators L^1_n and L^2_n and central charges c_1 and c_2, it is easy to see that state $c_2 L^1_{-m} - c_1 L^2_{-m}) |0\rangle$ is annihilated by all the positive modes of $L_n = L^1_n + L^2_n$. Hence all these states correspond to primary fields of the Virasoro algebra, but they become descendants if one adds the $h = 2$ operator $c_2 T^1(z) - c_1 T^2(z)$ (transforming like a conformal field with respect to $T(z)$) to the operator algebra. Obviously the only primary fields with respect to this extended algebra are the products of the primary fields of the original algebras, and they are finite in number if c_1 and c_2 are less than one. On the other hand, the unextended Virasoro algebra of the tensor product clearly has an infinite number of primary fields.

The simplest way to make modular invariant discrete series tensor product is to choose the same factors for the left- and right-moving sectors, and to make each left- and right-moving pair separately modular invariant. The modular invariants for elements of the discrete series have been classified completely [49], and just as for SU(2) Kac–Moody algebras have an A–D–E interpretation. However, when one considers tensor products this is certainly not the most general solution. There is no reason why one should choose the tensor product in a left-right symmetric way. Furthermore there is no reason why all factors should be separately modular invariant. This latter point is demonstrated quite convincingly if one takes as a building block the $c=\frac{1}{2}$ element (corresponding to the Ising model). The tensoring procedure is then equivalent to constructions with free fermions with periodic and anti-periodic boundary conditions, and this problem has a very large number of solutions for large c. The tensor product method has been used for constructing bosonic strings out of the discrete series of minimal CFT's in [59]. Gepner (paper 17) has applied a similar construction, using the discrete series of $N=2$ super-Virasoro representations, to heterotic string construction.

2.2.9. *The bosonic string spectrum*

The space–time sector of a bosonic string consists of d bosons X^μ, plus ghosts b and c. Quantization of a compactified bosonic string works in essentially the same way as for uncompactified strings. The ground state is $c(0)\,|0\rangle$, where $c(z)$ is a reparametrization ghost field and $|0\rangle$ the SL(2, C) invariant vacuum. The states in the theory are created by vertex operators of the form $V(z)\,\mathrm{e}^{ikX(z)}$, where V consists of fields of the internal sector, combined with polynomials of operators of the form $\partial^n X^\mu$ from the space–time sector. The physical states $c(0)V(0)\,\mathrm{e}^{ikX(0)}\,|0\rangle$ of the bosonic string must be annihilated by the BRST operator Q, which has the form [2]

$$Q=\oint\frac{\mathrm{d}z}{2\pi\mathrm{i}}\left[cT+bc\,\partial c+\tfrac{3}{2}\partial^2 c\right]. \tag{2.17}$$

Here T is the energy–momentum tensor T^{int} of the internal sector plus the energy–momentum tensor T^{st} of all d fields X^μ (but not the ghosts). Since $Q\,|0\rangle=0$ this is equivalent to requiring that Q commutes with the operators $cV\,\mathrm{e}^{ikX}$, which leads to the condition that $V\,\mathrm{e}^{ikX}$ must be conformal field of weight one (in the left-moving as well as the right-moving sector).

The BRST conditions on the physical states can be solved explicitly by choosing light-cone gauge, which leads to an expression of ∂X^- in terms of the transverse components X^i,

$$\partial X^-=\frac{1}{2p^+}\sum_{i=1}^{d-2}\partial X^i\,\partial X^i+\frac{1}{p^+}T^{\text{int}}, \tag{2.18}$$

valid for the non-zero modes of this expression. For the zero modes one gets the usual condition $L_0^{\text{st}}+L_0^{\text{int}}-1\,|\text{phys}\rangle=0$, which reduces in light-cone gauge to

$$\begin{aligned}&\left(L_0^{\text{int}}+\mathcal{N}-1\right)|\psi\rangle=\tfrac{1}{2}m_{\mathrm{R}}^2\,|\psi\rangle,\\&\left(\bar{L}_0^{\text{int}}+\bar{\mathcal{N}}-1\right)|\psi\rangle=\tfrac{1}{2}m_{\mathrm{L}}^2\,|\psi\rangle,\\&m^2=m_{\mathrm{L}}^2=m_{\mathrm{R}}^2,\end{aligned} \tag{2.19}$$

where

$$\mathcal{N} = \sum_{n=1}^{\infty} \alpha^i_{-n}\alpha^i_n .$$

The only way in which all this differs from corresponding expressions for uncompactified strings is that the sum on i runs only to $d-2$; the remaining terms ($i = d-1, \ldots, 24$) have been replaced by the internal energy–momentum tensor T^{int} and its modes L_n^{int}. A similar comment holds for the Lorentz algebra. In 26 dimensions one can express J^{i-} in terms of the transverse oscillators via (2.18). Below 26 dimensions one gets in the same way an expression for J^{i-} that involves also T^{int}, replacing some of the transverse oscillators.

A special feature of the space–time sector as opposed to the free bosons encountered in the internal sector is that it is completely determined by physical considerations. A flat space interpretation of the theory implies that every state can appear with any transverse momentum $k^i \in \mathbb{R}^{d-2}$. Hence the vertex operators $V(\boldsymbol{k}) = \mathrm{e}^{\mathrm{i}\boldsymbol{k}\cdot \boldsymbol{X}(\bar{z},z)}$ must be part of the set of mutually local operators in the theory for any value of k^i. The existence of these operators implies that also the operators $\partial_z X$ and $\partial_{\bar{z}} X$ must be present, since they appear in operator products. Hence one is inevitably led to the usual description of the space–time sector in terms of oscillators α^i_n and $\bar{\alpha}^i_n$. Just as for the internal sector, one can describe the primary fields of the space–time sector in terms of the $(\partial_z X^i, \partial_{\bar{z}} X^j)$ charges $(p^i_{\mathrm{L}}, p^j_{\mathrm{R}})$. The vertex operators $V(\boldsymbol{k})$ have $\boldsymbol{p}_{\mathrm{L}} = \boldsymbol{p}_{\mathrm{R}} = \boldsymbol{k}$. Locality of any other primary fields with respect to $V(\boldsymbol{k})$ for arbitrary $\boldsymbol{k}$ implies that $\boldsymbol{p}_{\mathrm{L}} = \boldsymbol{p}_{\mathrm{R}}$, so that nothing new can be added.

This demonstrates that the conventional way to construct the closed, oriented bosonic string and to combine it with the internal sector is indeed the only possible way. The Hilbert space of physical states is simply a tensor product of the internal CFT and a space–time CFT built with of $d-2$ left- and right-moving transverse oscillators. The contribution $\mathcal{P}^{\text{st}}$ to the partition function (2.5) can now be computed exactly as for the internal sector. It is simply the diagonal invariant of the $U(1)^{d-2}$ algebra, written in terms of characters (2.15):

$$\mathcal{P}^{\text{st}} = \int \frac{d^{d-2}k}{(2\pi)^{d-2}} \prod_i \chi_{k_i}(0\,|\,q)\chi_{k_i}(0\,|\,\bar{q})$$

$$= (2\pi\sqrt{2}\,\mathrm{Im}\,\tau)^{2-d}\eta(\tau)^{2-d}\eta(\bar{\tau})^{2-d},$$

which indeed is modular invariant.

The lowest state in the spectrum is the vacuum itself, which is a tachyon with $m^2 = -2$. We have seen earlier that this state is inevitable in a modular invariant theory, and we have chosen its multiplicity, M_{00}, equal to one. The space–time excitations of the vacuum yield a state $\alpha^i_{-1}\bar{\alpha}^j_{-1}$ of zero mass. This tensor product consists of a symmetric traceless tensor (the graviton), an anti-symmetric tensor (usually denoted as B_{ij}) and a trace (the dilaton). Their multiplicity is also equal to M_{00}, which implies first of all that they are present in any bosonic string theory, and furthermore motivates the choice $M_{00} = 1$.

The only other states of zero mass or less that can occur are scalars $|h, h\rangle$, with $h \le 1$, and vector bosons $\alpha^i_{-1}|1, 0\rangle$ or $\bar{\alpha}^i_{-1}|0, 1\rangle$. Every vector boson is automatically a gauge boson of a compact Lie algebra, which in general is some semi-simple algebra plus some U(1)'s. The generators of these gauge algebras are the zero modes of the currents that create the states $|0, 1\rangle$ and $|1, 0\rangle$. The other modes of these currents generate the corresponding Kac–Moody algebras. Conversely, any Kac–Moody symmetry that is realized on the internal sector manifests itself in the spectrum via gauge-bosons.

The scalars, which are tachyons if $h < 1$, can transform in any representation R which satisfies $C(R)/[C(A)+k] < h$, plus condition (2.12) for unitary highest weight representations (we consider simple algebras here; the generalization is obvious). Both conditions become irrelevant for $k \to \infty$, so that there is no fundamental limitation on the representations of massless scalars (however for large k the central charge of the KM algebra approaches dim $\mathscr{G}$, so that arbitrary large values of k are only possible for sufficiently small groups, and sufficiently small space–time dimensions). Of course c and h may receive contributions from additional factors of the internal sector, which with some cleverness may be tailored to complete their values to integers or whatever else one desires.

2.3. Fermionic strings

2.3.1. The bosonic string representation

In constructions of fermionic strings (i.e. heterotic and type-II) strings, one encounters two new problems: the ψ^{μ}, β, γ system, and the presence of an $N=1$ world-sheet supersymmetry. The first of these problems is quite easy to deal with. It implies the following general structure of the partition function for heterotic strings:

$$\mathscr{P}(\bar{\tau}, \tau) = \mathscr{P}^{\rm st} \sum_{\alpha} \mathscr{P}_{\alpha}^{\rm NSR}(\tau) \mathscr{P}_{\alpha}^{\rm ghost}(\tau) \mathscr{P}_{\alpha}^{\rm int}(\bar{\tau}, \tau).$$

Here $\mathscr{P}^{\rm st}$ is identical to the corresponding function for the bosonic string, $\mathscr{P}^{\rm NSR}$ represents the contribution of $d-2$ NSR-fermions ψ^{i}, $\mathscr{P}^{\rm ghost}$ contains the contribution of the remaining two fermions plus the β, γ ghosts, and $\mathscr{P}^{\rm int}$ represents the contribution of the internal sector. The summation is over all spin structures, and the general form of the expression is valid for any genus (with the period matrix denoted as τ). Our main interest is in the contribution of the internal sector. In general, this contribution may be different for different NSR spin structures. Under modular transformations these factors must transform into each other in such a way that the complete expression is invariant.

The main problem is the factor $\mathscr{P}^{\rm ghost}$. It contains not just the ghost and "longitudinal" fermion determinants, but also correlators of picture changing operators that appear due to the supermoduli integration. Fortunately we do not need to know an explicit expression for it, but only its transformation properties. It is reasonable to assume that those are the same for any fermionic string, and this assumption underlies any discussion of higher loop modular invariance that has appeared in the literature. Indeed, the ghost and fermion determinants are universal, and the picture changing operators, even though they appear to have some dependence on the internal sector via the supercurrent (discussed below), have always the same operator products and monodromy properties.

It is in fact convenient to replace $\mathscr{P}_{\alpha}^{\rm ghost}$ by something simpler that transforms in the same way. The most useful replacement is [65]

$$\mathscr{P}_{\alpha}^{\rm ghost} \to \mathscr{P}_{\alpha}^{\rm D_4} \mathscr{P}^{\rm E_8}, \tag{2.20}$$

where the first factor is the character of a level-1 $\rm D_4$ Kac–Moody algebra for spin structure α and the second factor is the character of $\rm E_8$, (level 1). The $\rm D_4$ character $\mathscr{P}_{\alpha}$ is simply the determinant of the Dirac operator on a Riemann surface of 8 Majorana–Weyl fermions with identical spin structure α. The E_8 character is the same for 16 fermions, summed over all 2^{2g} spin structures at genus g. These determinants have a simple expression in terms of θ-functions. Note that the conformal anomaly of the β, γ ghosts plus two fermions is 12, the same as that of a level-1 $D_4 \times E_8$ KM algebra. Hence this replacement preserves both conformal and modular invariance. It maps the non-unitary ghost system to a unitary conformal field theory. The resulting partition function satisfies all the consistency conditions to serve as the partition function of a bosonic string. The effect of (2.20) is to extend the SO($d-2$) KM algebra generated by the transverse NSR fermions to SO($d+6$), while keeping the same correlation with the spin structure labels of the internal sector. Conversely, any bosonic string theory with level-1 SO($d+6$) $\times E_8$ KM algebra (plus a realization of $N=1$ world-sheet supersymmetry, to be discussed below) corresponds to a heterotic string theory. To summarize, for *any* heterotic string there exists a bosonic string which faithfully represents its properties. We will call this the *bosonic representation* of a heterotic string, and it will often be convenient to work in this representation.

There is a similar map for type-II strings. The generic form of a type-II partition function is

$$\mathcal{P}(\bar{\tau}, \tau) = \mathcal{P}^{\mathrm{st}}(\bar{\tau}, \tau) \sum_{\alpha\beta} \mathcal{P}_{\alpha}^{\mathrm{NSR}}(\tau)_{\alpha}^{\mathrm{ghost}}(\tau) \mathcal{P}_{\beta}^{\mathrm{NSR}}(\bar{\tau})^{\mathrm{ghost}}(\bar{\tau}) \mathcal{P}_{\alpha\beta}^{\mathrm{int}}(\bar{\tau}, \tau).$$

Here one has the option of replacing one ghost system to obtain a heterotic string partition function, or both to obtain a bosonic string partition function.

Hence the classification problem of heterotic and type-II strings is contained in that of bosonic strings, in the following way:

$$\text{Bosonic strings} \supset \text{Heterotic strings} \supset \text{Type-II strings}. \tag{2.21}$$

We have discussed the conditions for modular invariance of bosonic strings in the previous section. Once the problem is understood there, one never has to worry about it again for fermionic strings (except to verify the aforementioned assumption about the transformation properties of $\mathcal{P}_{\alpha}^{\mathrm{ghost}}$).

2.3.2. World-sheet supersymmetry

The right-moving sector of heterotic strings, and both sectors of type-II strings should not just form representations of the Virasoro algebra, but of an $N=1$ super-Virasoro algebra. This is an extension of the Virasoro algebra by a current T_{F}, which is a conformal field of weight $\frac{3}{2}$. It must have the following operator product with itself:

$$T_{\mathrm{F}}(z) T_{\mathrm{F}}(w) = \frac{\frac{2}{3}c}{(z-w)^3} + \frac{2T(w)}{(z-w)}. \tag{2.22}$$

The form of T_{F} in the space–time sector is determined by Lorentz invariance, so that the general form of T_{F} is

$$T_{\mathrm{F}}(z) = \mathrm{i}\psi^{\mu}\, \partial X_{\mu} + T_{\mathrm{F}}^{\mathrm{ghost}} + T_{\mathrm{F}}^{\mathrm{int}}.$$

For the ghost part see [2]. Our main interest is in the internal part, which must satisfy (2.22) with $c = 15 - 3d/2$.

Since T_F is a half-integer spin operator with $(\bar{h}, h) = (0, \frac{3}{2})$, it cannot appear in the spectrum itself. Hence one may expect that it has non-trivial monodromy with respect to some field in the theory. The possible twists are restricted by two conditions: they must form an automorphism of (2.22), and T_F must have definite monodromy with respect to all primary fields in the theory. The only possible automorphism is $T_F \to -T_F$, so that the states on which T_F can act fall into two classes, just as in theories constructed with free fermions: those on which T_F acts without generating branch cuts form the Neveu–Schwarz (NS) sector, and those on which T_F has a square root branch cut form the Ramond (R) sector.

If one tensors several $N = 1$ algebras with supercurrents T_F^i and primary fields ϕ_{NS}^i or ϕ_R^i to make a new one with $T_F^{\text{total}} = \Sigma_i T_F^i$ the requirement of definite monodromy implies a restriction on primary fields of the super-Virasoro algebra of the tensor product: Products $\Pi_i \phi_{NS}^i$ of NS fields form the NS sector of the new algebra, and products $\Pi_i \phi_R^i$ form its Ramond sector. On mixed products T_F would give both integer and half-integer powers of $z - w$, so that such fields cannot be allowed. In particular one must pair the NS sectors of the space–time and internal CFT's with each other, and similarly for the R sectors.

The super-Virasoro generator T_F is integer-moded on R states and half-integer moded on NS states. The modes G_r of T_F have the following anti-commutator:

$$\{G_r, G_s\} = 2L_{r+s} + \frac{c}{3}(r^2 - \tfrac{1}{4})\delta_{r,-s}. \tag{2.23}$$

The Ramond sector has a zero-mode operator G_0 which satisfies $G_0^2 = L_0 - c/24$. For unitary representations this implies that in the Ramond sector $h \geq c/24$. This lower bound is saturated by all states that are annihilated by G_0. These states are called the Ramond ground states. There may not be such a state; if there is not supersymmetry is spontaneously broken. This does not mean that the resulting string theory is inconsistent, but it does imply that there are no massless fermions in the spectrum. Note that (2.23) also implies that the SL(2, C)-invariant vacuum $|0, 0\rangle$ is annihilated by $G_{-1/2}$.

Primary fields with respect to the super-Virasoro algebra must be primary with respect to the Virasoro algebra, and have at most a $(z - w)^{-3/2}$ singularity in their operator product with T_F. In the NS sector one gets then

$$\begin{aligned} T_F(z)\phi(w) &= \frac{1}{(z-w)}\phi'(w), \\ T_F(z)\phi'(w) &= \frac{2h}{(z-w)^2}\phi(w) + \frac{\partial_w \phi(w)}{z-w}. \end{aligned} \tag{2.24}$$

Here h is the conformal weight of ϕ (this equation does not hold for $h = 0$, i.e. $\phi = 1$). The additional field ϕ' appearing here is itself a Virasoro primary field of weight $h + \frac{1}{2}$. The effect of the superextension of the Virasoro algebra is thus to combine Virasoro primary fields into supermultiplets, whose members have weights h and $h + \frac{1}{2}$. It is often convenient to introduce an anti-commuting parameter θ and combine the two members of a supermultiplet into a superfield $\phi'(z) + \theta\phi(z)$. The supercurrent forms a supermultiplet with the energy–momentum tensor, of the form $T(z) + \theta T_F(z)$. This streamlines the calculation of operator products, because they can be written down for an entire supermultiplet. See, e.g., [2] for more details.

In the Ramond sector the super-Virasoro primary fields have a $(z - w)^{-3/2}$ singularity in their operator product with T_F, with the exception of the ground states, which have only a

$(z-w)^{-1/2}$ singularity. This is easily seen to imply that all excited states come in pairs that are mapped into each other by G_0. Only the ground states escape this pairing. This pairing plays an important rôle in establishing Lorentz invariance in the chiral sector of the theory. Lorentz invariance forces massive fermions to appear in pairs of opposite chirality. The existence of a chiral partner for every massive fermion is guaranteed by world-sheet supersymmetry, which implies a pairing of all excited states as explained a moment ago. The generator of world-sheet supersymmetry enters into the problem via the Lorentz generator L^{i-}, which contains a term $\int \psi^i \psi^- \propto \int \psi^i T_F$.

2.3.3. Physical states and picture changing

Just as for the bosonic string, the physical states of the fermionic string are subject to the BRST condition. The BRST charge of fermionic strings contains a few extra terms. First of all there is a term just like (2.17), but with an energy–momentum tensor T that also includes the NSR fields ψ^μ and the β, γ ghosts. There are two additional terms: one term of the form $\oint (\mathrm{d}z/2\pi\mathrm{i})\gamma(z)T_F(z)$, plus a pure ghost term that plays no rôle in the following.

The discussion of the consequences of the extra term involving the supercurrent is simplest if one bosonizes the $\beta\gamma$ ghost system as follows [2]:

$$\gamma(z) = \mathrm{e}^{\mathrm{i}\phi(z)}\eta(z); \qquad \beta(z) = \mathrm{e}^{-\mathrm{i}\phi(z)}\,\partial\xi(z).$$

Here ϕ is a boson with opposite metric, $\phi(z)\phi(w) = \log(z-w)$, and η and ξ form an auxiliary system whose properties do not matter in the following.

The canonical form of the vertex operators is now $\mathrm{e}^{-\mathrm{i}\phi}cV\,\mathrm{e}^{\mathrm{i}kX}$. The commutator with the first term in the BRST charge, of the form (2.17), vanishes as before if $\mathrm{e}^{-\mathrm{i}\phi}V\mathrm{e}^{\mathrm{i}kX}$ is a conformal field of weight 1 in the Virasoro algebra of all fields except b and c. Since the factor $\mathrm{e}^{-\mathrm{i}\phi}$ contributes $\frac{1}{2}$ to the weight, $V\,\mathrm{e}^{\mathrm{i}kX}$ must thus have weight $\frac{1}{2}$. The commutator with the second term in the BRST charge vanishes if and only if $V\,\mathrm{e}^{\mathrm{i}kX}$ is a super-Virasoro primary field, or in other words, the lower component of a supermultiplet. Both conditions can be solved explicitly by going to light-cone gauge. They lead to an expression of ∂X^- and ψ^- in terms of the light-cone fields. There is no essential difference between "compactified" and uncompactified strings here, except that the expressions for ψ^- and ∂X^- include fields from the internal sector.

There is a second simple solution to the BRST conditions. One may also consider vertex operators with vanishing ϕ-ghost charge, i.e. $cV'\,\mathrm{e}^{\mathrm{i}kX}$. Now one finds that $V'\mathrm{e}^{\mathrm{i}kX}$ must be the *highest* component of a super-Virasoro primary field with $h=\frac{1}{2}$ (in other words, it must satisfy the second line in (2.24)). In fact, NS vertex operators exist for any integral value of the ghost charge q, that is with any factor $\mathrm{e}^{\mathrm{i}q\phi}$, but the operators are simplest for $q=-1$ and $q=0$.

The ghost charge of a vertex operator $\mathcal{O}_q = \mathrm{e}^{\mathrm{i}q\phi}V_q\,\mathrm{e}^{\mathrm{i}kX}$ can be enlarged to $q+1$ by acting with the *picture changing operator*, which essentially means that one takes for $\mathcal{O}_{q+1}$ the finite terms in the operator product of $\mathrm{e}^{\mathrm{i}\phi}T_F$ and $\mathcal{O}_q$ (for a more precise description, see [2]). In passing from $q=-1$ to $q=0$ this operator acts on the vertex operator $V\,\mathrm{e}^{\mathrm{i}kX}$ exactly like a world-sheet supersymmetry transformation, which produces the other member, $V'\,\mathrm{e}^{\mathrm{i}kX}$, of the supermultiplet. Note that this operation can act very non-trivially on the internal sector. If one uses the same operation to increase q further the result is more complicated, and does not correspond anymore to a simple world-sheet supersymmetry transformation.

Both the $q=-1$ and $q=0$ pictures are needed to compute tree diagrams, since the total ghost charge of the external lines has to add up to -2 (and to $2(g-1)$ for a genus g Riemann surface). For determining the spectrum the $q=-1$ picture is simplest, since only in this picture the vertex operator $V\,e^{ikX}$ is a genuine super-Virasoro primary field. Hence the corresponding states are annihilated by *all* positive modes of T_F (in the $q=0$ picture they are not annihilated by $G_{1/2}$). This allows a simple solution of the constraints by going to light-cone gauge, as mentioned above. In the Ramond sector the analogous canonical choice is $q=-\frac{1}{2}$.

2.3.4. *Heterotic string spectra*

To enumerate all states in the spectrum one should fix a ϕ-ghost charge, and eliminate all states that do not fulfill the BRST condition. We have just seen that the simplest way to achieve that is to choose ghost charge -1 for the NS sector, and $-\frac{1}{2}$ for the R sector.

In the bosonic string representation introduced above this procedure can be performed extremely easily. Consider the $SO(d+6)$ KM algebra that is created from the $SO(d-2)$ light-cone algebra by the map (2.20). One simply decomposes $SO(d+6)$ to $SO(d-2)\times SO(8)$, and keeps only those states that are 8_v or 8_s ground states of the SO(8) Kac–Moody algebra (and that are also ground states of the additional E_8 factor in (2.20)). The states associated with 8_v transform like tensors of $SO(d-2)$, while those associated with 8_s transform like $SO(d-2)$ spinors (see paper 14 for a more detailed explanation). This procedure was first described in [66,67].

One can arrive at the same answer by computing the light-cone partition function. The functions $\mathscr{P}_\alpha^{\text{ghost}}$ can be calculated explicitly at genus 1. One finds

$$\mathscr{P}_3^{\text{ghost}}=-\mathscr{P}_4^{\text{ghost}}=-\mathscr{P}_2^{\text{ghost}}=\pm\mathscr{P}_1^{\text{ghost}},$$

where we have labelled the spin structures by θ-function indices. The relative signs between these functions are consistent with the rule derived above from the canonical ghost charge assignments: they imply that indeed the states associated with 8_v of SO(8) are the bosons that contribute to the one-loop partition function. Furthermore they ensure that space–time spinors contribute with an extra-sign relative to space–time tensors. Thus the correct spin-statistics relation is automatically obtained if one has the correct spin-statistics relation in the bosonic string.

The mass-formula for the fermionic string states is also most easily described in terms of their bosonic string representation: it is simply the bosonic string mass-formula (2.19) (where one should include the contribution $+\frac{1}{2}$ due to the 8_v and 8_s ground states).

To get states of mass zero or less one should have an $SO(d+6)$ KM representation with $h\leq 1$. For heterotic strings, the possibilities are as follows:

Singlet representation

Since it does not contain 8_v or 8_s, the singlet representation does not yield any physical states. In particular, the bosonic string tachyon does not appear in the heterotic string spectrum.

Adjoint representation

This yields a vector representation of $SO(d-2)$, with $m_R=0$. Since the state $|0,0\rangle$ is always a state in the bosonic string spectrum, one automatically has states $\bar{\alpha}^i_{-1}J^a_{-1}|0,0\rangle$, where J^a denotes the $SO(d+6)$ currents that generate the adjoint representation. In the heterotic string these states correspond – after pulling an 8_v of SO(8) out of the $SO(d+6)$ adjoint representation – to the graviton, the dilaton and the anti-symmetric tensor B^{ij}, which are thus as inevitable as they were in the bosonic string. If the bosonic string contains any other $(\bar{h}, h) = (1, 0)$ fields, they generate a KM algebra times some U(1) factors. In the heterotic string one has then states $J^a_{-1}|1, 0\rangle$, which yield the corresponding gauge bosons. Conversely, any massless vector boson that originate from the adjoint of SO$(d+6)$ implies the existence of a $(\bar{h}, h) = (1, 0)$ primary field, and hence all such vector bosons are gauge bosons of a compact gauge group, which in general is a product of simple Lie algebras and U(1)'s, denoted in the following as $\mathscr{G}_\mathrm{L}$.

Vector representation

The SO$(d+6)$ vector representation must appear in combination with primary fields of the internal sector whose h-values are $(\bar{h}, h) = (h + \frac{1}{2}, h)$. The corresponding physical state has mass $\frac{1}{2}m^2 = h - \frac{1}{2}$. All such states are space–time scalars, which are tachyonic for $h < \frac{1}{2}$. Tachyons need not appear; a sufficient (but not necessary) condition for their absence is space–time supersymmetry. Massless scalars are almost always present in the spectrum of heterotic strings below ten dimensions (but never in ten dimensions, since there is no internal sector to add $h = \frac{1}{2}$). In the heterotic string, this is the only way for getting massless scalars, apart from the dilaton (and in four dimensions $B_{\mu\nu}$, which is equivalent to a pseudo-scalar). If the left-moving sector has a KM algebra, these scalars are in some representation of the corresponding gauge group. The possible representations of massless scalars or tachyons are restricted (by the condition $h + \frac{1}{2} \leq 1$) in exactly the same way as for scalars and tachyons in the bosonic string (more about representations of scalars, and particularly the possibility for having scalars in the adjoint representation follows below).

There is another way for getting massless particles from the SO$(d+6)$ vector representation. It may happen that this representation appears in combination with a primary field $(\bar{h}, h) = (0, \frac{1}{2})$ of the internal sector. Although this does not satisfy the constraint $m_\mathrm{L} = m_\mathrm{R}$, one can get physical states by exciting the left-moving sector in one of two ways: by oscillators $\bar{\alpha}^i$, or by KM currents in the internal sector. The former kind of excitation gives rise to space–time vectors, the latter to scalars in the adjoint representation of the KM algebra of the left sector.

Thus we find here a second source for space–time vectors, in addition to the gauge bosons of the left-moving KM algebra. One would expect them to be gauge bosons as well (since that is the only way to make sense of massless vectors in field theory), and indeed they are: their gauge group $\mathscr{G}_\mathrm{R}$ is determined by the right-moving sector.

To see how this works let us represent the aforementioned $(0, \frac{1}{2})$ fields by N free fermions ψ^a, $a = 1, \ldots, N$, assuming that there are N such fields. Let us also denote the $(\bar{h}, h) = (0, \frac{1}{2})$ operators that generate the SO$(d+6)$ vector representation by ψ^A, $A = 1, \ldots, d+6$. Together these operators form new currents $\psi^a\psi^A$ of conformal weight (0, 1) that appear in addition to the KM currents of SO$(d+6)$. Since these new currents transform as SO$(d+6)$ vectors, the consequence is that SO$(d+6)$ is part of an SO$(d+6+N)$, $N > 0$, KM algebra. This enlarged algebra must have level 1, and the embedding of SO$(d+6)$ must have Dynkin index 1.

With respect to the super-Virasoro algebra of the right-moving sector the fermions ψ^a are primary, i.e. they define a $(0, \frac{1}{2})$ superfield. The higher component of this superfield are currents J^a. It is easy to work out the operator products of these fields using superconformal field theory. Normalizing the operators J^a and ψ^a so that they satisfy (2.11) and (2.16) respectively, one gets

$$J^a(z)\psi^b(w) = \frac{1}{z-w}\mathrm{i}f_{abc}\psi^c(w).$$

It follows that the currents J^a generate a KM algebra $\mathscr{G}_R$, and that the fermions ψ^a transform in the adjoint representation of this algebra. This system is called a super-Kac–Moody algebra.

The current J^a consists of two terms: one that acts on the N fermions ψ^a and one that acts on the rest of the internal sector. The former is completely determined by the fact that ψ^a transforms in the adjoint representation. Hence the embedding is $\mathscr{G}_R \subset SO(N) \subset SO(d+6+N)$, where the SO($N$) vector representation contains the adjoint of $\mathscr{G}_R$. The restriction of $\mathscr{G}_R$ to SO($d+6+N$) is therefore a KM algebra of level $C(A)$, the Dynkin index of the embedding ($C(A)$ is defined in section 2.2.4). The embedding of $\mathscr{G}_R$ in the rest of the internal sector is not subject to such a restriction. The possibilities for $\mathscr{G}_R$ and its (total) level are limited by the total available amount of central charge in the internal sector (see paper 23).

An important consequence of the existence of such $(0, \frac{1}{2})$ superfields ψ^a in a heterotic string is that one cannot get chiral fermions. Fermions originate from the spinor representations of SO($d+6$), and are chiral only if this algebra has two inequivalent spinor representations (which is only true if d is even), which belong to *different* representations of one of the gauge groups. If $SO(d+6) \subset SO(d+6+N)$ this cannot happen, since the SO($d+6+N$) spinors contain both chiralities of SO($d+6$) spinors (this argument rules out chiral fermions in $\mathscr{G}_L$ representations. They cannot be chiral with respect to $\mathscr{G}_R$ because $\mathscr{G}_R$ acts on the SO($d+6+N$) spinors via an embedding in the adjoint of SO(N), a real representation).

As we have seen above, one gets scalars in the adjoint representation of the left-moving KM algebra together with these vector bosons, however at the price of loosing chirality. Since adjoint scalars are needed for symmetry breaking in certain unification models, it would be interesting to know if one can get them in chiral theories. The answer is that nothing rules this out rigorously, but they have to be primary fields of the KM algebra: adjoint representations generated by the left-moving currents imply the existence of $(0, \frac{1}{2})$ fields ψ^a, with all the consequences described above. The only KM algebras that do not have adjoint representations as primary field are the ones with $k=1$. Thus only in this case (which however represents the bulk of the theories constructed so far) can adjoint scalars be ruled out rigorously.

The aforementioned problem with chirality can be avoided in type-II string theories. In those theories one can never get gauge bosons whose space–time index is due to a bosonic oscillator. The mechanism described here is the only source of gauge bosons. However, one can construct theories where such a gauge group comes only from one sector, and then the other sector may still yield chiral fermions.

Spinor representations

The spinor representations of SO($d+6$) yield space–time fermions. To get massless fermions one needs a contribution $(\bar{h}, h) = (1, h)$, where $h = (10-d)/16$, from the internal sector. The fields in the internal sector have Ramond boundary conditions, and it follows then that this value of h is precisely the smallest one possible (see (2.23)): world-sheet supersymmetry forbids fermionic tachyons. The gauge representations of massless fermions are restricted in a similar way as those of massless scalars. In heterotic strings these restrictions allow the standard model representations for gauge groups like SU(3) × SU(2) × U(1), SU(4) × SU(2) × SU(2), SU(5), SO(10) and E_6. In practice these groups appear frequently (although the presently available explorations are probably biased towards them), but usually in combination with other groups. Whenever these groups appear, the massless fermions are often in the standard model representation. Most interestingly, this representation usually occurs with a certain multiplicity, typically a number in the range 0 to 100, and some examples are known where it is 3 or 4.

2.3.5. Space–time supersymmetry

Analogous to massless scalars, there is the interesting possibility of having a (0, $(10-d)/16$) field appearing in combination with a spinor of SO($d+6$). Left excitations by $\bar{\alpha}^i_{-1}$ give rise to a gravitino; left excitations by currents $\bar{J}^a_{-1}$ yield spinors in the adjoint representation of the gauge groups: gaugino's.

Just as for scalars, the presence of such (0, $(10-d)/16$) fields in combination with the SO($d+6$) spinors implies the existence of extra (0, 1) currents in the right-moving sector of the bosonic string representation. Thus SO($d+6$) must be an index-1 embedding in a level-1 KM algebra such that some of the roots transform as SO($d+6$) spinors. It is elementary to find all the possibilities, and they are listed in table I (from [68]) (The even-dimensional part of this table appears in paper 14, where the relation between exceptional algebras and space–time supersymmetry was observed in the context of the covariant lattice construction.)

Supersymmetry in space–time can also be explained in terms of extended world-sheet supersymmetry. This description is completely equivalent to the characterization given in table I, but has the disadvantage that a separate discussion is needed for every dimension, and for every number of supersymmetries.

As an illustration, consider the most interesting case, $N=1$ supersymmetry in four dimensions. In the description of table I this corresponds to an embedding $\mathrm{SO}(10) \subset \mathrm{E}_6$, which implies the existence of an additional U(1) current ∂H in the theory, extending the Cartan sub-algebra of SO(10) to that of E_6. It is easy to see that the operator product of this current with the supercurrent T_F generates a new supercurrent T'_F. These two supercurrents generate an $N=2$ world-sheet super-Virasoro algebra (note that only one of these two supersymmetries is local; T'_F generates a global supersymmetry). The current ∂H is the U(1) generator in that algebra. Since the physical states it acts on are defined by E_6 weights, the U(1) charge is quantized in a definite way. This corresponds precisely to the "$N=2$ plus charge integrality" condition of [69].

The massless supermultiplets of $N=1$ supersymmetry are easily seen to correspond to the lowest non-trivial E_6 representations: The chiral multiplets correspond to the (27) and $(\overline{27})$, and the vector multiplet to the adjoint representation (78). For a more detailed discussion of multiplet structures in relation to exceptional group representations, and generalizations to extended supersymmetries see [54,70].

Looking at table I one may have the impression that on every line all entries except the first one are just torus compactifications. This is indeed true as far as the supermultiplet structure is concerned. However, this does not mean that for example an $N=4$ string theory in four dimensions (i.e. a theory of E_8 type) is necessarily a torus compactification (in a "stringy"

Table I
Complete classification of extended supersymmetry in heterotic string theories.

Type	Space–time dimension								
	10	9	8	7	6	5	4	3	2
E_8	1	1	1	2	2	4	4	8	(8,8)
E_7	–	–	–	–	1	2	2	4	(4,4)
E_6	–	–	–	–	–	–	1	2	(2,2)
F_4	–	–	–	–	–	–	–	1	(1,1)
Spin (N)	–	–	–	–	–	–	–	–	$(N-8,0)$ or $(0,N-8)$

sense) of a ten-dimensional theory. A counterexample to this tempting conjecture can be constructed as follows *: take the ten-dimensional $E_8 \times E_8$ ten-dimensional heterotic string, compactify it to four dimensions on the $(D_6)_L \times (D_6)_R$ Lorentzian self-dual lattice, and make a twist that interchanges the two E_8 factors, combined with a shift by a vector weight of $(D_6)_R$ (note the analogy with the construction of the E_8 string in paper 4). The resulting theory satisfies all consistency conditions, has $N=4$ space–time supersymmetry, but has a gauge group of rank 14 (namely E_8(level 2) $\times B_6$), whereas any torus compactification has a gauge group of rank 22.

One would expect a string theory with gravitons in its spectrum to be supersymmetric at all levels. For all supersymmetries in table I that are associated with exceptional algebras this is indeed easy to prove. One simply notes that physical states are obtained by projecting on the (8_v) and (8_s) representations of an SO(8) sub-algebra, and that SO(8) triality becomes an inner automorphism when SO(8) is embedded in an exceptional algebra. For more details see [68,54]. This implies vanishing of the one-loop cosmological constant for all superstrings. At higher loops less rigorous statements can be made at present, but one can derive identities, similar in structure to Riemann identities, that are satisfied by the internal sector partition functions of all superstrings bases on exceptional algebras [68]. At genus one, such identities have also been obtained in [71,72].

(2, 2) theories

A generic heterotic string has no left-moving supersymmetries and one local right-moving one. One can denote this situation as (0, 1) world-sheet supersymmetry. It has become customary to denote also the presence of additional global supersymmetries in this way. For example, the four-dimensional heterotic strings with $N=1$ space–time supersymmetry have an additional (global) world-sheet supersymmetry, and are therefore called (0, 2) theories. (Unfortunately, the notation that is used in the literature does not just indicate the number of global world-sheet supersymmetries, but refers also to the charge integrality condition. Many non-supersymmetric string theories have an $N=2$ world-sheet supersymmetry.)

One can also formally define global supersymmetries associated with the left-moving sector. The best way to do that is to think of heterotic strings as indicated in (2.21). A very small subset of the class of heterotic strings can be regarded as the image under the bosonic string map of a type-II string. Type-II strings are only consistent if they have at least a local (1, 1) supersymmetry. Since the internal sector of the type-II string is not affected by the map, this property is preserved. This defines a small class of heterotic strings with (1, 1) world-sheet supersymmetry, which always have a gauge group containing at least $SO(d+6) \times E_8$. The existence of a world-sheet supersymmetry, inherited from the type-II string, implies that the Ramond and Neveu–Schwarz states of $SO(d+6)$ are paired respectively with the Ramond and Neveu–Schwarz states of the internal sector of the type-II string.

If the heterotic string has $N=1$ space–time supersymmetry in four dimensions, the world-sheet supersymmetry is promoted to (1, 2). If furthermore the SO(10) gauge group is a sub-algebra of $(E_6)_L$, the heterotic string is said to be a (2, 2) theory. In that case the type-II theory of which it is the image has $N=2$ supersymmetry, with one gravitino coming from the left- and one from the right-moving sector. When the heterotic string is mapped to a bosonic string, such a bosonic string has an E_6 gauge group in the left- as well as the right-moving sector.

* With thanks to L. Dixon for a bitnet-discussion from which this example resulted.

It should be clear that nothing requires the left-moving supersymmetries to be present. However, when they are present, they lead to some interesting consequences. Because of the $N=1$ world-sheet supersymmetry on the left, one can consider the formal equivalent of picture changing. In the right-moving sector a vertex operator and its picture changed version describe the same physical state. However, on the left the equivalent of the picture changing operator is a seemingly innocuous chiral operator with $\bar{h}=2$. It has the form $\psi^a(\bar{z})T_F(\bar{z})$, where ψ^a is a fermion in the vector representation of SO($d+6$), and T_F is the supercurrent in the left-moving internal sector of the type-II string.

In the left-moving sector this "picture changing" operator takes physical states into different physical states. This is especially interesting if massless states are mapped to other massless states. In four dimensions it is easy to see that the adjoint of E_6 (the gauge bosons and gluino's in the spectrum) is mapped to massive states only. However, a massless (27) or $(\overline{27})$ is always mapped to a massless scalar.

In an $N=1$ heterotic string with $(E_6)_L$ gauge group, a left-moving (27) or $(\bar{2}7)$ with $m_L=0$ can only correspond to a physical state if it appears in combination with a chiral $N=1$ multiplet (i.e. a (27) or $(\bar{2}7)$ of $(E_6)_R$ in the bosonic string representation) of the right-moving sector with $m_R=0$. These massless chiral multiplets are the "families" and "anti-families", consisting of fermions and scalars in the (27) and $(\bar{2}7)$ of $(E_6)_L$. Because of "picture changing" in the left-moving sector, each such family or anti-family is accompanied by a massless chiral multiplet, which is a singlet of $(E_6)_L$.

The scalars in these supermultiplets are called the *moduli*, a name that originates from Calabi–Yau compactification. They correspond to continuous parameters of the space of conformal field theories. This means that if one gives a vacuum expectation value to any of these scalars one obtains another consistent (2, 2) string theory. Indeed, as one might expect, the potential of the moduli is flat to all orders in string perturbation theory. For reviews and references, see e.g. [31,73,74]. A discussion of interesting recent work on Kähler potentials of the moduli can be found in [75–77]. Although the discussion of continuous deformations of string theories (or conformal field theories) appears to be simplest for (2, 2) theories, such deformations exist for (0, 2) theories (see e.g. [78–80]).

Calabi–Yau compactifications provided the first examples of four-dimensional (2, 2) theories, and here the moduli have a simple physical interpretation: they correspond to zero modes of the metric, and parametrize the moduli space of deformations of the manifold. Gepner has conjectured that in fact all (2, 2) heterotic strings correspond to points in the moduli space of some Calabi–Yau manifold (paper 18). The existence of moduli supports this conjecture. In addition, many (2, 2) spectra have been identified with Calabi–Yau spectra, (see paper 18, [81–84]), while also the Yukawa couplings have been checked for one identification [85]. The partition function of (2, 2) models can of course be expressed in terms of characters of $N=2$ super-Virasoro algebras; see [86] for a review of this subject.

2.3.6. *Type-II spectra*

Almost all aspects of type-II spectra follow in a straightforward way from the discussion of heterotic strings given above. To get the massless states (and tachyons, if any are present) one simply combines the adjoint, vector and spinor representation of SO($d+6$) in all possible ways. The left–right combination of the adjoint representations yields the usual graviton, dilaton and $B_{\mu\nu}$, which are again inevitable. Combinations of a left-moving SO($d+6$) adjoint with a right-moving SO($d+6$) vector plus $(\bar{h}, h)=(0, \frac{1}{2})$ superfield yield gauge bosons (and analogously with left and right interchanged). The possible gauge groups are thus exactly the same as the ones that can be obtained from the *right*-moving sector of a heterotic string. Such gauge bosons do not have to be present, but if they are present in one chiral sector they prohibit chiral fermions from that sector. Consequently, the only chance of constructing chiral type-II strings in four dimensions (where a gauge group must be present in a chiral theory) is to have a gauge group in only one sector.

There is in fact another source of vector bosons in four dimensions, namely the tensor product of a left- and right-moving massless spinor. However, it can easily be shown that these vector bosons do not couple to any other field (including themselves) in the low-energy limit (see papers 22 and 23). In particular they are thus gauge bosons of a set of U(1)'s.

The possibilities for supergravity in type-II strings can be read off from table I by taking all possible combinations of extensions of $SO(d+6)$ on the left and the right. In four dimensions all possibilities can in fact be realized in a consistent string theory.

2.3.7. *Anomalies*

As a consequence of modular invariance, the fermion polygon anomalies of all chiral fermions in all theories discussed here factorize exactly in the same way as Green and Schwarz [87] discovered in ten dimensions:

$$A(F,\ R) = (\mathrm{Tr}\ F^2 - \mathrm{Tr}\ R^2)[X_{d-2}(F,\ R)], \tag{2.25}$$

where $A(F,\ R)$ denotes the usual character polynomial which parametrizes the anomaly [88] and X_{d-2} denotes a polynomial of order $(d-2)/2$ in the gauge and curvature two-forms F and R. The traces are in the vector representation, or normalized to the vector representation.

This result was proved in [89] for theories constructed out of free fermions or bosons. It was pointed out in [90] that it holds for all gauge symmetries corresponding to Kac–Moody algebras, because of the modular transformation properties of the Weyl–Kac character formula. Since the extension to additional U(1)'s is trivial, the factorization of the anomaly has now been established for all string theories discussed in this book.

The remaining anomaly is cancelled by the cooperation of two terms in the effective action, both involving the rank-2 anti-symmetric tensor $B_{\mu\nu}$. One, the Chern–Simons term, appears at string tree level and can be calculated quite straightforwardly for any string theory in any dimension. This term is always the same as in ten dimensions. The other term appears at one loop order in string perturbation theory, and depends non-trivially on the details of theory. It has been calculated in [91], but only for one kind of heterotic string construction, the covariant lattice construction of paper 14. Needless to say, the result confirms the expectation of Green and Schwarz, and the anomalies cancel. It should be easy to redo this calculation with abstract Kac–Moody generators instead of the Frenkel–Kac operators used in [91], in order to establish also this part of the calculation for the entire class of heterotic strings (anomalies in type-II theories can be discussed in an analogous way).

An interesting phenomenon related to anomaly cancellation occurs in four dimensions. Here the factor X_2 in (2.25) can only be $\mathrm{Tr}\, F$, and hence it can only be non-vanishing if some U(1) generator in the theory is not traceless. If this term is present, the corresponding anomaly cancelling term triggers a Higgs mechanism, so that the U(1) gauge bosons become massive by absorbing the $B_{\mu\nu}$ field (which is a scalar in four dimensions) [92,93]. In a supersymmetric theory, this is a first indication of supersymmetry breaking, and indeed it can be shown that in such theories supersymmetry is broken by a Fayet–Iliopoulos D-term [94]. At two loops this leads to a non-vanishing cosmological constant [95], and at still higher loops presumably to infinities. Hence four-dimensional theories with such non-traceless U(1)'s are not good string theories. This is in fact the only known example where a string theory that is supersymmetric at tree level develops problems in perturbation theory.

Fortunately there are many four-dimensional strings that do not have this problem. Obviously it does not occur for theories with only simple factors in their gauge groups. Less obvious is the fact that it cannot occur in any string theory that has a level-1 E_8 factor in its spectrum (which includes all (2, 2) theories). The reason is that modular invariance dictates that the traces in (2.25) must be over the entire gauge group, including the E_8 factor. But there is no non-trivial E_8 representation that can have a U(1) charge and still have $h = 1$.

3. The Birth of Superstrings

It is an inevitable consequence of the scope of this collection that most of the papers in it were written after 1984. This certainly does not do justice to older work, in which the foundation for all these developments was laid. It is unfortunately not possible to include all relevant older papers within the page limits of this volume, but it would be unfair to bypass this era completely.

Fermionic strings first appeared in the literature in 1971, with papers by Ramond and Neveu and Schwarz [96], who introduced world-sheet fermions with a space–time index. These two papers discuss the same two-dimensional theory, but with different boundary conditions for the fermions. At about the same time, world-sheet fermions were also discussed by Bardakci and Halpern [97].

A consistent interacting theory must have both Neveu–Schwarz and Ramond fermions, which give rise to space–time bosons and fermions respectively. Consistency of interacting strings imposes further constraints. These were first described by Gliozzi, Scherk and Olive (GSO) (paper 1), who observed that a "G-parity" projection on all the states in the spectrum solved simultaneously several known problems. They also observed that their projection made the theory supersymmetric in space–time (and hence tachyon-free), at least in the sense of having equal numbers of bosons and fermions at each level. With this paper the superstring had come to life, but no further work on it was done for about five years.

Although the GSO-projected theory appeared to be supersymmetric, one did not have a manifestly supersymmetric formulation. This problem was solved by Green and Schwarz (paper 2 and [98]). In paper 2 the two kinds of closed superstrings (type-IIA and type-IIB) are introduced. Perhaps this can be regarded as a first example of the notion of treating left- and right-movers independently, which plays a crucial rôle in heterotic strings. The latter were constructed in paper 3, which marks the end of the first era of superstrings, and the beginning of the second. Interacting heterotic strings are discussed in part II of this paper [99].

Another important line of development in early string history is that of modular invariance. The need for modular invariance in bosonic string loop amplitudes was realized by Shapiro [100]. In two remarkable papers [101] Nahm initiated the use of modular invariance as a restriction on new string theories (in the second of these papers he obtains in fact the partition function of the left-moving sector of the heterotic string). Modular invariance is also the key to the understanding of GSO projections, although it appears that this was not realized until recently (paper 6; see also [102]).

There are many other papers that introduced important new ideas. In [103] a mechanism for constructing $SU(N)$ algebras in terms of string variables is described. This is the first example of what later became known as the "Frenkel–Kac mechanism" [104] (see also [45]). There have been several attempts to reduce the number of dimensions in string theory. For example, in [105] it is observed that one can obtain four-dimensional strings by including in the bosonic string an internal sector consisting of 44 fermions. This is of course an important ingredient in the fermionic constructions that have become popular more recently, but not the only ingredient; the conditions due to modular invariance were not discussed until 1986. There was also work on compactification of the ten-dimensional theories (see e.g. [106]) and in particular on torus compactification [107]. Undoubtedly there have been many other relevant contributions that were forgotten, and were rediscovered more recently. In reading some of these older papers one gets the impression that many interesting ideas were not pursued more vigorously because they did not seem to help the application of string theory to hadron physics. Perhaps this is a lesson that is worth remembering.

Nuclear Physics B122 (1977) 253–290

SUPERSYMMETRY, SUPERGRAVITY THEORIES AND THE DUAL SPINOR MODEL

F. GLIOZZI ★ and J. SCHERK
Laboratoire de Physique Théorique de l'Ecole Normale Supérieure ★★

D. OLIVE
CERN, Geneva
and
The Niels Bohr Institute, Copenhagen

Received 14 January 1977

We study the connection between the dual spinor model and supersymmetry. We show that in the low-energy region, the dual spinor model yields a supersymmetric Yang-Mills theory with O(4) internal symmetry and a supergravity theory with O(4) internal symmetry.

1. Introduction

The non-renormalizability of the V–A theory of weak interactions has led to a long search for renormalizable theories. The problem was not solved by finding a new method to deal with the divergences of the V–A theories. Rather, it was solved by considering the V–A theory as a special limit ($M_W = \infty$) of a unified gauge theory of weak and electromagnetic interactions [1]. The success of this approach was due to the theoretical proof that the gauge theories are renormalizable [2] and to the experimental fact that the most striking predictions of these models, namely the existence of neutral currents, and of charm [3], turned out to be true.

The same techniques used to quantize gauge theories were then applied to Einstein's theory of gravity [4]. It was then found that one-loop diagrams in pure gravity were on-shell finite. However, as soon as interactions with matter fields were included, one obtained divergent, and non-renormalizable results [4].

It was then thought that the problem could be solved by making gravity more symmetric, thus including more fields in the gravitational multiplet. This led to the idea, and to the realization, of supergravity [5] $(2, \frac{3}{2})$ and of complex supergravity

★ Permanent address: Istituto di Fisica Nucleare, Sezione di Torino, Italy.
★★ Postal address: 24, rue Lhomond, 75231-Paris, France.

[6] $(2, \frac{3}{2}, \frac{3}{2}, 1)$. It is remarkable that in these models, one-loop on-shell amplitudes do turn out to be finite [7]. One can imagine more complex supersymmetric systems including the graviton. However, in this way one produces only a gravitational-like sector, as there is only one coupling constant, κ (the gravitational constant) present in such models. One is then led to introduce other multiplets of fields containing a Yang-Mills field [8–10], or scalar fields [10,11] in order to represent matter. In this case one recovers the faults of ordinary Einstein theory in interaction with matter. For instance, the locally supersymmetric system $(2, \frac{3}{2})$ coupled to $(1, \frac{1}{2})$ [8,9] turns out to give non-renormalizable one-loop divergences [12]. A possibility is that by making the theory even more symmetric, one will get convergent results even in the matter-gravity interactions.

Another line of thinking is to consider supergravity and its possible refinements as only a first step. For instance nothing guarantees that 2-loop diagrams in such a scheme will be finite [13]; interaction with matter may also not work. Then there would be an obvious parallelism between the V–A theory with coupling constant G_F and the gravitational theory with coupling constant $G_N \sim \kappa^2$ since both constants have the same dimensions. In the renormalizable gauge theories G_F is a phenomenological parameter, expressed as $G_F \sim g^2/M_W^2$ where g is the real dimensionless coupling constant of the theory and M_W^2 play the role of a cut-off. So one may ask whether one could find a similar expression for $G_N \sim (g^2)^n/\Lambda^2$ and identify g and Λ.

Dual models * provide an answer to this kind of question. The problem of obtaining a cut-off Λ for gravity is far from obvious, as there already is a gauge boson (the graviton) associated with gravity. In the dual model there is a sector (open strings) which is essentially a short distance modified Yang-Mills theory [15], the distance $\Lambda^{-1/2}$ being $\alpha'^{1/2}$, α' being the slope of the Regge trajectories. Then a sector (closed strings) containing a spin-2 massless graviton, is obtained as a bound state of the open-string sector, already from the one-loop diagrams. Then $n = 2$, and $\Lambda^{-1/2}$ is essentially of the order of Planck's length **.

As there are extra dimensions to be compactified *** Λ is in fact expressed both in terms of α' and of the volume associated with the compactified dimensions so that the model contains more than one parameter of the dimension of a length, thus giving a certain amount of freedom.

Further, we would like to show that dual models are in accordance with the program of supergravity. As there is only one dual model containing both bosons and fermions (the dual spinor model) [18], we consider its supersymmetric properties. It is already known that the dual spinor model can be obtained from a classical action, which is locally supersymmetric invariant on the two-dimensional surface span-

* For general reviews on dual models, see ref. [14].

** For the interpretation of dual models as theories of fundamental interactions rather than models of hadrons, see ref. [16].

*** For compactification of extra spacial dimensions in dual models or in field theory, see ref. [17].

ned by the string [19]. Here we give arguments why the spinor dual model of open strings should be globally supersymmetric in the 10-dimensional space-time in which the model is consistent. We show that in the limit where both the slope and the size of the 6-dimensional compact go to zero, the dual model of open strings turns into a hypersymmetric Yang-Mills theory (1 spin 1, 3 scalars, 3 pseudoscalars, 4 Majorana spin-$\frac{1}{2}$) with SU(4) symmetry. A model with the same field content was studied by Fayet [20]. Similar arguments apply for the sector of closed strings. In the same limit, we show that the dual model must yield a hypergravity theory with internal SU(4) symmetry, containing at least 1 spin-2, 3 spin-1 vectors, 3 spin-1 axial vectors, 2 scalars, 4 spin-$\frac{3}{2}$ (Majorana), 4 spin-$\frac{1}{2}$. The dual model *implies* the existence of this field theory which has not yet been written down. To make contact with already known theories, we suppress some of the components present in the dual model spectrum and obtain two known theories:

(a) supergravity in interaction with a massless supermultiplet $[(2, \frac{3}{2}), (\frac{1}{2}, 0)]$ [10,11];

(b) complex supergravity [6] $(2, \frac{3}{2}, \frac{3}{2}, 1)$.

Most of these results were announced in ref. [21]. The correspondence between supergravity and the dual spinor model was also conjectured by Schwarz [9]. The most important result of this study is to reveal a hierarchy of theories.

The hypersymmetric theories can be labelled with SU(N) in the case one takes all generators of the symmetry group, including those which contain γ^5, or by O(N) if those are excluded. By order of inclusion, we have the following list:

Theories of gravity. Einstein's theory; supergravity [O(1)] [5]; complex supergravity [O(2)]; O(4) supergravity (yet to be worked out); supergravity in 10 dimensions; the dual spinor model of closed strings (κ, α').

Theories of matter. Yang-Mills theory; supersymmetric Yang-Mills theory [22] [O(1)]; supersymmetric Yang-Mills with O(2) [22,23]; supersymmetric Yang-Mills with O(4), first presented here; supersymmetric Yang-Mills in 10 dimensions; the dual spinor model of open strings (g, α').

The final unification of matter and gravity appears only at the last stage, as closed strings are bound states of open strings.

As far as renormalizability is concerned, theories of gravity always give finite one-loop results. This is true even for the dual spinor model of closed strings, as the same arguments used by Shapiro [24] for the Virasoro-Shapiro model [25] must apply. The same topological arguments should prove that the N-loop diagrams are also finite for closed strings.

For open strings, one-loop diagrams give rise to a divergence. There are arguments that the divergence can be removed by a slope renormalization [26], but the subject is still under study.

In sect. 2, we discuss the spectrum of the dual spinor model and give arguments for its global supersymmetry in 10 dimensions. In sect. 3, we discuss its connection with the Yang-Mills theory with fermions in 10 dimensions, which is shown to be supersymmetric. Out of it we obtain the hypersymmetric Yang-Mills theory in 4 dimensions with internal O(4)/SU(4) symmetry. It appears that even if one is interested only in that model, the 10-dimensional notation is more compact and elegant. In sect. 4 we define and study the spectrum of the closed string sector of the dual spinor model. In sect. 5 we establish the correspondence between supergravity theories and the dual model of closed strings.

2. The spectrum of the dual spinor model

2.1. The spectrum of the Neveu-Schwarz model

The model is defined out of two sets of harmonic oscillators:

$$\text{Bose set:} \qquad [a_n^\mu, a_m^{\dagger\nu}]_- = \eta^{\mu\nu}\delta_{mn} \qquad n, m \text{ integers,} \tag{2.1}$$

$$\text{Fermi set:} \qquad [b_r^\mu, b_s^{\dagger\nu}]_+ = \eta^{\mu\nu}\delta_{rs} \qquad r, s \text{ half integers.} \tag{2.2}$$

$\eta^{\mu\nu} = (-{+}{+}{+})$, and $-$ and $+$ denote commutators and anticommutators, respectively.

The states $|\psi\rangle$ obtained by applying creation operators on the vacuum $|0\rangle$ do not form a Hilbert space because of the indefinite metric. However, one can define a sector of states, called the physical sector, which has semi-definite positive metric. They are defined by an infinite number of gauge conditions:

$$L_n|\phi\rangle = 0 \qquad n > 0\,, \tag{2.3}$$

$$G_r|\phi\rangle = 0 \qquad r > 0 \quad r \text{ half integer,} \tag{2.4}$$

$$(L_0 - \tfrac{1}{2})|\phi\rangle = 0\,, \tag{2.5}$$

The last equation being the mass shell condition. The gauge operators G_r, L_n form an infinite graded Lie algebra.

The mass shell condition gives a quantization condition for the masses:

$$\alpha' m^2 = n \qquad n = -\tfrac{1}{2}, 0, +\tfrac{1}{2}, 1 \ldots \text{etc.} \tag{2.6}$$

The ground state is thus a tachyon. More generally one can distinguish two sectors of states labelled by the eigenvalues of the operator $G = (-1)^{(\Sigma b_m^\dagger b_m - 1)}$.

The sector with $G = -1$ has $\alpha' m^2 = n$, $n = -\frac{1}{2}, \frac{1}{2}, \frac{3}{2}, \ldots$ and contains the tachyon, represented by $|0\rangle$;

The sector with $G = +1$ is tachyon free and has integrally quantized masses: $\alpha' m = n$, $n = 0, 1, 2, \ldots$.

The ground state of the $G = -1$ sector is a massless vector, represented by

$$b_{1/2}^{\dagger\mu}|0\rangle\,\epsilon_\mu(p)\,, \qquad p^2 = 0\,.$$

Because of the constraint (2.4) with $r = \frac{1}{2}$, $\epsilon_\mu(p)$ must satisfy the Lorentz condition: $p \cdot \epsilon = 0$. Strictly speaking, this is not a gauge condition as it can be derived from the free Maxwell action for on-shell photons (see ref. [10]). This equation decouples the time component of the photon polarization vector ϵ_μ and guarantees the absence of the negative metric states associated with it.

The physical states of the dual model contain amongst themselves some zero norm states: they are both "physical" (i.e. they satisfy eqs. (2.3)–(2.5) and spurious, i.e. can be written in the form

$$\sum_{n\geq 1} L_{-n}|\phi_n\rangle + \sum_{r\geq 1/2} G_{-r}|\phi_r\rangle) .$$

These states always decouple in the dual S-matrix because the external states are physical and the L_n, G_r operators have nice commutation relations with the vertices. In particular a very important property of the dual S matrix is that an external photon polarization vector can be subjected to an on-shell gauge transformation: $\epsilon_\mu(p) \to \epsilon_\mu(p) + \lambda p_\mu$ without affecting the S matrix. This is because the variation of the photon state is

$$p_\mu b^{\dagger\mu}_{1/2}|0\rangle\lambda = G_{-1/2}|0\rangle\lambda \tag{2.7}$$

and such a zero norm state is decoupled by the above arguments. Thus on-shell gauge invariance, together with the Lorentz condition proves that only the transverse components of the photon are coupled in the dual model, at all orders in α'. The number of physical degrees of freedom of the photon are then $D-2$.

All these statements hold for $D \leqslant 10$. For $D = 10$, the zero norm states are more numerous and one can show that not only is the photon transverse, but also all the states in the dual model. One can then reformulate the model in such a way that oscillators with only $D - 2 = 8$ Lorentz indices appear, so that the no-ghost theorem is trivial, but Lorentz invariance is less obvious (but true).

Finally if a U(N) group is introduced *via* Chan-Paton factors, all states of the NS model belong to the regular representation of U(N) and are N^2 degenerate.

2.2. *The spectrum of the Ramond model*

The model is defined out of two sets of oscillators:

$$\text{Bose set:} \qquad [a^\mu_n, a^{\dagger\nu}_m]_- = \eta^{\mu\nu}\delta_{mn} \qquad m, n \text{ integers,} \tag{2.8}$$

$$\text{Fermi set:} \qquad [d^\mu_n, d^{\dagger\nu}_m]_+ = \eta^{\mu\nu}\delta_{mn} \qquad m, n \text{ integers.} \tag{2.9}$$

The physical states satisfy an infinite number of gauge conditions:

$$F_n|\phi\rangle = L_n|\phi\rangle = 0 \qquad \text{for all } n \geqslant 0 . \tag{2.10}$$

The $n = 0$ equations give the mass-shell condition and a Dirac-like equation. The algebra of the F_n, L_n gauges differs only by its c-numbers from the Neveu-Schwarz case.

The masses are quantized through integers:

$$\alpha' m^2 = n , \qquad n = 0, 1, 2, \ldots . \tag{2.11}$$

In particular in such a model there is no tachyon. The ground state is $|0\rangle u$, where u is a (commuting) spinor satisfying the Dirac equation:

$$\not{p}u = 0 . \tag{2.12}$$

One can also define two sectors in the Ramond model, according to the eigenvalues of the operator

$$\tilde{\gamma}^{D+1} = \gamma^{D+1}(-1)^{\Sigma d_n^\dagger \cdot d_n} , \tag{2.13}$$

where $\gamma^{D+1} = \gamma^0\gamma^1 \ldots \gamma^{D-1}$. The Dirac γ matrices are conventionally defined to be in the representation of lowest dimension ($2^{D/2}$) of the Clifford algebra

$$[\gamma^\mu, \gamma^\nu]_- = 2\eta^{\mu\nu} \tag{2.14}$$

in D dimensions (one time, $D-1$ space dimensions, D assumed even). Then one can show that

$$(\tilde{\gamma}^{D+1})^2 = (-1)^{(D/2-1)} . \tag{2.15}$$

Left and right handed states can be defined for any even D.

Just as in the Neveu-Schwarz model, physical states are ghost free for $D \leqslant 10$, and for $D = 10$ only transverse (positive norm) states are coupled. The difference does not affect the ground-state spinor, which is not a gauge field.

Finally, under the U(N) group the fermions can belong to a different representation to that of the bosons. In the following, we shall see that assigning them to the regular representation of U(N) solves a problem one meets in computing fermion-fermion scattering.

2.3. Supersymmetry and the spectrum of the Neveu-Schwarz-Ramond model

2.3.1. Three problems of the NSR model

We want to review three problems which arise out of the calculations of the fermion-fermion scattering amplitudes, and how these three problems are solved by a supersymmetric assignment of the ground states.

(i) The first problem has to do with fermion number conservation. When computing the graphs of figs. 1(a), 1(b) depicting fermion-fermion scattering, one finds that the spectrum of mesonic states exchanged in the s channel (namely the Neveu-Schwarz model) is almost identical to the spectrum of bound states one finds by duality in the t channel. In fact the only difference is that the states with $G = -1$ exchange parity between s and t channels. If one wants to define a fermion number in the conventional way, one thus finds that for instance at $m = 0$ there are three photons, with fermion numbers 0 and ± 2. It is likely that if one could construct amplitudes with an arbitrary number of fermions, each state would appear infinitely de-

a b

Fig. 1. (a) The duality equation for fermion-antifermion scattering. (b) The duality equation for fermion-fermion scattering. The ~ sign indicates that the equality holds up to parity reversal of the $G = -1$ states.

generate with respect to the fermion number. This is clearly bad for a naive quark-gluon interpretation of that model.

A possible solution, proposed in ref. [17], was to quantize in the extra directions the momenta of the fermions in a way which would depend upon the fermion number: for example

$$p^4 = \frac{2\pi}{R}(n + \alpha Z) , \tag{2.16}$$

n being an integer, Z the fermion number, R an arbitrary radius, α an arbitrary and small parameter. This works well for small Z, but when $n + \alpha Z$ approaches an integer, the contribution of p^4 to the mass falls again to zero. Another solution, somewhat unnatural (but maybe consistent) is to force the model to conserve fermion numbers by considering only graphs of the type of fig. 1a rather than those of fig. 1b.

(ii) A second problem, related to the first is the "representation explosion". As s and t channels are degenerate one may expect (as is also based on duality arguments) that the same mesons will turn up as bound states of $2n$ fermions. If one wants to put the fermions in the spinorial representation of the U(N) group (for instance), an explosion of representations is bound to take place. The only known way to prevent this explosion is to put the spinor in the regular representation of the U(N) group, just like the bosons, and to use Chan-Paton factors. Then only the regular representation of U(N) appears as bound states, a well-known property of the Chan-Paton method.

(iii) The last problem is the strange doubling in parity of the $G = -1$ sector of the NS which distinguishes the two sides of a fermionic string. Because of this distinction, twisting fermionic strings is impossible and loops such as the one of fig. 2 is forbidden. Last but not least, the $G = -1$ sector now contains two tachyons, both of which are undesirable [27].

A nice way to solve the two first problems is to choose the fermions to be Majorana, and to belong to the regular representation of the U(N) group. This does not mean that fermion number conservation is to be abandoned, but rather that it has to be obtained in a different way, just as in ordinary supersymmetry [28].

The last problem is solved by making the ground state fermion left-handed. Since the sector of the NS model with $G = -1$ changes handedness, it is decoupled through

Fig. 2. The non-planar fermionic loop, which should give fermionic closed string bound states.

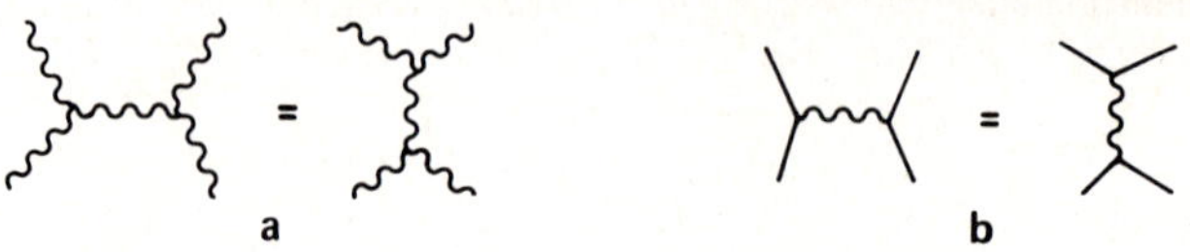

Fig. 3. (a) Duality property for boson-boson scattering. (b) Duality property for Majorana-Weyl fermion-fermion scattering.

such a choice and the tachyon disappears, and with it the problem of identifying the two sides of a fermionic string.

The correct model to be considered thus seems to be the $G = +1$ sector of the NS model together with the Weyl (left-handed) and Majorana (real) sector of the Ramond model. It seems indeed to be trouble-free (no tachyons, no ghosts).

In such a model fermion-fermion and boson-fermion scattering have exactly the same duality properties (see fig. 3).

It may look surprising to see the Weyl condition arising. On the other hand, it is worth noticing that in 4 dimensions for massless spinors, the Majorana and Weyl conditions are equivalent. In the real representation of the γ matrices a Majorana fermion is real. Let us consider a complex Weyl spinor defined by

$$\gamma^5 \nu(x) = i\nu(x) . \tag{2.17}$$

Decomposing $\nu(x)$ into Majorana (real) spinors one has

$$\nu(x) = \sqrt{\tfrac{1}{2}}(\lambda(x) + i\chi(x)) , \tag{2.18}$$

and

$$\chi = \gamma_5 \lambda \tag{2.19}$$

is equivalent to the Weyl condition. The Lagrangian of a "neutrino" field $\nu(x)$ is

$$\mathcal{L} = -\tfrac{1}{2} i\bar{\nu}(x)\gamma^\mu \overleftrightarrow{\partial}_\mu \nu(x) . \tag{2.20}$$

After a brief calculation using (2.19) it can be recast into

$$\mathcal{L} = -i\bar{\lambda}(x)\gamma^\mu \partial_\mu \lambda(x) , \tag{2.21}$$

which is the Lagrangian for a massless Majorana field $\lambda(x)$.

So for $D = 4$, the Majorana and Weyl conditions can be transformed into each other, and except for $m \neq 0$ one can use one formalism or the other. Note that for 4 dimensions imposing Majorana *and* Weyl on a spinor would imply that it vanishes.

2.3.2. Existence of a supersymmetric assignment of the NSR ground states

We now examine the problem of whether Majorana-Weyl spinors can exist at all. Assuming first that Majorana spinors exist for $D = 2d$, we see that since $(\gamma^{2d+1})^2 = (-1)^{d-1}$, in order for both conditions to be compatible, the eigenvalues of γ^{2d+1} must be ± 1 rather than $\pm i$, and therefore d must be *odd*.

We now look for which values of D Majorana spinors exist. We shall prove the following crucial theorem:

Theorem: Majorana spinors, and a real representation of the γ matrices, exist if and only if D is 2 or 4 modulo 8.

Throughout this section * denotes complex conjugation, † denotes hermitian conjugation and T the transposition.

Using Schur's lemma [29], since γ^μ and $\gamma^{*\mu}$ have the same algebra, and the representation of the γ^μ is of minimal dimension $2^{D/2}$, there must exist a matrix B such that

$$\gamma^{*\mu} = B\gamma^\mu B^{-1} . \qquad (2.22)$$

One can require further that

$$|\det B| = 1 . \qquad (2.23)$$

We note that B has involutive properties, namely,

$$\gamma^\mu = B^*\gamma^{\mu *}B^{*-1} , \qquad (2.24)$$

or

$$\gamma^{*\mu} = B^{*-1}\gamma^\mu B^* . \qquad (2.25)$$

Comparing equations we see that B^* and B^{-1} are proportional to each other. Using (2.23) one gets

$$B^*B = \epsilon I , \qquad |\epsilon| = 1 . \qquad (2.26)$$

Let us show that ϵ is real:

$$BB^* = \epsilon^* I , \qquad (2.27)$$

but multiplying eq. (2.26) by B and B^{-1} one gets

$$B(B^*B)B^{-1} = \epsilon I = \epsilon^* I .$$

So $\epsilon = \pm 1$. One can easily show that ϵ does not depend upon which representation of the γ matrices one uses. Therefore, it has an intrinsic significance, and depends only upon the space time metric $\eta^{\mu\nu}$.

ϵ is of crucial importance: if $\epsilon = -1$ no Majorana spinors exist, whilst for $\epsilon = +1$ they do exist.

To see this let us define Majorana spinors. It is easily seen that if ψ satisfies the Dirac equation, $B^{-1}\psi^*$ also satisfies it. Therefore, the charge conjugate of ψ is $B^{-1}\psi^* = \psi^c$. A majorana spinor is defined such that

$$\psi = \psi^c = B^{-1}\psi^* . \tag{2.28}$$

Now $\psi = B^{-1}\psi^*$ implies

$$\psi^* = B^{*-1}\psi = \epsilon B\psi , \tag{2.29}$$

and therefore we get the consistency condition

$$(1 - \epsilon)\psi = 0 , \tag{2.30}$$

which implies that if ψ does not vanish, ϵ must be +1. A more general definition of a Majorana spinor, i.e. $\psi = \lambda\psi^c$ does not affect this conclusion.

Until now, the signature of the metric, and the hermiticity properties of the γ matrices have not been used. We now bring them into the game by assuming one time dimension and $D - 1$ space dimensions; then $\gamma^{0\dagger} = -\gamma^0$ and $\gamma^{i\dagger} = \gamma^i$.

Then one has

$$\gamma^{\mu\dagger} = -\gamma^0\gamma^\mu(\gamma^0)^{-1} . \tag{2.31}$$

Defining the charge conjugation matrix C such that

$$\gamma^{\mu \mathrm{T}} = -C\gamma^\mu C^{-1} , \tag{2.32}$$

C can be computed in terms of the B matrix and γ^0 in two different ways:

$$\gamma^{\mu \mathrm{T}} = (\gamma^{\mu *})^\dagger = (\gamma^{\mu\dagger})^* . \tag{2.33}$$

Using the first way of writing $\gamma^{\mu \mathrm{T}}$ one obtains

$$\gamma^{\mu \mathrm{T}} = -(B^{\dagger -1}\gamma^0)\gamma^\mu(B^{\dagger -1}\gamma^0)^{-1} . \tag{2.34}$$

And using the second way one obtains

$$\gamma^{\mu \mathrm{T}} = -(B\gamma^0)\gamma^\mu(B\gamma^0)^{-1} . \tag{2.35}$$

This shows that the matrices B and $(B^\dagger)^{-1}$ are proportional to each other, thus

$BB^\dagger = \lambda$. Because of the normalization of B, $|\lambda| = 1$, and λ is positive. Thus $\lambda = 1$.

We recapitulate the results obtained so far: we have

$$BB^\dagger = B^\dagger B = I\,, \tag{2.36}$$

$$BB^* = \epsilon I\,. \tag{2.37}$$

This implies that

$$B = \epsilon B^{\mathrm{T}}\,, \tag{2.38}$$

so that B is either symmetric or antisymmetric.

Finally, we have

$$C = B\gamma^0\,, \tag{2.39}$$

$$C^{\mathrm{T}} = -\epsilon C\,. \tag{2.40}$$

We now go into the crucial phase of the proof which is the computation of ϵ [29]. We achieve this by counting the number of symmetric and antisymmetric matrices which form a complete basis for the product of γ matrices.

Let us denote by Γ^n $n = 0, 1, \ldots, D$ (assume here D even) the products of n γ matrices, completely antisymmetrized with respect to the indices of the γ matrices and normalized such that $(\Gamma^n)^2$ is either +1 or −1. (We shall not use this normalization condition here, but it will be important later on for Fierz transformation.) For a given n there are $\binom{D}{n}$ Γ^n matrices, and the total number of these matrices is 2^D. Since the γ matrices are $2^{D/2} \times 2^{D/2}$, this shows that the Γ^n matrices form a complete basis for any combination of sum or products of the γ matrices, and the independence of these matrices is also obvious.

Let us now consider $C\Gamma^n C^{-1}$. It is easy to see that this is equal to $\Gamma^{n\mathrm{T}}$, up to a sign which is given by $(-1)^n(-1)^{n(n-1)/2}$. The first factor comes from using (2.32) n times, and the second factors from reordering the n γ matrices inside Γ^n to get the correct ordering. In doing this reordering no contraction ever occurs as Γ^n is antisymmetrized with respect to the indices of the γ matrices inside. So we have

$$\Gamma^{n\mathrm{T}} = (-1)^n(-1)^{n(n-1)/2}C\Gamma^n C^{-1}\,, \tag{2.41}$$

or

$$(C\Gamma^n)^{\mathrm{T}} = \epsilon(-1)^{(n-1)(n-2)/2}(C\Gamma^n)\,. \tag{2.42}$$

So the matrices $C\Gamma^n$ are either symmetric or antisymmetric. This equation allows one to count the number of independent antisymmetric matrices in two different ways. On one hand we know that as the γ matrices are $2^{D/2}$ dimensional, a complete basis of antisymmetric matrices must contain $2^{D/2}(2^{D/2}-1)/2$ matrices. On the other hand, using eq. (2.42) the number of antisymmetric matrices can also be written as

$$N=\sum_{n=0}^{D}\tfrac{1}{2}(1-\epsilon(-1)^{(n-1)(n-2)/2})\binom{D}{n}=\tfrac{1}{2}(2^{D/2}(2^{D/2}-1))\,. \tag{2.43}$$

Thus we obtain an equation which determines ϵ:

$$\epsilon\sum_{n=0}^{D}(-1)^{(n-1)(n-2)/2}\binom{D}{n}=2^{D/2}\,. \tag{2.44}$$

The summation over n can be performed by noticing that

$$(-1)^{(n-1)(n-2)/2}=-\tfrac{1}{2}[(1+i)i^n+(1-i)(-i)^n] \tag{2.45}$$

and one finally obtains ϵ:

$$\epsilon=-\sqrt{2}\cos\frac{\pi}{4}(D+1)\,. \tag{2.46}$$

So

$$\epsilon=+1 \qquad \text{for } D=2,4\ [\text{mod } 8]\,, \tag{2.47}$$

$$\epsilon=-1 \qquad \text{for } D=6,8\ [\text{mod } 8]\,. \tag{2.48}$$

To complete our theorem, we now prove that when $\epsilon=+1$, a real representation of the γ matrices exists. Under a change of basis of the γ matrices $\gamma'^{\mu}=A\gamma^{\mu}A^{-1}$, B transforms as follows:

$$B'=A^*BA^{-1}\,. \tag{2.49}$$

For $\epsilon=+1$, B is symmetric and also unitary. Therefore if we decompose B into real and imaginary parts $B=B_1+iB_2$, B_1, B_2 are real, symmetric matrices, which satisfy because of the unitarity condition: $B_1^2+B_2^2=1$ and $[B_1,B_2]=0$. The last condition ensures that they can be simultaneously diagonalized. Thus B itself can be diagonalized, and since B is unitary it can be put in the form $B=\mathrm{diag}(e^{i\alpha_1},\ldots e^{i\alpha_D})$. Then choosing $A=\mathrm{diag}(e^{i\alpha_1/2},\ldots e^{i\alpha_D/2})$ brings B to the form $B=1$, which proves the existence of a real representation of the γ matrices. In such a representation the charge conjugation matrix C is just given by γ^0.

The requirement of the existence of Majorana-Weyl spinors gives thus the following condition on D: $D=2$ [mod 8]. The first non-trivial dimension (2 is somewhat trivial) is thus $D=10$, which is a nice way to recover the critical dimension of the NSR model.

We end this section by giving an explicit real representation of the γ matrices in 10 dimensions, which will be useful later on when making explicit the 4-dimensional content of a 10-dimensional theory. The 32-dimensional Dirac indices can be written as a direct product of a Dirac index running from 1 to 4 (ordinary space-time) and a Dirac index running from 1 to 8 ("internal" space). Thus the 32-dimensional γ matrices can be written as direct products of 4×4 ordinary Dirac matrices and of 8×8 ("internal symmetry" indices).

A real (Majorana) representation is then obtained, as given by

$$\text{(a)} \quad \gamma^\mu = \gamma^\mu \otimes \begin{pmatrix} I_4 & 0 \\ 0 & -I_4 \end{pmatrix}, \qquad \mu = 0, 1, 2, 3\,, \tag{2.50}$$

where γ^μ on the right are the 4×4 Dirac matrices in a real representation, I_4 the 4×4 identity matrix;

$$\text{(b)} \quad \gamma^{3+i} = \gamma^5 \otimes \begin{pmatrix} \rho_i & 0 \\ 0 & \rho_i' \end{pmatrix}, \qquad i = 1, 2, 3\,, \tag{2.51}$$

$$\rho_1 = \rho_1' \equiv \gamma^0\,, \tag{2.52}$$

$$\rho_2 = \rho_2' \equiv \gamma^5\,, \tag{2.53}$$

$$\rho_3 = -\rho_3' \equiv \gamma^0\gamma^5\,, \tag{2.54}$$

$$\text{(c)} \quad \gamma^{6+j} = I_4 \otimes \begin{pmatrix} 0 & \zeta_j \\ \zeta_j & 0 \end{pmatrix}, \qquad j = 1, 2, 3\,, \tag{2.55}$$

$$\zeta_1 \equiv \gamma^1\,, \qquad \zeta_2 \equiv \gamma^2\,, \qquad \zeta_3 \equiv \gamma^3\,. \tag{2.56}$$

When writing such equations such as (2.56), we only mean that the algebra of ζ^i and γ^i are identical. One can verify that this representation is (obviously) real and satisfies the correct transposition and anticommutation properties. The advantage of it is that the "extra" six γ matrices appear as three scalar and three pseudoscalar matrices in Dirac space. Computing $\gamma^{11} = \gamma^0\gamma^1 \ldots \gamma^9$ one obtains

$$\gamma^{11} = I_4 \otimes \begin{pmatrix} 0 & \rho_3 \\ -\rho_3 & 0 \end{pmatrix}. \tag{2.57}$$

A Dirac spinor in 10 dimensions can be written as

$$\psi = \begin{pmatrix} \psi_1 \\ \psi_2 \end{pmatrix}, \tag{2.58}$$

where ψ_1, ψ_2 are four Dirac spinors in four dimensions. A Majorana spinor in 10 dimensions is such that $\psi = \psi^*$ and is a collection of four Majorana spinors in four dimensions. Finally, a Weyl-Majorana spinor in 10 dimensions in this representation is given by $\psi = \gamma^{11}\psi$, i.e.:

$$\psi = \begin{pmatrix} \psi_1 \\ -\rho_3\psi_1 \end{pmatrix}, \tag{2.59}$$

$$\psi_1^* = \psi_1 . \tag{2.60}$$

This shows explicitly the existence, in 10 dimensions, of Majorana-Weyl spinors and also reveals that the 10-dimensional Weyl condition does not imply parity breaking in four dimensions. A Majorana-Weyl spinor in 10 dimensions simply appears as a collection of four four-dimensional Majorana spinors.

2.4. Counting the states of the NSR model

Having decided that the ground state fermion is a Majorana-Weyl fermion, and that it must belong also to the regular representation of U(N), we are left with the Majorana-Weyl-Ramond model and the $G = +1$ sector of the Neveu-Schwarz model. We shall show that this model has a good chance of being supersymmetric by showing that at each mass level, the number of physical states of the Bose and Fermi sectors are equal.

The quantization in masses for the model is now given by

$$\alpha' m^2 = n , \qquad n = 0, 1, 2 \ldots , \tag{2.61}$$

and the spectrum has no tachyon. As both Fermi and Bose sectors belong to the same representation of U(N), the factor N^2 will be dropped. We start with the ground state ($n = 0$). In the NS sector this is a "photon" with $D - 2 = 8$ physical degrees of freedom. In the R sector there is a spinor (32 components) subjected to two conditions, Majorana and Weyl, which decrease each by half the number of independent components. So the degeneracy $d(n)$ for $n = 0$ is 8 both for fermions and bosons.

We now compute the partition function of the Fermi sector for $D = 10$. Since we count only the coupled (i.e., transverse states), the partition function is obtained by using the counting operator made of transverse oscillators

$$R = \sum_{n=1}^{\infty} n(a_n^\dagger \cdot a_n + d_n^\dagger \cdot d_n) , \tag{2.62}$$

$$(a_n^\dagger \cdot a_n = \sum_{i=1}^{8} a_n^{\dagger i} a_n^i, \text{ etc.}) .$$

Since the ground state fermion is eight-fold degenerate, one has

$$f_{\mathrm{R}}(q) = \sum_{n=0}^{\infty} d_{\mathrm{R}}(n) q^{2n} = 8\,\mathrm{Tr}(q^{2R}) = 8 \prod_{m=1}^{\infty} (1 - q^{2m})^{-8} (1 + q^{2m})^{8} \,. \quad (2.63)$$

A similar computation for the $G = +1$ sector of the NS model gives with

$$R = \sum_{n=1}^{\infty} n\, a_n^{\dagger} \cdot a_n + \sum_{r=1/2}^{\infty} r\, b_r^{\dagger} \cdot b_r \,,$$

$$\begin{aligned} f_{\mathrm{NS}}(q) &= \sum_{n=0}^{\infty} d_{\mathrm{NS}}(q) q^{2n} \\ &= \frac{1}{q}\,\mathrm{Tr}[\tfrac{1}{2}(1+G) q^{2R}] \\ &= \prod_{n=1}^{\infty} (1 - q^{2n})^{-8} \frac{1}{2q} \{ \prod_{n=1}^{\infty} (1 + q^{2n-1})^{8} - \prod_{n=1}^{\infty} (1 - q^{2n-1})^{8} \} \,. \end{aligned} \quad (2.64)$$

The equality of these two expressions was known to Jacobi [30] who called it "*aequatio identica satis abstrusa*". For us its meaning is less abstruse as it means that $d_{\mathrm{NS}}(n) = d_{\mathrm{R}}(n)$ at each level, and therefore an important necessary condition for a linear realization of supersymmetry is satisfied.

A picture of the NSR spectrum is given in fig. 4.

To prove supersymmetry completely, it remains to define supersymmetric trans-

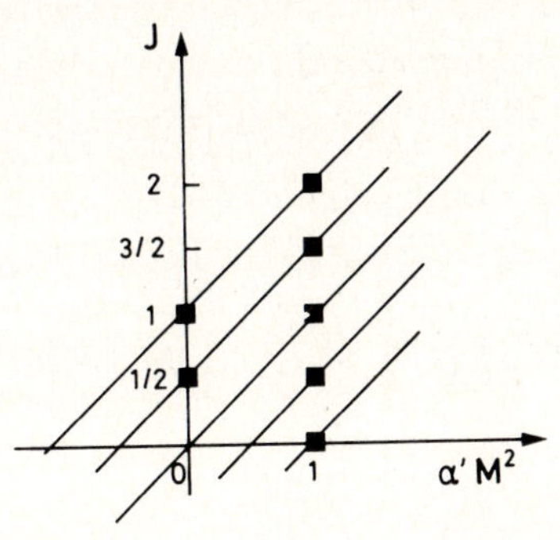

Fig. 4. The spectrum of the supersymmetric NSR model (open strings).

formations which exchange the NS and R states, and we plan to come back to this subject in a later publication.

3. Supersymmetric Yang-Mills theory and the dual spinor model

We shall not review the arguments [15] which lead one to conclude that in the $\alpha' = 0$ limit the dual spinor model on-shell amplitudes in the tree approximation are identical with those of the (10 dimensional) Yang-Mills theory. These are similar to those used in the Veneziano model to prove the connection with the Yang-Mills theory without fermions. So we shall simply list the main ingredients of the proof:

(a) The fact that in the $\alpha' = 0$ limit dual diagrams turn into a sum of Feynman diagrams, each Feynman diagram being obtained with the correct weight [31].

(b) The dimensionality of the couplings obtained when $\alpha' = 0$ (only the dual coupling constant g remains and it is identified with the Yang-Mills coupling constant).

(c) The on-shell gauge invariance of the dual S matrix ($\delta S = 0$ when $\delta\epsilon_\mu = \alpha p_\mu$) which implies the non-Abelian Yang-Mills invariance.

(d) The invariance of the dual S matrix under the U(N) group introduced by the Chan-Paton factors, the gauge group G of Yang-Mills being identified with U(N). Because of duality requirement the gauge coupling constant g' of the U(1) subgroup of U(N) is identical with g the gauge coupling constant of the SU(N).

The only non-trivial point which requires a bit of care is the first one. It actually requires that each external line of a Feynman graph can be obtained either from a twisted or untwisted line in a dual diagram. So it needs the definition of a fermion twist operator. For instance, the photon-fermion dual diagram of fig. 5a should be accompanied with the analogous diagram of fig. 5b. It is worthwhile to notice that it leads to problems when the $G = -1$ sector is included [27], but that the problems disappear when this sector is removed and one can obtain a consistent dual definition of the graph of fig. 5b.

So the theory one obtains in the zero slope limit is simply the Yang-Mills theory with fermions:

$$\mathcal{L} = \mathrm{Tr}(-\tfrac{1}{4}G_{\mu\nu}G^{\mu\nu} - \tfrac{1}{2}i\bar{\lambda}\gamma^\mu \mathcal{D}_\mu \lambda)\,, \tag{3.1}$$

$$G_{\mu\nu} = \partial_\mu A_\nu - \partial_\nu A_\mu + ig[A_\mu, A_\nu]\,, \tag{3.2}$$

$$A_\mu = A_\mu^0 \frac{1}{\sqrt{N}} + \sum_{i=1}^{N^2-1} A_\mu^i X^i\,, \tag{3.3}$$

$$\mathcal{D}_\mu \lambda = \partial_\mu \lambda + ig[A_\mu, \lambda]\,. \tag{3.4}$$

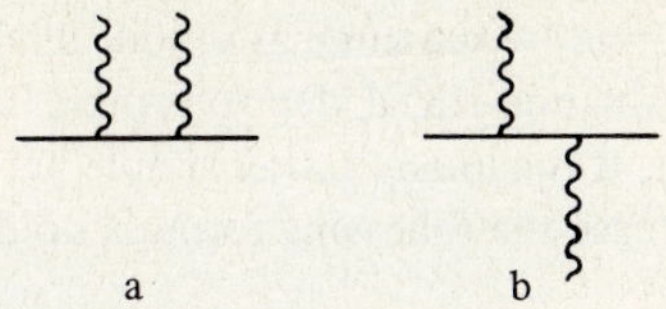

a b

Fig. 5. (a), (b) Two of the graphs which must contribute to boson-fermion scattering.

The X^i are the generators of SU(N) in the $N \times N$ representation, normalized such that $\mathrm{Tr}(X_i, X_j) = \delta_{ij}$. Note that the U(1) vector boson A^0_μ is in fact a free field.

All indices are 10 dimensional and the γ matrices are defined as in the previous section; the λ being 32 dimensional Majorana spinors:

$$\lambda^* = \lambda\,, \tag{3.5}$$

$$\gamma^{11}\lambda = \lambda\,. \tag{3.6}$$

We now show that the action $S = \int \mathrm{d}^{10}x\,\mathcal{L}$ is invariant under global supersymmetric transformations in 10 dimensions. One indeed checks that if ϵ is a constant anticommuting Majorana-Weyl spinor, δS vanishes under the following infinitesimal transformation:

$$\delta A_\mu = -i\bar{\epsilon}\gamma_\mu\lambda\,, \tag{3.7}$$

$$\delta\lambda = G_{\mu\nu}\sigma^{\mu\nu}\epsilon\,, \tag{3.8}$$

where

$$\sigma^{\mu\nu} = \tfrac{1}{4}[\gamma^\mu, \gamma^\nu]\,. \tag{3.9}$$

This transformation of the fields is the same as in four dimensions. The only difference in proving the vanishing of δS for four and 10 dimensions is that now one has to use the following identity:

$$\gamma^\mu\sigma^{\alpha\beta} = \tfrac{1}{2}[g^{\mu\alpha}\gamma^\beta - g^{\mu\beta}\gamma^\alpha - \frac{1}{7!}\epsilon^{\mu\alpha\beta\lambda_1\ldots\lambda_7}\gamma^{11}\gamma^{\lambda_1}\ldots\gamma^{\lambda_7}]\,, \tag{3.10}$$

where $\epsilon^{\mu_1\ldots\mu_{10}}$ is the completely antisymmetrized tensor in 10 dimensions.

One also verifies that the supersymmetric algebra closes on fields which satisfy the equations of motion:

$$[\delta_1, \delta_2]\lambda = 2i(\bar{\epsilon}_1\gamma^\nu\epsilon_2)\mathcal{D}_\nu\lambda\,, \tag{3.11}$$

$$[\delta_1, \delta_2]A_\mu = 2i(\bar{\epsilon}_1\gamma^\nu\epsilon_2)G_{\nu\mu}\,. \tag{3.12}$$

These commutation relations are the same as in four dimensions and therefore have the same interpretation. The non-trivial step to prove (3.12) is to use the 10-dimension Fierz transformation. It is actually rather simple because the spinors ϵ_1, ϵ_2 are both left handed. In that case the following formula holds:

$$(\bar{u}_1 \gamma_\nu u_2)\gamma^\nu u_3 = \tfrac{1}{2}(\bar{u}_1 \gamma_\alpha u_3)\gamma^\alpha u_2 - \tfrac{1}{4}(\bar{u}_1 \sigma_{\alpha\beta\gamma} u_3)\sigma^{\alpha\beta\gamma} u_2 , \quad (3.13)$$

if $\gamma_{11} u_i = u_i$; $i = 1, 2, 3$. $\sigma_{\alpha\beta\gamma}$ is the completely antisymmetrized product of 3γ matrices. The contribution of the second term vanishes in eq. (3.11) because of the Majorana property.

So the 10-dimensional Yang-Mills theory with fermions in the regular representation of the group is automatically supersymmetric provided that the fermions are Majorana-Weyl.

We now show that a very interesting four dimensional action can be obtained from this 10-dimensional formalism. In the field theory we can assume that the extra six coordinates, X^{3+i} ($i = 1, \ldots 6$) are compact. A simple compactification is obtained by introducing six radii, R_i such that $0 \leqslant x^{3+i} \leqslant R_i$; all fields being periodic in x^{3+i} between 0 and R_i. A Fourier expansion of the 10-dimensional fields reveals that each of them represents an infinite number of four-dimensional fields, of quantized masses given by $m^2(n_1, \ldots n_6) = 4\pi^2 \Sigma_1^6 n_i^2/R_i^2$. Assuming the R_i to be small, we take in the Fourier expansion only the mode corresponding to $n_1 = n_2 = \ldots = n_6 = 0$, i.e., we drop all the dependence of the fields upon the six extra coordinates. In this way, up to a rescaling of the fields and of the coupling constant g we obtain a four-dimensional local action with a finite number of massless fields.

This action can also be obtained by a direct zero slope limit of the dual spinor model in the following way [17]: quantize all momenta of the particles in the dual model by $p^{3+i} = 2\pi(n^i/R_i)$. The six n^i appear as six additive conserved quantum numbers. As α' goes to zero, let us also take R_i to zero with fixed ratios $\gamma_i = \alpha'/R_i^2$. Then one obtains the four-dimensional action we are going to write down.

The non-trivial character of the four-dimensional action we obtain in this way is due to the fact that the components A_{3+i} are independent of the x^{3+i}, but are non-vanishing, so that three pseudoscalar fields $B_i(x) = A_{3+i}(x)$ ($i = 1, 2, 3$) and three scalar fields $A_i(x) = A_{6+i}(x)$ ($i = 1, 2, 3$) appear associated with the vector field $A_\mu(x)$ ($\mu = 0, 1, 2, 3$). Further, the Majorana-Weyl spinor $\lambda(x)$ describes four Majorana spinors $\lambda_K(x)$ in four dimensions, the internal symmetry index K running from 1 to 4. The supersymmetry transformation parameter ϵ describes also four Majorana spinor parameters ϵ_K. So we obtain an action which contains one vector multiplet and three scalar multiplets, with only one coupling constant g, invariant under a supersymmetry transformation with an internal symmetry index. In fact, the action is not only supersymmetric, but hypersymmetric in the sense defined by Fayet [23]. It is straightforward to derive the action from eq. (3.1) together with the representation of the γ matrices written in sect. 2, and we shall only give the result where all symbols have a conventional, four-dimensional meaning:

$$\mathcal{L} = \mathrm{Tr}\{-\tfrac{1}{4}G_{\mu\nu}G^{\mu\nu} - \tfrac{1}{2}(\mathcal{D}_\mu A_i)^2 - \tfrac{1}{2}(\mathcal{D}_\mu B_i)^2 + \tfrac{1}{4}g^2([A_i, A_j]^2 + [B_i, B_j]^2 + 2[A_i, B_j]^2) - \tfrac{1}{2}i\bar{\lambda}_K \gamma^\mu \mathcal{D}_\mu \lambda_K + \tfrac{1}{2}g\bar{\lambda}_K[(\alpha^j_{Kl}A_j + \beta^j_{Kl}\gamma^5 B_j), \lambda_l]\} . \tag{3.14}$$

In this action, K, l indices run from 1 to 4, i and j indices from 1 to 3. The six matrices α^i and β^i satisfy the following algebra:

$$\{\alpha^i, \alpha^j\} = \{\beta^i, \beta^j\} = -2\delta^{ij} , \tag{3.15}$$

$$[\alpha^i, \beta^j] = 0 . \tag{3.16}$$

They are antisymmetric and real, and are related to the previously defined matrices ρ_i and ζ:

$$\beta_i = \rho^i , \qquad \alpha_j = -\rho_3\zeta^j . \tag{3.17}$$

An explicit representation of these matrices is given in terms of Pauli matrices:

$$\beta^1 = \begin{pmatrix} 0 & i\sigma^2 \\ i\sigma^2 & 0 \end{pmatrix}, \qquad \beta^2 = \begin{pmatrix} 0 & 1 \\ -1 & 0 \end{pmatrix}, \qquad \beta^3 = \begin{pmatrix} -i\sigma^2 & 0 \\ 0 & i\sigma^2 \end{pmatrix},$$

$$\alpha^1 = \begin{pmatrix} 0 & \sigma^1 \\ -\sigma^1 & 0 \end{pmatrix}, \qquad \alpha^2 = \begin{pmatrix} 0 & -\sigma^3 \\ \sigma^3 & 0 \end{pmatrix}, \qquad \alpha^3 = \begin{pmatrix} i\sigma_2 & 0 \\ 0 & i\sigma_2 \end{pmatrix}. \tag{3.18}$$

Before we discuss the full invariance of the Lagrangian, let us consider a few simple examples.

Let $A_i = B_i = 0$, and only one of the form λ_K non-vanishing, then we get

$$\mathcal{L} = \mathrm{Tr}(-\tfrac{1}{4}G_{\mu\nu}G^{\mu\nu} - \tfrac{1}{2}\bar{\lambda}\gamma^\mu \mathcal{D}_\mu \lambda) , \tag{3.19}$$

which is the supersymmetric (four-dimensional) non-Abelian vector multiplet action [22].

Let us choose $\lambda_3 = \lambda_4 = 0$; λ_1, λ_2 non-vanishing. Choose $A_1 = A_2 = B_1 = B_2 = 0$, $A_3 = A, B_3 = B$ non-vanishing. Using the explicit representation of the α, β matrices, we see that we obtain

$$\mathcal{L} = \mathrm{Tr}(-\tfrac{1}{4}G_{\mu\nu}G^{\mu\nu} - \tfrac{1}{2}(\mathcal{D}_\mu A)^2 - \tfrac{1}{2}(\mathcal{D}_\mu B)^2 + \tfrac{1}{2}g^2[A, B]^2 - \tfrac{1}{2}i\bar{\lambda}_1\gamma^\mu \mathcal{D}_\mu\lambda_1 - \tfrac{1}{2}i\bar{\lambda}_2\gamma^\mu \mathcal{D}_\mu\lambda_2 + g\bar{\lambda}_1[(A + \gamma^5 B), \lambda_2]) . \tag{3.20}$$

This is precisely the action for a vector multiplet in interaction with a massless scalar multiplet [22], both belonging to the regular representation of the gauge group. This action was shown to have a hidden SU(2) ⊗ U(1) symmetry [23] by Fayet, and the supersymmetric invariance with internal indices one derives from this was called by him "hypersymmetry".

We now return to our action of eq. (3.14) and list its invariances.

(a) Non-Abelian gauge invariance with respect to U(N)

$$\delta A_\mu = \partial_\mu \Lambda + ig[A_\mu, \Lambda]\,,$$

$$\delta A_j = ig[A_j, \Lambda]\,, \qquad j = 1, 2, 3\,.$$

Same for B_j, λ_K.

(b) Invariance under supersymmetric O(6) ~ SU(4) transformations, the supersymmetric transformation parameters being four Majorana spinors ϵ_K

$$\delta A_\mu = -i\bar{\epsilon}_K \gamma_\mu \lambda_K\,, \qquad K = 1, 2, 3, 4\,,$$

$$\delta A_i = -i\bar{\epsilon}_K (\alpha^i)_{Kl} \lambda_l\,,$$

$$\delta B_i = -i\bar{\epsilon}_K (\beta^i)_{Kl} \gamma^5 \lambda_l\,,$$

$$\delta \lambda_K = \{G_{\mu\nu}\sigma^{\mu\nu} + \gamma^\mu \mathcal{D}_\mu(\alpha^i A_i + \gamma_5 \beta^i B_i) + ig[B_i, B_j]\sigma^{ij}$$
$$+ ig[A_i, A_j]\sigma'^{ij} + 2ig[B_i, A_j]K^{ij}\gamma^5\}_{Kl}\epsilon_l\,,$$

where

$$\sigma^{ij} = \tfrac{1}{4}[\beta^i, \beta^j]\,, \qquad \sigma'^{ij} = \tfrac{1}{4}[\alpha^i, \alpha^j]\,, \qquad K^{ij} = \tfrac{1}{4}\{\alpha^i, \beta^j\}\,.$$

(c) Invariance under SO(6) ~ SU(4) transformations (or SU(2) ⊗ SU(2) if the γ^5 transformations are not considered)

$$\delta A_\mu = 0\,,$$

$$\delta \lambda_K = \tfrac{1}{2}[\sigma^{ij}\Lambda_{ij} + \sigma'^{ij}\Lambda'_{ij} + \gamma^5 K^{ij}\tilde{\Lambda}_{ij}]_{Kl}\lambda_l\,,$$

$$\delta A_i = \Lambda'_{ij} A_j + \tilde{\Lambda}_{il} B_l\,,$$

$$\delta B_i = \Lambda_{ij} B_j + \tilde{\Lambda}_{il} A_l\,.$$

It is obvious that to derive the results, the notation of (3.1) is both elegant and time saving. The O(6) invariance is trivially a consequence of the 10-dimensional

Lorentz invariance of (3.1) which is broken down to four-dimensional Lorentz invariance times O(6) by the ansatz that $\partial_{3+i} = 0$.

4. The spectrum of the closed strings of the dual spinor model

The equations of motion of a classical spinning string can be derived from a well defined action which exhibits local supersymmetry in the two-dimensional surface spanned by the string [19]. In a specific orthonormal gauge, these equations read:

$$\Box\phi^\mu = 0 , \tag{4.1}$$

$$\gamma^i \partial_i \psi^\mu = 0 , \tag{4.2}$$

where μ is the space-time index ($\mu = 0, 1, \ldots 9$), $i = 1, 2$ referring to the two-dimensional surface. The coordinates on the surface being $y^0 = \tau, y^1 = \sigma$. ψ^μ is a two-dimensional spinor

$$\psi^\mu = \begin{pmatrix} \psi_1^\mu \\ \psi_2^\mu \end{pmatrix} .$$

Using the representation of the γ matrices

$$\gamma^0 = \begin{pmatrix} 0 & 1 \\ -1 & 0 \end{pmatrix} , \qquad \gamma^1 = \begin{pmatrix} 0 & 1 \\ 1 & 0 \end{pmatrix} ,$$

these equations are easily solved as

$$\phi^\mu = \phi^\mu(\sigma + \tau) + \bar{\phi}^\mu(\sigma - \tau) , \tag{4.3}$$

$$\psi_1^\mu = \psi_1^\mu(\sigma + \tau) , \tag{4.4}$$

$$\psi_2^\mu = \psi_2^\mu(\sigma - \tau) . \tag{4.5}$$

We wish now to discuss the boundary conditions for a closed spinning string. For the closed string one must have

$$\phi^\mu(\sigma, \tau) = \phi^\mu(\sigma + 2\pi, \tau) , \tag{4.6}$$

as ϕ^μ represents the position of the string in space time. Thus in the expansion of ϕ^μ in Fourier series, one obtains two sets of modes, α_n^μ (expansion in $e^{-in(\tau+\sigma)}$), $\bar{\alpha}_n^\mu$ (expansion in $e^{-in(\tau-\sigma')}$) with α_0^μ being related to the momentum of the whole string.

For the fermionic fields ψ_1, ψ_2, there is however an arbitrariness. Looking at the gauge conditions, one can see that it would be inconsistent to identify $\psi_1(\sigma = 0)$ with $\psi_2(\sigma = 2\pi)$. However, there is a sign ambiguity in identifying $\psi_1(\sigma = 0)$ with $\psi_1(2\pi)$, and a similar one arises for ψ_2. Thus one obtains three possible cases:

$$\text{(a)}\quad \psi_1(\sigma = 0) = -\psi_1(\sigma = 2\pi)\,,\qquad \psi_2(\sigma = 0) = -\psi_2(\sigma = 2\pi)\,, \tag{4.7}$$

$$\text{(b)}\quad \psi_1(\sigma = 0) = -\psi_1(\sigma = 2\pi)\,,\qquad \psi_2(\sigma = 0) = +\psi_2(\sigma = 2\pi)\,, \tag{4.8}$$

$$\text{(c)}\quad \psi_1(\sigma = 0) = +\psi_1(\sigma = 2\pi)\,,\qquad \psi_2(\sigma = 0) = +\psi_2(\sigma = 2\pi)\,. \tag{4.9}$$

Expanding ψ_1, ψ_2 into a Fourier series:

$$\psi_1 = \sum_n c_n \exp[-in(\sigma + \tau)]\,, \tag{4.10}$$

$$\psi_2 = \sum_n \bar{c}_n \exp[-in(\tau - \sigma)]\,, \tag{4.11}$$

we see that in the first case both oscillators $c_n, \bar{c}_n$ are half-integrally moded, so that the model which results upon quantization is a doubling of the Neveu-Schwarz model. It has already been studied in detail in refs. [32] and [33], and it appears in loop diagrams as bound states of the $G = +1$ sector of the Neveu-Schwarz model. The particles it describes are bosonic and it contains a massless graviton.

In the second case, the c_n oscillators are of the Neveu-Schwarz type (half-integrally moded), but the $\bar{c}_n$ are of the Ramond type (integrally moded). This model describes fermions as the c_0^μ modes must be identified with the (32-dimensional) γ^μ matrices. It has not so far been studied and should, in principle, arise as a bound state of the interacting Neveu-Schwarz and Ramond models, through the graph of fig. 2 which has not yet been computed due to the difficulties in defining a fermion twist operator. Now that these conceptual difficulties are removed, it should become possible to compute this graph and to verify this conjecture. We are going to see that the ground state of this model contains a massless spin-$\frac{3}{2}$ particle.

Finally, in the third case we get a model obtained by doubling the Ramond model which should describe bosons in a Kemmer-Duffin-like formalism. However, this sector is known to be obtained in loop diagrams as bound states of the $G = -1$ sector of the Neveu-Schwarz model [32]. Since this sector is eliminated by the Weyl condition, the model with two sets of Ramond oscillators should not appear as a bound state in our supersymmetric dual model, and we shall disregard it.

We would like now to discuss model (b) and its spectrum. To make the discussion more transparent we will first recall how model (a) is defined and what its spectrum is.

The Neveu-Schwarz model was defined through the equations:

$$L_n|\phi\rangle = 0\,, \qquad n > 0\,, \qquad (L_0 - \tfrac{1}{2})|\phi\rangle = 0\,,$$

$$G_r|\phi\rangle = 0\,, \qquad r = \tfrac{1}{2}, \tfrac{3}{2}, \ldots\,. \tag{4.12}$$

These equations define a ghost-free sector of $D \leqslant 10$. Suppose now we want to "double" this model. We will get a ghost-free spectrum if we double all these conditions, i.e., assume that

$$L_n|\phi\rangle = \overline{L}_n|\phi\rangle = 0\,, \qquad n = 1, 2, \ldots\,,$$

$$G_r|\phi\rangle = \overline{G}_r|\phi\rangle = 0\,, \qquad r = \tfrac{1}{2}, \tfrac{3}{2}, \ldots\,,$$

$$(L_0 - \tfrac{1}{2})|\phi\rangle = (\overline{L}_0 - \tfrac{1}{2})|\phi\rangle = 0\,. \tag{4.13}$$

The last equation implies, in particular, that $(L_0 - \overline{L}_0)|\phi\rangle = 0$. As in the Neveu-Schwarz model, the ground state ($|0\rangle$) is a tachyon, with $\alpha' m^2 = -1$. The intercept of the tachyon trajectory is $\alpha(0) = 1$. One has

$$L_0 = \tfrac{1}{2}\alpha' p^2 + \sum n a_n^\dagger \cdot a_n + \sum r b_r^\dagger \cdot b_r\,, \tag{4.14}$$

$$\overline{L}_0 = \tfrac{1}{2}\alpha' p^2 + \sum n \overline{a}_n^\dagger \cdot \overline{a}_n + \sum r \overline{b}_r^\dagger \cdot \overline{b}_r\,. \tag{4.15}$$

As in the NS model one can define a conserved multiplicative quantum number G_p which distinguishes two sectors in this model:

$$G_p = (-1)^{\sum\limits_m b_m^\dagger b_m - 1} = (-1)^{\sum\limits_m \overline{b}_m^\dagger \overline{b}_m - 1}\,. \tag{4.16}$$

The last equality follows from the fact that for on-shell states $L_0 - \overline{L}_0 = 0$. G_p should not be confused with the G operator of the Neveu-Schwarz model for open strings.

So we see that we have two sectors in that model, one with $G_p = -1$ which has intercept 1 and contains a tachyon, and one with $G_p = +1$. The ground state of the $G_p = +1$ sector is represented by

$$b_{1/2}^{\dagger\mu}\, \overline{b}_{1/2}^{\dagger\nu}|0\rangle \epsilon_{\mu\nu}\,. \tag{4.17}$$

This state has zero mass. We see that because of the eqs. (4.13), $\epsilon_{\mu\nu}$ is subjected to two conditions:

$$p^{\mu} \epsilon_{\mu\nu} = p^{\nu} \epsilon_{\mu\nu} = 0 . \tag{4.18}$$

$\epsilon_{\mu\nu}$ is neither symmetric, nor antisymmetric, nor traceless. So it describes:

a massless antisymmetric tensor $A_{\mu\nu} = \frac{1}{2}(\epsilon_{\mu\nu} - \epsilon_{\nu\mu})$;

a massless scalar $\epsilon_{\mu\mu}$ (spin 0);

a massless symmetric tensor (spin 2): $e_{\mu\nu} = \frac{1}{2}(\epsilon_{\mu\nu} + \epsilon_{\nu\mu})$ ($\epsilon_{\mu\nu}$ traceless).

We have refrained from giving the spin of $A_{\mu\nu}$. In 10 dimensions it represents $8 \times \frac{7}{2} = 28$ physical degrees of freedom, while in 4 dimensions it represents $2 \times \frac{1}{2} = 1$ physical degree of freedom and can in fact be identified with a pseudoscalar, as we are going to see in a moment.

The transversality of $A_{\mu\nu}$ and $e_{\mu\nu}$ follows from the fact that if we replace $\epsilon_{\mu\nu}$ by $\epsilon_{\mu\nu} + p_{\mu}\zeta_{\nu} + p_{\nu}\bar{\zeta}_{\nu}$ where ζ_{ν} and $\bar{\zeta}_{\nu}$ are two vectors subject to the condition $p \cdot \zeta = p \cdot \bar{\zeta} = 0$, the variation of the dual S matrix will be zero as it amounts to replacing the massless tensor state by

$$G_{-1/2}\bar{b}^{\dagger\nu}_{1/2}|0\rangle\zeta^{\nu} + \bar{G}_{-1/2}b^{\dagger\nu}_{1/2}|0\rangle\bar{\zeta}^{\nu} , \tag{4.19}$$

and as this state is physical and spurious, it will be decoupled.

It would seem that for the symmetric part of $\epsilon_{\mu\nu}$, for instance, we get a gauge invariance which is less than general coordinate invariance since ζ^{μ} is constrained by the condition $p \cdot \zeta = 0$. But in fact the general equation derived from an on-shell spin-2 massless particle from the linearized Einstein action is $p^{\mu}e_{\mu\nu} = \frac{1}{2}p_{\nu}e_{\mu\mu}$. From eq. (4.18) we see that the dual model picks up a particular gauge, namely $e_{\mu\mu} = 0$ to describe the graviton, and this is the reason why our on-shell gauge invariance looks restricted. This point was discussed in detail in refs. [16], and it was concluded that in fact general coordinate invariance is implied by the invariance of the dual S matrix under the restricted gauge transformation. Similarly, for the antisymmetric part, it was shown [16] that it implies that the $A_{\mu\nu}$ field always appears through the gauge invariant, completely antisymmetric tensor:

$$F_{\mu\nu\rho} = \partial_{\mu}A_{\nu\rho} + \partial_{\nu}A_{\rho\mu} + \partial_{\rho}A_{\mu\nu} . \tag{4.19}$$

Now we look at model (b). Again we shall obtain a ghost-free spectrum for $D \leqslant 10$, if we define as physical states those which satisfy both the Ramond and the Neveu-Schwarz conditions for physical states, i.e.,

$$F_n|\phi\rangle = 0 \qquad n \geqslant 0 , \tag{4.20}$$

$$L_n|\phi\rangle = 0 \qquad n \geqslant 0 , \tag{4.21}$$

$$\bar{G}_r|\phi\rangle = 0 \qquad r = \tfrac{1}{2}, \tfrac{3}{2}, \dots , \tag{4.22}$$

$$\bar{L}_n|\phi\rangle = 0 \qquad n \geqslant 1\,, \tag{4.23}$$

$$(\bar{L}_0 - \tfrac{1}{2})|\phi\rangle = 0\,. \tag{4.24}$$

A surprising feature obtained by comparing eqs. (4.21) and (4.24) is that now

$$(L_0 - \bar{L}_0)|\phi\rangle = \tfrac{1}{2}|\phi\rangle \tag{4.25}$$

rather than zero. Thus the ground state in such a formalism is not the vacuum. Rather, it is given by

$$|\phi_0\rangle = \bar{b}^{\dagger\mu}_{1/2}|0\rangle\psi_\mu\,, \tag{4.26}$$

which satisfies

$$L_0|\phi_0\rangle = 0\,, \qquad (\bar{L}_0 - \tfrac{1}{2})|\phi_0\rangle = 0\,,$$

for $p^2 = 0$. Hence it describes a massless state. It is a spin-vector state, as the F operators contain γ matrices. The condition

$$F_0|\phi_0\rangle = 0 \tag{4.27}$$

implies

$$\not{p}\,\psi_\mu = 0\,, \tag{4.28}$$

and the condition

$$\bar{G}_{1/2}|\phi_0\rangle = 0 \tag{4.29}$$

implies

$$p^\mu\psi_\mu = 0\,. \tag{4.30}$$

So the ground state is a massless spin-vector state ψ_μ subject to the conditions (4.28) and (4.30).

Now we relate this result with what is known about the Rarita-Schwinger equation which describes spin-$\frac{3}{2}$ particles.

For $D = 4$, the equation reads

$$\epsilon^{\mu\nu\rho\sigma}\gamma_5\gamma_\rho\partial_\nu\psi_\sigma = (g^{\mu\nu}\gamma^\sigma - g^{\mu\sigma}\gamma^\nu + g^{\nu\sigma}\gamma^\mu - \gamma^\mu\gamma^\nu\gamma^\sigma)\partial_\nu\psi_\sigma$$
$$= 0\,. \tag{4.31}$$

For $D \neq 4$, we cannot use the first form of the RS equation as it contains explicitly the $\epsilon^{\mu\nu\rho\sigma}$ symbol. However, we can use the second form which is independent of the space-time dimension. In the second form, it is still true that the equation is invariant under $\psi_\sigma \to \psi_\sigma + \partial_\sigma \epsilon$ whatever D. In this way one can obtain the properties of a spin-$\frac{3}{2}$ particle for any space-time dimension. Going to the momentum representation we obtain, by contraction with γ^μ, a system of two equivalent equations:

$$p \cdot \psi = \not{p}\gamma \cdot \psi \,, \tag{4.32}$$

$$\not{p}\psi_\mu = p_\mu \gamma \cdot \psi \,. \tag{4.33}$$

Assume now that ψ_μ is on-shell ($p^2 = 0$). Multiplying (4.33) by $\not{p}$ we get

$$0 = p_\mu \not{p}\gamma \cdot \psi = p_\mu p \cdot \psi \,. \tag{4.34}$$

Hence if $p_\mu \neq 0$, the generalized on-shell Rarita-Schwinger equation reduces to

$$p \cdot \psi = 0 \,, \tag{4.35}$$

$$\not{p}\psi_\mu = p_\mu \gamma \cdot \psi \,. \tag{4.36}$$

This system is still invariant under $\psi_\mu \to \psi_\mu + \epsilon p_\mu$: therefore we can further specify a gauge, for instance, $\gamma \cdot \psi = 0$. Then we get three simple equations:

$$p \cdot \psi = \gamma \cdot \psi = \not{p}\psi_\mu = 0 \,. \tag{4.37}$$

Now we can compare these equations with those obtained for the ground state of the dual model. We see that in the dual model the condition $\gamma \cdot \psi = 0$ is missing. Therefore the ground state of this sector contains not only a spin-$\frac{3}{2}$ particle but also a spin-$\frac{1}{2}$ particle.

To separate more obviously the spin-$\frac{3}{2}$ and $\frac{1}{2}$ content present in the dual spinor ψ_μ one can use a method which was used in refs. [16] to separate the spin-2 and spin-0 content in $\epsilon_{\mu\nu}$. One introduces a momentum $\bar{p}^\mu$ conjugate to p^μ satisfying $p^2 = \bar{p}^2 = 0; p\bar{p} = 1$.

Then one defines the spin-$\frac{1}{2}$ part of ψ_μ as

$$\psi_\mu^{1/2} = \frac{1}{\sqrt{D-2}}[\gamma_\mu - p_\mu(\bar{p} \cdot \gamma)]\chi \,, \tag{4.38}$$

where χ is a Dirac spinor satisfying $\not{p}\chi = 0$, and notice that it satisfies the dual conditions $p^\mu \psi_\mu = 0; \not{p}\psi_\mu = 0$. Note that $\gamma \cdot \psi^{1/2} = (\sqrt{D-2})\chi$.

The spin-$\frac{3}{2}$ part of ψ_μ is defined as follows:

$$\psi_\mu^{3/2} = \psi_\mu - \tfrac{1}{2}p_\mu(\bar{p} \cdot \gamma)\gamma \cdot \psi \,, \tag{4.39}$$

where ψ_μ satisfies the on-shell Rarita-Schwinger equation:

$$p \cdot \psi = 0 , \qquad \not{p}\psi_\mu = p_\mu \gamma \cdot \psi , \tag{4.40}$$

one verifies easily that $\psi_\mu^{3/2}$ satisfies the dual equations, together with the additional condition

$$\gamma \cdot \psi^{3/2} = 0 . \tag{4.41}$$

In this way one can explicitly separate the content in spins $\frac{3}{2}, \frac{1}{2}$ of an amplitude with an arbitrary number of external states with spin $\frac{3}{2}$ or $\frac{1}{2}$. Further, it can be shown that the dependence of these amplitudes upon the momenta $\bar{p}$ will drop out so that the separation is in fact covariant.

Under a gauge transformation of $\psi \to \psi + p_\mu \epsilon$ where ϵ is an unrestricted spinor, $\psi_\mu^{3/2}$ transforms as

$$\delta\, \psi_\mu^{3/2} = p_\mu \epsilon' , \tag{4.42}$$

with $\epsilon' = \epsilon - \frac{1}{2}(\bar{p} \cdot \gamma)\not{p}\epsilon$. Since $p^2 = 0$, ϵ' satisfy $\not{p}\epsilon' = 0$. Therefore

$$\delta\,|\phi_0\rangle = p \cdot \bar{b}_{1/2}^{\dagger}|0\rangle\epsilon' = G_{-1/2}|0\rangle\epsilon' \tag{4.43}$$

is a state which is both spurious and physical, and it will be decoupled in the S matrix. This allows us to assert that in the dual model the spin-$\frac{3}{2}$ particle will also be transverse.

It is now easy to introduce an interaction between bosonic and fermionic closed strings. As far as defining a self-interaction of bosonic closed strings, this was already done in refs. [32] and [33]. So we only need define (in an analogous way to what one does in the NSR model) the vertices where a fermionic closed string line emits an arbitrary number of states of the bosonic closed strings. More specifically, we can require that the fermionic closed string line should emit an arbitrary number of massless states of the $G_p = +1$ sector of the bosonic closed string (graviton, scalar, antisymmetric tensor). Since the fermionic model is a product of the Neveu-Schwarz model in the $\mathcal{F}_2$ formalism by a Ramond model, the vertex for the emission of the tensor is constructed as a product of the Neveu-Schwarz vertex for a photon emission (in the $\mathcal{F}_2$ formalism), by the vertex of the Ramond model for a photon emission. The selection of which formalism R_1, or R_2 one makes for the Ramond part, depends upon the choice of the propagator. It seems most natural to use for propagator

$$P = \int_0^1 \mathrm{d}x\, x^{\bar{L}_0 - 1/2 + L_0 - 1} \int_0^{2\pi} \mathrm{d}\theta\, \mathrm{e}^{i\theta(L_0 - \bar{L}_0 - 1/2)}$$

and this selects the R_1 formalism for the fermions.

So the vertex for a tensor emission will be given by

$$V = \epsilon^{\mu\nu} V_\mu^{NS} V_\nu^{R} , \tag{4.44}$$

where $\epsilon^{\mu\nu}$ is subjected to the gauge conditions $p^\mu \epsilon_{\mu\nu} = p^\nu \epsilon_{\mu\nu} = 0$ and

$$V_\mu^{NS} = :(\bar{p}_\mu - \bar{H}_\mu \frac{p}{2} \cdot \bar{H}) \exp i \frac{p}{2} \cdot \bar{Q} : , \tag{4.45}$$

and

$$V_\mu^{R(1)} = :(p_\mu + \tfrac{1}{2}\Gamma_\mu \frac{p}{2} \cdot \Gamma) \exp i \frac{p}{2} \cdot Q :$$

$$= \{F_n, V_\mu^{R(2)}\} , \tag{4.46}$$

$$V_\mu^{R(2)} = :\Gamma_\mu \exp i \frac{p}{2} \cdot Q : . \tag{4.47}$$

The peculiarity of the R_1 formalism is that all vertices, except the last one in the chain, are $V_\mu^{(1)}$ vertices; the last one being $V_\mu^{(2)}$.

So we get the expression for the process $\frac{3}{2}$ or $\frac{1}{2} \to \frac{3}{2}$ or $\frac{1}{2} + N$ tensor states

$$A_N = \kappa^N \bar{\psi}_\alpha(q_1) \langle 0; q_1 | b_{1/2}^\alpha \epsilon^{\mu_1\nu_1} V_{\mu_1}^{NS} V_{\nu_1}^{R(2)} P$$

$$\times \epsilon^{\mu_2\nu_2} V_{\mu_2}^{NS} V_{\nu_2}^{R(1)} \ldots \epsilon^{\mu_N\nu_N} V_{\mu_N}^{NS} V_{\nu_N}^{R(1)} b_{1/2}^{\dagger\beta} | 0, q_2 \rangle \psi_\beta(q_2) , \tag{4.48}$$

where each x, θ integral is in fact extended to the whole complex $z = x e^{i\theta}$ plane to have non-planar duality.

We see that if the model is defined in this way, the $G_p = -1$ sector, which contains a tachyon, will never appear. Also we see that in such a process, the handedness of $\psi_\alpha(q_1)$, $\psi_\beta(q_2)$ is the same, and to define both to be left- or right-handed is consistent with our elimination of the $G_p = -1$ sector. Similarly as in the Ramond model, scattering of two spin vector states can also be defined and the same troubles would arise if the $G_p = -1$ sector was included. It is also easy to predict the relative handedness between the open and closed string fermions: since the Noether current of the 10-dimensional Yang-Mills is

$$J^\mu = \gamma^\alpha \gamma^\beta \gamma^\mu \, \mathrm{Tr}(\lambda F_{\alpha\beta}) . \tag{4.49}$$

As we need a non-vanishing coupling $\bar{\psi}_\mu J^\mu$, it follows that the closed string fermions ψ_α must have the same handedness as the open string fermions λ. A word of caution, however, is needed: when we decompose ψ_μ into its spin-$\frac{3}{2}$ and spin-$\frac{1}{2}$ content, we easily see in eq. (4.49) that ψ_μ and χ will have opposite handednesses. So what ap-

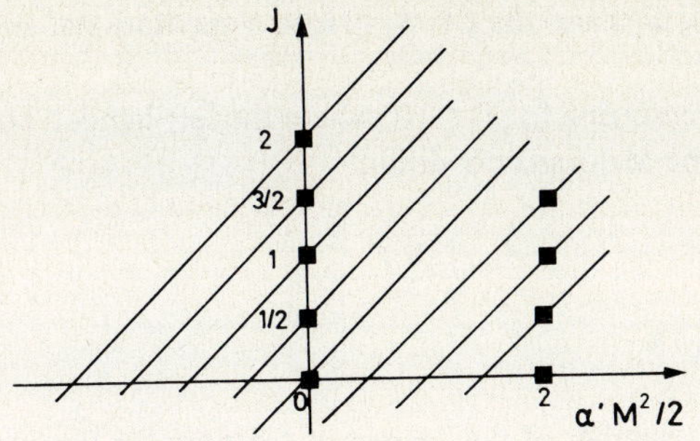

Fig. 6. The spectrum of the supersymmetric NSR model of closed strings.

pears as same handedness in the dual formalism for closed string will appear as opposite in the field theory: again this result is consistent with the generalization to 10 dimensions of a coupling of the type $\bar{\psi}_\mu \not{\partial} A \gamma^\mu \chi$ present in supergravity.

As far as quantum numbers are concerned, as we are dealing with closed string states, all are U(N) singlets, and the graviton and hemitrion do not carry any internal quantum numbers. Further as the fermion-fermion scattering will automatically have s, t, u singularities, it is impossible to assign (at least in a trivial way) a fermionic number to the fermionic closed string states and one is led to make them Majorana fermions.

To summarize our model of closed string states consists of:

(a) A doubled Neveu-Schwarz model, from which the $G_p = -1$ sector has been eliminated, which contains no tachyon nor ghosts, is bosonic, and where the ground state is a massless graviton, a massless scalar, and a massless antisymmetric tensor.

(b) A Neveu-Schwarz ⊗ Ramond model describing Majorana fermions, which are left-handed (in the dual notation), contains no tachyon nor ghosts and whose ground state is a massless hemitrion (spin $\frac{3}{2}$) and a massless scalar.

A pictorial description of this model is given in fig. 6.

Let us start counting the physical degrees of freedom in such a model for bosons and fermions. Let us first do it for $D = 4$, which can be obtained either by defining the model for $D = 4$ or doing it for $D = 10$, but putting to zero all unwanted components (in the first case one drops the left-handedness condition).

Graviton:	$\frac{1}{2}(D-1)(D-2)-1$	$= 2$ degrees of freedom:	e_μ^a fields ,
Dilaton:		1	A ,
Antisym. tensor:	$\frac{1}{2}(D-2)(D-3)$	$= 1$	$A_{\mu\nu} \sim B$,
Spin-$\frac{3}{2}$:	$(D-3)2^{D/2}\,\frac{1}{2}$	$= 2$	ψ_μ ,
Spin-$\frac{1}{2}$:	$2^{D/2}\,\frac{1}{2}$	$= 2$	χ .

We see that again the numbers of fermions and bosons are equal. Further, one can group together the (e_μ^a, ψ_μ) and the (A, B, χ) into a gravitational and a "matter" scalar multiplet.

Let us now do the counting for $D = 10$, where the left-handed condition is imposed in addition to the Majorana condition.

$$e_\mu^a : 35 \qquad A : 1 \qquad A_{\mu\nu} : 28\,,$$

$$\psi_\mu : 56 \qquad \chi : 8\,.$$

Again we get the same number of the degrees of freedom for bosons and fermions [34] but we can no longer split these fields into two multiplets, so that all of them appear necessary to define "pure" supergravity in 10 dimensions. If one can define a supergravity theory in 10 dimensions (which looks now very plausible) by using the same process as in the 10-dimensional Yang-Mills theory, one will get "ordinary" supergravity in interaction with a very large number of matter multiplets. For instance, each 10-dimensional ψ_μ will appear as four-dimensional $\psi_{\mu k}$ and a large number of spinors. Vector fields will appear from many sources, and one should recover as a particular case the recently discovered hypergravity with multiplets $(2, \frac{3}{2})$ and $(\frac{3}{2}, 1)$.

The general counting of fermions and boson states at each level will also work out if we notice that the fermionic closed strings also have their own G_p operator. This G_p operator is, in fact, the generalized γ^{11} operator: $\Gamma^{11} = \gamma^{11}(-1)^{\Sigma d_n^+ d_n}$. All states have $\Gamma^{11} = +1$, but massive states have both $\gamma^{11} = +1$ or -1. The identity one has to prove is then easily deduced from the Jacobi identity.

Finally, we note that we have made many assumptions which could, in principle, all be derived from the loop computations for open strings, and which should be, in principle, checked explicitly in a unified model of open and closed strings. In such a model the constant κ which appears in eq. (4.48) is in fact related to g through a relation of the type:

$$\kappa \sim g^2 \alpha'^{-1}\,. \tag{4.50}$$

In order to have a four-dimensional interpretation of this equality, one compactifies six dimensions. Let V_0 be the volume of the six-dimensional compact space. Then κ_4 and g_4 (the four-dimensional observable coupling constants) are given by

$$\kappa_4 = \kappa (V_0)^{-1/2}\,, \qquad g_4 = g(V_0)^{-1/2}\,. \tag{4.51}$$

So we obtain the relation

$$\kappa_4 \sim g_4^2 \frac{(V_0)^{1/2}}{\alpha'}\,. \tag{4.52}$$

If $V_0 \sim \alpha'^3$ we get $\kappa_4 \sim g_4^2 (\alpha')^{1/2}$, i.e., α' is of the order of Planck's length. On the other hand, if there is an arbitrariness in the scale of V_0, α' is more or less arbitrary. Note that eq. (4.52) is very similar to the expansion for the Fermi constant $\sqrt{G_F}$ in a unified theory of weak and electromagnetic interactions. Only the power of g differs, and this is because in our case the graviton appears as a bound state of photons.

5. Supergravity and the interacting closed strings of the dual spinor model

Now we consider the dual amplitudes for the scattering of closed string states. The external states are any of the ground states of the bosonic and fermionic closed string sectors, and depend upon the coupling constant κ of three closed strings and on α'. We ask whether in the limit $\alpha' \to 0$ these amplitudes can be recovered from a local action, and if this action has anything to do with supergravity. As we do not know (yet) the supergravity theory in 10 dimensions, we can restrict ourselves to establishing the connection with supergravity in four dimensions. To do this, we consider only the sector of the S matrix where the extra momenta are zero (after quantization), and further assume that all polarization tensors with extra indices vanish. As we have seen in Yang-Mills in 10 dimensions, in this case we must assume also that out of each of the four Majorana spinors which describe a 10-dimensional left-handed spinor, only one of them is non-vanishing:

$$\psi_{\mu\kappa} = \psi_\mu \delta_{\kappa 1} , \tag{5.1}$$

$$\chi_\kappa = \chi \delta_{\kappa 1} . \tag{5.2}$$

In this way we obtain a system of fields which are: e^a_μ (graviton), $A(x)$ (scalar), $A_{\mu\nu}(x)$ (pseudoscalar), $\psi_\mu(x)$ (spin $\frac{3}{2}$), $\chi(x)$ (spin $\frac{1}{2}$) where all symbols are four-dimensional. We want to show that in the limit $\alpha' = 0$, this subsector of the dual S matrix interacts just like supergravity with a scalar multiplet.

The subsector of that system which consists of Bose fields and its zero slope limit was studied in refs. [16]. It was shown that the zero slope limit was correctly given by the action:

$$S = \int d^4x\, e\left[-\frac{1}{2\kappa^2} R(e) - \tfrac{1}{2} g^{\mu\nu} \partial_\mu A \partial_\nu A - \tfrac{1}{12} F_{\mu\nu\rho} F^{\mu\nu\rho} f(A)\right] . \tag{5.3}$$

A few differences with respect to refs. [16] are that we now use a vierbein formalism to describe the spin-2 (which does not affect the purely Bose part of the theory), rather than the metric formalism. Also we have rescaled the action so as to agree with the conventions of refs. [10] and [11].

Now we explain why for $D = 4$ we can replace the $A_{\mu\nu}$ antisymmetric tensor by a pseudoscalar field $B(x)$. Let us define $B_\mu = \frac{1}{6}\epsilon_{\mu\alpha\beta\gamma} F^{\alpha\beta\gamma}$. As $F_{\alpha\beta\gamma}$ is completely antisymmetric, this formula can be inverted as $F_{\alpha\beta\gamma} = \epsilon_{\sigma\alpha\beta\gamma} B^\sigma$. So in the action we can replace $F_{\alpha\beta\gamma}$ by B^σ, provided that we take care of the fact that B_μ is constrained by the condition $D^\mu B_\mu = 0$. This we can do by adding to the action a Lagrange multiplier field $B(x)$. The action then becomes

$$S = \int d^4x e \left\{ -\frac{1}{2\kappa^2} R(e) - \tfrac{1}{2} g^{\mu\nu} \partial_\mu A \partial_\nu A + \tfrac{1}{2} B_\sigma B^\sigma f(A) - B^\mu \partial_\mu B \right\} . \qquad (5.4)$$

Using the equations of motion, we see that this is just the first-order formalism for a pseudoscalar field $B(x)$, the second-order formalism being obtained by replacing $B_\sigma f(A) = \partial_\sigma B$ in the action

$$S = \int d^4x e \left\{ -\frac{1}{2\kappa^2} R(e) - \tfrac{1}{2} g^{\mu\nu} \partial_\mu A \partial_\nu A - \tfrac{1}{2} g^{\mu\nu} \partial_\mu B \partial_\nu B f^{-1}(A) \right\} . \qquad (5.5)$$

So indeed we can replace the formalism with an antisymmetric tensor by a formalism with a pseudoscalar, as was already noticed in ref. [35]. More generally suppose that in the action, additional terms of the form

$$S_I = -\int d^4x e F_{\mu\nu\rho} J^{\mu\nu\rho} \qquad (5.6)$$

appeared with $J^{\mu\nu\rho}$ being a completely antisymmetric tensor depending on other fields. Then the second-order action we would get would be (dropping out the Einstein and kinematic term for the A field which are irrelevant for this problem)

$$S = \int d^4x e \{ -\tfrac{1}{2} g^{\mu\nu} \partial_\mu B \partial_\nu B f^{-1}(A) + \partial_\mu B J^\mu f^{-1}(A) - J_\mu J^\mu f^{-1}(A) \} , \qquad (5.7)$$

where

$$J_\mu = \epsilon_{\mu\alpha\beta\gamma} J^{\alpha\beta\gamma} \qquad (5.8)$$

So we see that if we have action, depending on $A_{\mu\nu}$, invariant under $\delta A_\mu = \partial_\mu \zeta_\nu - \partial_\nu \zeta_\mu$ such that only $F_{\alpha\beta\gamma}$ appears in the action and containing at most two derivatives, we can replace $A_{\mu\nu}$ by a pseudoscalar field $B(x)$. It is a pseudoscalar if $J^{\mu\nu\rho}$ is a true tensor, which will be the case. Further, we see that in the resulting action B appears only through its derivatives, and that new contact terms $(J_\mu J^\mu)$ can arise from the transition from one formalism to the other.

Similarly, in the dual S matrix formulation, one can also replace $A_{\mu\nu}$ by a pseudoscalar. The gauge invariance $\delta A_{\mu\nu} = p_\mu \zeta_\nu - p_\nu \zeta_\mu$ is quite different from the graviton gauge invariance as there are no self couplings of the antisymmetric tensor $A_{\mu\nu}$. In fact there is no non-trivial antisymmetric conserved current to which $A_{\mu\nu}$ can couple, so that an S matrix amplitude can always be rewritten in such a way that only $F_{\mu\nu\rho}(p)$ appears in the external states. As before we can replace $F_{\mu\nu\rho}$ by $B_\sigma(p) = \frac{1}{6} \epsilon_{\sigma\alpha\beta\gamma} F^{\alpha\beta\gamma}$. Now B_σ can be expressed as p_σ multiplied by a pseudoscalar quantity. Indeed in four dimensions we can introduce a system of four vectors:

$$p_\mu; \bar{p}_\mu; \epsilon^i_\mu \qquad i = 1, 2 , \qquad \text{such that} \quad p \cdot \bar{p} = 1 ,$$

$$p \cdot \epsilon^i = \bar{p} \cdot \epsilon^i = \epsilon^1 \cdot \epsilon^2 = 0 , \qquad p^2 = \bar{p}^2 = 0 , \qquad \epsilon_1^2 = \epsilon_2^2 = 1 .$$

$A_{\mu\nu}$ can be represented in this basis of vectors. Using the condition that $p^\mu A_{\mu\nu} = 0$ one gets

$$A_{\mu\nu} = A(\epsilon_\mu^1 \epsilon_\nu^2 - \epsilon_\nu^1 \epsilon_\mu^2) + B \sum_{i=1}^{2} (\epsilon_\mu^i p_\nu - \epsilon_\nu^i p_\mu) , \tag{5.9}$$

$$B_\sigma = 2\epsilon_{\sigma\alpha\beta\gamma} A \epsilon_1^\alpha \epsilon_2^\beta p^\gamma . \tag{5.10}$$

Expanding

$$B_\sigma = a p_\sigma + b \bar{p}_\sigma + \sum_{i=1,2} c_i \epsilon_\sigma^i , \tag{5.11}$$

one sees that

$$b = c_1 = c_2 = 0 ,$$

$$a = 2\epsilon_{\sigma\alpha\beta\gamma} \epsilon_1^\alpha \epsilon_2^\beta p^\gamma \bar{p}^\sigma , \tag{5.12}$$

and therefore

$$B_\sigma = p_\sigma (\epsilon^{\mu\alpha\beta\gamma} A^{\alpha\beta} p^\gamma \bar{p}^\mu) . \tag{5.13}$$

This result is rather important as it shows that in four dimensions a lot of couplings of the $A_{\mu\nu}$ field vanish on the mass shell. To give an example, in eq. (5.3) a function $f(A)$ was included because in general a coupling of order κ in $F_{\mu\nu\rho} F^{\mu\nu\rho} A$ is found, even on shell. However, in four dimensions this coupling vanishes on mass shell as it gives through the previous arguments a coupling proportional to $p^{(1)} \cdot p^{(2)} = 0$. So the function $f(A)$ can be got rid of (at least at first order in κ) by a field redefinition.

Now that the pseudoscalar character of $A_{\mu\nu}(x)$ is well established, we consider the zero slope limit of the dual model restricted to its four-dimensional sector. It is now obvious that one gets (up to canonical field redefinitions which do not affect the S matrix) the supergravity theory in interaction with the scalar multiplet written down in refs. [10] and [11]. This is because for $\alpha' = 0$ all the couplings have dimensions fixed by κ, and because we have both on-shell general covariance ($\epsilon_{\mu\nu} \to \epsilon_{\mu\nu} + p_\mu \zeta_\nu + p_\nu \zeta_\mu$) and local supersymmetry invariance ($\psi_\mu \to \psi_\mu + \epsilon p_\mu$), and the same system of local fields ($e_\mu^a, \psi_\mu, A, B, \chi$) as starting point. As explained in sect. 5 of ref. [10], on-shell gauge invariance of the S matrix is sufficient alone to generate all trilinear and higher couplings needed in the theory.

It is, however, both pedagogical and amusing to verify that the three-point couplings do agree. Also we shall learn about supergravity in 10 dimensions in doing so. As the three-point couplings for Bose fields have already been studied, we shall consider only the trilinear couplings where a fermionic line emits a boson. All these couplings are expressed by a single expression in the dual model:

$$V_3 = \kappa \overline{\psi}_\alpha(q_1)\langle 0, q_1 | b^\alpha_{1/2}\, \epsilon^{\mu\nu}(p)\, V^{\rm NS}_\mu\, V^{\rm R_2}_\nu\, b^{\dagger\beta}_{1/2} | 0, q_2\rangle \psi_\beta(q_2)\,, \tag{5.14}$$

$$V_3 = \kappa \overline{\psi}_\alpha(q_1)[\delta_{\alpha\mu} p^\beta + \delta_{\beta\mu} q_2^\alpha + \delta_{\alpha\beta} q_1^\mu]\gamma^\nu \psi_\beta(q_2)\epsilon_{\mu\nu}(p)\,. \tag{5.15}$$

The on-shell gauge invariance of this expression, both with respect to $\epsilon_{\mu\nu}$ and ψ_μ, is easy to verify. Note that the dependence on α' in V_3 disappears completely, so that the $\alpha' = 0$ limit is trivial on the trilinear couplings.

As ψ_μ represents both spin $\frac{3}{2}, \frac{1}{2}$ and $e_{\mu\nu}$ both spin 2 and two spin 0, the expression contains a lot of information. If we look at the vertex emission of a graviton ($\epsilon_{\mu\nu}$ symmetric and traceless) it is easy to verify that it couples to the energy momentum tensors of the spin $\frac{3}{2}$ and spin $\frac{1}{2}$ as it should. The graviton makes no transition between spin $\frac{3}{2}$ and spin $\frac{1}{2}$ for such a transition. Using the expression (4.38) for $\psi_\beta^{1/2}$, and the fact that $\psi^{3/2}\cdot\gamma = 0$, one has

$$V_3(\tfrac{3}{2};\tfrac{1}{2}) = \frac{\kappa}{\sqrt{D-2}}\overline{\psi}^{3/2}_\alpha(q_1)\gamma^\nu[\delta_{\alpha\mu} p\cdot\gamma + \gamma_\mu q_{2\alpha} + q_{1\mu}\gamma_\alpha]\chi\epsilon_{\mu\nu}(p)\,, \tag{5.16}$$

$$= \frac{\kappa}{\sqrt{D-2}}\overline{\psi}^{3/2}_\alpha(q_1)[2(q^{(1)}_\mu\delta_{\alpha\nu} - q^{(1)}_\nu\delta_{\alpha\mu} + \sigma^{\mu\nu}q_1^\alpha)$$

$$-\eta^{\mu\nu}p^\alpha]\chi\epsilon_{\mu\nu}\,, \tag{5.17}$$

In this expression the traceless symmetric part does not contribute. So only the antisymmetric tensor and the scalar contribute to the $\frac{3}{2}, \frac{1}{2}$ transition.

Before we look in greater detail into this transition we can look at the diagonal $(\frac{3}{2}, \frac{3}{2})$ or $(\frac{1}{2}, \frac{1}{2})$ processes. It is easy to check that the scalar A (i.e., the trace part of $\epsilon_{\mu\nu}$) gives vanishing contribution to these processes: therefore no term of the form $\overline{\psi}_\mu \ldots \psi_\nu A$ of $\overline{\chi} \ldots \chi A$ will appear in the action. The case of the antisymmetric part is much less obvious. If one looks, for instance, at a vertex like $A_{\mu\nu}\overline{\chi} \ldots \chi$, one finds it *a priori* non-vanishing and proportional to the expression

$$A_{\mu\nu}\overline{\chi}(q_1)(\gamma^\mu q_2^\nu - \gamma^\nu q_2^\mu)\chi(q_2)\,. \tag{5.18}$$

So a trilinear coupling of the form

$$A_{\mu\nu}\overline{\chi}(\gamma^\mu\partial_\nu - \gamma^\nu\partial_\mu - 2\sigma^{\mu\nu}\not\partial)\chi \tag{5.19}$$

should appear in the action. However, for a Majorana spinor χ, this expression is equivalent to

$$-\tfrac{1}{2}A_{\mu\nu}\partial_\rho[\bar{\chi}\gamma^\mu\gamma^\nu\gamma^\rho\chi] = \tfrac{1}{6}F_{\mu\nu\rho}(\bar{\chi}\gamma^\mu\gamma^\nu\gamma^\rho\chi) \tag{5.21}$$

by integrating by parts. The same manipulation can be made on the S matrix element which takes this form. In general ($D \neq 4$) this is non-vanishing even on-shell. For $D = 4$, however, it does, because it is proportional to $B_\sigma\bar{\chi}\gamma_5\gamma^\sigma\chi$ and on-shell, as we have seen B_σ is proportional to p_σ. Using then the Dirac equation for χ, $\bar{\chi}$, one obtains zero on-mass shell. The same result holds after a more complicated computation for the $\frac{3}{2}, \frac{3}{2}, A_{\mu\nu}$ vertex: it does vanish on-shell, but only for $D = 4$.

Therefore the only non-vanishing couplings, apart from the gravitational couplings, are the $\frac{3}{2} - \frac{1}{2} - 0^+$ and $\frac{3}{2} - \frac{1}{2} - 0^-$ couplings. The first one is very easily seen to be proportional to $\bar{\psi}_\mu \not{\partial} A\gamma^\mu\chi$. For the second one, one has to use manipulations similar to the previous one, but one obtains a non-zero result, namely a $\bar{\psi}_\mu\gamma_5\not{\partial} B\gamma^\mu\chi$ coupling. A careful study reveals that also the magnitudes of the coupling agree with those of refs. [10] and [11].

A non-trivial check of our results is that the supergravity action in four dimensions, in interaction with a massless scalar multiplet, can be written in such a way that B appears only through its derivative. From the dual model we know that a consistent formalism in four dimensions would be to replace the scalar multiplet (A, B, χ) by an $(A, A_{\mu\nu}, \chi)$ multiplet. The gauge invariance $\delta A_{\mu\nu} = \partial_\mu\zeta_\nu - \partial_\nu\zeta_\mu$ would imply, through the reasoning of the beginning of this section, that in the (A, B, χ) formalism only $\partial_\sigma B$ appears.

Looking at the action first written in ref. [10] and proved to be exact to all orders in ref. [11], we see that there is only one offending term in the action which does not contain a derivative on B. It is (we make a temporary return to the notations and conventions of ref. [10] to avoid notational confusion):

$$-\tfrac{1}{8}i\kappa^2 B\partial_\rho A[e\bar{\chi}\gamma_5\gamma^\rho\chi - \epsilon^{\mu\nu\rho\sigma}\bar{\psi}_\mu\gamma_\nu\psi_\sigma]\,. \tag{5.22}$$

Integrating by parts, it gives a $\partial_\rho B$ term, and the still offending term

$$\tfrac{1}{4}i\kappa^2 AB[e\bar{\chi}\gamma_5\gamma^\mu D_\mu\chi - \epsilon^{\mu\nu\rho\sigma}\bar{\psi}_\mu\gamma_\nu D_\rho\psi_\sigma]\,. \tag{5.23}$$

This term can be grouped with the kinetic terms of χ and ψ_μ and now reads

$$-\tfrac{1}{2}\bar{\chi}(1 - \tfrac{1}{2}i\kappa^2\gamma_5 AB)\gamma^\mu D_\mu\chi - \tfrac{1}{2}\epsilon^{\mu\nu\rho\sigma}\bar{\psi}_\mu(1 + \tfrac{1}{2}i\kappa^2\gamma_5 AB)\gamma_\nu D_\rho\psi_\sigma\,. \tag{5.24}$$

We can now get rid of it through a field redefinition. Setting:

$$\chi' = \exp(-\tfrac{1}{4}i\kappa^2\gamma_5 AB)\chi\,, \tag{5.25}$$

$$\psi'_\mu = \exp(\tfrac{1}{4}i\kappa^2\gamma_5 AB)\psi_\mu\,, \tag{5.26}$$

one sees that one obtains for these four terms the expression

$$-\tfrac{1}{2}\bar{\chi}'\gamma^\mu D_\mu \chi' - \tfrac{1}{2}\epsilon^{\mu\nu\rho\sigma}\bar{\psi}'_\mu \gamma_\nu D_\rho \psi'_\sigma \, . \tag{5.27}$$

Further, all other terms in the action turn into their counterparts. Apart from the Noether current, all bilinears appearing in the action are of the form $\bar{\psi}_\mu \gamma^\alpha \psi_\nu$ or $\bar{\psi}_\mu \gamma^\alpha \gamma^5 \psi_\nu$, or $\bar{\chi}\gamma^\alpha\gamma^5\chi$. For instance

$$\bar{\psi}_\mu \gamma^\alpha \psi_\nu = \bar{\psi}'_\mu \exp(\tfrac{1}{4} i\kappa^2 \gamma_5 AB)\gamma^\alpha \exp(\tfrac{1}{4} i\kappa^2 \gamma_5 AB)\psi'_\nu = \bar{\psi}'_\mu \gamma^\alpha \psi'_\nu \, .$$

The only exception is the Noether current where an even number of γ's appear. This is, however, just what is needed, as χ and ψ transform with the opposite sign in the exponential. So the supergravity in interaction with matter can indeed be written in the following way:

$$\mathcal{L} = \mathcal{L}^{\mathrm{G}}(e^a_\mu, \psi'_\mu) + \mathcal{L}^{\mathrm{M}} \, ,$$

$$\begin{aligned}\mathcal{L}^{\mathrm{M}} = & -\tfrac{1}{2} e g^{\mu\nu}(\partial_\mu A \partial_\nu A + \partial_\mu B \partial_\nu B) + \tfrac{1}{2} e\kappa \bar{\psi}'(\partial\!\!\!/ A + i\gamma_5 \partial\!\!\!/ B)\gamma^\mu \chi' \\ & - \tfrac{1}{2} e\bar{\chi}'\gamma^\mu D_\mu \chi' + \tfrac{1}{32}\kappa^2(\bar{\chi}'\gamma_5\gamma_\tau\chi')[\epsilon^{\mu ab\tau}\bar{\psi}'_\mu \gamma_a \psi'_b - 2e(\bar{\psi}'_\alpha \gamma_5 \gamma^\tau \psi'^{\alpha})] \\ & + \tfrac{1}{4} i\kappa^2 A \partial_\beta B[e\bar{\chi}'\gamma_5\gamma^\beta\chi' - \epsilon^{\alpha\beta\mu\tau}\bar{\psi}'_\alpha \gamma_\tau \psi'_\mu] - \tfrac{1}{16} e\kappa^2(\bar{\chi}'\chi')^2 \, . \end{aligned} \tag{5.28}$$

Using eqs. (5.3)–(5.5), it is now trivial to go from the (A, B, χ) formalism to the $(A, A_{\mu\nu}, \chi)$ formalism. Note that the existence of that formalism, deduced from the dual model, gives an interesting insight as to why in the $m = 0$ case no infinite series occurred in the scalar multiplet action.

Now we notice that to obtain the $(\frac{3}{2}, 2); (\frac{1}{2}, 0)$ supergravity theory we have enormously restricted the dual spectrum at zero mass. The O(2) supergravity theory [6] can also be obtained out of the dual S matrix. It contains one spin 2, two spin $\frac{3}{2}$ and one spin 1. The spin 2 is simply obtained out of the vierbein components e^a_μ of the 10-dimensional theory with $0 \leqslant \mu, a \leqslant 3$. The 2 spin $\frac{3}{2}$ are obtained by setting to zero only 2 of the 4 Majorana spin $\frac{3}{2}$ which compose the 10-dimensional spin $\frac{3}{2}$: $\psi = (\psi^{(1)}_\mu, \psi^{(2)}_\mu, 0, 0)$. The vector field A_μ is harder to obtain as it arises from various sources. It can come out of extra components of the vierbein, namely e^6_μ, e^9_μ which are seen to make transitions between $\psi^{(1)}_\mu, \psi^{(2)}_\mu$, from the vertex of eq. (5.15). Actually, the computation of this vertex reveals that these components of the vierbein have the same Noether couplings as in ref. [6]. Similarly, the $A_{\mu 6}$ or $A_{\mu 9}$ fields are also candidates to contribute to A_μ. $F_{\mu\nu 9} = \partial_\mu A_{\nu 9} - \partial_\nu A_{\mu 9}$ is indeed a Maxwell tensor and one can verify that $A_{\mu 6}, A_{\mu 9}$ couple again to $\psi^{(1)}_\rho, \psi^{(2)}_\rho$ with the same Noether coupling as in ref. [6]. So the A_μ field is in fact a combination of those four fields. Once the Noether coupling is correctly obtained, the previous arguments imply that the quartic couplings are the same as in field theory.

By restricting less and less the dual spectrum one will obtain a model with 3 spin-$\frac{3}{2}$, 3 spin-1 and one spin-$\frac{1}{2}$ [34], and internal symmetry O(4). Next a model with 4 spin-$\frac{3}{2}$, 6 vectors, 4 spin-$\frac{1}{2}$, and two spin-zero, and internal symmetry O(4). In this model which forms an irreducible representation of O(4), the degrees of freedom inherent in the 10-dimensional model are not yet exhausted, as in principle we can have 6 vectors $e_{\mu i}$ and 6 vectors $A_{\mu i}$. So the largest model one can in principle obtain out of 10 dimensions will be O(4) invariant but will contain several representations of O(4) supergravity. Counting reveals that it will contain 6 times the representation of O(4) supersymmetry with $J_{\max} = 1$, which itself contains one spin-1, four spin-$\frac{1}{2}$ and six spin-0. Even though these models have not yet been constructed, the dual theory is seen to imply their existence.

References

[1] S. Weinberg, Phys. Rev. Letters 19 (1967) 1264;
A. Salam, Elementary particle theory, ed. N. Svartholm (Almquist, Stockholm, 1968).

[2] G. 't Hooft, Nucl. Phys. B35 (1971) 167.

[3] S. Glashow, J. Iliopoulos and L. Maiani, Phys. Rev. D2 (1970) 1285.

[4] G. 't Hooft and M. Veltman, Ann. Inst. H. Poincaré 20 (1974) 69;
S. Deser, H.-S. Tsao and P. Van Nieuwenhuizen, Phys. Rev. D10 (1974) 3337;
P. Van Nieuwenhuizen, Stony Brook report No. ITP-SB-75-46 (1973).

[5] D.Z. Freedman, P. Van Nieuwenhuizen and S. Ferrara, Phys. Rev. D13 (1976) 3214;
S. Deser and B. Zumino, Phys. Letters 62B (1976) 335.

[6] S. Ferrara and P. Van Nieuwenhuizen, ITP-SB-76-48 preprint (1976), to be published.

[7] M.T. Grisaru, P. Van Nieuwenhuizen and J.A.M. Vermaseren, ITP-SB-76/42 preprint (1976), to be published.

[8] S. Ferrara, P. Van Nieuwenhuizen and J. Scherk, Phys. Rev. Letters 37 (1976) 1035.

[9] D.Z. Freedman and J.H. Schwarz, ITP-SB-76-41 preprint (1976), to be published.

[10] S. Ferrara, F. Gliozzi, J. Scherk and P. Van Nieuwenhuizen, Nucl. Phys. B117 (1976) 333.

[11] S. Ferrara, D.Z. Freedman, P. Van Nieuwenhuizen, P. Breitenlohner, F. Gliozzi and J. Scherk, ITP-SB-76-46 preprint (1976), to be published.

[12] P. Van Nieuwenhuizen and J.A.M. Vermaseren, ITP-SB-76-44 preprint (1976), to be published.

[13] P. Van Nieuwenhuizen and C.C. Wu, ITP-SB-76-38 preprint (1976).

[14] Dual Theory, ed. M. Jacob (North-Holland, Amsterdam, 1974);
P.H. Frampton, Dual resonance models (Benjamin, 1974);
J. Scherk, Rev. Mod. Phys. 47 (1975) 123.

[15] A. Neveu and J. Scherk, Nucl. Phys. B36 (1972) 155.

[16] J. Scherk and J.H. Schwarz, Nucl. Phys. B81 (1974) 118;
T. Yoneya, Nuovo Cimento Letters 8 (1973) 951;
T. Yoneya, Progr. Theor. Phys. 51 (1974) 1907;
J. Scherk and J.H. Schwarz, Phys. Letters 52B (1974) 347;
J. Scherk and J.H. Schwarz, Phys. Letters 57B (1975) 463;
J. Scherk, Proc. Int. Conf. on non-local quantum field theory, Alushta, 1976.

[17] E. Cremmer and J. Scherk, Nucl. Phys. B103 (1976) 399; B108 (1976) 409; B118 (1977) 61.

[18] A. Neveu and J.H. Schwarz, Nucl. Phys. B31 (1971) 86; Phys. Rev. D4 (1971) 1109;
P. Ramond, Phys. Rev. D3 (1971) 2415.

[19] L. Brink, P. di Vecchia and P. Howe, Göteborg preprint (1976);
S. Deser and B. Zumino, Phys. Letters 65B (1976) 369.
P.A. Collins and R.W. Tucker, Phys. Letters 64B (1976) 207.
[20] P. Fayet, private communication.
[21] F. Gliozzi, J. Scherk and D. Olive, Phys. Letters 65B (1976) 282.
[22] S. Ferrara and B. Zumino, Nucl. Phys. B79 (1974) 413;
A. Salam and J. Strathdee, Phys. Letters 51B (1974) 353.
[23] P. Fayet, Nucl. Phys. B113 (1976) 135.
[24] J. Shapiro, Phys. Rev. D5 (1972) 1945.
[25] M.A. Virasoro, Phys. Rev. 177 (1969) 2309;
J.A. Shapiro, Phys. Letters 33B (1970) 361.
[26] D. Ebert, H.J. Otto and V.N. Pervushin, presented at Int. Conf. on non-local field theory, Alushta, 1976.
[27] L. Brink and D. Olive, Aspen Talk, 1974, unpublished.
[28] P. Fayet and S. Ferrara, PTENS 76/11 (1976), to be published in Phys. Reports, and references therein.
[29] W. Pauli, Ann. Inst. Poincaré 6 (1936) 109;
K.M. Case, Phys. Rev. 97 (1954) 810;
A. Pais, J. Math. Phys. 3 (1962) 1135.
[30] C.G.J. Jacobi, Fundamenta, Königsberg (1829).
[31] J. Scherk, Nucl. Phys. B31 (1971) 222;
N. Nakanishi, Progr. Theor. Phys. 48 (1972) 355.
[32] J. Schwarz, in Dual Theory.
[33] L. Clavelli and J. Shapiro, Nucl. Phys. B57 (1973) 490.
[34] R. Haag, I. Łopuszanski and M. Sohnius, Nucl. Phys. B88 (1976) 257.
[35] E. Cremmer and J. Scherk, Nucl. Phys. B72 (1974) 117.
[36] S. Ferrara, J. Scherk and B. Zumino, PTENS 76/23, to be published in Phys. Letters B;
D.Z. Freedman, ITP-SB-76-58 preprint.

SUPERSYMMETRICAL STRING THEORIES ☆

Michael B. GREEN [1] and John H. SCHWARZ
California Institute of Technology, Pasadena, CA 91125, USA

Received 11 December 1981

A supersymmetrical light-cone-gauge string action is presented. It provides a basis for understanding the previously-studied supersymmetrical dual string theory as well as two new closed-string theories that have extended supersymmetry in ten dimensions, corresponding to $N = 8$ supersymmetry in four dimensions.

In several recent papers [1–3], we have developed a new oscillator description of the supersymmetrical dual string theory [4]. It has proved to be superior to older formulations for understanding the ten-dimensional supersymmetry of the theory and for explicitly calculating scattering amplitudes. This letter describes a light-cone-gauge string action from which this oscillator formalism can be derived. The string action also leads to the discovery of two new closed-string theories with extended supersymmetry in ten dimensions (corresponding to $N = 8$ supersymmetry in four dimensions). Unlike any other known dual string theory, these closed-string theories turn out to be completely free of divergences in the one-loop approximation (which is all that has been checked so far).

The conventional (Veneziano) string model can be based on an action principle for the string coordinates $X^{\mu}(\sigma, \tau)$ [5] ‡1. The action can be written as a σ–τ reparametrization-invariant expression by using an auxiliary two-dimensional metric tensor. The local invariance allows one to choose an orthonormal gauge in which the light-cone component $X^{-}(\sigma, \tau)$ may be regarded as a dependent variable [7]. The remaining gauge freedom (plus partial use of equations of motion) allows one to set

$$X^{+}(\sigma, \tau) = x^{+} + 2\alpha' p^{+} \tau, \tag{1}$$

where x^{+} and p^{+} are constants. Once this is done, the string dynamics resides entirely in the transverse coordinates $X^{i}(\sigma, \tau)$. A description of the supersymmetrical string theories requires fermionic degrees of freedom, $S^{Aa}(\sigma, \tau)$, as well. The index $A = 1, 2$ describes a spinor in the σ–τ space, whereas the index a is 32-valued, as appropriate for a spinor in 10-dimensional spacetime. S^{1a} and S^{2a} are Majorana–Weyl spinors in ten dimensions. For much of the analysis it is not necessary to specify whether S^{1} and S^{2} have the same or opposite handedness, but ultimately both cases will turn out to be of interest. Corresponding to eq. (1), the fermionic field is required to satisfy the gauge condition

$$(\gamma^{+})^{ab} S^{Ab} = 0. \tag{2}$$

The postulated two-dimensional string action in the light-cone gauge is ‡2

☆ Work supported in part by the US Department of Energy under Contract No. DE-AC03-81-ER40050.

[1] On leave from Queen Mary College, University of London, England.

‡1 As customary, σ is a spacelike variable that runs from 0 to π along the string, while τ is a timelike evolution parameter. These are represented collectively by a two-vector index α. μ is a vector index for the physical spacetime, 26-valued for the Veneziano model and 10-valued for the supersymmetrical models. For a collection of review articles, see ref. [6].

‡2 For the two-dimensional Dirac matrices, the Majorana representation $\rho^{0} = \begin{pmatrix} 0 & -i \\ i & 0 \end{pmatrix}$, $\rho^{1} = \begin{pmatrix} 0 & i \\ i & 0 \end{pmatrix}$ is convenient. We also define $\bar{S}^{Aa} = S^{\dagger Bb} (\gamma^{0})^{ba} (\rho^{0})^{BA}$, $\rho_{3} = \rho^{0}\rho^{1} = \mathrm{diag}(1, -1)$ A spacelike metric is used for both two-vectors and ten-vectors. We henceforth set $\alpha' = 1/2$.

$$\int d\tau \int_0^\pi d\sigma[-(4\pi\alpha')^{-1}\partial_\alpha X^i \partial^\alpha X^i + (i/4\pi)\bar{S}\gamma^- \rho^\alpha \partial_\alpha S]. \tag{3}$$

A covariant gauge-invariant action from which eqs. (1)–(3) can be deduced is not yet known. The equations of motion implied by (3), subject to boundary conditions to be discussed, are

$$\partial^\alpha \partial_\alpha X^i = 0 \quad \text{and} \quad \rho^\alpha \partial_\alpha S = 0. \tag{4}$$

This action is invariant under the supersymmetry transformations

$$\delta X^i = (p^+)^{-1/2} \bar{\epsilon}\gamma^i S, \tag{5a}$$

$$\delta S = i(p^+)^{-1/2}\gamma_- \gamma_\mu (\rho \cdot \partial X^\mu)\epsilon, \tag{5b}$$

again assuming appropriate boundary conditions. These transformations may be contrasted with those in an earlier formulation in which ϵ was a spinor only in the two-dimensional sense [8]. The parameters ϵ^{Aa} are Majorana–Weyl in $d = 10$, and may either be regarded as parametrizing two supersymmetries in ten dimensions or 16 supersymmetries in two dimensions. They may even depend on σ and τ provided that

$$(\partial/\partial\tau - \rho_3 \partial/\partial\sigma)\epsilon = 0, \tag{6}$$

but we are primarily interested in the case of constant ϵ. Commuting two global supersymmetry transformations, one finds (using $\rho \cdot \partial S = 0$) that

$$[\delta_1, \delta_2] X^i = \xi^\alpha \partial_\alpha X^i + a^i, \qquad [\delta_1, \delta_2] S = \xi^\alpha \partial_\alpha S, \tag{7a,b}$$

with

$$\xi^\alpha = -(2i/p^+)\bar{\epsilon}^{(1)}\gamma_- \rho^\alpha \epsilon^{(2)}, \tag{8}$$

$$a^i = -2i\bar{\epsilon}^{(1)}\rho^0 \gamma^i \epsilon^{(2)}, \tag{9}$$

which may be interpreted as translations of (σ, τ) and X^i, respectively. The supersymmetry charges obtained from eqs. (3) and (5) by the Noether method are

$$Q^{Aa} = \frac{i}{\pi}(p^+)^{-1/2} \int_0^\pi [\gamma_\mu (\rho \cdot \partial X^\mu)\rho^0 S]^{Aa} d\sigma. \tag{10}$$

Boundary conditions needed to derive the equations of motion (4) must be considered carefully. In the case of open strings, it is necessary that S^1 and S^2 have the same handedness and

$$\partial X^i(0, \tau)/\partial\sigma = \partial X^i(\pi, \tau)/\partial\sigma = 0, \tag{11}$$

$$S^{1a}(0, \tau) = S^{2a}(0, \tau), \qquad S^{1a}(\pi, \tau) = S^{2a}(\pi, \tau). \tag{12a, b}$$

Possible minus signs in eq. (12) are excluded by requiring invariance under global supersymmetry transformations. In this case, one also needs $\epsilon^{1a} = -\epsilon^{2a}$, so that there is, in fact, just one ten-dimensional supersymmetry invariance in the open-string case. Eqs. (4), (11), (12) lead to the mode expansions

$$X^i = x^i + p^i\tau + i \sum_{n\neq 0} \frac{1}{n} \alpha_n^i \cos n\sigma\, e^{-in\tau}, \tag{13}$$

$$S^{1a} = \sum_{n=-\infty}^{\infty} S_n^a \exp[-in(\tau - \sigma)], \tag{14a}$$

$$S^{2a} = \sum_{n=-\infty}^{\infty} S_n^a \exp[-in(\tau + \sigma)], \tag{14b}$$

Substituting these expansions in eq. (10), one finds that Q^{1a} and Q^{2a} are not separately conserved (they depend on τ) but that

$$Q^a = \tfrac{1}{2}(Q^{1a} + Q^{2a})$$

$$= i(p^+)^{1/2}(\gamma_+ S_0)^a + i(p^+)^{-1/2} \sum_{-\infty}^{\infty} (\gamma_i S_{-n})^a \alpha_n^i \tag{15}$$

is conserved.

Canonical quantization gives

$$[\alpha_m^i, \alpha_n^j] = m\delta_{m+n,0}\delta^{ij}, \tag{16}$$

$$\{S_m^a, \bar{S}_n^b\} = (\gamma^+ h)^{ab}\delta_{m+n,0}, \tag{17}$$

$$[\alpha_m^i, S_n^a] = 0, \tag{18}$$

where h represents a Weyl projection operator $\frac{1}{2}(1 \pm \gamma_{11})$. Eqs. (15)–(18) were discovered in ref. [1], where it is proved that

$$\{Q^a, \bar{Q}^b\} = -2(h\gamma \cdot p)^{ab}, \tag{19}$$

provided that one imposes the mass-shell condition

$$p^- = (1/2p^+)(p^i p^i + N), \tag{20}$$

$$N = \sum_{n=1}^{\infty} (\alpha_{-n}^i \alpha_n^i + \tfrac{1}{2} n \bar{S}_{-n}\gamma^- S_n). \tag{21}$$

Ref. [1] gives formulas for the Lorentz generators $J^{\mu\nu}$ in terms of the α_n^i and the S_n^a and proves that they have the correct algebra when eq. (20) is satisfied. A proof that Q^a transforms as a spinor in the ten-dimensional sense is also given there. Lorentz covariance is not at all obvious in the light-cone-gauge formalism [the action in eq. (3) only has manifest covariance in the eight-dimensional transverse space]. However, functional integral methods analogous to those developed for the conventional model [9] can be used to obtain covariant S-matrix elements from the light-cone-gauge action. The results are the same as obtained by the operator methods in our previous works [1–3].

Turning attention to closed strings, we replace the boundary conditions in eqs. (11) and (12) with the requirement that X^i and S^{Aa} are periodic in σ with period π. This, together with eq. (4), leads to the closed-string mode expansions

$$X^i(\sigma,\tau) = x^i + p^i\tau + \frac{\mathrm{i}}{2}\sum_{n\neq 0}\frac{1}{n}\{\alpha_n^i \exp[-2\mathrm{i}n(\tau-\sigma)] + \tilde{\alpha}_n^i \exp[-2\mathrm{i}n(\tau+\sigma)]\}, \tag{22}$$

$$S^{1a} = \sum_{-\infty}^{\infty} S_n^a \exp[-2\mathrm{i}n(\tau-\sigma)], \tag{23a}$$

$$S^{2a} = \sum_{-\infty}^{\infty} \tilde{S}_n^a \exp[-2\mathrm{i}n(\tau+\sigma)]. \tag{23b}$$

The canonical commutation relations give eqs. (16)–(18) for the operators with tildes as well as the ones without. Also

$$[\alpha_m^i, \tilde{\alpha}_n^j] = [\alpha_m^i, \tilde{S}_n^a] = [\tilde{\alpha}_m^i, S_n^a] = \{S_m^a, \tilde{S}_n^b\} = 0. \tag{24}$$

The handedness of S^1 and S^2 may be independently assigned. Substituting the mode expansions (22) and (23) into eq. (10) gives

$$Q^{1a} = \mathrm{i}(p^+)^{1/2}(\gamma_+ S_0)^a + 2\mathrm{i}(p^+)^{-1/2}\sum_{-\infty}^{\infty}(\gamma_i S_{-n})^a \alpha_n^i, \tag{25a}$$

$$Q^{2a} = \mathrm{i}(p^+)^{1/2}(\gamma_+ \tilde{S}_0)^a + 2\mathrm{i}(p^+)^{-1/2}\sum_{-\infty}^{\infty}(\gamma_i \tilde{S}_{-n})^a \tilde{\alpha}_n^i. \tag{25b}$$

These are separately conserved and give the algebra of extended supersymmetry in ten dimensions

$$\{Q^{Aa}, \bar{Q}^{Bb}\} = -2(h^A \gamma\cdot p)^{ab}(\rho^0)^{AB}. \tag{26}$$

The mass-shell condition in eq. (20) is replaced for closed strings by the conditions

$$L_0 = \tfrac{1}{8}p^2 + N = 0, \qquad \tilde{L}_0 = \tfrac{1}{8}p^2 + \tilde{N} = 0, \tag{27a,b}$$

where N is given in eq. (21) and $\tilde{N}$ is the corresponding expression with $\tilde{\alpha}$ and $\tilde{S}$. (The closed-string Regge slope is $\frac{1}{4}$—half of the open-string slope.) Eq. (27) is required both for the closure of the Lorentz algebra and for the validity of the supersymmetry algebra in eq. (26). One can define a truncation to a closed-string theory with simple supersymmetry by restricting the physical spectrum to states that are symmetrical under the interchanges $\alpha_m^i \leftrightarrow \tilde{\alpha}_m^i$ and $S_m^a \leftrightarrow \tilde{S}_m^a$. (This requires that S_m^a and $\tilde{S}_m^a$ have the same handedness, $h^1 = h^2$.) In this case $Q = \frac{1}{2}(Q^1 + Q^2)$ is the only remaining supersymmetry. This truncated theory is of importance because it describes the closed strings that interact with the open strings describes above, whereas untruncated closed-string models cannot interact with open strings (just as $N = 4$ Yang–Mills cannot couple to $N = 8$ supergravity).

The 256 massless physical closed-string states may be described by ket vectors $|ij\rangle$, $|ia\rangle$, $|ai\rangle$, $|ab\rangle$ where the first label refers to the S_0 space and the second to $\tilde{S}_0$ space.

$$|ai\rangle = \tfrac{1}{8}\mathrm{i}(\gamma_j S_0)^a|ji\rangle, \qquad |ia\rangle = \tfrac{1}{8}\mathrm{i}(\gamma_j \tilde{S}_0)^a|ij\rangle, \tag{28a,b}$$

$$|ab\rangle = -\tfrac{1}{64}(\gamma_i S_0)^a(\gamma_j \tilde{S}_0)^b|ij\rangle. \tag{28c}$$

The supersymmetry and Lorentz transformation rules for these states can be deduced from eqs. (17) and (25), as well as the Lorentz generator representations, which have the form given in ref. [1] with identical extra terms formed from oscillators with tildes. The spin content of this multiplet is precisely that of extended supergravity in ten dimensions. There are two distinct cases. If the S and $\tilde{S}$ oscillators have opposite handedness then the two gravitinos contained in $|ai\rangle$ and $|ia\rangle$ have opposite handedness and the states $|ab\rangle$ correspond to a vector and third-rank antisymmetric tensor field. This corresponds to the spin content obtained by reduction of eleven-dimensional supergravity to ten dimensions. If S and $\tilde{S}$ have the same handedness, then the two gravitinos have the same handedness and the states $|ab\rangle$ correspond to a scalar, second-rank antisymmetric tensor, and fourth-rank antisymmetric self-dual tensor. This is clearly a distinct alternative and cannot be obtained by reduction from a higher dimension. Nonetheless, the two theories are extremely close in their structure. It is amusing that even though the spin content of the two theories is different in ten dimensions, it coincides on reduction to any lower dimension.

The interacting closed-string theories can be described in the operator formalism in terms of a propagator for states satisfying eq. (27):

$$\Delta = \int_{|z|<1} d^2z\, z^{L_0-1}\bar{z}^{\tilde{L}_0-1}, \tag{29}$$

and vertices describing the emission of massless states. The vertex formulas are given as direct products of open-string vertices [2] in the tilde and no tilde spaces. More specifically, representing the open-string vertex for emitting a massless vector with polarization vector ζ^μ as $\zeta^\mu V^B_\mu$ and the vertex for emitting a massless fermion with Dirac spinor u^a as $\bar{u}^a V^F_a$, the closed string vertices have the form $\zeta^{\mu\nu} V^B_\mu \tilde{V}^B_\nu$, $\bar{u}_1^{\mu a} V^B_\mu \tilde{V}^F_a$, $\bar{u}_2^{a\mu} V^F_a \tilde{V}^B_\mu$, and $\bar{\eta}^{ab} V^F_a \tilde{V}^F_b$. Using the vertex formulas given in ref. [2], tree and loop amplitudes for specific processes are straightforward to calculate. One finds, for example, that one-loop graphs with three or fewer external (on mass-shell) lines are identically zero for the same reasons as in the open-string case [3], namely the vanishing of zero-mode traces.

The one-loop contribution to the N-particle connected S-matrix is given by a single Feynman graph in the case of closed strings. For the case of any four massless external states, the one-loop contribution consists of a kinematical factor involving the polarization vectors and spinors as well as the momenta of the external states times the integral

$$\int \frac{d^2\tau}{(\mathrm{Im}\,\tau)^5} \int \prod_{i=1}^{3} d^2\nu_i \prod_{1\leqslant i<j\leqslant 4} \{\exp[-\pi(\mathrm{Im}\,\nu_{ji})^2/\mathrm{Im}\,\tau] \times |\vartheta_1(\nu_{ji}|\tau)|\}^{(1/2)k_i\cdot k_j}. \tag{30}$$

In this integral ϑ_1 is a Jacobi elliptic function, $\nu_{ji} = \nu_j - \nu_i$ and $\nu_4 = \tau$. The ν_i integration regions are restricted by

$$0 \leqslant \mathrm{Im}\,\nu_i \leqslant \mathrm{Im}\,\tau, \qquad -\tfrac{1}{2} \leqslant \mathrm{Re}\,\nu_i \leqslant \tfrac{1}{2}. \tag{31}$$

To determine the appropriate τ integration region, one notes that eq. (30) may be written in the form $\int d^2\tau\, (\mathrm{Im}\,\tau)^{-2} F(\tau)$, where F is an automorphic function. That is, it is invariant under the modular group

$$\tau \to (a\tau + b)/(c\tau + d), \tag{32}$$

where a, b, c and d are integers satisfying $ad - bc = 1$. Therefore, to avoid multiple counting, one must restrict the τ integration to the "fundamental" region [10]

$$-\tfrac{1}{2} \leqslant \mathrm{Re}\,\tau \leqslant \tfrac{1}{2}, \qquad \mathrm{Im}\,\tau \geqslant 0, \qquad |\tau| \geqslant 1. \tag{33}$$

It is easy to see that the one-loop integral with these limits has the expected poles and branch cuts and is otherwise completely free from divergences. This contrasts, in particular, with the conventional Shapiro–Virasoro model in which there is a divergence associated with the corner where $\nu_1, \nu_2, \nu_3 \to \tau$ [10]. A massless scalar particle at zero momentum in the propagator connecting a tadpole to a tree graph is responsible for the divergence. This does not occur in the supersymmetrical theory, because the supersymmetry algebra makes loops with three or fewer on-shell massless states vanish identically.

In conclusion, we have found a string action that explains the oscillator description of the theory of open and closed strings with simple supersymmetry in ten dimensions. It also leads to the discovery of two new closed-string theories, both of which correspond to $N = 8$ supergravity in four dimensions in the limit that the radii of six compactified dimensions and the slope parameter α' are shrunk to zero. If it should turn out that $N = 8$ supergravity is not finite to all orders in perturbation theory, but that these closed-string theories are, then they may provide the means to make sense of the quantum theory.

We are grateful to Lars Brink for helpful discussions.

References

[1] M.B. Green and J.H. Schwarz, Nucl. Phys. B181 (1981) 502.

[2] M.B. Green and J.H. Schwarz, Caltech preprint CALT-68-872.

[3] M.B. Green and J.H. Schwarz, Caltech preprint CALT-68-873.

[4] A. Neveu and J.H. Schwarz, Nucl. Phys. B31 (1971) 86; Phys. Rev. D4 (1971) 1109;
P. Ramond, Phys. Rev. D3 (1971) 2415.

[5] Y. Nambu, Proc. Intern. Conf. on Symmetries and quark models (Wayne State University, 1979) (Gordon and Breach, New York, 1970), p. 269;
L. Susskind, Nuovo Cimento 69A (1970) 457.

[6] M. Jacob, ed., Dual models (North-Holland, Amsterdam, 1974);
J. Scherk, Rev. Mod. Phys. 47 (1975) 123.
[7] P. Goddard, J. Goldstone, C. Rebbi and C.B. Thorn, Nucl. Phys. B56 (1973) 109.
[8] J.-L. Gervais and B. Sakita, Nucl. Phys. B34 (1971) 477.
[9] S. Mandelstam, Nucl. Phys. B64 (1973) 205.
[10] J. Shapiro, Phys. Rev. D5 (1972) 1945.

Nuclear Physics B256 (1985) 253-284

HETEROTIC STRING THEORY*
(I). The free heterotic string

David J. GROSS, Jeffrey A. HARVEY, Emil MARTINEC and Ryan ROHM

Joseph Henry Laboratories, Princeton University, Princeton, New Jersey 08544, USA

Received 26 February 1985

A new theory of closed orientable superstrings is constructed as a chiral combination of the closed $D=26$ bosonic and $D=10$ fermionic strings. The resulting ten-dimensional theory is $N=1$ supersymmetric, Lorentz invariant, and free of tachyons. An alternate formulation where sixteen of the bosonic coordinates are replaced by fermionic coordinates is presented and used to construct a manifestly Lorentz covariant and reparametrization invariant action for the theory. The theory is consistent only for gauge group $\mathrm{Spin}(32)/\mathrm{Z}_2$ or $\mathrm{E}_8 \times \mathrm{E}_8$.

1. Introduction

Contemporary attempts to construct unified field theories have largely focused on the search for new and hidden symmetries of nature. The most promising of these are based on the beautiful supersymmetric extension of Poincaré invariance and on the revival of the ideas of Kaluza and Klein of hidden, compact spatial dimensions. These ideas have the potential of providing a framework for a truly unified theory of gravity and matter, which can provide an explanation for the known low-energy gauge theory of matter and predict its full particle content. Within the framework of ordinary field theory, however, these attempts have reached an impasse. First, they do not provide a cure for the nonrenormalizability of quantum gravity. Therefore new physics must appear at the Planck length. Even if one ignores this problem and focuses on the low-energy structure of such theories, it appears to be impossible to construct realistic theories without a great loss of predictive power. In order to generate the observed spectrum of chiral fermions it appears necessary to retreat from the most ambitious Kaluza-Klein program, which would uniquely determine the low-energy gauge group as isometries of some compact space, and introduce gauge fields by hand [1]. Finally no realistic and compelling model has emerged.

String theories offer a way out of this impasse. They are based on an enormous increase of the number of degrees of freedom, with a corresponding enlargement of the symmetry group of nature. It has been clear for some time that the ten-dimensional superstring provides a consistent, finite theory which has as its low-energy manifestation ten-dimensional supergravity [2]. However, in order to generate

* Supported in part by NSF grant PHY-80-19754, and DOE contract DE-AC02-76ER03072.

the observed chiral fermions one must start with a chiral string theory which has potential anomalies. Until recently the only known anomaly-free chiral superstring theory was that of orientable, type II, $N=2$ supersymmetric closed strings [3]. This ten-dimensional theory does not contain Yang–Mills interactions; it is unlikely that they, together with the known fermions, could emerge upon compactification. The recent discovery that nonorientable, type I, open and closed superstrings with $N=1$ supersymmetry are finite and anomaly free if the gauge group is SO(32) [4] has increased the phenomenological prospects of unified string theories.

The anomaly cancellation mechanism of Green and Schwarz can be understood in terms of the low-energy field theory that emerges from the superstring, which is a slightly modified version of $D=10$ supergravity. The dangerous Lorentz and Yang–Mills chiral anomalies cancel if and only if the gauge group of the theory is SO(32) or $E_8 \times E_8$. The ordinary superstring theory cannot incorporate $E_8 \times E_8$. However the apparent correspondence between the low-energy limit of anomaly-free superstring theories and anomaly-free supergravity theories suggests the existence of a new kind of string theory whose low-energy limit would contain an $E_8 \times E_8$ gauge group. This is the motivation that led to the construction of the heterotic string. It is of more than academic interest to construct such a theory, since the phenomenological prospects for an $E_8 \times E_8$ theory appear to be much brighter.

In the standard formulation of string theory gauge groups are introduced by attaching (Chan–Paton) labels to the edges of open strings [5]. This procedure can only yield, in a consistent theory, the gauge groups SO(N) and Sp($2N$); it cannot produce $E_8 \times E_8$ [6]. Alternatively one might hope to obtain a gauge group from the compactification of a closed string theory, à la Kaluza–Klein. Such a possibility was suggested to us by the construction due to Frenkel and Kac of representations of Kac–Moody algebras in terms of string vertex operators [7]. The heterotic string theory is a theory of closed strings, in which the compactification of sixteen internal dimensions produces the gauge group SO(32) or $E_8 \times E_8$*.

Another motivation behind the construction of the heterotic string was the search for new consistent string theories. Previously known string theories are the bosonic theory (the Veneziano model) in 26 dimensions and the fermionic, superstring theory in ten dimensions. The new string theory is constructed as a chiral hybrid of these. The possibility of doing this for closed strings becomes evident once one realizes that the physical degrees of freedom of closed strings are two-dimensional massless fields – the transverse coordinates (and their supersymmetric partners) of the world sheet of a propagating string. These fields can be separated into right and left movers, even in the presence of string interactions.

For the consistency of the string theory it is necessary and sufficient that each sector be separately consistent. Thus the heterotic string is constructed as a combina-

* The observation that rank 16 is equal to the difference of the 26 dimensions of the bosonic string theory and the 10 of the superstring was made by Freund [8].

tion of the right-moving coordinates of the ten-dimensional superstring and the left-moving coordinates of the 26-dimensional bosonic string. This yields a theory which is Lorentz invariant in $D=10$ and which has appealing features not found in either theory separately*.

In this paper we shall construct in some detail the free heterotic string. In sect. 2 we present an operator construction using light-cone gauge quantization. Much of this is simply putting together the bosonic and fermionic string components. The new feature is the treatment of the 16 internal, left-moving, bosonic coordinates. We show that these must, for consistency, be compactified on a particular torus. This compactification leads to a set of 496 massless vector bosons. Some of these (16) can be thought of as ordinary Kaluza–Klein gauge bosons corresponding to the 16 isometries of the torus; the others (480) are massless solitons. Together they fill out the adjoint representation of $G=E_8\times E_8$ or $G=\mathrm{Spin}(32)/Z_2$. We show that the theory is free of tachyons and exhibit the massless states and the first massive excitations. We demonstrate that the theory is Lorentz invariant, $N=1$ supersymmetric and that all the states form unitary representations of the gauge group G.

In sect. 3 we present an alternative formulation of the heterotic string theory in terms of internal fermionic coordinates. This amounts to a "fermionization" of the internal coordinates, with minor complications due to the finite length of the closed string. This formulation has some advantages, particularly in setting up a manifestly covariant formalism. In sect. 4 we show that one can describe the heterotic string in a covariant fashion, with no anomalies. Sect. 5 consists of some concluding remarks.

In the following paper [9] we shall discuss the interacting heterotic string, construct the vertex operators for the emission of massless strings, calculate various tree and loop amplitudes, derive the low-energy field theoretic limit of the theory, and verify that it is finite.

2. Construction of the free heterotic string

In this section we shall present the operator formulation of the free heterotic string in light-cone gauge. Before proceeding let us review the philosophy behind the construction.

Noninteracting strings are constructed by first quantization of an action given by the invariant area of the world sheet swept out by the string, or by its supersymmetric generalization. The reparametrization invariance of the classical action permits one to choose a "gauge" in which the timelike parameter of the world sheet, τ, is identified with light-cone time, $X^+=\sqrt{\tfrac{1}{2}}\,(X^0+X^9)$. In this light-cone gauge the theory reduces to a two-dimensional free field theory of the physical degrees of

* Thus the name "heterotic".

freedom (the 24 transverse coordinates, X^i, of the bosonic string or the 8 transverse coordinates, X^i, and their fermionic partners, of the superstring) subject to constraints. This procedure is only valid, however, in the critical dimension of 26 for the bosonic string and 10 for the superstring. In other space-time dimensions the existence of anomalies implies that the conformal degree of freedom of the two-dimensional metric does not decouple [10]. If this degree of freedom is ignored then there is a breakdown of world-sheet reparametrization invariance [11].

In the critical dimension the physical degrees of freedom, being massless two-dimensional free fields, can be decomposed into right and left movers, i.e. functions of $\tau-\sigma$ and $\tau+\sigma$ respectively. If we consider only closed strings then the right- and left-moving modes never mix. The separation of these modes will persist even in the presence of interactions, as long as we consider orientable world surfaces. This is because interactions between closed strings are constructed, order by order in perturbation theory, by modifying the topology of the two-dimensional world sheet on which the strings propagate. In terms of the first quantized two-dimensional theory, however, no interaction is introduced; the right and left movers still propagate independently. Thus there is no obstacle to treating the right movers and the left movers differently, as long as each sector is separately consistent. This is the idea behind the construction of the heterotic string, which combines the right movers of the superstring with the left movers of the bosonic string. It is necessarily a theory of closed orientable strings, since open string boundary conditions would mix right and left movers, and since one can clearly define the orientation of such a string.

In this section we shall construct the heterotic string in light-cone gauge and show that it is indeed consistent, Lorentz invariant, supersymmetric and free of tachyons. The only subtlety arises in the quantization and compactification of the sixteen extra left-moving bosonic coordinates. An alternate version of the heterotic string (see sect. 3) can be given in which these coordinates are replaced by 32 fermionic coordinates. This formulation allows one to construct a manifestly covariant form of the action, as will be shown in sect. 4.

2.1. THE PHYSICAL DEGREES OF FREEDOM

The physical degrees of freedom of the right-moving sector of the fermionic superstring consists of 8 transverse coordinates $X^i(\tau-\sigma)$ $(i=1\cdots8)$ and 8 Majorana-Weyl fermionic coordinates $S^a(\tau-\sigma)$ [12]. The physical degrees of freedom of the left-moving sector of the bosonic string consist of 24 transverse coordinates, $X^i(\tau+\sigma)$ and $X^I(\tau+\sigma)$ $(i=1\cdots8,\ I=1\cdots16)$. Together they comprise the physical degrees of freedom of the heterotic string. The eight transverse right- and left-moving coordinates combine with the longitudinal light-cone coordinates $X^{\pm}(\tau,\sigma)$ to describe the position of the string embedded in ten-dimensional flat space-time.

The dynamics of the free heterotic string is governed by the light-cone gauge action ($\alpha = 1, 2$)

$$S = -\int d\tau \int_0^\pi d\sigma \frac{1}{4\pi\alpha'} [\partial_\alpha X^i \partial^\alpha X^i + \partial_\alpha X^I \partial^\alpha X^I + i\bar{S}\gamma^- (\partial_\tau + \partial_\sigma) S] , \qquad (2.1)$$

where S is a ten-dimensional Majorana–Weyl spinor, with components S^a, i.e.

$$\gamma^+ S = hS = 0 ,$$
$$h \equiv \tfrac{1}{2}(1 + \gamma_{11}) . \qquad (2.2)$$

(From here on, unless otherwise noted, we shall work in units in which the string tension $T_0 = 1/2\pi\alpha' = 1/\pi$ and follow the notation of ref. [12].) The coordinates X^i will contain both right movers, $X^i(\tau - \sigma)$, and left movers, $X^i(\tau + \sigma)$; however, we must impose a constraint restricting X^I to consist of left-moving modes alone. This can be achieved by demanding that

$$\Phi^I \equiv (\partial_\tau - \partial_\sigma) X^I = 0, \qquad I = 1 \cdots 16 . \qquad (2.3)$$

Alternatively, following [13], we can add to the action a term proportional to $\lambda^I [(\partial_\tau - \partial_\sigma) X^I]^2$, which has the same effect. Siegel has shown that the resulting action possesses a local gauge symmetry which allows the Lagrange multipliers λ^I to be gauged away, leaving one with the constraint of eq. (2.3).

The above action is invariant under a supersymmetry transformation that relates right-moving bosons and fermions, while the left-moving bosons remain unchanged:

$$\delta X^i = \frac{1}{\sqrt{p^+}} \bar{\varepsilon} \gamma^i S ,$$

$$\delta S = \frac{i}{\sqrt{p^+}} \gamma_- \gamma_\mu (\partial_\tau - \partial_\sigma) X^\mu \varepsilon . \qquad (2.4)$$

where ε is a right-moving Majorana–Weyl light-cone spinor. In terms of two-dimensional field theory we have $N = \frac{1}{2}$ supersymmetry.

Canonical quantization of the transverse and fermonic coordinates is straightforward. Since we are dealing with closed strings, the fields are periodic functions of $0 \leqslant \sigma \leqslant \pi$ and can be expanded as

$$X^i(\tau - \sigma) = \tfrac{1}{2} x^i + \tfrac{1}{2} p^i (\tau - \sigma) + \tfrac{1}{2} i \sum_{n \neq 0} \frac{\alpha_n^i}{n} e^{-2in(\tau - \sigma)} ,$$

$$X^i(\tau + \sigma) = \tfrac{1}{2} x^i + \tfrac{1}{2} p^i (\tau + \sigma) + \tfrac{1}{2} i \sum_{n \neq 0} \frac{\tilde{\alpha}_n^i}{n} e^{-2in(\tau + \sigma)} \qquad (2.5)$$

$$S^a(\tau - \sigma) = \sum_{n = -\infty}^{+\infty} S_n^a \, e^{-2in(\tau - \sigma)} , \qquad (2.6)$$

where

$$[x^i, p^j] = i\delta^{ij}, \qquad [\alpha^i_n, \alpha^j_m] = [\tilde{\alpha}^i_n, \tilde{\alpha}^j_m] = n\delta_{n+m,0}\delta^{ij},$$
$$[\alpha^n_i, \tilde{\alpha}^m_j] = 0,$$
$$\{S^a_n, \bar{S}^b_m\} = (\gamma^+ h)^{ab}\delta_{n+m,0}. \tag{2.7}$$

In the case of the left-moving internal coordinates, X^I, we must introduce Dirac brackets to take into account the second-class constraints $\Phi^I = 0$. Thus the Poisson bracket of X^I with its canonical momentum, $P^I = (1/\pi)\partial_\tau X^I$, which is canonically given by

$$\{X^I(\sigma, \tau), P^J(\sigma', \tau)\} = \delta(\sigma - \sigma')\delta^{IJ} \tag{2.8}$$

is modified to yield the Dirac bracket

$$\{X^I(\sigma, \tau), P^J(\sigma', \tau)\}_{\mathrm{DB}} = \{X^I(\sigma, \tau), P^J(\sigma', \tau)\} - \int \mathrm{d}\sigma''\, \mathrm{d}\sigma''' \{X^I(\sigma, \tau), \Phi^K(\sigma'', \tau)\}$$
$$\times C_{KK'}(\sigma'', \sigma''')\{\Phi^{K'}(\sigma''', \tau), P^J(\sigma', \tau)\}, \tag{2.9}$$

where $C_{KK'}(\sigma, \sigma')$ is the inverse of the Poisson bracket of $\Phi^K(\sigma)$ and $\Phi^K(\sigma')$. Since $C_{KK'}(\sigma, \sigma') = \delta^{KK'}(1/2\pi\partial_\sigma)\delta(\sigma - \sigma')$, one finds that

$$\{X^I(\sigma, \tau), P^J(\sigma', \tau)\}_{\mathrm{DB}} = \tfrac{1}{2}\delta(\sigma - \sigma')\delta^{IJ}. \tag{2.10}$$

Therefore X^I has the expansion

$$X^I(\tau + \sigma) = x^I + p^I(\tau + \sigma) + \tfrac{1}{2}i \sum \frac{\tilde{\alpha}^I_n}{n} \mathrm{e}^{-2in(\tau+\sigma)}, \tag{2.11}$$

where upon quantization

$$[\tilde{\alpha}^I_n, \tilde{\alpha}^J_m] = n\delta_{n+m,0}\delta^{IJ}, \tag{2.12}$$

$$[x^I, p^J] = \tfrac{1}{2}i\delta^{IJ}. \tag{2.13}$$

The factor of $\frac{1}{2}$ in the commutator of x^I and p^I, the center-of-mass position and momenta in the internal space, arises due to the constraint that ensures that X^I is a function of $\tau + \sigma$. It means that $2p^I$, $I = 1 \cdots 16$ are the generators of translations in the internal space. The allowed values of p^I will be determined, as we shall see below, by the structure of the internal sixteen-dimensional space.

In light-cone gauge $X^+(\tau, \sigma) = x^+ + p^+\tau$ and X^- is determined by solving the constraints that result from choosing a conformally flat metric on the world sheet [11]. If we expand $X^-(\tau, \sigma)$ as

$$X^-(\tau, \sigma) = x^- + p^-\tau + \tfrac{1}{2}i \sum_{n \neq 0} \frac{1}{n}(\alpha^-_n \mathrm{e}^{-2in(\tau-\sigma)} + \tilde{\alpha}^-_n \mathrm{e}^{-2in(\tau+\sigma)}), \tag{2.14}$$

then α_n^- is given, as in the fermionic string, as $(n \neq 0)$

$$\alpha_n^- = \frac{1}{p^+} \sum_m \alpha_m^i \alpha_{n-m}^i + \frac{1}{2p^+} \sum_m (m - \tfrac{1}{2}n) \bar{S}_{n-m} \gamma^- S_m , \tag{2.15}$$

and $\tilde{\alpha}_n^-$ is constructed, as in the bosonic string, as $(n \neq 0)$

$$\tilde{\alpha}_n^- = \frac{1}{p^+} \sum_m (\tilde{\alpha}_m^i \tilde{\alpha}_{n-m}^i + \tilde{\alpha}_m^I \tilde{\alpha}_{n-m}^I) . \tag{2.16}$$

(Note, in these formulas $\alpha_0^i = \tilde{\alpha}_0^i = \frac{1}{2}p^i$, $\tilde{\alpha}_0^I = p^I$.)

These same constraints determine p^-, the generator of τ-translations conjugate to X^+, and thereby the mass operator $m^2 = 2p^+p^- - (p^i)^2$ of the string

$$\tfrac{1}{4}(\text{mass})^2 = N + (\tilde{N} - 1) + \tfrac{1}{2} \sum_{I=1}^{16} (p^I)^2 , \tag{2.17}$$

where $N(\tilde{N})$ are the normal-ordered number operators for the right (left) movers

$$N = \sum_{n=1}^{\infty} (\alpha_{-n}^i \alpha_n^i + \tfrac{1}{2} n \bar{S}_{-n} \gamma^- S_n) ,$$

$$\tilde{N} = \sum_{n=1}^{\infty} (\tilde{\alpha}_{-n}^i \tilde{\alpha}_n^i + \tilde{\alpha}_{-n}^I \tilde{\alpha}_n^I) . \tag{2.18}$$

The subtraction of -1 in (2.17) is due to the normal ordering of $\tilde{N}$, which is unnecessary in the case of the right movers due to fermion-boson cancellations. This subtraction can also be seen to be necessary to ensure ten-dimensional Lorentz invariance. Finally the factor of $\frac{1}{2}(p^I)^2$ comes from the internal momentum and winding number (see below) of the left movers.

In addition there is a further constraint that requires

$$N = \tilde{N} - 1 + \tfrac{1}{2} \sum_{I=1}^{16} (p^I)^2 . \tag{2.19}$$

This constraint has a simple physical explanation. Since there is no distinguished point on a closed string, we are free to shift the origin of the σ-coordinate by an arbitrary amount Δ. This is achieved by the unitary operator

$$U(\Delta) \equiv e^{2i\Delta(N - \tilde{N} + 1 - \frac{1}{2}\sum_I (p^I)^2)} , \tag{2.20}$$

which (recall that $[x^I, p^J] = \frac{1}{2} i \delta^{IJ}$) satisfies $U(\Delta) F(\tau, \sigma) U^+(\Delta) = F(\tau, \sigma + \Delta)$, where F can be X^i, X^I or S^a. The operator $U(\Delta)$ must therefore equal the identity operator on the space of physical states, which must therefore satisfy eq. (2.19). Again the subtraction constant, -1, is due to the normal ordering of $\tilde{N}$ and is necessary for Lorentz invariance. In sect. 4 these constraints will be rederived starting from a covariant formulation of the heterotic string.

2.2. COMPACTIFICATION OF THE INTERNAL DIMENSIONS

The extra 16 left-moving coordinates of the heterotic string can be interpreted as parametrizing an "internal" compact space T. It is unlikely that T could be curved without producing an inconsistent string theory, since then the resulting two-dimensional field theory of X^I would be an interacting nonlinear σ-model with inevitable conformal anomalies. Therefore we consider only flat compact manifolds, in other words T is a sixteen-dimensional torus*.

Since closed strings contain gravity in their low-energy limit one expects that a compactified closed string theory will contain massless vectors associated with the isometries of the compact space. In the case of a flat 16-dimensional torus this would yield the gauge bosons of $[U(1)]^{16}$ A remarkable feature of closed string theories is that for special choices of the compact space there will exist additional massless vector mesons. These are in fact massless solitons of the closed string theory. They combine with the Kaluza-Klein gauge bosons to fill out the adjoint representation of a simple Lie group whose rank equals the dimension of T. In the case of the heterotic string the structure of T is so severely limited that only two choices are consistent. These produce the gauge mesons of $\mathrm{Spin}(32)/Z_2$ or $E_8 \times E_8$!

To illustrate this phenomenon consider the simple example of the bosonic (26-dimensional) string, with one coordinate compactified to a circle (say $0 \leq X^{24} \leq 2\pi R$). The coordinates X^i, $i = 1 \cdots 23$, all have the normal mode expansion of eq. (2.5). However the expansion of $X^{24}(\tau, \sigma)$ is modified. First the momentum, p^{24}, in the compact dimension is quantized in units of $1/R$. In addition string theories give rise to a new phenomenon which has no counterpart in field theories of point particles, namely the existence of solitons which are consequences of the multiply connected structure of configuration space [15]. The function $X^{24}(\tau, \sigma)$ maps the circle $0 \leq \sigma \leq \pi$ onto the circle $0 \leq X^{24} \leq 2\pi R$, and such maps fall into homotopy classes characterized by a winding number L that counts how many times X^{24} wraps around the circle. Thus

$$X^{24}(\tau, \sigma) = x^{24} + p^{24}\tau + 2LR\sigma + \tfrac{1}{2}i \sum_{n \neq 0} \frac{1}{n} (\alpha_n^{24}\, e^{-2in(\tau-\sigma)} + \tilde{\alpha}_n^{24}\, e^{-2in(\tau+\sigma)})$$

$$p^{24} = \frac{M}{R}; \quad M, L \text{ are integers}. \tag{2.21}$$

String configurations with nonvanishing, integer L are solitons. They must be included in the spectrum of interacting closed string theories, since a string with $L = 0$ can split to form two strings with $L = +1$ and $L = -1$ (see fig. 1). Such solitons will always arise whenever a closed string propagates on a nonsimply connected

* For consistent compactification of supersymmetric coordinates there is strong evidence that it suffices for T to be Ricci flat [14]. The only known consistent possibility for compactification of purely bosonic coordinates is for T to be a flat torus.

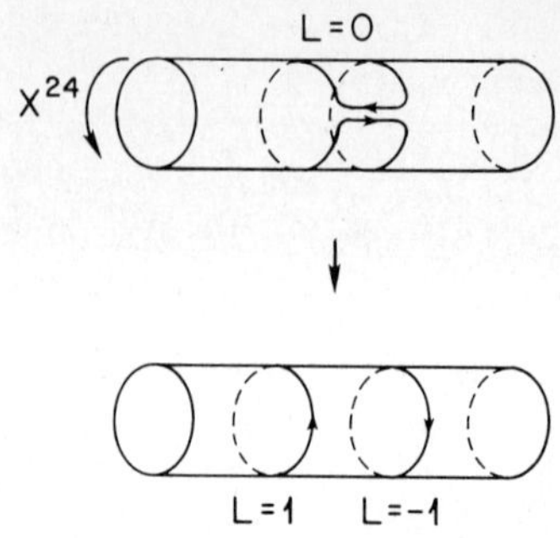

Fig. 1. A string with winding number $L=0$ splits into two strings of winding numbers $L=\pm 1$.

manifold. The excitations of the string with nonzero momentum (winding number) will have masses proportional to $1/R^2$ (R^2). The mass operator for the standard, 26-dimensional, closed string is given by

$$\tfrac{1}{4}(\text{mass})^2 = N + \tilde{N} - 2 + \frac{M^2}{4R^2} + L^2R^2 \,, \tag{2.22}$$

where N and $\tilde{N}$ are the number operators for the left and right movers; the factor of $M^2/4R^2 = \frac{1}{4}(p^{24})^2$ arises from the momentum in the compact dimension, and the term L^2R^2 is the energy required to wrap the string around the circle.

The physical states of the closed string satisfy a constraint which for an uncompactified string is simply $N=\tilde{N}$. This constraint implies that the unitary operator $U(\Delta) = \exp[2i(N-\tilde{N})\Delta]$, which shifts σ by an amount Δ, is equal to the identity operator on physical states, thus ensuring the absence of a distinguished point on the string. If the string contains a winding number term, $2LR\sigma$, one must modify $U(\Delta)$ so as to shift this as well. Thus

$$U(\Delta) = \exp[2i(N - \tilde{N} - p^{24}RL)\Delta] \tag{2.23}$$

and the physical states must satisfy

$$N = \tilde{N} + ML \,. \tag{2.24}$$

Let us now examine the massless spectrum of the effective 25-dimensional theory. It consists of the transverse components of the graviton, antisymmetric tensor and dilaton, $\tilde{\alpha}^i_{-1}\alpha^j_{-1}|0\rangle$, $1 \le i, j \le 23$; the transverse components of two vector mesons, $\tilde{\alpha}^i_{-1}\alpha^{24}_{-1}|0\rangle$ and $\alpha^i_{-1}\tilde{\alpha}^{24}_{-1}|0\rangle$. These correspond to the gauge bosons expected from the usual Kaluza-Klein compactification on a circle with isometry $U(1)$. However for the particular choice of the radius of the circle

$$R = \sqrt{\alpha'} = \sqrt{\tfrac{1}{2}} \,, \tag{2.25}$$

the solitons with winding numbers $L=\pm 1$ and momenta $M=\pm 1$ (with $N-\tilde{N}=ML$) are also massless. Thus there are four more massless vectors $\alpha^i_{-1}|M=L=\pm 1\rangle$ and $\tilde{\alpha}^i_{-1}|M=-L=\pm 1\rangle$.

These six massless vectors form the adjoint representation of SU(2) × SU(2), with the diagonal generators having eigenvalues $\frac{1}{2}(M+L)$ and $\frac{1}{2}(M-L)$ respectively. This construction is a physical realization of the work of Frenkel and Kac on representations of Kac–Moody Lie algebras [7]. Their work suggests that this construction can be generalized to produce massless vector mesons in the adjoint representation of any simply laced Lie group G, i.e. a Lie group whose root vectors all have the same length. This restricts G to be SU(n), SO($2n$), E_6, E_7, E_8 or direct products of these. The method employs closed strings moving on a torus such that the lattice of allowed momenta coincides with the root lattice of the group. In other words the string moves on the maximal torus of the group G. However the quantization differs from the usual quantization of closed bosonic strings.

First consider compactification of the usual closed bosonic string on a D-dimensional torus, which may be thought of as R^D modulo a lattice Γ generated by D independent basis vectors e_i^I ($i=1\cdots D$). We shall choose the length of the e_i to be equal to $\sqrt{2}$ and identify the center-of-mass coordinate x^I with its translation by $2\pi R_i$ in the direction of e_i, so that

$$x^I \equiv x^I + \sqrt{2}\pi \sum_{i=1}^{D} n_i R_i e_i^I \,, \tag{2.26}$$

with n_i being integers. The mode expansion for the internal coordinates is

$$X^I(\sigma,\tau) = x^I + p^I\tau + 2L^I\sigma + \tfrac{1}{2}i \sum_{n\neq 0} \frac{1}{n}\left(\alpha_n^I\, e^{-2in(\tau-\sigma)} + \tilde{\alpha}_n^I\, e^{-2in(\tau-\sigma)}\right). \tag{2.27}$$

For a torus R^D/Γ the allowed center-of-mass momenta p^I lie on the dual lattice $\tilde{\Gamma}$, generated by e_i^{*I} ($i=1\cdots D$) defined by

$$\sum_{I=1}^{D} e_i^I e_j^{*I} = \delta_{ij} \,. \tag{2.28}$$

Since p^I generates translations of x^I we must have

$$p^I = \sqrt{2} \sum_{i=1}^{D} \frac{m_i}{R_i} e_i^{*I} \tag{2.29}$$

for integer m_i. The allowed winding numbers are

$$L^I = \sqrt{\tfrac{1}{2}} \sum_{i=1}^{D} R_i n_i e_i^I \tag{2.30}$$

for integer n_i. The constraint (2.24) is generalized to

$$N = \tilde{N} + \sum_I p^I L^I = \tilde{N} + \sum_i m_i n_i \,, \tag{2.31}$$

and the mass operator is now

$$\begin{aligned}\tfrac{1}{4}(\text{mass})^2 &= N+\tilde{N}-2+\sum_I (\tfrac{1}{4}p^I p^I + L^I L^I)\\ &= N+\tilde{N}-2+\sum_{i,j}\left(\frac{1}{2}\frac{m_i}{R_i}\,g_{ij}\,\frac{m_j}{R_j}+\tfrac{1}{2}n_i R_i g^*_{ij} n_j R_j\right),\end{aligned} \tag{2.32}$$

where the metric $g_{ij}=\sum_{I=1}^{D} e_i^I e_j^I$ is the Cartan matrix for the Lie group with maximal torus R^D/Γ and $g^*_{ij}=\sum_{I=1}^{D} e_i^{*I} e_j^{*I}=(g_{ij})^{-1}$ is the dual metric. With our normalization g_{ij} is integer valued and $g_{ii}=2$ for simply laced Lie groups.

As before we obtain $2D$ massless vector mesons $\tilde{\alpha}^i_{-1}\alpha^I_{-1}|0\rangle$ and $\alpha^i_{-1}\tilde{\alpha}^I_{-1}|0\rangle$ as expected since the isometry group of the torus is $\mathrm{U}(1)^D$. However the constraints make it clear that it is not possible to obtain the additional massless vectors with $N-\tilde{N}=\pm 1$ needed to complete the adjoint representation of a simple nonabelian group of rank $D>1$ for any real R_i. The resolution of this difficulty provides a key to the construction of the heterotic string.

We return to the mode expansion (2.27) and write $X^I(\sigma,\tau)=X^I(\tau-\sigma)+\tilde{X}^I(\tau+\sigma)$ with

$$\begin{aligned} X^I(\tau-\sigma) &= \tfrac{1}{2}x^I+(\tfrac{1}{2}p^I-L^I)(\tau-\sigma)+\tfrac{1}{2}i\sum_{n\neq 0}\frac{1}{n}\alpha^I_n\,\mathrm{e}^{-2in(\tau-\sigma)},\\ \tilde{X}^I(\tau+\sigma) &= \tfrac{1}{2}x^I+(\tfrac{1}{2}p^I+L^I)(\tau+\sigma)+\tfrac{1}{2}i\sum_{n\neq 0}\frac{1}{n}\tilde{\alpha}^I_n\,\mathrm{e}^{-2in(\tau+\sigma)}.\end{aligned} \tag{2.33}$$

Note that the left and right oscillators are treated independently but the center-of-mass coordinates and momenta are *not*. For ordinary space-time coordinates with $L=0$ this is clearly necessary. However for an internal nonsimply-connected space we are free to regard X^I and $\tilde{X}^I$ as *completely independent* massless two-dimensional fields with expansions

$$\begin{aligned} X^I(\tau-\sigma) &= x^I+p^I(\tau-\sigma)+\text{oscillators},\\ \tilde{X}^I(\tau+\sigma) &= \tilde{x}^I+\tilde{p}^I(\tau+\sigma)+\text{oscillators},\end{aligned} \tag{2.34}$$

and $[x^I,\tilde{p}^J]=[\tilde{x}^I,p^J]=0$, but $[x^I,p^J]=[\tilde{x}^I,\tilde{p}^J]=\tfrac{1}{2}i\delta^{IJ}$ as in (2.13). Now $2p^I(2\tilde{p}^I)$ generates translations of $x^I(\tilde{x}^I)$ so the allowed momenta are

$$p^I=\sqrt{\tfrac{1}{2}}\sum_{i=1}^{D}\frac{m_i}{R_i}e_i^{*I} \tag{2.35}$$

and similarly for $\tilde{p}^I$.

The left- and right-moving sectors are now related only by the constraint which follows from invariance under shifts in σ which becomes

$$N+\tfrac{1}{2}\sum_I (p^I)^2=\tilde{N}+\tfrac{1}{2}\sum_I(\tilde{p}^I)^2. \tag{2.36}$$

The mass operator is now

$$\tfrac{1}{4}(\text{mass})^2 = N + \tilde{N} - 2 + \tfrac{1}{2}\sum_I (p^I p^I + \tilde{p}^I \tilde{p}^I)\,. \tag{2.37}$$

Choosing the radius to be

$$R_i = R = \sqrt{\tfrac{1}{2}}\,(=\sqrt{\alpha'})\,, \tag{2.38}$$

we now find $2D$ U(1) gauge bosons as before but in addition there are massless vector meson states with $N - \tilde{N} = +1\ (-1)$ and $(\tilde{p}^I)^2 = 2((p^I)^2 = 2)$. These complete the adjoint representation of G×G, where G is a simply-laced Lie group with rank $\mathrm{G} = D$.

The heterotic string is constructed by utilizing the left-moving part of the above construction with $D = 16$ to produce gauge bosons in the adjoint representation of $\mathrm{E}_8 \times \mathrm{E}_8$ or Spin(32)/Z_2. This choice is dictated by the following argument. Since X^I is to be a function of $\tau + \sigma$ only, the allowed momenta (2.35) must coincide with an allowed winding number (2.30). Thus the allowed states must have momenta which lie on the intersection of Γ and $\tilde{\Gamma}$. Furthermore, in order to satisfy the constraint, eq. (2.19), the square of the allowed momenta must be even integers. Finally, the lattice must be self-dual, i.e. Γ must equal $\tilde{\Gamma}$ (equivalently $\det g_{ij} = 1$). This last requirement is necessary to have a consistent theory of *interacting* closed strings. We shall discuss this in detail in the following paper [9]. For the moment we simply note that the (euclidean) two-dimensional world sheet which describes a one-loop closed string amplitude is a torus. This torus is clearly symmetric under the interchange of σ and τ. To achieve this symmetry, necessary for global reparametrization invariance, the coefficients of τ and σ in X^I must span all the states of the *same* lattice. All of these requirements can be satisfied if and only if all the radii are determined as in (2.38) and the lattice is integral, even and self-dual. This means that the metric $g_{ij} = \sum_{I=1}^{16} e_i^I e_j^I$ is integer valued, g_{ii} is even and $\Gamma = \tilde{\Gamma}$. For this reason the obvious torus, given by $\mathrm{R}^{16}/\mathrm{Z}^{16}$ is unacceptable. Z^{16} is self-dual if one chooses the basis vectors to be of length one – but then it is not even. The sublattice consisting only of even points is not self-dual.

Even, self-dual lattices are extremely rare. In fact, they only exist in $8n$ dimensions. In sixteen dimensions there are two such lattices, $\Gamma_8 \times \Gamma_8$ and Γ_{16}. The first is the direct product of two Γ_8, where Γ_8 is the root lattice of E_8, and g_{ij} is the Cartan matrix of E_8:

$$g_{ij}^{\mathrm{E}_8} = \left|\begin{array}{cccccccc} 2 & -1 & & & & & & \\ -1 & 2 & -1 & & & & & \\ & -1 & 2 & -1 & & & & \\ & & -1 & 2 & -1 & & & \\ & & & -1 & 2 & -1 & & -1 \\ & & & & -1 & 2 & -1 & \\ & & & & & -1 & 2 & 0 \\ & & & & -1 & & 0 & 2 \end{array}\right|\,. \tag{2.39}$$

The construction of Γ_{16} is slightly more complicated. The basis vectors for the SO(32) root lattice can be written in terms of a set of orthonormal vectors $\{u_i\}$, $i=1\cdots 16$ as $e_i=u_i-u_{i+1}$, $i=1\cdots 15$, $e_{16}=u_{15}+u_{16}$. This root lattice is not self-dual since $\det g_{ij}=4$. However a self-dual lattice can be constructed from the root lattice by adding points that are multiples of *one* of the spinor weights of Spin(32). We can choose the spinor weight to be $s_1=\frac{1}{2}(u_1-u_2+u_3-u_4+\cdots+u_{15}-u_{16})$, $s_1^2=4$ and take the basis vectors for Γ_{16} to be s_1 and e_i, $i=2\cdots 16$. Since e_1 is an integer linear combination of the basis vectors, Γ_{16} contains all the points of the root lattice of SO(32) plus additional points that correspond to spinor weights of Spin(32)/Z_2. Note that this is not the same as SO(32). The center of Spin(32) is $Z_2\times Z_2$. Removing the diagonal combination of the two Z_2 factors eliminates all spinor representations, leaving SO(32). Removing one Z_2 eliminates one of the two spinor representations of Spin(32) and all representations in the same Z_2 conjugacy class as the spinor (including the vector).

On either of these tori x^I is given by eq. (2.26) where the allowed momenta are

$$p^I=\sum_{i=1}^{16} n_i e_i^I \qquad (n_i=\text{integer})\,, \tag{2.40}$$

and n_i also measures the winding number in the ith direction around the torus. The values of $(p_I)^2$ are all even integers, the minimal value is 2. Each of these sixteen-dimensional self-dual lattices has precisely 480 vectors of length squared 2. These will yield, in the heterotic string theory, 480 massless vector bosons. These 480 solitons are clearly labeled by the weights of the adjoint representations of G = $E_8\times E_8$ or Spin(32)/Z_2. They combine with the 16 massless vectors with zero weights (the Kaluza–Klein gauge bosons) to fill out the adjoint representation of G. We shall see below that all the states of the heterotic string will form representations of G.

2.3. THE SPECTRUM

The physical states of the heterotic string are simply direct products of Fock space states $(|\rangle_R\times|\rangle_L)$ of the right-moving fermionic string and the left-moving bosonic string, subject to the constraint (2.19). The right-moving ground state is annihilated by α_n^i, S_n^a $(n>0)$ and N. It forms an irreducible representation of the zero mode oscillators S_0^a and consists of 8 bosonic states $|i\rangle_R$ and 8 fermionic states $|a\rangle_R$, which satisfy

$$|a\rangle_R=\tfrac{1}{8}i(\gamma_i S_0)^a|i\rangle_R\,,\qquad |i\rangle_R=\tfrac{1}{16}i(\bar{S}_0\{\gamma_i,\gamma_+\})^a|a\rangle_R\,. \tag{2.41}$$

The left-handed ground state is annihilated by $\tilde{\alpha}_n^i$, $\tilde{\alpha}_n^I$ $(n>0)$, $\tilde{N}$ and p^I. By itself it is a tachyon, but it is removed from the physical spectrum by the constraint which requires that $\tilde{N}+\frac{1}{2}(p^I)^2=N+1>0$.

The spectrum of physical states contains only nonnegative masses due to the constraint. In fact

$$(\text{mass})^2 = 8N \geq 0\,,$$
$$N = \tilde{N} + \tfrac{1}{2}(p^I)^2 - 1\,. \tag{2.42}$$

The ground state is therefore massless and consists of the direct product of $|i\rangle_R$ or $|a\rangle_R$ with $\tilde{\alpha}^j_{-1}|0\rangle_L$, $\tilde{\alpha}^I_{-1}|0\rangle_L$ or $|p^I\rangle_L$ (where $(p^I)^2=2$). All these states have $N=\tilde{N}+\frac{1}{2}(p^I)^2-1=0$. As expected they form irreducible 10-dimensional, massless, supermultiplets. The states $|i$ or $a\rangle_R \times \tilde{\alpha}^j_{-1}|0\rangle_L$ form the irreducible $N=1$, $D=10$ supergravity multiplet. Note that the antisymmetric tensor appears here in the antisymmetric product of two SO(8) vectors as compared to the type I closed string theory where it appears in the antisymmetric product of two SO(8) spinors. The states $|i\rangle_R \times \tilde{\alpha}^I_{-1}|0\rangle_L$ and $|a\rangle_R \times \tilde{\alpha}^I_{-1}|0\rangle_L$ together with the 16×480 states $|i$ or $a\rangle_R \times |p^I, (p^I)^2=2\rangle_L$ form an irreducible $N=1$, $D=10$ super-Yang–Mills multiplet of G (where $G = E_8\times E_8$ or Spin $(32)/Z_2$).

Although the spectrum of the massless states of the heterotic string (with $G=\text{Spin}(32)/Z_2$) coincides with that of the type I superstring with gauge group SO(32), the difference between these two theories is apparent in the structure of the massive levels. In the heterotic string we will obtain states with arbitrary p^I = weight vector of G, and thus states which are in arbitrarily large representations of G. The type I open string theory contains only the antisymmetric tensor (adjoint) and the symmetric tensor representations of SO(32).

We shall exhibit explicitly the structure of the first massive level with $N=1$, $(\tilde{N}, (p^I)^2) = (2,0)$, $(1,2)$ or $(0,4)$, and $(\text{mass})^2=8$. Already at this level there are 18 883 584 physical degrees of freedom! The left- and right-moving sectors separately can be assembled into SO(9) multiplets, since the Lorentz and supersymmetry generators act separately on each. Among the right movers we have 128 bosonic states $\alpha^i_{-1}|j\rangle_R$ and $S^a_{-1}|b\rangle_R$ which fill out the SO(9) representations $\square\square$ and $\begin{smallmatrix}\square\\\square\\\square\end{smallmatrix}$ of dimensions 44 and 84 respectively. In addition there are the fermion states, $\alpha^i_{-1}|a\rangle_R$ and $S^a_{-1}|i\rangle_R$ which form a single irreducible SO(9) representation of dimension 128.

The left-moving sector consists of the SO(8) scalars $|p^I, (p^I)^2=4\rangle_L$, $\tilde{\alpha}^I_{-1}|p^I, (p^I)^2=2\rangle_L$, $\tilde{\alpha}^I_{-1}\tilde{\alpha}^J_{-1}|0\rangle_L$ and $\tilde{\alpha}^I_{-2}|0\rangle_L$; the SO(8) vectors $\tilde{\alpha}^i_{-1}|p^I, (p^I)^2=2\rangle_L$, $\tilde{\alpha}^i_{-2}|0\rangle_L$, $\tilde{\alpha}^i_{-1}\tilde{\alpha}^I_{-1}|0\rangle_L$ and the SO(8) symmetric tensors $\tilde{\alpha}^i_{-1}\tilde{\alpha}^j_{-1}|0\rangle_L$. To count the number of these states we use the fact that the number of points of length squared $2m$ on the lattices Γ_{16} or $\Gamma_8\times\Gamma_8$ is given by $480\,\sigma_7(m)$, where $\sigma_7(m)$ is the sum of the seventh powers of the divisors of m [16]. Using this formula we find 69 752 SO(8) singlets which can be assembled into irreducible $E_8\times E_8$ representations as $(3875,1)+(1,3875)+(248,248)+(248,1)+(1,248)+(1,1)+(1,1)$. For $\text{Spin}(32)/Z_2$ the irreducible representations are $\begin{smallmatrix}\square\\\square\\\square\\\square\end{smallmatrix}$ (dimension 35 960), $\begin{smallmatrix}\square\\\square\end{smallmatrix}$ (496), $\square\square$ (527), the spinor of dimension (32 768) and one singlet (1). Note the appearance at the first massive level of the SO(32) spinor. This state cannot decay and might pose cosmological problems. Similarly the 497 SO(8) vectors transform as the adjoint representation of G plus a singlet while the SO(8) symmetric tensor is a singlet under G.

These states may be easily assembled into SO(9) multiplets, whereby the G singlets are in the SO(9) representation $\underline{\underline{44}}+\underline{1}$ and the G adjoint representations form a SO(9) vector. The other G representations are SO(9) singlets. Finally the complete SO(9)×G multiplets are obtained by taking the SO(9) tensor product of the representations $(\square\!\square+\begin{smallmatrix}\square\\\square\end{smallmatrix})$ occurring in the right-moving sector with the representations $(\square\!\square+\square+\cdot)$ occurring in the left-moving sector. As we shall show below the physical states of the heterotic string form a representation of the super $D=10$ Poincaré group and of the algebra of $G=E_8\times E_8$ or $Spin(32)/Z_2$. This guarantees that a similar grouping of the states exists at each mass level.

The number of states of a string theory increases dramatically as a function of mass. In fact the degeneracy of the levels with mass M increases exponentially as $d(M)\sim\exp(\bar{\beta}M)$. This leads to a "limiting temperature" for the noninteracting string of $kT_L=1/\bar{\beta}$, since at this temperature the specific heat diverges [17]. Of course interactions might modify this conclusion. The value of $\bar{\beta}$ is easily determined by calculating the "partition" function $P(x)$ which, in the case of the standard bosonic closed string in 26 dimensions, is given by

$$P_B(x)=\left\{\prod_{n=1}^{\infty}(1-x^n)\right\}^{-24}=\sum_{N=0}^{\infty}P_B(N)x^N. \tag{2.43}$$

It is easy to see that $P_B(N)$ counts the number of string states with right-moving oscillator number N. Since for the closed string $N=\tilde{N}$ and $M^2=4(N+\tilde{N}-2)=8(N-1)$ we have that $d_B(M)=P_B^2\,(\frac{1}{8}M^2+1)$. Similarly the partition function of the fermionic superstring in 10 dimensions is given by

$$P_F(x)=\left[\frac{\prod_{n=1}^{\infty}(1+x^n)}{\prod_{n=1}^{\infty}(1-x^n)}\right]^8=\sum_{N=0}P_F(N)x^N. \tag{2.44}$$

Here however $M^2=4(N+\tilde{N})=8N$ and the ground state is 256-fold degenerate so that $d_F(M)=256P_F^2\,(\frac{1}{8}M^2)$. Since the heterotic string is a hybrid of the bosonic and fermionic strings and $M^2=4(N+\tilde{N}-1+\frac{1}{2}(p^I)^2)=8N$ we have

$$d_M(M)=16P_F(\tfrac{1}{8}M^2)P'_B(\tfrac{1}{8}M^2), \tag{2.45}$$

where $P'_B(x)$ is the partition function of the left movers

$$P'_B(x)=\frac{1}{x}\left\{\prod_{n=1}^{\infty}(1-x^n)\right\}^{-24}\left(1+\sum_{m=1,2,\dots}480\sigma_7(m)x^m\right)=\sum_{\tilde{N}=-1}P'_B(\tilde{N})x^{\tilde{N}}. \tag{2.46}$$

The large-N behavior of the $P(N)$ can be determined from the behavior of $P(x)$ as $x \to 1$. Using the formulae

$$\prod_{n=1}^{\infty} (1 \pm x^n) = \exp \sum_{n=1}^{\infty} \ln (1 \pm x^n) \cdot \underset{x \sim 1}{\simeq} \exp - \sum_{1}^{\infty} \frac{(\mp)^k}{k^2} \frac{1}{(1-x)},$$

we deduce that

$$P_B(x) \sim P'_B(x) \underset{x \to 1}{\simeq} \exp \frac{4\pi^2}{1-x}, \qquad P_F(x) \sim \exp \frac{2\pi^2}{1-x}.$$

This translates into exponential growth of $P(N)$, which behaves as $P_B(N) \sim P'_B(N) \sim \exp (4\pi\sqrt{N})$; $P_F(N) \sim \exp (2\sqrt{2}\pi\sqrt{N})$. In terms of the "limiting temperature" we then have

$$\bar{\beta}_B = \sqrt{8}\pi = (4\pi\sqrt{\alpha'}),$$
$$\bar{\beta}_F = 2\pi = (2\sqrt{2}\pi\sqrt{\alpha'}), \tag{2.47}$$

for the standard bosonic and fermionic strings. The limiting temperature of the heterotic string lies in between these

$$\bar{\beta}_H = \tfrac{1}{2}(\bar{\beta}_B + \bar{\beta}_F) = (1+\sqrt{2})\pi = (2+\sqrt{2})\pi\sqrt{\alpha'}. \tag{2.48}$$

2.4. SYMMETRIES

It is straightforward to verify that the heterotic string is Lorentz invariant and $N=1$ supersymmetric in 10 dimensions. This is because the generators of these transformations act separately on the right and left movers. Since these are formulated in their critical dimensions of 10 and 26 respectively, the would-be anomalies cancel.

Thus the heterotic string contains one supersymmetry, generated by

$$Q^a = i\sqrt{p^+}\,(\gamma^+ S_0)^a + 2i\frac{1}{\sqrt{p^+}}\sum_n (\gamma_i S_{-n})^a \alpha^i_n, \tag{2.49}$$

which acts on the right-moving sector. The demonstration that

$$\{Q^a, Q^b\} = -2(h\gamma^\mu P_\mu)^{ab} \tag{2.50}$$

is identical to the demonstration of the right-moving supersymmetry of the type II closed superstring theory [12] (half of its $N=2$ supersymmetry). Similarly the generators of the 10-dimensional Lorentz group, $J^{\mu\nu}$, can be written as

$$J^{\mu\nu} = l^{\mu\nu} + J^{\mu\nu}_R + J^{\mu\nu}_L, \tag{2.51}$$

where

$$l^{\mu\nu} = x^\mu p^\nu - x^\nu p^\mu, \tag{2.52}$$

and $J_{\mathrm{R}}^{\mu\nu}(J_{\mathrm{L}}^{\mu\nu})$ are the Lorentz generators on the oscillator coordinates of the right-moving (left-moving) sector of the fermionic (bosonic) strings respectively. For example, $J_{\mathrm{L}}^{ij}=-i\sum_{n=1}^{\infty}(1/n)(\tilde{\alpha}_{-n}^{i}\tilde{\alpha}_{n}^{j}-\tilde{\alpha}_{-n}^{j}\alpha_{n}^{i})$, $J_{\mathrm{L}}^{+-}=J_{\mathrm{L}}^{-+}=0$, $J_{\mathrm{L}}^{i-}=-i\sum_{n=1}^{\infty}(1/n)(\tilde{\alpha}_{-n}^{i}\tilde{\alpha}_{n}^{-}-\tilde{\alpha}_{-n}^{-}\tilde{\alpha}_{n}^{i})$. The demonstration that the $J^{\mu\nu}$ satisfy the 10-dimensional Lorentz algebra is identical to the demonstration of the Lorentz invariance of the 10-dimensional superstring [12] and the 26-dimensional bosonic string [11]. Note that these demonstrations do require the normal ordering subtractions of $\tilde{N}$ that appear in eqs. (2.17) and (2.19).

We shall now show that the states of the heterotic string form representations of the group $\mathrm{G}=\mathrm{E}_8\times\mathrm{E}_8$ or $\mathrm{Spin}(32)/\mathrm{Z}_2$. We are guided by the work of Frenkel and Kac, and Segal who used string theory operators to construct representations of Kac-Moody algebras [7]. To this end we construct an operator $E(K^I)$ which acts on the left-moving states of the heterotic string; $E(K^I)$ represents the generators of G that translate states on the weight lattice by a root vector $K^I((K^I)^2=2)$. In order to understand the construction of Frenkel and Kac it is useful to consider a hamiltonian treatment of the left-moving sector of the heterotic string viewed as a two-dimensional field theory with solitons. The fields $X^I(\sigma)$ are maps from the circle S^1 parametrized by $0\leqslant\sigma\leqslant\pi$ onto the 16-dimensional torus T. The configuration space can thus be thought of as the loop space $\Omega\mathrm{T}$ of maps $\mathrm{S}^1\to\mathrm{T}$. This loop space has the structure of a group under pointwise multiplication of maps and falls into disconnected components labelled by the winding numbers of the torus, i.e. $\pi_0(\Omega\mathrm{T})=\pi_1(\mathrm{T})$. In quantizing the theory we consider unitary operators $U(f_1)$ which when acting on the vacuum create states corresponding to the map $f_1:\mathrm{S}^1\to\mathrm{T}$, $|f_1\rangle=U(f_1)|0\rangle$. If we consider two maps f_1, f_2 and their composition f_1f_2 then quantum mechanically we are only interested in rays in a Hilbert space so we only require

$$U(f_1)U(f_2)=\varepsilon(f_1,f_2)U(f_1f_2)\,, \tag{2.53}$$

with $|\varepsilon(f_1,f_2)|=1$. The phase factor ε is required by associativity to satisfy

$$\varepsilon(f_1,f_2)\varepsilon(f_1f_2,f_3)=\varepsilon(f_1,f_2f_3)\varepsilon(f_2,f_3)\,, \tag{2.54}$$

which is just the statement that ε is a two-cocycle. ε is nontrivial if it cannot be factored in the form $\mathrm{e}^{i\phi(f_1)}\mathrm{e}^{i\phi(f_2)}\mathrm{e}^{-i\phi(f_1f_2)}$ and reabsorbed in the definition of U. In this case the operators U give a nontrivial projective representation of the group $\Omega\mathrm{T}$. These representations have been considered previously in the physics literature in connection with quantization of solitons in theories such as the sine-Gordon model [7] and the Skyrme model [18]. General conditions for the existence of nontrivial projective representations are given in a classic paper of Bargmann [19].

We are interested in representing the group of translations on the weight lattice. In order to reproduce the correct group commutation relations we must consider a nontrivial projective representation of this group. Points on the weight lattice correspond to allowed winding numbers (i.e. elements of $\pi_0(\Omega\mathrm{T})$) with addition of lattice vectors corresponding to group multiplication of elements in disconnected components of ΩT. For lattice vectors K^I, L^I, M^I the two-cocycle condition is thus

$$\varepsilon(K, L)\varepsilon(K+L, M) = \varepsilon(L, M)\varepsilon(K, L+M) . \tag{2.55}$$

For our purposes it is convenient to modify the translation operator directly and to consider the operators

$$E(K^I) = \int \frac{\mathrm{d}z}{2\pi i z} :\mathrm{e}^{2iK^I \cdot X^I(z)}: C(K^I) , \tag{2.56}$$

with $(K^I)^2 = 2$, $z = \mathrm{e}^{2i(\tau+\sigma)}$, and the contour runs around the unit circle. These operators contain the usual translation operator $\mathrm{e}^{2iK^I X^I}$ which translates the internal momenta (= winding numbers) by K^I. The additional term $C(K^I)$ can be viewed as an operator 1-cocycle which will be chosen so that the 480 $E(K^I)$ with $(K^I)^2 = 2$ plus the 16 operators p^I satisfy the Lie algebra of G (in the Chevalley basis).

Using the properties of harmonic oscillator coherent states it is straightforward to verify that

$$:\mathrm{e}^{2iK^I X^I(z)}::\mathrm{e}^{2iL^I X^I(w)}: = :\mathrm{e}^{2i(K\cdot X(z) + L\cdot X(w))}: (z-w)^{K\cdot L} (wz)^{-K\cdot L/2} \quad \text{for} \quad |w| < |z| . \tag{2.57}$$

Under the interchange $(K, z) \leftrightarrow (L, w)$ the RHS acquires a factor of $(-)^{K\cdot L}$. We therefore demand that $C(K)C(L) = (-1)^{K\cdot L} C(L)C(K)$ in order that the E have simple commutation relations. The analog of eq. (2.53) is just

$$C(K)C(L) = \varepsilon(K, L)C(K+L). \tag{2.58}$$

In addition to the two-cocycle condition (2.55) $\varepsilon(K, L)$ may be chosen to satisfy [7]

$$\begin{gathered} \varepsilon(K, L)\varepsilon(L, K) = (-1)^{K\cdot L} \\ \varepsilon(K, 0) = -\varepsilon(K, -K) = 1 . \end{gathered} \tag{2.59}$$

Now, if we compute the commutator of $E(K)$ and $E(L)$ (K and L are root vectors), the integral over $(z-w)^{K\cdot L}$ will yield a nonzero result only if $K \cdot L = -1$ (and therefore $K + L$ is another root vector since $(K+L)^2 = 2+2-2 = 2$), or if $K \cdot L = -2$ which requires $K = -L$. This is a consequence of the fact that

$$\begin{aligned} :\mathrm{e}^{2iK\cdot X(z)}: = \exp\left[+ \sum_{n=1}^{\infty} K \cdot \tilde{\alpha}_{-n}(z^n/n) \right] \exp[+2iK \cdot x + K \cdot p \ln z] \\ \times \exp\left[- \sum_{n=1}^{\infty} K \cdot \tilde{\alpha}_n(z^{-n}/n) \right] \end{aligned} \tag{2.60}$$

is a meromorphic function of z (recall that $K \cdot p$ is integer valued on the lattice of allowed momenta).

Therefore one easily shows that

$$[E(K), E(L)] = \begin{cases} \varepsilon(K, L)E(K+L)\,, & (K+L)^2 = 2\,, \\ K^I \cdot p^I\,, & K+L=0\,, \\ 0\,, & \text{otherwise}\,, \end{cases} \tag{2.61}$$

$$[p^I, E(K)] = K^I.$$

These are precisely the commutators of the generators of G, as long as the $\varepsilon(K, L)$ are chosen to equal the structure constants (± 1 in this basis) of the group G. Frenkel and Kac show that $C(K)$ and $\varepsilon(K, L)$ can be constructed to satisfy these properties. In fact $\varepsilon(K, L)$ is defined for arbitrary K and L on the lattice by extending the structure constants (defined for $K^2 = L^2 = (K+L)^2 = 2$) to the whole lattice using (2.59) and bilinearity.

Therefore we have an explicit representation of the generators of G on the Fock space of the left movers of the heterotic string. Since the right movers are G-singlets and the constraint, eq. (2.19), is G-invariant, it follows that the free heterotic string is G-invariant.

3. Fermionic representation

There is an alternate description of the internal degrees of freedom of the heterotic string (which are responsible for the Yang-Mills interactions) in terms of chiral, anticommuting, Lorentz-scalar coordinates. These fermionic coordinates are related to the bosonic coordinates discussed above by the well-studied bosonization of two-dimensional fermion fields, with minor complications due to the finite length of the string. For purposes of describing the spectrum and interactions of the string the two formulations are equivalent. They correspond to alternate constructions of a particular representation of the affine Lie algebra based on the gauge group of the theory, whose properties determine the structure of the string in the light-cone gauge*.

We shall now present the construction of the heterotic string in terms of internal fermionic coordinates, beginning with the model with Spin(32)/Z_2 gauge group. The construction of the $E_8 \times E_8$ model requires subtle modifications. The construction parallels that of the superstring in terms of the Neveu-Schwarz [21] and Ramond [22] models. Here we introduce a set of 32 real, anticommuting, left-moving coordinates $\psi^i(\tau+\sigma)$ which transform as the fundamental representation of SO(32). Regarded as two-dimensional fields on the string world-sheet the ψ^i are Majorana-Weyl fermions, however they transform as ten-dimensional Lorentz scalars. The light-cone gauge action of these fields is simply $(1/2\pi)\int \mathrm{d}\tau \int_0^\pi \mathrm{d}\sigma\, \psi^i\, (\partial/\partial\tau - \partial/\partial\sigma)\psi^i$, yielding the equation of motion $(\partial/\partial\tau - \partial/\partial\sigma)\psi^i = 0$.

* The use of fermionic variables to create representations of an internal symmetry group is not new; Bardacki and Halpern introduced this notion [20] as a possible way of incorporating hadronic symmetries in dual resonance models.

At this stage we have the option of choosing periodic or antiperiodic boundary conditions for ψ^i. As in the case of the "old superstring", we find it necessary to include in the Hilbert space of string states two sectors corresponding to the two possible boundary conditions. This is required to have a consistent, unitary description of string loop amplitudes. We have then one sector, analogous to the Ramond model, in which the fermions obey periodic boundary conditions in σ and thus have a Fourier expansion

$$\psi^i(\tau+\sigma) = \sum_{n=-\infty}^{+\infty} \psi_n^i \, e^{-2in(\tau+\sigma)}, \qquad n \text{ integer}. \tag{3.1}$$

Upon quantization, the oscillators ψ_n^i obey

$$\{\psi_n^i, \psi_m^j\} = \delta^{ij}\delta_{n+m,0}. \tag{3.2}$$

The other (Neveu-Schwarz) sector contains fermions which obey antiperiodic boundary conditions in σ, so that the integer frequencies n in eqs. (3.1) and (3.2) are shifted by $\frac{1}{2}$. The choice of boundary condition determines the ground state in each sector. In the antiperiodic case the ground state is an SO(32) singlet state annihilated by all ψ_n^i $(n>0)$. In the periodic sector the zero frequency oscillators, ψ_0^i, form a Clifford algebra and therefore cannot annihilate the vacuum. Hence we must take the ground state to be the appropriate spinor representation of SO(32).

The hamiltonian describing the contribution of the left-moving bosonic (the eight transverse coordinates $\tilde{X}^i(\tau+\sigma)$) and fermionic modes to the mass operator of the string is

$$\tilde{N} = \tilde{N}_B + \tilde{N}_F,$$
$$\tilde{N}_F = \sum_{n>0} n\psi_{-n}^i\psi_n^i - C_{A,P}. \tag{3.3}$$

$C_A(C_P)$ is the normal ordering constant in the case of antiperiodic (periodic) fermions respectively. This subtraction, which includes the contribution from the spatial oscillations as well, is uniquely determined by the requirement of 10-dimensional Lorentz invariance. A straightforward calculation along the lines of ref. [12] provides the values

$$C_A = 1, \qquad C_P = -1. \tag{3.4}$$

We then obtain the spectrum pictured in fig. 2 (disregarding the spatial oscillations). The two sets of states so constructed each form irreducible Fock space representations of the respective sets of oscillators. However, as previously mentioned, we are actually interested in constructing representations of the affine Lie algebra based on SO(32), whose generators M_n^{ij} are bilinear in the fermions:

$$M_n^{ij} = \tfrac{1}{2} \sum_{m=-\infty}^{+\infty} \psi_{n-m}^i \psi_m^j \tag{3.5}$$

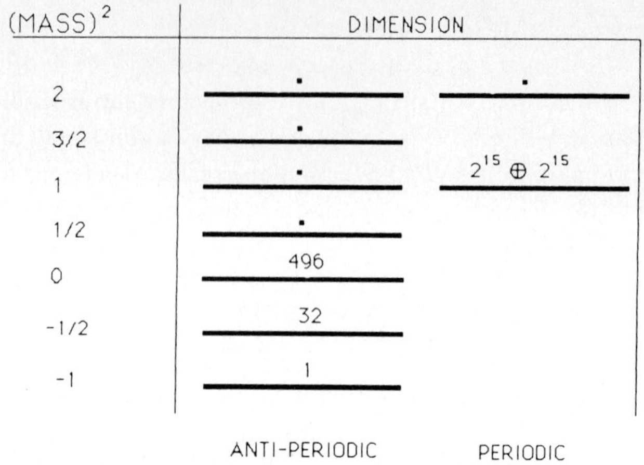

Fig. 2. Spectrum of states for the group Spin(32)/Z_2.

(n = integer or half-integer)

$$[T_1 \cdot M_m, T_2 \cdot M_n] = [T_1, T_2] \cdot M_{n+m} + \tfrac{1}{2} n \operatorname{Tr}(T_1 T_2) \delta_{m+n,0}, \tag{3.6}$$

where the T_{ij} are 32×32 antisymmetric matrices and $T \cdot M_n = T^{ij} M_n^{ij}$. Consequently the two fermion Fock space representations are each reducible, splitting into representations of the affine Lie algebra with even or odd "fermion number". (These correspond to the four conjugacy classes of SO(32) representations, whose characteristic members are singlet, spinor, spinor′ and vector, respectively.)

For the purpose of constructing the heterotic string it is necessary to exploit this reducibility by projecting onto the subspace of states which in the antiperiodic sector contain an even number of fermions (i.e. the integer mass levels), and in the periodic sector satisfy a Weyl projection with $\frac{1}{2}(1+\bar{\gamma})$ where $\bar{\gamma} = \psi_0^1 \psi_0^2 \cdots \psi_0^{32}$ on their SO(32) spinor index. The necessity of this truncation of the Hilbert space stems from the fact that the right-moving half of the theory contains only integer mass levels, whereas the unreduced spectrum of left movers has half-integer modes. Such states could never satisfy the constraint $N = \tilde{N}$, which is necessary to have a consistent spectrum, and so the half-integer levels must be removed. Since these correspond to the vector ($\underline{32}$) conjugacy class of SO(32) we must, for consistency, also discard half of the states in the periodic sector, since [spinor]×[spinor]=[vector].

The resulting theory does not contain any states of the vector or spinor types. Consequently the gauge group is Spin(32)/Z_2, and the oscillators (ψ_n^i) used to construct this group are not quite a representation of this group. This circumstance is familiar from the construction of superstring; although the unreduced Neveu-Schwarz and Ramond models were themselves consistent (at least at tree level), a truncation of this sort was necessary in order to find a supersymmetry generator which exchanged bosonic (Neveu-Schwarz) and fermionic (Ramond) sectors.

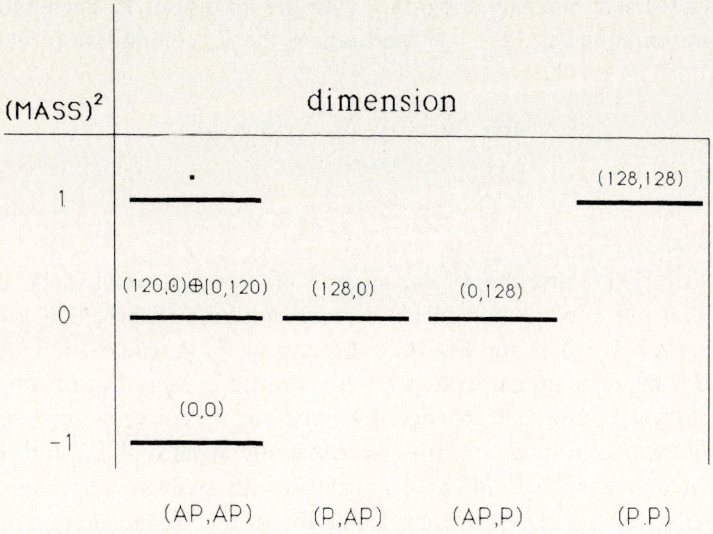

Fig. 3. Spectrum of states for the group $E_8 \times E_8$.

We now turn to the fermionic description of the $E_8 \times E_8$ heterotic string. Remarkably, it requires only a simple modification of the procedure used to obtain the $Spin(32)/Z_2$ theory, although demonstrating that the correct model is obtained in this way is rather subtle. We proceed by defining two sets of 16 fermionic coordinates ψ^i and $\tilde{\psi}^i$ $(i = 1 \cdots 16)$. These are quantized as before, with the boundary conditions and projections being imposed separately on each set of fermionic coordinates. We now have four sectors, depending on which boundary conditions are applied to ψ^i or $\tilde{\psi}^i$. We keep only those states with an even number of ψ^i excitations, independent of the boundary conditions applied to $\tilde{\psi}$ and of the $\tilde{\psi}$ internal quantum numbers. We then obtain states which are explicit representations of $SO(16) \times SO(16)$, which are restricted in the same way that we restricted the SO(32) states.

The mass spectrum of the theory can be calculated if we include the correct normal-ordering constant. In the two sectors whose boundary conditions match we obtain the same result as before; in the two sectors where one set of fermions has antiperiodic boundary conditions and the other has periodic boundary conditions the normal-ordering constant is zero. Then the low-lying levels have the mass spectrum and $SO(16) \times SO(16)$ quantum numbers depicted in fig. 3 (again ignoring spatial excitations).

The massless states in the (AP, AP) sector are in the adjoint representation of $SO(16) \times SO(16)$, as is necessary in order for the heterotic string to contain Yang–Mills interactions. However, there are also massless states in the $\underline{128}$ spinor representation of each SO(16). Since these also become gauge bosons (and gauginos) when we attach the right-moving modes, the theory must contain a gauge group larger than $SO(16) \times SO(16)$. The resolution of this problem lies in the fact that E_8

contains SO(16) as a maximal subgroup, with the adjoint of E_8 decomposing into SO(16) representations as $\underline{120}+\underline{128}$, and where the E_8 commutation relations are represented as $(i, j, \ldots = 1 \cdots 16)$

$$[M_{ij}, M_{kl}] = (\delta_{ik} M_{jl} + \text{perms.}) ,$$
$$[M_{ij}, S^a] = \tfrac{1}{4}(\Sigma^{ij})_{ab} S^b ,$$
$$[S^a, S^b] = (\Sigma^{ij})_{ab} M^{ij} , \tag{3.7}$$

where $\Sigma^{ij} = \frac{1}{2}i\,[\Gamma^i, \Gamma^j]$ and the Γ^i are gamma matrices for SO(16). We therefore conjecture that our construction yields a representation of the affine Lie algebra based on $E_8 \times E_8$, in which the SO(16) subgroup of E_8 is realized linearly on the operators M^{ij}; the remaining operators S^a are realized in a nonlinear fashion. These generators, however, must interchange the Fock spaces corresponding to the two kinds of boundary conditions in order to realize the algebra of E_8, and no simple expression involving the ψ^i could accomplish this. An analogous problem arises in the construction of the supersymmetry operator in the Neveu-Schwarz-Ramond realization of the superstring, which also is required to exchange the bosonic and fermionic Fock spaces. In that case, unfortunately, the supersymmetry operator did not have a simple expression in terms of oscillators of either sector. The superstring solution was to adopt a representation of the string coordinates for which the supersymmetry generators have a natural action. The analogous procedure in our case would be to abandon the fermionic coordinates altogether and use the algebra generated by $\{M_n^{ij}, S_n^a\}$ - the affine Lie algebra $\hat{E}_8$. However, the representations of this algebra are difficult to construct directly, so much so that mathematicians were led to use dual models as a means of constructing them [7, 23]. For practical purposes it is easier to calculate using the fermionic coordinates, wherein the SO(16) invariance is manifest and can be extended by gauge invariance to the full E_8 invariant expression.

The question remains whether we are sure that this construction actually produces a representation of the affine Lie algebra $(\hat{E}_8 \times \hat{E}_8)$. We are, but we will only present an outline of the argument here; for a more rigorous assurance we refer the reader to ref. [23] where it is shown that the representation of $\hat{E}_8$ which is of interest to us can be decomposed into $\widehat{\mathrm{SO}}(16)$ representations in precisely the combination found in our model, where the "basic" representation of $\hat{E}_8$ is the sum of the "basic" representation of $\widehat{\mathrm{SO}}(16)$ and the "spinor" representation of $\widehat{\mathrm{SO}}(16)$. This decomposition is based on an automorphism of the $\hat{E}_8$ Lie algebra, which in the notation of (3.7) leaves M_{ij} unchanged and takes S^a into $-S^a$. This automorphism can be extended trivially to the affine Lie algebra $\hat{E}_8$, and the commutation relations are unchanged. The representations of $\hat{E}_8$ can be decomposed into representations of the subalgebra of $\hat{E}_8$ which commutes with the automorphism, namely, the generators of $\widehat{\mathrm{SO}}(16)$; the states which are even under the automorphism are created by an even number of S_m^a, and those odd under the automorphism are created by an odd number of S^a_m. It is straightforward to show that the low-lying states of the $\hat{\mathrm{E}}_8$ representation, when decomposed this way, correspond to those of the $\widehat{\mathrm{SO}}(16)$ representations we constructed. There is also the circumstantial evidence that the number of states at each level of the $\hat{\mathrm{E}}_8$ "basic" representation match the number of states in the $\widehat{\mathrm{SO}}(16)$ construction, as seen by comparing the partition function $\mathrm{ch(G)} = \mathrm{Tr}\, q^{\tilde{N}_F}$ for the two cases (ch(G) is called the character function by the authors of ref. [7])

$$\begin{aligned}\mathrm{ch}\,(\hat{\mathrm{E}}_{8(\mathrm{basic})}) &= \frac{\left(1+240\sum_{n=1}^{\infty}\sigma_3(n)q^n\right)}{\prod_{n=1}^{\infty}(1-q^m)^8} \\ &= \mathrm{ch}\,(\widehat{\mathrm{SO}}(16)) + \mathrm{ch}\,(\widehat{\mathrm{SO}}(16)_{(1/2)}) \\ &= \frac{1}{2}\left(\prod_{n=1}^{\infty}(1-q^{n-1/2})^{16} + \prod_{n=1}^{\infty}(1+q^{n-1/2})^{16}\right) + 128\prod_{n=1}^{\infty}(1+q^n)^{16}, \qquad (3.8)\end{aligned}$$

where $\sigma_3(n)$ is the sum of the third powers of the divisors of n (a nontrivial identity, whose proof is based on the theory of modular forms [16, 24]). We conclude that our fermionic construction is indeed a representation of $\hat{\mathrm{E}}_8 \times \hat{\mathrm{E}}_8$; the generators of the subalgebra SO(16) × SO(16) are realized linearly in this construction, but the other generators unfortunately do not have a simple or useful realization in terms of the fermionic coordinates.

It is helpful to examine this result in relation to what we found in the bosonic realization. In that case, only the generators in the Cartan subalgebra were realized linearly, although the nonlinearly-realized operators also have a useful form. This is consistent with our expectations of the standard bosonization procedure, which gives linear realizations for the bosonic fields of only the flavor-diagonal fermionic currents. In this respect, both of the representations could be improved by turning to the third realization mentioned in sect. 4, which corresponds to a nonabelian bosonization of the fermionic theory; however this third realization is not practical for the kind of calculations currently contemplated.

From this point on the construction of the heterotic string proceeds as before, except that the internal bosonic operators must be replaced with the corresponding fermionic operators.

4. Covariant quantization

Thus far we have exhibited an apparently consistent light-cone quantization of the heterotic string. However, the light-cone formalism receives its justification as a geometric theory of surfaces by virtue of its being the gauge restriction of a reparametrization-invariant theory. Maintaining reparametrization invariance in the quantum theory is an essential requirement for the consistency of the theory; it is responsible for the critical dimensions 10 and 26. In this section we will describe both the path integral and operator formalisms for covariant quantization; the key point is that a proper BRS quantization shows that all potential anomalies cancel between the Faddeev–Popov ghosts and the string fields.

The path integral approach begins with a covariant action for the heterotic string. In principle, we would like an action which is manifestly group and supersymmetry invariant. The appropriate covariant framework for the $E_8\times E_8$ and $Spin(32)/Z_2$ currents is a nonlinear sigma model on the group manifold, with a Wess–Zumino term [25]. Of course the sigma model fields have to be constrained to be left moving. Similarly, a covariant action for the superstring exists [26] which may be thought of as a nonlinear sigma model on superspace (the super-translation group manifold) with a WZ term [27]. An arbitrary element of the super-translation group may be written as $h=\exp\{i[X\cdot P+\theta\cdot Q]\}$, where X_μ ($\mu=0\cdots 9$) and θ^a (a ten-dimensional Majorana–Weyl spinor) are the superspace coordinates and P and Q generate translations and supersymmetry transformations, respectively. In terms of the derivatives

$$h^{-1}(x)\partial_\alpha h(x)=(\partial_\alpha X^\mu-i\bar\theta\gamma^\mu\partial_\alpha\theta)P_\mu+\partial_\alpha\theta\cdot Q\,, \tag{4.1}$$

the covariant superstring action may be written

$$S_{\rm string}=\int d^2\xi\, e\tfrac{1}{2}\,{\rm tr}\,[(h^{-1}e_a^\alpha\partial_\alpha h)(h^{-1}e^{a\beta}\partial_\beta h)]+\Gamma_h\,. \tag{4.2}$$

The e_a^α define a local tangent space frame in terms of the coordinates ξ^α which parametrize the world sheet ($e=\det e_\alpha^a$). We choose the arbitrary normalization of the generators ${\rm tr}\,(P_\mu P_\nu)=\eta_{\mu\nu}$; also ${\rm tr}\,(Q_aQ_b)=0$. Γ_h is the Wess–Zumino action [25] which cannot be written as a local functional of the fields, but can be constructed in terms of the derivatives (4.1) as a three-dimensional integral when the world surface has spherical topology

$$\Gamma_h=\int_{M,\partial M=S^2} d^3\xi\,\varepsilon^{\alpha\beta\gamma}\,{\rm tr}\,[(h^{-1}\partial_\alpha h)(h^{-1}\partial_\beta h)(h^{-1}\partial_\gamma h)\,. \tag{4.3}$$

Γ_h is nonvanishing due to the relation ${\rm tr}\,[Q_aQ_bP^\mu]=(\gamma^\mu C^{-1})_{ab}$.

In the case of the heterotic string we want (4.2) to describe the dynamics of the "right" movers, where "right" is defined in terms of the local frame. We therefore add to the action the term

$$S_{\rm constraint}=\int d^2\xi\, e\lambda^{++}(h^{-1}e_+^\alpha\partial_\alpha h)(h^{-1}e_+^\beta\partial_\beta h) \tag{4.4}$$

(here $\xi^\pm=\tau\pm\sigma$ are two-dimensional light-cone coordinates). This indeed enforces the constraint that the superstring coordinates are only right moving. Similarly, to describe the left-moving coordinates of the heterotic string, we introduce $g=\exp\{i[X\cdot P+\phi^aT^a]\}$ where T^a are the generators of $G=E_8\times E_8$ or $Spin(32)/Z_2$ and ϕ^a are the local coordinates on G. The dynamics of the left movers is then described by the same action (4.2), together with the constraint (4.4), but now with all +'s replaced by −'s. This ensures that g is only a function of left-moving coordinates (in the local frame).

However, the covariant superspace action is difficult to quantize due to the presence of certain second-class constraints [26], and the sigma-model approach is unwieldy for practical calculations. The heterotic string is no worse in this regard than the ordinary superstring, with the possible exception of the question of reconciling the local fermionic symmetry of (4.2) [26] with the new constraints (4.4).

Thus although it would be nice to have a manifestly reparametrization invariant, Lorentz invariant, group invariant, and supersymmetric formalism, such an approach is not yet practical. Two options are then available: first, one can give up manifest Lorentz and reparametrization invariance and construct the theory on the light cone as we have done in sects. 2 and 3. Here supersymmetry is manifest, but full Lorentz invariance is not and must be checked. The light-cone approach is the one most suitable for simple calculations, and we will employ it in the following paper on interactions [9]. The second alternative is to give up manifest supersymmetry while retaining Lorentz and reparametrization invariance; it is this approach that we will now discuss.

We have seen that the group theory may be represented in terms of left-moving, Lorentz scalar, bosons or fermions. Their action may be written in a manifestly covariant form as

$$S_{\mathrm{G,fermion}} = \int \mathrm{d}^2\xi \, e[\tfrac{1}{2} i\psi^I e^\alpha_+ \partial_\alpha \psi^I] \,,$$

or

$$S_{\mathrm{G,boson}} = \int \mathrm{d}^2\xi \, e[\tfrac{1}{2}(e^\alpha_a \partial_\alpha X^I)^2 + \lambda^{--}(e^\alpha_- \partial_\alpha X^I)^2] \,. \tag{4.5}$$

Here $\psi^I (I = 1 \cdots 32)$ are the internal fermionic coordinates, $X^I (I = 1 \cdots 16)$ are the internal boson coordinates, and λ^{--} is a Lagrange multiplier which, as in (4.4), eliminates the right-moving bosons [13].

The ordinary superstring has a similar covariant formulation in terms of the Neveu-Schwarz-Ramond model. Here one describes the fermions by fields in the vector representation of O(9, 1). This is precisely analogous to the representation of the heterotic internal degrees of freedom given in sect. 3, where (say for G = Spin(32)/Z_2) one employs fermion fields ψ^I, in the vector representation of O(32). As we have explained, the states of the system will depend on the choice of boundary conditions. Periodic (antiperiodic) boundary conditions on the fermion fields will yield states which are O(9, 1) spinors (tensors). Space-time supersymmetry interchanges the two sectors and will not be manifest. In the case of the heterotic string the fermionic fields are further constrained to be right-moving, and thus the total space-time action is [28]

$$S_{\text{space-time}} = \int \mathrm{d}^2\xi\, e[\tfrac{1}{2}(e_a^\alpha \partial_\alpha X^\mu)^2 - \tfrac{1}{2}i\psi^\mu \rho^- e_-^\alpha \partial_\alpha \psi^\mu + \tfrac{1}{2}i(\zeta_\alpha \rho^\beta \rho^\alpha \psi^\mu)\partial_\beta X^\mu]\,. \quad (4.6)$$

This action involves the space-time coordinates X^μ, O(9, 1) fermions ψ^μ, the zweibein e_a^α and its right-handed gravitino partner ζ_α (ρ^α are two-dimensional γ-matrices). In addition to local reparametrization invariance, this action has an "$N=\frac{1}{2}$" local supersymmetry

$$\begin{aligned}
\delta e_-^\alpha &= i\varepsilon\rho_- \xi^\alpha\,,\\
\delta\zeta_\alpha &= -2\nabla_\alpha \varepsilon\,,\\
\delta X^\mu &= i\varepsilon\psi^\mu\,,\\
\delta\psi^\mu &= [\partial_\alpha X^\mu + \tfrac{1}{2}i(\zeta_\alpha \psi^\mu)]\rho^\alpha \varepsilon\,, \qquad (4.7)
\end{aligned}$$

with a right-handed spinor parameter that satisfies $\rho^-\varepsilon = \nabla_+\varepsilon = 0$.

Quantization is most easily performed in the superconformal gauge

$$e_a^\alpha = \mathrm{e}^\phi \delta_a^\alpha\,, \qquad \zeta^\alpha = \rho^\alpha \eta\,. \quad (4.8)$$

With this choice the total heterotic string action, (4.5)+(4.6), describes a set of free fields (we choose the fermionic form of S_G for simplicity):

$$S_{\text{heterotic}} = \int \mathrm{d}^2\xi[\tfrac{1}{2}(\partial_\alpha X^\mu)^2 + \tfrac{1}{2}i\psi^\mu \rho^- \partial_- \psi^\mu + \tfrac{1}{2}i\psi^I \rho^+ \partial_+ \psi^I] \quad (4.9)$$

since the conformal degrees of freedom ϕ and η decouple classically. This action still has a residual invariance under conformal reparametrizations (and their right-handed two-dimensional superpartners)

$$\xi^\pm \to f(\xi^\pm)\,, \quad (4.10)$$

which preserve the gauge (4.8). These transformations are generated by the components of the two-dimensional energy-momentum tensor and superconformal current

$$\begin{aligned}
T_{++} &= (\partial_+ X^\mu)^2 + i\psi^\mu \rho_+ \partial_+ \psi^\mu\,,\\
T_{--} &= (\partial_- X^\mu)^2 + i\psi^I \rho_- \partial_- X^I\,,\\
S_+ &= i(\partial_+ X^\mu)\psi^\mu\,. \qquad (4.11)
\end{aligned}$$

The (super)conformal symmetries allow us to classically specify the further gauge choices

$$X^+ = x^+ + p^+\tau\,, \qquad \psi^+ = 0\,. \quad (4.12)$$

This procedure can be justified quantum mechanically in a hamiltonian framework [29]. The constraints $T_{++} = T_{--} = S_+ = 0$ enable us to solve for X^- and ψ^- in terms of the physical degrees of freedom X^i, ψ^i ($i=1\cdots 8$), ψ^I ($I=1\cdots 32$); this leaves only global τ and σ translations, generated by

$$H=\tfrac{1}{2}\int(\dot{X}^2+X'^2+i\psi^i\cdot\psi^{i\prime}+i\psi^I\psi^{I\prime})\,\mathrm{d}^2\xi\,,$$

$$P=\int(\dot{X}\cdot X'+i\psi^i\cdot\psi^{i\prime}-i\psi^I\psi^{I\prime})\,\mathrm{d}^2\xi\,, \tag{4.13}$$

as residual invariances. Finally we arrive at the light-cone gauge action

$$S=\int \mathrm{d}^2\xi[\tfrac{1}{2}(\partial_\alpha X^i)^2+\tfrac{1}{2}i\psi^i\rho^-\partial_-\psi^i+\tfrac{1}{2}i\psi^I\rho^+\partial_+\psi^I]\,. \tag{4.14}$$

Eqs. (4.13) and (4.14) are just the Neveu-Schwarz-Ramond version of the light-cone gauge heterotic string action (2.1) and constraints (2.17) and (2.19). Thus we see that one may derive the light-cone heterotic string from a manifestly reparametrization invariant action. Checking that Lorentz invariance is maintained upon quantization is the easiest way to ensure that no anomalies of reparametrization invariance have appeared.

At this point we are more interested in maintaining manifest Lorentz invariance as well as checking two-dimensional coordinate invariance explicitly. Hence we will not choose the light-cone gauge (4.12), but rather explore the quantum mechanical consequences of (4.8). In the path integral over the action, factoring out the gauge group volume via the choice (4.8) involves computing a jacobian measure factor from the fluctuations of the metric induced by infinitesimal reparametrizations

$$\delta g_{\alpha\beta}=\nabla_{(\alpha}\delta\xi_{\beta)}\,. \tag{4.15}$$

This jacobian, as well as the corresponding factor arising from fixing the $N=\frac{1}{2}$ supersymmetry, may be represented via standard manipulations [30] as a path integral over ghost fields. The ghost action is

$$S_{\text{ghost}}=\int \mathrm{d}^2\xi[b^{++}\nabla_+c_++b^{--}\nabla_-c_-+\beta^-\nabla_-\gamma]\,, \tag{4.16}$$

where b, c are the reparametrization ghosts and β, γ are the local supersymmetry ghosts.

Evaluating the path integral in the gauge (4.8) reduces to calculating a series of functional determinants of differential operators of various conformal spins. The conformal spin of a quantity is defined by the (possibly half-integral) phase $\mathrm{e}^{in\theta}$ it acquires under a two-dimensional rotation by angle θ. The covariant derivative $\nabla_n=\mathrm{e}^{-(n-1)\phi}\partial_+\,\mathrm{e}^{n\phi}$ maps the tensor space of spin n into the tensor space of spin $(n+1)$. The effective action consists of the combination

$$\ln S_{\text{eff}}=\det{}^{-10/2}[\Delta_0=\nabla_0^+\nabla_0+\nabla_0\nabla_0^+]\det{}^{10}(\nabla_{1/2}^+)\det{}^{32}(\nabla_{-1/2})$$

$$\times\det(\nabla_{3/2}^+)\det(\nabla_1)\det(\nabla_{-1}^+)\,. \tag{4.17}$$

Because of the chiral asymmetry of this expression there is an ambiguity in the phases of each of these determinants (except Δ_0). The last two are hermitian conjugates and therefore do not contribute to the phase anomaly. The remaining spin-$\frac{1}{2}$ and spin-$\frac{3}{2}$ anomalies have been computed by Alvarez-Gaumé and Witten [3] with the result that 22($=32-10$) spin-$\frac{1}{2}$ determinants precisely cancel the anomaly of a spin-$\frac{3}{2}$ determinant*. Note that if we had worked with the bosonic representation of G, we could calculate the anomaly of a left-moving boson by regarding it as an antisymmetric 0-index tensor. One finds as expected that one such boson has the same gravitational anomaly as two Majorana–Weyl fermions. Thus we find that the heterotic string is anomaly free.

In addition to the potential phase anomaly, there is also a possibility of an anomaly in the modulus of the expression (4.17) since conformal invariance is an anomalous symmetry. The modulus may be regulated in any number of ways [31] and the anomaly again cancels in the critical dimension, 10 for the right movers and 26 for the left movers.

One may also study the anomaly cancellation in an operator approach [30]. There, analytic reparametrizations are generated by the stress-energy tensor T_{++}, anti-analytic ones by T_{--}; conformal invariance is expressed by the tracelessness condition $T_a{}^a=0$. The Fourier components

$$L_n = \int \mathrm{d}\xi\, \mathrm{e}^{in\xi} T_{++}(\xi)\,,$$

$$\tilde{L}_n = \int \mathrm{d}\xi\, \mathrm{e}^{in\xi} T_{--}(\xi)\,, \tag{4.18}$$

generate the algebra of reparametrizations (the Virasoro algebra)

$$[L_n, L_m] = (n-m)L_{n+m} + \tfrac{1}{12}c(n^3-n)\delta_{n,-m}\,,$$

$$[\tilde{L}_n, \tilde{L}_m] = (n-m)\tilde{L}_{n+m} + \tfrac{1}{12}\tilde{c}(n^3-n)\delta_{n,-m}\,, \tag{4.19}$$

which has, in general, a c-number anomaly given by the second term. When the stress tensor includes all the relevant string and ghost field contributions, one finds $c=\tilde{c}=0$. This is the operator equivalent of the anomaly cancellation found in the path integral approach.

It is this striking cancellation of anomalies that results in string theories being the only known examples of quantum field theories embodying quantum-mechanical reparametrization invariance; it is this geometric fact that accounts for their remarkable properties.

* We should note that the number 23 in sect. 12 of [3] has had the contribution of a single fermion subtracted.

5. Conclusions

In this paper we have constructed a new type of quantum string theory, the heterotic string, an amalgamation of the old fermionic and bosonic strings. We have shown that, at the noninteracting level, the heterotic string theory satisfies all the usual consistency requirements. Thus the formal reparametrization invariance, embodied in the geometrical formulation of the action, suffers no anomalies upon first quantization. Equivalently the light-cone gauge (unitary) formalism preserves the Lorentz invariance and ($N=1$) supersymmetry of the theory. The spectrum of the free heterotic string is ghost and tachyon free, with a massless ground state consisting of $D=10$, $N=1$, supergravity and super-Yang-Mills gauge bosons and their fermionic partners. In the following paper [9] we shall show that interactions can be consistently introduced and that the theory is one-loop finite. It is certainly true that the formulation of the heterotic string appears somewhat awkward and contrived. This is not a shortcoming of our theory; rather it is indicative of the present level of understanding of all quantum string theories, which leaves much to be desired. Many of the most remarkable features of these theories emerge without a full comprehension of their origin. Most mysterious are the general coordinate invariance and local gauge symmetries of string theories, whose appearance lacks a geometrical explanation. This suggests that there exists a more profound formulation of string theory, in which these features would be manifest. In such a formulation the heterotic theory might appear more natural. It certainly occupies a natural niche between the 26-dimensional closed bosonic string theory and the 10-dimensional, closed fermionic string theory.

The special feature of the heterotic string theory is the appearance of a (almost) uniquely determined local gauge symmetry group $G=E_8\times E_8$ or $Spin(32)/Z_2$. Heterotic strings are closed, so that there is no place to attach charges as one does in the case of open strings. Instead the gauge group arises via a Kaluza–Klein-type mechanism stemming from the compactification of 16 of the internal coordinates of the string. These are required, by consistency of the theory, to lie on a particular 16-dimensional torus. This compactification produces 16 gauge bosons associated with the isometries of the torus, but in addition it yields 480 massless solitons which wind around the torus. Together they fill out the adjoint representation of G, and all the physical states of the heterotic string form unitary representations of G. Nowhere do we require anomaly cancellation. Instead the gauge groups $E_8\times E_8$ and $Spin(32)/Z_2$ arise from other considerations. In fact we can now understand why $D=10$, $N=1$ supergravity is anomaly free only if one chooses these gauge groups. These anomaly-free field theories are simply the low-energy limit of the heterotic string (as we shall see explicitly in the following paper [9]).

Among the ill-understood features of quantum string theories is their high degree of uniqueness. The heterotic string provides us with two more entries (depending on whether $G=E_8\times E_8$ or $Spin(32)/Z_2$) to an extremely short list of consistent string theories. Are there any more? We do not know the complete answer to this question. However, the only new theories that we know how to construct, using the ideas that underlie the heterotic theory, are two-dimensional. Since the heterotic strings are formed by combining left and right movers of separately consistent closed string theories, the only additional theories of this type that we know are based on the $D=2$ fermionic string theory of ref. [32]. The right-moving sector of this theory could be combined with the left-moving sectors of the $D=10$ fermionic string or the $D=26$ bosonic string to yield new two-dimensional string theories. In the latter case we would again compactify the 24 internal dimensions of the string on a torus based on a 24-dimensional self-dual lattice, of which there are exactly 24 types (including the famous Leech lattice). It would be very amusing to construct these theories.

From a practical point of view the heterotic string, particularly in its $E_8 \times E_8$ manifestation, provides a very promising candidate for a unified theory of gravity and matter. Recent investigations have shown that this theory possesses features which may solve many long-standing problems in particle physics. Among these are: compactifications with zero cosmological constant of 6 of the 10 space-time dimensions to a four-dimensional theory with an unbroken $E_8 \times E_8'$ symmetry and some number of chiral fermion families of E_6 [33], a natural mechanism for breaking E_6 down to an acceptable low-energy gauge group [33], a mechanism for splitting weak Higgs doublets from color Higgs triplets [34], and a mechanism for supersymmetry breaking with zero cosmological constant in the "hidden" E_8' sector [35]. Although much work remains to be done there seem to be no insuperable obstacles to deriving all of known physics from the $E_8 \times E_8$ heterotic string.

References

[1] E. Witten, 1983 Shelter Island Conf. (MIT, 1984) to appear

[2] M.B. Green and J.H. Schwarz, Nucl. Phys. B181 (1981) 502; B198 (1982) 252, 441; Phys. Lett. 109B (1982) 444;
M.B. Green, J.H. Schwarz and L. Brink, Nucl. Phys. B198 (1982) 474

[3] L. Alvarez-Gaumé and E. Witten, Nucl. Phys. B234 (1983) 269

[4] M.B. Green and J.H. Schwarz, CALT-68-1182, CALT-68-1194

[5] J. Paton and H.M. Chan, Nucl. Phys. B10 (1969) 519

[6] J.H. Schwarz, Proc. Johns Hopkins Workshop, Florence (1982);
N. Marcus and A. Sagnotti, Phys. Lett. 119B (1982) 97

[7] I.B. Frenkel and V.G. Kac, Inv. Math. 62 (1980) 23;
G. Segal, Comm. Math. Phys. 80 (1982) 301;
I.B. Frenkel, Lectures in Applied Math, Vol. 21 (1985) AMS.
P. Goddard and D. Olive, Proc. Conf. on vertex operators in mathematics and physics, Publications of the Mathematical Sciences Research Institute no. 3 (Springer, Berlin, 1984).

[8] P.G.O. Freund, ITP preprint (1984)

[9] D.J. Gross, J.A. Harvey, E. Martinec and R. Rohm, Heterotic string theory. (II). The interacting heterotic string, Princeton preprint (1985).

[10] A.M. Polyakov, Phys. Lett. 103B (1981) 207

[11] P. Goddard, J. Goldstone, C. Rebbi and C. Thorn, Nucl. Phys. B56 (1973) 109
[12] J.H. Schwarz, Phys. Reports 89 (1982) 223;
M.B. Green, Surveys in high energy physics 3 (1983) 127;
L. Brink, Superstrings, CERN preprint-H4006 (1984)
[13] W. Siegel, Nucl. Phys. B238 (1984) 307
[14] L. Alvarez-Gaumé and D.Z. Freedman, Phys. Rev. D22 (1980) 846;
Comm. Math. Phys. 80 (1981) 282
[15] E. Cremmer and J. Scherk, Nucl. Phys. B103 (1976) 399
[16] J.P. Serre, A course in arithmetic (Springer, New York, 1973)
[17] K. Huan and S. Weinberg, Phys. Rev. Lett. 25 (1970) 895
[18] S.G. Rajeev, Phys. Rev. D29 (1984) 2944
[19] V. Bargmann, Ann. Math. 59 (1954) 1
[20] K. Bardakci and M.B. Halpern, Phys. Rev. D3 (1971) 2493
[21] A. Neveu and J.H. Schwarz, Nucl. Phys. B31 (1971) 86; Phys. Rev. D4 (1971) 1109
[22] P. Ramond, Phys. Rev. D3 (1971) 2415
[23] J. Leponsky and R.L. Wilson, Comm. Math. Phys. 62 (1978), 43;
V. Kac, D. Kazhdau, J. Lepowsky and R.L. Wilson, Adv. in Math. 42 (1981) 83;
I.B. Frenkel, J. Lepowsky and A. Meurman, An E_8 approach to F_1, Proc. 1982 Montreal Conf. on Finite group theory, ed. J. McKay, Springer Lecture Notes in Mathematics (1984); A moonshine module for the monster, Proc. Conf. on Vertex operators in mathematics and physics, Publications of the Mathematical Sciences Research Institute no. 3 (Springer, Berlin, 1984)
[24] A. Erdelyi et al., Higher transcendental functions, vol. 2 (McGraw-Hill, New York, 1953) ch. 13
[25] E. Witten, Comm. Math. Phys. 92 (1984) 451
[26] M. Green and J. Schwarz, Nucl. Phys. B243 (1984) 285
[27] M. Henneaux and L. Mezincescu, Texas preprint UTTG-26-84;
E. Martinec, unpublished
[28] L. Brink, P. Di Vecchia and P. Howe, Phys. Lett. 65B (1976) 471;
S. Deser and B. Zumino, Phys., Lett. 65B (1976) 369
[29] E. Martinec, Phys. Rev. D30 (1984) 1195
[30] D. Friedan, Proc. 1982 Les Houches Summer School, ed. J.B. Zuber and R. Stora (North-Holland, Amsterdam, 1984)
[31] A.M. Polyakov, Phys. Lett. 103B (1981) 207, 211;
P. Di Vecchia, B. Durhuus, P. Olesen and J.L. Peterson, Nucl. Phys. B198 (1982) 157; B207 (1982) 77;
O. Alvarez, Nucl. Phys. B216 (1983) 125;
E. Martinec, Phys. Rev. D28 (1983) 2604;
K. Fujikawa, Phys. Rev. D25 (1982) 2584
[32] M. Ademollo et al., Phys. Lett. 62B (1976) 105; Nucl. Phys. B111 (1976) 77
[33] P. Candelas, G.T. Horowitz, A. Strominger and E. Witten, Princeton preprint (1984), Nucl. Phys. B, to be published
[34] E. Witten, Princeton preprint (1985)
[35] M. Dine, R. Rohm, N. Seiberg and E. Witten, in preparation

4. Non-Supersymmetric Ten-Dimensional Strings

The papers collected in this chapter are of interest not so much because of what is constructed, but because of how it is constructed. After the appearance of paper 3, it was taken for granted that all possibilities in ten dimensions were exhausted. There was also a widespread belief that fermionic string theories are necessarily supersymmetric. This turned out to be wrong, and an important lesson was learned, which led to an explosion of four-dimensional strings not much thereafter.

One thing that was understood properly for the first time was how to sum over spin structures in the most general way. In particular, it became clear that such summations do not necessarily have to be done for left- and right-movers separately. The relation between modular invariance, spin statistics and spin-structure summation is beautifully explained in paper 6. As an application, these authors construct two new, until then unknown, type-II string theories with tachyons (and hence no supersymmetry).

New heterotic strings were constructed roughly simultaneously in papers 4 and 5. This includes the $O(16) \times O(16)$ strings, the most well-behaved one of the new members of the family. Although it is the only one free of tachyons, its one-loop cosmological constant does not vanish, and its perturbation series is therefore almost certainly ill-behaved at higher loop orders. There are six other, tachyonic, string theories known in ten dimensions. They were first constructed by an orbifold-like procedure in paper 4, and were then also obtained by a systematic exploration of spin-structure summation in paper 7.

Altogether we know now nine heterotic strings and four type-II strings in ten dimensions. It was shown in paper 8 that all but one of these theories can be derived from even self-dual Euclidean lattices: the two 16-dimensional even self-dual lattices yields the type-II strings, and the 24-dimensional Niemeier lattices yield eight of the nine heterotic strings. The missing one has a rank-8 gauge group (E_8), and does not admit a lattice description.

The table in paper 8 provides a nice summary of all the heterotic theories, and for completeness we give here a similar description of the E_8 string. As was shown in [108], the left-moving sector of this theory can be described as a level-2 E_8 Kac–Moody algebra tensored with an Ising model. Both these CFT's have three primary fields: for E_8 they are identified by the singlet, the 248 and 3875; for the Ising model one has the well-known possibilities $h = 0$, $h = \frac{1}{16}$ and $h = \frac{1}{2}$. We label these primary fields respectively by $l = 0$, 2 and 1, for E_8 as well as for the Ising model, and denote their partition functions as $\mathscr{P}^l_{E_8}$ or $\mathscr{P}^l_I$. To specify the theory, we simply have to indicate which combinations of these fields belong to the spin structures of the right-moving NSR-model, or, in the bosonic string representation, to the conjugacy classes of the right-moving D_8 factor. The result is

$$
\begin{aligned}
\mathscr{P}_0 &= \mathscr{P}^0_{E_8}\mathscr{P}^0_I + \mathscr{P}^1_{E_8}\mathscr{P}^1_I, \\
\mathscr{P}_v &= \mathscr{P}^0_{E_8}\mathscr{P}^1_I + \mathscr{P}^1_{E_8}\mathscr{P}^0_I, \\
\mathscr{P}_s &= \mathscr{P}_c = \mathscr{P}^2_{E_8}\mathscr{P}^2.
\end{aligned}
\tag{4.1}
$$

Here $\mathscr{P}_x$ is the left-moving internal partition function that belongs to the conjugacy class (x) of the right-moving D_8 factor.

Despite the fact that systematic studies have been done within the framework of certain constructions, there is as yet no rigorous proof that the known cases exhaust all possibilities. Nevertheless this looks like one of the simplest classification problems. One simply has to find four characters of $c = 16$ CFT's that transform into each other exactly like D_8 characters, and have the right h-values to satisfy level matching. A still simpler characterization is the one used in paper 8, generalized from lattices to arbitrary CFT's: one can obtain all ten-dimensional heterotic strings from a list of modular-invariant $c = 24$ theories that contain a level-1 D_8 Kac–Moody algebra. Such a left-moving D_8 algebra can be traded for a right-moving one without affecting modular invariance. At present 25 modular $c = 24$ theories are known: the 24 Niemeier lattices, plus the internal sector of the E_8 string, combined with a *left*-moving D_8 factor. It would be extremely interesting – if only for mathematical reasons – to know the complete list of modular invariant, chiral CFT's with $c = 24$ (note that $c = 24$ is the smallest non-zero value of c for which a chiral CFT can be modular invariant). If anything can be classified completely, it should be this.

A rigorous classification exists only in two cases: the ten-dimensional type-II strings (for which one merely has to find all modular-invariant combinations of a left- and right-moving level-1 D_8 algebra) and the supersymmetric heterotic strings (where anomaly cancellation, a manifestation of modular invariance, uniquely selects only two gauge groups of rank not exceeding 16).

Nuclear Physics B274 (1986) 93–105
North-Holland, Amsterdam

STRING THEORIES IN TEN DIMENSIONS WITHOUT SPACETIME SUPERSYMMETRY

L.J. DIXON[1] and J.A. HARVEY[2]

Joseph Henry Laboratories, Princeton University, Princeton, NJ 08544, USA

Received 5 March 1986

We analyze the possibilities for new modular invariant ten-dimensional string theories without spacetime supersymmetry which are twisted versions of the closed superstring or of the heterotic string. The most interesting possibility is a tachyon-free theory with gauge group $O(16) \otimes O(16)$, chiral fermions, and a positive one-loop contribution to the cosmological constant.

The anomaly cancellation mechanism discovered by Green and Schwarz [1] and the construction of the heterotic string [2] have prompted many discussions of the relationship between various string theories as well as the possibility of new types of string theories [3, 4]. It is tempting to interpret the presence of a tachyonic state in the twenty-six dimensional bosonic string as a signal of vacuum instability and hence to search for a stable vacuum which would be tachyon-free. It has been proposed [4] that ten-dimensional superstrings might emerge as the correct vacuum in such an approach. Although there are many intriguing relations between the bosonic string and superstring theories, no compelling dynamical mechanism has yet appeared.

The presence of a tachyon in the bosonic string, as well as divergences associated with dilaton tadpoles [5], are indications of vacuum instability at either the classical or quantum level, but they do not indicate any fundamental inconsistency in the theory. It would be interesting to know whether there are any other string theories with the same degree of consistency as the twenty-six dimensional closed bosonic string; that is, Lorentz invariant, modular invariant, interacting string theories. In this letter we will present several such theories. Although they are related in a rather simple fashion to previously known theories, they do not seem to have been discussed previously.

The first model we will consider is related to the ten-dimensional closed superstring. We will work in light-cone gauge and with the old Ramond-Neveu-Schwarz

[1] Supported by a Charlotte Elizabeth Procter fellowship.
[2] Supported by NSF Grant PHY80-19754.

0550-3213/86/$03.50 © Elsevier Science Publishers B.V.
(North-Holland Physics Publishing Division)

formalism for the spinning string. The physical degrees of freedom of this theory are the transverse bosonic coordinates $X^i(\sigma,\tau)$, $i=1,\dots,8$ transforming as the $\mathbf{8}_v$ representation of SO(8), with the mode expansion

$$X^i(\sigma,\tau)=x^i+p^i\tau+\tfrac{1}{2}i\sum_{n\neq 0}\frac{\alpha_n^i}{n}\mathrm{e}^{-2in(\tau-\sigma)}+\tfrac{1}{2}i\sum_{n\neq 0}\frac{\tilde{\alpha}_n^i}{n}\mathrm{e}^{-2in(\tau+\sigma)},\tag{1}$$

and the fermionic coordinates $\psi^i(\tau-\sigma)$ and $\tilde{\psi}^i(\tau+\sigma)$, also transforming as vectors under SO(8). Consider first the (NS, NS) sector with antiperiodic boundary conditions for both left- and right-moving fermionic fields. The mode expansions for ψ^i and $\tilde{\psi}^i$ are then

$$\psi^i(\tau-\sigma)=\sum_{n\in Z+1/2}b_n^i\mathrm{e}^{-2in(\tau-\sigma)},$$

$$\tilde{\psi}^i(\tau+\sigma)=\sum_{n\in Z+1/2}\tilde{b}_n^i\mathrm{e}^{-2in(\tau+\sigma)}.\tag{2}$$

The $(\text{mass})^2$ operator, in units where $\alpha'=\frac{1}{2}$, is given by

$$\tfrac{1}{4}(\text{mass})^2=\left(L_0-\tfrac{1}{2}\right)+\left(\bar{L}_0-\tfrac{1}{2}\right),\tag{3}$$

where

$$L_0=\sum_{n=1}^{\infty}\alpha_{-n}^i\alpha_n^i+\sum_{n=1/2}^{\infty}nb_{-n}^ib_n^i,$$

$$\bar{L}_0=\sum_{n=1}^{\infty}\tilde{\alpha}_{-n}^i\tilde{\alpha}_n^i+\sum_{n=1/2}^{\infty}n\tilde{b}_{-n}^i\tilde{b}_n^i,\tag{4}$$

and physical states are subject to the constraint $L_0=\bar{L}_0$ which follows from invariance under rigid shifts in σ. The $-\frac{1}{2}$ terms added to L_0 and $\bar{L}_0$ in the $(\text{mass})^2$ operator arise from normal ordering and are required for closure of the Lorentz algebra in light-cone gauge.

As it stands this theory is inconsistent due to a lack of modular invariance. To see this consider the string path integral continued to euclidean space $(\sigma,t)=(\sigma,i\tau)$ for a world sheet with the topology of a torus. The metric on the torus, modulo conformal equivalence, depends on a single complex modular parameter τ with $\operatorname{Im}\tau>0$. The modular group is generated by the transformations

$$\tau\to\tau+1,$$

$$\tau\to-\frac{1}{\tau},\tag{5}$$

which correspond respectively to cutting the torus along a line of constant t and reconnecting with a 2π twist, and to interchanging σ and t directions. The

Neveu-Schwarz model corresponds to (σ, t) boundary conditions of $(-1,-1)$ on both sets of fermionic fields ψ^i and $\tilde{\psi}^i$. This model is clearly invariant under $\tau \to -1/\tau$, but not under $\tau \to \tau+1$, which takes $(-1,-1)$ to $(-1,+1)$. This can easily be checked by explicit calculation. For example, the one-loop contribution to the cosmological constant in this theory is easily calculated along the lines of [6, 7] to be

$$\Lambda_{\text{1-loop}} \sim -\int \frac{d^2\tau}{(\operatorname{Im}\tau)^2} (\operatorname{Im}\tau)^{-4} \left| \frac{\prod_{n=1}^{\infty}(1+z^{n-1/2})}{\prod_{n=1}^{\infty}(1-z^n)} \right|^{16}$$

$$\sim -\int \frac{d^2\tau}{(\operatorname{Im}\tau)^2} (\operatorname{Im}\tau)^{-4} \left| \frac{\vartheta_3(0,\tau)}{\vartheta_1'(0,\tau)} \right|^{8}, \tag{6}$$

where $z = e^{2\pi i\tau}$ and the Jacobi theta functions ϑ_i are described in [8]. Using the transformation properties of the theta functions one has

$$\left| \frac{\vartheta_3(0,\tau+1)}{\vartheta_1'(0,\tau+1)} \right|^8 = \left| \frac{\vartheta_4(0,\tau)}{\vartheta_1'(0,\tau)} \right|^8,$$

$$\left| \frac{\vartheta_3(0,-1/\tau)}{\vartheta_1'(0,-1/\tau)} \right|^8 = |\tau|^{-8} \left| \frac{\vartheta_3(0,\tau)}{\vartheta_1'(0,\tau)} \right|^8,$$

$$(\operatorname{Im}(-1/\tau))^{-4} = |\tau|^8 (\operatorname{Im}\tau)^{-4}, \tag{7}$$

so the integrand is invariant under $\tau \to -1/\tau$ but not under $\tau \to \tau+1$.

In order to ensure modular invariance it is necessary to sum over a modular invariant combination of boundary conditions on the fermionic fields ψ^i and $\tilde{\psi}^i$. The usual closed superstring corresponds to summing over the boundary conditions $(-1,-1)$, $(+1,-1)$, $(-1,+1)$ and $(+1,+1)$ independently for ψ^i and $\tilde{\psi}^i$. In terms of a hamiltonian formulation, the sum over t boundary conditions for fields antiperiodic in σ corresponds to the GSO projection [9] onto states with $G=(-1)^F=1$ and $\tilde{G}=(-1)^{\tilde{F}}=1$, where $F=\sum_{n=1/2}^{\infty} b_{-n}^i b_n^i - 1$ and $\tilde{F}=\sum_{n=1/2}^{\infty} \tilde{b}_{-n}^i \tilde{b}_n^i - 1$ are the world-sheet fermion numbers for ψ^i and $\tilde{\psi}^i$. The -1 in the definition of F, or equivalently the assignment of fermion number -1 to the tachyonic vacuum, is required for modular invariance. The GSO projection in the (NS, NS) sector destroys invariance under $\tau \to -1/\tau$. To restore this invariance it is necessary to include "twisted" sectors where the string closes only up to an element of the $Z_2 \otimes Z_2$ group with elements $\{1, (-1)^F, (-1)^{\tilde{F}}, (-1)^{F+\tilde{F}}\}$. These are just the (NS, NS), (R, NS), (NS, R), and (R, R) sectors of the closed string, where in the Ramond (R) sectors the fermion fields are periodic in σ with the mode expansions

$$\psi^i = \sum_{m\in Z} d^i_m e^{-2im(\tau-\sigma)},$$

$$\tilde{\psi}^i = \sum_{m\in Z} \tilde{d}^i_m e^{-2im(\tau+\sigma)}. \tag{8}$$

The operators L_0 and $\bar{L}_0$ in the Ramond sectors are given by

$$L_0 = \sum_{m=1}^{\infty} \alpha^i_{-m}\alpha^i_m + \sum_{m=1}^{\infty} m d^i_{-m} d^i_m,$$

$$\bar{L}_0 = \sum_{m=1}^{\infty} \tilde{\alpha}^i_{-m}\tilde{\alpha}^i_m + \sum_{m=1}^{\infty} m \tilde{d}^i_{-m} \tilde{d}^i_m. \tag{9}$$

The normal-ordering constant, which enters into the $(\text{mass})^2$ operator and into the constraint, is zero for the Ramond sectors. In these sectors $(-1)^F$ and $(-1)^{\tilde{F}}$, which must anticommute with right-moving and left-moving fermionic fields respectively, are given by

$$(-1)^F = \pm d^1_0 \dots d^8_0 (-1)^{\sum_{m=1}^{\infty} d^i_{-m} d^i_m},$$

$$(-1)^{\tilde{F}} = \pm \tilde{d}^1_0 \dots \tilde{d}^8_0 (-1)^{\sum_{m=1}^{\infty} \tilde{d}^i_{-m} \tilde{d}^i_m}. \tag{10}$$

Choosing the same sign in the definition of $(-1)^F$ and $(-1)^{\tilde{F}}$ in the Ramond sector yields the chiral type IIB superstring, with the massless fields in the (R, R) sector transforming as the SO(8) representations contained in $\mathbf{8}_s \otimes \mathbf{8}_s$. Choosing the opposite sign yields the type IIA superstring, with the massless fields in the (R, R) sector transforming as $\mathbf{8}_s \otimes \mathbf{8}_{s'}$.

The closed string theory originally considered by GSO [9] and recently reconsidered in [10, 11] actually consists only of the (NS, NS) and (NS, R) sectors. While this gives a supersymmetric spectrum, since the right-movers alone are supersymmetric, it does not result in a modular invariant theory. This is clear from the (σ, t) boundary conditions and can also be checked by an explicit one-loop calculation [11].

There is one further modular invariant combination of boundary conditions, which consists of summing over the *same* boundary conditions for left-movers and right-movers. That is, we project the (NS, NS) sector onto states with $(-1)^{F+\tilde{F}} = 1$ and add the "twisted" (R, R) sector, again projected onto states with $(-1)^{F+\tilde{F}} = 1$. There are actually two such theories, corresponding to the relative sign in the definitions of $(-1)^F$ and $(-1)^{\tilde{F}}$ in the (R, R) sector, just as in the case of the closed superstring. Each theory contains only spacetime bosons. However, the bosonic states are not a subset of the bosonic states of the superstring, since we are now including states with $(-1)^F=(-1)^{\tilde{F}}=-1$ which were previously discarded. In particular, the tachyon in the (NS, NS) sector is now physical. The partition function for the two theories is the same. In the projected (NS, NS) sector it is given by

$$\frac{|A(z)|^2+|B(z)|^2}{2|z||f(z)|^{16}}. \tag{11}$$

Here we have defined the functions

$$A(z)\equiv\prod_{n=1}^{\infty}(1+z^{n-1/2})^8=\vartheta_3^4/f(z)^4,$$

$$B(z)\equiv\prod_{n=1}^{\infty}(1-z^{n-1/2})^8=\vartheta_4^4/f(z)^4,$$

$$C(z)\equiv 8\prod_{n=1}^{\infty}(1+z^n)^8=\vartheta_2^4/\left(2z^{1/2}f(z)^4\right),$$

$$f(z)^3\equiv\prod_{n=1}^{\infty}(1-z^n)^3=\vartheta_1'/2z^{1/8}, \tag{12}$$

using the infinite-product representations of the Jacobi theta functions and the standard abbreviation $\vartheta_i\equiv\vartheta_i(0,\tau)$. The partition function for the projected (R, R) sector is given by

$$\frac{|C(z)|^2}{2|f(z)|^{16}}. \tag{13}$$

Adding (11) to (13) and using (12) we find the total partition function to be

$$\frac{2^7\left(|\vartheta_2|^8+|\vartheta_3|^8+|\vartheta_4|^8\right)}{|\vartheta_1'|^8}, \tag{14}$$

which when combined with a factor $(\operatorname{Im}\tau)^{-4}$ from the integration measure is clearly invariant under modular transformations.

So far our analysis has relied on the old Ramond-Neveu-Schwarz formalism. Given the equivalence between this formalism and the manifestly supersymmetric Green-Schwarz formalism, it is natural to ask whether this theory can be obtained within the latter formalism. In this formalism the fields ψ^i and $\tilde{\psi}^i$ are replaced by periodic fields S^a and $\tilde{S}^a$ transforming as the spinor $\mathbf{8}_{s'}$ of SO(8) (or as the $\mathbf{8}_{s'}$ and $\mathbf{8}_s$ in the case of the type IIA superstring). The irreducible representation of the S_0^a zero mode algebra is sixteen-dimensional and consists of an $\mathbf{8}_v$ plus an $\mathbf{8}_s$ of SO(8). The $\tilde{S}_0^a$ algebra is treated similarly. The only symmetry of this theory which leaves invariant ten-dimensional Minkowski space is a 2π spatial rotation, say $e^{2\pi i J_{12}}$, or in other words $(-1)^{F_S}$, where F_S is spacetime (as opposed to world-sheet) fermion number. Projecting onto states with $(-1)^{F_S}=1$ removes all the fermionic states and leaves all the bosonic states of the superstring. To preserve modular invariance we must include the twisted sector where the string closes only up to a transformation by $(-1)^{F_S}$. In other words, we must include the sector where S^a and $\tilde{S}^a$ are both antiperiodic, projected onto states with $(-1)^{F_S}=1$. In this sector the normal-ordering addition to L_0 and $\bar{L}_0$ is $-\frac{1}{2}$ for the same reason as it was in the (NS, NS) sector above. So the lowest state is a singlet tachyon with $\frac{1}{4}(\text{mass})^2=-1$, the next states consist of

$$\tilde{S}^a_{-1/2}|0\rangle_L \otimes S^b_{-1/2}|0\rangle_R$$

and are massless, and so on. It is easy to see that this procedure simply reproduces the previous spectrum of states.

So far we have only discussed the free spectrum of this theory. Lorentz invariance and the inclusion of interactions follow in a straightforward way, since in the Ramond-Neveu-Schwarz formalism we have only changed the projection relating the left- and right-hand sectors which were previously consistent. Because this theory has a tachyon and no fermions it is unlikely to be of much interest except as a toy model and we will not discuss it further.

It is natural to ask whether there are any analogous new theories which can be constructed from the heterotic string. Rather than looking at various combinations of $(-1)^F$ projections in the fermionic formulation of the heterotic string, we for the moment use the Green-Schwarz formalism for the right-moving fermionic degrees of freedom and the bosonic Frenkel-Kač construction of the $E_8 \otimes E_8$ or Spin(32)/Z_2 affine Lie algebra. We then need to consider the possible twists that leave invariant ten-dimensional Minkowski space.

One possibility is to again twist by a 2π spatial rotation. However, in this case it is easy to check that one simply recovers the heterotic string, with the spacetime fermions now appearing in the twisted sector. We can also twist by an $E_8 \otimes E_8$ or Spin(32)/Z_2 gauge transformation, but it can be shown [12] that this can only give back one or the other heterotic string. (For certain twists it is possible to start with $E_8 \otimes E_8$ and obtain Spin(32)/Z_2 and vice versa.) The only new possibility is to accompany a 2π spatial rotation with an order two twist of the $E_8 \otimes E_8$ or Spin(32)/Z_2 affine Lie algebra.

We thus consider Z_2 twists of the form

$$e^{2\pi i J_{12}}|_{\text{spacetime}} \cdot \gamma_\delta , \tag{15}$$

where γ_δ is of order two acting in $E_8 \otimes E_8$ or Spin(32)/Z_2. We can choose γ_δ to lie in a Cartan subalgebra, in which case we can represent γ_δ as a translation (or "shift") by a vector $\pi\delta$ on the $E_8 \otimes E_8$ or Spin(32)/Z_2 maximal torus; i.e. $\gamma_\delta = \exp(2\pi i p^I \delta^I)$, where p^I is the center of mass momentum operator for the string coordinate X^I on the maximal torus. Since $\gamma_\delta^2 = 1$, 2δ must be a lattice vector. In the twisted sector the string closes only up to a 2π spatial rotation plus a translation of the coordinates of the $E_8 \otimes E_8$ or Spin(32)/Z_2 torus by $\pi\delta$; i.e. $X^I(\pi) = X^I(0) + \pi\delta^I$. Thus in the twisted sector the momenta p^I lie on a "shifted" lattice: $p \in L + \delta$, where $L = \Gamma_8 \otimes \Gamma_8'$ or Γ_{16}, Γ_8 is the root lattice of E_8, and Γ_{16} is the weight lattice of Spin(32)/Z_2. Hence the left-moving contribution to the (mass)^2 operator is shifted by $\frac{1}{2}\delta^2$. In order to have matching contributions to the (mass)^2 from left- and right-movers, as required by modular invariance [12], δ^2 must be an integer. Also, shift vectors δ which differ by a lattice vector are clearly equivalent, so we can reduce the length of δ by subtracting off lattice vectors. It turns out [13] that by this procedure we can arrange that $\delta^2 \leqslant 2$ in the case of Γ_{16}, and $\delta = (\delta_1, \delta_1')$ with $\delta_1^2 \leqslant 1$, $\delta_1'^2 \leqslant 1$ in the case of $\Gamma_8 \otimes \Gamma_8'$. We thus conclude that $\delta^2 = 0$, 1 or 2.

If $\delta^2 = 0$, then γ_δ acts as the identity and we just get back the heterotic string. For $\delta^2 = 1$ there are two inequivalent choices for δ for $E_8 \otimes E_8$ and three for Spin(32)/Z_2. (Shift vectors which are related by some automorphism of the lattice correspond to conjugate, and hence equivalent, gauge transformations.) A convenient choice for the roots of E_8 is given by

$$\begin{aligned} &\pm e_i \pm e_j, && i \neq j, \\ &\tfrac{1}{2}(\pm e_1 \pm e_2 \pm \cdots \pm e_8) && \text{(even number of + signs)}, \end{aligned} \tag{16}$$

where the e_i, $i = 1, \ldots, 8$, are orthonormal vectors. The roots of SO(32) are given by $\pm e_i \pm e_j$, $i \neq j$, where the e_i, $i = 1, \ldots, 16$, are orthonormal vectors. The weight lattice of Spin(32)/Z_2 is generated by integer combinations of these roots as well as one of the spinor weights of Spin(32), say $\frac{1}{2}(e_1 + \cdots + e_{16})$. Using these bases, and using powers to denote repeated entries, the inequivalent shift vectors δ of length one are given by

$$\begin{aligned} &\delta = (1, 0^7; 0^8), \quad \left(\tfrac{1}{2}, \tfrac{1}{2}, 0^6; \tfrac{1}{2}, \tfrac{1}{2}, 0^6\right), && E_8 \otimes E_8, \\ &\delta = (1, 0^{15}), \quad \left(\left(\tfrac{1}{2}\right)^4, 0^{12}\right), \quad \left(\left(\tfrac{1}{4}\right)^{16}\right), && \text{Spin}(32)/Z_2. \end{aligned} \tag{17}$$

For $\delta^2 = 2$ there is a unique choice for either $E_8 \otimes E_8$ or Spin(32)/Z_2, given by

$$\begin{aligned} &\delta = (1, 0^7; 1, 0^7), && E_8 \otimes E_8, \\ &\delta = \left(\left(\tfrac{1}{2}\right)^8, 0^8\right), && \text{Spin}(32)/Z_2. \end{aligned} \tag{18}$$

The theories with $\delta^2 = 1$ all have physical tachyons occurring in the twisted sector. (Recall that the right-handed vacuum is at $-\frac{1}{2}$ in this sector, which matches the left-handed vacuum at $-1 + \frac{1}{2}\delta^2 = -\frac{1}{2}$.) Note that the tachyons are not gauge singlets. We will not analyze these theories in much detail. However, we should at least determine the spectrum of massless states to verify that they correspond to anomaly-free ten-dimensional theories. In each case we find the bosonic part of the ten-dimensional supergravity multiplet, $g_{\mu\nu}$, $B_{\mu\nu}$, ϕ, coming from the untwisted sector. In addition we find the following massless states. For $\delta = (1, 0^7; 0^8)$ in $E_8 \otimes E_8$: gauge bosons in the adjoint representation of $O(16) \otimes E_8$ and fermions in the $\mathbf{8}_s$ of SO(8) and the $(\mathbf{128}, \mathbf{1})$ of $O(16) \otimes E_8$ from the untwisted sector, and fermions in the $\mathbf{8}_{s'}$ of SO(8) and the $(\mathbf{128'}, \mathbf{1})$ of $O(16) \otimes E_8$ from the twisted sector. For $\delta = (\frac{1}{2}, \frac{1}{2}, 0^6; \frac{1}{2}, \frac{1}{2}, 0^6)$ in $E_8 \otimes E_8$: gauge bosons in the adjoint representation of $E_7 \otimes SU(2) \otimes E_7 \otimes SU(2)$ and $\mathbf{8}_s$ fermions in the $(\mathbf{56}, \mathbf{2}; \mathbf{1}, \mathbf{1}) \oplus (\mathbf{1}, \mathbf{1}; \mathbf{56}, \mathbf{2})$ representation of the gauge group from the untwisted sector, and $\mathbf{8}_{s'}$ fermions in the $(\mathbf{56}, \mathbf{1}; \mathbf{1}, \mathbf{2}) \oplus (\mathbf{1}, \mathbf{2}; \mathbf{56}, \mathbf{1})$ representation from the twisted sector. For $\delta = (1, 0^{15})$ in $\mathrm{Spin}(32)/Z_2$: gauge bosons in the adjoint representation of SO(32) and no massless fermions. For $\delta = ((\frac{1}{2})^4, 0^{12})$ in $\mathrm{Spin}(32)/Z_2$: gauge bosons in the adjoint representation of $SO(24) \otimes SO(8)$ and $\mathbf{8}_s$ fermions in the $(\mathbf{24}, \mathbf{8}_v)$ representation of the gauge group from the untwisted sector, and $\mathbf{8}_{s'}$ fermions in the $(\mathbf{24}, \mathbf{8}_{s'})$ representation of the gauge group from the twisted sector. For $\delta = ((\frac{1}{4})^{16})$ in $\mathrm{Spin}(32)/Z_2$: gauge bosons in the adjoint representation of $SU(16) \otimes U(1)$ and $\mathbf{8}_s$ fermions in the $(\mathbf{120}, +\mathbf{2}) \oplus (\overline{\mathbf{120}}, -\mathbf{2})$ representation of the gauge group from the untwisted sector, and $\mathbf{8}_{s'}$ fermions in the $(\mathbf{120}, -\mathbf{2}) \oplus (\overline{\mathbf{120}}, +\mathbf{2})$ representation from the twisted sector. All of these theories are anomaly-free, either because they are non-chiral as in the third case, or because the anomalies cancel trivially using the relations $\mathrm{Tr}_{\mathbf{8}_v} F^{2n} = \mathrm{Tr}_{\mathbf{8}_s} F^{2n} = \mathrm{Tr}_{\mathbf{8}_{s'}} F^{2n}$ for SO(8) and $\mathrm{Tr}_{\mathbf{128}} F^{2n} = \mathrm{Tr}_{\mathbf{128'}} F^{2n}$ for SO(16), or using the Green-Schwarz anomaly cancellation mechanism [1] for $(E_7 \otimes SU(2))^2$ and for $SU(16) \otimes U(1)$. There is one other possible set of Z_2 twists, which cannot be represented by shift vectors because they involve the outer automorphism of $E_8 \otimes E_8$, which exchanges the two E_8's. These cases also can be shown to have physical tachyons, and we will not discuss them further here.

Finally we come to the most interesting case which is $\delta^2 = 2$. This case leads to a tachyon-free theory since in the twisted sector the right-handed vacuum at $-\frac{1}{2}$ does not match the left-handed vacuum at $-1 + \delta^2 = 0$. So in the untwisted sector the $L_0 = \bar{L}_0$ constraint projects out the tachyon from the bosonic string, since the superstring vacuum is at level zero; whereas in the twisted sector this constraint projects out the tachyon from the twisted superstring, since the twisted bosonic string vacuum is at level zero! It turns out that starting with either $E_8 \otimes E_8$ or $\mathrm{Spin}(32)/Z_2$ yields the same theory, so we will just analyze the theory starting from $E_8 \otimes E_8$. In addition, although the spectrum is not hard to work out in this "shifted" bosonic formulation, it is even simpler in the fermionic formulation of $E_8 \otimes E_8$. In this formulation the shift by $\pi\delta$ in $E_8 \otimes E_8$ is equivalent to a 2π rotation in each O(16) factor in the $O(16) \otimes O(16)'$ subgroup of $E_8 \otimes E_8$. (See ref. [12] for details.) So the twist (15) becomes

$$R = \mathrm{e}^{2\pi i J_{12}}\big|_{\text{spacetime}} \mathrm{e}^{2\pi i j_{12}} \mathrm{e}^{2\pi i j'_{12}}, \tag{19}$$

where j_{12} and j'_{12} are generations of rotations in O(16) and O(16)′ respectively.

In the fermionic formulation of $E_8 \otimes E_8$ the fermions χ^I and χ'^I transform as $(\mathbf{16}, \mathbf{1}') \oplus (\mathbf{1}, \mathbf{16}')$ under $O(16) \otimes O(16)'$, so in the twisted sector the fermions obey the same boundary conditions as before. However, the fermion number of the vacuum changes by 1 for each set of fermions, so the $(-1)^F$ and $(-1)^{F'}$ projections are changed; i.e. in the twisted sector we keep the states which previously had $(-1)^F = (-1)^{F'} = -1$. Among the left-moving states with $\frac{1}{4}(\text{mass})^2 \leqslant 0$ this leaves only the massless states $\chi^I_{-1/2}\chi'^J_{-1/2}|0\rangle_L$, from the (NS, NS) sector. (States in the (R, R) sector start at $\frac{1}{4}(\text{mass})^2 = 1$; states in the (NS, R) and (R, NS) sectors start at $\frac{1}{4}(\text{mass})^2 = 0$, but these states have the wrong eigenvalue for either $(-1)^F$ or $(-1)^{F'}$.) To form physical states we must now combine left- and right-moving states at the same mass level and project onto $R = R_L R_R = 1$, where $R_{L(R)}$ is the left-(right-)moving contribution to R. In the untwisted sector, those states with $R_L = 1$ will combine with the bosonic right-moving states ($R_R = 1$) to form physical bosons; those states with $R_L = -1$ will combine with the fermionic right-moving states ($R_R = -1$) to form physical fermions. In the twisted sector, the eigenvalues of R_L and R_R are a bit peculiar: bosonic right-moving states have $R_R = -1$ and fermionic right-moving states have $R_R = 1$, and the contributions to R_L from the tensor and spinor representations of O(16) and O(16)′ are similarly reversed. Putting these results together, we find that the massless states consist of the bosonic part of the supergravity multiplet $(g_{\mu\nu}, B_{\mu\nu}, \phi)$, gauge bosons in the adjoint representation of $O(16) \otimes O(16)'$, $\mathbf{8}_s$ fermions in the $(\mathbf{128}, \mathbf{1}) \oplus (\mathbf{1}, \mathbf{128})$ representation of the gauge group, and $\mathbf{8}_{s'}$ fermions in the $(\mathbf{16}, \mathbf{16})$ representation.

We will show below that this string theory is modular invariant. One then expects that the low-energy ten-dimensional field theory should be anomaly free. To see that this is the case first note that pure gravitational anomalies cancel because there are 256 $\mathbf{8}_s$ and 256 $\mathbf{8}_{s'}$ fermions, albeit in different representations of $O(16) \otimes O(16)$. To check the cancellation of pure gauge anomalies and mixed anomalies, we need to relate traces in the spinor **128** representation of O(16) to those in the vector **16**:

$$\begin{aligned}
\mathrm{Tr}_{\mathbf{128}} F^6 &= 16\,\mathrm{tr}\,F^6 + \tfrac{15}{4}(\mathrm{tr}\,F^2)^3 - 15\,\mathrm{tr}\,F^2\,\mathrm{tr}\,F^4 ,\\
\mathrm{Tr}_{\mathbf{128}} F^4 &= 6(\mathrm{tr}\,F^2)^2 - 8\,\mathrm{tr}\,F^4 ,\\
\mathrm{Tr}_{\mathbf{128}} F^2 &= 16\,\mathrm{tr}\,F^2 .
\end{aligned} \tag{20}$$

Here we denote traces in the vector **16** representation simply by tr. We will also denote by Tr the signed trace over the fermionic representations which are present at the massless level, $\mathrm{Tr} \equiv \mathrm{Tr}_{(\mathbf{128},\mathbf{1})} + \mathrm{Tr}_{(\mathbf{1},\mathbf{128})} - \mathrm{Tr}_{(\mathbf{16},\mathbf{16})}$. The trace relations (20) may be derived from trace relations for the adjoint representation of E_8, along with relations between the vector and adjoint traces for O(16) (both sets of which may be found in ref. [1]), by decomposing the adjoint of E_8 into the adjoint plus spinor of

O(16). Using (20), we can write the twelve-form [14] representing the pure gauge anomaly as

$$\tfrac{1}{15}\mathrm{Tr}\,F^6=\tfrac{1}{4}\left[\left(\mathrm{tr}\,F_1^2\right)^3+\left(\mathrm{tr}\,F_2^2\right)^3\right]-\left(\mathrm{tr}\,F_1^2+\mathrm{tr}\,F_2^2\right)\left(\mathrm{tr}\,F_1^4+\mathrm{tr}\,F_2^4\right), \tag{21}$$

and the twelve-forms for the mixed anomalies as

$$\tfrac{1}{24}\mathrm{Tr}\,F^4\mathrm{tr}\,R^2=\left\{\tfrac{1}{4}\left[\left(\mathrm{tr}\,F_1^2\right)^2+\left(\mathrm{tr}\,F_2^2\right)^2-\mathrm{tr}\,F_1^2\mathrm{tr}\,F_2^2\right]-\mathrm{tr}\,F_1^4-\mathrm{tr}\,F_2^4\right\}\mathrm{tr}\,R^2, \tag{22}$$

$$\tfrac{1}{30}\mathrm{Tr}\,F^2\left[\tfrac{1}{8}\mathrm{tr}\,R^4+\tfrac{5}{32}\left(\mathrm{tr}\,R^2\right)^2\right]=0. \tag{23}$$

Here F_1 and F_2 are field strengths in the two O(16) factors, $\mathrm{tr}\,R^{2n}$ is a trace in the vector representation of SO(9,1), and the relative normalization of the three terms is correct. Note the absence of $\mathrm{tr}\,F_i^6$ terms, which makes it possible to cancel the anomalies with local counterterms. From (21), (22) and (23) we can extract the actual anomalies along the lines of ref. [1]; they are given by the expression

$$\begin{aligned}G_{\mathrm{f}}\sim\int\Big\{&\tfrac{1}{4}\left[\omega^1_{2\mathrm{Y},1}\left(\mathrm{tr}\,F_1^2\right)^2+\omega^1_{2\mathrm{Y},2}\left(\mathrm{tr}\,F_2^2\right)^2\right]-\tfrac{1}{3}\left(\omega^1_{2\mathrm{Y},1}+\omega^1_{2\mathrm{Y},2}\right)\left(\mathrm{tr}\,F_1^4+\mathrm{tr}\,F_2^4\right)\\&-\tfrac{2}{3}\left(\omega^1_{6\mathrm{Y},1}+\omega^1_{6\mathrm{Y},2}\right)\left(\mathrm{tr}\,F_1^2+\mathrm{tr}\,F_2^2\right)\\&+\left[\tfrac{1}{6}\left(\omega^1_{2\mathrm{Y},1}\mathrm{tr}\,F_1^2+\omega^1_{2\mathrm{Y},2}\mathrm{tr}\,F_2^2\right)-\tfrac{1}{12}\left(\omega^1_{2\mathrm{Y},1}\mathrm{tr}\,F_2^2+\omega^1_{2\mathrm{Y},2}\mathrm{tr}\,F_1^2\right)\right.\\&\qquad\left.-\tfrac{2}{3}\left(\omega^1_{6\mathrm{Y},1}+\omega^1_{6\mathrm{Y},2}\right)\right]\mathrm{tr}\,R^2\\&+\tfrac{1}{3}\omega^1_{2\mathrm{L}}\left[\tfrac{1}{4}\left[\left(\mathrm{tr}\,F_1^2\right)^2+\left(\mathrm{tr}\,F_2^2\right)^2-\mathrm{tr}\,F_1^2\mathrm{tr}\,F_2^2\right]-\mathrm{tr}\,F_1^4-\mathrm{tr}\,F_2^4\right]\Big\}.\end{aligned} \tag{24}$$

The $2n$-forms $\omega^1_{2n\mathrm{Y(L)}}$ can be defined via the gauge variations of the Yang-Mills(Lorentz) Chern-Simons forms, $\delta\omega^0_{(2n+1)\mathrm{Y(L)}}=-\mathrm{d}\omega^1_{2n\mathrm{Y(L)}}$. The Chern-Simons forms can in turn be defined via $\mathrm{d}\omega^0_{(2n+1)\mathrm{Y}}=\mathrm{tr}\,F^{n+1}$, $\mathrm{d}\omega^0_{(2n+1)\mathrm{L}}=\mathrm{tr}\,R^{n+1}$. Taking the antisymmetric tensor field $B_{\mu\nu}$ to transform as

$$\delta B=-\xi\left(\omega^1_{2\mathrm{Y},1}+\omega^1_{2\mathrm{Y},2}\right)-\omega^1_{2\mathrm{L}} \tag{25}$$

(modulo exact forms), with $\xi=1$, the gauge variation of the counterterm

$$\begin{aligned}S_{\mathrm{c.t.}}\sim\int\Big\{&B\left[\tfrac{1}{4}\left[\left(\mathrm{tr}\,F_1^2\right)^2+\left(\mathrm{tr}\,F_2^2\right)^2-\mathrm{tr}\,F_1^2\mathrm{tr}\,F_2^2\right]-\mathrm{tr}\,F_1^4-\mathrm{tr}\,F_2^4\right]\\&-\tfrac{2}{3}\left(\omega^0_{3\mathrm{Y},1}+\omega^0_{3\mathrm{Y},2}\right)\left(\omega^0_{7\mathrm{Y},1}+\omega^0_{7\mathrm{Y},2}\right)-\tfrac{1}{4}\omega^0_{3\mathrm{Y},1}\omega^0_{3\mathrm{Y},2}\left(\mathrm{tr}\,F_1^2-\mathrm{tr}\,F_2^2\right)\\&+\omega^0_{3\mathrm{L}}\left[\tfrac{1}{6}\left(\omega^0_{3\mathrm{Y},1}\mathrm{tr}\,F_1^2+\omega^0_{3\mathrm{Y},2}\mathrm{tr}\,F_2^2\right)-\tfrac{1}{12}\left(\omega^0_{3\mathrm{Y},1}\mathrm{tr}\,F_2^2+\omega^0_{3\mathrm{Y},2}\mathrm{tr}\,F_1^2\right)\right.\\&\qquad\left.-\tfrac{2}{3}\left(\omega^0_{7\mathrm{Y},1}+\omega^0_{7\mathrm{Y},2}\right)\right]\Big\}\end{aligned} \tag{26}$$

cancels precisely the anomalous gauge variation (24) due to the fermions.

Finally we wish to show that this theory is in fact modular invariant. There are three sets of fermionic fields in the heterotic string, the spacetime fermions ψ^i and the $O(16)\otimes O(16)'$ fermions χ^I and χ'^I, each with either NS or R boundary conditions. One can check that this $O(16)\otimes O(16)'$ theory has in the untwisted sector the following four triples of (spacetime, O(16), O(16)′) boundary conditions: (NS, NS, NS), (NS, R, R), (R, NS, R), and (R, R, NS), all with the usual $(-1)^F$ projections. In addition the twisted sector contains the remaining four triples: (R, R, R), (R, NS, NS), (NS, R, NS), and (NS, NS, R), all with the *opposite* set of states surviving the $(-1)^F$ projections (because the fermion number of each of the three Fock space vacua has changed by 1). With this fact in hand it is straightforward to evaluate the partition functions for the spacetime bosons and for the spacetime fermions. Using the definitions (12) they are given respectively by

$$P_B(z,\bar z)=\frac{1}{2z^{1/2}|f(z)|^{16}}\Big\{[A(z)-B(z)]\Big[4\bar z C(\bar z)^4+\frac{1}{4\bar z}\big(A(\bar z)^2+B(\bar z)^2\big)^2\Big]$$
$$+[A(z)+B(z)]\big[2C(\bar z)^2\big(A(\bar z)^2-B(\bar z)^2\big)\big]\Big\},\quad(27)$$

$$P_F(z,\bar z)=\frac{C(z)}{|f(z)|^{16}}\Big\{2C(\bar z)^2\big[A(\bar z)^2+B(\bar z)^2\big]+4\bar z C(\bar z)^4$$
$$+\frac{1}{4\bar z}\big[A(\bar z)^2-B(\bar z)^2\big]^2\Big\}.\quad(28)$$

Use of the identities $\vartheta_2^4+\vartheta_4^4=\vartheta_3^4$ and $\vartheta_1'=\vartheta_2\vartheta_3\vartheta_4$ results in considerable simplification.

We can evaluate for example the one-loop contribution to the cosmological constant which is given [6, 7] by

$$\Lambda_{\text{1-loop}}\sim-\int_F\frac{d^2\tau}{(\operatorname{Im}\tau)^2}(\operatorname{Im}\tau)^{-4}\big[P_B(z,\bar z)-P_F(z,\bar z)\big],\quad(29)$$

with $z=e^{2\pi i\tau}$ and the integration region restricted to a fundamental domain F of the modular group, e.g. $-\frac{1}{2}\leqslant\operatorname{Re}\tau\leqslant\frac{1}{2}$, $|\tau|\geqslant1$, $\operatorname{Im}\tau>0$. Using (27), (28) and (12) we find

$$\Lambda_{\text{1-loop}} \sim -\int_F \frac{d^2\tau}{(\operatorname{Im}\tau)^2} \frac{1}{(\operatorname{Im}\tau)^4 \vartheta_1'^4(\tau)} \left\{ \frac{\vartheta_2^4(\tau)}{\vartheta_2^8(\bar{\tau})} + \frac{\vartheta_4^4(\tau)}{\vartheta_4^8(\bar{\tau})} - \frac{\vartheta_3^4(\tau)}{\vartheta_3^8(\bar{\tau})} \right\}. \quad (30)$$

The integrand is clearly invariant under modular transformations. What is not so clear is the sign or magnitude of $\Lambda_{\text{1-loop}}$. Although we know of no simple method for evaluating integrals of this type, the absence of a tachyon ensures the absence of divergences in the integral, and the existence of convergent power series expansions for the theta functions throughout the fundamental domain makes numerical evaluation of the integral straightforward but tedious. We have found that in fact (30) is *positive*, in contrast to the negative contribution in the bosonic string [7] or in the models analyzed by Rohm [6].

This theory appears to be the unique modular-invariant, tachyon-free string theory in ten dimensions without spacetime supersymmetry. At the one-loop level the presence of a nonzero cosmological constant will lead to a dilaton tadpole and hence to quantum instability. It is not clear whether one can shift the vacuum to find a perturbatively stable vacuum also without spacetime supersymmetry, but the prospects appear brighter than for the bosonic string, given the absence of a tachyon in the free spectrum.

It should also be noted that the ten-dimensional theories considered here bear a close resemblance to the models considered by Rohm [6], in which propagation of the closed superstring and also of the heterotic string on a "twisted torus" was studied. A twisted torus results from identifying points in euclidean space under a group containing not just pure translations (which would yield an ordinary torus), but also elements which are combined translations and rotations. If we compactify the first theory we have considered here (which is purely bosonic) on a cubic six-torus and let the lengths of the sides of the torus approach zero (relative to $\sqrt{\alpha'}$), we find that the spectrum approaches that for the closed superstring on a twisted torus (with the twist $\theta = 2\pi$ as defined in [6]), in the limit of vanishing size for the twisted torus. In the case of the heterotic string on a twisted torus, we can also accompany the above-mentioned combined translations and rotations by gauge transformations. If we choose one of the Z_2 gauge transformations which we have listed above, we can find a similar correspondence of the spectrum for the heterotic string on a twisted torus with the spectrum for one of the above ten-dimensional theories compactified on an ordinary torus, again in the limit of vanishing size for the two tori.

Finally, we have also found tachyon-free theories with some of the spacetime dimensions compactified via orbifolds in a way that breaks spacetime supersymmetry. These theories are classically stable in the orbifold limit. If it is possible to go away from the orbifold limit by blowing up the orbifold singularities, then there should exist conformally invariant nonlinear sigma models with non-Kähler target spaces. These theories, as well as the numerical evaluation of $\Lambda_{1\text{-loop}}$ for the above theory, will be discussed elsewhere [15].

Some of the ten-dimensional tachyonic theories described here have also been considered by N. Seiberg and E. Witten [16]. Related results have been obtained by C. Vafa [16].

We would like to thank J. Bagger and F. MacKintosh for enlightening conversations.

Note added in proof. The possibility of a consistent bosonic string theory in ten-dimensions was suggested by C. Thorn at the November 1985 Santa Fe meeting on string theory. The tachyon-free $O(16) \otimes O(16)$ model has also been discussed by Alvarez-Gaumé, Ginsparg, Moore, and Vafa [17]. Kawai, Lewellen, and Tye [18] have given a classification of modular invariant ten-dimensional strings using a fermionic representation. Their results agree with ours and include a tachyonic E_8 model which is the unique model obtained using the outer automorphism of $E_8 \otimes E_8$ mentioned in the text.

References

[1] M.B. Green and J.H. Schwarz, Phys. Lett. 144B (1984) 117
[2] D. Gross, J. Harvey, E. Martinec and R. Rohm, Nucl. Phys. B256 (1985) 253, B267 (1986) 75
[3] P. Freund, Phys. Lett. 151B (1985) 387;
J. Thierry-Mieg, Phys. Lett. 156B (1985) 199;
C. Vafa and E. Witten, Phys. Lett. 159B (1985) 265
[4] A. Casher, F. Englert, H. Nicolai and A. Taormina, Phys. Lett. 162B (1985) 121
[5] M. Ademollo et al., Nucl. Phys. B94 (1975) 221
[6] R. Rohm, Nucl. Phys. B237 (1984) 553; Princeton University PhD thesis, unpublished
[7] J. Polchinski, University of Texas preprint UTTG-13-85
[8] A. Erdelyi et al., Higher transcendental functions, vol. 2 (McGraw-Hill, New York, 1953) ch. 13
[9] F. Gliozzi, J. Scherk and D. Olive, Phys. Lett. 166B (1976) 282; Nucl. Phys. B122 (1977) 253
[10] L Castellani, Phys. Lett. 166B (1986) 54;
L. Castellani, R. D'Auria, F. Gliozzi and S. Sciuto, Torino preprint (1985)
[11] R. Bluhm and L. Dolan, Rockefeller preprint (1986)
[12] L. Dixon, J. Harvey, C. Vafa and E. Witten, Princeton preprint (February, 1986)
[13] H.M. S. Coxeter, Integral cayley numbers, *in* Twelve geometric essays (Southern Illinois University Press, 1968)
[14] L. Alvarez-Gaumé and E. Witten, Nucl. Phys. B234 (1983) 269;
B. Zumino, Y.S. Wu and A. Zee, Nucl. Phys. B239 (1984) 477
[15] L. Dixon, J. Harvey and F. MacKintosh, to appear
[16] E. Witten, private communication
[17] L. Alvarez-Gaumé, P. Ginsparg, G. Moore and C. Vafa, HUTP 86/A013
[18] H. Kawai, D.C. Lewellen and S.-H.H. Tye, CLNS 68/732

AN O(16) × O(16) HETEROTIC STRING

L. ALVAREZ-GAUMÉ, P. GINSPARG, G. MOORE and C. VAFA
Lyman Laboratory of Physics, Harvard University, Cambridge, MA 02138, USA

Received 5 March 1986

By modifying the definition of the fermion number projection in the $E_8 \times E_8$ heterotic superstring theory in a manner consistent with modular invariance and factorization, a tachyon-free string theory is obtained without spacetime supersymmetry. The low energy limit is a ten-dimensional anomaly-free chiral SO(16) × SO(16) gauge theory coupled to gravity.

Some effort has been expended recently to elucidate the properties of superstring theories [‡1]. Following the discovery of the anomaly cancellation [2] in the low energy limit of O(32) open superstrings, and as well in the $E_8 \times E_8$ field theory, the development of heterotic strings [3] provided new possibilities [4] for making contact between strings and phenomenology. It has recently been shown that the introduction of torsion can dramatically alter the physical consequences of heterotic string theories [5]. In this letter we will use similar modifications of the path integral to construct a new heterotic string with O(16) × O(16) gauge symmetry which is tachyon-free, has no spacetime supersymmetry, and has an anomaly-free chiral ten-dimensional field theory as its low energy limit. Instead of introducing a background $B_{\mu\nu}$ as in ref. [5], we modify the definition of the $(-1)^{F_i}$ projections for the three sets of fermions in the standard construction of the $E_8 \times E_8$ heterotic string. Such modifications of $(-1)^{F_i}$ are not arbitrary and are subject to powerful constraints arising from the demand that the theory have modular invariant factorizing amplitudes. We will see that these constraints allow no non-trivial modifications of the O(32) string and allow a unique modification of the $E_8 \times E_8$ string [‡2]. We shall also show that the field theory limit of this string theory is a new ten-dimensional chiral field theory free of gauge and gravitational anomalies.

[‡1] For recent reviews see ref. [1].
[‡2] See the note added for a discussion of other theories which may be similarly modified.

We start by recalling the degrees of freedom of the $E_8 \times E_8$ heterotic string in the light-cone gauge. In addition to the 8 left- and right-moving bosonic degrees of freedom, there are 8 right-moving Neveu–Schwarz–Ramond (NSR) fermions $\psi^{\mu=1,8}$ transforming as the fundamental representation of O(8), the transverse part of the Lorentz group, together with two groups of 16 left-moving fermions, $\psi^{i=1,16}$ and $\tilde{\psi}^{i=1,16}$, taken to transform as (16, 1) and (1, 16) of O(16) × O(16). In the hamiltonian formalism, we can specify separately P or A sectors for each of the three groups of fermions, corresponding to their having periodic or antiperiodic boundary conditions along the spatial direction of the string. This gives a total of 8 sectors (∗; ∗, ∗) labelled by their boundary conditions (∗ = P, A). For example, (A; P, P) refers to the NS sector, with the two sets of O(16) gauge fermions taken in their periodic sectors. In this formalism, we project onto the physical Hilbert space of the theory by retaining only those states corresponding to a given eigenvalue of a fermion parity operator $(-1)^{F_i}$ chosen for each sector.

In the $E_8 \times E_8$ heterotic string, $(-1)^F$ projections are chosen so that those right-movers in the (A; ∗, ∗) sectors give a massless vector 8_v of O(8), while those in the (P; ∗, ∗) sectors give a massless Majorana–Weyl spinor 8_s of O(8). The left-movers in the (∗; A, A) sectors give massless states in the (120, 1) ⊕ (1, 120) of O(16) × O(16), together with an 8_v of O(8) coming from the bosonic degrees of freedom x^μ. The (∗; A, P) and (∗; P, A) sectors give, respectively, massless states

0370-2693/86/$ 03.50 © Elsevier Science Publishers B.V.
(North-Holland Physics Publishing Division)

transforming as (1, 128) and (128, 1) of O(16) × O(16), and the (*; P, P) sector contains no massless states. Assembled, we find that the massless states correspond to $(8_v \oplus 8_s, 1)_R \otimes [(8_v, 1) \oplus (1, \mathrm{Adj}(E_8 \times E_8))]_L$, the spectrum of $N = 1$ supergravity coupled to $E_8 \times E_8$ super Yang–Mills theory. We now discuss the extent to which we are free to alter this construction by defining new $(-1)^F$ projections for fermions in each sector.

In the path integral formulation of the theory, the $(-1)^F$ projections correspond at l-loop order to choosing phases for the sum over different spin structures on a Riemann surface of genus l. At one-loop other, a spin structure is determined by the boundary conditions along each of the two homology cycles on the torus. The different boundary conditions about one cycle (say the σ_1 direction) correspond to the different sectors discussed above, and the different boundary conditions about the other cycle (the σ_2 direction) correspond to insertion of $(-1)^F$ factors. In the path integral formulation, the sum over spin structures thus has the effect of enforcing the $(-1)^F$ projections, and the choice of signs multiplying the string measure for different spin structures corresponds to the choice of the definition of $(-1)^F$. We thus need to consider the possible consistent choices of signs in path integral measures.

In ref. [6], it was pointed out that the constraints of modular invariance and factorization of higher loop amplitudes can be used to obtain restrictions on the possible choices of phases. The idea is that amplitudes from two different spin structures on a genus-2 surface which are related by the action of the symplectic modular group can be factorized into products of one-loop amplitudes which are (one-loop) modular inequivalent. Factorization and modular invariance thus impose important constraints on the products of those phases attached at the one-loop level. It turns out that no additional constraints arise from considering factorization of higher than two-loop diagrams. This approach has been further explored in the context of modding out a string theory [‡3] by a symmetry group G in ref. [7], where it was shown that the possible choices of phases consistent to all loop orders of G abelian are classified by $H^2(G, U(1))$ (the second group cohomology with U(1) coefficients [8]) [‡4]. In the case at hand, it was suggested in ref. [7] to take $G = Z_2 \times Z_2 \times Z_2$, corresponding to modding out by the three different $(-1)^F$'s. In fact [‡5] only the relative $(-1)^F$'s are relevant, and taking $G = Z_2 \times Z_2$ is sufficient to probe all physically distinct situations. Further insight into this from the viewpoint of theta characteristics may be found in ref. [9].

To exemplify this we introduce some notation: let S^a denote a spin structure on the Riemann surface. On a surface of genus $g = 1$ (one-loop), there are four choices for S^a: (AA), (AP), (PA) and (PP), representing the choices of boundary conditions for fermions ψ around the two non-trivial cycles corresponding respectively to σ_2 and σ_1. While the different spin structures at one loop can be characterized by specifying these boundary conditions, for higher genus surfaces this is not quite correct since there is no longer a canonical flat reference bundle. It remains correct, however, that the *difference* between two spin structures on an arbitrary genus surface is represented by a ±1 choice for each cycle, insofar as *changing* boundary conditions by ± signs effects the transition between any two spin structures. We shall use α to denote the Z_2 group operation which takes us from one spin structure to another by changing appropriate boundary conditions.

In the heterotic string, the spin structures $S^a_{(1)}$, $S^a_{(2)}$, and $S^a_{(3)}$ for the three groups of fermions may be specified independently. We shall represent the contribution to the path integral from the measures for these sectors symbolically as $\Sigma_{a,b,c}\, S^a_{(1)} S^b_{(2)} S^c_{(3)}$, where the summation sign incorporates a canonical choice of relative phases for the different sectors. Consider first the expression in which all three spin structures are identified: $\Sigma_a\, S^a_{(1)} S^a_{(2)} S^a_{(3)}$. In the path integral measure for the O(32) heterotic string, we have instead $\Sigma_{a,b}\, S^a_{(1)} S^b_{(2)} S^b_{(3)}$, with the sum running over two independent spin structures a and b, and with the equal spin structures $S^b_{(2),(3)}$ corresponding to the spin structure for the 32 gauge fermions. We can view this theory as obtained from the former expression by modding out by a Z_2 defined to act as α on the second and third groups of fermions. Symbolically, we represent

[‡3] Modding out a string theory means to throw away certain Hilbert space sectors and add on new twisted sectors.

[‡4] The question of other consistent choices of phases for G non-abelian was left open.

[‡5] We are grateful to E. Witten for pointing this out.

the effect of this action by

$$\sum_a S^a_{(1)} S^a_{(2)} S^a_{(3)} \to \sum_{a,b} S^a_{(1)} S^b_{(2)} S^b_{(3)} . \tag{1}$$

Applying the analysis of ref. [7], we find that the inequivalent choices for the phases of the O(32) heterotic string are given by $H^2(Z_2, U(1))$, which is trivial. There are consequently no non-trivial choices for the phases of this theory. In the $E_8 \times E_8$ theory, we introduce two Z_2's corresponding to two independent Z_2 operations α and β acting independently on the second and third groups of fermions respectively. Modding out by $Z_2 \times Z_2$ then has the effect

$$\sum_a S^a_{(1)} S^a_{(2)} S^a_{(3)} \to \sum_{a,b,c} S^a_{(1)} S^b_{(2)} S^c_{(3)} , \tag{2}$$

and the inequivalent phase choices are classified by

$$H^2(Z_2 \times Z_2, U(1)) = Z_2 . \tag{3}$$

(The only other changes of phase in eqs. (1), (2) correspond to changing the relative sign of the diagrams in which the ith group of fermions has an even or odd spin structure. These sign changes correspond to physically irrelevant changes in the chiralities of the ground state spinors.) This method of altering a heterotic string thus leads to only one new possibility, which we now investigate.

The four elements of the group $Z_2 \times Z_2$ generated by α and β may be denoted $\alpha^{k_1}\beta^{k_2}$, with k_1 and k_2 equal to 0 or 1. Let

$$S^b_{(2)} = S^a_{(2)} A^{ab} , \quad S^b_{(3)} = S^a_{(3)} B^{ab} , \tag{4}$$

where A^{ab} and B^{ab}, respectively, act to change the spin structure S^a to spin structure S^b for the second and third groups of fermions by changing appropriate boundary conditions. On a surface of genus g, A and B may be considered as $2g$ component objects $(\alpha^{k_j^{ab}})$ and $(\beta^{l_j^{ab}})$, with k_j^{ab} and l_j^{ab} ($j = 1, ..., 2g$) equal to 0 or 1 depending on whether or not there is a change in boundary conditions between spin structures S^a and S^b around the jth cycle. At one loop, this corresponds to

$$A^{ab} = (\alpha^{k_1^{ab}}, \alpha^{k_2^{ab}}) , \qquad B^{ab} = (\beta^{l_1^{ab}}, \beta^{l_2^{ab}}) . \tag{5}$$

Relative to the $E_8 \times E_8$ heterotic string, we now introduce new phases ϵ according to

$$\sum_{a,b,c} S^a_{(1)} S^b_{(2)} S^c_{(3)}$$

$$\to \sum_{a,b,c} S^a_{(1)} S^b_{(2)} S^c_{(3)} \, \epsilon(\alpha^{k_2^{ab}} \beta^{l_2^{ac}}, \alpha^{k_1^{ab}} \beta^{l_1^{ac}}) . \tag{6}$$

The constraints of modular invariance and factorization to all loop orders require that the phases ϵ satisfy [7]

$$\epsilon(h_1, h_2, h_3) = \epsilon(h_1, h_3)\, \epsilon(h_2, h_3) ,$$

$$\epsilon(h_1, h_2)\, \epsilon(h_2, h_1) = 1 , \quad \epsilon(h_1, h_1) = 1 , \tag{7}$$

where $h_i \in Z_2 \times Z_2$. These properties allow all values of ϵ to be determined in terms of the value of $\epsilon(\alpha, \beta)$ as

$$\epsilon(\alpha^{k_2^{ab}} \beta^{l_2^{ac}}, \alpha^{k_1^{ab}} \beta^{l_1^{ac}}) = \epsilon(\alpha, \beta)^{k_2^{ab} l_1^{ac} - l_2^{ac} k_1^{ab}} . \tag{8}$$

The non-trivial modification of the standard $E_8 \times E_8$ theory corresponds to the choice $\epsilon(\alpha, \beta) = -1$.

We can now compare the hamiltonian with the path integral formalism to determine the massless states in the theory, coming from the Hilbert space sectors with non-positive zero-point energies. These are the (*; A, A), (*; A, P), and (*; P, A) sectors. We may determine the changes in the operator $(-1)^{F_i}$, $i = 1, 2, 3$ for the three sets of fermions by considering the phases associated to various diagrams in the path integral. For example, consider the massless bosons. To determine $(-1)^{F_1}$ in the (A; A, A) sector we examine the diagram $(P; A, A)_{\sigma_2}(A; A, A)_{\sigma_1}$. This notation means that the first set of fermions have spin structure PA while the second and third sets have spin structure AA. The contribution of this diagram to the path integral is the same as

$$\mathrm{Tr}_{(A;A,A)} (-1)^{F_1} q^{H_L} \bar{q}^{H_R} ,$$

where H_L and H_R are the left- and right-moving hamiltonians, and $q = \exp(2\pi i \tau)$ (τ is the modular parameter on the torus). The diagram also has a spin structure corresponding, by (5) and (6), to the action of the pair of $Z_2 \times Z_2$ group elements $(\alpha^{k_2^{ab}} \beta^{l_2^{ac}}, \alpha^{k_1^{ab}} \beta^{l_1^{ac}}) = (\alpha\beta, 1)$. But from (8), we see that $\epsilon(\alpha\beta, 1) = 1$, so modifying the partition function by the phase $\epsilon(\alpha\beta, 1)$ leaves $(-1)^{F_1}$ unchanged in this sector. Similarly, by considering $(A; P, A)_{\sigma_2}(A; A, A)_{\sigma_1}$ and $(A; A, P)_{\sigma_2}(A; A, A)_{\sigma_1}$, we see that all $(-1)^{F_i}$ are unchanged in the (A; A, A) sector and we recover the familiar $(8_v, 1, 1)$

$\otimes[(8_v,1,1)\oplus(1,120,1)\oplus(1,1,120)]$ states of the $E_8\times E_8$ heterotic string. An identical argument shows that the $(-1)^{F_i}$ are also unchanged in the (P; P, P) sector, where the states are massive anyway.

In the (A; P, A) sector (which contributes particles in the (128, 1) of O(16) × O(16) to the $E_8\times E_8$ string), for example, the two theories begin to differ. In the new theory, the diagram $(A;A,P)_{\sigma_2}(A;P,A)_{\sigma_1}$ occurs with phase $\epsilon(\beta,\alpha)=-1$ in the path integral so the definition of $(-1)^{F_3}$ in the (A; P, A) sector is changed. In the heterotic string, $(-1)^{F_3}$ on the massless left-moving (A) ground state is +1, so in the modified theory this state is projected out and there are no massless (128, 1) states. An identical argument applies to the (A; A, P) sector for $(-1)^{F_2}$, removing the (1, 128)'s. The sectors (A; P, A) and (A; A, P) instead give rise to $\frac{1}{2}$-integral bosonic states starting at $m^2=\frac{1}{2}$. Analyzing all other sectors similarly, we can summarize the $(-1)^F$'s changed with respect to the $E_8\times E_8$ theory as follows:

in sectors (P; A, A) and (A; P, P):

$$(-1)^{F_2},(-1)^{F_3}\ \text{are changed}\ ,$$

in sectors (P; P, A) and (A; A, P):

$$(-1)^{F_1},(-1)^{F_2}\ \text{are changed}\ ,$$

and in sectors (P; A, P) and (A; P, A):

$$(-1)^{F_1},(-1)^{F_3}\ \text{are changed}\ .$$

It is easy now to verify that the spectrum is free of tachyons, and to determine the massless spectrum of the theory.

The massless bosons come from the (A; A, A) sector and transform under O(8) × O(16) × O(16) as

$$(8_v,1,1)_R\otimes[(8_v,1,1)\oplus(1,120,1)\oplus(1,1,120)]_L\ , \tag{9a}$$

comprising the adjoint bosons of an O(16) × O(16) gauge group together with a graviton, dilaton, and antisymmetric tensor field. There are massless fermions from the (P; A, A) sector which transform as

$$(8_s,16,16)\ , \tag{9b}$$

and also massless fermions from the (P; A, P) and (P; P, A) sectors which transform as

$$(8'_s,1,128')\ \text{and}\ (8'_s,128',1)\ . \tag{9c}$$

Note that the chiralities $(8'_s)$ of the spinor representations in (9c) are opposite those of the $E_8\times E_8$ string. This arises from a change in $(-1)^{F_1}$ in the associated Ramond sector.

We now wish to show that the massless sector of the string theory we have constructed corresponds to an anomaly free ten-dimensional field theory. From (9a)–(9c), this low energy field theory consists of the SO(16) × SO(16) gauge fields A_μ, the graviton $g_{\mu\nu}$, the antisymmetric tensor field $B_{\mu\nu}$, the dilaton ϕ, and positive and negative chirality Majórana–Weyl spin-$\frac{1}{2}$ fermions transforming, respectively, in the (16, 16) and $(128',1)\oplus(1,128')$ representations of SO(16) × SO(16). We write the field strength for the gauge field coupled to the positive chirality fermions as $F_+=F_1\otimes\mathbf{1}+\mathbf{1}\otimes F_2$, where F_i, $i=1,2$, are taken in the fundamental (16) representation of the two SO(16)'s (and $\mathbf{1}$ is the 16 × 16 unit matrix). For the negative chirality fermions, we write $F_-=\hat F_1\oplus\hat F_2$ where $\hat F_i$, $i=1,2$ are taken in the spinor (128′) representation of the SO(16)'s. Note that the dimensions of the positive and negative chirality representations satisfy $d_{r_+}=d_{r_-}=256$.

For a single SO(16), the traces of generators $\hat F$ in the spinor representation can be usefully expressed in terms of traces of generators F in the fundamental representation as

$$\operatorname{tr}\hat F^6=16\operatorname{tr}F^6-15\operatorname{tr}F^4\operatorname{tr}F^2+\tfrac{15}{4}(\operatorname{tr}F^2)^3\ ,$$

$$\operatorname{tr}\hat F^4=-8\operatorname{tr}F^4+6(\operatorname{tr}F^2)^2\ ,$$

$$\operatorname{tr}\hat F^2=16\operatorname{tr}F^2 \tag{10}$$

(these can be derived, for example, using formulas from the appendix of ref. [10]). For the SO(16) × SO(16) representations above, these identities imply

$$\operatorname{tr}F_+^6-\operatorname{tr}F_-^6=15(\operatorname{tr}F^2)X_8\ ,$$

$$\operatorname{tr}F_+^4-\operatorname{tr}F_-^4=24X_8\ ,$$

$$\operatorname{tr}F_+^2-\operatorname{tr}F_-^2=0\ , \tag{11}$$

where the gauge invariant 8-form X_8 is given by

$$X_8\equiv\operatorname{tr}F_1^4+\operatorname{tr}F_2^4-\tfrac{1}{4}[(\operatorname{tr}F_1^2)^2+(\operatorname{tr}F_2^2)^2-\operatorname{tr}F_1^2\operatorname{tr}F_2^2]\ ,$$

and

$$\operatorname{tr}F^2\equiv\operatorname{tr}F_1^2+\operatorname{tr}F_2^2\ ,$$

henceforth denotes the trace written in terms of the gauge field strength $F = F_1 \oplus F_2$ in the fundamental $(16,1) \oplus (1,16)$ representation of $SO(16) \times SO(16)$.

The gauge and gravitational anomalies for this ten-dimensional theory are characterized by a 12-form I_{12}, formed by taking the difference between the 12-form contributions from $\frac{1}{2}\hat{A}(M)\,\mathrm{ch}(F_\pm)$ for the positive and negative chirality (Majorana) sectors of the theory. Using the results of ref. [11] (see also ref. [10])

$$\tfrac{1}{2}\hat{A}(M)\,\mathrm{ch}(F_\pm) = \tfrac{1}{2}\{1 + [1/(4\pi)^2]\tfrac{1}{12}\,\mathrm{tr}\,R^2 + \ldots\}$$
$$\times \{d_{r_\pm} + [\mathrm{i}^2/2(2\pi)^2]\,\mathrm{tr}\,F_\pm^2 + [\mathrm{i}^4/4!(2\pi)^4]\,\mathrm{tr}\,F_\pm^4$$
$$+ [\mathrm{i}^6/6!(2\pi)^6]\,\mathrm{tr}\,F_\pm^6 + \ldots\},$$

together with the relations (11), we find that I_{12} takes the factorized form

$$I_{12} = [1/96(2\pi)^6](\mathrm{tr}\,R^2 - \mathrm{tr}\,F^2)X_8\,, \qquad (12)$$

where the overall constant is irrelevant. We may now proceed as in refs. [2,12]. We first define the Chern–Simons forms Q_3 satisfying $\mathrm{d}Q_{3\mathrm{Y}} = \mathrm{tr}\,F^2$, $\mathrm{d}Q_{3\mathrm{L}} = \mathrm{tr}\,R^2$, and $\delta Q_3 = \mathrm{d}Q_2^1$. Since X_8 is also a closed gauge-invariant form, we can transgress to a 6-form X_6^1, defined via $X_8 = \mathrm{d}X_7$, $\delta^{\mathrm{Y}}X_7 = \mathrm{d}X_6^1$, where δ^{Y} denotes the Yang–Mills gauge variation. A similar transgression from I_{12} gives the anomalous variation of the effective action

$$\delta\Gamma = [\mathrm{i}/96(2\pi)^5]$$
$$\times [\tfrac{1}{3}(Q_{2\mathrm{L}}^1 - Q_{2\mathrm{Y}}^1)X_8 + \tfrac{2}{3}(\mathrm{tr}\,R^2 - \mathrm{tr}\,F^2)X_6^1]\,, \qquad (13)$$

with the relative coefficient between the two terms determined by Bose symmetry in the external gauge fields. We now take the gauge invariant field strength for the antisymmetric tensor field to be $H = \mathrm{d}B + Q_{3\mathrm{L}} - Q_{3\mathrm{Y}}$. The gauge and Lorentz variations of B are given by

$$\delta^{\mathrm{Y}}B = Q_{2\mathrm{Y}}^1\,, \quad \delta^{\mathrm{L}}B = -Q_{2\mathrm{L}}^1\,,$$

and we see that the anomalous variation (13) of the effective action is cancelled by the counterterm

$$C = [\mathrm{i}/96(2\pi)^5][BX_8 - \tfrac{2}{3}(Q_{3\mathrm{L}} - Q_{3\mathrm{Y}})X_7]\,. \qquad (14)$$

It is interesting to note some differences between the anomaly cancellation we have demonstrated here and that of ref. [2]. In the supersymmetric case treated in ref. [2], the removal of the pure gravitational anomaly required that the dimension of the gauge group be 496. Since our theory is not supersymmetric we have neither the contribution from a spin-$\frac{3}{2}$ gravitino field nor a relation between the number of spin-$\frac{1}{2}$ fermions and the dimension of the gauge group. Instead, the pure gravitational anomaly cancels in (12) simply because there are an equal number of positive and negative chirality fermions ($2 \times 128 = 16^2$). Similarly, the mixed gauge-gravitational contributions $(\mathrm{tr}\,R^2)^2 \times \mathrm{tr}\,F_\pm^2$ and $\mathrm{tr}\,R^4\,\mathrm{tr}\,F_\pm^2$ cancel identically here due to the third of the relations (11). The curious way in which the numerical coincidences embodied in (10) conspire to allow the anomalies to be cancelled by (14) is indeed tantalizing.

Finally, let us address the question of the cosmological constant and the degeneracies of the spectrum of the theory. By conformal invariance, Λ vanishes at zero-loop order. In the one-loop vacuum amplitude, one must sum various powers of the determinant of the (chiral) Dirac operator for the different spin structures on the torus. In this sum only even spin structures (namely (AA), (AP), (PA)) contribute because the odd spin structure (PP) has a zero mode [‡6]. Therefore, if we split the gauge fermions into two groups of sixteen, only five distinct modular orbits contribute. These orbits are characterized as follows: There is one orbit in which all three spin structures are identical. There are three orbits in which two spin structures are identical while the third is different. There is one orbit in which all three spin structures are distinct. By convention the first class does not have a phase change: $\epsilon(1,1) = 1$. In the second class, for which two spin structures are identical and the third is different, each orbit has a phase change of the form $\epsilon(\alpha,\alpha)$, $\epsilon(\beta,\beta)$, $\epsilon(\alpha,1)$, $\epsilon(1,\alpha)$, $\epsilon(\beta,1)$, or $\epsilon(1,\beta)$. By (8), we see that all of these are +1. In the remaining orbit where all spin structures are distinct, the phase change is of the form $\epsilon(\alpha,\beta)$, $\epsilon(\alpha,\alpha\beta)$, $\epsilon(\beta,\alpha)$, $\epsilon(\beta,\alpha\beta)$, $\epsilon(\alpha\beta,\alpha)$, or $\epsilon(\alpha\beta,\beta)$. Again, using (8) we see that these phases are all given by $\epsilon(\alpha,\beta)$.

The $O(16) \times O(16)$ string is defined by choosing $\epsilon(\alpha,\beta) = -1$. We thus find that the sign of a single modular orbit in the expression for the cosmological

‡6 In the path integral one does *not* take det′ since there is no physical reason (such as spacetime translational invariance) for dividing out by the zero mode.

constant is changed. Since the one-loop cosmological constant vanishes in the $E_8 \times E_8$ heterotic string, it is easily seen to be given in the $O(16) \times O(16)$ string by

$$\Lambda = \frac{T^5}{4(2\pi)^5} \int_F \frac{d^2\tau}{(\tau_2)^2} \frac{1}{(\tau_2)^4} \frac{1}{|\eta(\tau)|^{16}}$$

$$\times \left[\left(\frac{\vartheta[^0_1](0|\tau)}{\eta(\tau)} \right)^8 \left(\frac{\vartheta[^1_0](0|\tau)}{\eta(\tau)} \right)^8 \overline{\left(\frac{\vartheta[^0_0](0|\tau)}{\eta(\tau)} \right)^4} \right.$$

$$- \left(\frac{\vartheta[^0_0](0|\tau)}{\eta(\tau)} \right)^8 \left(\frac{\vartheta[^0_1](0|\tau)}{\eta(\tau)} \right)^8 \overline{\left(\frac{\vartheta[^1_0](0|\tau)}{\eta(\tau)} \right)^4}$$

$$\left. - \left(\frac{\vartheta[^1_0](0|\tau)}{\eta(\tau)} \right)^8 \left(\frac{\vartheta[^0_0](0|\tau)}{\eta(\tau)} \right)^8 \overline{\left(\frac{\vartheta[^0_1](0|\tau)}{\eta(\tau)} \right)^4} \right], \tag{15}$$

where

$$\eta(\tau) = q^{1/24} \prod_{n=1}^{\infty} (1 - q^n) \qquad (q = \exp(2\pi i \tau))$$

is the Dedekind eta function, $\tau = \tau_1 + i\tau_2$ is again the modular parameter, and the ϑ's are the Jacobi theta functions [13] with characteristics. T is the string tension, normalized so that the tachyon of the closed bosonic string has $m^2 = -8\pi T$ (and $T = 1/2\pi\alpha'$). The above expression is modular invariant as expected. To avoid overcounting we restrict the integration region to the fundamental domain of SL(2, Z) as usual [14]. Using the Jacobi triple product identity

$$2\eta^3(\tau) = \vartheta[^0_0](0|\tau)\, \vartheta[^0_1](0|\tau)\, \vartheta[^1_0](0|\tau)$$

in (15), we can infer that the difference of the boson and fermion partition functions satisfies

$$\sum_{n \in Z/2} [d_B(n) - d_F(n)] \exp(-4\pi m_n^2 \tau_2)$$

$$= -2^7 \int_{-1/2}^{1/2} d\tau_1 \frac{1}{\Pi(1-q^n)^8 (1-\bar{q}^n)^8}$$

$$\times \left(\frac{\Pi(1+\bar{q}^{n-1/2})^8}{\bar{q}^{1/2}\Pi(1+q^{n-1/2})^{16}} \right.$$

$$\left. - \frac{1}{16q} \frac{\Pi(1+\bar{q}^n)^8}{\Pi(1+q^n)^{16}} - \frac{\Pi(1-\bar{q}^{n-1/2})^8}{\bar{q}^{1/2}\Pi(1-q^{n-1/2})^{16}} \right), \tag{16}$$

where $d_B(n)$ and $d_F(n)$ are the degeneracies of the bosonic and fermionic states and the infinite products run from $n = 1$ to $n = \infty$. In particular, since there are no tachyons in the theory, Λ should be finite. Indeed one finds for large τ_2 that (16) can be expanded as

$$-2^7 [\tfrac{33}{2} - 1152(\bar{q}q)^{1/2} + 36828(\bar{q}q)$$

$$- 771840(\bar{q}q)^{3/2} + ...] .$$

The integrand of (15) thus goes as $(1/\tau_2)^6$ and the expression for Λ is finite.

We have expanded the infinite products in (16) and evaluated (15) numerically with the result

$$\Lambda \approx +T^5 [2^6/(2\pi)^5](5.67) \approx +0.0371 T^5 .$$

(The terms $q^{n/2}\bar{q}^{m/2}$ in the expansion contribute according to their "level number" $m + n$. Although the degeneracies have not yet approached their asymptotic behaviour by level 14, we believe the above result to be reliable since the contribution from this level is down by 8 orders of magnitude, and further levels are each successively suppressed by more than an order of magnitude.) Since $\Lambda > 0$ we expect that the dilaton one-point function is not zero. This indicates [4] that ten-dimensional flat space is not an appropriate vacuum for the theory. Considering the vacuum energy $V(R)$ as a function of the compactification scale R, we see that $V(R \to \infty) \to \Lambda$. The positive sign of Λ thus suggests that for fixed dilaton expectation value, the vacuum energy could decrease to a minimum value at a finite compactification scale (if zero, this would give a lower dimensional theory with zero cosmological constant and broken supersymmetry). One possibility for the naive lower dimensional phenomenology of such a model could be assessed just as for the $E_8 \times E_8$ case in ref. [4]. Since the 128 of SO(16) decomposes as $(16, 4) \oplus (\overline{16}, \bar{4})$ of $SO(10) \times SO(6)$, identifying the spin connection of a six-dimensional internal manifold with the SO(6) part of the gauge connection would give a four-dimensional world having SO(10) grand unification with families of chiral fermions in 16's, and a shadow SO(16) sector.

In conclusion we would like to emphasize that all heterotic string theories are intimately related. For example, the close relation between the O(32) and the $E_8 \times E_8$ theories is explained in refs. [5,15]. Similarly, the $E_8 \times E_8$ and $O(16) \times O(16)$ theories are obtained from one another by a mere change of a few signs. We believe it is remarkable that such a simple modification of the string path integral leads to such profound alterations of the physics.

We would like to thank N. Seiberg and E. Witten for many valuable discussions and suggestions. We also thank Jürg Kramer and E. de Shalit for discussions on modular forms. This work was supported in part by NSF contract PHY-82-15249. G.M. and C.V. are also supported by the Harvard Society of Fellows, and L. A.-G. by the A.P. Sloan foundation.

Note added. Similar modifications of the heterotic string have been considered by Seiberg and Witten in ref. [6], and the O(16) × O(16) model presented here is one of the models discussed by Dixon and Harvey in ref. [16]. Some of the other models of refs. [6,16] (which are tachyonic, but otherwise consistent) may be obtained by starting from theories with partitionings of spin structures different from those of the standard heterotic strings, and then following the procedure (here implemented only for the standard partitionings) of introducing possible new non-trivial phases. We begin with the theory $\Sigma_a S^a_{(1)}S^a_{(2)}S^a_{(3)}$ mentioned before eq. (1), defined by taking the 8 NSR right-moving fermions and 32 left-moving gauge fermions all to have the same spin structure. This is a tachyonic O(32) theory, which we shall denote by $\mathcal{U}$. If instead of acting with the Z_2 transformation (decoupling the identification of left–right spin structures) which gives the standard O(32) heterotic string of eq. (1), we act with the Z_2 which gives $\Sigma_{a,b} S^a_{(1)} \times S^a_{(2)}S^b_{(3)}$, (dividing the gauge fermions into two groups of 16, one of which has the same spin structure as the 8 NSR fermions, and the other independent), we get the O(16) × E_8 theory of refs. [6,16]. Acting on either of these with the other Z_2 gives a sum over three independent spin structures (for the group of 8 and two groups of 16 fermions as in eq. (2)), and we get the standard $E_8 \times E_8$ theory or, since two independent choices of phases appear at this stage, the O(16) × O(16) theory. But modular invariance also allows other partitionings of the fermions, as long as the groups each have size a multiple of 8 [17], and other theories may be constructed. Taking the gauge fermions in groups of 8 and 24, for example, and summing over three independent spin structures again allows two non-trivial choices of phase, one of which reproduces the standard O(32) theory, and the other of which gives the O(8) × O(24) theory of ref. [16].

Most generally, we see that the way to reproduce all known and possibly new heterotic models is to mod out the $\mathcal{U}$ theory by a subgroup G of its symmetry group. In the light-cone gauge, the $\mathcal{U}$ theory has a symmetry group SO(8) × SO(32) × Z_2 (the Z_2 changes the relative spin structures of left- and right-movers). If we wish to preserve the space–time symmetries, i.e. to work with ten-dimensional Minkowski space, we consider any theory $\mathcal{U}$/G, with G a discrete subgroup of SO(32) × Z_2. If such a $\mathcal{U}$/G theory is consistent, i.e. modular invariant to all orders (to one-loop order the necessary and sufficient condition for modular invariance is left–right level matching [7, 15]), then $\mathcal{U}$/G or theories related by inequivalent phase choices (corresponding to the elements of H^2(G, U(1))) might emerge as new interesting (tachyon-free) theories.

References

[1] J.H. Schwarz, Superstrings, Vols. I, II (World Scientific, Singapore, 1985);
M.B. Green and D.J. Gross, Proc. Unified string Workshop (Santa Barbara, CA, 1985).

[2] M.B. Green and J.H. Schwarz, Phys. Lett. B 149 (1984) 117.

[3] D.J. Gross, J.A. Harvey, E. Martinec and R. Rohm, Phys. Rev. Lett. 54 (1985) 502; Nucl. Phys. B 256 (1985) 253, and Nucl. Phys. B, to be published.

[4] P. Candelas, G. Horowitz, A. Strominger and E. Witten, Nucl. Phys. B 285 (1985) 48.

[5] K.S. Narain, Ruttherford Lab preprint;
K.S. Narain, H. Sarmadi and E. Witten, in preparation;
E. Witten, unpublished.

[6] N. Seiberg and E. Witten, Spin structures in string theory, Princeton preprint.

[7] C. Vafa, Modular invariance and discrete torsion on orbifolds, Harvard preprint HUTP-86/A011.

[8] K. Brown, Cohomology of groups (Springer, Berlin, 1982).

[9] L. Alvarez-Gaumé, G. Moore and C. Vafa, Theta functions, modular invariance, and strings, Harvard preprint HUTP-86/A017.

[10] L. Alvarez-Gaumé and P. Ginsparg, Ann. Phys. 161 (1985) 423.

[11] L. Alvarez-Gaumé and E. Witten, Nucl. Phys. B 234 (1984) 269.

[12] M.B. Green, J.H. Schwarz and P.C. West, Nucl. Phys. B 254 (1984) 327.

[13] D. Mumford, Tata Lectures on Theta, Vols. I, II (Birkhäuser, Basel, 1983).

[14] J. Shapiro, Phys. Rev. D 5 (1972) 1945;
R. Rohm, Nucl. Phys. B 237 (1984) 553;
J. Polchinski, Evaluation of the one loop string path integral, Comm. Math. Phys., to be published.
[15] L. Dixon, J. Harvey, C. Vafa and E. Witten, Strings on orbifolds II, Princeton preprint.
[16] L. Dixon and J. Harvey, String theories in ten dimensions without spacetime supersymmetry, Princeton preprint.
[17] E. Witten, Global anomalies in string theory, in: Geometry, anomalies, topology, eds. W.A. Bardeen and A.R. White (World Scientific, Singapore, 1985).

Nuclear Physics B276 (1986) 272–290
North-Holland, Amsterdam

SPIN STRUCTURES IN STRING THEORY

N. SEIBERG

Department of Physics, Weizmann Institute of Science, Rehovot 76100, Israel and Institute for Advanced Study, Princeton, NJ 08540, USA

E. WITTEN

Joseph Henry Laboratories, Princeton University, Princeton NJ 08544, USA

Received 11 March 1986

The effect of modular transformations which change the spin structure of the string world-sheet is discussed. Invariance of string perturbation theory under such transformations forces us to sum over different spin structures with well defined coefficients. The sum over spin structures amounts to performing the GSO projections. We explain why the GSO projection is compatible with unitarity; that is, why wrong G-parity states are not produced in pairs. On the contrary, we show that unitarity requires either the GSO projection, which gives a supersymmetric spectrum, or another projection which removes the massless spin-$\frac{3}{2}$ particle from the spectrum. This is in agreement with the familiar idea that a consistent theory with a massless spin-$\frac{3}{2}$ particle must be supersymmetric. These considerations lead us to find four new string theories in ten dimensions. Although they all have tachyons, they are otherwise consistent, modular invariant and unitary. One of these theories is particularly interesting since it has anomaly free chiral fermions in ten dimensions.

1. Introduction

String theory [1] is described in perturbation theory in terms of a two-dimensional field theory on the string world-sheet. The fields in the field theory are the coordinates of the string as well as some auxiliary fields. In the simplest case the only coordinates are X^{μ} which are two-dimensional scalars. In other cases two-dimensional spinors are also present. The two-dimensional metric then has to be supplemented with more information – the spin structure. The metric tells us how two-dimensional vectors are parallel transported. When a vector is transported along a closed loop it is rotated by an O(2) matrix. When a spinor is transported along the same loop, it is rotated by the square root of this matrix. Clearly there is a sign ambiguity in taking this square root. For contractible loops this sign ambiguity can be resolved by demanding continuity. For non-contractible loops, the ambiguity can be resolved and two different rules for parallel transporting spinors exist for the

0550-3213/86/$03.50

same metric. We should therefore specify the rule for each non-contractible loop. This information is the spin structure.

In field theory, there is no tunneling between different spin structures and therefore there is no need to sum over them. In string theory the situation is different. Just as one sums over topologies of the world-sheet there is no reason not to sum over its spin structures. In fact, we will show here that this is necessary for the consistency of the theory. One interesting question related to this sum is how to determine the relative weights of the different terms. Clearly, two-dimensional field theory considerations cannot fix them. We will show that they are almost uniquely determined by other requirements such as modular invariance and unitarity.

Before we sum over the spin structures we should make sure that each term separately is well-defined and does not suffer from anomalies. The absence of two-dimensional gravitational anomalies places strong constraints on the possible theories. At the one-loop level, invariance under global reparametrizations is one reason that the gauge group of the heterotic superstring must be $E_8 \times E_8$ or Spin(32) [2]. Global reparametrization anomalies were analyzed to all orders in (3). It was shown there that for the heterotic string no such anomalies arise at any order of perturbation theory. In the fermionic formulation of the heterotic string three kinds of two-dimensional fermions are present: 32 left-moving Majorana-Weyl spinors ψ^I which lead to the internal quantum numbers, 10 right-moving Majorana-Weyl spinors λ^μ and a spin-$\frac{3}{2}$ field. In the superconformal gauge the fermion integration is quadratic and one finds the fermion determinants*

$$S_{\text{het}}(\alpha, \beta, \gamma, g) = (\det{}_\alpha iD_+)^5 (\det{}_\alpha iR_+)^{-1/2} (\det{}_\beta iD_-)^8 (\det{}_\gamma D_-)^8, \quad (1.1)$$

where $D_\pm = \frac{1}{2}D(1 \pm \rho^3)$ are the chiral Dirac operations and R is the Rarita-Schwinger operator. g is the two-dimensional metric. α, β and γ are the spin structures of the fermions (the 32 left-moving fermions were split into two groups of 16 fermions each). According to ref. [3] this expression is invariant under reparametrization transformations that do not change the spin structure. For type II theories, in which both the left-moving and the right-moving sectors have two-dimensional supersymmetry, the result of the fermion integration is

$$S_{\text{II}}(\alpha, \beta, g) = (\det{}_\alpha iD_+)^5 (\det{}_\alpha iR_+)^{-1/2} (\det{}_\beta iD_-)^5 (\det{}_\beta iR_-)^{-1/2}. \quad (1.2)$$

It is easy to show by taking ratios of expressions of the type of eq. (1.1) that eq. (1.2) is also invariant under diffeomorphisms that leave the spin structures intact; alternatively, this can be deduced from the arguments of ref. [3].

* Ref. [3] uses a different definition of the determinant (they are normalized as determinants for Majorana fermions, while here we use Dirac fermions) and therefore the exponents here are smaller than there by a factor of two.

Once one has ensured that the measure for a given set of spin structures α, β, γ is invariant under diffeomorphisms that leave α, β, γ fixed, it is still necessary to consider diffeomorphisms that map α, β, γ to, say, α', β', γ'. This forces us to sum over spin structures in order to achieve modular invariance, but otherwise leads to no new constraints, for reasons that we will review below.

At tree level these considerations do not really arise. A tree graph is topologically a sphere. Since the fundamental group of the sphere is trivial, there are no non-contractible loops and there is only one spin structure. In sect. 2 we analyze the first non-trivial case – the one-loop graph, which is topologically a torus. It has four different spin structures. Modular invariance fixes the precise linear combination of them up to one coefficient. Unitarity demands that its value is ± 1. The physical interpretation of this constant is clarified and the freedom in its sign leads to the possibility of constructing the two type II theories. This sum over spin structures is equivalent to the Gliozzi, Scherk and Olive (GSO) projection [4], as we will see. We will find important differences between two-dimensional spinors which lead to internal quantum numbers, and two-dimensional spinors which lead to fermions in space-time. Sect. 3 continues the same analysis to all orders. Using unitarity all coefficients in the sum over spin structures are uniquely determined (in terms of the arbitrary coefficient in the one-loop graph). We point out a relation between these coefficients and the action of the modular group on spin structures of higher genus surfaces. After examining the consistency of known theories these considerations lead us in sect. 4 to find four new string theories by identifying the spin structure of the left-movers with that of the right-movers. The critical dimension of these theories is ten but they do not have space-time supersymmetry. All these theories contain gravity and they are manifestly modular invariant and unitary. Unfortunately they all suffer from tachyons in their spectra. One of these four theories is particularly interesting since it has chiral fermions in an anomaly-free representation of the gauge group O(16).

In what follows, we include for the sake of completeness many observations that may be known to workers in the field but are difficult or impossible to find in the published literature.

2. One-loop

The topology of the world-sheet for the one-loop graphs is that of a torus. By a two-dimensional reparametrization and a Weyl transformation its metric can be brought to a simple form [5]

$$g_{ab} = \delta_{ab}. \tag{2.1}$$

With such a flat metric, the world-sheet becomes a parallelogram with opposite sides identified (fig. 1). It can be understood to consist of the complex z plane with

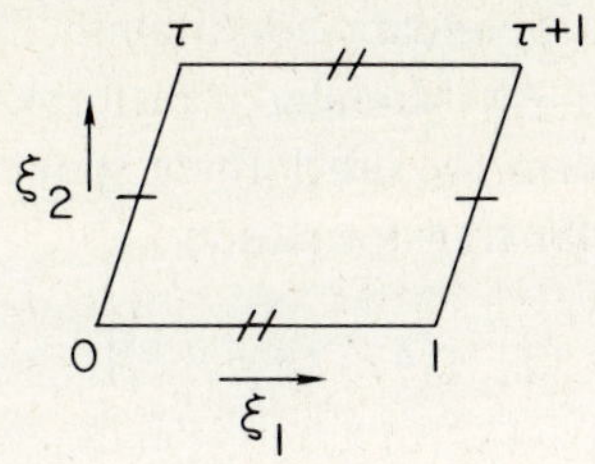

Fig. 1. A one-loop world-sheet can be obtained by gluing together opposite sides of the parallelogram indicated here. The parameters ξ_1 and ξ_2 run in the two independent directions on the parallelogram.

an identification of points $z \sim z+1 \sim z+\tau$; here τ is an arbitrary complex number (whose imaginary part can be taken to be positive by suitable rotation and scaling of the parallelogram in the complex plane). The sum over surfaces becomes, then, an integral over τ.

As is well known, the choice of eq. (2.1) does not completely fix the gauge. There is an infinite set of reparametrizations which preserve the gauge choice. These are the large diffeomorphisms

$$(\xi_1, \xi_2) \to (k\xi_1 + l\xi_2, m\xi_1 + n\xi_2), \tag{2.2}$$

with k, l, m, n integers and $kn - lm = 1$. Under such a transformation

$$\tau \to (n\tau + l)/(m\tau + k). \tag{2.3}$$

These are called modular transformations and τ is known as the modular parameter. To avoid over-counting, the integral over τ should be restricted to the fundamental region

$$-\tfrac{1}{2} < \mathrm{Re}\,\tau < \tfrac{1}{2}, \quad \mathrm{Im}\,\tau > 0, \quad |\tau| > 1. \tag{2.4}$$

When two-dimensional spinors are present one should specify their spin structure. The torus has two independent non-contractible loops; one of them winds around the ξ_1 direction in fig. 1 and the other around the ξ_2 direction. We need a rule for parallel transporting spinors around these two loops; such a rule is called a spin structure. Since the torus is flat, a vector parallel transported around a non-contractible loop returns to its original value. As a vector can be made from two spinors by $v^i = \bar{\psi}\gamma^i\psi$, so if vectors remain unchanged under parallel transport, spinors can only change by a factor of ± 1. The two possible signs correspond to periodic or anti-periodic boundary conditions on the fermions. Likewise, spinors change by a plus or minus sign under parallel transport around a non-contractible loop in the ξ_2 direction. We will denote the spin structure by these two signs – one for each non-contractible loop. A plus sign corresponds to periodic boundary conditions and a minus sign to anti-periodic boundary conditions. The four spin structures are $(++)$, $(+-)$, $(-+)$ and $(--)$. Modular transformations leave the $(++)$ spin structure invariant but transform the other three spin structures into one another. For instance, under the modular transformation

$$(\xi_1, \xi_2) \to (\xi_2, -\xi_1), \tag{2.5}$$

$(+-)$ and $(-+)$ are exchanged while $(--)$ is left invariant. Under

$$(\xi_1, \xi_2) \to (\xi_1, \xi_1 + \xi_2), \tag{2.6}$$

$(--)$ and $(+-)$ are exchanged while $(-+)$ is left invariant.

As mentioned above S_{het} and S_{II} are invariant under modular transformations which do not change the spin structure. These functional integrals are therefore well defined and describe sensible two-dimensional field theories. This is not enough for string theory where we demand invariance under every reparametrization. Since the modular group is a subgroup of the reparametrization group we should insist on modular invariance and make sure that the perturbation expansion is invariant under modular transformations which change the spin structure. The $(++)$ spin structure is modular invariant and therefore so is the functional integral based on it. The other three spin structures are not modular invariant and if we want to include any of them in the sum over surfaces, we should include them all. The relative normalizations of these three functional integrals cannot be determined by two-dimensional field theoretic considerations since each functional integral corresponds to a well-defined field theory just by itself. They can be determined by demanding that the sum be modular invariant. As suggested in ref. [3], we can simply *define* the functional integral for one spin structure and metric in terms of the functional integral for another spin structure and metric:

$$S_{\text{het}}(\alpha, \beta, \gamma, g) = S_{\text{het}}(\alpha', \beta', \gamma', g'), \tag{2.7}$$

$$S_{\text{II}}(\alpha, \beta, g) = S_{\text{II}}(\alpha', \beta', g'), \tag{2.8}$$

where α', β', γ' and g' are obtained from α, β, γ and g by a modular transformation*. This procedure determines the normalization of the functional integral for the three spin structures which are related by the modular group – $(--)$, $(+-)$ and

* One may wonder if there is a non-trivial issue of consistency to be faced in making this definition. Indeed if Φ and $\tilde{\Phi}$ are two different modular transformations that map spin structures α, β, γ into α', β', γ', we should ask whether we get the same answers by using Φ or $\tilde{\Phi}$ to compare $S(\alpha', \beta', \gamma')$ to $S(\alpha, \beta, \gamma)$. As was discussed in ref. [3], we will not run into any problem if (for each (α, β, γ)), $S(\alpha, \beta, \gamma)$ is invariant under modular transformations that leave α, β, and γ fixed. Indeed, in the situation under discussion, $\tilde{\Phi}^{-1}\Phi$ leaves α, β, and γ fixed, so writing $\Phi = \tilde{\Phi} \cdot (\tilde{\Phi}^{-1}\Phi)$, we see that if $\tilde{\Phi}^{-1}\Phi$ leaves $S(\alpha, \beta, \gamma)$ invariant, then Φ and $\tilde{\Phi}$ are equivalent ways to transform α, β, γ to α', β', γ'.

$(-+)$, in terms of the normalization of one of them, say for $(--)$. It does not fix the normalization of the functional integral for the fourth one $(++)$. Using these normalizations the sum of S_{het} (or S_{II}) over the different spin structures is guaranteed to be invariant both under modular transformations which change or do not change the spin structure.

Let us compute the fermion determinant for a group of eight left-moving real fermions, all with the same spin structure. This object suffers from a perturbative anomaly. But if we think of it as a factor in an anomaly free expression as S_{het} or S_{II}, we can fix the gauge of eq. (2.1) and consider $A(\alpha, \tau) = (\text{Det}_\alpha D_-)^4$ where τ is the modular parameter. The path integral on a torus computes the propagation of a string through a "time" τ, so it is conveniently written (up to a normalization constant which we will have to determine) as $\text{tr}\, e^{i\tau H}\Lambda$ where H is the appropriate hamiltonian (the Ramond or Neveu-Schwarz hamiltonian in the case of periodic or anti-periodic boundary conditions in the ξ_1 direction), while $\Lambda = 1$ for anti-periodic boundary conditions in the ξ_2 direction, and $\Lambda = (-1)^F$ (the operator that counts fermions modulo two) for periodic ones. Explicitly, the Ramond and Neveu-Schwarz hamiltonians are

$$H_{\text{NS}} = 2\pi\left(\sum_{r=1/2}^{\infty} r b_r^{i\dagger} b_r^i - \tfrac{1}{6}\right) \tag{2.9}$$

$$H_{\text{R}} = 2\pi\left(\sum_{n=1}^{\infty} n d_n^{i\dagger} d_n^i + \tfrac{1}{3}\right) \tag{2.10}$$

(notice the normal ordering constants). If, thus, we denote the path integral for $(--)$ spin structure as $A((--), \tau)$, and likewise for the others, then the world-sheet path integrals are

$$A((--), \tau) = \eta_{(--)} \text{Tr}\exp(i\tau H_{\text{NS}}), \tag{2.11}$$

$$A((+-), \tau) = \eta_{(+-)} \text{Tr}\exp(i\tau H_{\text{R}}), \tag{2.12}$$

$$A((-+), \tau) = \eta_{(-+)} \text{Tr}\exp(i\tau H_{\text{NS}})(-1)^F, \tag{2.13}$$

$$A((++), \tau) = \eta_{(++)} \text{Tr}\exp(i\tau H_{\text{R}})(-1)^F, \tag{2.14}$$

where the η's are normalization constants which we will have to determine.

We arbitrarily define $\eta_{(--)} = 1$ and find

$$A((--),\tau) = q^{-1/3}\prod_{n=1}^{\infty}(1+q^{2n-1})^8 = (2\pi)^{4/3}\frac{\theta_3(0,\tau)^4}{\theta_1'(0,\tau)^{4/3}},$$

$$q = \exp(i\pi\tau). \tag{2.15}$$

Clearly, there is a cube root of unity ambiguity in this expression. It is cancelled by the other determinants in the amplitude. The other η's are then found by requiring modular invariance or in other words by imposing eqs. (2.7)–(2.8). Using formulae from ref. [6], we find

$$A((+-),\tau) = A((--),\tau') = (2\pi)^{4/3}\frac{\theta_3(0,\tau')^4}{\theta_1'(0,\tau')^{4/3}} = -(2\pi)^{4/3}\frac{\theta_2(0,\tau)^4}{\theta_1'(0,\tau)^{4/3}}, \tag{2.16}$$

where $\tau' = \tau/(\tau+1)$. Since

$$\mathrm{Tr}\exp(i\tau H_{\mathrm{R}}) = 16q^{2/3}\prod_{n=1}^{\infty}(1+q^{2n})^8 = (2\pi)^{4/3}\frac{\theta_2(0,\tau)^4}{\theta_1'(0,\tau)^{4/3}}, \tag{2.17}$$

$\eta_{(+-)} = -1$. Similarily,

$$A((-+),\tau) = A((+-),\tau') = -(2\pi)^{4/3}\frac{\theta_2(0,\tau')^4}{\theta_1'(0,\tau')^{4/3}} = -(2\pi)^{4/3}\frac{\theta_4(0,\tau)^4}{\theta_1'(0,\tau)^{4/3}}, \tag{2.18}$$

where $\tau' = -1/\tau$. Since

$$\mathrm{Tr}\exp(i\tau H_{\mathrm{NS}})(-1)^F = q^{-1/3}\prod_{n=1}^{\infty}(1-q^{2n-1})^8 = (2\pi)^{4/3}\frac{\theta_4(0,\tau)^4}{\theta_1'(0,\tau)^{4/3}}, \tag{2.19}$$

$\eta_{(-+)} = -1$. Clearly, $\eta(++)$ cannot be determined this way but we readily find

$$A((++),\tau) = \eta_{(++)}\mathrm{Tr}\exp(i\tau H_{\mathrm{R}})(-1)^F = 0. \tag{2.20}$$

The vanishing of $A((++),\tau)$ can also be understood as a consequence of a harmonic spinor (a fermion zero mode) on the torus with periodic boundary conditions.

The sum $A = A((--),\tau) + A((-+),\tau) + A((+-),\tau) + A((++),\tau)$ is modular invariant. In fact, using the identity [6]

$$\theta_2(0,\tau)^4 - \theta_3(0,\tau)^4 + \theta_4(0,\tau)^4 = 0 \tag{2.21}$$

and eq. (2.20) we see that it vanishes.

This sum over spin structure has a simple physical interpretation. Using eqs. (2.11)–(2.14) and the values of the normalization constants η this sum turns out to be

$$A = \mathrm{Tr}\exp(i\tau H_{\mathrm{NS}})\tfrac{1}{2}\left(1+(-1)^{F+1}\right) - \mathrm{Tr}\exp(i\tau H_{\mathrm{R}})\tfrac{1}{2}\left(1-\eta_{(++)}(-1)^{F}\right). \quad (2.22)$$

We see that in the Neveu-Schwarz sector the sum amounts to projection on the states with $(-1)^F = -1$. Modular invariance forces us to perform a well-defined projection. This is one of the two GSO projections [4]. The relative sign between the contributions of the Neveu-Schwarz and the Ramond sectors ($\eta_{(+-)} = -1$) is also interesting. It means that the contribution of the states in the Ramond sector to the one-loop graph has the opposite sign to that of the states in the Neveu-Schwarz sector. Equivalently, these two sectors have opposite statistics and the spinors of the Ramond sector are indeed fermions. $\eta_{(++)}$ cannot be determined by modular invariance. The sum A can be interpreted as a sum of $\exp(i\tau H)$ over states only for $\eta_{(++)} = 0, \pm 1$. We will later show that unitarity excludes the case $\eta_{(++)} = 0$. $\eta_{(++)} = +1$ (or -1) corresponds to a projection on $(-1)^F = -1$ (or $+1$) in the Ramond sector. This is the second GSO projection.

For 16 two-dimensional Majorana-Weyl spinors, all with the same spin structure, the situation is somewhat different. Here we find the square of the determinants for eight fermions and therefore $\eta_{(+-)} = \eta_{(-+)} = +1$. This means that the projection in the Neveu-Schwarz sector is on states with $(-1)^F = +1$ and that the states in the Ramond sector satisfy Bose statistics. Since the Ramond sector consists of states with spinor quantum numbers, they cannot be interpreted as fermions and they must be spinors of an internal symmetry – not of the Lorentz group. The sum over the four spin structures

$$A = (2\pi)^{8/3}\frac{\theta_2(0,\tau)^8 + \theta_3(0,\tau)^8 + \theta_4(0,\tau)^8}{\theta_1'(0,\tau)^{8/3}} \quad (2.23)$$

is indeed modular invariant [6] (up to a cube root of unity ambiguity which is not present in the amplitude (1.1)).

All this is consistent with what we know of the various string theories. Let us consider first the heterotic string. On the torus we can pick the light-cone gauge. The determinant of the Rarita-Schwinger ghost cancels the determinant of two of the right-moving spinors. Since there are eight right-moving spinors, we must project on states with $(-1)^F = -1$ in the Neveu-Schwarz sector and on a half of the Ramond sector. There is no essential difference whether we pick $\eta_{(++)} = +1$ or -1. $(-1)^F$ has a factor of γ^9 in the light cone (or γ^{11} in a covariant treatment) and the difference between the two choices of $\eta_{(++)}$ can be absorbed into a redefinition of γ^9 (or γ^{11}), namely into redefinition of what we mean by left-handed or right-handed chirality. This difference is clearly not observable and the same theory is obtained whether we pick $\eta_{(++)} = +1$ or -1. We also learn that the states in the Ramond sector satisfy Fermi statistics and as space time spinors, the relation between the spin and the statistics is preserved. The spectrum of the theory has space-time supersymmetry in this case and indeed the one-loop contribution to the cosmological constant vanishes because $A = 0$.

There are 32 left-moving spinors in the heterotic string. They lead to the internal symmetry [3]. When they are grouped into two groups of 16 spinors in each, the symmetry is $E_8 \times E_8$. The analysis above shows that unlike the case for the right-movers, here one should project in the Neveu-Schwarz sector on $(-1)^F = +1$. Furthermore, as the fermions come in groups of 16, the states in the Ramond and in the Neveu-Schwarz sector satisfy the same statistics and the states in the Ramond sector are spinors of an internal symmetry rather than of space-time. As for the right-moving spinors the sign of $\eta_{(++)}$ is not observable; it can be absorbed into a redefinition of γ^{17}. The situation is similar when all the left-moving spinors have the same spin structure. The resulting symmetry is $\mathrm{Spin}(32)/Z_2$, the projection in the Neveu-Schwarz sector is on states with $(-1)^F = +1$, the Ramond sector yields spinors of an internal symmetry and the value of $\eta_{(++)}$ is not observable.

As proved in ref. [3] S_{het} is invariant under modular transformations which do not change the spin structure even when the left-moving spinors are grouped in groups of 8 (or 24) fermions each. If we sum over the spin structures, we obtain a modular invariant expression. However, such a theory cannot be given a consistent space time interpretation. The phases dictated by the modular group are such that the states in the Ramond and the Neveu-Schwarz sectors satisfy opposite statistics. These spinors do not lead in this case to an internal symmetry and the space-time interpretation is ruined.

It is straightforward to carry out a similar analysis for the type II theory where both the left-movers and the right-movers are supersymmetric. As stated above S_{II} is invariant under modular transformations which do not change the spin structures. Invariance under modular transformation is obtained when we sum over the spin structures. As for the right-movers in the heterotic string, we should project on $(-1)^F = -1$ in each Neveu-Schwarz sector (separately for the left-movers and for the right-movers). Since there are 8 fermions in each group, the states in the Ramond sector have the opposite statistics relative to the states in the Neveu-Schwarz sector. Space-time bosons are obtained when both the left- and right-movers have the same boundary conditions (Neveu-Schwarz or Ramond). Space-time fermions are obtained when the left- and right-movers are subject to different boundary conditions. Unlike the heterotic string, here the sign of $\eta_{(++)}$ is meaningful. There are two such constants – one for the left-movers and one for the right-movers. One of them, say $\eta_{(++)}$ for the left-movers, can be set to $+1$ by redefining the chirality – the sign of γ^9 (or γ^{11} if the light-cone gauge is not picked). Once this choice was made there is no more freedom for the other $\eta_{(++)}$ and its value is observable. If this constant is $+1$, the projections in the Ramond sector of the left-movers and of the right-movers are on the same chirality and we obtain the chiral theory IIB. If on the other hand, $\eta_{(++)} = -1$ the two projections are on different chirality and this corresponds to the non-chiral theory IIA.

In conclusion, we have seen that modular invariance and unitarity force us to sum over the spin structures with well-defined coefficients. This is equivalent to performing the GSO projections. If there are eight spinors in a group, the states in the Ramond and in the Neveu-Schwarz sector satisfy opposite statistics and the states in the Ramond sector must be interpreted as space-time spinors. In this case the projection in the Neveu-Schwarz sector is on states with $(-1)^F = -1$. For groups of sixteen spinors, the Neveu-Schwarz and the Ramond sectors satisfy the same statistics and therefore the states should be interpreted as having quantum numbers of an internal symmetry. The projection in the Neveu-Schwarz sector should be on states with $(-1)^F = +1$. One constant, the coefficient of the $(++)$ spin structure, cannot be determined by modular invariance. Unitarity will force this coefficient to be ± 1 (unless one makes some more elaborate modifications of the GSO projection that we will consider later). In some cases the sign of $\eta_{(++)}$ is not observable, but in the type II theories we can use it to construct both the type IIA and the type IIB theories.

3. Higher orders

In higher orders in the loop expansion the world-sheet is a higher genus Reimann surface. To determine its spin structures we should first find all non-contractible loops. Perhaps the easiest way to do that is by viewing the genus L surface as made by gluing L tori to one another (fig. 2). There are two non-contractible loops on each of these tori. These $2L$ loops generate all the non-contractible loops–all other non-contractible loops can be obtained by deforming and by joining several of these $2L$ loops. When a two-dimensional vector is parallel transported around any of these loops it is rotated by an O(2) matrix. When a two-dimensional spinor is transported along the same loop, it is rotated by the square root of this matrix. The sign ambiguity in this square root leads to the various spin structures. Since these $2L$ loops generate all the non-contractible loops, we should specify the rule for parallel transporting spinors along these loops and this determines the rule for all loops. There are two choices for this sign for each of the $2L$ loops and hence there are 2^{2L} different spin structures.

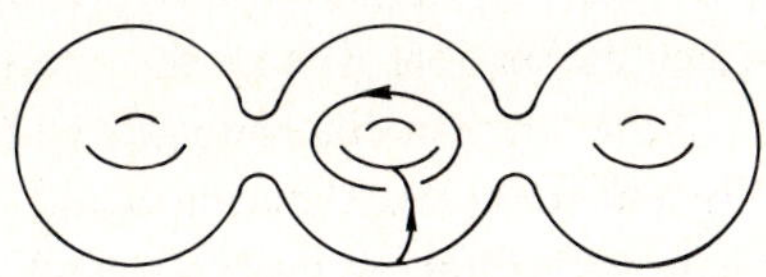

Fig. 2. An L-loop graph and two of its $2L$ non-contractible loops which generate all the other non-contractible loops.

If we think about the one-loop case, it is clear that with $(++)$ boundary conditions, it is possible to have a constant spinor which is a zero mode of the Dirac operator. For, say, positive chirality spinors there is precisely one such zero mode. For boundary conditions other than $(++)$ there is no constant spinor on the torus, and no Dirac zero modes. This situation generalizes to higher genus surfaces in the following way. On any Riemann surface with a given spin structure the number of positive chirality zero modes of the Dirac operator is a topological invariant modulo two. On a genus L surface, there are $2^{L-1}(2^L+1)$ spin structures for which the Dirac operator (on fermions of definite chirality) has an even number of zero modes, while there are $2^{L-1}(2^L-1)$ of them for which the number of zero modes is odd [7]. Evidently, the number of Dirac zero modes modulo two is a modular invariant, so these two groups transform separately under modular transformations. Surprisingly, it seems to have been proved only recently (partly under stimulus from string theory!) that these two groups each transform irreducibly under modular transformations so that there are no other modular invariants characterizing a spin structure [8]. In fact, the precise form of the action of the modular transformations of spin structures has been worked out in ref. [8] for arbitrary genus.

Let us refer to spin structures for which the number of Dirac zero modes (of positive chirality) is even or odd as even and odd spin structures, respectively. It can be shown that the number of chiral Dirac zero modes modulo two is additive when we glue together two Riemann surfaces. (This follows from the fact that the number of chiral zero modes modulo two is a bordism invariant.) Therefore, to make an even spin structure on a genus-two surface, we must glue together two genus-one surfaces which each have even spin structures (there are $3 \cdot 3 = 9$ ways to do this) or which each have odd spin structures (there is $1 \cdot 1 = 1$ choice). In all we have $9 + 1 = 10$ even spin structures on a genus-two surface. On the other hand, an odd spin structure in genus two is made by gluing an odd spin structure on genus one to an even one or vice-versa. There are $3 \cdot 1 + 1 \cdot 3 = 6$ possibilities. In a similar way, one can go on to prove inductively (as a consequence of the additivity modulo two of the number of chiral zero modes) that the number of even spin structures on genus L is $2^{L-1}(2^L+1)$. The fact that there are ten even spin structures for genus two and that these transform irreducibly under modular transformations will be crucial in what follows.

Our experience from the one-loop case tells us that for invariance under the full reparametrization group we have to sum over different spin structures. The precise coefficients of the different terms can be determined by the procedure used for the one-loop graph in the previous section. Using the modular group one can fix the relative normalizations of the fermion determinants within each sector. Just as in the one-loop case one coefficient cannot be determined this way – this is the relative weight of the two different sectors. Other considerations should be used to fix it.

There is another, more physical way to determine the weights of the different spin structures. They can all be determined by demanding that the theory be unitary.

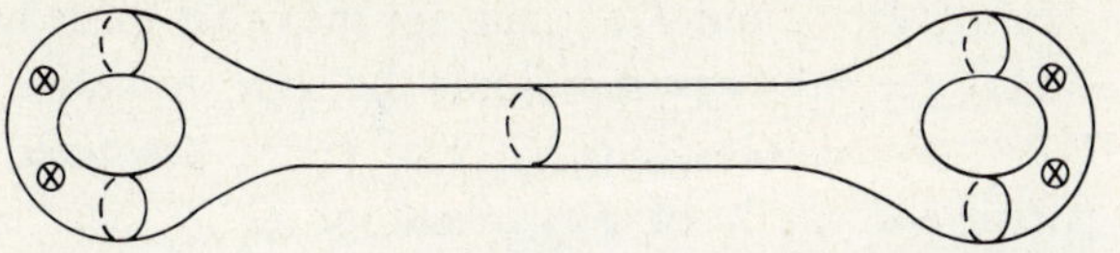

Fig. 3. A two-loop graph can be deformed to two-tori connected by an almost everywhere flat tube. If the vertex operators do not introduce cuts, all the spinors are anti-periodic around the flat tube.

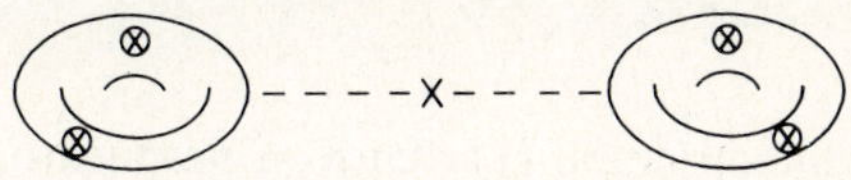

Fig. 4. The two-loop graph factorizes near the pole to a product of a propagator and two one-loop graphs.

Consider first a two-loop graph with some external legs (fig. 3). For simplicity we consider the case where the vertex operators for the external states do not generate cuts in the surface. They correspond to states for which all two-dimensional spinors are in the Neveu-Schwarz sector. Using the reparametrization freedom, the graph can be deformed to two widely separated tori connected by a tube which is almost everywhere flat. Since all external legs are bosonic (their vertex operators are periodic), the two-dimensional spinors are anti-periodic around the flat tube*. If the external momenta are such that the state in the tube is close to its mass-shell, unitarity demands that the graph factorizes to a product of two one-loop graphs and a propagator (fig. 4)**. Since all two-dimensional spinors are anti-periodic around the tube, the vertex operator for the corresponding state is periodic and the two one-loop graphs are without cuts.

The two-loop graph is given by a linear combination of functional integrals with different spin structures. Since we have already determined the coefficients for the different spin structures of the one-loop graph, we know how to evaluate the two-loop graph close to the pole. This provides us with enough information to fix the normalizations of the functional integrals for the spin structures of the double torus.

Let us focus on one group of two-dimensional spinors, say one of the two groups of the left-moving spinors in the heterotic string. The one-loop graph with the external legs is given by a sum of four terms corresponding to the four spin structures for this group of spinors. Each term is itself given by a sum over the spin structures of the other spinors. Denote these four contributions for one of the

* The fact that it is *antiperiodic* boundary conditions in the space direction that correspond to *bosonic* vertex operations might be counterintuitive, but follows from the fact that the NS states are bosons.

** There might be extra divergences in this case corresponding to mass renormalization [9]. These will not affect our argument.

two-tori by $F_{(++)}$, $F_{(-+)}$, $F_{(+-)}$ and $F_{(--)}$ and for the other torus by $G_{(++)}$, $G_{(-+)}$, $G_{(+-)}$ and $G_{(--)}$ where the subscript represents the spin structure. The normalizations of $F_{(-+)}$, $F_{(+-)}$, $F_{(--)}$, $G_{(-+)}$, $G_{(+-)}$ and $G_{(--)}$ are fixed by a modular invariance. The coefficient $\eta_{(++)}$ of $F_{(++)}$ and of $G_{(++)}$ was not determined uniquely. The amplitude of one-torus is therefore

$$F_{(-+)}+F_{(+-)}+F_{(--)}+\eta_{(++)}F_{(++)}. \tag{3.1}$$

The other one is

$$G_{(-+)}+G_{(+-)}+G_{(--)}+\eta_{(++)}G_{(++)}. \tag{3.2}$$

Close to the pole the two-loop graph is given by a propagator times

$$\begin{aligned}&\left(F_{(-+)}+F_{(+-)}+F_{(--)}+\eta_{(++)}F_{(++)}\right)\cdot\left(G_{(-+)}+G_{(+-)}+G_{(--)}+\eta_{(++)}G_{(++)}\right)\\ &\quad=\left(F_{(-+)}+F_{(+-)}+F_{(--)}\right)\cdot\left(G_{(-+)}+G_{(+-)}+G_{(--)}\right)+\eta^2_{(++)}F_{(++)}G_{(++)}+\\ &\qquad\eta_{(++)}\left(F_{(++)}\cdot\left(G_{(-+)}+G_{(+-)}+G_{(--)}\right)+\left(F_{(-+)}+F_{(+-)}+F_{(--)}\right)\cdot G_{(++)}\right).\end{aligned} \tag{3.3}$$

We should obtain the same result by evaluating the two-loop graph and by summing over its spin structures. The sixteen terms that we have found should equal the contribution from the sixteen different spin structures of the double torus. As mentioned above, these sixteen spin structures are split into two irreducible representations of the modular group. One of them consists of ten spin structures and the other irreducible representation has six elements. If in the above equation, we choose $\eta_{(++)}=0$, we find that only nine of the spin structures of the double torus contribute to the two-loop graph. Such a result cannot be modular invariant, since the two modular invariant combinations contain ten and six spin structures, respectively. Hence $\eta_{(++)}$ cannot vanish. We saw in section 2 that $\eta_{(++)}=0$ or ± 1. We therefore conclude that $\eta_{(++)}=\pm 1$.

With this value of $\eta_{(++)}$ the sixteen terms in eq. (3.3) split into two sectors. The contribution of one sector is independent of $\eta_{(++)}$. The contribution of the other sector is proportional to $\eta_{(++)}$. There are 10 spin structures in the first sector and 6 spin structures in the second sector. It is natural to conjecture that these two sectors correspond to the two irreducible representations of the modular group and that the relative phases found by unitarity are equivalent to those obtained by the modular transformations. The situation at two loops is thus analogous to the one loop case. There are two sectors of spin structures. The relative coefficients within each sector can be determined by the modular group and unitarity fixes the relative coefficient between the two sectors to be equal to the phase $\eta_{(++)}$ of the one-loop graph.

A similar analysis can be easily carried out to all orders in the loop expansion by factorizing an L-loop graph on a pole between two sub-graphs one of them having L_1 loops and the other with $L - L_1$ loops. We can determine the phases of the contributions of the various spin structures at the L-loop level in terms of those at lower levels. Using such an induction, we learn that the pattern observed at the one- and two-loop levels persists to all higher orders. There are two sectors. One of them has $2^{L-1}(2^L + 1)$ spin structures and the other has $2^{L-1}(2^L - 1)$ spin structures. The contribution of the second sector is again proportional to $\eta_{(++)}$ and that of the first sector is independent of $\eta_{(++)}$. We conclude that all the coefficients in the perturbation expansion are determined up to the coupling constant and the phase $\eta_{(++)}$.

The observation that unitarity forces $\eta_{(++)} = \pm 1$ has a very interesting qualitative significance. These are the values of $\eta_{(++)}$ for which one has a supersymmetric spectrum by the GSO mechanism. The other value of $\eta_{(++)}$ that made sense at the one loop level was $\eta_{(++)} = 0$. At $\eta_{(++)} = 0$ we have a spectrum with a massless spin-$\frac{3}{2}$ particle (of both chiralities) but no supersymmetry. A consistent theory with a massless spin-$\frac{3}{2}$ particle, on the other hand, should have the spin-$\frac{3}{2}$ particle coupled to a conserved current, and therefore should be supersymmetric. It has thus been felt for many years that consistency of interacting string theory must require either the GSO projection or some other modification. In the present framework, the way that this works out is that unitarity requires $\eta_{(++)} = 1$, so that the GSO projection is forced on us. We will see in the last section some other ways to obtain unitary theories; they will illustrate the fact that one can avoid the GSO projection only via some other modification of the theory which eliminates the massless spin-$\frac{3}{2}$ particle.

We can also conveniently answer another question which frequently arises. Why is the GSO projection compatible with unitarity? Why are not the states of wrong G-parity, which are removed by the GSO projection, created in pairs in loop diagrams? Since we have claimed that the GSO projection is *required* by unitarity (unless some more substantial modification is made) an answer to the question of why it is *allowed* must be implicit in our discussion. Indeed, the prescription "sum over all spin structures" amounts to making a GSO projection in every possible channel at once. If, as in fig. 4, we "cut" any one line in a string diagram, then the sum over possible boundary conditions on the line that has been cut amounts to the fact that the cut line may be either a Ramond or a Neveu-Schwarz state (either a fermion or a boson), while the sum over possible boundary conditions in the "orthogonal" direction is the GSO projection. The fact that the GSO projection is *compatible* with unitarity (wrong G-parity states are not created in pairs) is thus, as it should be, a corollary of the same sort of reasoning which shows that the GSO projection is *required* by unitarity.

We now move on to other issues. The fermion determinant for the torus with spin structure $(++)$ vanishes. This can be understood as a result of the existence of a zero mode of the Dirac operation for these boundary conditions. This fact makes it possible for the coefficient of this determinant to be unobservable in some cases both for 16 and for 8 two-dimensional spinors in a group. The generalization of this fact to higher orders is based on the fact that spin structures of odd type (those with the coefficient $\eta_{(++)}$) are those for which the chiral Dirac operator always has at least one zero mode (in fact, it always has an odd number of them). Therefore, each fermion determinant in this sector vanishes. Again, this fact makes it possible for $\eta_{(++)}$ to be unobservable in some cases.

We did not prove that the normalizations of the functional integrals for the different spin structures as determined by unitarity coincide with those determined by the modular group. An explicit verification of this conjecture will clearly be very interesting. We once again observe here the deep connection between modular invariance and unitarity.

4. New theories

In the previous sections we saw that for the consistency of the theory we have to sum over the spin structures with well-defined coefficients. Here we will consider new possibilities to sum over the spin structures in a consistent way. The resulting theories seem to be different from the known existing theories. We cannot prove however, that they are not different vacua of these theories.

The basic idea is very simple. In all theories discussed above the spinors are split into groups of eight or sixteen and we sum over the spin structures of the various groups. Avoiding anomalies and preserving a space time interpretation in which particles that appear in partition functions with negative signs are fermions prevent us from splitting the world-sheet spinors into smaller groups. We may try, however, to combine the spinors into fewer and larger groups and thus construct new theories.

Consider first the type II theories. Here we sum $S_{\mathrm{II}}(\alpha, \beta, g)$ over α and β. One of the two $\eta_{(++)}$'s can be set to $+1$ and the other $\eta_{(++)}$ is observable. These two theories (with the two possible values of the observable $\eta_{(++)}$) can be modified by summing over the spin structures subject to the condition $\alpha = \beta$. It is easy to perform some consistency checks for the two resulting theories. They are clearly Weyl invariant in ten dimensions. Each functional integral is well-defined and does not suffer from anomalies. The sum over the spin structures with the right coefficients guarantees modular invariance and unitarity.

It is straightforward to find the spectrum of these theories in the light-cone gauge. The condition $\alpha = \beta$ tells us that these theories have only two sectors – either both the left-movers and the right-movers satisfy anti-periodic boundary conditions (Neveu-Schwarz) or they both satisfy periodic boundary conditions (Ramond). This is unlike the standard type II theories where there are two more sectors where the left-movers and the right-movers satisfy opposite boundary conditions. These two sectors of the usual type II theories lead to space-time fermions. As they do not exist in our theories we have only space-time bosons. Modular invariance tells us to project in the Neveu-Schwarz sector on states with $(-1)^F = +1$ where F is the sum of the fermion number for the left-movers and for the right-movers. The relative sign between the Ramond and the Neveu-Schwarz sectors shows that the states in the Ramond sector are bosons (as in the type II theories). The two possible values of the observable $\eta_{(++)}$ correspond to two different projections in the Ramond sector – on $(-1)^F = -1$ or $+1$. These two possibilities lead to two different spectra. Finally the lack of a preferred point of the string leads to the constraint that the contribution to the mass square from the left-movers equals that from the right-movers.

Using these constraints we find the spectra of the two theories. The content of the Neveu-Schwarz sector is identical in both of them. The lowest state $|0\rangle$ is a tachyon with $\frac{1}{2}\alpha'(\text{mass})^2 = -1$. (Notice that in the type II theories this state is projected out of the spectrum since there we also impose $(-1)^{F_L} = -1$ where F_L is the number of left-moving fermions.) The next level is massless. It contains a graviton, an anti-symmetric tensor and a scalar. Using standard notation, these are the states

$$b^i_{-1/2}\tilde{b}^j_{-1/2}|0\rangle. \tag{4.1}$$

The next level is at $\frac{1}{2}\alpha'(\text{mass})^2 = +1$.

The two theories differ in their Ramond sectors. Before projecting on well-defined fermion number states they both have 256 massless ground states which are cross-products of the left-moving ground states and the right-moving ground states. They are

$$|8_s + 8_c\rangle_L \times |8_s + 8_c\rangle_R, \tag{4.2}$$

where 8_c and 8_s are the two spinors of O(8). The projector eliminates half of these states. The two different ways to project correspond to the two different values of $\eta_{(++)}$ and they lead to two different theories. In one of these theories we are left with the 128 states

$$|8_s\rangle_L \times |8_c\rangle_R + |8_c\rangle_L \times |8_s\rangle_R. \tag{4.3}$$

They transform under the Lorentz group as a complex vector (16 states) and a complex three-index anti-symmetric tensor (112 states). The other possibility is to keep the other 128 states.

$$|8_s\rangle_L \times |8_s\rangle_R + |8_c\rangle_L \times |8_c\rangle_R. \tag{4.4}$$

They transform under the Lorentz group as a complex scalar (2 states), a complex two-index anti-symmetric tensor (56 states) and a real four-index anti-symmetric tensor. (Notice that the four-index field is not self-dual and therefore does not lead to anomalies.)

Two more theories can be constructed out of the heterotic string by identifying the spin structures of the left-movers and the right-movers. If we sum $S_{\rm het}(\alpha, \beta, \gamma, g)$ subject to the condition $\alpha = \beta = \gamma$ the theory has only two sectors. In the Ramond sector all two-dimensional spinors are periodic and all the states are space-time fermions and transform as spinor representations of the gauge group Spin(32). The states of the Neveu-Schwarz sector are space-time bosons. This theory trivially passes the consistency checks mentioned above. The relative sign between the contribution of the Neveu-Schwarz states and the Ramond states to the one-loop graph is such that the relation between the spin and the statistics is preserved. Modular invariance tells us to project on states with $(-1)^F = -1$ (F is the total two-dimensional fermion number) in the Neveu-Schwarz sector.

The fastest way to find the spectrum is to use the light-cone gauge. We start by considering the spectrum in the Neveu-Schwarz sector. The lowest state is a tachyon with $\frac{1}{2}\alpha'(\text{mass})^2 = -1$. It is a Lorentz scalar and a Spin(32) vector. It is given by

$$b^I_{-1/2}|0\rangle_{\rm L} \times |0\rangle_{\rm R}, \tag{4.5}$$

where the index I labels the 32 left-moving spinors. The states at the next mass level are massless. Here we find the graviton, the anti-symmetric tensor and the dilaton as

$$\alpha^j_{-1}|0\rangle_{\rm L} \times \tilde{b}^i_{-1/2}|0\rangle_{\rm R} \tag{4.6}$$

and the Spin(32) gauge bosons

$$b^I_{-1/2}b^J_{-1/2}|0\rangle_{\rm L} \times \tilde{b}^i_{-1/2}|0\rangle_{\rm R}. \tag{4.7}$$

The lowest states in the Ramond sector are at $\frac{1}{2}\alpha'(\text{mass})^2 = +2$. Before imposing the projector these are

$$|S_1 + S_2\rangle_{\rm L} \times \left(\tilde{d}^i_{-1} \text{ or } \tilde{\alpha}^i_{-1}\right)|8_{\rm c} + 8_{\rm s}\rangle_{\rm R}, \tag{4.8}$$

where S_1 and S_2 are the two spinors of Spin(32). Here the sign of $\eta_{(++)}$ is not observable and can be absorbed into a redefinition of γ^{33}. After the projection we are left with the states

$$|S_1\rangle_{\rm L} \times \left(\tilde{d}^i_{-1}|8_{\rm c}\rangle_{\rm R} + \tilde{\alpha}^i_{-1}|8_{\rm s}\rangle_{\rm R}\right) + |S_2\rangle_{\rm L} \times \left(\tilde{d}^i_{-1}|8_{\rm s}\rangle_{\rm R} + \tilde{\alpha}^i_{-1}|8_{\rm c}\rangle_{\rm R}\right). \tag{4.9}$$

These are two Spin(32) spinors which transform as the 128 dimensional representation of O(9). Notice that unlike the usual heterotic string theory, here we find all the representations of Spin(32) in the spectrum.

Our fourth theory is obtained by summing $S_{\rm het}(\alpha, \beta, \gamma, g)$ subject to the condition $\alpha = \beta$. As the other theories, this one also passes the simple consistency checks mentioned above. The spectrum is found similarly to other theories. Since there are two independent groups of two-dimensional spinors and each of them can satisfy either periodic or anti-periodic boundary conditions, the theory has four sectors. One group of spinors – the sixteen left-movers (with spin structure γ) lead to the internal symmetry E_8. We project in its Neveu-Schwarz sector on $(-1)^F = +1$. In the Neveu-Schwarz sector of the other group of spinors we have to project on $(-1)^F = -1$. Here the states in the Ramond sector are space-time fermions.

Consider first the sector where all spinors are anti-periodic. Denote by B_r^I $(I = 1,\dots,16)$ the oscillators of the sixteen spinors in the first group (whose spin structure is not identified with that of the right-movers) and by b_r^N $(N = 1,\dots,16)$ the oscillators of the other sixteen left-moving spinors. The lowest states

$$b^N_{-1/2}|0\rangle_L \times |0\rangle_R \tag{4.10}$$

are at $\frac{1}{2}\alpha'(\text{mass})^2 = -1$. They form a Spin(16) vector. The next level is massless. It consists of the graviton, the anti-symmetric tensor and the dilaton

$$\alpha^i_{-1}|0\rangle_L \times \tilde{b}^i_{-1/2}|0\rangle_R . \tag{4.11}$$

The other massless states in this sector are the gauge bosons

$$B^I_{-1/2}B^J_{-1/2}|0\rangle_L \times \tilde{b}^i_{-1/2}|0\rangle_R , \tag{4.12}$$

$$b^N_{-1/2}b^M_{-1/2}|0\rangle_L \times \tilde{b}^i_{-1/2}|0\rangle_R . \tag{4.13}$$

In the second sector the spinors in the first group are periodic and those in the second group are anti-periodic. The lowest states here are massless. These are the other 128 gauge bosons needed to complete the adjoint representation of E_8. In the third sector the spinors in the second group are periodic and those in the first group are anti-periodic. The lowest states in this sector are massless. After imposing the projection operator (the sign of $\eta_{(++)}$ is not observable in this case) we find the states

$$|S_1\rangle_L \times |8_c\rangle_R + |S_2\rangle_L \times |8_s\rangle_R , \tag{4.14}$$

where S_1 and S_2 are the two spinors of Spin(16). All the states in the fourth sector where all the two-dimensional spinors are periodic are massive. The gauge symmetry of this theory is $E_8 \times$ Spin(16). The lowest state is a tachyon which is a Spin(16) vector. At the massless level we find apart from the gravity multiplet and the gauge bosons, chiral fermions which transform as spinors of Spin(16). The right-handed fermions transform as the positive chirality spinor of Spin(16) which we may denote as the 128, while the left-handed fermions transform as the negative chirality spinor of SO(16), which we may call the 128′. It is interesting to note that this set of chiral fermions is an anomaly-free combination considered earlier [10].

What is the significance of the models that we have formulated in this section? As they have tachyons, they are unlikely to be the basis for physical models, even though the last has a more or less reasonable gauge group and fermion representation. They do, however, illustrate an important theoretical point which we have alluded to in the previous section. One thing that the four models that we have just formulated have in common is that none of them has a massless spin-$\frac{3}{2}$ particle. This is quite compatible with the view that a constant (unitary) theory must have either the GSO projection (to achieve supersymmetry) or a related projection (like those considered above) to eliminate the massless spin-$\frac{3}{2}$ particle. A consistent theory in which there is a massless spin-$\frac{3}{2}$ particle must be supersymmetric. Technically, there may be an alternative to this in which the massless spin-$\frac{3}{2}$ particle decouples at very low energy, but this does not seem to be an option in string theory because prior to the GSO or other projection the massless spin-$\frac{3}{2}$ particle does not decouple in this way.

The work of N. S. was supported in part by DOE contract DE-AC02-76ER02220 and that of E. W. by NSF grant PHY80-19754. We would like to thank V. Kaplunovsky for assistance.

After completion of this paper we became aware of other related work. Vafa [11] uses a technique similar to that of sect. 3 to determine the phases in theories based on orbifolds. Modular invariant combinations of partition functions somewhat similar to those in sect. 4, but more intricate, have been constructed by Dixon and Harvey [12], who found and studied in detail a tachyon free model, and by Alvarez-Gaumé et al. [13]. Narain [14] also studied similar theories.

References

[1] J.H. Schwarz, Phys. Reports 89 (1982) 223;
M.B. Green, Surv. High Energy Phys. 3 (1982) 127

[2] D.J. Gross, J.A. Harvey, E. Martinec and R. Rohm, Phys. Rev. Lett. 54 (1985) 502; Nucl. Phys. B256 (1985) 253, B267 (1986) 75

[3] E. Witten, *in* Anomalies, geometry, topology, ed. W. Bardeen and A. White (World Scientific, Singapore, 1985)

[4] F. Gliozzi, J. Scherk and D.I. Olive, Phys. Lett. 65B (1976) 282; Nucl. Phys. B122 (1977) 253

[5] M. Schiffer and D.C. Spencer, Functionals of finite Riemann surfaces (Princeton Univ. Press, 1954)

[6] A. Erdelyi et al., Higher transcendental functions (McGraw-Hill, New York, 1953)

[7] M.F. Atiyah, Ann. scient. Ec. Norm. Sup. 4 se. 4 (1971) 47

[8] R. Lee and E. Miller, to appear

[9] S. Weinberg, Texas preprint UTTG-22-85

[10] E. Witten, *in* Proc. 1983 Shelter Island Conference, ed N. Khuri *et al.* (MIT Press, 1985)

[11] C. Vafa, Harvard preprint (1986)

[12] L. Dixon and J. Harvey, in preparation

[13] L. Alvarez-Gaumé, P. Ginsparg, G. Moore, and C. Vafa, in preparation

[14] K.S. Narain, Phys. Lett. B169 (1986) 41

Classification of closed-fermionic-string models

H. Kawai, D. C. Lewellen, and S. -H. H. Tye
Newman Laboratory of Nuclear Studies, Cornell University, Ithaca, New York 14853
(Received 20 May 1986)

A classification of ten-dimensional closed-fermionic-string models with world-sheet superconformal invariance, modular invariance, and physical states with proper space-time statistics is presented. This classification is based on an analysis of the spin structures of the Neveu-Schwarz-Ramond fermions and the fermionic variables used in representing the internal-symmetry space.

I. INTRODUCTION

One of the most remarkable features of superstring theories is their near uniqueness; only a very few models appear to be mathematically consistent. Indeed at one time superstrings garnered little interest because it was believed that any such theories with pretensions of being physically realistic unified models must contain gauge and gravitational anomalies.[1] The discovery by Green and Schwarz[2] of at least one exception to this rule, the SO(32) type-I model, opened the floodgates of string research—in particular, the search for other consistent theories. This search met with one notable success. A clever hybrid construction of closed-string theories by Gross, Harvey, Martinec, and Rohm[3] led to two heterotic string models, with the gauge groups $\mathrm{Spin}(32)/Z_2$ or $E_8 \times E_8$ apparently uniquely singled out by the requirement of modular invariance of one-loop amplitudes.

A consistent fermionic string theory should be invariant under reparametrizations, superconformal transformations, and two-dimensional (world-sheet) Lorentz and supersymmetry transformations. In general these symmetries are spoiled by anomalies at the quantum level. The absence of such anomalies impose stringent constraints on string-model building; for example, the vanishing of conformal and superconformal anomalies requires the space-time dimension to be 10 and, if the string possesses an internal gauge symmetry described by fermionic fields, the number of such internal fermion fields must be 32. Modular transformations are discrete reparametrizations (or equivalently global Lorentz transformations) and so modular invariance is included in the requirements listed above. In calculating modular invariant loop amplitudes, the integration over moduli (which describe inequivalent surfaces) should be restricted to the fundamental domain of the modular group. This restriction is the key to the ultraviolet finiteness of string theory so that modular invariance is indeed a necessary requirement of any sensible closed-string theory.

Recently several groups[4–9] have realized that using a fermionic representation for the internal-symmetry group of a closed string and considering sometimes intricate correlations of the spin structures appearing in one-loop amplitudes, new models may be constructed which are also one-loop modular invariant. Some of these, in particular, a tachyon-free model with gauge group $SO(16)\times SO(16)$,[5,6] are possible unified theories; others, however, must be discarded because they have statistics inappropriate for a physical theory; i.e., either the space-time fermions and bosons have the same statistics, or different sectors of the excitations in the internal space appear with opposite statistics and so cannot be sensibly interpreted as giving rise to an internal-symmetry group. In the present work we systematically consider the possibilities for such constructions with periodic or antiperiodic boundary conditions for the fermion fields, finding closed-fermionic-string spin structures which give modular invariant one-loop amplitudes and a projection onto the subspace of physical states which have the appropriate spacetime statistics for a physically sensible theory; i.e., Fermi-Dirac statistics for space-time fermions and Bose-Einstein statistics for space-time bosons and internal-symmetry degrees of freedom. The work of Seiberg and Witten[4] clearly suggests that these requirements, one-loop modular invariance, and proper space-time statistics, are enough to ensure modular invariance of multiloop amplitudes. Our results are summarized in Table I, which we believe probably represents a complete list of possible ten-dimensional closed-fermionic-string models for which the fermionic fields representing the internal-symmetry space are noninteracting and satisfy either periodic or antiperiodic boundary conditions. The gauge group identification for the tachyonic models is tentative, since we have not checked the complete spectra and/or constructed the corresponding vertex operators. Each of the models also has a graviton multiplet (or supermultiplet if space-time supersymmetry is present) which is not included in Table I. The first 12 models are known (see, in particular, the elegant work of Dixon and Harvey[5] in which some of these models were discovered by modifying the bosonic Frenkel-Kac construction of the internal-symmetry group), while the final model is new. Unlike the known models with internal-symmetry groups, this new model has a gauge group of rank 8 rather than 16. A more de-

TABLE I. Ten-dimensional modular-invariant closed-fermionic-string theories with proper space-time statistics. The references are GS, Green and Schwarz (Ref. 12); SW, Seiberg and Witten (Ref. 4); GHMR, Gross, Harvey, Martinec, and Rohm (Ref. 3); DH, Dixon and Harvey (Ref. 5); AGMV, Alvarez-Gaume, Ginsparg, Moore, and Vafa (Ref. 6).

Identification	Number of gauge bosons	Number of massless fermions[a]	Number of tachyons	Space-time SUSY	Chiral	Reference
Type IIA		0	0	Yes	No	GS
Type IIB		0	0	Yes	Yes	GS
Type A		0	1	No	No	SW
Type B		0	1	No	No	SW
SO(32) heterotic	496	496	0	Yes	Yes	GHMR
$E_8\times E_8$ heterotic	496	496	0	Yes	Yes	GHMR
$O(16)\times O(16)$	240	512	0	No	Yes	DH, AGMV
SO(32)	496	0	32	No	No	DH, SW
$O(16)\times E_8$	368	256	16	No	Yes	DH, SW
$O(8)\times O(24)$	304	384	8	No	Yes	DH
$(E_7\times SU_2)^2$	272	448	4	No	Yes	DH
U(16)	256	480	2	No	Yes	DH
E_8	248	496	1	No	No	

[a]Not counting those in the graviton (super)multiplet.

tailed examination of the spectra and properties of these models will be considered elsewhere.

This paper is organized as follows. In Sec. II we collect some general results concerning modular invariance and spin structure which provide the starting point for the discussion following. In Sec. III we find modular invariant, physically sensible spin structures and consider the spectra of these allowed models in Sec. IV. Section V concludes with some overall discussions of our work. Appendix A contains a theorem key to the argument in Sec. III, and Appendix B includes some details extending and clarifying the main discussion.

II. MODULAR INVARIANCE AND SPIN STRUCTURE

The possibilities for closed-fermionic-string theories are severely limited by the symmetries required for their consistency. This is perhaps best exemplified by the heterotic string action which, while it can be rewritten in many forms, is essentially unique and in some sense the most general such action that one can consider. Employing the Neveu-Schwarz-Ramond (NSR) formalism[10] for the space-time fermions, as we will throughout the present work, the covariant first quantized action is[3,11]

$$S(e,\psi,\lambda,X,\chi)=\int d^2\xi e\left[\tfrac{1}{2}g^{mn}\partial_m X^\mu\partial_n X^\mu+\frac{i}{2}\lambda_R^\mu\gamma^m\partial_m\lambda_R^\mu+\frac{i}{2}\psi_{nR}\gamma^m\gamma^n\lambda_R^\mu\left[\partial_m X^\mu-\frac{i}{4}\psi_{mR}\lambda_R^\mu\right]+\frac{i}{2}\chi_L^J\gamma^m\partial_m\chi_L^J\right]. \tag{2.1}$$

In this covariant formalism the zweibein e_m^a and the metric tensor $g_{mn}=e_m^a e_{an}$ appear in the above form to guarantee reparametrization invariance, which is required to eliminate the timelike modes of the string coordinate X^μ. In similar fashion the elimination of the timelike fermionic modes necessitates two-dimensional local supersymmetry which in turn requires the gravitino ψ. In the sector of right moving fields $(X_R^\mu,\lambda_R^\mu,\psi_{nR})$ we must have $\mu=1,2,\ldots,D=10$ in order for the superconformal anomaly to vanish. In the bosonic left-moving sector, the vanishing of the conformal anomaly requires 26 dimensions; here 16 of these dimensions are compactified to produce an internal-symmetry group given in a fermionic representation by the χ_L^J, $J=1,\ldots,32$. If both right- and left-moving sectors are NSR strings the resulting theory is a type-II superstring; if both right- and left-moving sectors are bosonic, one has a pure bosonic theory with an internal-symmetry group. Apart from these modifications (and the possibilities for world-sheet fermion interactions or nontrivial background fields), (2.1) appears to uniquely specify possible closed-string theories. This is not the case, however, because we have yet to specify the boundary conditions on the fermions, i.e., the spin structures. At this level the only constraint that (2.1) places on these is that, due to the form of the interactions [specifically the third term in (2.1)] the right-moving gravitino and the ten right-moving fermions, λ_R^μ, must have the same spin structure. *A priori* the 32 χ_L^J are not so limited, but by examining yet another symmetry, modular invariance of one-loop amplitudes, we will find that all of the spin structures must be intricately related.

To calculate a string one-loop amplitude we must sum over world sheets with the topology of a torus. In the conformal gauge all dependence on the world-sheet metric of the torus is contained in the complex modular parameter τ which parametrizes inequivalent tori. A consistent string theory should not depend on how we choose to parametrize its world sheet. At the one-loop level this requires that amplitudes be invariant under the modular group which consists of discrete reparametrizations (or equivalently global Lorentz transformations) and is generated by the transformations

$$\tau \to \frac{-1}{\tau}, \tag{2.2a}$$

$$\tau \to \tau+1 . \tag{2.2b}$$

Essentially (2.2a) interchanges the σ and t directions on the torus, while (2.2b) corresponds to cutting the torus at some constant t slice, twisting one end through 2π, and reconnecting. The path integrals of the bosonic and fermionic degrees of freedom on the torus must give functions of τ which are invariant under (2.2). The bosonic integration is already modular invariant if the Regge intercept is the usual one chosen to ensure unitarity; our concern here is with the fermionic sector. The fermion path integral depends on what spin structures we choose for the two noncontractable loops (with winding number one) on the torus, i.e., whether the fermions are chosen to have periodic or antiperiodic boundary conditions when traversing the σ or t directions on the torus. More complicated boundary conditions, e.g., some phase other than ± 1, will be considered in a future paper.[13] Under modular transformations, functions of different spin structures may be mapped into each other;[4–9] thus a modular-invariant path integral must, in general, include a sum over spin structures.

To be explicit, let us introduce a notation which will prove useful later on. Z_b^a is the contribution to the vacuum loop path integral from integrating out a single left-moving fermionic degree of freedom. The superscript indicates the spin structure in the σ direction on the torus, $a=0$ (1) for antiperiodic (periodic) boundary conditions corresponding to Neveu-Schwarz (Ramond) sectors. Similarly the subscript b indicates the spin structure in the t direction. The phases of these partition functions may be chosen such that under modular transformations they become,[9] for $\tau \to -1/\tau$,

$$\begin{aligned} &Z_1^1 \to Z_1^1 , \\ &Z_0^0 \to Z_0^0 , \\ &Z_0^1 \to Z_1^0 , \\ &Z_1^0 \to Z_0^1 ; \end{aligned} \tag{2.3a}$$

for $\tau \to \tau+1$,

$$\begin{aligned} &Z_1^1 \to e^{i\pi/12} Z_1^1 , \\ &Z_0^0 \to e^{i\pi/12} Z_1^0 , \\ &Z_0^1 \to e^{i\pi/12} Z_0^1 , \\ &Z_1^0 \to e^{i\pi/12 - i\pi/4} Z_0^0 . \end{aligned} \tag{2.3b}$$

Right-moving fermionic degrees of freedom give results which are the complex conjugates of those in (2.3); notice that as a necessary prerequisite for modular invariance the overall factor of $e^{i\pi/12}$ in (2.3b) must cancel between the left- and right-moving degrees of freedom, giving a constraint on the total number of fermionic modes:

$$N_R - N_L = 0 \quad (\text{mod} 24) . \tag{2.4}$$

Actually (2.4) follows rigorously from (2.3) only if none of the fermions have spin structure $\binom{1}{1}$. This is because Z_1^1 in fact vanishes for a vacuum loop due to the presence of a fermion zero mode; thus if one or more fermions have spin structure $\binom{1}{1}$ then the product of fermion determinants will vanish and modular invariance at this level is trivially satisfied [the transformations for Z_1^1 in (2.3) are hence merely symbolic]. Even so we can make some demands on the allowed spin structure in these sectors by considering the factorization of multiple-loop scattering amplitudes where Z_1^1 need not vanish yet the entire amplitude must still be modular invariant. In addition, we will see in Sec. III and Appendix A that proper space-time statistics requires the presence of sectors of the theory where no fermions have spin structure $\binom{1}{1}$; thus (2.4) will hold. Since the number of fermions must be the same for all spin structures (2.4) is true in all sectors; it follows that, for a consistent theory, the transformation for Z_1^1 given in (2.3b) should hold even for the appropriate scattering amplitudes. The known consistency of the NSR model[10,12] shows that this is in fact the case at least for fermions in groups of eight with the same spin structure.

Earlier we alluded to three general possibilities for closed-string theories; we can check that all of these satisfy (2.4). For type-II and purely bosonic models,[12] where the number of left- and right-moving fermionic modes is the same, this is trivially satisfied. For heterotic-type theories in the light-cone gauge, the gravitino and two fermionic modes drop out leaving 8 right-moving fermions which, with the 32 left-moving fermions, also satisfies (2.4). The methods we will present in Sec. III work equally well for all three types of models; however, we will restrict ourselves to the last case with 8 right-moving and 32 left-moving fermionic modes. The possibilities for type-II theories have been exhausted elsewhere[4] and we merely include the results in Table I. As for the purely bosonic models with internal-symmetry groups we simply mention in passing that from our analysis it is clear that all of these models are tachyonic. In general (leaving aside the possibilities for nontrivial background fields, etc.) the string in $D=4n+2$ space-time dimensions will have $8(6-n)$ left-moving fermions and an equal number of right-moving fermions in the internal-symmetry space.

In Sec. III we will determine the spin structures consistent with (2.3); for now we limit ourselves to deriving one general constraint which follows from (2.3) and (2.4). Applying (2.3b) twice gives (dropping the overall factor which cancels between left- and right-moving sectors)

$$\begin{aligned} &Z_1^1 \to Z_1^1 , \quad Z_0^0 \to e^{-i\pi/4} Z_0^0 , \\ &Z_0^1 \to Z_0^1, \quad Z_1^0 \to e^{-i\pi/4} Z_1^0 . \end{aligned} \tag{2.5}$$

Thus modular invariance requires that in any given contribution to the path integral the number of fermions with spin structure $\binom{0}{0}$ and $\binom{0}{1}$ [denote these by $\Gamma\binom{0}{0}$ and $\Gamma\binom{0}{1}$, respectively] must satisfy,

$$\Gamma\begin{bmatrix}0\\0\end{bmatrix}+\Gamma\begin{bmatrix}0\\1\end{bmatrix}=0 \pmod 8 . \tag{2.6a}$$

Using (2.3) this can be extended to

$$\Gamma\begin{bmatrix}0\\0\end{bmatrix}+\Gamma\begin{bmatrix}1\\0\end{bmatrix}=0 \pmod 8 , \tag{2.6b}$$

$$\Gamma\begin{bmatrix}1\\0\end{bmatrix}+\Gamma\begin{bmatrix}0\\1\end{bmatrix}=0 \pmod 8 . \tag{2.6c}$$

If in some sector $\Gamma\binom{1}{1}=0 \pmod 8$ then (2.6) taken with (2.4) implies that (recall that the 8 right-handed fermions must already have the same spin structure)

$$\Gamma\begin{bmatrix}0\\0\end{bmatrix},\ \Gamma\begin{bmatrix}1\\0\end{bmatrix},\ \Gamma\begin{bmatrix}0\\1\end{bmatrix}=0 \pmod 8$$

$$\text{if } \Gamma\begin{bmatrix}1\\1\end{bmatrix}=0 \pmod 8 . \tag{2.7}$$

In Sec. III we will in fact restrict ourselves to this case:

$$\Gamma\begin{bmatrix}0\\0\end{bmatrix},\ \Gamma\begin{bmatrix}0\\1\end{bmatrix},\ \Gamma\begin{bmatrix}1\\0\end{bmatrix},\ \Gamma\begin{bmatrix}1\\1\end{bmatrix}=0 \pmod 8 . \tag{2.8}$$

The remaining possibility permitted by (2.6), that of spin structures identified in groups of four, will be considered elsewhere.[13]

One final piece of formalism will prove necessary for the discussion of Sec. III. We represent the results of fermion path integrals in Hamiltonian form as a sum over states,[4,12]

$$Z_b^a=\phi_b^a\,\mathrm{Tr}[e^{i\tau\hat{H}_a}(-1)^{b\hat{N}_a}] , \tag{2.9}$$

where

$$\hat{H}_0=\hat{H}_{\rm NS}=2\pi\left[\sum_{r=1/2}^{\infty} r b_r^\dagger b_r-\frac{1}{48}\right] , \tag{2.10a}$$

$$\hat{H}_1=\hat{H}_R=2\pi\left[\sum_{n=1}^{\infty} n d_n^\dagger d_n+\frac{1}{24}\right] , \tag{2.10b}$$

$$\hat{N}_0=\hat{N}_{\rm NS}=\sum_{r=1/2}^{\infty} b_r^\dagger b_r , \tag{2.11a}$$

$$(-1)^{\hat{N}_1}=\pm d_0(-1)^{\sum_{n=1}^{\infty}d_n^\dagger d_n} , \tag{2.11b}$$

and ϕ_b^a is chosen so that (2.3) holds. For our purposes we need only that $(\phi_b^a)^8=(-1)^{(a+b)}$. The Neveu-Schwarz and Ramond Hamiltonians in (2.10) are for single fermionic degrees of freedom. All we need to know about the bosonic contributions for the present work is that for the combined eight bosonic degrees of freedom the contribution to the normal-ordering constant is $-\frac{1}{3}$. Strictly speaking, Eq. (2.11b) is correct only when we consider products of an even number of fermions, i.e., $\prod_{i=1}^{2n}(-1)^{\hat{N}_1}$; for a single fermion it is simply a formal expression whose precise meaning is explained in Appendix B. The $\pm$ in (2.11b) corresponds to left or right chirality in the Ramond sector and may be absorbed into the representation chosen for d_0.

III. CLASSIFICATION OF SPIN STRUCTURES

The most general possible contribution to the one-loop string path integral from the fermionic degrees of freedom is a sum of products of fermion determinants with arbitrary coefficients over all possible combinations of spin structures:

$$Z=\sum_{\{\mathbf{a},\mathbf{b}\}} C_{b_0b_1\cdots b_{32}}^{a_0a_1\cdots a_{32}} Z_{b_0b_1\cdots b_{32}}^{a_0a_1\cdots a_{32}} . \tag{3.1}$$

Here we use a natural generalization of the notation introduced in the preceding section; $Z_{b_0b_1\cdots b_{32}}^{a_0a_1\cdots a_{32}}$ is a product of the fermion determinant from the right-moving modes with spin structure $\binom{a_0}{b_0}$ and the 32 fermion determinants from the left-moving modes with spin structures $\binom{a_i}{b_i}$. Often we will employ an obvious vector notation, e.g., $Z_{\mathbf{b}}^{\mathbf{a}}$. The sum in (3.1) runs over all combinations of a_i and b_j equal to 0 or 1. Explicitly we have

$$Z_{\mathbf{b}}^{\mathbf{a}}=(-1)^{a_0+b_0+(1/8)\sum_{i=1}^{32}(a_i+b_i)}\,\mathrm{Tr}\left[\exp\left[i\tau\sum_{j=1}^{32}\hat{H}_{a_j}-i\bar{\tau}\hat{H}_{a_0}\right](-1)^{\mathbf{b}\cdot\hat{\mathbf{N}}_{\mathbf{a}}}\right] . \tag{3.2}$$

The components of $\hat{\mathbf{N}}_{\mathbf{a}}$ are the appropriate fermion number operators; for the right-moving sector $\hat{H}_{a_0}$ and $\hat{N}_{a_0}$ are the eight fermion analogs of the single fermion Hamiltonian and number operators of (2.10) and (2.11):

$$\hat{H}_{a_0=0}=2\pi\left[\sum_{r=1/2}^{\infty}\sum_{i=1}^{8} r b_r^{\dagger i} b_r^i-\frac{1}{6}\right] , \tag{3.3a}$$

$$\hat{H}_{a_0=1}=2\pi\left[\sum_{n=1}^{\infty}\sum_{i=1}^{8} n d_n^{\dagger i} d_n^i+\frac{1}{3}\right] , \tag{3.3b}$$

$$\hat{N}_{a_0=0}=\sum_{r,i} b_r^{\dagger i} b_r^i , \tag{3.4a}$$

$$(-1)^{\hat{N}_{a_0=1}}=\pm d_0^1 d_0^2\cdots d_0^8(-1)^{\sum_{n,i}d_n^{\dagger i}d_n^i} . \tag{3.4b}$$

The factor of $\frac{1}{8}$ in (3.2) causes no problems since from (2.8)

$$C_{\mathbf{b}}^{\mathbf{a}}=0 \text{ unless } \Gamma\begin{bmatrix}a\\b\end{bmatrix}=0 \pmod 8 \tag{3.5}$$

for any particular choice of $a,b\in Z_2$.

The phase in (3.2) is chosen so that under modular transformations the different $Z^{\mathrm{a}}_{\mathrm{b}}$ transform into each other without picking up any additional phases. For Z to be modular invariant it is necessary and sufficient that the coefficients of terms in (3.1) related by modular transformations be identified. In our notation this condition is [cf. (2.3)]

$$C^{\mathrm{a}}_{\mathrm{b}}=C^{\mathrm{b}}_{\mathrm{a}} \tag{3.6a}$$

for invariance under (2.3a)

$$C^{\mathrm{a}}_{\mathrm{b}}=C^{\mathrm{a}}_{\mathbf{1}+\mathrm{a}+\mathrm{b}} \tag{3.6b}$$

for invariance under (2.3b), where $\mathbf{1}$ is the 33-dimensional vector with all components equal to one, and in such manipulations the subscripts and superscripts are elements of Z_2.

For (3.1) to be interpreted as a sum of $e^{i\tau H}$ over string states the $C^{\mathrm{a}}_{\mathrm{b}}$ must be chosen so that in each physical sector we obtain a sensible projection operator, i.e.,

$$\sum_{\{\mathrm{b}\}}(-1)^{(1/8)\sum_{i=1}^{32}(a_i+b_i)+\mathrm{b}\cdot\mathrm{n}}C^{\mathrm{a}}_{\mathrm{b}}=\eta^{\mathrm{a}}_{\mathrm{n}}=0 \text{ or } 1 . \tag{3.7}$$

For a given collection of fermion numbers (n) a string state either appears in the sum of $e^{i\tau H_{\mathrm{a}}}$ contributing to Z ($\eta^{\mathrm{a}}_{\mathrm{n}}=1$) or it does not ($\eta^{\mathrm{a}}_{\mathrm{n}}=0$). The sign accompanying the term in the sum is determined by requiring the states to have physically sensible statistics. In particular, in writing (3.7) we must define $\mathbf{n}$ to be the eigenvalue of $\hat{\mathbf{N}}$ (mod 2) except for the zeroth component n_0, which differs from the eigenvalue of $\hat{N}_0$ by 1; thus, the phase factor in (3.7) differs from that in (3.2) by a factor of $(-1)^{a_0}$ which ensures the desired statistics. Space-time bosons (Neveu-Schwarz sector, $a_0=0$) contribute to loop amplitudes with a positive sign while space-time fermions (Ramond sector, $a_0=1$) have the opposite statistics, contributing with a negative sign. On the other hand, for the left-moving degrees of freedom the states in the two sectors $a_i=0$ or 1 contribute to the loop amplitude with the same sign, i.e., have the same statistics, as is appropriate for an internal-symmetry group.

Our present goal is to solve (3.7) subject to the condition (3.6). A general solution to (3.7) is easy enough to obtain; multiplying through by $(-1)^{\mathrm{b}'\cdot\mathrm{n}}$ and summing over all $n_i=0,1$ one finds

$$(-1)^{(1/8)\sum_1^{32}(a_i+b_i')}C^{\mathrm{a}}_{\mathrm{b}'}=2^{-33}\sum_{\{\mathrm{n}\}}(-1)^{\mathrm{b}'\cdot\mathrm{n}}\eta^{\mathrm{a}}_{\mathrm{n}} . \tag{3.8}$$

It remains to determine for which choices of $\eta^{\mathrm{a}}_{\mathrm{n}}$ (3.6) holds. Consider first the pure Neveu-Schwarz sector, $\mathbf{a}=\mathbf{0}$. Equation (3.8) becomes (dropping the primes on $\mathbf{b}$)

$$(-1)^{(1/8)\sum_{i=1}^{32}b_i}C^{0}_{\mathrm{b}}=2^{-33}\sum_{\{\mathrm{n}\}}(-1)^{\mathrm{b}\cdot\mathrm{n}}\eta^{0}_{\mathrm{n}} . \tag{3.9}$$

From (3.6b) we know, in particular, that $C^0_0=C^0_1$ which implies that

$$\sum_{\{\mathrm{n}\}}[1-(-1)^{\sum_{i=0}^{32}n_i}]\eta^{0}_{\mathrm{n}}=0 . \tag{3.10}$$

In other words, η^0_{n} vanishes unless

$$\sum_{i=0}^{32}n_i=0 \pmod 2 . \tag{3.11}$$

From (3.6a) we require $C^0_{\mathrm{b}}=C^{\mathrm{b}}_0$ which with (3.8) implies

$$\sum_{\{\mathrm{n}\}}(-1)^{\mathrm{b}\cdot\mathrm{n}}\eta^{0}_{\mathrm{n}}=\sum_{\{\mathrm{n}\}}\eta^{\mathrm{b}}_{\mathrm{n}}\geq 0 \text{ for all } \mathbf{b} . \tag{3.12}$$

Equations (3.5), (3.11), and (3.12) determine the allowed choices for η^0_{n}. Equation (3.9) then determines those choices for $\mathbf{b}$ which give nonvanishing C^0_{b}. It turns out that these choices for $\mathbf{b}$ form a vector space over the field Z_2 (this is shown explicitly in Appendix A). Knowing this we can reconsider our equations, restricting $\mathbf{b}$ to an r-dimensional vector space, with basis vectors $\mathbf{V}^l$,

$$\mathbf{b}=\beta_1\mathbf{V}^1+\cdots+\beta_r\mathbf{V}^r\equiv\boldsymbol{\beta}\cdot V . \tag{3.13}$$

Here and throughout the following discussion, boldface greek letters denote r-dimensional vectors and boldface roman letters 33-dimensional vectors. The only exception is V which is an $r\times 33$ matrix so that $\mathbf{V}^l$ and $\mathbf{V}_i$ are 33- and r-dimensional vectors, respectively ($l=1,\ldots,r$; $i=0,\ldots,32$).

Given (3.13), Eqs. (3.12), (3.7), and (3.6) guarantee that $\eta^{\mathrm{a}}_{\mathrm{n}},C^{\mathrm{a}}_{\mathrm{b}},C^{\mathrm{b}}_{\mathrm{a}}$ are zero unless

$$\mathbf{a}=\boldsymbol{\alpha}\cdot V , \quad \mathbf{b}=\boldsymbol{\beta}\cdot V ; \quad \alpha_l,\beta_l\in Z_2 . \tag{3.14}$$

With these restrictions, Eq. (3.7) becomes

$$\sum_{\{\boldsymbol{\beta}\}}(-1)^{(1/8)\sum_{i=1}^{32}(\boldsymbol{\alpha}+\boldsymbol{\beta})\cdot\mathbf{V}_i+\boldsymbol{\beta}\cdot\boldsymbol{\gamma}}C^{\boldsymbol{\alpha}\cdot V}_{\boldsymbol{\beta}\cdot V}=\eta^{\boldsymbol{\alpha}\cdot V}_{\boldsymbol{\gamma}}=0 \text{ or } 1 ,$$
$$\gamma^l\equiv\mathbf{V}^l\cdot\mathbf{n}\in Z_2 , \tag{3.15}$$

and the analog of (3.8) is

$$(-1)^{(1/8)\sum_{i=1}^{32}(\boldsymbol{\alpha}+\boldsymbol{\beta})\cdot\mathbf{V}_i}C^{\boldsymbol{\alpha}\cdot V}_{\boldsymbol{\beta}\cdot V}=2^{-r}\sum_{\{\boldsymbol{\gamma}\}}(-1)^{\boldsymbol{\beta}\cdot\boldsymbol{\gamma}}\eta^{\boldsymbol{\alpha}\cdot V}_{\boldsymbol{\gamma}} . \tag{3.16}$$

Now $|C^{\boldsymbol{\alpha}\cdot V}_0|=2^{-r}$ from Eq. (A12) in Appendix A, so that for each $\boldsymbol{\alpha}$ one and only one of the $\eta^{\boldsymbol{\alpha}\cdot V}_{\boldsymbol{\gamma}}$ is nonzero. With this in mind (3.16) becomes

$$(-1)^{(1/8)\sum_{i=1}^{32}(\boldsymbol{\alpha}+\boldsymbol{\beta})\cdot\mathbf{V}_i}C^{\boldsymbol{\alpha}\cdot V}_{\boldsymbol{\beta}\cdot V}=2^{-r}(-1)^{\boldsymbol{\beta}\cdot\boldsymbol{\gamma}(\boldsymbol{\alpha})} , \tag{3.17}$$

where $\boldsymbol{\gamma}$ now depends on $\boldsymbol{\alpha}$.

Before enforcing modular invariance to restrict $\boldsymbol{\gamma}(\boldsymbol{\alpha})$, let us determine which choices of basis vectors $\mathbf{V}^l$ are consistent with our requirements. From (3.17) it is clear that $C^{\boldsymbol{\alpha}\cdot V}_{\boldsymbol{\beta}\cdot V}$ is nonzero for any choice of $\boldsymbol{\alpha}$ and $\boldsymbol{\beta}$, so (3.5) places a severe constraint on all but the zeroth component of the basis vectors $\mathbf{V}^l$. Let $\{\mathbf{U}^m\}$ be a basis for the 32-dimensional subspace obtained by eliminating the zeroth component from the vector space spanned by $\{\mathbf{V}^l\}$. We must require that any vector in the subspace have 0 (mod 8) nonzero components and furthermore any two vectors must share a multiple of 8 nonzero components. We may now easily catalog all of the allowed choices for $\{\mathbf{U}^m\}$. Since all of the 32 fermions in the internal-symmetry space are equivalent we need not consider bases that differ only by reordering of the components. From the analysis of Appendix A we know that each basis must contain the vector with all components equal to one, $\mathbf{U}^1=\mathbf{1}$; thus this is the only possible one-dimensional basis. For two-dimensional subspaces we have two choices; we can add a vector with 8 components equal to 1 ($\overline{\mathbf{U}}_i^2=1$ for $i\le 8$; $=0$ for $i>8$) or one with 16 nonzero components ($\mathbf{U}_i^2=1$ for $i\le 16$; $=0$ for $i>16$). The vector with 24 nonzero components is contained in the first subspace ($\mathbf{U}^1+\overline{\mathbf{U}}^2$). For three-dimensional subspaces we can have two "8-vectors" with no ones in common or two "16-vectors" with eight ones in common. Any allowed sub-basis with an 8- and 16-vector is equivalent to a sub-basis with two nonoverlapping 8-vectors. Following such considerations one can easily find the nine distinct bases listed in Table II. Any collection of more than three 8-vectors or five 16-vectors with allowed overlaps are not independent. For the higher-dimensional bases one must consider mutual overlaps between three or more vectors, but a careful check shows that only those subspaces spanned by the vectors in Table II meet our requirements.

The allowed 33-dimensional basis vectors, $\mathbf{V}^l$, are obtained by adding a zeroth component to the $\mathbf{U}^m$ vectors with value 0 and/or 1. From the analysis of Appendix A it is clear that $\mathbf{V}^1=\mathbf{1}$ must always appear. Two distinct cases now arise: either the vector space includes the vector with components $\mathbf{V}_i^0=\delta_{i,0}$ in which case all vectors in the space can have their zeroth component equal to both 0 and 1 (i.e., the zeroth component and the 32-dimensional vectors completely factor) or it does not, in which case the basis vectors $\mathbf{V}^l$ $(l>1)$ coincide with the $\mathbf{U}^l$ with the zeroth component added with value *either* 0 *or* 1 (if both appeared, the sum of the two vectors would be $\mathbf{V}^0$).

Now we return to Eq. (3.17). Knowing from the above discussion that any two vectors in the vector space spanned by $\mathbf{U}^l$ have overlap=0 (mod 8) we can rewrite (3.17) as

$$C^{\alpha\cdot V}_{\beta\cdot V}=2^{-r}(-1)^{\beta\cdot\gamma(\alpha)+(\alpha+\beta)\cdot\rho}\,,\qquad \rho\equiv\left[\tfrac{1}{8}\sum_{i=1}^{32}\mathbf{V}_i\right]\ (\text{mod}2)\,. \tag{3.18}$$

All that remains to be done is to determine which choices of $\gamma(\boldsymbol{\alpha})$ are consistent with modular invariance, which, in the present notation, is the demand that [cf. (3.6)]

$$C^{\alpha\cdot V}_{\beta\cdot V}=C^{\beta\cdot V}_{\alpha\cdot V}\,, \tag{3.19a}$$

$$C^{\alpha\cdot V}_{\beta\cdot V}=C^{\alpha\cdot V}_{(\alpha+\beta)\cdot V+V^1}\,. \tag{3.19b}$$

Applying (3.19a) to (3.18) gives

$$\boldsymbol{\beta}\cdot\gamma(\boldsymbol{\alpha})=\boldsymbol{\alpha}\cdot\gamma(\boldsymbol{\beta})\,. \tag{3.20}$$

This equation must hold for all values of α_l and β_m, for example, choosing $\beta_m=\delta_{lm}$ (3.20) becomes

$$\gamma_l(\boldsymbol{\alpha})=\boldsymbol{\alpha}\cdot\gamma(\beta_l=1)\,. \tag{3.21}$$

In other words, γ_l is a linear function of the α_l's and (3.18) may be written

$$C^{\alpha\cdot V}_{\beta\cdot V}=2^{-r}(-1)^{\sum_{l,m}\beta_l k_{lm}\alpha_m+(\alpha+\beta)\cdot\rho}\,, \tag{3.22}$$

where k_{lm} take values in Z_2 and are independent of α_l and β_l. Now (3.19a) simply requires that

$$k_{lm}=k_{ml}\,. \tag{3.23}$$

Equation (3.19b) applied to (3.22) gives

$$\sum_{l,m}\alpha_l k_{lm}\cdot\alpha_m+\sum_l k_{1l}\alpha_l+\boldsymbol{\alpha}\cdot\rho+\rho_1=0\ (\text{mod}2)\quad\text{for all }\boldsymbol{\alpha}\,. \tag{3.24}$$

This reduces to

$$k_{ll}+k_{1l}+\rho_l=0\ (\text{mod}2)\,, \tag{3.25}$$

where we have used the fact that $\alpha_l^2=\alpha_l$, $\mathbf{V}^1=\mathbf{1}$, and $\rho_1=0$. Inserting (3.22) and (3.2) into (3.1) gives finally

$$Z=2^{-r}\sum_{\{\alpha,\beta\}}(-1)^{\alpha\cdot V_0}\,\text{Tr}\left[\exp\left[i\tau\sum_{j=1}^{32}\hat{H}_{\alpha\cdot V_j}-i\bar{\tau}\hat{H}_{\alpha\cdot V_0}\right](-1)^{\sum_l\beta_l(V_0^l+\sum_m k_{lm}\alpha_m+\sum_i V_i^l\hat{N}_i)}\right]\,. \tag{3.26}$$

For each given choice of the set of basis vectors $\mathbf{V}^l$ and each given set $\{k_{lm}\}$ that is consistent with (3.23) and (3.25), we have a fermionic string model. Its spectrum is generated by the various choices of α_l, β_l, and n_i that are allowed by the constraints.

IV. CLASSIFICATION OF CLOSED-FERMIONIC-STRING MODELS

Equation (3.26) along with (3.23), (3.25), and Table II represent a complete catalog of physically interpretable, closed-fermionic-string spin structures satisfying (2.8). In this section we will demonstrate how to extract from this information the possibilities for string models, in particular, their spectrum of physical states. A given spin structure serves to project out a particular collection of string states. Explicitly, Eq. (3.26) implies that in the sector with Hamiltonian $\sum_{j=0}^{32}\hat{H}_{\alpha\cdot V_j}$ the physical spectrum includes those states whose fermion numbers n_k satisfy

$$V_0^l+\sum_m k_{lm}\alpha_m+\sum_i V_i^l n_i=0\ \ (\text{mod}2)\,. \tag{4.1}$$

The (mass)^2 of a given state depends on its fermion numbers, and on the (mass)^2 of the ground state of the physical sector (represented by $\boldsymbol{\alpha}\cdot V$) which is given by a sum of normal-ordering constants [cf. (2.10)]. Figure 1 gives the ground-state (mass)^2 levels for each sector we need to consider, with each R or N representing eight fermions in a Ramond or Neveu-Schwarz sector. In addition to satisfying (4.1) a physical state must have the same mass in both its right- and left-moving modes as a consequence of in-

TABLE II. Allowed subspace bases. All basis vector components not equal to one vanish.

Basis	Basis vectors U^m
1	U^1 ($U_i^1=1$; $i=1,\ldots,32$)
2	U^1, U^2 ($U_i^2=1$; $i=1,\ldots,16$)
3	U^1, U^2, U^3 ($U_i^3=1$; $i=1,\ldots,8, 17,\ldots,24$)
4	U^1, U^2, U^3, U^4 ($U_i^4=1$; $i=1,\ldots,4, 9,\ldots,12, 17,\ldots,20, 25,\ldots,28$)
5	U^1, U^2, U^3, U^4, U^5, ($U_i^5=1$; $i=4n+1, 4n+2$; $n=0,\ldots,7$)
6	U^1, U^2, U^3, U^4, U^5, U^6 ($U_i^6=1$; $i=$odd)
7	U^1, $\overline{U}^2$ ($\overline{U}_i^2=1$; $i=1,\ldots,8$)
8	U^1, $\overline{U}^2$, $\overline{U}^3$ ($\overline{U}_i^3=1$; $i=9,\ldots,16$)
9	U^1, $\overline{U}^2$, $\overline{U}^3$, $\overline{U}^4$ ($\overline{U}_i^4=1$; $i=17,\ldots,24$)

variance of the theory under shifts in the world-sheet parameter σ. This is enforced by the integration of (3.26) (multiplied by the corresponding contribution of the bosonic fields to the path integral) over τ in the fundamental domain; the result vanishes unless the right $(\text{mass})^2$ and left $(\text{mass})^2$ are equal.

A priori each possible combination of choices of V from Table II and **k** satisfying (3.23) and (3.25) represents a potential new string model. In practice, however, we have found that out of this large number of models only a few are physically distinct; a given gauge group can have many different fermionic representations.

Let us consider a particular example, basis 7 from Table II, and determine the spectrum of tachyons and gauge bosons for the models it represents. First consider the case where the basis of 33-dimensional vectors does not contain $\mathbf{V}^0$. Then (3.23) and (3.25) give

$$\begin{aligned} &k_{11}=0 \text{ or } 1 ,\\ &k_{12}=k_{21}=1+k_{22} . \end{aligned} \tag{4.2}$$

Equation (4.1) becomes

$$\begin{aligned} &1+k_{11}\alpha_1+(1+k_{22})\alpha_2=n_0+N_8+N_{24} \quad (\text{mod}2) ,\\ &V_0^2+k_{22}(\alpha_2+\alpha_1)+\alpha_1=N_8+V_0^2 n_0 \quad (\text{mod}2) , \end{aligned} \tag{4.3}$$

$$N_8\equiv\sum_{i=1}^{8} n_i,\quad N_{24}\equiv\sum_{i=9}^{32} n_i .$$

Notice that in this case the spin structures for the first 8 left-moving fermions are always identified, as are the spin structures for the last 24; the physical states will naturally be grouped in representations of $SO(8)_R\times SO(8)_L \times SO(24)_L$.

Consider first the $N_R(NNNN)_L$ sector (using the notation of Fig. 1). The only way to build this sector from our vector space is to take $\alpha_1=\alpha_2=0$ so that $\boldsymbol{\alpha}\cdot V$ is the 33-dimensional zero vector. Any tachyon present must have $n_0=0$, $N_8+N_{24}=1$. Equation (4.3) then tells us that the tachyons in this sector fall into two possible representations of $SO(8)_R\times SO(8)_L\times SO(24)_L$ depending on the value of V_0^2, $(1,1,24_v)$ or $(1,8_v,1)$ for $V_0^2=0$ or 1, respectively. The massless states in this sector have $n_0=1$, $N_8+N_{24}=2$. From (4.3) the gauge bosons in this sector for both $V_0^2=0$ and 1 are in the representation $(8_v,28,1)+(8_v,1,276)$ since we can have two fermions in either the first 8 or the last 24 modes of the internal space, but not split between the two. The only other bosonic states with $(\text{mass})^2$ less than or equal to zero fall into the $N_R(RNNN)_L$ sector. If $V_0^2=1$ then the vector space does not contain this sector and the only nonpositive mass states are those given above. If $V_0^2=0$ then $\alpha_1=0$, $\alpha_2=1$ are the necessary choice (giving $\boldsymbol{\alpha}\cdot V$ with 8 ones among the 32 left movers, the other components vanishing). Tachyons are present if $n_0=N_{24}=0$ is allowed. N_8 can be either 0 or 1. Since the first 8 left-moving fermions fall into a Ramond sector this just determines which of the two eight-dimensional SO(8) spinor representations the ground state transforms as. From (4.3) one indeed finds tachyons, in the representation $(1,8_s,1)$. $N_{24}=n_0=1$ satisfies (4.3) for $\alpha_1=0$, $\alpha_2=1$ so one finds massless bosons in the representation $(8_v,8_s,24_v)$. Taken together with the results from the pure Neveu-Schwarz sector we find for $V_0^2=0$ that the gauge bosons fill out the adjoint representation of SO(32) while the tachyons form an SO(32) vector. $V_0^2=1$ gives the $SO(8)\times SO(24)$ tachyonic model. In similar fashion one can obtain the massless fermionic states of these models by examining the $R_R(NNNN)_L$ and $R_R(RNNN)_L$ sectors (only the latter appears in the present case and only for $V_0^2=1$).

Adding $\mathbf{V}^0$ to the vector space considered above we can, in a completely analogous manner, find the physical spectrum of another set of models. Here we obtain the same vector space for either choice of V_0^2, but one still finds two different models distinguished now by the value of k_{02}. For $k_{02}=1$ we again find the tachyonic $SO(8)\times SO(24)$ model, though built with a quite different spin structure from the case considered above. If $k_{02}=0$ the model is the tachyon-free SO(32) heterotic string.

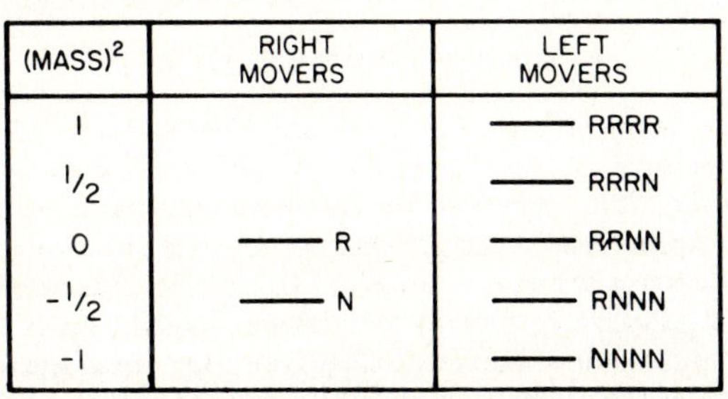

FIG. 1. Ground-state $(\text{mass})^2$ level structure for left- and right-moving modes. R, Ramond (periodic); N, Neveu-Schwarz (antiperiodic).

Bases 8 and 9 in Table II require somewhat more effort to analyze, but the manipulations are completely straightforward. Cases 1 through 6, those bases built from "16-vectors," are easier to work out. In these cases all $\rho_l=0$ and the value of V_0^l, $l>1$, may be chosen to be zero, since $\mathbf{V}^{l'}=\mathbf{V}^1+\mathbf{V}^l$ also appears in the vector space and has all the symmetry properties of $\mathbf{V}^l$ but with $V_0^{l'}=1$. Given this, Eqs. (3.23) and (3.25) become

$$k_{11}=0 \text{ or } 1\ , \qquad k_{1l}=k_{l1}=k_{ll}\ . \tag{4.4}$$

In turn (4.1) reduces to

$$\delta_{l,1}+\delta_{l,0}+\sum_m k_{lm}\alpha_m=\sum_j V_j^l n_j \pmod 2\ . \tag{4.5}$$

The spectra of these models can be considered in detail just as outlined in the example above; here we restrict ourselves to counting the number of tachyons and gauge bosons. It is not difficult to consider bases 1 through 6 simultaneously. For this purpose define p to be the number of basis 16-vectors for the given case. Then $h=2^{5-p}$ is the "block size" in which all the spin structures must be identified, and $q=16/h$ is the number of such blocks in a group of 16. Consider first the case where $\mathbf{V}^0$ is not included in the vector space. The only sector in which tachyons can appear for these models is $N_R(NNNN)_L$ (all $\alpha_l=0$), $n_0=0$ and $\sum_{i=1}^{32} n_i=1$ [$N_R(RNNN)_L$ cannot be obtained with 16-vectors]. With these restrictions, Eq. (4.5) implies that the only n_i which can be nonzero are those such that $V_i^l=0$ for all $l=2,\ldots,p+1$. The number of these, and hence the number of tachyons, is easily seen to be h.

For the massless gauge bosons we must examine two sectors, $N_R(NNNN)_L$ and $N_R(RRNN)_L$, all others give only massive states or do not appear from our vector space. In the former case we need all $\alpha_l=0$, $n_0=1$, and $\sum_{i=1}^{32} n_i=2$. To satisfy (4.5) two $n_i(i\neq 0)$ can be nonzero only if they appear in the same block, thus the total number of gauge bosons in this sector is $qh(h-1)$.

Any nonvanishing combination of α_l, $l\neq 1$, along with $\alpha_1=0$ gives spin structure in the $N_R(RRNN)_L$ sector. Since all such choices are equivalent we may count the number of gauge bosons for the case $\alpha_l=\delta_{l,2}$ and multiply by the number of combinations, $(2q-1)$, to obtain the total number of gauge bosons. For massless bosons we need $n_0=1$ and $\sum_{i=17}^{32} n_i=0$. The first 16 fermions are broken into q blocks of size h, each in the Ramond sector. The total fermion number in each block determines the chirality of the ground-state spinor representation in that block. To get the number of gauge bosons we must multiply a factor of the dimension of the spinor representation $(2^{(h/2)-1})$ for each of the q blocks by the number of fermion number combinations allowed by (4.5). Now (4.5) places p constraints on the total of q fermion numbers (one for each block) leaving 2^{q-p} allowed combinations; thus the number of gauge bosons for $\alpha_2=1$ is $(2^{(h/2)-1})^q 2^{q-p}$. Putting our results together, the total number of gauge bosons for each model is

$$qh(h-1)+(2q-1)(2^{(h/2)-1})^q 2^{q-p}=240+2^{8-p} \tag{4.6}$$

which gives, in order $p=0, 1,\ldots,5$, the last six entries of Table I. Strictly speaking for case 6, where h is one and the dimension of the spinor representation given above is $1/\sqrt{2}$, one should be more careful and directly consider the algebra of the Ramond zero modes, but one finds that (4.6) is in fact the correct result (this is shown in Appendix B).

Massless space-time fermions in these models arise only from the $R_R(RRNN)_L$ sector. The counting of these states is essentially the same as the counting of gauge bosons in this sector given above. The only effective difference is that n_0 may be either 0 or 1 so that the number of massless fermions (beyond those in the graviton multiplet) is twice the result given above,

$$(2q-1)(2^{(h/2)-1})^q 2^{q-p+1}=2^{9-p}(2q-1)\ . \tag{4.7}$$

Adding $\mathbf{V}^0$ to the vector spaces considered above gives from (4.5) an additional equation:

$$n_0=1+\sum_l k_{0l}\alpha_l\ . \tag{4.8}$$

For $\alpha_l=0$ this gives $n_0=1$; i.e., there are no tachyons allowed in the physical spectra of these theories. Upon examination one finds two distinct possibilities in these cases depending on whether all of the k_{0l} vanish or some are nonzero. In either case the counting for massless states in the $N_R(NNNN)_L$ sector proceeds exactly as before. The difference arises in the number of allowed combinations of values of α_l which give the $N_R(RRNN)_L$ sector. If $k_{0l}=0$ for all l then the number of such combinations is twice that found above so that the total number of gauge bosons is

$$qh(h-1)+(2q-1)(2^{(h/2)-1})^q 2^{q-p+1}\ , \tag{4.9a}$$

which turns out to be 496 for all five cases. If any of the k_{0l} are nonzero then regardless of the particular choice of k_{0l}, there are $(2q-2)$ combinations of α_l which produce the $N_R(RRNN)_L$ sector and so a total of

$$qh(h-1)+(2q-2)(2^{(h/2)-1})^q 2^{q-p} \tag{4.9b}$$

gauge bosons. Remarkably this gives a value of 240 for all the cases at hand, producing many copies of the SO(16)×SO(16) model.[5,6] Note as well that bases 1 through 6 in Table II are sufficient to generate all of the models in Table I (aside from the type-II superstrings). Bases 7, 8, and 9 generate only a subset, not including the last three models.

V. DISCUSSION AND REMARKS

To maintain manifest covariance in the path-integral formulation of closed-fermionic strings, it is convenient to use the NSR formalism for the space-time fermions and to express the internal-symmetry group in terms of world-sheet fermionic variables. Invariance under world-sheet superdiffeomorphisms, which include both reparametrization and local supersymmetry invariance, is necessary to remove the timelike components of the string superfield X^μ. The measures needed in the definition of the functional integrals, in general, break world-sheet superconformal and Lorentz invariances. This is the origin of superconformal and Lorentz anomalies. The absence of such anomalies impose stringent constraints on the physically acceptable models. Superconformal invariance requires the space-time dimension to be $D=10$, and the number of Majorana-Weyl fermions in the internal-symmetry space (for models with gauge symmetry) to be 32. The absence of global Lorentz anomalies implies that string models must be modular invariant.[9] In fact, modular invariance is the key ingredient to the ultraviolet finiteness property of string theories. With superdiffeomorphism, superconformal, and Lorentz invariances, all potential sources of anomalies are removed. Hence the resulting string models should be completely anomaly-free. In particular, space-time gravitational and gauge anomalies are expected to be absent in these string models.

To obtain a sensible fermionic-string model, the space-time fermions must have Fermi-Dirac statistics while all internal-symmetry degrees of freedom and space-time bosons must have Bose-Einstein statistics. This condition plus one-loop modular invariance impose stringent constraints on physically interesting string models. As pointed out by Seiberg and Witten[4] and others, these conditions may also be sufficient to guarantee multiloop modular invariance. In this case our classification of ten-dimensional closed-fermionic-string models is complete provided we only consider periodic or antiperiodic boundary conditions for the fermionic fields, (2.8) is satisfied, and the world-sheet fermions in the internal symmetry space remain noninteracting. While we have not yet exhausted the possibilities for more complicated boundary conditions for the fermionic fields, our results so far suggest that Table I probably includes all models.[13]

Some of the models in our classification do not have space-time supersymmetry and some have tachyons. The absence of space-time supersymmetry implies that a nonzero cosmological constant Λ will be present at the quantum level, as, for example, has been explicitly calculated for the SO(16)$\times$SO(16) model where Λ is finite and positive.[5,6] This clearly indicates that the dilaton tadpole is not zero. The presence of tachyons, as well as the presence of nonzero dilaton one-point functions, are clear signs of vacuum instability at either the classical or quantum level. Since strings that are stable and remain in $D=10$ are physically undesirable, one may go so far as to consider the presence of tachyons and/or a dilaton tadpole as virtues rather than fatal flaws. It will be most interesting to analyze the nonsupersymmetric string models to see if there exist compactifications that result in effectively $D=4$, $N=1$ supersymmetric models with realistic gauge groups and fermion contents.

Dixon and Harvey[5] have given a simple, elegant classification of fermionic string models based on the bosonic formulation of the internal-symmetry space. Their approach, which classifies inequivalent shift vectors on the group lattice, generates all the models except the E_8 model. This is not surprising since in our analysis all of the models with the exception of the E_8 one have spin structures of pairs of fermions identified so that the fermionic variables can be easily bosonized. The E_8 model involves only single fermions and so does not readily emerge from a bosonic formulation. Dixon and Harvey observe, however, that there is one other possible set of models involving the outer automorphism of $E_8 \otimes E_8$, which exchanges the two E_8's. The E_8 model generated by the spin-structure method may belong to this set. It will be interesting to understand the relation between these two different but complementary approaches.

Finally it should be pointed out that new closed-fermionic-string models in fewer than ten dimensions may be constructed by an extension of the method described in this paper. It should be obvious to the reader that our analysis is a generalization of the method of Gliozzi, Scherk, and Olive.[14] A further generalization allows us to construct, among other models, $N=1$ supersymmetric chiral string models in four space-time dimensions. The results will be presented elsewhere.[13]

Note added. Lance Dixon and Jeffrey Harvey informed us that they have also constructed the E_8 model from their approach. We thank them for useful discussions.

ACKNOWLEDGMENTS

The authors would like to thank P. Ginsparg for useful discussions. This work was supported by the National Science Foundation.

APPENDIX A

Here we demonstrate that in order to satisfy Eqs. (3.9), (3.11), and (3.12) the values of $\mathbf{b}$ for which $C_{\mathbf{b}}^0$ is nonzero must form a linear vector space over the field Z_2.

First note that if only one of the $\eta_{\mathbf{n}}^0$ is one then (3.12) implies that $\mathbf{b}\cdot\mathbf{n}=0$ for all $\mathbf{b}$, i.e., $\mathbf{n}=0$. Similarly if only $\eta_{\mathbf{n}_1}^0$ and $\eta_{\mathbf{n}_2}^0$ are nonzero then one of them again must be η_0^0 and Eq. (3.9) becomes

$$(-1)^{(1/8)\sum_{i=1}^{32} b_i} C_{\mathbf{b}}^0 = 1+(-1)^{\mathbf{b}\cdot\mathbf{n}_1} . \tag{A1}$$

In other words,

$$C_{\mathbf{b}}^0=0 \quad \text{unless } \mathbf{b}\cdot\mathbf{n}_1=0 \ (\text{mod}2) . \tag{A2}$$

We have a linear equation for the allowed $\mathbf{b}$ over the field Z_2.

Now if some number M of the $\eta_{\mathbf{n}_l}^0$ are equal to one then (3.12) becomes

$$\sum_{l=1}^{M} (-1)^{\mathbf{b}\cdot\mathbf{n}_l} \geq 0 \quad \text{for all } \mathbf{b} . \tag{A3}$$

In general only some number s of the exponents in (A3) will be linearly independent functions of the b_i (mod2); designate these as B_j, $j=1,\ldots,s$. We wish to show that the left-hand side of (A3) (which we will call F_s) must be of the form

$$F_s(B_1,\ldots,B_s)=\prod_{j=1}^{s}[1+(-1)^{B_j}] . \tag{A4}$$

If this is the case then $C_{\mathbf{b}}^0$ will be nonvanishing if and only if $\mathbf{b}$ satisfies

$$B_1=B_2=\cdots=B_s=0 \ (\text{mod}2) . \tag{A5}$$

This is our desired result, the allowed $\mathbf{b}$ being the vector space of solutions of the system (A5) of s-independent linear homogeneous equations on Z_2.

Our proof of (A4) proceeds by induction. From (A1) it is clear that (A4) holds for $s=1$. Assume that it holds for $s=N-1$. For $s=N$ (A3) can be written

$$\overline{G}(B_1,\ldots,B_{N-1})+(-1)^{B_N}G(B_1,\ldots,B_{N-1})\geq 0 \quad \text{for all } B_j=0 \text{ or } 1 . \tag{A6}$$

Comparing (A6) for $B_N=0$ and 1 we see that $\overline{G}$ satisfies

$$\overline{G}(B_1,\ldots,B_{N-1})\geq 0 \ \text{ for all } B_j,\ j=1,\ldots,N-1, \tag{A7}$$

So $\overline{G}$ is the solution to (A3) for $S=N-1$ which by assumption takes the form F_{N-1}. Now if $B_1=1$ then F_{N-1} vanishes for every choice of B_j, $j=2,\ldots,N-1$, and (A6) gives

$$(-1)^{B_N}G(1,B_2,\ldots,B_{N-1})\geq 0 \quad \text{for all } B_j,\ j=2,\ldots,N . \tag{A8}$$

For this to be true for $B_N=0$ or 1 we must have

$$G(1,B_2,\ldots,B_{N-1})=0 \quad \text{for all } B_j,\ j=2,\ldots,N-1 . \tag{A9}$$

But we can always write

$$G=G_0(B_2,\ldots,B_{N-1})+(-1)^{B_1}G_1(B_2,\ldots,B_{N-1}) , \tag{A10}$$

so that (A9) implies $G_0=G_1$ or

$$G=[1+(-1)^{B_1}]G_0(B_2,\ldots,B_{N-1}) . \tag{A11}$$

There was, however, nothing special about B_1 so we must in fact have $G=F_{N-1}$ so that the left-hand side of (A6) becomes F_N, the desired result.

Finally we note two other results assumed in Sec. III which follow immediately from the present analysis. First from (3.11) if $\eta_{\mathbf{n}}^0$ is nonzero then $\mathbf{n}$ has an even number of nonzero components so that the vector $\mathbf{b}$ with all components equal to one is always present in the vector space of allowed solutions of (A5). Second it follows from (3.9) and (A4) that

$$|C_0^{\mathbf{b}}|=|C_{\mathbf{b}}^0|=2^{-r} , \tag{A12}$$

if $\mathbf{b}$ is in the vector space of solutions of (A5) (and vanishes otherwise).

APPENDIX B

Here we derive the factor $(2^{(h/2)-1})^q(2^{q-p})=2^{8-p}$ (for $p=1,2,\ldots,5$) that appears in Eq. (4.6) and in the process clarify our use of formal expressions such as (2.11b). For the purpose of constructing the internal-symmetry group with fermionic variables, we use the reducibility property of the Ramond sector by employing Weyl projections of the 32 internal Majorana-Weyl fermions. For the basis vector U^2, the Weyl projection is $\frac{1}{2}(1+\Gamma)$, where

$$\Gamma_1=\Gamma(p=1)=\gamma_1\gamma_2\cdots\gamma_{16} . \tag{B1}$$

Here γ_i is the zero mode of the Ramond fermion $\chi_i[d_0^i$ in the notation of (2.11)]. This means the $\Gamma(p=1)=1$ states are physical. The dimension of the spinor representation of SO(16) is 128, which is the number of states for the $p=1$ case. For $p=2$,

$$\Gamma_2=\gamma_1\gamma_2\cdots\gamma_8=1 . \tag{B2}$$

Equations (B1) and (B2) together imply

$$\Gamma_2'=\gamma_9\gamma_{10}\cdots\gamma_{16}=1 . \tag{B3}$$

The spinor representation of SO(8)$\times$SO(8) is $(8_s,8_s)$ which has 64 states. For $p=3$, we have

$$\Gamma_3=\gamma_1\gamma_2\gamma_3\gamma_4\gamma_9\gamma_{10}\gamma_{11}\gamma_{12}=1 . \tag{B4}$$

Comparing with (B2), we see that

$$\Gamma_3'=\gamma_1\gamma_2\gamma_3\gamma_4=a=\pm 1 . \tag{B5}$$

Comparing with (B2) and (B3), we find that

$$\gamma_5\gamma_6\gamma_7\gamma_8=\gamma_9\gamma_{10}\gamma_{11}\gamma_{12}=\gamma_{13}\gamma_{14}\gamma_{15}\gamma_{16}=a . \tag{B6}$$

Since the dimension of the spinor representation of SO(4) is 2, the number of states is $2(2^4)=32$ where the first factor comes from the two choices of a. For $p=4$, we have

$$\Gamma_4=\gamma_1\gamma_2\gamma_5\gamma_6\gamma_9\gamma_{10}\gamma_{13}\gamma_{14}=1 . \tag{B7}$$

We can freely choose

$$\begin{aligned} \gamma_1\gamma_2&=b , \\ \gamma_5\gamma_6&=c , \\ \gamma_9\gamma_{10}&=d , \end{aligned} \tag{B8}$$

where b, c, $d=\pm 1$. Then (B1) to (B7) give

$$\begin{aligned} \gamma_3\gamma_4&=ab , \\ \gamma_7\gamma_8&=ac , \\ \gamma_{11}\gamma_{12}&=ad , \\ \gamma_{13}\gamma_{14}&=bcd , \\ \gamma_{15}\gamma_{16}&=abcd . \end{aligned} \tag{B9}$$

Since the spinor representation of SO(2) is 1, the number of states is given by (choices of a, b, c, and d) $2^4=16$. Finally we come to the $p=5$ case, where

$$\Gamma_5=\gamma_1\gamma_3\gamma_5\gamma_7\gamma_9\gamma_{11}\gamma_{13}\gamma_{15}=1 . \tag{B10}$$

In contrast with the earlier cases, $\gamma_{2i-1}\gamma_{2i}$ $(i=1,2,\ldots,8)$ from the $p=4$ case anticommutes with Γ_5:

$$\{\gamma_{2i-1}\gamma_{2i},\Gamma_5\}=0 \tag{B11}$$

while

$$[\gamma_{4l+1}\gamma_{4l+2}\gamma_{4l+3}\gamma_{4l+4},\Gamma_5]=0 \tag{B12}$$

for $l=0,1,2,3$. For example, let us consider the eigenstates of $\gamma_1\gamma_2$,

$$\gamma_1\gamma_2|\pm\rangle=\pm|\pm\rangle \tag{B13}$$

then

$$\gamma_1\gamma_2\Gamma_5|\pm\rangle=-\Gamma_5\gamma_1\gamma_2|\pm\rangle=\mp\Gamma_5|\pm\rangle ,$$

that is

$$\Gamma_5|\pm\rangle=|\mp\rangle . \tag{B14}$$

Then $(1/\sqrt{2})(|+\rangle+|-\rangle)$ is an eigenstate of Γ_5 with eigenvalue 1. In general

$$\frac{1}{\sqrt{2}}(|b,ab,c,ac,d,ad,bcd,abcd\rangle+|-b,-ab,-c,-ac,-d,-ad,-bcd,-abcd\rangle) \tag{B15}$$

is an eigenstate of Γ_5. A simple counting gives $\frac{1}{2}(2^4)=8$ states. Therefore the $(1/\sqrt{2})$ dimensional spinor representation can be interpreted as this projection of one component eigenstates of $\gamma_{2i-1}\gamma_{2i}$ to an eigenstate of Γ_5.

---

[1]L. Alvarez-Gaume and E. Witten, Nucl. Phys. **B234**, 269 (1983).

[2]M.B. Green and J.H. Schwarz, Phys. Lett. **149B**, 117 (1984); **151B**, 21 (1985).

[3]D.J. Gross, J.A. Harvey, E. Martinec, and R. Rohm, Phys. Rev. Lett. **54**, 502 (1985); Nucl. Phys. **B256**, 253 (1985).

[4]N. Seiberg and E. Witten, Nucl. Phys. **B276**, 272 (1986).

[5]L.J. Dixon and J.A. Harvey, Nucl. Phys. **B274**, 93 (1986). See also L. Dixon, J. Harvey, C. Vafa, and E. Witten, *ibid.* **B261**, 678 (1985); **B274**, 285 (1986).

[6]L. Alvarez-Gaume, P. Ginsparg, G. Moore, and C. Vafa, Phys. Lett. **171B**, 155 (1986).

[7]L. Alvarez-Gaume, G. Moore, and C. Vafa, Harvard University Report No. HUTP-86/A017, 1986 (unpublished).

[8]E. D'Hoker and D.H. Phong, Columbia University Report No. CU-TP-340, 1986 (unpublished); W. Nahm, Report No. Bonn-HE-86-01, 1986 (unpublished).

[9]S. Chaudhuri, H. Kawai, and S.-H.H. Tye, Cornell University Report No. CLNS-86/723, 1986 (unpublished).

[10]A. Neveu and J.H. Schwarz, Nucl. Phys. **B31**, 86 (1971); P. Ramond, Phys. Rev. D 3, 2415 (1971).

[11]L. Brink, P. DiVecchia, and P. Howe, Phys. Lett. **65B**, 471 (1976); S. Deser and B. Zumino, *ibid.* **65B**, 369 (1976).

[12]For a review, see, e.g., J.H. Schwarz, Phys. Rep. **89**, 223 (1982); M.B. Green, Surv. High Energy Phys. 3, 127 (1983).

[13]H. Kawai, D.C. Lewellen, and S.-H.H. Tye, Phys. Rev. Lett. **57**, 1832 (1986).

[14]F. Gliozzi, J. Scherk, and D. Olive, Nucl. Phys. **B122**, 253 (1977).

TEN-DIMENSIONAL HETEROTIC STRINGS FROM NIEMEIER LATTICES

W. LERCHE[a], D. LÜST[b,1,2] and A.N. SCHELLEKENS[a]
[a] *CERN, CH-1211 Geneva 23, Switzerland*
[b] *CALTECH, Pasadena, CA 91125, USA*

Received 5 September 1986

It is shown how all known ten-dimensional heterotic string theories with rank-16 gauge groups can be derived in a direct way from Niemeiers classification of euclidean even self-dual lattices in 24 dimensions. This result is based on a recent construction of heterotic strings from odd self-dual lorentzian lattices. A natural embedding of all these theories in the 26-dimensional closed bosonic string is suggested by this construction.

1. Introduction. Although the consistency requirements which string theories have to satisfy are quite restrictive, it has become clear that there are more solutions than one originally expected [1–4]. At the same time, it has become clear that many string theories are closely related. In particular, it has been shown in ref. [5] that a certain class of heterotic strings with rank-16 gauge group is related to odd lorentzian self-dual lattices. The uniqueness of those lattices guarantees that any such lattice can be Lorentz rotated into any other.

Although the possibility of making Lorentz rotations suggests a continuous infinity of new ten-dimensional theories (as in ref. [6]), there is actually only a discrete set of theories that makes physical sense, as we will explain below. The purpose of this paper is to construct all lattices corresponding to sensible ten-dimensional heterotic string theories. There is no need to consider GSO-projections or lattice shifts at all. We begin with a brief summary of the results obtained in ref. [5].

2. Odd self-dual lattices. The heterotic strings considered in ref. [5] are constructed out of the usual ten-dimensional left- and right-moving bosonic coordinates X^μ plus 16 left- and 6 right-moving bosons, compactified on a lattice $\Gamma_{16;5,1}$, with signature $(16(-), 5(+), -)$. The first 16 bosons produce a gauge group G by the Frenkel–Kac mechanism. The last six produce covariant Neveu–Schwartz or Ramond theories, and can be thought of as a bosonization of NS or R world sheet fermions and ghosts [7]. We can write the vectors on the lattice as $\boldsymbol{w}=(\boldsymbol{\lambda}_{\rm L},\boldsymbol{\lambda}_{\rm R},q)$, where $\boldsymbol{\lambda}_{\rm L}$ is a 16-dimensional weight vector of G, $\boldsymbol{\lambda}_{\rm R}$ a five-dimensional weight of the Wick rotated Lorentz group SO(10), and q the ghost charge. The lattice contains far too many states, because classes of states whose ghost charges differ by an integer are in fact equivalent. The light-cone partition function for the physical states (not including oscillators) can be obtained by dividing the partition function of $\Gamma_{16;5,1}$ by the partition function of the even lorentzian self-dual lattice $\Gamma_{1,1}$. These partition functions are given by [5]

$$\theta_{16;5,1}(\tau,\bar\tau)=\sum_{w\in\Gamma_{16;5,1}}\exp(-\pi i\bar\tau\lambda_{\rm L}^2)\times\exp[\pi i\tau(\lambda_{\rm R}^2-q^2-2q-2)]\exp(2\pi iq)\,,$$

$$\theta_{1,1}(\tau)=\sum_{w\in\Gamma_{1,1}}\exp(\pi i\tau w^2)\,. \tag{2.1}$$

Notice that $\theta_{1,1}$ depends only on τ(not $\bar\tau$), because it only describes right-movers. Although these two functions are separately ill-defined because of the

[1] Supported by a scholarship of the Max Kade Foundation, New York.

[2] Address after October 1, 1986: Max-Planck-Institut für Physik und Astrophysik, Föhringer Ring 6, D-8000 Munich 40, Fed. Rep. Germany.

lorentzian metric for the right-movers, their ratio is just the light-cone partition function. After dividing by $\theta_{1,1}$, the states remaining in the light-cone partition function are those for which the last two entries of $\boldsymbol{w}$ have fixed values [8]:

$$(...0,\,-1) \quad \text{or} \quad (...,1/2,\,-1/2)\,. \tag{2.2}$$

Thus we can select the physical states out of the lattice by keeping only states with these last two entries. The masses of these states are given by $m^2 = m_L{}^2/2 + m_R{}^2/2$,

$$\tfrac{1}{8}m_L^2 = \tfrac{1}{2}\lambda_L^2 + N_L - 1\,,$$

$$\tfrac{1}{8}m_R^2 = \tfrac{1}{2}\boldsymbol{u}^2 + N_R - \tfrac{1}{2}\,, \tag{2.3}$$

where $\boldsymbol{u}$ contains the first four entries of $\boldsymbol{\lambda}_R$, and we have included the occupation numbers N_L and N_R of the bosonic oscillator modes. Furthermore, physical states have to satisfy $m_L^2 = m_R^2$.

To check modular invariance of these theories it suffices to consider the transformations $\tau \to \tau + 1$ and $\tau \to -1/\tau$. The partition function is invariant under $\tau \to \tau + 1$ if [5]

$$\lambda_L^2 - \lambda_R^2 + q^2 + 2q = 0 \bmod 2\,. \tag{2.4}$$

For the second transformation one finds that the partition function transforms properly if $\Gamma_{16;5,1}$ is self-dual, apart from a phase factor $\exp(4\pi i q)$, which is irrelevant only if q is integer or half-integer.

It is thus clear that the theory is not modular invariant for arbitrary SO(17,5) Lorentz transformations of $\boldsymbol{w}$. Now, to impose sensible space–time properties we furthermore require that q is integer if $\boldsymbol{\lambda}_R$ is a vector weight of SO(10), and half-integer if $\boldsymbol{\lambda}_R$ is a spinor weight. The factor $\exp(2\pi i q)$ in (2.1) ensures then that space–time bosons and fermions contribute with a relative minus sign to the partition function, so that the spin-statistics theorem is satisfied. This, together with (2.4), is all we need to know to classify the sensible solutions.

3. Classification. We can write the SO(10) weight $\boldsymbol{\lambda}_R$ as $(\boldsymbol{u}, p)$, where $\boldsymbol{u}$ is a SO(8) weight and p is an integer or half-integer. Because of the spin-statistics relation mentioned above, p and q are either both integer or half-integer. Since a self-dual lattice contains all vectors having integer dot product with all lattice vectors, if must certainly contain $(\boldsymbol{\lambda}_L, \boldsymbol{u}, p, q) = (0,\,0,\pm 1,\pm 1)$. By adding linear combinations of these vectors we can always restrict p and q to have the values (0, 0), (0, −1), (1/2, −1/2) and (1/2, 1/2). These vectors can be regarded as representatives of "conjugacy classes" of the lattice $\Gamma_{1,1}$. We denote these classes as (0), (v), (s) and (c), respectively. Notice that because of the physical state selection rule (2.2) only classes (v) and (s) contain physical states. In exactly the same way one can see that the lattice must contain the vectors $(0, \boldsymbol{r}, 0, 0)$ of the SO(8) root lattice, so that we can restrict $\boldsymbol{u}$ to be a representative of the SO(8) conjugacy classes, which we again denote as (0), (v), (s) and (c). The lattice thus consists of the following sets of vectors

$$(\Delta_{00}, 0, 0)\,, \quad (\Delta_{0v}, 0, v)\,,$$

$$(\Delta_{v0}, v, 0)\,, \quad (\Delta_{vv}, v, v)\,,$$

$$(\Delta_{sc}, s, c)\,, \quad (\Delta_{ss}, s, s)\,,$$

$$(\Delta_{cc}, c, c)\,, \quad (\Delta_{cs}, c, s)\,. \tag{3.1}$$

Here $\Delta_{\alpha\beta}$ denotes a set of vectors $\boldsymbol{\lambda}_L$ associated with the SO(8) conjugacy class α and the $\Gamma_{1,1}$ conjugacy class β.

It is now easy to see that (2.4) is satisfied if and only if Δ_{v0} and Δ_{0v} contain only vectors of odd length and all other $\Delta_{\alpha\beta}$ contain only vectors of even length. In this case, (3.1) defines indeed an odd, integral lattice: for example, (Δ_{ss}, s, s) contains only vectors of odd length.

To construct a self-dual lattice $\Gamma_{16;5,1}$ with the required properties we have to find a combination of sets of vectors $\Delta_{\alpha\beta}$ with:

1. $(\Delta_{\alpha\beta}, \Delta_{\gamma\delta}) - (\alpha, \gamma) - \langle \beta, \delta \rangle \in \mathbb{Z}$.

2. Δ_{0v}, Δ_{v0} odd, all others even.

3. If $\boldsymbol{w} = (\boldsymbol{\lambda}_L, \boldsymbol{u}, s)$ and $(\boldsymbol{\lambda}_L, \Delta_{\alpha\beta}) - (\boldsymbol{u}, \alpha) - \langle \boldsymbol{s}, \beta \rangle \in \mathbb{Z}$

   for all α and β, then $w \in \Gamma_{16;5,1}$. (3.2)

Here (,) denotes an euclidean and ⟨ , ⟩ a lorentzian dot product with metric (1, −1). The crucial observation is now that in $\Gamma_{1,1}$ the conjugacy classes have the following mutual dot products:

$$\langle 0, \alpha \rangle \in \mathbb{Z} \quad (\alpha = 0, v, c, s) ,$$

$$\langle v, s \rangle \in \mathbb{Z} + 1/2 ,$$

$$\langle v, c \rangle \in \mathbb{Z} + 1/2 ,$$

$$\langle s, c \rangle \in \mathbb{Z} + 1/2 . \qquad (3.3)$$

These are precisely the same as the corresponding euclidean dot products of the SO(8) conjugacy classes. This is not so for the lengths of the vectors. Here we find for $\Gamma_{1,1}$ and SO(8), respectively,

$$\langle 0, 0 \rangle \in 2\mathbb{Z} , \quad (0, 0) \in 2\mathbb{Z} ,$$

$$\langle v, v \rangle \in 2\mathbb{Z} + 1 , \quad (v, v) \in 2\mathbb{Z} + 1 ,$$

$$\langle s, s \rangle \in 2\mathbb{Z} , \quad (s, s) \in 2\mathbb{Z} + 1 ,$$

$$\langle c, c \rangle \in 2\mathbb{Z} , \quad (c, c) \in 2\mathbb{Z} + 1 . \qquad (3.4)$$

We can use these facts to relate the problem of satisfying (3.2) to a problem already solved in the mathematical literature. Suppose one has a lattice (3.1) obeying all three conditions. Now regard the third component of $(\Delta_{\alpha\beta}, \alpha, \beta)$ not as a $\Gamma_{1,1}$ weight, but as a SO(8) weight in the "same" conjugacy class. Furthermore, replace the metric by an euclidean one on the whole lattice. Then the resulting lattice Γ_{24} is an even self-dual lattice of dimension 24. It is even, because in going from $\Gamma_{1,1}$ to SO(8) all weights change in precisely the right way. It is integral, because all three terms in (3.2) remain unchanged, and since they are integer or half-integer, the relative signs do not matter. It is self-dual, because if $\boldsymbol{v} \notin \Gamma_{24}$ has integral dot products with all $\boldsymbol{v} \in \Gamma_{24}$, one can easily construct a vector $\boldsymbol{w} \notin \Gamma_{16;5,1}$ with integral dot product with all $\boldsymbol{w} \in \Gamma_{16;5,1}$, contradicting the fact that $\Gamma_{16;5,1}$ was self-dual.

All 24-dimensional even euclidean self-dual lattices have been classified by Niemeier [9]. To use these lattices for our purpose we must decompose each vector $\boldsymbol{w}$ on such a lattice as $(\boldsymbol{\lambda}, \boldsymbol{\alpha}, \boldsymbol{\beta})$, where $\boldsymbol{\alpha}$ and $\boldsymbol{\beta}$ are SO(8) weights and $\boldsymbol{\lambda}$ is a weight of the remaining symmetry (gauge) group. This is only possible if the group G associated with the Niemeier lattice contains SO(8)×SO(8) in a regular embedding. In particular, this rules out all embeddings in SU(N). Furthermore, only 8 of the 16 conjugacy classes of SO(8)×SO(8) should appear because of

Table 1
Contents of the four sectors.

SO(16)	SO(8) ×SO(8)	Physical	$m_R^2/8$	Spin
(0)	(0, 0)	no		
	(v, v)	yes	0, 1, 2,...	odd-rank tensor
(v)	$(v, 0)$	no		
	$(0, v)$	yes	– 1/2, 1/2,...	even-rank tensor
(s)	(c, c)	no		
	(s, s)	yes	0, 1, 2,...	left spinor
(c)	(s, c)	no		
	(c, c)	yes	0, 1, 2,...	right spinor

the spin-statistics relation. This rules out any embedding of the two SO(8)'s in separate simple factors of G, for the following reason. Every embedding of SO(8) produces at least the conjugacy class (0). If in addition it contains (s) or (c), then combination with the second SO(8) is certain to produce the inacceptable combinations $(0, s)$ or $(0, c)$. Hence the only acceptable SO(8) decompositions of a simple factor of G are (0) or $(0+v)$. However, embedding two SO(8) factors in this way, we would get a lattice consisting only of vectors $(\Delta_{00}, 0, 0)$, $(\Delta_{0v}, 0, v)$, $(\Delta_{v0}, v, 0)$ and (Δ_{vv}, v, v). But such a lattice cannot be self-dual: since Δ_{00} and Δ_{vv} have even vectors and Δ_{0v} and Δ_{v0} only odd ones, the vectors $(\Delta_{vv}, 0, v)$, $(\Delta_{v0}, 0, 0)$, etc. are examples of vectors which are not on the lattice, but have integer dot products with it.

Thus the only remaining possibility is the embedding SO($2N$) ⊃ SO($2N$–16)×SO(16) or E_8 ⊃ SO(16), with SO(16) ⊃ SO(8)×SO(8). Consequently, the conjugacy classes of the two SO(8) are correlated, and the lattice can be written in the form $(\Delta_0, 0) + (\Delta_v, v) + (\Delta_s, s) + (\Delta_c, c)$, where the second entry denotes a SO(16) conjugacy class. Comparing with (3.1) we have, therefore

$$\Delta_{00} = \Delta_{vv} = \Delta_0 , \quad \Delta_{0v} = \Delta_{v0} = \Delta_v ,$$

$$\Delta_{ss}\Delta_{cc} = \Delta_s , \quad \Delta_{sc} = \Delta_{cs} = \Delta_c ,$$

Table 1 summarizes the content of the various sectors. The combination $\Delta = \Delta_0 + \Delta_v + \Delta_s + \Delta_c$ must itself form a lattice, and Δ_0 must contain the dual of that lattice. If Δ is (part of) a weight lattice of a group G, Δ_0 contains the root lattice, because of $\Delta_0 = \Delta_{vv}$ the presence of massless G gauge bosons is automatically guaranteed. The sector (v) is the only one that may contain tachyons; this is the case if Δ_v has vectors of length 1. The massless states in the other sectors are always due to vectors of length 2. A theory has chiral fermions if $\Delta_s \neq \Delta_c$ and Δ_s or Δ_c has vectors of length 2. Absence of anomalies is guaranteed by modular invariance [10].

To find all relevant possibilities for Γ_{24}, we consider the Niemeier lattices [9]. Each one of these is characterized by the root lattice of some semi-simple algebra of rank 24, together with certain weights. All such root lattices containing $D_8 \sim SO(16)$ are given in the first column of table 2. The second column gives the conjugacy classes of the additional weights. These are easily obtained from the "glue vectors" tabulated in ref. [11]. The D_8 sublattices, the remaining gauge groups G_{het} and the conjugacy classes of G_{het} appearing in $\Delta_0,...,\Delta_c$ can be read off directly from the first two columns, and are presented in the other columns. We get precisely all the theories obtained with the bosonic construction of ref. [2] and no more. It is a priori clear (since we can obtain only rank-16 groups) that the fermionically constructed E_8 theory of ref. [4] (suggested in ref. [2] and perhaps identical to the heterotic string constructed in ref. [12]) does not fit in our framework.

It is straightforward to apply this method to type II superstrings. One starts with a lorentzian odd self-dual lattice $\Gamma_{5,1;5,1}$, which can be mapped onto an euclidean even self-dual lattice Γ_{16}. This lattice can only be $E_8 \times E_8$ or D_{16}. From $E_8 \times E_8$ one obtains the type IIA and type IIB superstrings [‡1], and from D_{16} the type A and B theories constructed fermionically in ref. [14].

The above construction has an obvious relation to the one presented in refs. [15,16] for superstrings out of the bosonic string. To make this more precise, write a Niemeier lattice Γ_{24} as "$D_8 \times G_{het}$", and define a new lorentzian lattice $\Gamma_{16;16} = (G_{het})_L \times (D_8 \times E_8)_R$, with metric $(16(+1), 16(-1))$. Since D_8 has only integer length weights, this is an even lorentzian self-dual lattice, on which the bosonic string can be compactified. The construction presented in ref. [16] is immediately applicable to this lattice: one removes states belonging to the extra E_8, decomposes SO(16) to $SO(8)_1 \times SO(8)_2$ and keeps only states with fixed weights $(0, 0, 0, -1)$ and $(1/2, 1/2, 1/2, -1/2)$ of $SO(8)_2$, and no $SO(8)_2$ excitations [16]. This is precisely the physical state selection rule (2.2) we described earlier for the $\Gamma_{1,1}$ lattice. It obviously would be interesting to make the map from $SO(8)_2$

‡1 The open superstring related to $\Gamma_{0;5,1}$[13] can be mapped uniquely onto $\Gamma_8 = E_8$.

Table 2
Ten-dimensional heterotic strings.

Niemeier lattice		Heterotic string, SO(16)-sector				
roots	weights	algebra	(0)	(v)	(s)	(c)
$E_8{}^3$		$E_8 \times E_8$	$(0+s, 0+s)$	–	$(0+s, 0+s)$	–
$E_8 \times D_{16}$	$(0, s)$	D_{16}	$(0+s)$	–	$(0+s)$	–
$E_8 \times D_{16}$	$(0, s)$	$E_8 \times D_8$	$(0+s, 0)$	$(0+s, v)$	$(0+s, s)$	$(0+s, c)$
D_{24}	(s)	D_{16}	(0)	(v)	(s)	(c)
$D_{12} \times D_{12}$	$[v, s]$ [a)]	$D_4 \times D_{12}$	$(0, 0)$	$(v, 0)$	(s, v)	(s, c)
	(c, c)		(v, s)	$(0, s)$	(c, c)	(c, v)
$D_{10} \times E_7{}^2$	$(s, 1, 0)$	$D_2 \times E_7{}^2$	$(0, 0, 0)$	$(v, 0, 0)$	$(s, 1, 0)$	$(c, 1, 0)$
	$(c, 0, 1)$		$(v, 1, 1)$	$(0, 1, 1)$	$(c, 0, 1)$	$(s, 0, 1)$
	$(v, 1, 1)$					
$D_9 \times A_{15}$	$(s, 4k+2)$	$D_1 \times A_{15}$	$(0, 4k)$	$(v, 4k)$	$(s, 4k+2)$	$(c, 4k+2)$
	$(0, 4k)$					
$D_8{}^3$	$[s, v, v]$	$D_8 \times D_8$	$(0, 0)$	(s, v)	(v, v)	$(c, 0)$
	$[c, 0, c]$		(c, c)	(v, s)	(s, s)	$(0, c)$
	(s, s, s)					

a) In column 2, square brackets indicate cyclic permutation. For E_7, 1 denotes the conjugacy class containing the (56); for A_{15}, an entry m denotes the class containing tensors of rank $m (k=0,...,4)$.

to $\Gamma_{1,1}$ more precise, since it would open the way to a covariant approach to the embedding of all superstrings in the bosonic string. In any case, as embeddings in the bosonic string all theories in table 2 are clearly on the same footing. It is attractive to think of all of them as "vacua" of the bosonic string.

It is also possible to apply these ideas to compactified $(10-d)$-dimensional strings, by considering odd self-dual lattices $\Gamma_{16+d;5+d,1}$, which can be mapped onto even self-dual lorentzian lattices $\Gamma_{16+d;8+d}$. Unfortunately, the lattice classification theorems are not of much help here. Because the left and right vectors need not have integral lengths, we cannot map, in general, this lattice onto an even euclidean lattice. In particular, for $d=6$ this is manifestly impossible, since the resulting lattice would have dimension 36, and there are no such euclidean even self-dual lattices. However, an interesting subset of six-dimensional theories can be related to 32-dimensional euclidean even self-dual lattices. It is known that there are huge numbers of such lattices, but for that reason they have not been classified.

References

[1] D. Gross, J. Harvey, E. Martinec and R. Rohm, Nucl. Phys. B 256 (1985) 253.
[2] L. Dixon and J. Harvey, Nucl. Phys. B 274 (1986) 93.
[3] L. Alvarez-Gaumé, P. Ginsparg, G. Moore and C. Vafa, Phys. Lett. B 171 (1986) 155.
[4] H. Kawai, D. Lewellen and S. Tye, Classification of closed fermionic string models, Cornell preprint CNLS 86/732.
[5] W. Lerche and D. Lüst, Covariant heterotic strings and odd self-dual lattices, Caltech preprint CALT-68-1376 (1986).
[6] K. Narain, Phys. Lett. B 169 (1986) 41.
[7] D. Friedan, E. Martinec and S. Shenker, Nucl. Phys. B 271 (1986) 93.
[8] V. Kostelecký, O. Lechtenfeld, W. Lerche, S. Samuel and S. Watamura, in preparation.
[9] H. Niemeier, J. Number Theor. 5 (1973) 142.
[10] A. Schellekens and N. Warner, Phys. Lett. B 177 (1985) 317; Anomaly cancellation and self-dual lattices, preprint CERN-TH 4465/86, Phys. Lett. B, to be published; Anomalies, characters and strings, preprint CERN-TH 4529/86, in preparation
[11] J. Conway and M. Sloane, J. Number Theor. 15 (1982) 83.
[12] D. Benett, M. Brene, L. Mizrachi and H. Nielsen, Phys. Lett. B 178 (1986) 179.
[13] J. Cohn, D. Friedan, Z. Qiu and S. Shenker, Covariant quantization of supersymmetric string theories: the spin field of the Ramond–Neveu–Schwartz model, Chicago preprint EFI 85-90 (October 1985, revised February 1986).
[14] N. Seiberg and E. Witten, Spin structures in string theory, IAS preprint (February 1986).
[15] A. Casher, F. Englert, H. Nicolai and A. Taormina, Phys. Lett. B 162 (1985) 121.
[16] F. Englert, H. Nicolai and A. Schellekens, Nucl. Phys. B 274 (1986) 315.

* * *

W. Lerche, D. Lüst and A.N. Schellekens, Ten-dimensional heterotic strings from Niemeier lattices, Phys. Lett. B 181 (1986) 71.

The entry for $D_1 \times A_{15}$ in table 2 on page 74 is incorrect. It should be as follows:

Niemeier lattice			Heterotic string, SO(16)-sector			
roots	weights [a]	algebra	(0) [a]	(v) [a]	(s) [a]	(c) [a]
$D_9 \times A_{15}$	$(0, 8k)$ $(s, 8k+2)$ $(v, 8k+4)$ $(c, 8k+6)$	$D_1 \times A_{15}$	$(0, 8k)$ $(v, 8k+4)$	$(v, 8k)$ $(0, 8k+4)$	$(s, 8k+2)$ $(c, 8k+6)$	$(c, 8k+2)$ $(s, 8k+6)$

[a] k is either zero or one.

* * *

5. Four-Dimensional Heterotic Strings

The historical development of four-dimensional string constructions has partly been in inverse logical order: the most difficult approach was discovered first, and since then simpler and simpler constructions were found. This somewhat strange course of events fits rather well in a theory that was derived from a scattering amplitude rather than the other way around.

In 1984 essentially the only known way of "compactifying" string theories was torus compactification. This was not a very attractive option. The most appealing feature of heterotic strings was that they have a chiral spectrum in ten dimensions, while avoiding anomalies. These nice features are ruined by torus compactification, which leads always to a non-chiral spectrum.

The search for a solution to this problem did not last very long, since one could build on experience with field theory compactifications, gathered mainly during the decade before 1984. The interest in compactification of higher dimensions had several different origins. Some of the work in this area can be traced back to the first wave of string theory, which confronted many people for the first time with the notion of higher dimensions. Another reason for studying compactification was the 11-dimensional supergravity theory, which to some seemed a good candidate for a "theory of everything". One believed to have a good chance to realize the dream of Kaluza and Klein of obtaining all interactions from higher-dimensional gravity. However, one quickly faced the problem of obtaining chiral fermions in four dimensions. This is a lost cause from the very beginning in an 11-dimensional theory, but turns out to be impossible in any field theory without any other interactions than gravity [109].

The most promising approaches from the point of view of chirality used additional gauge bosons. In these models, chiral fermions could be obtained only if the gauge group was topologically non-trivial – a monopole or instanton-like configuration (see e.g. [109–112]) – and a generic way of achieving this was known: one could choose the classical gauge field equal to the spin-connection of the manifold. However, the idea of using extra gauge fields was against the Kaluza–Klein philosophy, and nothing seemed to be gained by it. Actually, something *was* gained. In [109] Witten showed that compactification of a ten-dimensional anomaly-free SO(16) gauge theory on a six-manifold with Euler number χ can give a four-dimensional theory with χ generations of 16's of SO(10) – an acceptable unification model. Thus at least one could hope to get an understanding of the replication of fermion generations in this way. This idea turned out to be remarkably close to what was realized in string theory a few years later.

With all this experience from field theory available, it was natural to study compactifications of the field theory limit of string theory. An early exploration in this direction was made in [92] for the SO(32) theory, with the encouraging result that four-dimensional chiral theories were possible. In [113] the field theory limit of the $E_8 \times E_8$ heterotic string was studied, and compactified with the additional requirement that the result should have a surviving $N = 1$ supersymmetry in four dimensions. It turns out that such a compactification is possible if the manifold is a Ricci-flat Kähler manifold (a Calabi–Yau (CY) manifold), with a classical gauge field equal to the spin-connection. The gauge field must have very particular embedding in $E_8 \times E_8$ to satisfy the famous $\mathrm{Tr}\, R^2 - \mathrm{Tr}\, F^2 = 0$ condition [92] and this embedding was found to break $E_8 \times E_8$ to $E_6 \times E_8$. These theories have $\frac{1}{2}\chi$ generations in the (27) of E_6, which is an acceptable though not ideal unification group. Some other papers discussing field theory compactifications of string theories are for example [93,114–116].

Although some attention was paid in [113] to string consistency conditions, this paper is about field theory compactifications and not about string construction. Nevertheless it has survived remarkably well, and it is now a widely held belief that the Calabi–Yau compactifications do correspond to consistent string theories. Unlike the theories collected in this volume, general Calabi–Yau theories are not exactly solvable with any known method, and one can for example not compute the massive mode spectrum or arbitrary correlation functions.

The desire to have something more manageable led to the idea of considering limits of CY manifolds in which the curvature is concentrated in a finite number of singular points. It was realized that such *orbifolds* do lead to exactly solvable theories, which are exact solutions to the string equations of motion. In their first paper, the authors seem somewhat ambivalent about their true intentions (whether to regard orbifold compactifications as a poor man's approach to Calabi–Yau compactification, or as consistent string compactifications in their own right). Indeed, paper 9 pays a lot of attention to the process of repairing ("blowing up") singularities in order to get a smooth manifold.

The emphasis shifted not much later towards the point of view that orbifold compactification yields perfectly consistent string theories even if the singularities are not removed, and even if they cannot be regarded as singular limits of any smooth manifold. Indeed, the whole idea of thinking of compactification on six-manifolds began to loose more and more ground.

By the time that the sequel to the original orbifold paper (paper 10) appeared, Narain had published paper 13, which was the first construction of string theories directly in four dimensions (actually there were of course such attempts in the early days of string theory, but none that dealt properly with the constraints due to conformal and modular invariance). However, in a subsequent paper Narain, Sarmadi and Witten [57] showed that this construction could also be described as a conventional torus compactification in which additional background fields, the internal components of the graviton, the anti-symmetric tensor B_{IJ} and the vector bosons A_I^a have constant values.

By now it had become clear that the consistency conditions of string theory were nowhere nearly as restrictive as they were once thought to be. In early 1986 it was realized that not only there were many possibilities in four dimensions, but that also in ten dimensions some string theories had been overlooked (see chapter 4). The more general summations over fermionic spin structures that were discovered in ten dimensions soon found an application to four dimensions in the work of Kawai, Tye and Lewellen [117], extensively explained in paper 15).

Several other papers were inspired by the new developments in ten dimensions. Lerche and Lüst [118] had the idea of using a lattice description of the NSR model plus ghosts [2,119], and combining it with the left-moving lattice (this is called the *covariant lattice construction*). They realized that by doing this the ten-dimensional heterotic strings could be put on equal footing with the $O(16)\times O(16)$ string, and that both could be described by odd self-dual lattices satisfying some additional requirements. They speculated that by enlarging the lattice to include internal bosons and fermions one should be able to do the same in four dimensions – an idea that ultimately turned out to be correct, as was shown in paper 14.

Towards the end of 1986 it had become clear to many others that it was easy to construct four-dimensional strings. Antoniadis, Bachas and Kounnas found a different way to formulate fermionic string construction (see also section 2.2.7). There was also news from the orbifold front. Mueller and Witten [120] considered twists that are automorphism only for special values of the radius of the compactification torus, in an attempt to fix the radius. Ibañez, Nilles and Quevedo (paper 11) found an ingenious combination of shifts and twists that allowed them to break the gauge group to something resembling the standard model, while reducing at the same time the number of generations. A rather general formulation of orbifold constructions, covering all the ones mentioned above, was found by Narain, Sarmadi and Vafa (paper 12). These authors allowed the left- and right-moving sectors of the string to be twisted separately, an idea that does not have a space–time interpretation. However, by that time the irrelevance of such an interpretation had become clear.

Confronted with so many different ways of constructing new string theories, one had to suspect that there are many overlaps between them. It is now clear that this is indeed true, although the precise regions of overlap are still not known. The similarities and differences between the three approaches (orbifolds, free fermions and free bosons (lattices)) can be described as follows.

As pointed out in section 2.3.1, the conditions for modular invariance of a d-dimensional heterotic string are identical to those of a d-dimensional bosonic string with a right-moving level-1 $SO(d+6)\times E_8$ KM algebra. In fermionic constructions one can describe this algebra in terms of free fermions, and one does the same with the internal sector of the heterotic string. In this way one obtains a bosonic string whose internal sector is described entirely in terms of free fermions, and to which the discussion of section 2.2.7 applies directly. Similarly, in bosonic constructions one describes the KM algebra by a lattice (this can be done only if d is even) and incorporates it into an even self-dual Lorentzian lattice $\Gamma_{26-d;26-d}$. In asymmetric orbifold constructions one starts with a torus compactified ten-dimensional heterotic string, whose bosonic string representation has the form $\Gamma_{26-d;10-d}\times(E_8\times E_8)_R$, where $\Gamma_{22-d;10-d}$ is an even self-dual Lorentzian lattice. Then one applies shifts and/or twists to the lattice Γ and one of the E_8 factors, in order to break the latter to $SO(d+6)$ (or an exceptional algebra containing it, if one wants to get a supersymmetric theory).

As far as modular invariance is concerned, no further condition is needed beyond the ones of bosonic strings. A new constraint one has to deal with is world-sheet supersymmetry. In orbifold constructions this is solved by making sure that world-sheet supersymmetry of the ten-dimensional theory is never broken. If one describes the E_8 factor representing the NSR model by fermions, and twist the $10-d$ internal fermions in exactly the same way as the internal bosons, the supercurrent $\psi^{i+}\,\partial X_{i-}$ is local with respect to all the twisted states in the NS sector (and it has the required square root branch cut in the R sector). In fermionic and lattice constructions however one does not start from an already consistent superstring theory, and one has to build supercurrents directly in four dimensions.

A general method that applies to all constructions is to write down the most general expression of the correct conformal weight in terms of the primary fields of the theory. In practice this is usually not feasible, and one chooses simple, but not completely general ansätze. The following expressions are used in the three constructions:

$$\psi^{i+}\,\partial X_{i-}+\text{c.c.} \qquad \left(i=1,\ldots,\frac{10-d}{2}\right) \qquad \text{(orbifolds)},$$

$$\frac{\mathrm{i}}{6}g_{abc}\psi^a\psi^b\psi^c \qquad (a=1,\ldots,30-3d) \qquad \text{(free fermions)},$$

$$\sum_{l}\boldsymbol{B}(\boldsymbol{l})\cdot\partial\boldsymbol{X}\,\mathrm{e}^{\mathrm{i}\boldsymbol{l}\cdot\boldsymbol{X}}+\sum_{l}A(\boldsymbol{t})\,\mathrm{e}^{\mathrm{i}\boldsymbol{t}\cdot\boldsymbol{X}} \qquad (\boldsymbol{l},\,\boldsymbol{t}\in\mathbb{R}^{15-3d/2}) \qquad \text{(free bosons)},$$

$$(\boldsymbol{l}^2=1,\ \boldsymbol{t}^2=2,\ \boldsymbol{l}\cdot\boldsymbol{B}(\boldsymbol{l})=0).$$

The operator product of T_F with itself can now be used to determine the coefficients. This operator product is automatically satisfied for the orbifold supercurrent, and for fermionic supercurrents the result is that g_{abc} have to be structure constants of a semi-simple Lie algebra

[121] (which differ from the structure constants f_{abc} introduced earlier by a normalization factor: $g_{abc} = \sqrt{2/C(A)}\, f_{abc}$).

The general solution for bosonic supercurrents is not yet known. The expression shown above was first considered in [122] and papers 21 and 22, where a few special solutions were found. One way to get a solution is to bosonize the fermionic expression for the supercurrent. A far larger set of solutions was found in [123] by means of an orbifold-inspired procedure.

Given a supercurrent, one has to construct a conformal field theory with respect to which it has well-defined monodromy. As explained above, this is automatic in the orbifold procedure. In fermionic constructions it implies "triplet constraints" on the boundary conditions of fermions, taken in groups of three. In the covariant lattice construction it implies that all vectors of the form $((v), \boldsymbol{l})$ and $((v), \boldsymbol{t})$, where (v) denotes the vector conjugacy class of $SO(d+6)$, have to be present on the lattice (see paper 23).

In paper 14 the problem was approached in a different way. The starting point was the observation that in chiral string theories one can expect to have problems with massive excitations (which should not be chiral). (This concern about massive chiral states was a result of earlier work on anomaly cancellation [89], where this question comes up if one tries to define a chiral partition function.) Of course one expects massive chiral states to be absent precisely because of world-sheet supersymmetry, which is needed for Lorentz invariance. Chirality provided a simple way to get a handle on that problem. In paper 14 conditions for absence of massive chiral states were formulated for lattices built out of D_1 factors. The result could be interpreted as a bosonic version of the triplet constraint of [117].

It was thus immediately clear that lattice constructions with the supercurrent of paper 14 can be written in terms of free fermions. In fact, as explained in [63], all conformal field theories bases on even self-dual lattices can be written in terms of fermions with Thirring interactions, or free fermions with in general complicated boundary conditions. However, this does not mean that all lattice constructions of heterotic string theories can be obtained within the framework of the fermionic constructions that are presently known. The point is that the form of the bosonic supercurrent is the most general one under the assumption that one has a lattice theory: the presence of c bosonic currents ∂X, where c is the conformal anomaly of the internal sector. On the other hand, the trilinear expression $g_{abc}\psi^a\psi^b\psi^c$ is not the most general supercurrent for a fermionic theory. To construct the most general one, one would also have to take spin fields (and additional twist fields if one uses arbitrary complex boundary conditions) into consideration (the same remark applies also to the standard orbifold supercurrent). Indeed, many of the supercurrents found in [123] do not have a representation in terms of fermion trilinears.

On the other hand, there are many theories that can be obtained with fermionic constructions, but not with lattice constructions. This is manifestly so if the rank of the gauge group is smaller than $26-d$.

There is also a big overlap between orbifolds and the other two constructions. All presently known lattice theories use one of the supercurrents obtained in [123]. Those supercurrents were derived by writing Weyl-twisted orbifolds in terms of lattices, and it is thus a tautology to say that all presently known lattice theories can be written as orbifolds. It would be very interesting to know if there are entirely different supercurrents that do not have such an interpretation. The large overlap between fermionic and lattice constructions implies obviously that many fermionic theories are equivalent to orbifolds as well.

No one expected the proliferation of new string theories to end here, and indeed it has not. One of the most exciting new ideas has been the construction of the internal sector as a tensor product of other CFT's, and in particular those in the discrete series. In principle one could use the $N=0$ discrete series for this purpose. Indeed, this possibility has been studied for bosonic strings in [59]. To get fermionic strings, one merely has to build a supercurrent out of the primary fields in the theory, and then construct a modular invariant that contains it as a chiral field. This has indeed been done for the first member of the discrete series, the Ising model. Fermionic constructions with periodic and anti-periodic boundary conditions are nothing but tensor constructions of Ising models. The success of fermionic constructions suggests that this procedure might work for other elements of the discrete series as well, but this has not been done so far.

One can simplify the work considerably by starting with $N=1$ building blocks, which already have a supercurrent built in. If one wishes to build a space–time supersymmetric theory, one can look for a combination of primary fields that can serve as a gravitino. Again this task can be simplified by starting with building blocks that have an $N=2$ supersymmetry. By imposing a charge quantization condition on the U(1) charge in the $N=2$ tensor algebra, one automatically obtains a gravitino in the spectrum. This procedure has been worked out by Gepner (paper 17) (classifications of four-dimensional solutions can be found in [82,83]. The use of $N=2$ building blocks in this procedure is a convenience, not a necessity. World-sheet supersymmetry plus the presence of a graviton (i.e. an E_6 root in the notation of table I) imply automatically the existence of an $N=2$ algebra.

There are numerous other constructions that have appeared in the literature more recently. In [124] four-dimensional string theories are constructed in the Green–Schwarz formalism (almost all other papers use the NSR formalism); in [125] group manifolds are used to construct lower-dimensional strings; an interesting recent development is the work by Kazama and Suzuki [126], who use the coset construction [21] to construct new solvable $N=2$ models that can be used for building (2, 2) models. There are so many other relevant contributions that it is unfortunately impossible to even list them all. An undoubtedly incomplete list (with apologies to those authors whose work is overlooked here) is [127–136]. For a review of some of these recent developments see [137].

Although from the point of view of string construction one appears to be building unrelated new string theories in each of these constructions, it has become popular to think of them as different "vacua" of one generic heterotic string. Unfortunately it has not been possible so far to define this statement more precisely, but it is supported by the observation that seemingly different string theories turn out to be connected to each other when they are compactified to lower dimensions (see e.g. [57,52,138]). Of course this is very attractive from a philosophical point of view, since it would restore the notion of a unique theory, but in practice it does not help much. One still has to understand the gigantic space of ground states to be able to make progress with true string phenomenology.

Nuclear Physics B261 (1985) 678-686

STRINGS ON ORBIFOLDS*

L. DIXON[1], J.A. HARVEY, C. VAFA and E. WITTEN

Joseph Henry Laboratories, Princeton University, Princeton, NJ 08540, USA

Received 26 June 1985

String propagation on the quotient of a flat torus by a discrete group is considered. We obtain an exactly soluble and more or less realistic method of string compactification.

The most thoroughly understood example of superstring compactification is compactification on a flat torus [1]. This example is exactly soluble and obeys all of the equations of string theory. It is also clearly a significant example since it is one way to understand the construction of the heterotic string theory [2]. For compactification of physical dimensions from ten to four, the flat torus is not suitable since it leads in many ways to unappealing phenomenology. More realistic compactification schemes involve non-trivial supersymmetric non-linear sigma models, and the realistic ones seem to be rather complicated mathematically. The purpose of this paper is to describe a scheme of compactification of extra physical dimensions which is almost as simple as standard toroidal compactification and far more realistic.

The main phenomenological question involves the realization of spacetime and gauge symmetry breaking. In the string theory a way to achieve symmetry breaking is to twist the boundary conditions in such a way that there is no net charge corresponding to the broken symmetries on the string world sheet. Such a procedure has been used for supersymmetry [3] and gauge symmetry [4] breaking and it is natural to attempt to do so for Lorentz symmetries as well [5]. The idea of using twisted boundary conditions to break Lorentz symmetries is that spacetime coordinates $x^\mu(\sigma)$ should not be periodic functions of σ but periodic up to a Lorentz transformation. Modular invariance puts severe constraints on such attempts, since a string sector with twisted boundary conditions in the σ-direction is related by modular transformations to sectors with various boundary conditions in the σ- and τ-directions. What will be described below is one consistent framework in which the twisted boundary conditions are related to the construction of a new manifold (or at least, a new orbifold, as we will describe later). The method of constructing new manifolds (or orbifolds) that we will employ is a classical geometrical method that we will implement directly in string theory.

* Research supported in part by NSF Grant PHY80-19754.

[1] Supported by an NSF Graduate Student Fellowship.

A classical example, which has been described in the physics literature [6], involves the K3 manifold. Consider the four-dimensional torus T with periodic coordinates x_i, $i = 1 \cdots 4$, obeying $x_i \approx x_i + 1$. Let g be the transformation $g(x_i) = -x_i$. Since $g^2 = 1$, g generates the group Z_2. If g acted freely (without fixed points), then the quotient space $\mathrm{X} = \mathrm{T}/\mathrm{Z}_2$ would be a smooth, non-singular manifold. Actually, g does not act freely. There are sixteen fixed points - the points in which each coordinate x_i is an integer or half-integer. Because of this, X is not a smooth manifold but a so-called orbifold - a manifold with singularities that correspond to those obtained by dividing a smooth manifold by the non-free action of a discrete group. In this case, the singularities can be smoothed out or resolved by "blowing up" the fixed points. The resulting smooth manifold is a rather subtle and interesting one called the K3 surface. The singularities encountered in the examples we will consider below can likewise be resolved, but this will not be our main concern. Our main concern will be to treat the string propagation directly on the singular orbifold. As we will discuss, this is exactly soluble. (In resolving singularities, one must introduce free parameters corresponding to the "size" of the blown-up fixed points. When these parameters are tiny, the exactly soluble problem of string propagation on the orbifold is an arbitrarily good approximation to propagation on the smooth manifold, if that is what is desired.)

The K3 manifold is in many ways unique. For instance, it is the only four-dimensional manifold of SU(2) holonomy. The construction that leads to K3 has an analogue in real dimension six(complex dimension three) that does not have the same degree of uniqueness. The simplest example was constructed by mathematicians and was described in [4], where it was called Z. More general examples have been constructed recently in [7]. Although the discussion readily generalizes to the other cases, we will illustrate our ideas by discussing string propagation on the Z manifold. To review the construction of the Z manifold, let z_i, $i = 1, 2, 3$ be three complex variables. For each i, let T_i be the torus defined by $z_i \approx z_i + 1 \approx z_i + \mathrm{e}^{i\pi/3}$, and let α be the transformation $z_i \to z_i\, \mathrm{e}^{2i\pi/3}$ acting on the product manifold $\mathrm{T} = \mathrm{T}_1 \times \mathrm{T}_2 \times \mathrm{T}_3$. Then α generates a group G isomorphic to Z_3. There are twenty seven fixed points of α, namely the points in which each z_i is an integer multiple of $\sqrt{\frac{1}{3}}\, \mathrm{e}^{i\pi/6}$. Because of the existence of these twenty seven fixed points, the quotient space $\mathrm{Z} = \mathrm{T}/\mathrm{G}$ is not a smooth manifold but an orbifold with twenty seven isolated singularities. We wish to describe string propagation on this orbifold.

To describe the quantum mechanics of a point particle propagating on Z, one approach would be to begin with the Hilbert space H_0 of wave functions for a particle propagating on T and then restrict ourselves to the Hilbert space H of G-invariant wave functions. The particle wave functions on Z can be identified with G-invariant wave functions on T. Actually, in general some choices must be made to define what we mean by the action of G on H_0. If the particle under study has some internal quantum numbers, so that the wave function is actually a section of some vector bundle V over T, it is necessary to "lift" the action of G from T to V, and a choice of such a lifting will determine what we mean by the G-invariant subspace H. Differently put, in general we may wish to choose G to act on the internal quantum numbers as well as on the space-time manifold T, and how G acts on the internal quantum numbers will influence what is meant by the G-invariant subspace of Hilbert space.

Following this logic for strings, the first step to describe string propagation on Z is to construct the Hilbert space H_0 of string states on T and project this onto the G-invariant subspace. As in the particle case, we must choose a G action on the internal degrees of freedom of the string, and in the case of the $E_8 \times E_8$ heterotic theory, this means that we must pick a homomorphism of G into $E_8 \times E_8$. Although other possibilities might be interesting, we will pick here a simple choice which is motivated by the idea of embedding the spin connection in the gauge group [4], and which also turns out to make sense in terms of string propagation on the orbifold. Picking an $E_6 \times SU(3)$ subgroup of the first E_8, let β be the generator of the center of SU(3). Letting $\gamma = \alpha\beta$, and defining the action of γ on space-time fermions so that $\gamma^3 = 1$ (the other possibility would have been $\gamma^3 = (-1)^F$), γ generates a Z_3 group which we will also call G. Thus, one ingredient in constructing the Hilbert space that describes string propagation on Z is to construct the G-invariant subspace of the string Hilbert space for propagation on T.

In contrast to the point-particle case, this is not the end of the story. If G acted freely, dividing T by the G action to obtain Z would introduce new sectors in the string propagation - winding sectors in which the string wraps around a non-contractible loop in Z. Mathematically, these are sectors in which the coordinates $x^i(\sigma)$ of the string obey not $x^i(\sigma + 2\pi) = x^i(\sigma)$ but $x^i(\sigma + 2\pi) = gx^i(\sigma)$ for some $g \in G$. In the case at hand, G does not act freely, but we can hardly expect this extra complication to eliminate the need to consider the sectors (which, as we will see, no longer deserve to be called winding sectors) in which the string is closed only modulo a G transformation. We will call the new sectors twisted sectors. Including the twisted sectors is necessary if x and gx are to be considered equivalent space-time points. Like the untwisted sector, the twisted ones must be projected onto G-invariant subspaces if we are interested in propagation on $T/G \approx Z$. (A precise argument for the need to include and project the twisted sectors will be given later on the basis of modular invariance.)

Returning to the Z manifold, there are three sectors - the untwisted sector and the sectors twisted by γ to γ^{-1}. We will discuss first the untwisted sector. To construct this sector, one simply formulates string propagation on the ordinary torus T and projects onto γ invariant states. Massless modes on the torus are made by combining massless modes of the right-movers (which have the degrees of freedom of the open superstring) with massless modes of the left-movers (which carry $E_8 \times E_8$ as well as space-time bosonic degrees of freedom). A right-moving mode with eigenvalue λ of γ must be combined with a left-moving mode of eigenvalue λ^{-1}. Here λ may be any cube root of unity (since $\gamma^3 = 1$). For right-movers the massless modes consist of a standard super-Maxwell multiplet whose relevant quantum numbers (λ and the four-dimensional helicity) are

$$\begin{aligned} \lambda = 1: &\quad (1, \tfrac{1}{2}) + (-\tfrac{1}{2}, -1)\,, \\ \lambda = e^{2i\pi/3}: &\quad (\tfrac{1}{2}, 0) + (\tfrac{1}{2}, 0) + (\tfrac{1}{2}, 0)\,, \\ \lambda = e^{-2i\pi/3}: &\quad (0, -\tfrac{1}{2}) + (0, -\tfrac{1}{2}) + (0, -\tfrac{1}{2})\,. \end{aligned} \tag{1}$$

For left-movers, in addition to λ and the four-dimensional helicity, massless states must be labeled by the unbroken gauge quantum numbers $E_6 \times SU(3) \times E_8$, this being the subgroup of the gauge group that commutes with γ. The massless states are as follows ($E_6 \times SU(3) \times E_8$ quantum numbers are indicated in brackets and helicities in parentheses):

$$\begin{gathered} \lambda = 1: \\ ([1,1,1], (1) + (-1));\ ([78,1,1] + [1,8,1], (0));\ ([1,1,248], (0))\,, \\ \lambda = e^{-2i\pi/3}: \\ ([1,1,1], (0) + (0) + (0));\ ([27,3,1], (0))\,, \\ \lambda = e^{2i\pi/3}: \\ ([1,1,1], (0) + (0) + (0));\ ([\overline{27}, \bar{3}, 1], (0)]\,. \end{gathered} \tag{2}$$

Coupling together states from the above equations to make $\lambda = 1$, the allowed states are as follows. The gauge singlets are the supergravity multiplet $(2, \frac{3}{2})$ and $(-\frac{3}{2}, -2)$ and ten matter multiplets $(\frac{1}{2}, 0)$ and $(0, -\frac{1}{2})$. In addition there are massless gauge supermultiplets of unbroken $E_6 \times SU(3) \times E_8$. And there are massless chiral matter fields with gauge quantum numbers. The states with helicity $(\frac{1}{2}, 0)$ are three copies of $[27, 3]$ under $E_6 \times SU(3)$. Their antiparticles, of helicity $(0, -\frac{1}{2})$, transform as $[\overline{27}, \bar{3}]$. Viewed just in terms of E_6, there are nine chiral generations in the untwisted sector.

Let us consider the twisted sector corresponding to twisting by γ. There are some new features in this sector. For the string to have a massless state, it must be capable of shrinking to zero size. To do so it must sit at one of the fixed points of γ. In the case under study, the fixed points of γ are isolated, and a zero-size string sitting at one fixed point cannot be continuously deformed into a zero-size string sitting at another fixed point. Expansion about strings sitting at different fixed points leads to disjoint sectors in the string propagation. In the Z manifold, as there are 27 fixed points, we obtain 27 sectors corresponding to the expansion of the string about each of these fixed points. These sectors lead to isomorphic physics, as they are related by obvious symmetries (translations on the underlying torus). It is therefore adequate to concentrate on any one of the twenty seven sectors.

In the twisted sector, the quanta of oscillators are not all integral. To enumerate the physical degrees of freedom we will use the light cone gauge, with manifestly supersymmetric spacetime fermions. Let x_3, x_4 be the uncompactified transverse oscillators, and let x_i, $i=5,\dots,10$ be the compactified transverse oscillators. Then x_3 and x_4 are quantized with integral quanta of oscillation, but the other x_i are quantized with quanta that are integers plus or minus one third (plus one third for three of the x_i and minus one third for the other three). As we have chosen γ to respect space-time supersymmetry (since it lies in an SU(3) subgroup of the rotation group which leaves invariant one supersymmetry), the space-time fermions are quantized in the same way. In particular, only two of the space-time fermions (the ones that are quantized with integral oscillators) have zero modes. Quantization of these zero modes causes the ground state of the right-movers to be a doublet with $\lambda=1$ and helicity $(\frac{1}{2}, 0)$. The discussion of the left-movers in the twisted sector is more complicated. A relatively simple approach involves realizing an $O(16)\times O(16)$ subgroup of the gauge group linearly on thirty two fermions. Actually, the first O(16) is broken to $O(10)\times SU(3)$ by the twisting. It is particularly easy to find and count the states that are spinors of this O(10). They come from a sector in which the ten fermions realizing this O(10) have integral quanta of oscillation and therefore have zero modes (and where the sixteen fermions of the other O(16) are antiperiodic). The six other fermions of the broken O(16) have quanta of oscillation that are not integral, three being shifted by $\frac{1}{3}$ and three by $-\frac{1}{3}$. Quantization of the ten zero modes gives a spinor of O(10) (of definite "chirality") which has no additional left-moving quantum numbers since there are no other zero modes to be quantized. Inclusion of the sector corresponding to a minus sign change in the boundary conditions of the sixteen fermions realizing the broken O(16) can be seen to fill this out into a 27 of E_6, and to generate as well massless E_6 singlets transforming as three triplets of SU(3). (These states all have $\lambda=1$.) The spectrum just described comes from expanding about one of the twenty seven fixed points of γ. All told, we get from the sector twisted by γ twenty seven positive chirality matter multiplets transforming as the [27, 1] of $E_6\times SU(3)$ and eighty one such multiplets transforming as [1, 3]. Happily, little further effort is required to analyze the states obtained by twisting with γ^{-1}. They are just the antiparticles of the states just mentioned.

Let us collect these results and count the number of chiral multiplets obtained by combining all the pieces. Under E_6, there are a total of thirty six "families" (the number of families is defined as the net number of positive chirality 27's minus negative chirality 27's). Nine positive chirality 27's (transforming as three triplets of the "flavor" group SU(3)) came from the untwisted quantization, while twenty seven positive chirality 27's come from the sector twisted by γ. No positive chirality

27's come from the sector twisted by γ^{-1}, and no negative chirality 27's come from any sector. In addition, there are a large number of E_6 singlet massless chiral fields.

Let us compare these results with what one learns by studying the same manifold by field theoretic means [4]. In that framework, embedding the spin connection in the gauge group, the number of generations is predicted to be one-half the Euler character of the manifold. The Euler characteristic is 72 for the Z manifold, so the number of generations is predicted to be 36 - the same number we have just found. This suggests that the string propagation on the orbifold with α embedded in the gauge group in the manner we considered is a limiting case of string propagation on the smooth manifold with spin connection embedded in the gauge group. Some differences between our results here and results in [4] correspond to the fact that the unbroken SU(3) in our construction disappears if the fixed points are blown up and smoothed with spin connection embedded in the gauge group. Therefore, the 36 chiral generations were not in [4] classified as representations of a family SU(3). Also, the number of massless E_6 singlets that appear in the field theoretic analysis, although large, is less than we have found here, presumably because some states that are kept massless by SU(3) but not by any topological considerations get mass if SU(3) is broken.

The procedure just given for counting the number of generations that appear in quantization on the Z orbifold was fairly explicit but also fairly long-winded. We will now consider an efficient path integral method for computing the number of generations on an arbitrary orbifold. Thus, we start with an arbitrary smooth manifold M, with a discrete symmetry group G. In the case of the Z manifold, M was a flat torus, but for present purposes we can be more general. Let K be the orbifold M/G. If the spin connection of M (and the G action on tangent vectors of M) is embedded in the gauge group, then the number of generations is one half the Euler characteristic of K, so what we really want is an efficient method to calculate the Euler characteristic of an orbifold.

In fact, the Euler characteristic of a manifold M is naturally viewed as the index $\mathrm{Tr}\,(-1)^F$ of the 1+1 dimensional supersymmetric non-linear sigma model [8]. (If the spin connection is embedded in the gauge group, this is the field theory that describes string propagation on M.) This can be computed as a path integral [9]:

$$\mathrm{Tr}\,(-1)^F = \int \mathrm{d}x^i(\sigma,\tau)\,\mathrm{d}\psi^i(\sigma,\tau)\exp[-I]\,, \tag{3}$$

where $\psi^i(\sigma,\tau)$ obey periodic boundary conditions in both sigma and tau, and I is the action of the supersymmetric non-linear sigma model.

To study the orbifold, one must introduce twisted boundary conditions. In general, one could try to pick any two elements h and g of G and require $x^i(\sigma+2\pi,\tau)=gx^i(\sigma,\tau)$, $x^i(\sigma,\tau+2\pi)=hx^i(\sigma,\tau)$. It turns out that we will only have need for the case in which g and h commute with each other. Indeed, the boundary conditions just stated are more or less inconsistent otherwise.

Let $A(g, h)$ denote the value of the integral in (3) with boundary conditions twisted by g and h. We will describe how to compute the $A(g, h)$ and how to express the Euler characteristic of the orbifold in terms of them. Computing the $A(g, h)$ is particularly simple. Let $\mathrm{M}(g, h)$ be the subspace of M consisting of points that are invariant under both g and h, and let $\chi(g, h)$ be the Euler characteristic of $\mathrm{M}(g, h)$. Then as we will now argue, $A(g, h) = \chi(g, h)$. In fact, as $A(g, h)$ is a topological invariant, we may as well compute it by going to weak coupling (very large radius for M). The path integral is then dominated by paths of very nearly zero action, which means that the x^i must be (nearly) independent of sigma and tau. This is compatible with the boundary conditions only if x^i is invariant under g, h, so x^i must lie in (or very near) $\mathrm{M}(g, h)$. The bosonic and fermionic oscillations transverse to $\mathrm{M}(g, h)$ give determinants which cancel in the usual supersymmetric fashion, so the path integral in (3) reduces to the path integral for a similar integral for a supersymmetric non-linear sigma model on $\mathrm{M}(g, h)$. By the usual analysis of $\mathrm{Tr}\,(-1)^F$, the value of the integral is $\chi(g, h)$.

Now that we have calculated the $A(g, h)$, we want to use them to determine the Euler characteristic of the orbifold. Suppose that we want to discuss strings that obey untwisted boundary conditions in sigma, so $g = 1$. A winding by h in the tau direction then amounts to studying not $\mathrm{Tr}\,(-1)^F$ but $\mathrm{Tr}\, h(-1)^F$. (Our above arguments show that $\mathrm{Tr}\, h(-1)^F = \chi(1, h)$, an assertion which is known as the Lefschetz fixed-point theorem.) Projecting onto invariant states is carried out by the projection operator $Q = (1/N) \sum h$, where N is the number of elements in G and the sum in $\sum h$ runs over all elements in G. The contribution of strings untwisted in sigma to the Euler characteristic of the orbifold is thus $\mathrm{Tr}\, Q(-1)^F = (1/N) \sum \chi(1, h)$. (In the case of the Z manifold, this would give $\frac{1}{3}(0+27+27) = 18$, in agreement with the fact that we previously found nine generations from this sector.)

In field theory, we would stop here. In string theory, we cannot stop here; we must go on to discuss the contributions of sectors twisted in sigma as well as tau. The concrete reason for why we must do this is as follows. A modular transformation $(\sigma, \tau) \to (k\sigma + l\tau, m\sigma + n\tau)$, k, l, m, n integers, $kn - lm = 1$ will turn a sector twisted in tau only into one twisted both in sigma and tau. Modular invariance is of course needed for consistency of the theory of closed strings. At first sight one might think that modular invariance would force us to consider arbitrary pairs (g, h), but that is not so. If g and h commute with each other, they are mapped into the commuting pair $(g^k h^l, g^m h^n)$, so modular invariance permits us to restrict attention to commuting pairs (g, h). This is just as well, since the twisted boundary conditions do not make too much sense unless g and h commute. These facts suggest the following formula for the Euler characteristic of K:

$$\chi(\mathrm{K}) = \frac{1}{N} \sum_{gh=hg} \chi(g, h) . \tag{4}$$

This formula is modular-invariant, receives the proper contribution from $g = 1$, and only involves commuting pairs. To see that it is also correct requires more thought. To construct the required Hilbert space, one begins with all possible sigma windings by group elements g; denote the corresponding Hilbert spaces as H_g. One then tries to make a G-invariant Hilbert space by inserting a projection operator. One expects the projection operator to be a symmetry of the Hilbert space, but in fact the individual H_g are not invariant under all of G. A given H_g only has a lesser symmetry group G_g consisting of elements that commute with g, which we may call the "little group" (or centralizer) of g. An element h of G that does not commute with g turns H_g into $H_{h^{-1}gh}$. The way to form a G-invariant and modular-invariant Hilbert space is to sum over all possible H_g and project each H_g onto states invariant under its own little group - by including all possible tau windings that commute with g. This corresponds to the formula in (4). It is again easy to check this formula for the Z manifold; there are eight non-zero terms in (4), each contributing $(\frac{1}{3})\times 27=9$, so that $\chi(Z)=72$. As this agreement shows, (4) properly includes corrections to the Euler characteristic coming from blowing up the fixed points. It is remarkable that in this simple way string theory "knows" what happens when the singularities are resolved by blowing up the fixed points. Field theory does not lead in this way to the proper formula, since it omits the contributions of the twisted sectors.

The agreement between the string theory calculation of the Euler characteristic and the correct value is not an accident of the Z manifold but a general feature, at least for manifolds of SU(3) holonomy. If one divides a manifold M by the action of a discrete group G, singularities are created if G does not act freely. The theory of how to resolve these singularities is rather complicated in general, but under favorable conditions there are simple prescriptions, which were summarized in [10]. In these cases, (4) gives a value of the Euler characteristic which agrees with that obtained in more standard ways. For instance, if G is Z_3 and N is the fixed point set of G in M, then the quotient $X=M/G$ has (after blowing up fixed points) the Euler characteristic [10] $\chi(X)=\frac{1}{3}(\chi(M)+8\chi(N))$, which is easily seen to agree with (4). It does not appear that eq. (4) is known to mathematicians, and as this formula makes sense in situations in which a standard procedure for resolving the singularities does not exist, (4) may be a clue to a more general procedure.

We have given the Z manifold as an example in which the string propagation is exactly soluble, but the considerations are more general. The general idea is to begin with a flat torus T with symmetry group G, and study the quotient T/G. If G is embedded in a suitable way in E_8, one is studying an idealized limit of "embedding the spin connection in the gauge group." It is possible to study more general embeddings, but we have found that problems typically arise [11] from a mismatch of left- and right-moving energy levels. The simplest consistent examples occur when G is a subgroup of SU(3) leading to an unbroken $N=1$ supersymmetry in four dimensions. Again, one can consider more general examples; one typically encounters tachyons unless group elements not in SU(3) act freely [11] and even then there may be one-loop instabilities [3], but it may be that some examples work. Even though no smooth Ricci-flat manifolds with O(6) holonomy are known [12], we can still consider string theory on oribifolds with broken supersymmetry. We will discuss these and other matters related to the phenomenology of string theories on orbifolds more fully in [11].

We would like to thank J. Lepowsky for many discussions on twisted Kac–Moody algebras, and A. Strominger and T. Killingback for discussions of generalizations of the Z manifold. Furthermore, C.V. is greatly indebted to C. Hodgson for many enlightening conversations.

References

[1] M.B. Green, J.H. Schwarz and L. Brink, Nucl. Phys. B198 (1982) 474
[2] D.J. Gross, J.A. Harvey, E. Martinec and R. Rohm, Phys. Rev. Lett. 54 (1985) 502, Princeton preprints (1985)
[3] R. Rohm, Nucl. Phys. B237 (1984) 553
[4] P. Candelas, G. Horowitz, A. Strominger and E. Witten, Nucl. Phys. B258 (1985) 46
[5] C. Vafa and E. Witten, Princeton preprint (1985)
[6] G.W. Gibbons and C.N. Pope, Comm. Math. Phys. 66 (1979) 291;
D.N. Page, Phys. Lett. 80B (1978) 55
[7] A. Strominger, I.A.S. preprint (1985)
[8] E. Witten, Nucl. Phys. B202 (1982) 253
[9] S. Cecotti and L. Girardello, Phys. Lett. 110B (1982) 39;
L. Alvarez-Gaumé, Comm. Math. Phys. 90 (1983) 161;
D. Friedan and P. Windey, Nucl. Phys. B235 [FS11] (1984) 395;
E. Witten, unpublished
[10] A. Strominger and E. Witten, I.A.S. preprint, 1985
[11] L. Dixon, J. Harvey, C. Vafa, and E. Witten, to appear
[12] S.T. Yau, private communication

Nuclear Physics B274 (1986) 285–314
North-Holland, Amsterdam

STRINGS ON ORBIFOLDS (II)

L. DIXON,[1†] J. HARVEY,[2†] C. VAFA[3] and E. WITTEN[2†]

Received 12 March 1986

A general framework for analyzing string propagation on the quotient of a manifold by a discrete group is presented. The conditions necessary for modular invariance and absence of tachyons are discussed. Models with four or fewer generations which correspond to the standard embedding of the spin connection in the gauge group are analyzed. We all discuss some generalizations which may lead to more phenomenologically promising models.

1. Introduction

Superstring theories are serious candidates for unified theories of all known interactions [1–3]. However if they are to make contact with the real world it will be necessary to understand how the enormous symmetry of these theories is broken down to something like that which we observe at low energies. In a previous paper [4] we discussed a simple but powerful method of achieving symmetry breaking in string theories via twisted boundary conditions. Depending on how the twists act on the spacetime and internal degrees of freedom this general approach can describe toroidal compactification, symmetry breaking by Wilson lines, a singular limit of compactification on certain Calabi-Yau spaces, and may even be used to derive one heterotic string theory from the other! In addition to its generality this method has the advantage of allowing explicit calculations of tree and loop string amplitudes with realistic symmetry breaking.

In this paper we will present a general framework for discussing strings with twisted boundary conditions and explore some of the phenomenological consequences of this approach to symmetry breaking. Twisted boundary conditions naturally arise when one considers string propagation on an orbifold. We will define an orbifold to be the object one obtains by dividing a manifold by the action of a discrete group G. (More generally an orbifold is locally, but not necessarily globally,

† Joseph Henry Laboratories, Princeton University, Princeton, NJ 08544, USA.

1 Supported by a Charlotte Elizabeth Proctor Fellowship.
2 Supported by NSF Grant PHY80-19754.

3 Junior Fellow, Harvard Society of Fellows, Lyman Laboratory of Physics, Harvard University, Cambridge, MA 02138, USA. Supported by NSF Grant PHY82-15249.

such an object.) If G acts freely (without fixed points) we simply obtain another manifold. When G has fixed points the orbifold has singularities which in certain cases appear to be innocuous in string theory. In general one must also define the action of G on the internal degrees of freedom of the string. When G acts freely, choosing a non-trivial action for G on the gauge degrees of freedom is equivalent to symmetry breaking by Wilson lines [3, 5]. When G has fixed points we obtain a discrete version of identifying the spin connection with the gauge connection [3, 6] or its generalizations [7]. In certain cases one can also contemplate modding out the string theory by a discrete group which acts only on the internal gauge degrees of freedom. In the heterotic string this allows one to "derive" the $\mathrm{Spin}(32)/\mathrm{Z}_2$ theory from the $\mathrm{E}_8 \times \mathrm{E}_8$ one and vice versa in a way which bears a certain resemblance to the construction of the monster group presented in [8].

The outline of this paper is as follows. In sect. 2 we discuss some general features of the discrete groups used to construct orbifolds when we take the original manifold to be euclidean space. We discuss the possible actions of these groups on the internal degrees of freedom and derive an expression for the fundamental group of an orbifold. Sect. 3 is devoted to a description of the Hilbert space of strings on orbifolds. In sect. 4 we discuss the constraints on G necessary to obtain a consistent string theory and discuss various possibilities for embedding G in $\mathrm{E}_8 \times \mathrm{E}_8$. In sect. 5 we consider models where G is embedded in a standard SO(6) subgroup of E_8. In the case of a six-dimensional orbifold it turns out that the point group P consisting of the rotations in G must lie in an SU(3) subgroup of SO(6) in order to avoid the presence of tachyons in the string theory. (There is one way to avoid the SU(3) requirement, which is mentioned in this section.) The discrete subgroups of SU(3) which can lead to small numbers of generations are then classified. We show that it is possible to obtain two or four generations but not three in this approach. Sect. 6 is devoted to some brief comments on interactions for twisted strings and to a general discussion of our results. In order to flesh out the discussion in the main text we present some explicit examples of the construction of the twisted sectors of string Hilbert space in the appendix. We also show how one can go between the $\mathrm{E}_8 \times \mathrm{E}_8$ and $\mathrm{Spin}(32)/\mathrm{Z}_2$ heterotic strings by dividing by an involution and including the appropriate twisted sector. The techniques we present here seem to be most powerful in the heterotic string where the gauge interactions are a manifestation of the underlying affine Lie algebra. Since the $\mathrm{E}_8 \times \mathrm{E}_8$ heterotic string appears to be the most promising phenomenologically most of our discussion will focus on this theory.

2. Space groups and point groups

In this section we will discuss some general features of the discrete groups used to obtain an orbifold from a manifold when we take the manifold to be euclidean space. Other possibilities may be of interest, such as orbifolds constructed from

Calabi-Yau spaces, but these will be much less tractable. The discrete group we pick should preserve the metric of the space, and thus for euclidean space will be a discrete subgroup of the euclidean group consisting of translations and rotations. For example, consider a d-dimensional torus T^d. It could be equivalently viewed as the d-dimensional euclidean space R^d divided by the action of a d-dimensional discrete translation group Z^d:

$$T^d = R^d/Z^d. \tag{1}$$

We would like to generalize this definition of the torus by enlarging the discrete translation group to include some discrete rotations and/or rotations accompanied by translations. This generalized group is called a space group. The space group is the restriction to spacetime of our discrete group G; that is, it ignores the action of G on the gauge group.

A d-dimensional space group S consists of elements made out of pairs of rotations θ and translations v, represented by $g = (\theta, v)$. The action of the group elements on the d-dimensional vector x is given by

$$gx = \theta x + v. \tag{2}$$

Dividing by S means that we think of x and gx as the same point. In other words each point is identified with its orbit under the action of the group (hence the name orbifold). It is straightforward to check the following relations:

$$(\theta, v)(\omega, u) = (\theta\omega, v + \theta u),$$

$$(\theta, v)^{-1} = (\theta^{-1}, -\theta^{-1}v),$$

$$(\theta, u)(1, v)(\theta, u)^{-1} = (1, \theta v). \tag{3}$$

The subgroup Λ of S consisting of pure translations (shifts), i.e., elements of the form $(1, v)$, forms a d-dimensional discrete translation group Z^d referred to as the lattice of S, and defines a torus by $T^d = R^d/\Lambda$. We also consider what is called a point group P, which is a subgroup of $O(d)$ consisting of rotations θ such that (θ, v) is an element of S for some v. P is the restriction to $O(d)$ of S, which is in turn the restriction to $E(d)$ (the d-dimensional euclidean group of rotations and translations) of G.

For each element θ of P, there corresponds a unique, up to translations in Λ, shift vector v with (θ, v) in S. To see the uniqueness, consider two elements of S, (θ, v) and (θ, u). Then we have

$$(\theta, v)(\theta, u)^{-1} = (\theta, v)(\theta^{-1}, -\theta^{-1}u) = (1, v - u). \tag{4}$$

So we see that $v-u$ is an element of Λ. This implies that P has a well-defined action on the torus T^d which we denote by $\bar{\mathrm{P}}$. (We do this to remind ourselves that unlike P, $\bar{\mathrm{P}}$ may be accompanied by shifts.) In fact $\bar{\mathrm{P}}$ can be identified with the quotient $\bar{\mathrm{P}} = \mathrm{S}/\Lambda$.

This brings us to two equivalent ways of describing orbifolds. We can either start with euclidean space and divide by the space group, or equivalently consider the corresponding torus and divide by the point group. That is

$$\Omega = \mathrm{R}^d/\mathrm{S} = \frac{\mathrm{R}^d/\Lambda}{\bar{\mathrm{P}}} = \mathrm{T}^d/\bar{\mathrm{P}}. \tag{5}$$

P can be thought of as the holonomy group of Ω: if $x = (\theta, v)y$ then not only are x and y identified, but also the tangent vectors at x are identified with the tangent vectors at y rotated by θ. Now if we connect y to x by a path, which is a closed loop in Ω, and parallel transport a vector along it, the vector undergoes no rotation (the torus is flat), but compared to the vector at x it differs by a rotation θ. This shows that the holonomy group P of the orbifold is a discrete group, rather than a continuous Lie group as in the case of a smooth manifold. P is not an arbitrary discrete subgroup, but must act crystallographically on the lattice Λ. Eq. (3) shows that if v is an element of Λ then so is θv for each θ of P. Choosing a basis for the lattice in which all the vectors of Λ are represented by integer column vectors implies that P should have an integer representation (a representation where the entries of the matrices are integers). This places a strong restriction on what P can be.

An orbifold is in general not a manifold because the space group may not act freely on the euclidean space; that is, it may have fixed points. At these fixed points the orbifold fails to be a manifold. (Some lines emanating from the fixed point are identified with each other.) In certain cases it is known how to repair these singularities by removing these points and gluing in manifolds with the appropriate boundary. This is called blowing up and resolving the singularities. This smoothes out the holonomy group of the orbifold into a Lie group containing P as a discrete subgroup. Even though one might think that there could be difficulties if the singularities are not resolved, we have not encountered any. Also, the manifolds required to repair the singularities are not known in every case of interest.

The space groups for up to four dimensions have been classified [23] (the three-dimensional space groups have been studied in physics in connection with crystallography). The case of interest for compactification of superstrings is the six-dimensional space groups, which have not been classified. We will discuss some general features of strings propagating on six-dimensional orbifolds which do not require the full classification of space groups. If we want to preserve $N=1$

supersymmetry in four dimensions the point group (holonomy group) P should be a discrete subgroup of SU(3). If we do not want to preserve supersymmetry the holonomy group could be a discrete subgroup of SO(6), but when supersymmetry is broken, as will be discussed later, one must be careful if one wants to avoid tachyons. In addition, once supersymmetry is broken the cosmological constant will generically receive large contributions at the one-loop level.

So far we have discussed the action of the group G on the space-time degrees of freedom. We could also allow the group to act on the gauge degrees of freedom. If S consists only of translations, $S = \Lambda$, then $R^6/S = T^6$ and $\pi_1(T^6) = Z^6$. Choosing the action of Λ on the gauge degrees of freedom means associating a gauge transformation with a translation around a closed loop on T^6, i.e. choosing a set of Wilson lines for the six generators of $\pi_1(T^6)$. For the $E_8 \times E_8$ heterotic string we can represent the $E_8 \times E_8$ degrees of freedom using sixteen left-moving bosons X^I parametrizing the maximal torus of $E_8 \times E_8$. The action of Λ can then be simply represented as translations on the torus by shift vectors δ^I, one for each generator of Λ.

For a general space group, if we want to identify the spin connection with the gauge connection, we should pick an SO(6) subgroup of E_8 and consider $g = (\theta, v)$ acting on the string degrees of freedom representing E_8. If we represent E_8 through sixteen fermions ψ transforming as the **16** of O(16), then viewing O(6) as the subgroup of O(16) under which the **16** transforms as a **6** plus singlets, we set $g\psi = \theta\psi$. If we represent E_8 in terms of the torus, it will be difficult to describe the action of the space group on it if the point group is non-abelian. However if the point group is abelian, i.e. it sits in a Cartan subalgebra of O(6), the action of each element in the space group on the E_8 torus is given by a shift. In particular if we consider J_{12}, J_{34}, J_{56} to be the orthonormal generators of a Cartan subgroup of O(6), and P_1, P_2, P_3 to be generators of translation in three of the coordinates of the torus, they are identified respectively (up to normalizations). If we use the orthonormal basis vectors for the lattice vectors of E_8 which are described in the appendix, then the spatial rotation $\exp(2\pi i(aJ_{12} + bJ_{34} + cJ_{56}))$ is identified as a shift by $(a, b, c, 0, 0, 0, 0, 0)$ on the group torus. In some cases the shift on the torus is equivalent to an SO(8) rotation of the E_8 root lattice. This corresponds to different, but equivalent, ways of embedding O(6) in E_8. These matters and the string Hilbert space for each of these cases will be discussed in the appendix with some specific examples. We may also consider different actions of the space group on the gauge group which do not correspond to identifying the spin connection with the gauge connection. These possibilities will be discussed in sect. 4.

For each point group there correspond many inequivalent space groups depending on which lattice we pick and how the shift vectors have been chosen. However, many features of the string theory on orbifolds can be deduced merely from the choice of the point group. For example, many point groups can be ruled out if we are interested in low numbers of generations. Since the discrete subgroups of SU(3)

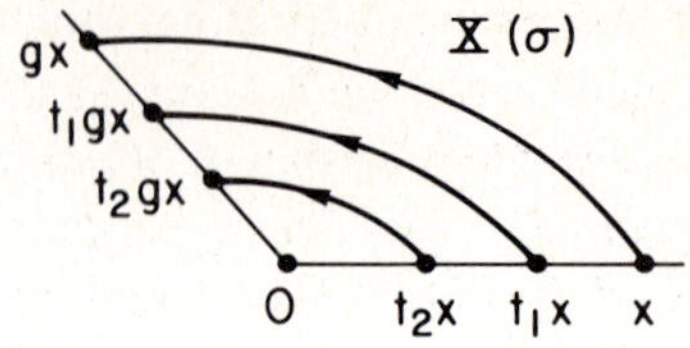

Fig. 1. The homotopy $f(t,\sigma) = tX(\sigma)$, $t \in [0,1]$, $\sigma \in [0, 2\pi]$, shows that the loop $X(\sigma)$ is contractible if $X(2\pi) = gX(0)$ and g has a fixed point.

have already been classified we will discuss the possibilities of obtaining low numbers of generations in the case of unbroken supersymmetry. In the case of SO(6) the maximal discrete subgroups have been classified [30], but a full classification of their subgroups does not seem to have been made.

If we choose to identify the spin connection with the gauge connection, we may still pick those elements of S which generate the fundamental group of the orbifold to act non-trivially on the remaining unbroken gauge degrees of freedom in order to achieve more realistic symmetry breaking. Thus we wish to determine the fundamental group of an orbifold. Consider euclidean space divided by the space group S. Since all strictly periodic loops in R^d are contractible, the only possibly nontrivial loops are those which are periodic up to a transformation in S. It might therefore appear that the fundamental group of the orbifold is simply S. This is not true: if there are elements g which have fixed points (or fixed surfaces), then all loops periodic up to g may be shrunk to a point. To see this note that for g to have a fixed point it must be a rotation about that point. That means the two ends of the loop are related by a rotation about the fixed point, and so are the lines joining the endpoints to the fixed point. So if we take the origin as our fixed point, as in fig. 1, we have $gx = \theta x$ for some rotation θ. Now we can contract the loop onto the fixed point in such a way that the loop maintains its shape but is shrunk in size; i.e. $X(\sigma) \to tX(\sigma)$. The endpoints will still be related by the same rotation θ and the loop will remain a closed loop under this contraction. The whole loop will contract to the fixed point and is therefore trivial.

Suppose g_1 and g_2 are two group elements which have fixed points. So the loops periodic up to g_1 and g_2 are contractible. Furthermore, the loops periodic up to $g_1 g_2$ are also contractible, because we can deform the loop periodic up to $g_1 g_2$ into two loops, one periodic up to g_1 and the other up to g_2. Since each of these loops is contractible, we see that the whole loop is contractible. In this way we see that any product of elements which have fixed points corresponds to a contractible loop. Let F denote the subgroup of S generated by elements which have fixed points. Our modified guess, which turns out to be the correct answer, is that the fundamental group of the orbifold is S/F. (This can be proven using the Seifert-Van Kampen theorem [9].) (It is easy to check that F is a normal subgroup of S, so that S/F is indeed a group.)

3. Hilbert space of strings on orbifolds

To describe the string Hilbert space on an orbifold we start by considering string propagation on the manifold before dividing by the group action. (In our case the manifold would be either R^d or T^d.) To each element g in the group, there corresponds an operator $\bar{g}$ acting on the Hilbert space of the string. If we want x and gx to correspond to the same point in the orbifold we should consider the subspace of the Hilbert space invariant under the action of $\bar{g}$, i.e. $\bar{g}$ should act as the identity operator for the orbifold.

For example, consider starting from R^d and dividing by the discrete translation group Z^d to obtain the torus T^d. To each translation v in Z^d corresponds the translation operator $e^{iP\cdot v}$ which acts on the string Hilbert space of R^d. We should project onto the subspace invariant under these actions, which amounts to requiring that the eigenvalues of the momentum operator lie on the lattice dual to the lattice defined by Z^d. This subspace defines the string Hilbert space for T^d.

If we were describing open strings that would be all that is necessary. For closed strings, however, the story continues: the usual boundary condition for closed strings is $X(2\pi) = X(0)$ (or antiperiodic for fermions in some sectors). This is now not the only possibility because for the orbifold X and gX are the same points and we should therefore require only that $X(2\pi) = gX(0)$ for some g in our group. This implies that for each g in the group we get a sector H_g of the Hilbert space in which the boundary conditions are changed to periodicity up to g transformation. These we shall call twisted sectors.

In the example above this implies that we get sectors H_v in which $X(2\pi) = X(0) + v$ where v is a translation in Z^d. These sectors are just the usual winding sectors in which the string wraps around the torus. We still have to project onto the g invariant subspace, which again amounts to requiring that momenta lie on the dual lattice. Also, the projection $L_0 = \bar{L}_0$ which ensures that the string is invariant under $\sigma \to \sigma + \Delta$ (and which we discuss in the next section) is slightly modified from what it was in the untwisted sector. These twisted sectors must clearly be included since they can be pair produced in interactions of untwisted strings. They are also necessary to ensure modular invariance.

In the general case the projection of the twisted sector onto the group invariant subspace is a little more tricky. This is because if the group we are dividing by is not abelian then some group elements take one twisted sector to another. Suppose we consider a state in the Hilbert space H_g, which means that $X(2\pi) = gX(0)$. If we act on this state by some other group element h, the initial point of the string is taken to the point $hX(0)$ and the final point is taken to $hX(2\pi) = hgX(0) = hgh^{-1}(hX(0))$. So we see that the string is periodic up to hgh^{-1} and therefore is in the sector $H_{hgh^{-1}}$. If g and h do not commute this is different from H_g. In other words under the action of the group, sectors which are in a given conjugacy class of the group mix. (Two elements g_1 and g_2 are conjugate if $g_1 = hg_2h^{-1}$ for some h in the group.) So to form group invariant states we will have to take a state in the sector

H_g, project it onto the invariant subspace of the centralizer, or little group, of g (the subgroup of elements commuting with g, which therefore take H_g to itself), and then take the sum of corresponding states from sectors in the same conjugacy class, each of which is projected onto its little group's invariant subspace. So in a sense, we form states for each conjugacy class in the group. As was argued in [4], this construction is necessary to form a G invariant and modular invariant string Hilbert space.

To illustrate these ideas consider the case when the manifold is R^1, the real line, and the group is the space group $(\pm 1, n)$, where n is an integer. In other words x is identified with $\pm x + n$. This group is not abelian and the elements of the form $(+1, n)$ and $(+1, -n)$ are conjugate. So in forming group invariant states, we have to take states from $H_{(+1,n)} + H_{(+1,-n)}$. Also the elements $(-1, n)$ and $(-1, n+2k)$ are conjugate so we get two additional conjugacy classes, corresponding to whether n is even or odd.

As discussed in the last section, we could equivalently construct this orbifold starting from the torus instead and dividing by the point group. In this case the point group is Z_2. (The circle is defined by $x \sim x+1$ and the nontrivial element of the point group acts by taking x to $-x$.) So we get two sectors. The untwisted sector has further subsectors corresponding to winding states. These are the $(+1, n)$ sectors described above. Projecting the untwisted sector on to the Z_2-invariant states implies that we have to take states from the sum of each pair of opposite windings, because the nontrivial element in Z_2 takes a sector with winding number n to a sector with winding number $-n$. This corresponds to the fact that $(+1, n)$ and $(+1, -n)$ are conjugate in the other picture. In the twisted sector we get two subsectors. This comes about from the quantization of the center of mass of the string. To see this consider the point limit of the string, i.e. take $X(\sigma) = X$ independent of σ. But in the twisted sector this is consistent with the boundary condition $X(0) = -X(2\pi) + n$ only if $X = 0$ or $\frac{1}{2}$. By taking the center of mass to lie at either of these two points we get two subsectors of the twisted sector. These two subsectors correspond to the two conjugacy classes $(-1, 2k)$ and $(-1, 2k+1)$ discussed in the other picture. In this example the fundamental group is trivial, because $(-1, 1)$ and $(-1, 0)$ generate the space group and both of these elements have fixed points.

In the general case the point group will be much more complicated than Z_2 and it may be non-abelian, in which case we should take our states from each conjugacy class of the point group as discussed earlier. Of course each conjugacy class may contain subsectors corresponding to various choices for the center of mass coordinate, or in other words corresponding to the fact that there may be more than one conjugacy class in the space group for a given conjugacy class in the point group.

This general framework can also be used to describe the Ramond-Neveu-Schwarz formulation of the superstring. For simplicity consider only the right-moving modes. Then in addition to the space-time coordinates one has (in light-cone gauge) fermionic degrees of freedom $\psi^i(\sigma-\tau)$ transforming as an $\mathbf{8}_v$ under SO(8). In the Neveu-Schwarz sector the ψ^i obey anti-periodic boundary conditions, the ground state is a tachyon, the first excited state, $\psi^i_{-1/2}|0\rangle$, is a massless vector and so on. One then considers the Z_2 group generated by the operator $(-1)^F$ which anticommutes with fermionic fields and commutes with bosonic fields. Projecting onto Z_2-invariant states throws out the tachyon and all states at half-integer levels. To preserve modular invariance we must include the "twisted" or Ramond sector with $\psi^i(\pi)=-(-1)^F\psi^i(0)=\psi^i(0)$ and again project onto Z_2-invariant states. This is of course just the GSO [24] projection which ensures a supersymmetric spectrum of states. A similar description applies to the fermionic formulation of the gauge degrees of freedom of the heterotic string. In fact one of the motivations in constructing orbifolds was to generalize this twisting of fermionic fields to twisting of bosonic fields.

4. Constraints from modular invariance

In a closed string theory physical states must be invariant under global shifts in the space-like coordinate of the string world-sheet, $\sigma\to\sigma+\Delta$, reflecting the lack of a preferred point on the string. Since $L_0-\bar{L}_0$ is the generator of infinitesimal shifts of σ and the $(\text{mass})^2$ operator is proportional to $L_0+\bar{L}_0$ this implies that physical states receive equal contributions to their $(\text{mass})^2$ from left- and right-moving modes. In the heterotic string this projects out the tachyon from the left-moving bosonic string and results in an infinite tower of massive physical states. When we include twisted sectors of string propagation we must ensure that the theory remains tachyon-free and that L_0 and $\bar{L}_0$ eigenvalues continue to match. In the heterotic string these constraints are non-trivial. Consider a twist of order n acting on the space-time and/or gauge degrees of freedom. In the twisted sectors the mode numbers of the left- and right-moving oscillators will be shifted by an amount $\eta=r/n$ $(0\leqslant\eta<1)$ with r an integer. For a right- (left-) moving bosonic oscillator with moding shifted by η the normal ordering contribution to L_0 $(\bar{L}_0)$ is given by $f(\eta)=\frac{1}{2}\Sigma_{n\geqslant 0}(n+\eta)$. This can be computed using zeta function regularization. Alternatively, if we only want the η dependence there is a short cut. We note that $f(\eta+1)=f(\eta)-\frac{1}{2}\eta$ because the sum for $f(\eta+1)$ starts at $\eta+1$ and thus misses the first η. Given also an a priori knowledge that $f(\eta)$ is a polynomial in η, this equation can be solved to give $f(\eta)=\text{const}+\frac{1}{4}\eta(1-\eta)$. (The constant equals $-\frac{1}{24}$ as can be deduced in various ways.) For fermionic oscillators we get the same contribution with the opposite sign.

To derive the level matching conditions for L_0 and $\bar{L}_0$ we will work with Ramond-Neveu-Schwarz fermions for the superstring and use the fermionic formulation of $E_8\times E_8$. Consider a sector twisted by a group element g of finite order n, $g^n=1$, with its action on space-time degrees of freedom embedded in O(8) and its action on $E_8\times E_8$ embedded in an $O(16)\times O(16)$ subgroup. Diagonalize the action of g on the vector representation of these orthogonal groups by choosing a complex basis. That is, let g have eigenvalues $e^{2\pi i r_a/n}$ in the **4** of $U(4) \subset O(8)$ and eigenvalues $e^{2\pi i p_i/n}$ and $e^{2\pi i q_j/n}$ in the **8** of the two U(8) subgroups of $O(16) \times O(16)$. Although the fermions transform under the vector representation of these orthogonal groups, we also need to ensure that g has order n in the spinor representations. Remember that the adjoint of E_8 contains the double-valued spinor of SO(16) as well as the adjoint of SO(16). Thus in order that g has a sensible action in Spin(8) and in $E_8 \times E_8$ we must have

$$\sum r_a = 0 \bmod 2,$$

$$\sum p_i = 0 \bmod 2,$$

$$\sum q_i = 0 \bmod 2. \tag{6}$$

For n even these conditions can also be derived directly from the requirement that the L_0 and $\bar{L}_0$ levels match, when the $(-1)^F$ projections are properly included. For n odd they are not independent constraints and are always satisfied by an appropriate choice of the complex basis.

In the Ramond sector, with all fermions periodic before twisting, the normal ordering contribution to L_0 cancels between bosons and fermions and the contribution to $\bar{L}_0$ is

$$-1 + \tfrac{1}{2}\sum_a \frac{r_a}{n}\left(1 - \frac{r_a}{n}\right) - \tfrac{1}{2}\sum_i \frac{p_i}{n}\left(1 - \frac{p_i}{n}\right) - \tfrac{1}{2}\sum_j \frac{q_j}{n}\left(1 - \frac{q_j}{n}\right). \tag{7}$$

Since left- and right-handed oscillators have modings which are multiples of $1/n$, the existence of states with $L_0 = \bar{L}_0$ is equivalent to the requirement

$$\tfrac{1}{2}\left(\sum r_a - \sum p_i - \sum q_i\right) - \frac{1}{2n}\left(\sum r_a^2 - \sum p_i^2 - \sum q_i^2\right) = m, \tag{8}$$

with m integer. For n odd this is satisfied if

$$\sum r_a^2 - \sum p_i^2 - \sum q_i^2 = 0 \bmod n, \qquad n \text{ odd}, \tag{9}$$

while for n even we must require

$$\sum r_a^2 - \sum p_i^2 - \sum q_i^2 = 0 \bmod 2n, \qquad n \text{ even}. \tag{10}$$

Level matching in the Neveu-Schwarz sectors where some or all of the various groups of fermions start with anti-periodic boundary conditions is also ensured by (6), (9), (10).

The conditions (6), (9), (10) are necessary to ensure modular invariance. To argue that they are necessary recall that the construction of the string Hilbert space described in sect. 3 guarantees invariance under $\tau \to -1/\tau$ which interchanges σ and t boundary conditions. Invariance under $\tau \to \tau + 1$ requires not only that L_0 and $\bar{L}_0$ match for physical states, but that the states which survive the $(-1)^F$ projections and the projection onto group invariant states have *integer* eigenvalues for $L_0 - \bar{L}_0$. To see this recall that the propagator in a closed string theory has the form

$$\Delta = \int \frac{d^2 z}{|z|^2} z^{L_0} \bar{z}^{\bar{L}_0}, \tag{11}$$

with $z = e^{2\pi i \tau}$. The $\mathrm{Re}\,\tau$ piece of the propagator projects onto states satisfying $L_0 = \bar{L}_0$ when the integration is restricted to a fundamental region $-\frac{1}{2} < \mathrm{Re}\,\tau < \frac{1}{2}$, but only if $L_0 - \bar{L}_0$ has integer eigenvalues. But in a sector twisted by g, $\tau \to \tau + 1$ measures the g charge of each state since $\tau \to \tau + 1$ changes the (σ, t) boundary conditions from (g, h) to (g, gh). Therefore the projection onto group invariant states should ensure invariance under $\tau \to \tau + 1$.

A careful check that the conditions (6), (9), (10) are sufficient for modular invariance can be made by explicit evaluation [10] of the determinants appearing in the measure for the heterotic string [2] when the world sheet has the topology of a torus. As argued in [11] one must check invariance of the measure under modular transformations that leave the (g, h) boundary conditions fixed. A special case of the condition (10) was discussed in [11] where it was interpreted as a special case of a general requirement that the first Pontryagin classes of the $E_8 \times E_8$ bundle V and the tangent bundle T of the space-time manifold should be equal even including torsion. A similar interpretation should be possible here once the cohomology of orbifolds is properly defined [10]. If the quadratic conditions (9, 10) imply equality of the first Pontryagin classes then it is certainly plausible that the linear conditions (6) correspond to the vanishing of the Z_2 parts of the first Chern classes, or more precisely the second Stiefel-Whitney classes $w_2(V)$ and $w_2(T)$, which are precisely the obstructions to the existence of globally defined spinors.

We will now discuss examples of orbifolds which are consistent with the requirement of modular invariance. The first example corresponds to a discrete version of embedding the spin connection in the gauge group as in [3]. If we start with a six-torus then the point group will be a subgroup of SO(6); we wish to accompany the twist of the space-time coordinates by a corresponding twist in the SO(6) subgroup of E_8. To see that this ensures a tower of physical states we note that the L_0 and $\bar{L}_0$ eigenvalues match before twisting, but the twisting cannot affect this matching since the twist acts on the eight left-handed bosons the same as on the eight right-handed bosons, and on six of the left-handed O(16) fermions the same as on the eight right-handed space-time fermions, due to the identification of the spin and gauge connections. In the untwisted case the physical states all lie at integer (rather than half-integer) levels due to the $(-1)^F$ projections. The fermion number of the vacuum may change with respect to the untwisted case (for an example see the appendix), but it will change in the same way for left- and right-handed vacuum states. Thus this construction always leads to matching eigenvalues for L_0 and $\bar{L}_0$. The Z-orbifold discussed in [4] is an example of an orbifold with this standard embedding of the spin connection in the gauge group. It is possible to make a fairly complete classification of such orbifolds with small numbers of generations. We will discuss the possibilities in sect. 5. There are however more general possibilities which presumably correspond to orbifold limits of the theories discussed in [7]. These should allow one to find explicit examples of orbifolds which may solve various phenomenological problems along the lines discussed in [7]. Here we will be content to illustrate some of the possibilities by examining generalizations of the Z-orbifold.

One possibility is to choose a non-standard embedding of the point group in $E_8 \times E_8$. In the case of the Z-orbifold there are in fact very few such generalizations, so we will enumerate them here. Since the point group is abelian, $P = Z_3$, we can use shifts in the bosonic lattice representation of E_8 to describe how P acts on the gauge group, as discussed in the appendix. Denote the shift vector by $v = (v_1, v_2)$, where v_1 (v_2) is a shift vector for the first (second) E_8. Then the vectors $3v_i$ should both be E_8 lattice vectors, so that the spacetime rotation $\exp(\frac{2}{3}\pi i(J_{12} + J_{34} + 2J_{56}))$, which is of order 3, is accompanied by a gauge transformation of order 3 in $E_8 \times E_8$.

The bosonic formulation is convenient here because we can see that two shift vectors v_1 and v_1', say, which differ by an E_8 lattice vector must have *equivalent* actions on E_8. Also, if v_1 can be rotated into v_1' by an automorphism of the E_8 lattice, then v_1 and v_1' correspond to conjugate elements in E_8, and so in our Z_3 case they will result in identical patterns of symmetry breaking. These equivalences, plus the constraint (9) from modular invariance (with $n = 3$), leave only five distinct possibilities, as we will now see. When we rewrite (9) for the case of the Z-orbifold, and in terms of the bosonic formulation of the gauge degrees of freedom, it tells us that the (length)^2 of the shift vector, v^2, should be $\frac{2}{3}$ times some integer. But we also know that $3v_i$ is a lattice vector in the even E_8 lattice, so v_i^2 is $\frac{2}{9}$ times some integer, for each i. It can be shown [28] that any point in the eight-dimensional space containing the E_8 lattice is within the distance 1 from some lattice point. This result is important for us because it means that we can quite generally take the shift vector v_i to have $v_i^2 \leqslant 1$, by subtracting off an appropriate lattice vector from some longer but equivalent shift vector v_i^0. Then the only allowed combinations of lengths for the shift vectors v_i (modulo interchange of v_1 and v_2) are: (i) $v_1^2 = v_2^2 = 0$; (ii) $v_1^2 = \frac{2}{3}$, $v_2^2 = 0$; (iii) $v_1^2 = \frac{2}{9}$, $v_2^2 = \frac{4}{9}$; (iv) $v_1^2 = v_2^2 = \frac{2}{3}$; and (v) $v_1^2 = \frac{8}{9}$, $v_2^2 = \frac{4}{9}$. It turns out that these lengths lead to *unique* shift vectors v_i in each case (modulo the equivalences under automorphisms of the E_8 lattice discussed above).

We will now briefly describe these five possibilities. Case (i) is just the identity transformation, so it leads to gauge bosons in the adjoint representation of $E_8 \times E_8$

and of course no chiral fermions. In case (ii) we may write the shift vectors as $v_1 = (\frac{1}{3}, \frac{1}{3}, \frac{2}{3}, 0, \ldots, 0)$, $v_2 = 0$. So this Z_3 gauge transformation is $\exp(\frac{2}{3}\pi i(J_{12} + J_{34} + 2J_{56}))$ in O(16) and hence describes the "standard" identification of the spin connection with the gauge connection which was used in [4]. In case (iii) we may take $v_1 = (\frac{1}{3}, \frac{1}{3}, 0, \ldots, 0)$, $v_2 = (\frac{2}{3}, 0, \ldots, 0)$, which splits the action of Z_3 between the two E_8 factors as $\exp(\frac{2}{3}\pi i(J_{12} + J_{34} + 2J'_{12}))$ in $O(16) \times O(16)'$. Here the gauge symmetry is broken to $E_7 \times U(1) \times SO(14)' \times U(1)'$. Case (iv) embeds the spin connection once into each E_8 and so breaks $E_8 \times E'_8$ down to $E_6 \times SU(3) \times E'_6 \times SU(3)'$. This case maintains the symmetry between the two E_8 factors, as suggested in [7]. Finally, the shift vectors in case (v) may be written as $v_1 = (\frac{2}{3}, \frac{1}{3}, \frac{1}{3}, \frac{1}{3}, \frac{1}{3}, 0, 0, 0)$, $v_2 = (\frac{2}{3}, 0, \ldots, 0)$, and the gauge symmetry is broken to $SU(9) \times SO(14)' \times U(1)'$. For the Z-orbifold the point group is abelian and as a result the above cases do not lead to models that are phenomenologically promising. For non-abelian point groups the conditions (6, 9, 10) will be more difficult to satisfy but should also make it possible to obtain models where the rank of the unbroken gauge group is quite small.

Another possibility is to embed the space group nontrivially in $E_8 \times E_8$, rather than just the point group. Equivalently, we start with a torus with Wilson lines and then divide by the point group. Concretely, consider elements of the point group $(\alpha, 0)$ and generators of the translation group $(1, e_i)$. Choosing a homomorphism of the space group into $E_8 \times E_8$ means that we have group elements $(\alpha, 0, \beta), (1, e_i, \gamma_i)$ where β and γ_i are $E_8 \times E_8$ elements which have the same generators and relations as the corresponding space group. Thus $[e_i, e_j] = 0$ implies $[\gamma_i, \gamma_j] = 0$, and if α is a rotation of order n which fixes no component of e_i then

$$[(\alpha, 0, \beta)(1, e_i, \gamma_i)]^n = (\alpha, \alpha e_i, \beta\gamma_i)^n = (1, 0, (\beta\gamma_i)^n). \tag{12}$$

Thus in addition to the condition $\beta^n = 1$ which follows from the fact $\alpha^n = 1$ there is a condition on the Wilson lines that $(\beta\gamma_i)^n = 1$ for $E_8 \times E_8$ elements β which are associated with rotations that leave no components of e_i fixed.

As an example we again take the Z-orbifold. We start with a six-torus $T^6 = T_1 \times T_2 \times T_3$ where T_i is the torus defined by $z_i \sim z_i + e_{i1}$, $z_i \sim z_i + e_{i2}$ with $e_{i1} = 1$, $e_{i2} = e^{2\pi i/3}$. The point group is generated by α: $z_i \to z_i e^{2\pi i/3}$. Choosing an embedding of β in $E_8 \times E_8$ we have group elements $(\alpha, 0, \beta)$ with $\alpha^3 = \beta^3 = 1$. Embedding the translation group gives elements $(1, e_{i1}, \gamma_{i1}), (1, e_{i2}, \gamma_{i2})$ where the γ's commute with each other. Since α rotates all the e_i's we have the condition $(\beta\gamma_{i1})^3 = (\beta\gamma_{i2})^3 = 1$. In addition since $e_{i2} = \alpha e_{i1}$ we have $(1, e_{i2}, \gamma_{i2})(\alpha, 0, \beta) = (\alpha, e_{i2}, \gamma_{i2}\beta) = (\alpha, \alpha e_{i1}, \beta\gamma_{i1})$ so that $\gamma_{i2} = \beta\gamma_{i1}\beta^{-1}$. So in this example the most general possibility is to choose three commuting $E_8 \times E_8$ elements γ_{i1} and an $E_8 \times E_8$ element β with $\beta^3 = (\beta\gamma_{i1})^3 = 1$ subject to the constraints (6, 9, 10). Note that in the twisted sector the 27 fixed points are fixed points of the same element of the point group but correspond to different elements of the space group. Thus the different fixed points are associated with different gauge transformations in general. Models of this kind have also been considered by Strominger [12], from the point of view of the effective field theory.

It is also interesting to note that the presence of Wilson lines before dividing by the point group can lead to fractionally charged states as in [13] even though the fundamental group of the resulting orbifold may be trivial.

This procedure breaks $E_8 \times E_8$ down to the subgroup commuting with both the embedded point group elements β and the "Wilson lines" γ_i. This makes it rather easy to obtain explicit models where E_6 is broken down further to SO(10) or SU(5). This general approach contains both the non-standard embeddings described earlier, as well as true Wilson lines if the orbifold has a non-trivial fundamental group, and should make it possible to find models which solve many of the earlier phenomenological problems of superstring theories.

Finally we should mention that the embedding of the space group we described is a special case of an even more general class of possibilities. Narain [14] has shown that it is possible to classify modular invariant toroidal compactifications of the heterotic string down to four dimensions by a classification of self-dual lorentzian lattices with signature $(+1^{22}, -1^{6})$. It has also been shown [15] that these are equivalent to ordinary toroidal compactification with constant expectation values for gauge fields in the Cartan subalgebra $\langle A_i^I \rangle$ and antisymmetric tensor fields $\langle B_{ij} \rangle$. The models we have considered have $\langle A_i^I \rangle \neq 0$ but $\langle B_{ij} \rangle = 0$. There may be other models of interest which start with both $\langle A_i^I \rangle \neq 0$ and $\langle B_{ij} \rangle \neq 0$.

5. Models with the standard embedding of the point group

In this section we will give a fairly complete classification of models where the point group is embedded in the standard way in a single SO(6) subgroup of one E_8 factor. As argued earlier this will lead to level matching of L_0 and $\bar{L}_0$ in twisted sectors, hence to a modular invariant string theory. With this standard embedding, many choices of point group elements θ are unacceptable because they result in physical tachyons in some twisted sector. If we choose a diagonal basis for θ we can represent $\theta \in SO(6)$ by $\theta = \exp(2\pi i(aJ_{12} + bJ_{34} + cJ_{56}))$. We will show that $\pm a \pm b \pm c$ must be zero for some choice of signs in order for the theory to be free of tachyons. Because the eight eigenvalues of θ acting on the spinor of SO(8) are $e^{\pi i(\pm a \pm b \pm c)}$, there must then be at least two zero modes for the right-handed SO(8) spinor fermions in the Green-Schwarz formalism. So at least one four-dimensional supersymmetry will be preserved.

To show that $\pm a \pm b \pm c$ must be zero, we can first classify all the SO(6) elements which can act crystallographically, i.e. as automorphisms of six-dimensional lattices. Such elements must have order $m = 2$ through 10, 12, 14, 15, 18, 20, 24 or 30. (If we take into account the action of the elements on spinors as well the order may be doubled.) This is the six-dimensional analog of the fact that three-dimensional

TABLE 1
Acceptable point group elements $\theta = \exp 2\pi i(aJ_{12} + bJ_{34} + cJ_{56})$

$(\lvert a\rvert, \lvert b\rvert, \lvert c\rvert)$	Order	$\chi(F_\theta)$	$(\lvert a\rvert, \lvert b\rvert, \lvert c\rvert)$	Order	$\chi(F_\theta)$
$(0, \frac{1}{2}, \frac{1}{2})$	2	0	$(\frac{1}{6}, \frac{1}{3}, \frac{1}{2})$	6	12
$(0, \frac{1}{3}, \frac{1}{3})$	3	0	$(\frac{1}{7}, \frac{2}{7}, \frac{3}{7})$	7	7
$(\frac{1}{3}, \frac{1}{3}, \frac{2}{3})$	3	27	$(\frac{1}{8}, \frac{1}{4}, \frac{3}{8})$	8	4
$(0, \frac{1}{4}, \frac{1}{4})$	4	0	$(\frac{1}{8}, \frac{3}{8}, \frac{1}{2})$	8	8
$(\frac{1}{4}, \frac{1}{4}, \frac{1}{2})$	4	16	$(\frac{1}{12}, \frac{1}{3}, \frac{5}{12})$	12	3
$(0, \frac{1}{6}, \frac{1}{6})$	6	0	$(\frac{1}{12}, \frac{5}{12}, \frac{1}{2})$	12	4
$(\frac{1}{6}, \frac{1}{6}, \frac{1}{3})$	6	3			

crystals have only 2-, 3-, 4- or 6-fold rotational axes of symmetry. We do not have to check all elements with these orders; an element θ is unacceptable if the sector twisted by θ^n contains a tachyon. We then only need to check some ten elements with $\pm a \pm b \pm c$ not equal to zero. In order to determine if there is a tachyon we must compute the normal ordering contribution to the (mass)^2 operator for both the left- and right-moving modes. Using the results from sect. 4 we find that all elements with $\pm a \pm b \pm c \neq 0$ give rise to tachyons.

There remain thirteen classes of acceptable SO(6) elements which can act crystallographically without producing tachyons. These elements are listed in table 1 along with the number of fixed points for each element. The number of fixed points is easily determined using the Lefschetz fixed point theorem. A discrete symmetry θ maps the pth cohomology class H^p of the torus into itself. If we represent the action of θ on H^p by a $\binom{6}{p} \times \binom{6}{p}$ matrix θ_p and denote by F_θ the fixed point set of θ, then the Lefschetz fixed point theorem gives the Euler character $\chi(F_\theta)$ as

$$\chi(F_\theta) = \sum_{p=0}^{6} (-1)^p \operatorname{Tr} \theta_p = \det(1-\theta), \tag{13}$$

where the determinant is in the vector **6** representation of SO(6). If the fixed point set of θ consists only of points, i.e. has dimension zero, then $\chi(F_\theta)$ just counts the number of fixed points. If $\chi(F_\theta) = 0$ then θ has fixed tori, or perhaps no fixed points if it is accompanied by appropriate shifts v.

It may be possible to use one of the "unacceptable" elements if it is accompanied by a shift which causes it to have no fixed points. In this case the potential tachyon state which is in the twisted sector has an additional contribution to its energy since it must wind a finite distance around the orbifold. If the radius of the orbifold is sufficiently large then all states will have positive (mass)^2. However, since supersymmetry is broken the vacuum energy will depend on the orbifold radius and typically the orbifold will shrink until a tachyon appears. This situation was first studied by Rohm [16]. It is also possible in some cases to choose non-standard embeddings of the point group which shift the twisted left-moving ground states to positive or zero $(\text{mass})^2$, resulting in tachyon-free theories [29].

Neglecting the above exceptions, each element θ of P preserves a four-dimensional supersymmetry; that is, θ fixes some component of the spinor **4** of SO(6), or equivalently the vector **4** of SU(4). We will now show that the entire point group P also preserves a four-dimensional supersymmetry, i.e. P is a discrete subgroup of SU(3). Since we are interested in the action of P on spinors we consider the Lefschetz number

$$L(\mathscr{P}) = \sum_{i=0}^{4} (-1)^i \mathrm{Tr}_i \mathscr{P}, \tag{14}$$

where $\mathscr{P} = (1/|\mathrm{P}|)\Sigma_{\theta \in \mathrm{P}} \theta$ is a projection operator onto the P-invariant subspace of R^6 and $\mathrm{Tr}_i \mathscr{P}$ denotes the trace of $\mathscr{P}$ in the i times antisymmetrized tensor representation of SU(4). Now

$$\sum_{i=0}^{4} (-1)^i \mathrm{Tr}_i \theta = \det{}_1(1-\theta) = 0, \tag{15}$$

since in the **4** of SU(4) θ has at least one eigenvalue of unity, so $L(\mathscr{P}) = 0$. Using $\mathrm{Tr}_0 \mathscr{P} = \mathrm{Tr}_4 \mathscr{P} = 1$ for the trace over a singlet and $\mathrm{Tr}_1 \mathscr{P} = \mathrm{Tr}_3 \mathscr{P}$ for the trace over the **4** and $\bar{\mathbf{4}}$ we see that $0 = 2(1 - \mathrm{Tr}_1 \mathscr{P}) + \mathrm{Tr}_2 \mathscr{P}$. Now Tr_i is just the dimension of the invariant subspace. So if $\mathscr{P}$ has no invariant subspace acting on R^6, i.e. $\mathrm{Tr}_2 \mathscr{P} = 0$, then $\mathrm{Tr}_1 \mathscr{P} = 1$ and $\mathscr{P}$ fixes some component of the **4** of SU(4) or $\mathrm{P} \subseteq \mathrm{SU}(3)$ as claimed. If $\mathrm{Tr}_2 \mathscr{P} > 0$ then $\mathrm{Tr}_1 \mathscr{P} > 1$ and P lies in a proper Lie subgroup of SU(3).

To classify models generated by orbifolds with the standard embedding of the point group we therefore need only the classification of discrete subgroups of SU(3) [17]. Here we will consider only models with four or fewer generations. The number of generations is given by one half the Euler characteristic χ where [4]

$$\chi = \frac{1}{|\mathrm{P}|} \sum_{g,h \in \mathrm{P}} \chi_{g,h}. \tag{16}$$

Here the sum is over commuting pairs of elements, $|\mathrm{P}|$ is the order of the point group P, and $\chi_{g,h}$ is the Euler character of the intersection of the fixed point sets of g and h. This definition agrees with the Euler character of the manifold obtained by repairing the singularities of the orbifold in the cases where it is known how to do that. The "field theory" contribution to χ,

$$\chi_{\mathrm{ft}} = \frac{1}{|\mathrm{P}|} \sum_h \chi_{1,h}, \tag{17}$$

gives the strict Euler number of the orbifold as determined by simplicial decomposition.

The discrete subgroups of SU(3) can be divided into those which are contained in $SU(2) \times U(1)/Z_2$ and those which are not. The former will involve tensor products of the cyclic groups Z_k with the discrete subgroups of $SU(2)/Z_2$ and SU(2), which consist of Z_l, the dihedral group D_l, the point symmetry groups of the tetrahedron (T), cube (O) and the icosahedron (I), and their double covers. The remaining discrete subgroups of SU(3), like those of SU(2), form a few infinite sequences plus a finite number of "exceptional" groups. Since we want the subgroup to act as a symmetry of some six-dimensional lattice, all the elements must be of the type listed in table 1. This reduces the possible subgroups to a finite number to which (16) can be applied.

It is convenient to reduce the number of terms in the formula for the Euler character by rewriting it as a sum over the conjugacy classes of the point group. Since the fixed point set of the element $g_1 g g_1^{-1}$ is just the fixed point set of g rotated and translated by g_1, it is easy to see that $\chi_{g_1 g g_1^{-1}, g_1 h g_1^{-1}} = \chi_{g,h}$. Thus in the formula (16), the terms in the sum over elements g which come from a single conjugacy class are all equal. The number of elements in the conjugacy class of g (denoted by $\{g\}$) is $|G|/|C(g)|$, where $C(g)$ is the centralizer of g (the elements commuting with g). So the Euler character is

$$\chi = \sum_{\{g\}} \frac{1}{|C(g)|} \sum_{h \in C(g)} \chi_{g,h} . \tag{18}$$

One can generally simplify eq. (18) quite a bit further without knowing anything about the space group S or the lattice Λ. First of all, if e is the identity element of P, $\chi_{e,g} = \chi(F_g) = \det(1-g)$ is independent of the shift vector used in g, and is listed in table 1 for all the acceptable types of elements. Since x is a fixed point of gh if it is a fixed point of g and a fixed point of h, one sees that $\chi_{g,h} = \chi_{g,gh}$. Similarly, $\chi_{g,h} = \chi_{g^{-1},h}$, since the fixed point sets of g and g^{-1} are identical. Using these facts and table 1, one can evaluate all terms of the form χ_{g^m, g^n} in eq. (18). Usually this information alone is sufficient to show that four or fewer generations ($\chi \leqslant 8$) cannot be obtained using the given point group.

One finds that only four point groups P can yield four or fewer generations. Writing the group elements here as SU(3) matrices acting on three complex coordinates for R^6, we define the following group elements:

$$u = \begin{pmatrix} 0 & 1 & 0 \\ 0 & 0 & 1 \\ 1 & 0 & 0 \end{pmatrix}, \qquad \theta = \begin{pmatrix} 0 & -1 & 0 \\ 1 & 0 & 0 \\ 0 & 0 & 1 \end{pmatrix},$$

$r = \mathrm{diag}(1, -1, -1)$, $r' = uru^{-1} = \mathrm{diag}(-1, -1, 1)$, $\omega = \mathrm{diag}(e^{2\pi i/3}, e^{2\pi i/3}, e^{2\pi i/3})$, and $a = \mathrm{diag}(1, e^{2\pi i/3}, e^{-2\pi i/3})$. Let P be the 12-element tetrahedral group T generated by r and u. Then $\chi=\frac{1}{2}\chi_{r,r'}$. If P is the 24-element octahedral group O generated by r, u and θ, then $\chi=\frac{3}{4}\chi_{r\theta,r'}+\frac{1}{4}\chi_{r,r'}$. If P is the 8-element dihedral group D_4 generated by r and θ, then $\chi=\frac{3}{4}\chi_{r,r'}+\frac{3}{4}\chi_{r\theta,r'}$. And if P is the 27-element SU(3) "dihedral" group $\Delta(3\cdot 3^2)$ generated by ω, a and u, then $\chi=8+\frac{16}{9}(\chi_{\omega,a}+\chi_{\omega,u}+\chi_{\omega,au}+\chi_{\omega,a^2u})$. At this point, one must choose a lattice for P to act on, and a shift v for each element θ, in order to define the group $\overline{\mathrm{P}}$. The shifts must respect the point group structure. For example, if $\theta^n=1$ in P and $g=(\theta,v)$, then $g^n=1$ in $\overline{\mathrm{P}}$, so g^n is a pure translation by a lattice vector.

Consider first the case $\mathrm{P}=\Delta(3\cdot 3^2)$. P is a symmetry group for the root lattice of E_6 in particular. This lattice is generated by identifying $z_i \sim z_i+1 \sim z_i+e^{2\pi i/3}$ for each of three complex coordinates z_i, plus $(z_1,z_2,z_3)\sim(z_1,z_2,z_3)+(d,d,d)$, where $d=\sqrt{\frac{1}{3}}\,e^{i\pi/6}$. To get four generations ($\chi=8$) we require $\chi_{\omega,u}=\chi_{\omega,au}=\chi_{\omega,a^2u}=0$. One solution is to arrange the shifts v_a and v_u for a and u so that a, u, au and a^2u have no fixed points. On the E_6 root lattice, a consistent choice is $v_a=(f,f,f-d)$, $v_u=(f,f,f+d)$, where $f=\frac{1}{3}e^{i\pi/3}$. One can calculate the fundamental group by noting that all elements of the space group S which have fixed points have the form $(\omega^{\pm1},l)$ where l is any lattice vector. Modding out of S the group F generated by these elements leaves the fundamental group $Z_3\times Z_3$. The above choice of lattice and shifts which gives four generations is not unique. However, the fundamental group is always a subgroup of $Z_3\times Z_3$ for $\mathrm{P}=\Delta(3\cdot 3^2)$, because $(\omega^{\pm1},l)$ always has fixed points.

For the three remaining cases (P = T, O or D_4), we need to determine the possible values for $\chi_{r,r'}$, where $r=(\mathrm{diag}(1,-1,-1),v_r)$, $r'=(\mathrm{diag}(-1,-1,1),v_{r'})$, with v_r and $v_{r'}$ denoting possible shifts. In fact the possible values for $\chi_{r,r'}$ do not depend on the shifts. If $\chi_{r,r'}\neq 0$, then r and r' must each fix two-tori, which intersect at a finite number of points. We can choose our coordinate system so that one intersection is at the origin, which corresponds to setting v_r and $v_{r'}$ to zero. Next it is convenient to write r using a basis in which all lattice vectors are represented by integer column vectors. This basis may be chosen so that r has the form $\mathrm{diag}(1,1,-1,-1,-1,-1)$, $\mathrm{diag}(\sigma_1,1,-1,-1,-1)$ or $\mathrm{diag}(\sigma_1,\sigma_1,-1,-1)$, where $\sigma_1=\begin{pmatrix}0&1\\1&0\end{pmatrix}$. These forms have 16, 8 and 4 fixed tori, respectively. Similarly, one may classify the possible forms for the pair of r and r'. There are several inequivalent possibilities, but they all result in $\chi_{r,r'}=64$, 32, 16 or 8 (or zero, if shifts are included). (The fact that $\chi_{r,r'}\geqslant 8$ is needed to show that a few point groups which we have not mentioned here cannot yield four or fewer generations.)

It might appear that $\chi_{r,r'}=8$ or $\chi_{r\theta,r'}=8$ would allow one to construct a three-generation model based on the point group O or D_4. There is a unique form for r and r' which gives $\chi_{r,r'}=8$: $r=\mathrm{diag}(\sigma_1,\sigma_1,-1,-1)$, $r'=\mathrm{diag}(-1,-1,-\sigma_1,-\sigma_1)$. However, this form turns out to be incompatible with enlarging the symmetry group of the lattice from $Z_2\times Z_2$ (generated by r and r') to D_4 by adding the element θ! The same argument shows that $\chi_{r\theta,r'}\neq 8$ as well. Since D_4 is a subgroup of O, O also fails to give three generations, and in fact no three-generation model preserving supersymmetry is possible via orbifolds with the standard embedding of the point group in E_8.

The above form for r and r' does allow the symmetry group of the lattice to contain T, making two generation models possible. For example, the six-dimensional cubic lattice may be used, with the basis $\{(1, \pm i, 0), (i, 0, \pm 1), (0, 1, \pm i)\}$ (written in complex coordinates). The fundamental group turns out to be trivial for these models.

Four-generation models using T arise from taking $\chi_{r,r'} = 16$. (One such model was constructed by Strominger [26].) There are seven inequivalent forms for r and r' which give $\chi_{r,r'} = 16$, but only three of them allow T to be a symmetry group of the lattice. These three forms are represented by the following lattices: (i) two copies of the three-dimensional body-centered cubic lattice (BCC $\oplus$ BCC), with basis $\{(1,1,1), (1,-1,-1), (-1,-1,1), (i,i,i), (i,-i,-i), (-i,-i,i)\}$; (ii) the direct sum of a body-centered cubic lattice with a face-centered cubic lattice (BCC $\oplus$ FCC), with basis $\{(1,1,1), (1,-1,-1), (-1,-1,1), (i,i,0), (i,0,i), (0,i,i)\}$; and (iii) two copies of the face-centered cubic lattice (FCC $\oplus$ FCC), with basis $\{(1,1,0), (1,0,1), (0,1,1), (i,i,0), (i,0,i), (0,i,i)\}$. The fundamental group is $Z_2 \times Z_2$ in case (i), Z_2 in case (ii), and trivial in case (iii): One can show that the group structure of T implies that the shift accompanying u must be chosen so that u has fixed points as well as r. Since u and r generate T, the group F of elements in the space group S which have fixed points must include each element of T with *some* shift. In this case, a nontrivial fundamental group arises only if there are lattice vectors which are not in F. Since $(r,0)$ and $(1,l)(r,0)$ both have fixed points if the lattice vector l has the form $(0, l_2, l_3)$, the element $(1,l)$ must be in F. Similarly all translations by lattice vectors of the form $(l_1, l_2, 0)$ and $(l_1, 0, l_3)$ are in F. In case (iii) this implies that F contains the lattice basis and hence the entire lattice, so the fundamental group is trivial. In case (ii) $(1,1,1)$ is the single generator of the fundamental group (modulo F); and in case (i) $(1,1,1)$ and (i,i,i) are the generators.

Finally, one may construct two- and four-generation models based on the point group O. To get two generations one needs $\chi_{r,r'} = 16$. Since O contains T, one is restricted to the same three forms for r and r' mentioned in the previous paragraph. In fact, each of the above three lattices has O as a symmetry group, but in case (i) one cannot arrange $\chi_{r\theta,r'} = 0$ so one cannot get two generations. Cases (ii) and (iii) do lead to two-generation models, both with trivial fundamental group. In the case of four generations with point group O, $\chi_{r,r'} = 32$ is required. There are three inequivalent forms for r and r' which give $\chi_{r,r'} = 32$, but only two of them allow O as a symmetry group, and of these two only one allows a choice of shifts so that $\chi_{r\theta,r'} = 0$. The remaining case is represented by the direct sum of a face-centered cubic lattice with a cubic lattice (FCC $\oplus$ cubic), with basis $\{(1,1,0), (1,0,1), (0,1,1), (i,0,0), (0,i,0), (0,0,i)\}$. The fundamental group is again trivial. This exhausts the possible models with four or fewer generations based on this class of orbifolds.

6. Discussion

One view of string compactification on orbifolds is that it provides a singular limit of compactification on certain Calabi-Yau spaces which is amenable to direct analysis at the string theory level. It thus provides a useful testing ground for general topological arguments which are applied to compactification on Calabi-Yau spaces. As an example it has recently been shown that Yukawa couplings in the four-dimensional effective field theory can be computed in terms of the cohomology ring of the Calabi-Yau space [18, 19]. One of the puzzling results of [19] was that in certain cases the normalization matrix for kinetic energy terms of some of the massless modes diverged in the orbifold limit. In particular for the Z-manifold discussed in [3] there are 27 non-contractible four-surfaces B_i located at the bolts which come from repairing the singularities at the 27 fixed points of the Z-orbifold. The massless modes associated with the B_i have a normalization which vanishes in the orbifold limit, in which the volume of the B_i goes to zero. However, one might expect that string effects would effectively cut off the size of the B_i at the scale of the string tension, leading to finite Yukawa couplings. In string theory the massless modes associated with the B_i are just the massless modes in the twisted sector of string propagation.

Thus one can calculate Yukawa couplings directly in the string theory by calculating correlation functions of the appropriate twisted vertex operators. The operators for the emission of states belonging to the untwisted sector are relatively simple to construct [25] because they act within a single sector of the string Hilbert space. For example, if the string shown in fig. 2 is twisted by g, the vertex operators for the 3 untwisted states are just the usual vertex operators (e.g. $:e^{ik\cdot X}:$ for a tachyon in the Veneziano model), except that the operators X, etc., are now operators in the twisted sector Hilbert space H_g, so they contain fractionally moded oscillators in their mode expansion. The entire calculation can be done in H_g.

In contrast, fig. 3 shows an amplitude which cannot be calculated within a single sector of the Hilbert space, due to the presence of more than 2 external states from the twisted sector. However, the calculation is still tractable [20, 21]. If we choose a complex basis for the operators X^i, ψ^i, etc., in order to diagonalize g and h, we find that these operators acquire phases when they are carried around the locations z_i of the twisted vertex operators on the world-sheet (the complex z-plane). That is, $X(e^{2\pi i}(z-z_i)) = e^{i\phi_i} X(z-z_i)$, where $e^{in_i\phi_i} = 1$ if n_i is the order of the twist for

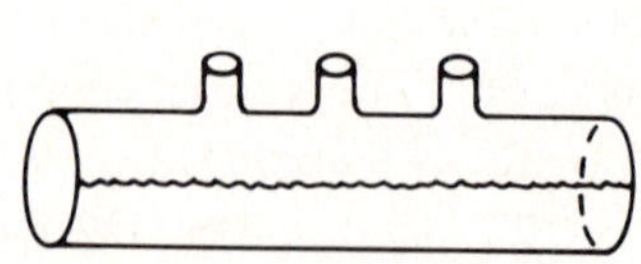

Fig. 2. Emission of three untwisted strings from a string twisted by g. The cut signifies that the string is not closed on the torus before dividing by g.

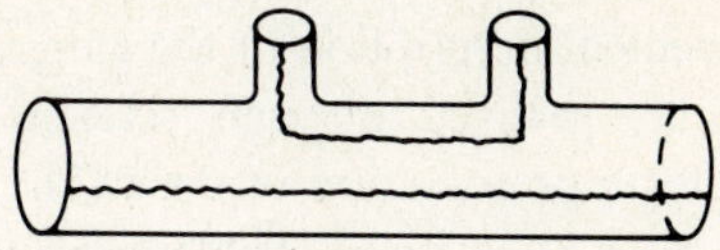

Fig. 3. Emission of a string twisted by h and a string twisted by h^{-1} from a string twisted by g.

state i. So correlation functions can be calculated in the z-plane with nth root branch cuts emanating from the locations of the twisted vertex operators. These calculations indicate that as expected Yukawa couplings are all finite when calculated directly in string theory [20, 21]. Since there is no difficulty in defining the normalization of states in the string Hilbert space, the vanishing of the normalization of the corresponding field theory states presumably can also be understood in terms of renormalization of D-terms in the appropriate supersymmetric non-linear sigma model.

The similarity in the mathematical treatment of symmetry breaking by Wilson lines, compactification on orbifolds, and symmetry rearrangement by twists acting purely on the gauge degrees of freedom (see the appendix) makes it tempting to speculate that orbifolds and twisted strings are in some sense more fundamental than the more general framework of compactification on Calabi-Yau spaces. Even if this possibility is not borne out it is quite possible that dynamical effects may drive a class of Calabi-Yau spaces toward the orbifold limit. Orbifolds have the advantage that they allow very explicit construction of quasi-realistic string compactifications. Although the models analyzed in detail in sect. 5 do not appear particularly promising from a phenomenological point of view, it should be possible to find much more promising models along the lines suggested in sect. 4.

Special orbifolds like the Z-orbifold can be related to smooth manifolds of SU(3) holonomy obtained by a classical operation of blowing up fixed points. One of the initial surprises in the theory of string propagation on orbifolds was the fact that string propagation on orbifolds makes perfect sense even for orbifolds not related to a smooth manifold by known operations. Recently, however, generalizations of the "blowing up" constructions have been found [27], which seem to apply whenever the point group is a subgroup of SU(3). It is conceivable that one of the lessons of orbifolds will turn out to be that the moduli space of conformally invariant sigma models is "better" than that of the corresponding manifolds, and that the conformal sigma models remain smooth even in limits (such as the orbifold limit) in which the corresponding manifolds become singular.

Appendix

In this appendix we give some examples of the construction of the string Hilbert space in the twisted sectors. We will concentrate mainly on the $E_8 \times E_8$ gauge degrees of freedom of the heterotic string.

In general the most convenient description of the gauge degrees of freedom is the fermionic formulation of E_8 realized through sixteen left-handed fermions ψ^I transforming as the **16** vector representation of the O(16) maximal subgroup of E_8. If the point group is abelian, it is as simple to describe the twisted sector through shifts in the bosonic lattice representation of E_8. In certain cases it is also possible to describe a non-abelian group of twists in the bosonic formulation if the group lies in the Weyl group of E_8; i.e. acts as an automorphism of the root lattice. This third construction bears the closest resemblance to the way twists on the spatial degrees of freedom are described, in which case the E_8 lattice is replaced by the spatial lattice defining the six-torus T^6. We will explain these three equivalent descriptions of the twisted gauge sectors through two examples in each case. First however we shall discuss briefly what is meant by a twisted affine Lie algebra.

Suppose we have a Lie algebra of a group with an automorphism ω. An automorphism of a Lie algebra is a linear invertible transformation that preserves the commutation relations, i.e.

$$[\omega T^i, \omega T^j] = \omega([T^i, T^j]). \tag{A.1}$$

An automorphism is said to be of order n if n is the smallest integer such that $\omega^n = 1$. We choose a basis for the Lie algebra in such a way that ω acts diagonally. (Note that in this basis the T^i are not hermitian.) The eigenvalues of ω are nth roots of unity, i.e. $e^{2\pi i\eta}$ with $\eta = k/n$ where k is an integer. We will denote the twist corresponding to the generator T^i by η^i.

The untwisted affine Lie algebra is defined by

$$[T^i_m, T^j_n] = f^{ijk} T^k_{m+n} + cm\delta_{m+n,0} \mathrm{Tr}\, T^i T^j \tag{A.2}$$

and the affine Lie algebra twisted by ω is defined by

$$[T^i_{m+\eta^i}, T^j_{n+\eta^j}] = f^{ijk} T^k_{m+n+\eta^k} + c(m+\eta^i)\delta_{m+n+\eta^i+\eta^j,0} \mathrm{Tr}\, T^i T^j, \tag{A.3}$$

where $\eta^k = \eta^i + \eta^j$ for nonzero f^{ijk}. This follows because ω preserves the commutation relations.

The affine Lie algebra has a simple interpretation in terms of currents on the string. The algebra corresponds to the usual current commutation relations with an additional Schwinger term, written in terms of the Fourier components. The untwisted algebra corresponds to periódic boundary conditions for the currents on the string, whereas the twisted algebra corresponds to twisting the boundary conditions:

$$T^i(2\pi) = e^{2\pi i\eta^i} T^i(0). \tag{A.4}$$

The currents $T^i(\sigma)$ may be represented in the E_8 case by either eight bosons or sixteen fermions [2].

In string theories $T^i_{n+\eta^i}$ are represented by operators which raise the eigenvalue of the hamiltonian (the $(\text{mass})^2$) by $n+\eta^i$. To discuss the gauge quantum numbers of various states we have to find generators which preserve the mass eigenvalue, i.e. that commute with the hamiltonian. These are the zero modes of generators which are invariant under the action of ω. So under the action of ω the gauge symmetry is broken to the subgroup which commutes with ω. What we will be describing through twisted string Hilbert spaces of the gauge group is simply the realization of this twisted affine Lie algebra. If the automorphism ω is inner (that is, if it can be written as a group element) then the twisted affine Lie algebra is in fact isomorphic to the untwisted algebra. However the isomorphism does not commute with the hamiltonian so the two constructions do not lead to isomorphic physics.

Let us fix an O(16) maximal subgroup of E_8 and pick an orthonormal Cartan subalgebra of it whose elements we denote by $J_{i,i+1}$ with i running over odd numbers from 1 to 15. We will consider two twistings, the first one by $\omega_1 = e^{2i\pi J_{12}}$ and the second by $\omega_2 = e^{i\pi(J_{12}+J_{34})}$. Both of these twists are of order two; i.e. the square of both equals one. They correspond to the breakings

$$E_8 \to O(16)\,,$$
$$E_8 \to E_7 \times SU(2)\,, \tag{A.5}$$

respectively. (If we instead identify J_{12} and J_{34} with space-time rotation generators, the first one leaves the space-time bosonic degrees of freedom untouched, while multiplying the fermionic degrees of freedom by -1. The second one acts on the coordinates by $x^i \to -x^i$ for i from 1 to 4, and leaves the other space-time coordinates invariant.) We will now describe the construction of the Hilbert space for these two twistings in three ways, by twisting fermions, by shifting on the lattice, and by rotations on the lattice.

Let us first review the untwisted fermionic construction of E_8 in the heterotic string. For the fermionic representation of E_8 without the twisting we have two sectors, one periodic and one antiperiodic. In both of these sectors we have to project onto a positive $(-1)^F$ subspace. From the first sector we obtain as the ground state a spinor **128** of O(16) which is obtained by representing the Clifford algebra of the fermionic zero modes. The ground state of the second sector is shifted down from the first sector because the contribution of zero-point energies of oscillators with different boundary conditions is different. Recall that the normal ordering term for a bosonic (fermionic) oscillator twisted by $0 \leqslant \eta < 1$ is $(-)$ $(-\frac{1}{24}+\frac{1}{4}\eta(1-\eta))$. Thus in the antiperiodic sector the ground state energy is lowered by $16(\frac{1}{16})=1$. The ground state is a singlet. The first excited states correspond to $\psi^I_{-1/2}$ acting on the ground state, which gives a **16** of O(16). If we assign the ground state to be invariant under $(-1)^F$ this excited state is odd under $(-1)^F$ and will be projected out. The next state corresponds to $\psi^I_{-1/2}\psi^J_{-1/2}$ acting on the ground state and will form the adjoint **120** of O(16), and furthermore is even under $(-1)^F$. The mass level of this excited state matches the ground state of the periodic sector which is a spinor of O(16). The two sectors together will therefore give the adjoint of E_8: **248** = **120** + **128**.

In the twisted constructions we still have two sectors, but with modified boundary conditions. In the first sector $\psi^I(2\pi) = \omega\psi^I(0)\omega^{-1} = R^I_J(\omega)\psi^J(0)$. (Here R is the rotation matrix corresponding to ω.)

Let us consider twisting by ω_1. In this case the rotation matrix is just the identity, because the fermions are in the vector representation of O(16). So it seems that we do not change the boundary conditions in each sector and that we should get the same oscillators in each sector. However, what is different now is that the ground states behave differently under $(-1)^F$. In particular the ground state of the antiperiodic sector which was even under $(-1)^F$ is now odd under it. (To see this one can continuously change the boundary conditions and note that one fermionic state crosses the ground state in the process, thus changing the fermion number of the vacuum.) What we end up doing, therefore, is keeping the states which we projected out in the usual construction of E_8. So in the antiperiodic sector the first state which is even under $(-1)^F$ is the **16** of O(16). In the periodic sector we get the spinor of O(16) as the ground state, except with the opposite chirality. Furthermore we have no corresponding state from the antiperiodic sector to complete it to the adjoint of E_8. The symmetry is broken to O(16) and the states form representations of O(16) and not of E_8 (see fig. 4). Note that charge fractionalization occurs in the twisted sector: The **16** of O(16) is in the "wrong" conjugacy class for the embedding $E_8 \supset O(16)$.

Now consider twisting by ω_2. In this case $\psi^1, \psi^2, \psi^3, \psi^4$ change their boundary conditions. In the periodic sector they become antiperiodic and in the antiperiodic sector they become periodic. The rest of the fermions will have the same boundary conditions as before. In the first sector we have 12 integral modes and 4 half-integral modes. In the second sector we have 12 half-integral modes and 4 integral modes. The second sector is shifted down by $\frac{1}{2}$ relative to the first sector. The ground state of the first sector is a **32** spinor of O(12) which one gets by

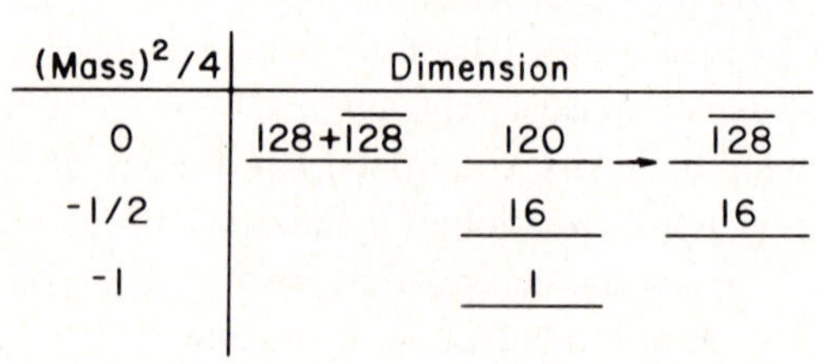

Fig. 4. Spectrum of low-lying states for $E_8 \to O(16)$ symmetry breaking. States are labeled by O(16) quantum numbers, the states to the right of the arrow indicate states that survive the modified $(-1)^F$ projection.

$(\text{Mass})^2/4$	Dimension			
$-1/4$	(32,1)+(32′,1)	(12,2)+(12,2′)	→	(56,1)
$-3/4$		(1,2)+(1,2′)		(1,2)

Fig. 5. Spectrum of low-lying states for $E_8 \to E_7 \times SU(2)$ symmetry breaking. States are labeled by $SO(12) \times SO(4)$ to the left of the arrow, to the right the states surviving the $(-1)^F$ projection are reassembled into $E_7 \times SU(2)$ multiplets.

representing the 12 zero modes and projecting onto a definite chirality. The ground state of the second sector is a **2** spinor of O(4) of a definite chirality. The first excited state of the second sector is obtained by acting with $\psi^5_{-1/2}, \dots,$ or $\psi^{16}_{-1/2}$ on the spinor of O(4) of the opposite chirality, so we get two vector representations of O(12). This state is at the same mass level as the ground state of the first sector and together they form **56** = 32 + 2(12). So putting the two sectors together we see that the ground state is doubly degenerate, corresponding to the doublet of SU(2), and the first excited state is 56-fold degenerate, corresponding to the fundamental representation of E_7. E_8 is broken to $E_7 \times SU(2)$ and so each level forms a representation of $E_7 \times SU(2)$ (see fig. 5). Again charge fractionalization occurs, as neither the **(1, 2)** nor the **(56, 1)** of $E_7 \times SU(2)$ appear in the decomposition of any E_8 representation.

We will now see how the twisted sector can be realized in the lattice formulation of E_8. We shall first review the bosonic construction of E_8 in the untwisted heterotic string. Recall that the root lattice of E_8 in an orthonormal basis consists of vectors of the form $(n_1, \dots, n_8)$ and $(n_1 + \frac{1}{2}, \dots, n_8 + \frac{1}{2})$ with n_i being integers satisfying the constraint that $\Sigma n_i =$ even. The mass eigenstates of the E_8 are given by various winding states on the lattice, acted upon by the 8 bosonic oscillators which correspond to excitations of these wound states. The mass levels are given by $\frac{1}{2} W^2 + N$ where N is the number of excited oscillators and W is the vector whose components give the winding numbers in different directions. The ground state has winding number zero and is a singlet of E_8. The next level is obtained by taking one of the 240 states with $\frac{1}{2} W^2 = 1$, or by letting one of the 8 bosonic oscillators α^i_{-1} act on the ground state. Together they form the **248** adjoint of E_8.

Now consider twisting the boundary conditions. Since we have chosen ω_1 and ω_2 to lie in the Cartan subalgebra of E_8, they commute with the Cartan subalgebra currents $\partial_{\bar{z}} X$ where X denotes the coordinates of the E_8 torus. So ω_i can only shift the coordinates by some constant. In the case of ω_1 we have

$$\omega_1 X \omega_1^{-1} = X + v_1 , \qquad (A.6)$$

where $v_1 = (1, 0, 0, 0, 0, 0, 0, 0)$. The boundary condition $X(2\pi) = X(0) + v_1$ (up to a vector in the lattice of E_8) implies that the winding vector now lies on the E_8 lattice

$+v_1$; i.e., $W = W_0 + v_1$ where W_0 is a vector in the E_8 lattice. The ground state is now 16-fold degenerate due to the 16 choices for W_0, $(-1,0,\ldots,\pm1,0,\ldots)$, $(-2,0,\ldots)$ and $(0,0,\ldots)$, which give $W=(0,\ldots,\pm1,0,\ldots)$, so that $\frac{1}{2}W^2=\frac{1}{2}$. These states correspond to the vector representation of O(16). Similarly one sees that the next state corresponds to the spinor **128** of O(16), with the opposite chirality to the one in the untwisted E_8. These are just the states obtained by twisting in the fermionic representation (see fig. 4). If we had calculated the contribution of the zero point energies of the oscillators they would have also agreed.

Similarly twisting by ω_2 corresponds to the shift $v_2 = (\frac{1}{2},\frac{1}{2},0,0,0,0,0,0)$. The ground state is now doubly degenerate due to the two choices $(0,0,\ldots)$ and $(-1,-1,0,\ldots)$ for W_0. These correspond to the doublet of SU(2). The next level is similarly seen to be 56-fold degenerate and corresponds to the fundamental representation of E_7. These results agree with what we obtained for the fermionic representation (see fig. 5). The equivalence of shifting in the lattice and twisting in the fermionic representation can be checked by considering the partition function in either picture, and seeing that they agree. The equality of the two partition functions is basically a consequence of

$$\frac{q^{-\eta^2/2}\Sigma q^{(n+\eta)^2/2}}{\Pi(1-q^n)} = \prod_{n\geqslant 0}\left(1+q^{n+\eta+1/2}\right)\left(1+q^{n-\eta+1/2}\right), \qquad (A.7)$$

which follows from the infinite product representations for the Jacobi theta functions.

In the above cases it turns out that we can realize the twist in yet a different, but equivalent way on the root lattice of E_8, by choosing ω_1 and ω_2 *not* to lie in the Cartan subalgebra. In the Cartan-Weyl basis for a Lie algebra, the generators are divided into the generators H_i of the Cartan subalgebra, and the "raising and lowering" operators E_α (α is a root), with the commutation relations $[H_i, H_j]=0$, $[H_i, E_\alpha]=\alpha_i E_\alpha$, $[E_\alpha, E_{-\alpha}]=\alpha\cdot H$, etc. So if $\omega_\alpha = \exp[\frac{1}{2}i\pi(E_\alpha+E_{-\alpha})]$, then $\omega_\alpha H_i \omega_\alpha^{-1} = H_i - \alpha_i \alpha\cdot H$, and ω_α represents a reflection in the hyperplane orthogonal to the root α on the E_8 root lattice. The ω_α generate the Weyl group of E_8. We may represent the twist ω_1 as a product of ω_α's: $\omega_1 = \prod_{j=1}^{8}\omega_{\alpha_j}$, where the eight $\alpha_j = (1,\pm1,0,\ldots,0)$, $(0,0,1,\pm1,0,\ldots,0)$, $(0,\ldots,0,1,\pm1,0,0)$, $(0,\ldots,0,1,\pm1)$. Then $\omega_1^2=1$ and $\omega_1 H_i \omega_1^{-1} = -H_i$. The action of ω_1 on the coordinates of the E_8 torus is therefore

$$\omega_1 X \omega_1^{-1} = -X,$$

i.e. ω_1 inverts all the coordinates. The boundary condition $X(2\pi) = -X(0)$ (up to an E_8 lattice vector) prohibits all winding states, and shifts the bosonic oscillator mode numbers to the half-integers. Had we been discussing bosons with both left- and right-moving modes, there would have been an overall factor of 256 in the counting of states, corresponding to the center of mass of the string sitting at one of the 2^8 solutions of $X^i = -X^i$ (modulo the E_8 lattice). However, the E_8 bosons have only left-moving modes and one must work not with the E_8 lattice Γ_8 but with a projective representation of it, $\bar{\Gamma}_8$. Thus rather than simply solving $X^i = -X^i$ we should find a faithful representation of the group $\overline{\Gamma_8/2\Gamma_8}$. The lowest dimensional faithful representation of this group has dimension $16 = 2^4$. So in effect we have to take the square root of the number of fixed points. This is also required to maintain modular invariance. The ground state forms a **16** of O(16) and the next state, obtained by exciting the ground state with each of the eight oscillators $\alpha^i_{-1/2}$, forms a **128** spinor. The partition function

$$16q^{-1/2}\prod_{n>0}\left(1-q^{n-1/2}\right)^{-8} = 16q^{-1/2}\prod_{n>0}\left(1+q^{n/2}\right)^{8} \tag{A.8}$$

is equal to that found in the two previous constructions, as may be shown by squaring the "aequatio identico satis abstrusa" of Jacobi,

$$\tfrac{1}{2}\left(\prod_{n>0}\left(1+q^{n-1/2}\right)^8 - \prod_{n>0}\left(1-q^{n-1/2}\right)^8\right) = 8q^{1/2}\prod_{n>0}\left(1+q^{n}\right)^8, \tag{A.9}$$

and substituting.

In the second case we may write

$$\omega_2 = \prod_{i=1}^{4}\exp\left[\tfrac{1}{2}i\pi\left(E_{\alpha_i}+E_{-\alpha_i}\right)\right], \tag{A.10}$$

where the product runs over only the first four of the roots α_j used in ω_1. Let "a" denote indices from 1 to 4, and "b" indices from 5 to 8. Then

$$\omega_2 H_a \omega_2^{-1} = -H_a, \qquad \omega_2 H_b \omega_2^{-1} = +H_b. \tag{A.11}$$

The appropriate boundary conditions on $X(\sigma)$ are

$$X^a(\pi) = -X^a(0) + \sum_{k=1}^{8} n_k e_k^a,$$

$$X^b(\pi) = +X^b(0) + \sum_{k=1}^{8} n_k e_k^b, \tag{A.12}$$

where n_k are integers and e_k are the E_8 basis vectors. Expansion of

$$X^i(\sigma,\tau) = x^i + W^i(\tau+\sigma) + \tfrac{1}{2}i\sum_{n\neq 0}\alpha^i_n e^{-2in(\tau+\sigma)} \tag{A.13}$$

shows that four of the oscillators have half-integral modes and four have integral modes. We also find that $W^a = 0$, $W^b = \Sigma_{k=1}^{8} n_k e_k^b$, $x^a = 2\pi \frac{1}{2} \Sigma_{k=1}^{8} n_k e_k^a (\mathrm{mod}\ \mathrm{E}_8$ lattice) and x^b is unrestricted. So the winding vectors W^i must lie on the lattice formed by projecting the E_8 lattice onto the subspace left invariant by the automorphism ω_2. In this case the lattice is the root lattice of F_4, which has a basis $\{(0,\dots,0,1,0,0,0),(0,\dots,0,0,1,-1,0),(0,\dots,0,0,0,1,-1),(0,\dots,0,\frac{1}{2},\frac{1}{2},\frac{1}{2},\frac{1}{2})\}$. The location of the set of allowed x^a's depends on the winding vector chosen, but the set always consists of four distinct four-tori. (For example, for the ground state, $W^i = 0$, the tori pass through the origin, $(\frac{1}{2},0,0,0,\dots)$, $(0,\frac{1}{2},0,0,\dots)$ and $(0,0,\frac{1}{2},0,\dots)$.) Again, since only left-moving bosons are present, the overall degeneracy factor is $\sqrt{4} = 2$. The ground state is a doublet of SU(2). The next excited level has $2 \times 4 = 8$ components obtained by exciting the ground state with each of the four oscillators $\alpha^a_{-1/2}$, and $2 \times 24 = 48$ components obtained by using each of the 24 points with $(\text{length})^2 = 1$ on the F_4 root lattice as a winding vector. This level forms the **56** of E_7.

The partition function is

$$2q^{-3/4}\theta_{F_4}(q)\prod_{n>0}(1-q^n)^{-4}(1-q^{n-1/2})^{-4}, \tag{A.14}$$

where

$$\theta_{F_4}(q) = \sum_{W \in \Lambda_{F_4}} q^{W^2/2} = \left[\sum_{n \in Z} q^{n^2/2}\right]^4 + \left[\sum_{n\in Z} q^{(n+1/2)^2/2}\right]^4 \tag{A.15}$$

is the theta function for the F_4 root lattice. The identity (25) can be used to show that the partition function again equals that found in the two previous constructions.

We have only described here the representation of two simple twists on $\mathrm{E}_8 \times \mathrm{E}_8$. In general the twists will be more complicated and will be accompanied by shifts or twists on the six-dimensional compact space in order to describe Wilson lines or to embed the spin connection in the gauge group as described in the text. Also we must project onto invariant states in each twisted sector as described in sect. 3.

One can also contemplate dividing the theory by a twist acting only on the internal degrees of freedom representing $\mathrm{E}_8 \times \mathrm{E}_8$ or $\mathrm{Spin}(32)/\mathrm{Z}_2$. As we will now show this allows one to view the $\mathrm{E}_8 \times \mathrm{E}_8$ and $\mathrm{Spin}(32)/\mathrm{Z}_2$ heterotic strings as simply twisted versions of each other.

Consider the $\mathrm{E}_8 \times \mathrm{E}_8'$ theory twisted by the element $\theta = \omega_1\omega_1'$, where ω_1 acts on the first E_8 and ω_1' on the second, and ω_1 is the twist described earlier. In the fermionic formulation this twist does not change the boundary conditions on the fermions, but only the $(-1)^F$ projection. We have 4 sectors corresponding to

$(\text{Mass})^2/4$	Dimension		
0	(120, 1)	(1, 120)	(16, 16)
-1/2	(16, 1)	(1, 16)	
-1	(1, 1)		

Fig. 6. Spectrum of low-lying states in the (NS, NS) sector in the fermionic formulation of $E_8 \times E_8$ before projection with $(-1)^F$. The states are labeled by $O(16) \times O(16)$ quantum numbers.

periodic (P) or antiperiodic (AP) boundary conditions for the fermions representing E_8 and E_8'. In the (AP, AP) sectors before projecting with $(-1)^F$ the spectrum is as shown in fig. 6. In terms of $E_8 \times E_8' \supset SO(16) \times SO(16)'$, the usual (untwisted) $(-1)^F$ projection keeps the $(\mathbf{120}, \mathbf{1}) + (\mathbf{1}, \mathbf{120}')$, which combine with the $(\mathbf{128}, \mathbf{1})$ and $(\mathbf{1}, \mathbf{128}')$ from the (P, AP) and (AP, P) sectors to form the adjoint of $E_8 \times E_8$.

When we consider the theory twisted by θ we must project onto θ-invariant states in the untwisted sector and then onto the θ-invariant states in the twisted sector. In the untwisted sector the $(-1)^F$ projection is unchanged but we must project onto states with $\theta = 1$. θ has eigenvalue $+1$ on the adjoint of SO(16) and eigenvalue -1 on the spinor or spinor′ of SO(16). So the $\theta = 1$ projection keeps the $(\mathbf{120}, \mathbf{1}) + (\mathbf{1}, \mathbf{120}')$ massless states from the (AP, AP) sector, but projects out the $(\mathbf{128}, \mathbf{1})$ and $(\mathbf{1}, \mathbf{128}')$ from the (P, AP) and (AP, P) sectors. In the twisted sector the $(-1)^F$ projection is changed, since the vacuum fermion number changes by one for each set of 16 fermions. The θ projection acts as before, so in this sector the massless states are just the $(\mathbf{16}, \mathbf{16}')$ coming from the (AP, AP) sector. Thus at the massless level we have $(\mathbf{120}, \mathbf{1}) + (\mathbf{1}, \mathbf{120}') + (\mathbf{16}, \mathbf{16}')$, which is just the $SO(16) \times SO(16)'$ content of the adjoint of SO(32). This construction maintains modular invariance and thus, by the uniqueness [22] of the modular form of weight 8, $\theta_{\Gamma_{16}}(q) = [\theta_{\Gamma_8}(q)]^2$, the partition function for this twisted construction must agree with the partition function for $\text{Spin}(32)/Z_2$ (which is the same as that for $E_8 \times E_8$). (This is also easy to verify directly.) Thus we see that the basic plus spinor representations of the SO(32) affine Lie algebra that appear in the $\text{Spin}(32)/Z_2$ heterotic string can be viewed as simply a sum of untwisted and twisted representations of the $E_8 \times E_8$ affine Lie algebra with the appropriate projections. A similar construction allows one to obtain $E_8 \times E_8$ from $\text{Spin}(32)/Z_2$. In fact any twist acting only on $E_8 \times E_8$ or $\text{Spin}(32)/Z_2$ that satisfies the conditions for modular invariance (6, 9, 10) will by uniqueness reproduce the spectrum of states of either $E_8 \times E_8$ or $\text{Spin}(32)/Z_2$. The close relation between these theories in ten dimensions and the fact that they are continuously related upon toroidal compactification [14, 15] strongly suggest that they should be viewed as different vacua of an underlying fundamental second-quantized theory.

References

[1] J.H. Schwarz, Phys. Reports 84 (1982) 223;
M.B. Green, Surveys in High Energy Phys. 3 (1982) 127;
M.B. Green and J.H. Schwarz, Phys. Lett. 144B (1984) 117
[2] D.J. Gross, J.A. Harvey, E. Martinec and R. Rohm, Phys. Rev. Lett. 54 (1985) 502; Nucl. Phys. B256 (1985) 253, B267 (1986) 75
[3] P. Candelas, G. Horowitz, A. Strominger and E. Witten, Nucl. Phys. B258 (1985) 46
[4] L. Dixon, J.A. Harvey, C. Vafa and E. Witten, Nucl. Phys. B261 (1985) 651
[5] Y. Hosotani, Phys. Lett. 129B (1984) 193
[6] E. Witten, Proc. 1983 Shelter Island Conference, ed. N. Khuri et al. (MIT Press, 1985)
[7] E. Witten, Nucl. Phys. B268 (1986) 79
[8] I.B. Frenkel, J. Lepowsky and A. Meurman, *in* Vertex operators in mathematics and physics, Publications of the Mathematical Sciences Research Institute No. 3 (Springer, Berlin, 1984)
[9] H. Seifert and W. Threlfall, A textbook of topology (Academic Press, New York, 1980)
[10] C. Vafa, to appear
[11] E. Witten, *in* Anomalies, geometry and topology, ed. A. White (World Scientific, 1985)
[12] A. Strominger, Santa Barbara preprint (1985)
[13] X.-G. Wen and E. Witten, Nucl. Phys. B261 (1985) 651
[14] K. Narain, Phys. Lett. 169B (1986) 41
[15] K. Narain, M.H. Sarmadi, and E. Witten, Rutherford Lab. preprint (1986)
[16] R. Rohm, Nucl. Phys. B237 (1984) 553
[17] M.W. Fairbairn, T. Fulton and W.H. Klink, J. Math. Phys. 5 (1964) 1038
[18] A. Strominger and E. Witten, Comm. Math. Phys. 101 (1985) 341
[19] A. Strominger, Nucl. Phys. B274 (1986) 253
[20] L. Dixon, D. Friedan, E. Martinec and S. Shenker, to appear
[21] S. Hamidi and C. Vafa, to appear
[22] J.P. Serre, A course in arithmetic (Springer, New York, 1973)
[23] R.L.E. Schwarzenberger, N-dimensional crystallography (Pitman, San Francisco, 1980)
[24] F. Gliozzi, J. Scherk and D. Olive, Nucl. Phys. B122 (1977) 253
[25] J. Lepowsky, preprint (1985);
V.G. Kac and D.H. Peterson, *in* Anomalies, geometry and topology, ed. A. White (World Scientific, 1985)
[26] A. Strominger, preprint NSF-ITP-85-105 (1985)
[27] S.-S. Roan and S.-T. Yau, UCSD preprint (1985)
[28] H.M.S. Coxeter, Twelve geometric essays (Southern Illinois Univ. Press, 1968)
[29] L. Dixon and J. Harvey, Nucl. Phys. B274 (1986) 93
L. Dixon, J. Harvey, and F. MacKintosh, to appear
[30] W. Plesken and M. Pohst, Math. Comp. 31 (1977) 552

ORBIFOLDS AND WILSON LINES

L.E. IBÁÑEZ, H.P. NILLES and F. QUEVEDO
CERN, CH-1211 Geneva 23, Switzerland

Received 12 December 1986

We study the consequences of the presence of gauge background fields on the torus underlying orbifold compactification of the heterotic string. It is pointed out that such Wilson lines provide us with a mechanism for controlling the number of chiral matter states both from twisted and untwisted sectors, as well as breaking the symmetry group. Starting from the Z orbifold, we can construct a variety of four-dimensional string models with three families of quarks and leptons and different gauge groups such as E_6, $SU(3)^3$, $SU(6)\times U(1)$ or $SU(5)\times[SU(2)\times U(1)]^2$.

1. Introduction. A clarification of an eventual connection between superstring theories and physics necessitates a better understanding of four-dimensional string theories. Compactification of higher-dimensional string theories on a torus $T^n = R^n/\Lambda$ is well·understood [1] but seems to be unrealistic because of the absence of chiral fermions. The consideration of orbifolds [2,3] (or, more precisely, of twisted tori $O^n = T^n/P$ [‡1]) seems to be more promising in this respect, still sharing the conceptual simplicity of torus compactification [4,5]. In this direction the work has mainly been concentrated on the choice of an abelian point group with standard embedding in the gauge group $E_8\times E_8$ of the heterotic string [3]. Whilst this leads to encouraging results concerning the breakdown of E_8 and the existence of chiral fermions, it usually leads to an embarrassing proliferation of matter fields. Also, one example with nonstandard embedding has been worked out [6].

In this paper we consider the inclusion of background gauge fields corresponding to noncontractible loops on T^n in the framework of orbifold compactification. Such Wilson lines can be equivalently understood as the embedding of the full space group S into the gauge group of the high-dimensional string theory. We show, as might have been expected [3], that this leads to more flexibility with respect to gauge symmetry breakdown. However, it will also be shown, as might not have been expected from the beginning, that such Wilson lines provide us with a mechanism for controlling the number of matter multiplets. In the framework of the Z orbifold [2] we will see that this method allows us to easily construct models with three generations of quarks and leptons with semi-realistic gauge groups such as E_6, $SU(6)\times U(1)$, $SU(3)^3$, and $SU(5)\times[SU(2)\times U(1)]^2$. In a certain way our work is complementary to the attempts to classify all possible four-dimensional string models on the basis of modular invariant partition functions [7,8]. Although here a very large number of such models can be classified, a mechanism for selecting the interesting ones is still lacking.

Let us therefore consider the motion of closed strings on the twisted torus $O^n = T^n/P \cong R^n/S$. Consistency requires that the action of P leaves the torus T^n invariant. We denote the two-dimensional coordinates by $\sigma = \sigma_1 + \tau\sigma_2$, where τ is the modular parameter. We then observe the existence of several sectors of closed strings denoted by (g, h), where g and h are elements of P:

$$X(\sigma_1+\pi, \sigma_2) = hX(\sigma_1, \sigma_2),$$
$$X(\sigma_1, \sigma_2+\pi) = gX(\sigma_1, \sigma_2)\ , \tag{1}$$

and g and h have to commute for the boundary conditions to be compatible. The contribution of the $(g,$

[‡1] P is a discrete group called the point group. Equivalently, we can consider O^n/S, where the space group S contains P as well as the shifts Λ of the lattice defining the torus T^n.

0370-2693/87/$ 03.50 © Elsevier Science Publishers B.V.
(North-Holland Physics Publishing Division)

h) sector to the partition function is given by Tr $g \times e^{2\pi i\tau H}$, with H being the hamiltonian corresponding to the h-twist. The sector $h=1$ is called the untwisted sector; it corresponds to strings already closed on T^n. Because of the division of T^n by P, we can keep only those states in the Hilbert space of strings moving on T^n that are invariant under the action of P. This projection is actually given by the sum of the contributions of the $(g, 1)$ sectors with $g \in P$ [‡2]. The untwisted sector contains, in general as massless states, the graviton, possibly the gravitinos, the gauge bosons, and possibly also matter fields.

The sectors with $h \neq 1$ are called twisted sectors. Applying a modular transformation such as $\tau \to -1/\tau$ on the untwisted sector shows that they have to be included. These are the string configurations that are not closed on T^n but only on T^n/P. The number of these twisted sectors can be computed either by the consideration of conjugacy classes of P or by counting the fixed points of T^n under the action of P. The last statement can easily be seen by expanding $X(\sigma)$, and requiring $X(\pi)=hX(0)$ modulo a lattice vector. This needs the centre-of-mass coordinate to coincide with a fixed point. For the oscillators this generally implies the presence of noninteger modes depending on the order of P. In any of the twisted sectors, we again have to sum over all the sectors (g, h) with $gh=hg$ projecting on the P-invariant Hilbert space.

So much for the action of P on the space–time degrees of freedom. Discussing the heterotic string, we could have the twist P accompanied by a gauge transformation in, for example, the $E_8 \times E_8'$ gauge group. We then take a subgroup $P' \subset E_8 \times E_8$ homomorphic to P and act with the twist $P \times P'$. In the bosonic formulation, such an embedding can always be realized as transformations on the $E_8 \times E_8$ lattice. In the case of an abelian point group, the action of P' can always be chosen to be a shift vector v^I, $I=1, ..., 16$, in the $E_8 \times E_8$ lattice accompanying the space–time twist. These shifts cannot be chosen arbitrarily, but they have to satisfy the group law and also restrictions from modular invariance leading to left–right level matching [3, 9]. The modular projection that previously led to the P-invariant Hilbert space now selects states that are invariant under the diagonal subgroup of $P \times P'$. In such a situation the gauge group $E_8 \times E_8$ will be broken down to the subgroup that commutes with P'. The twisted sectors contain only matter fields, whilst the graviton and the gauge bosons come exclusively from the untwisted sector.

[‡2] This is the generalization of, for example, the GSO projection in the spinning string obtained by a summation over spin structures.

2. Wilson lines. We would now like to add constant background gauge fields (Wilson lines) corresponding to the noncontractible loops on T^n. In the interpretation $O^n = R^n/S$, this is nothing else than the embedding of the full space group S into the gauge group [3]. Contrary to the case of torus compactification, these background gauge field values have to satisfy certain constraints because of the nontrivial group structure of S. In particular, the gauge P' and Wilson-line W transformations must fulfil $(P'W)^n = 1$ for an order n moding [3]. As a consequence, the values of the Wilson lines are quantized, and it turns out that in general we cannot consider all the n possible values independently.

A nontrivial consequence of the presence of Wilson lines shows up in the discussion of the twisted sectors. They are identified by the fixed points of certain elements of S. Whilst in the previous case only the subgroup $P \subset S$ was relevant, we now have to consider the full space group in order to properly treat the embedding in the twisted sectors. In group-theoretic language, this comes from the fact that members belonging to the same conjugacy class of the point group might belong to different conjugacy classes of the space group. We have also to take into account that the space group is non-abelian. This fact leads to complications, but in this paper we restrict ourselves to an abelian embedding in the gauge group, and the non-abelian structure of S will only affect the projections in the twisted sectors.

Let us now discuss the untwisted sector, which can be understood from torus compactification. We split the string into left and right movers, $X = \frac{1}{2}(X_L + X_R)$:

$$X_L^A = q^A + 2L^A u + \tfrac{1}{2}i\sum_n \frac{\alpha_n^A}{n} e^{-2inu},$$

$$X_R^i = \tilde{q}^i + 2\tilde{L}^i v + \tfrac{1}{2}i\sum_n \frac{\tilde{\alpha}_n^i}{n} e^{-2inv}, \qquad (2)$$

with $i=1, ..., 6$ and $A=(i,I)$, $I=1, ..., 16$, for the heterotic string, in an obvious notation. The lattice (L^A, $\tilde{L}^i$) for the compactified dimensions [1] should be lorentzian, even, and self-dual, with signature (22, 6). The mass spectrum can be obtained from

$$\tfrac{1}{4}m_L^2=N_L-1+\tfrac{1}{2}L^2, \quad \tfrac{1}{4}m_R^2=N_R+\tfrac{1}{2}\tilde{L}^2 , \tag{3}$$

with the constraint $m_L^2=m_R^2$. The invariant Wilson line is given by

$$\int_i A_\mu^I \, dx^\mu \equiv 2\pi A_\mu^I e_i^\mu = 2\pi a_i^I .$$

The vectors L including gauge background fields a_j^I are given by [‡3]

$$L_1=n^i R_i e_i+\frac{m_i}{2R_i}e^{*i}+\frac{a_i^I}{2R_i}e^{*i}[p^I-\tfrac{1}{2}a_j^I n^j] ,$$

$$\tilde{L}=L-2n^i R_i e_i ,$$

$$L_2^I=p^I-a_i^I n^i , \tag{4}$$

with $L=(L_1, L_2^I)$. The vectors e_i ($i=1, ..., 6$) define the lattice of the six-torus with corresponding radii R_i and we have chosen $(e_i^2)=2$, whilst the e^{*i} define the dual lattice $e^{*i}e_j=\delta_j^i$; n and m are integers corresponding to the winding numbers and the quantized momenta on the torus. Finally, the p^I define the extra-dimensional lattice, e.g. $E_8\times E_8$ root lattice. To illustrate these formulae, let us consider the Z orbifold [2], where the six-dimensional lattice consists of three copies of the SU(3) root lattice. In cartesian coordinates we have $e_1=(\sqrt{2}, 0; 0, 0; 0, 0)$, $e_2=(-1/\sqrt{2}, \sqrt{3}/\sqrt{2}; 0,0; 0,0)$, and similar expressions for the remaining e_i. The point group $P=Z_3$ consists of 120° rotations of the SU(3) lattices simultaneously. It is useful to consider complex coordinates and represent the six real coordinates by three complex ones with lattice vectors such as (1, 0, 0) and ($e^{2\pi i/3}$, 0, 0). There are 27 fixed points of the Z_3 transformation given by $(p_1, p_2, p_3)d$, with $d=\sqrt{1/3}\,e^{i\pi/6}$ and $p_\alpha=0,\pm 1$. The group law in S gives restrictions for the choice of the background values a_i^I, in that $3a_i^I$ has to be a vector in the $E_8\times E_8$ lattice. Furthermore, here one can only choose three independent values, since $a_1=a_2$, $a_3=a_4$ and $a_5=a_6$ have to be satisfied [3]. This is because $Pe_i=e_{i+1}$ for $i=1,3,5$.

[‡3] Inclusion of a background value for the antisymmetric field B_{ij} does not change our results except for particular values of the compactification radii. Models with enlarged gauge groups have been considered in ref. [10].

In the untwisted sector, one starts with the string as it moves on T^6, and then one has to project onto the S-invariant states in the Hilbert space, corresponding to the sum of the contributions from the (g, 1) sectors. For $P=Z_3$ this leads to the well-known projections. Including the a_i gives new restrictions concerning a selection of special values of $a_i^I p^I$. To illustrate this, consider the massless sector which requires $\tilde{L}=0$ and in turn implies $n^i=0$ in the given example. Since m_i are integers, we observe that we have to project onto integer $a_i^I p^I$, as can be derived from formula (4). This is equivalent to projecting onto states which are Wilson-line singlets and generally leads to a further breakdown of the gauge group.

For the discussion of the twisted sectors, we should remember that there are 27 fixed points. In the usual Z orbifold all these 27 sectors lead to identical physics [2]. The oscillators corresponding to the six compact dimensions have third-integral moding and the twist $h=\theta=e^{2\pi i/3}$ is embedded in the gauge group, leading to a shift in the $E_8\times E_8$ lattice $p^I\to p^I+v^I$. Again one has to project onto invariant states, summing over the (g, θ) sectors. This gives the states of positive helicity; the ones with negative helicity are obtained from the θ^{-1} sectors.

Inclusion of the Wilson lines leads to a completely different situation, since the different twisted sectors no longer show a universal behaviour. This comes from the fact that the fixed points belong to different elements of the space group. The fixed point (0, 0, 0)d, for example, is a fixed point of the θ transformation; here we have to consider the lattice p^I+v^I in the left-handed sector. The fixed point (d, 0, 0), however, is not a fixed point of θ alone; a shift by e_1 on the SU(3) lattice has to accompany this rotation. In this twisted sector the embedding in the gauge group requires the shift $p^I\to p^I+v^I+a_i^I$, and with nontrivial a_i^I different twisted sectors lead to different spectra. The twisted sectors can therefore be labelled by $h=(\theta, e_i)$ corresponding to a rotation and a shift. In each of these sectors we have to perform the projection on invariant states, and here we have to take into account the non-abelian structure of the space group. In a given sector twisted by h, we have to sum the contributions of the sectors (g, h), not for all $g \in S$ but only for those where g commutes with h.

The various shifts v^I and a_i^I have to satisfy further restrictions from modular invariance such as left–right level matching, and this will be discussed in the examples given below. Here we should state explicitly that in all our examples we have embedded P in an abelian way in the gauge group. Together with the fact that we use only inner automorphisms acting on the lattices, this implies that we cannot reduce the rank of the gauge group. The rank can, however, be lowered by an explicit consideration of these complications.

3. Examples. All the above points may be conveniently illustrated by considering the existence of a specific Wilson line on the Z orbifold. There are four nontrivial embeddings of the point group operation θ into the $E_8 \times E_8$ gauge group consistent with modular invariance constraints [3]. These are represented by the four different shift vectors v^I shown in table 1. The corresponding unbroken gauge groups were given in ref. [3]. We also give here the matter superfield content from untwisted and twisted sectors. One can add Wilson lines a_i^I on any of the four possible embeddings (i.e. on top of any of the modings generated by the four v^I's). Only the standard embedding 1 in the table has been studied in detail in the literature [2], and it is also the only one which contains complete standard quark–lepton generations in the twisted sector. One version of the second embedding in the table has also been studied [6]. Later we will show that all the noncanonical embeddings can lead to interesting models when one adds Wilson lines. As we remarked above, modular invariance restricts the possible Wilson-line shifts a^I in a given twisted sector of the theory [3, 9]. In the Z orbifold case, one has the condition

$$\sum_{a=1}^{3} 9r_a^2 = 6 = \sum_I 9(v^I + a^I)^2 \mod 3\,, \tag{5}$$

where the r_a's indicate the shift in the space–time sector. A simple way to find a^I's satisfying this equation is to choose them equal to any of the shifts of table 1 but without any overlap with the v^I shift of the moding. However, this is not a necessary condition (see example 1 of tables 2 and 3). Notice that since different twisted sectors may get different a^I shifts, one has to check that condition (5) is respected in all the twisted sectors.

Let us consider one example in detail. Take the usual canonical embedding with the P' moding generated by the shift $v^I = (\frac{1}{3}, \frac{1}{3}, \frac{2}{3}, ...)$ and add a Wilson line on one of the six compactified dimensions $a_1^I = (0, 0, ..., \frac{1}{3}, \frac{1}{3})(\frac{2}{3}, ...)$, in an obvious notation. Since $\theta e_1 = e_2$, we have necessarily to add another (equal) Wilson line on the other dimension, $a_2^I = a_1^I \equiv a^I$. We assume that the other four dimensions have no Wilson line, $a_3^I = a_4^I = a_5^I = a_6^I = 0$. Notice that $3a^I$ is a vector in the $E_8 \times E_8'$ lattice, as it should

Table 1
Models without Wilson lines. [All U(1) quantum numbers have been omitted.]

Example	Shift	Gauge group	Untwisted sector	Twisted sector
1	$(\frac{1}{3}\frac{1}{3}\frac{2}{3}\ ...)(0\ ...)$	$E_6 \times SU(3) \times E_8'$	3[(27,3), 1′]	27[(27,1),1′] + 81[(1,$\bar{3}$),1′]
2	$(\frac{1}{3}\frac{1}{3}\frac{2}{3}\ ...)(\frac{1}{3}\frac{1}{3}\frac{2}{3}...)$	$[E_6 \times SU(3)]^2$	3[27,3)(1,1)′ + (1,1)(27,3)′]	27[(1,$\bar{3}$),(1,$\bar{3}$)′]
3	$(\frac{1}{3}\frac{1}{3}...)(\frac{2}{3}\ ...)$	$[E_7 \times U(1)] \times [SO(14) \times U(1)]'$	3(56,1′) + 3(1,1′) + 3(1,14′) + 3(1,64′)	27[(1,1′) + (1,14′)] + 81(1,1′)
4	$(\frac{1}{3}\frac{1}{3}\frac{1}{3}\frac{1}{3}\frac{2}{3}...)(\frac{2}{3}...)$	$SU(9) \times SO(14)' \times U(1)'$	3(84,1′) + 3(1,14′) + 3(1,64′)	27($\bar{9}$,1′)

be. It is interesting to remark that a Wilson line such as $(0, 0, ..., \frac{1}{3}, \frac{1}{3})(0, ..., 0)$ would have induced a non-modular invariant result, which is why we included an action of the a^I shift on E_8'. Let us start by considering the massless sector of the *untwisted sector*. The states in the left-handed sector, before adding the Wilson line, can be classified according to their eigenvalue with respect to $P' = \exp(2\pi i p^I v^I)$. There are the 16 U(1)'s corresponding to $\alpha^I|0\rangle_L$ and the states which complete the adjoint of $SU(3) \times E_6 \times E_8'$, all of these with $P'=1$. Then there are states with quantum numbers (**3**, **27**) and $P'=\alpha^2$ ($\alpha = e^{2\pi i/3}$), and states with quantum numbers ($\bar{\mathbf{3}}$, $\overline{\mathbf{27}}$) and $P'=\alpha$. After addition of the Wilson lines, the GSO-like projection will project on states ($P_I^2=2$ roots) which satisfy $p^I a^I = m$, m=integer. In the $P'=1$ sector these states will correspond to the E_8 roots $(0, 0, 0, ..., \pm 1, ..., \pm 1, 0, 0), \pm(0, 0, ..., 1, -1), \pm(1, -1, 0, ...), \pm(1, 0, 1, ...), \pm(0, 1, 1, 0, ...), \pm(-\frac{1}{2}, -\frac{1}{2}, \frac{1}{2}, \pm\frac{1}{2}, \pm\frac{1}{2}, \pm\frac{1}{2}, \frac{1}{2}, -\frac{1}{2}), \pm(-\frac{1}{2}, -\frac{1}{2}, \frac{1}{2}, \pm\frac{1}{2}, \pm\frac{1}{2}, \pm\frac{1}{2}, -\frac{1}{2}, \frac{1}{2})$. These correspond to the non-vanishing roots of an $SU(3) \times SU(6) \times U(1)$ group. The second E_8' gives in the $P'=1$ sector the gauge group $SO(14)' \times U(1)'$. All these $P'=1$ left movers combine with the $P=1$ right-handed ground states from $|i\rangle_R$, $|a\rangle_R$ to form gauge bosons and gauginos with gauge group $SU(3) \times SU(6) \times U(1) \times SO(14)' \times U(1)'$.

Doing a similar exercise with the $P'=\alpha^2$ untwisted sector, one easily verifies that from the original (**3**, **27**) states there survive only the $p^I a^I = m$ projected states which transform as a (**3**, **15**, 0) under $SU(3) \times SU(6) \times U(1)$. There are no states with $SO(14)' \times U(1)'$ quantum numbers. The surviving $P'=\alpha^2$ left movers combine with the three $P=\alpha$ right-moving states $|i\rangle_R$, $|a\rangle_R$ [$i,a \in 3$ of $SU(3) \subset SU(6)$] to form $P \times P' \times$ Wilson-line singlets. The $P'=\alpha$ sectors give us the antiparticles. Thus as a whole we obtain from the untwisted sector (apart from the supergravity and "dilatonic" particles) three chiral superfields transforming as (**3**, **15**, 0) of $SU(3) \times SU(6) \times U(1) \times SO(14)' \times U(1)'$.

Let us now consider the twisted sectors. They are associated with string modes which expand around any of the 27 fixed points of the Z orbifold. Since there are Wilson lines in one of three complex dimensions of the orbifold, the 27 fixed points are split into three groups of nine fixed points according to the string boundary conditions. The first nine fixed points, which we denote by TT (trivial twisted), are the ones with coordinates $(0, m, n)d$, $(m, n = 0, \pm 1)$. The bosonic string coordinates have the usual closed-string boundary conditions: $x^I(\pi) = x^I(0) + 2\pi v^I$; $x^I(\pi) = \theta x^I(0)$ + (lattice vectors without Wilson lines). The states associated with these nine fixed points do not feel the presence of the Wilson line; thus the zero modes satisfy the usual Z orbifold condition

$$\tfrac{1}{2}(p^I + v^I)^2 + \tilde{N} - \tfrac{2}{3} = 0\,. \tag{6}$$

The states of these sectors are also the usual ones of the Z-orbifold twisted sector. For $\tilde{N}=0$ one has $(p^I + v^I)^2 = \frac{4}{3}$, and the nine states transforming as [(**1**, **15**) + 2(**1**, $\bar{\mathbf{6}}$)] under $SU(3) \times SU(6) \times U(1)$ will be massless. For $\tilde{N} = \frac{1}{3}$ one has $(p^I + v^I)^2 = \frac{2}{3}$, and the $9 \times 3 = 27$ states transforming as ($\bar{\mathbf{3}}$, **1**) are massless. All these states survive under the relevant projection in this case, which is on $P \times P'$ invariant states.

The second set of nine fixed points, which we denote by NTT-1 (nontrivial twisting 1), are those with coordinates $(+1, m, n)d$, $(m, n = 0, \pm 1)$. The bosonic string coordinates have the closed-string boundary conditions: $x^I(\pi) = x^I(0) + 2\pi(v^I + a^I)$; $x^i(\pi) = \theta x^i(0) + 2\pi e_1^i$ + (shifts without Wilson lines), since there is a Wilson line a^I associated with the lattice vector e_1^i, and all these nine fixed points require a shift by e_1^i to be really fixed points. The associated twisted sectors do feel the presence of the Wilson lines, and the massless modes have to satisfy

$$\tfrac{1}{2}(p^I + v^I + a^I)^2 + \tilde{N} - \tfrac{2}{3} = 0\,. \tag{7}$$

There are only solutions to this with $\tilde{N}=0$, and they correspond to elements of the $(p^I + v^I + a^I)$ shifted lattice, with p^I being $(0, 0, ..., 0)$; $(-1, 0, -1, 0, ...)$; $(0, -1, -1, 0, ...)$; $(0, 0, -1, ..., -1, 0)$; $(0, 0, -1, ..., 0, -1)$; $(-\frac{1}{2}, -\frac{1}{2}, -\frac{1}{2}, \pm\frac{1}{2}, \pm\frac{1}{2}, \pm\frac{1}{2}, -\frac{1}{2}, -\frac{1}{2})$. Taking into account the chirality of the twisted right-handed vacuum, the quantum numbers of these states correspond to a [($\bar{\mathbf{3}}$, **1**) + (**1**, $\bar{\mathbf{6}}$)] under $SU(3) \times SU(6) \times U(1)$. No massless states with $SO(14)' \times U(1)'$ quantum numbers appear. All these states survive the GSO-like projection, which in this case projects onto states with

$$\exp\{2\pi i[(p^I + v^I + a^I)(v^I + a^I)]\} = \alpha, \quad \alpha = e^{2\pi i/3}$$

Finally, the third set of nine fixed points, which we

denote by NTT-2 (nontrivial twisting 2), are those with coordinates $(-1, m, n)d$, $(m, n=0,\pm 1)$. The bosonic string coordinates have the closed-string boundary conditions: $x^I(\pi)=x^I(0)+2\pi(v^I-a^I)$; $x^i(\pi)=\theta x^i(0)-2\pi\, e_1^i+$ (shifts without Wilson lines). The massless modes have to satisfy

$$\tfrac{1}{2}(p^I+v^I-a^I)^2+\tilde{N}-\tfrac{2}{3}=0\,. \tag{8}$$

One can easily check that they correspond to a $[(\bar{\mathbf{3}}, \mathbf{1})+(\mathbf{1}, \bar{\mathbf{6}})]$, although in this case the $\bar{\mathbf{6}}$ has a U(1) charge opposite to the one in the NTT-1 sector. Again all these states survive projection. In summary, the matter content of this model with respect to $SU(3)\times SU(6)\times U(1)$ is

$$\mathrm{U}\to 3(\mathbf{3}, \mathbf{15}, 0)\,,$$

$$\mathrm{TT}\to 9[(\mathbf{15}, 0)+(\mathbf{1}, \bar{\mathbf{6}}, a)+(\mathbf{1}, \bar{\mathbf{6}}, -a)] +27(\bar{\mathbf{3}}, \mathbf{1}, 0)\,,$$

$$\text{NTT-1}\to 9[(\bar{\mathbf{3}}, \mathbf{1}, 0)+(\mathbf{1}, \bar{\mathbf{6}}, a)]\,,$$

$$\text{NTT-2}\to 9[(\bar{\mathbf{3}}, \mathbf{1}, 0)+(\mathbf{1}, \bar{\mathbf{6}}, -a)]\,. \tag{9}$$

One gets the antiparticles from the sector twisted by θ^{-1} and $(-v^I)$. Notice how all four sectors are interrelated in order to obtain an anomaly-free result. This model contains nine families from the TT sector, which feel only the $SU(6)\times U(1)$ interactions. The other three sectors give rise to nine further families, which feel also the SU(3) interactions. The gauge group $SO(14)'\times U(1)'$ is hidden. This example is not particularly interesting from the phenomenological point of view, but we have chosen it here because it has all the features of a general model based on the Z orbifold with a Wilson line.

One can obtain more interesting models by using different combinations of point-group embedding (v^I's) and Wilson lines (a^I's). Some examples are shown in table 2. Thus, for instance, one may get rid of the untwisted matter by putting some part of the Wilson line on top of the shift v^I. That is done in example 1 of table 2 which leads to a model with nine E_6 families (all from the TT sector) and a hidden $E_6'\times SU(3)^2$. Very often the addition of a Wilson line has the effect (in the case of the canonical v^I) of eliminating the families which otherwise would have

Table 2
Models with one Wilson line. [All U(1) quantum numbers have been omitted.]

Example	Shift	Wilson line	Gauge group	Untwisted sector	Twisted sector	Comments
1	$(\frac13\frac13\frac23\,0\ldots0)(0\ldots)$	$(\frac13\frac13\frac23\,0\ldots)(\frac13\frac13\frac23\ldots)$	$[E_6\times SU(3)]^2$	Nothing	$9(27,1)+27(1,\bar{3})+27(1',\bar{3}')+9(\overline{27}',1')+9(1,3,1',\bar{3}')$	Nine E_6 families + …
2	$(\frac13\frac13\frac23\,0\ldots)(0\ldots)$	$(0\ldots\frac13\frac13\frac23)(\frac13\frac13\frac23\ldots)$	$[SU(3)]^4\times E_6'\times SU(3)'$	$3[(3,3,3,1)]$	$9[(1,3,3,1)+(1,\bar{3},1,3)(1,1,\bar{3},\bar{3})]+27(\bar{3},1,1,1)+9[(1,\bar{3},1,1)(\bar{3},1)']+9[(1,1,\bar{3},1)(\bar{3},1)']$	Nine $SU(3)^4$ families + …
3	$(\frac13\frac13\frac23\,0\ldots)(0\ldots)$	$(0\ldots\frac13\frac13)(\frac23\ldots)$	$[SU(3)\times SU(6)\times U(1)]\times[SO(14)\times U(1)]'$	$3[(3,15)]$	$9[1,15+\bar{6}+\bar{6})]+27(\bar{3},1)+9[(1,\bar{6}+\bar{6})+(\bar{3}+\bar{3},1)]$	Eighteen $SU(6)\times SU(3)\times U(1)$ families
4	$(\frac13\frac13\frac23\,0\ldots)(0\ldots)$.	$(0\ldots\frac13\frac13\frac13\frac13\frac23)(\frac23\ldots)$	$[SU(6)\times SU(3)\times U(1)]\times[SO(14)\times U(1)]'$	$3[(3,\bar{6})]$	$9[1,15+\bar{6}+\bar{6})]+27(\bar{3},1)+9[(3,1)+(1,6)]+9(1,1)+$ hidden matter	Nine $SU(6)\times SU(3)\times U(1)$ families + …
5	$(\frac13\frac13 0\ldots)(\frac23\,0\ldots)$	$(0\,0\,\frac23\ldots)(0\,\frac13\frac13\ldots)$	$[SO(10)\times SU(2)\times U(1)^2]^2$	$3[(16',2')+(10',1')]+3[(10,2)+(1,1)]$	$9(1,10')+$ singlets and doublets	Six $SO(10)\times SU(2)'\times U(1)'^2$ families + …
6	$(\frac13\frac13\frac13\frac13\frac23\ldots)(\frac23\ldots)$	$(0\ldots\frac23)(0\,\frac13\frac13\ldots)$	$[SU(5)\times SU(2)\times U(1)^2]\times[SO(10)'\times SU(2)'\times U(1)'^2]$	$3(\overline{10},1,1)+3(5,2,1)+3(5,1,2)+$ hidden matter	$9[(\bar{5},1,1)]+$ singlets and doublets	Three $SU(5)\times SU(2)^2\times U(1)^2$ families

Table 3
Models with two Wilson lines. [All U(1) quantum numbers have been omitted.]

Example	Shift	Wilson line	Gauge group	Untwisted sector	Twisted sector	Comments
1	$(\frac{1}{3}\frac{1}{3}\frac{2}{3}\,0\ldots)(0\ldots)$	$a_1=(\frac{1}{3}\frac{1}{3}\frac{2}{3}\,0\ldots)$ $(\frac{1}{3}\frac{1}{3}\frac{2}{3}\ldots)$ $a_3=(\frac{1}{3}\frac{1}{3}\frac{2}{3})$ $(\ldots\frac{1}{3}\frac{1}{3}\frac{2}{3})$	$[E_6\times SU(3)]\times[SU(3)']^4$	Nothing	3[(1,27)] + singlets and triplets	Three E_6 families
2	$(\frac{1}{3}\frac{1}{3}\frac{2}{3}\ldots)(0\ldots)$	$a_1=(\ldots\frac{1}{3}\frac{1}{3}\frac{2}{3})$ $(\frac{1}{3}\frac{1}{3}\frac{2}{3}\ldots)$ $a_3=(\frac{1}{3}\frac{1}{3}\frac{2}{3}\ldots\frac{1}{3}\frac{1}{3}\frac{2}{3})$ $(\ldots\frac{1}{3}\frac{1}{3}\frac{2}{3})$	$[SU(3)]^8$	Nothing	$3[(1,3,3,1)+(1,\bar{3},1,3)+(1,1,\bar{3},\bar{3})]+\ldots$	Three $[SU(3)]^3$ families + ...
3	$(\frac{1}{3}\frac{1}{3}\frac{2}{3}\ldots)(0\ldots)$	$a_1=(\ldots\frac{1}{3}\frac{1}{3})$ $(\frac{2}{3}\ldots)$ $a_3=(\frac{1}{3}\frac{1}{3}\frac{2}{3}\ldots)$ $(\ldots\frac{1}{3}\frac{1}{3}\frac{2}{3})$	$[SU(6)\times SU(3)\times U(1)]\times[SU(3)\times SU(2)^2\times U(1)^4]$	Nothing	$3[(15+\bar{6}+\bar{6},1)]+\ldots$	Three SU(6) × U(1) families + ...

appeared in the NTT-1 and NTT-2 sectors. This happens in examples 1 to 4 in table 2. Thus the addition of Wilson lines is an efficient way of controlling the number of generations of a given orbifold. It is then easy to obtain models with three families of "quarks" and "leptons". This is the case of example 6 in the table. It starts from a "noncanonical" v^I shift originating an SU(9), which is broken by the Wilson line down to an $SU(5)\times[SU(2)\times U(1)]^2$ model with three SU(5) families all originating in the untwisted sector (the twisted sector does not usually contain interesting matter for noncanonical v^I's).

More three-family models may be obtained by adding an appropriate second Wilson line on another complex dimension, e.g. $a_3^I=a_4^I\equiv\tilde{a}^I$. This should be done taking care that the usual modular invariance conditions are respected. In this case the gauge group is broken down to the subgroup commuting with both Wilson lines. The twenty-seven fixed points split into nine sectors of three fixed points each. If one starts from a canonical v^I, a second Wilson line can easily convert a model with nine generations (from the TT sector) into a model with three generations, since the TT sector has now only three fixed points. This is what happens in the examples shown in table 3. There are no families from the untwisted sector, since some part of the Wilson lines overlap with v^I, and then those families are projected out. All generations in the twisted sectors are also eliminated by the Wilson lines except for those in the TT sector. Since that sector contains three fixed points [e.g. $(0,\,0,\,m)$, $m=0,\pm1$], we get three generations. Obviously we can build many models with different gauge groups by adding two Wilson lines around two complex dimensions and considering different embeddings, and the examples shown in the table are just a few possibilities. It is rather remarkable that such a large variety of models (and generations) can be obtained just by adding Wilson lines to the simplest supersymmetric orbifold, the Z orbifold.

In principle one can obtain models which look more like the Standard Model, such as $SU(3)\times SU(2)\times U(1)^n$, but this does not guarantee that other phenomenological problems, such as neutrino masses and too fast proton decay, are solved. Furthermore, it is not clear that the additional U(1)'s present would not cause additional phenomenological problems. In this context it would be useful to study the effect of Wilson lines on non-abelian orbifolds in which the additional U(1)'s could be absent.

There is another interesting point related to U(1)'s in this type of orbifold construction. Very often the U(1)'s obtained upon compactification are anomalous. This is true not only for models with additional Wilson lines but also for the usual noncanonical embeddings of table 1 without any Wilson lines. This fact is not new and has recently appeared in different contexts [8, 11, 12]. Since the theories we are discussing are modular invariant, one suspects that some mechanism, probably related to the anomaly cancellation mechanism of Green and Schwarz, must be present to take care of anomalies. In fact the ten-dimensional "Wess–Zumino" counterterm [13] $B \wedge \mathrm{Tr}(F \wedge F \wedge F \wedge F)$ yields, upon compactification [11], a one-loop four-dimensional coupling proportional to $B_{\mu\nu}\mathrm{Tr}(F_{\varrho\sigma})\,\epsilon^{\mu\nu\varrho\sigma}$, F being the field strength of the anomalous U(1). After a duality transformation, this term acquires the aspect of the usual Goldstone boson to gauge boson coupling, and the anomalous U(1) gauge boson becomes massive – hence harmless. Since this mechanism seems to involve just one chiral superfield, it seems to generate a SUSY-breaking D-term [‡4]. This is not necessarily a problem as long as the "observed" matter does not have anomalous U(1) quantum numbers.

4. Conclusions. Let us conclude with a few comments. We have shown how the addition of Wilson lines on orbifold tori is an efficient method for controlling the unbroken gauge group as well as the number of generations left after compactification. Thus, for example, orbifolds with a large Euler characteristic may yield realistic models if appropriate Wilson lines are present. In this context it would be interesting to see the connection between orbifold-like compactifications and the four-dimensional string models discussed in refs. [7] and [8]. Pre-existing Wilson lines could also be useful in the left–right asymmetric orbifolds recently considered by Narain, Sarmadi and Vafa [14]. The orbifold-type constructions seem to have the advantage that one has a certain intuitive control on the gauge interactions and number of generations to be expected from a given model. In the present paper we have studied, in some detail, the case of the Z orbifold, and we have shown how this very simple example can lead to interesting three-family models with gauge groups such as E_6, $SU(3)_L \times SU(3)_R \times SU(3)_c$, $SU(6) \times U(1)$, $SU(5) \times [SU(2) \times U(1)]^2$, etc. These do not exhaust the list of possibilities, and it would be interesting to search for examples closer to the observed Standard Model.

‡4 The existence of this mechanism seems to be known by a number of people. We thank L. Alvarez-Gaumé, J. Distler, W. Lerche and N. Schellekens for discussions about these points.

To our knowledge these seem to be the first systematic sources of orbifold models with three generations in the literature [‡5]. It is amusing that, in all the models discussed above, the fact that one model has three generations is directly or indirectly related to the fact that there are three compactified complex extra dimensions.

We acknowledge discussions with L. Alvarez-Gaumé, W. Lerche, K. Narain and N. Schellekens.

‡5 See also the example of ref. [6].

References

[1] K.S. Narain, Phys. Lett. B 169 (1986) 41;
K.S. Narain, M.H. Sarmadi and E. Witten, A note on toroidal compactification of heterotic string theory, Rutherford preprint RAL-86-022 (1986);
P. Ginsparg, Comment on toroidal compactification of heterotic superstrings, Harvard preprint HUTP-86/A053 (1986).

[2] L. Dixon, J.A. Harvey, C. Vafa and E. Witten, Nucl. Phys. B 261 (1985) 678.

[3] L. Dixon, J.A. Harvey, C. Vafa and E. Witten, Nucl. Phys. B 274 (1986) 286.

[4] S. Hamidi and C. Vafa, Interactions on orbifolds, Harvard–CalTech preprint HUTP-86/A041, CALT-86-1349 (1986).

[5] L. Dixon, D. Friedan, E. Martinec and S. Shenker, The conformal field theory of orbifolds, Princeton–Chicago preprint EFI86-42 (1986).

[6] D.-X. Li, Direct compactification of heterotic strings, modular invariance, and three families of chiral fermions, McGill University preprint (1986).

[7] H. Kawai, D.C. Lewellen and S.-H.H. Tye, Construction of fermionic string models in four dimensions, Cornell preprint CLNS86/751 (1986).

[8] W. Lerche, D. Lüst and A.N. Schellekens, Chiral four-dimenional heterotic strings from self-dual lattices, preprint CERN-TH 4590/86 (1986).

[9] C. Vafa, Nucl. Phys. B 273 (1986) 592.

[10] M. Mueller and E. Witten, Phys. Lett. B 182 (1986) 28.

[11] E. Witten, Phys. Lett. B 149 (1984) 351.

[12] K. Pilch and A.N. Schellekens, Nucl. Phys. B 259 (1985) 637.

[13] M. Green and J. Schwarz, Phys. Lett. B 149 (1984) 117.

[14] K.S. Narain, private communication;
K.S. Narain, M.H. Sarmadi and C. Vafa, in preparation.

Nuclear Physics B288 (1987) 551–577
North-Holland, Amsterdam

ASYMMETRIC ORBIFOLDS

K.S. NARAIN and M.H. SARMADI

Rutherford Appleton Laboratory, Chilton, Didcot, OX11 0QX, UK

C. VAFA

Lyman Laboratory of Physics, Harvard University, Cambridge, MA 02138, USA

Received 3 February 1987

We describe string theories in which the left-moving and right-moving degrees of freedom live on different orbifolds.

1. Introduction

In the study of string theories, some of the most interesting issues relate to the applicability or inapplicability of our intuition gained from dealing with quantum field theories. This intuition has been at the root of some recent developments in string theory. For example, the fact that point particles and their interactions are most elegantly described by a quantum field theory, motivates the study of string field theories. This has led to extending the point particle symmetries such as general coordinate invariance to an infinite dimensional symmetry of the string theory.

On the other hand there are a number of examples showing how our naive intuition fails. For instance, the field theory of particles gives no clue as to why compactifying on special tori, such as is the case for heterotic strings [1] and their toroidal compactifications to lower dimensions [2], should give rise to gauge invariances besides the ones expected from the Kaluza-Klein picture. These are "stringy" effects which cannot be explained from a point particle viewpoint. The winding of strings around the non-trivial loops of the tori is responsible for giving rise to gauge degrees of freedom. Since this involves extended states, it is not expected from a point particle point of view. Another counter-intuitive aspect of heterotic string (in its bosonic formulation) is that the left-movers and right-movers live on different spaces. Left-movers live on the product of 10-dimensional Minkowski space with one of two choices of 16-dimensional tori, and the right-movers live on the 10-dimensional Minkowski space. The 10-dimensional

0550-3213/87/$03.50

Minkowski space is viewed as common to both right-movers and left-movers.

Another example where our naive point particle intuition fails is for strings propagating on orbifolds [3]. Orbifolds, which generally are not manifolds, would give problems with the unitarity of point particle theory (due to the existence of points where orbifolds fail to be manifolds). However, the string propagation on them is not only consistent, but in cases where they are formed from tori, is natural and rather simple to describe. Again, the phenomenon is "stringy" and the unitarity is rescued due to the existence of "twisted sectors" (strings closed up to a group action).

In this paper we shall combine these two "stringy" phenomena, and describe another "stringy" phenomenon in which, roughly speaking, the left-movers of a string theory live on one orbifold, and the right-movers live on another orbifold. We shall call such spaces *asymmetric orbifolds*. In view of the fact that the separation of left- and right-movers is already done in heterotic strings and that orbifolds are consistent, it seems somewhat surprising that these two ideas had not been combined before*. The main obstacle is to overcome the "intuition" that the 10-dimensional space-time is somehow fundamentally "different" from the 16 internal dimensions (as is suggested by the field theory view of the massless degrees of freedom), and unlike the 16 internal coordinates should be treated symmetrically between the left-movers and right-movers.

In sect. 2 we describe the basic setup and discuss the importance of modular invariance for asymmetric orbifolds. In sect. 3 we construct the Hilbert space of an asymmetric orbifold. This is done by imbedding an asymmetric orbifold in a symmetric orbifold. In sect. 4 we consider some examples of asymmetric orbifolds as well as describe the relation of some asymmetric orbifolds to Lorentz rotations on odd self-dual lattices. And finally in sect. 5 we present our conclusion. In appendices A and B we explain some mathematical ideas which are useful in the study of asymmetric orbifolds (appendix A discusses the decomposition of a self-dual lattice to two orthogonal subspaces and appendix B describes some group theory relevant to the description of vertex operators in an asymmetric orbifold).

Ideas which can be interpreted as asymmetric orbifolds have already been discussed in the fermionic formulation in ref. [4]. Our work was motivated in part to understand the relation of these approaches to orbifolds.

2. Asymmetric orbifolds

In the construction of toroidal orbifolds [3], one starts from a torus, and considers string propagation on the quotient of the torus by an isometry subgroup of it. This results in a theory in which the left- and right-moving bosonic space-time coordinates are treated on the same footing. In view of the fact that the heterotic string is

* This was done in a limited way in ref. [3] (as discussed in sects. 3 and 4 of this paper).

asymmetric in the bosonic formulation, it is natural to ask whether one can continue this asymmetry all the way to four dimensions. In particular, we wish to investigate the extent to which this separation of left-right moving degrees of freedom of an orbifold for ten-dimensional space-time coordinates can be done.

As has been shown in ref. [2] one can describe the momenta of left-movers and right-movers of a string theory in such a way that the combined left-right momentum gives rise to a self-dual even lorentzian lattice. These were interpreted physically as turning on Wilson lines and anti-symmetric tensor fields on toroidally compactified string theories. If we consider a d-dimensional toroidal compactification of the heterotic string, the lattice we obtain is an even self-dual lattice with signature $((16+d)(+), d(-))$, which we denote by $\Gamma^{p,q}$ with $p=16+d$ and $q=d$. This lattice is unique, but to describe the physics we have to choose two orthogonal subspaces of dimensions $16+d$ and d, which have definite signatures $(+)$ and $(-)$ respectively. When we write $\Gamma^{p,q}$ we shall mean not just the lattice but also a choice of such a decomposition. Two different choices of such a decomposition are related by an $\mathrm{SO}(p,q)$ rotation on the lattice. The momenta of the left-movers and the right-movers are obtained by projecting the lattice vectors on the two subspaces respectively. We thus write an arbitrary element of the lattice in the form $(P_\mathrm{L}, P_\mathrm{R})$, where P_L is a $16+d$ dimensional vector and P_R is d-dimensional.

We may now construct an orbifold by considering automorphisms of this space which do not necessarily treat the left-right spaces symmetrically in any sense. The orbifold construction in ref. [3] corresponds to considering $\Gamma^{p,q}=\Gamma^{16,0}+d\Gamma^{1,1}$, and modding out by groups which act symmetrically between left and right on the $d\Gamma^{1,1}$. Now we will allow the more general possibility of a left-right asymmetric action on $d\Gamma^{1,1}$. However not every automorphism of the lattice can serve as a starting point to construct an orbifold: when we mod out the Hilbert space by an automorphism, it must not mix up the left- and right-moving Hilbert spaces, otherwise it would not be a symmetry of the string theory. So consider a group element denoted by $g=(\theta_\mathrm{L}, a_\mathrm{L}; \theta_\mathrm{R}, a_\mathrm{R})$. Then g is defined to act on the momentum (winding) degrees of freedom of the Hilbert space by

$$g|P_\mathrm{L}; P_\mathrm{R}\rangle = \exp(2\pi i(P_L\cdot a_L - P_R\cdot a_R))|\theta_\mathrm{L}P_\mathrm{L}; \theta_\mathrm{R}P_\mathrm{R}\rangle, \tag{2.1}$$

where θ_L and θ_R are $16+d$ and d-dimensional rotations and a_L and a_R are $(16+d)$ and d-dimensional shift vectors respectively. The action of g on the bosonic oscillator is also clear, it is simply given by the action of the rotation matrices θ_L and θ_R on the oscillators. The action on the NSR fermions is determined by the requirement of preserving world sheet supesymmetry, and so is simply given by θ_R rotation (the necessity of preserving world sheet supersymmetry has been discussed in ref. [4]). Note that the action on the gauge degrees of freedom is already contained in the action of it on the lattice, because we are realizing the gauge degrees of freedom as part of the lattice. So g completely specifies the action on all string degrees of freedom (apart from the possibility of projecting out the different G parities for the NSR fermions).

It is easy to see that for two group elements the multiplication rule is

$$(\theta_L, a_L; \theta_R, a_R)(\omega_L, b_L; \omega_R, b_R) = (\theta_L\omega_L, a_L\omega_L + b_L; \theta_R\omega_R, a_R\omega_R + b_R)$$

(where the shift vectors are written as row vectors). Note that (a_L, a_R) is well defined up to an addition of a vector in the $\Gamma^{p,q}$ lattice, because the action of g given by (2.1) does not change by such an addition (this is because $\Gamma^{p,q}$ is an integral lattice).

To describe all the string states, we have to introduce the twisted sectors, i.e. string states which close only up to g action. Here is the main place we face a difficulty: It is not easy to construct the twisted Hilbert space for an asymmetric orbifold. The reason is that the operation of modding out by g cannot be described geometrically as a string propagating on an orbifold, and we lose all our geometrical intuition in constructing the twisted sectors. In other words (2.1) is defined in the momentum (winding) space and does not give rise to a sensible action on the x coordinates. In particular in the left-right symmetric orbifolds the fixed points of the g transformation play an important role in describing the twisted sector, but now the fixed points which are common between left and right degrees of freedom lose their meaning, because the action of group element g may have a different number of fixed points for left- or right-moving degrees of freedom.

Not all is lost, however. Modular transformations come to the rescue to give a hint about the structure of the twisted Hilbert space, even before constructing it! Let us consider the partition function of the identity sector twisted by g, i.e.,

$$\mathrm{Tr}\, g q^{H_L} \bar{q}^{H_R}. \tag{2.2}$$

This corresponds in the path integral formulation to twisting the boundary condition in the time direction by g and having periodic boundary conditions in σ direction. To get the partition function for the string in the twisted sector g, we simply consider the modular transformation which exchanges the space and time of the torus, that is the transformation $\tau \to -1/\tau$. Let us focus on the contribution of bosons to (2.2). There are two types of bosonic contributions, one is the lattice contribution, the other is the oscillator contribution. Let us consider the subspace I of vectors in $\Gamma^{p,q}$ which are left invariant under $\theta = (\theta_L; \theta_R)$ rotation. Then it is clear that only such states contribute to (2.2) because the trace kills all the states not left invariant under g (up to phases). The contribution of the bosonic oscillators is also easily computed. Now under $\tau \to -1/\tau$ the partition function is seen to correspond to the momenta lying on $I^* + a^*$ where I^* is the lattice dual to I, and a^* is the orthogonal projection of $a = (a_L; a_R)$ onto I, i.e. $a = a^* + b$ where

$a^* \in I$ and $b \cdot I = 0$. Similarly the modular transformation of the oscillator contribution to (2.2) corresponds to the partition function of fractionally moded harmonic oscillators, depending on the eigenvalues of θ_L and θ_R. However, there are also two overall numerical factors. One is the inverse of the volume of the fundamental cell of I (which comes from the Poisson resummation) $1/v = 1/\sqrt{|I^*/I|}$ (where I is a subspace of I^* by integrality of the lorentzian lattice, and $|I^*/I|$ denotes the index of I in I^*). The other factor comes from the oscillators and is given by $\sqrt{\det(1-\theta_L)\det(1-\theta_R)} = \sqrt{\det(1-\theta)}$ (the determinant is over the eigenvalues of θ not equal to one). The simplest way to show this is to remember that in the symmetric case, the contribution of symmetrically twisted oscillators (with no invariant subspace) under modular transformations is equal to the number of fixed points of θ which by the Lifshitz fixed point theorem is $\det(1-\theta)$. So altogether we get the factor

$$D = \sqrt{\frac{\det(1-\theta)}{|I^*/I|}} . \tag{2.3}$$

This should correspond to the degeneracy of the bosonic ground state of the twisted sector (had we included the NSR contribution to the partition function we would have obtained an additional factor which would give the degeneracy of the fermionic ground state, which is always an integer). Being the degeneracy of the bosonic ground state, D should be an integer. But that is far from being obvious. Note that D is entirely determined from the group action on the lattice alone. So whether or not D is an integer is a question which could be posed independently of string theory, and just from the point of view of lattices and their automorphisms. In fact there seem to be two puzzles: one is why the expression under the square root is an integer rather than a fraction, and secondly why it is a perfect square. The answer to the first puzzle turns out to be easy and one can show by a very simple argument that the expression under the square root is an integer (see the next section). What is surprising is that D always turns out to be an integer! This fact is intimately connected to the self-duality of $\Gamma^{p,q}$ but unfortunately does not seem to have a simple proof. We will discuss a proof of it [5] in the next section (with the details given in appendix B) involving the representation theory of finite groups!

But we are still not finished with the construction of the Hilbert space. We should interpret D as the "number of fixed points" of g in some space. We could, for constructing the Hilbert space, apply a tensor in a space of dimension D, but we would not know a priori how the vertex operators act on them, and thus we would be unable to compute interactions. In the next section when we give the proof that D is an integer we will see that the proof already tells us how the vertex operators act on the fixed point Hilbert space.

To clarify the physical meaning of the ground state, we will consider in the next section doubling the degrees of freedom of the orbifold to obtain a left-right symmetric orbifold, and imbedding the Hilbert space of the asymmetric orbifold in the Hilbert space of the symmetric orbifold, and interpret D as the square root of the number of fixed points of g acting on the symmetric orbifold.

Before going to the next section let us note that modular invariance puts restrictions on g. Suppose $g^n = 1$. The necessary and sufficient condition for modular invariance at one loop (in the fermionic form) is the level matching conditions for the left-right Hilbert spaces [6]

$$n(E_{\rm L} - E_{\rm R}) = 0 \bmod 1\,, \tag{2.4}$$

where $E_{\rm L}$ and $E_{\rm R}$ are the energies of any two states of the left-moving and right-moving Hilbert spaces (in the bosonic form the necessity of (2.4) follows from invariance under the modular transformation $\tau \to \tau + n$ (or $2n$) [6,3] and the sufficiency at one loop can probably be shown by arguments similar to ref. [6]). Let $e^{2\pi i r_i/n}$ and $e^{2\pi i \tilde{r}_s/n}$ denote the complex eigenvalues of $\theta_{\rm L}$ and $\theta_{\rm R}$ respectively. Then we find the condition (2.4) for the g sector implies that for n odd

$$(na^*)^2 = \sum r_i^2 \bmod n\,; \tag{2.5}$$

for n even, eq. (2.5) is equality modulo $2n$ and in addition we get two mod-two conditions

$$\sum \tilde{r}_s = 0 \bmod 2\,, \qquad v \cdot \theta^{n/2} v = 0 \bmod 2 \tag{2.6}$$

for all $v \in \Gamma^{p,q}$. In checking level matching one has to remember that in the bosonic form it does not suffice to check the condition (2.4) only for the ground states, as it would be in the fermionic form, because here, there are winding (momentum) states, which do not necessarily have a fraction of n contribution to the energies, and so we will have to check (2.4) for the ground state of all winding (momentum) states. In fact checking level matching for all winding (momentum) states introduces only one additional condition which is the second of the mod 2 conditions in (2.6). In these conditions $\theta_{\rm R}$ appears only (if n is even) in the mod 2 conditions. The reason for this is that in the Ramond sector by world sheet supersymmetry the zero point energies of the bosons and fermions cancel and so in applying (2.4) to the ground states only the left-moving ground state energy contributes (the linear mod 2 condition for the eigenvalues of $\theta_{\rm R}$ in the n even case, is as a result of an imperfect cancellation between bosonic and fermionic zero point energies in the NS sector). Note that $g^n = 1$ implies that $na^* \in I$, and so the condition (2.5) is a condition between integers (since $(na^*)^2$ is even, from the analogue of (2.5) in the case n is even it follows that $\Sigma r_i^2 = \Sigma r_i = 0 \bmod 2$).

In general if we mod out the $\Gamma^{p,q}$ theory by an isometry group H, the condition (2.5) holds for each element of finite order. If H is abelian it has been recently shown [7], that modular invariance at one loop suffices to prove it for all loops (the discussion in ref. [7] is in the fermionic form, and to apply those arguments we should fermionize all the string degrees of freedom, which one can always do for arbitrary radii of torus [8]). If the orbifold group is non-abelian, however, it is shown there that there are in general new conditions for modular invariance. All these conditions turn out to be expressible in terms of the vanishing of particular elements in the group cohomology of H, depending on the representation of H. Note that unlike the symmetric orbifold where the space group is not a finite group, in this formulation the asymmetric orbifold group is always finite, as it has absorbed the non-finite part of the space group in the choice of the lorentzian lattice.

3. Hilbert space construction

Asymmetric structures have already been encountered in the construction of heterotic strings. So we will review the construction of the Hilbert space in that case to guide us in the construction of the Hilbert space for the asymmetric orbifolds. For the heterotic string (in its bosonic formulation) one starts by considering both the left- and right-moving degrees of freedom of the string on a lattice and by appropriate truncation of the degrees of freedom, one considers only the left-moving degrees of freedom in sixteen dimensions. More precisely, we consider the torus defined by $\frac{1}{2}(\Gamma^8+\Gamma^8)$ (for the $E_8\times E_8$ theory) and turn on an appropriate $B_{\mu\nu}$ field (the explicit form of B for cases of this type is described in ref. [9]). Then the left and right momenta are given by (P_L, P_R) where P_L and P_R each belong to the $E_8\times E_8$ root lattice. In addition there are oscillators for both left-movers and right-movers, which we use to construct the Hilbert space. The standard heterotic string is obtained by restricting to momenta of the form $(P_L, 0)$ and using only the left-moving oscillators to construct the Hilbert space.

This suggests that we should construct an asymmetric orbifold in a similar way by restricting to an appropriate subspace of a symmetric orbifold. This is the approach that we will follow. We will first review the Hilbert space construction for left-right symmetric orbifolds. In particular for simplicity we will focus on bosonic degrees of freedom.

In a left-right symmetric orbifold, the group acts directly on the coordinates, which do not by themselves have a left or right index on them, i.e. the left- and right-moving degrees of freedom are treated the same way. So suppose the underlying space is a torus, defined by a lattice Γ. This means that the coordinate x is identified with $x+v$ for all $v\in\Gamma$. Consider a group element g which is an automorphism of Γ, and acts on the coordinates by

$$gx=\theta x\,, \tag{3.1}$$

where θ is a rotation matrix, which takes the lattice to itself. For simplicity we have considered the case where g is not accompanied by shift vectors. The construction we give below easily generalizes to the case where there are shifts. Let us construct the twisted Hilbert space corresponding to the strings twisted by g, i.e. strings which satisfy

$$X(2\pi) = \theta X(0) + v\,, \tag{3.2}$$

where we have parameterized the string by an angle running from 0 to 2π. The oscillator content of the twisted Hilbert space is determined by the eigenvalues of g, and is straightforward to construct. The more interesting piece comes from the zero modes, which characterize the winding states and the center of mass position of the twisted string.

To determine the center of mass we should consider constant loops, which implies that we should find fixed points of the transformation (3.1). Let I denote the subspace of Γ left invariant by θ, and let N denote the subspace of Γ which is orthogonal to I. Each choice of center of mass determines an element of the lattice v from (3.2), that is, $(1-\theta)X_{\text{cm}} = v$, and we shall denote the corresponding state by $|v\rangle$. It is clear that $v \in N$, because for every $w \in I$, $(1-\theta)w = 0$, and so it follows that $v \cdot w = 0$, which implies that v is in N. Furthermore, any vector in N gives rise to a choice of the X_{cm}. However, two different v's which differ by $(1-\theta)u$ for $u \in \Gamma$ define the same center of mass, because that simply corresponds to shifting X_{cm} by u which is a latter vector. Note that $(1-\theta)\Gamma$ is a sublattice of N, as follows from what we said above. So we see that the inequivalent choices for the center of mass are given by the index of the quotient

$$\left|\frac{N}{(1-\theta)\Gamma}\right| . \tag{3.3}$$

In addition, we will have winding states for each $v \in \Gamma$. Except that the components of v in the N direction have already been considered in the oscillator expansion in the twisted Hilbert space. In order to form orthogonal Hilbert spaces, we should consider the orthogonal projection of Γ onto I, which we call $\tilde{I}$. So for each vector in $\tilde{I}$ we get a winding state in the twisted sector over the Hilbert space of each fixed point (with the center of mass of each shifted by the component of v parallel to N). In addition we still have momenta, but now the motion in the N direction is prohibited by the boundary conditions. And the motion of the particle is only allowed in the I direction, so the momentum belongs to the dual of I, namely I^*.

Having described the structure of the twisted Hilbert space, we would like to compare it to what we found from modular transformation of the untwisted partition function for the asymmetric orbifolds in the last section. First we write D which is supposed to be the degeneracy of the ground state as follows. As discussed in appendix A, since $\Gamma^{p,q}$ is a self-dual lattice we have $I^*/I = N^*/N$. In addition

$\det(1-\theta)=|N^*/(1-\theta)N^*|$ (note that $(1-\theta)N^*$ is a sublattice of N^*, and as discussed in appendix A, N^* (I^*) could be obtained by an orthogonal projection of $\Gamma^{p,q}$ on N (I), and so $(1-\theta)N^*=(1-\theta)\Gamma^{p,q}$). So we see that

$$D=\sqrt{\frac{|N^*/(1-\theta)N^*|}{|N^*/N|}}=\sqrt{\left|\frac{N}{(1-\theta)\Gamma^{p,q}}\right|}\,. \tag{3.4}$$

So we see the reason why D^2 is an integer: it is simply the index of a sublattice $((1-\theta)\Gamma^{p,q})$ in a lattice (N). We notice that D^2 resembles the number of inequivalent choices of the center of mass of a symmetric orbifold (3.3). This suggests that we should somehow consider a symmetric orbifold, instead of an asymmetric orbifold, and take the "square root of the number of fixed points".

The first difficulty in doing this is that the lattice $\Gamma^{p,q}$ in (3.4) is not a euclidean lattice, and cannot be identified with Γ in eq. (3.3).In order to form a symmetric orbifold, we consider the lattice $\Gamma^{p,q}$, but with a *euclidean* signature, which, to avoid confusion, we shall call $\tilde{\Gamma}^{p,q}$. We consider a symmetric orbifold whose windings are allowed to be on $\frac{1}{2}\tilde{\Gamma}^{p,q}$ and the momenta lie on the dual lattice. $\tilde{\Gamma}^{p,q}$ is not self-dual with the euclidean signature. However, using the self-duality of $\Gamma^{p,q}$ one can easily see that for each vector $(p_1,p_2)\in\tilde{\Gamma}^{p,q}$ the vector $(p_1,-p_2)$ is on the dual lattice, and this characterizes the dual lattice. In addition we will turn on a certain B field, similar to the E_8 case (the mod two reduction of the metric in a lattice basis). The general structure of the momenta for the asymmetric orbifold is [2]

$$(P_L,P_R)=\left(\tfrac{1}{2}P-BW+W;\tfrac{1}{2}P-BW-W\right), \tag{3.5}$$

where

$$P=2(k_1,-k_2),\qquad W=\tfrac{1}{2}(k_3,k_4), \tag{3.6}$$

and

$$(k_1,k_2),(k_3,k_4)\in\tilde{\Gamma}^{p,q}.$$

The B field is chosen so that if e_i generate $\tilde{\Gamma}^{p,q}$, then

$$(e_iBe_j)=e_i\cdot e_j\qquad \mathrm{mod}\,2,$$

where the dot product on the right is with respect to the *lorentzian* signature (it is easy to see that such a B can always be defined). With this choice of B one can use the formula (3.5) to obtain the left-right momenta of the states of the symmetric orbifold. This is a lattice $\Gamma^{p+q,p+q}$ which is generated by vectors of the form

$$(k_1,0;0,-k_2),(0,k_2;-k_1,0),$$

where $(k_1,k_2)\in\tilde{\Gamma}^{p,q}$. Now to obtain the asymmetric Hilbert space (in the untwisted sector), we should restrict to momenta which are of the form $(k_1, 0; 0, -k_2)$, and use only the first p left-moving oscillator excitations, and the last q right-moving oscillator excitations.

Now we will move on to the construction of twisted sectors. Any symmetry of $\Gamma^{p,q}$ which can be used to construct an asymmetric orbifold, can also be used to construct a symmetric orbifold out of $\tilde{\Gamma}^{p,q}$. This easily follows, because any symmetry of $\Gamma^{p,q}$ which does not mix the left and right movers is also a symmetry of $\tilde{\Gamma}^{p,q}$ since it sits in $SO(p)\times SO(q)$. So if (x_1, x_2) denote the coordinates of the torus generated by $\frac{1}{2}\tilde{\Gamma}^{p,q}$, we can mod out by

$$(x_1, x_2) \to (\theta_L x_1, \theta_R x_2).$$

In the twisted sector of the symmetric orbifold the ground state will have a degeneracy given by eq. (3.3) with Γ replaced by $\tilde{\Gamma}^{p,q}$, which is therefore exactly the square of (3.4) (it is not too difficult to see that the N's in the two formulas (3.3) and (3.4) are the same (up to a factor of $\frac{1}{2}$)). So we will have to restrict to a "square root of the fixed point Hilbert space" of the symmetric orbifold to describe asymmetric orbifold states in the twisted sector. It is also not too difficult to see that the momenta and windings of the symmetric orbifold in the twisted sector are generated by vectors of the form $(k_1, 0; 0, -k_2)$ and $(0, k_2; -k_1, 0)$ with $(k_1, k_2) \in I^*$. So restricting to the vectors $(k_1, 0; 0, -k_2)$, we obtain the momenta in the twisted sector of the asymmetric orbifold, which gives a partition function as predicted from modular transformations discussed in the last section.

So we see that the main subtlety in the construction of the Hilbert space of left-right asymmetric orbifolds, starting from a symmetric one, occurs in the twisted sector where one has to take the square root of the ground state of a left-right symmetric orbifold. In ref. [3] it was suggested in the context of twisting the root lattice of a left-moving E_8, that the fixed points of a symmetric orbifold should be thought of as representing a tensor product structure of left-movers times that of the right-movers, and restricting to left-movers would mean restricting to the left-moving part of the tensor product. We will now discuss how taking this "square root" of the Hilbert space of the fixed point set is accomplished, and why one is *forced* to do that from the point of view of interactions, as well as generalize it for an arbitrary asymmetric orbifold.

Structures which we shall interpret as asymmetric orbifolds have already come up in the study of twisted affine Lie algebras [5] and their vertex operator constructions will be of great importance for the interactions on asymmetric orbifolds. We shall make use of these constructions below.

As we mentioned above, to understand the structure of the twisted Hilbert space of an asymmetric orbifold we will have to consider interactions on orbifolds. Interactions on orbifolds have been considered in ref. [10]. Let us concentrate on the couplings between two twisted states with an untwisted state. These are among the simplest couplings to describe, as one can compute them without constructing twist operators. To compute them one considers

$$\langle t_{\mathrm{f}}| V_{\mathrm{u}} |t_{\mathrm{i}}\rangle,$$

where $|t_{\mathrm{i}}\rangle$ and $|t_{\mathrm{f}}\rangle$ label two twisted states and V_{u} denotes the vertex for the emission of an untwisted state, from a twisted sector. Note that the emission of an untwisted state keeps one within the same twisted sector. Let us ask whether there is any way that in a symmetric orbifold, one can start from the twisted string expanded about one fixed point, to get to another string expanded about a different fixed point. The answer to this question is yes [10], we simply have to emit an untwisted winding state. The vertex operator would be the usual vertex operator written in terms of the twisted string variables, except that it is accompanied, in addition, by a finite dimensional matrix T which acts on the space of fixed points. We can label the fixed points by vectors in the lattice N (modulo $(1-\theta)\Gamma$), and with that labeling emission of winding state w acts on the fixed points by $T_w|v\rangle = |v+w\rangle$. There are other states in the identity sector, and in particular, the momentum states. We could, more generally consider the emission of both momentum and winding states. Emission of momentum states does not change the fixed points, so that $T_{p,w}|v\rangle = |v+w\rangle$.

However, the action of T on the fixed points is not completely determined by which fixed point gets mapped to which fixed points. We will also have to define the *phase* of the matrix elements of T. In the above paragraph we had assumed that T has no non-trivial phases. It turns out, as we shall see below, that if the matrix T simply shifts the fixed point set, without introducing phases, there would be an inconsistency with the symmetry structure required in a closed string amplitude. In particular we will be *forced* to include phases in the matrix elements of T for consistency of the closed string theory. We must allow the more general possibility of

$$T_{p,w}|v\rangle = \mathrm{e}^{i\alpha(p,w,v)}|w+v\rangle$$

for some α. The order of emission of states should not change the amplitude in a closed string theory, because there is no intrinsic meaning to any ordering of points on a Riemann surface without boundaries. To find the phases in the matrix elements of T, we will have to investigate the order independence of the amplitudes. Non-trivial phases of the type we shall encounter have already been encountered in the construction of the vertex operators for affine Lie algebras [11], and are used for the emission of gauge particles in the bosonic formulation of the heterotic string [1]. There it was seen that one should properly define the vertex operators for the emission of momentum states taking into account a cocycle structure described in ref. [11]. The introduction of the cocycles was necessary to insure that the vertex operators have the correct commutation relations. Or, equivalently, the order in which the gauge particles are emitted should not affect the amplitude in a closed string theory. This means that when computing an amplitude, we must have

$$\langle \dots V(p_1, z_1) V(p_2, z_2) \dots \rangle = \langle \dots V(p_2, z_2) V(p_1, z_1) \dots \rangle ,$$

but if one uses the standard tachyon emission vertex, the left-hand side gives a factor proportional to $(z_1 - z_2)^{p_1 \cdot p_2}$, whereas the right-hand side gives $(z_2 - z_1)^{p_1 \cdot p_2}$. The ratio between these two is $(-1)^{p_1 \cdot p_2}$. So for the above equality to hold it seems that we should modify the vertex operators by multiplying them by operators e_{p_i} such that

$$e_{p_1} e_{p_2} = (-1)^{p_1 \cdot p_2} e_{p_2} e_{p_1} .$$

But we do not have to introduce new operators in the theory. We simply have to note that the vertex for the emission of a gauge particle shifts the momentum, by the momentum of the gauge particle, but again, as is the case for shift matrices T, the phases are not determined just by the momentum conservation, and one can include phases in the definition of the vertex operators so as to avoid the problem mentioned above.

To ensure appropriate commutation relations or the vertex operator representation of twisted affine algebras similar cocycles have been found to be necessary [5]. Interpreted from the physical viewpoint this means that we need to accompany the vertex operators for the emission of untwisted states from twisted sectors by operators whose commutation relation is given by a cocycle factor. Let n be the order of the automorphism θ, and let $\omega = e^{2\pi i/n}$. Suppose we are considering the emission of untwisted states with momenta $p_1 = (p_{1L}, p_{1R})$ and $p_2 = (p_{2L}, p_{2R})$ (note that here the word "momenta" refers both to momenta and windings). Let $X = X_L + X_R$ denote the string variable, and suppose we are in the sector twisted by θ, which means that $X(2\pi) = \theta(X(0))$ (this notation means that $X_L(2\pi) = \theta_L X_L(0)$ and $X_R(2\pi) = \theta_R X_R(0)$). We note that to each vertex operator is associated a matrix T_p which acts on the center of mass of the twisted string. So the vertex operator is given by $T_p e^{ip \cdot X}$ (where again the left and right labels are implicit). The condition found in ref. [5] is that for proper commutation relations of the vertex operator we must have

$$T_{p_1} T_{p_2} = C(p_1, p_2) T_{p_2} T_{p_1} , \tag{3.7}$$

where

$$C(p_1, p_2) = \prod_{k=1}^{n} (-\omega^k)^{(\theta^k p_1) \cdot p_2} \tag{3.8}$$

(note that the first reference in [5] deals with a general even integral lattice of any signature which includes the case of interest to us). This result can be easily derived by considering the amplitude for the emission of two untwisted states in two different orders from the twisted sector (using the formula for the relevant Green function discussed in ref. [10]). Note that by (3.7) we must have $C(p_1, p_2) = C(p_2, p_1)^{-1}$. It is easy to see that (3.8) indeed satisfies this condition (in the case that n is even, we have to use the second condition in (2.6) to show this).

Now we are finally in a position to discuss the taking of the square root of the fixed point set. Let us concentrate on emission of untwisted states which does not introduce winding states, and only acts on the ground states of the twisted sector. For this purpose, we consider all the vectors of the type $p = (k_1, 0; 0, -k_2)$ such that $(k_1, k_2) \in N$. Associated to these are the matrices T_p which act only on the fixed point Hilbert space. These matrices are associated to emission of untwisted states with momentum p of the asymmetric orbifold from the ground state of the twisted sector. Now if we consider all such matrices T_p, they generate a non-abelian group G acting on the fixed point Hilbert space. Eq. (3.7) shows the non-abelian structure of G (restricted to the subspace N, C simplifies to $C(p_1, p_2) = \prod_{k=1}^{n}(\omega^k)^{(\theta^k p_1)\cdot p_2}$). The matrices associated to two p's which differ by $(1-\theta)\Gamma^{p,q}$ act on the fixed point states the same way, and are equal. To see this, note that fixed point states are characterized by vectors in N modulo vectors in $(1-\theta)\Gamma^{p,q}$, so two such T's will shift the fixed points in an identical manner. The only way they could have differed is in their phase structure. But using eq. (3.8) one can show that they satisfy identical commutation relations with other T's. This follows from the fact that $C(p, *) = 1$ for all $p \in (1-\theta)\Gamma^{p,q}$. So the group G has $D^2 \cdot n$ elements where D is given by (3.4), and n is the order of automorphism. This is because each element in G corresponds to a vector in $N/(1-\theta)\Gamma^{p,q}$, multiplied by an nth root of unity ω^k. The nth root of unity is in G because of the relations (3.7) and (3.8). More properly speaking we should think of G as an abstract group which has a faithful representation furnished by T's and nth roots of unity.

It is important to ask whether this representation of G is irreducible or not. If it were reducible, it would mean that by taking appropriate linear combinations of states in the Hilbert space of the fixed point set to form irreducible representations, the operators for the emission of untwisted states will only mix them within the same irreducible representation of G. This would mean that the emission of untwisted states keeps the state in the given irreducible representation of G. It turns out that in fact no interactions, not just the emission of untwisted states, can mix states in different irreducible representations of G, and therefore, the physics decouples for each irreducible representation. To illustrate the reason for this, suppose that there were a non-vanishing amplitude involving two twisted states from two different irreducible representations of G. Then, we can reconstruct the amplitude (by duality), in such a way that the two twisted states are connected by a state in the untwisted sector, which is impossible because the vertex operators of the untwisted sector do not mix different irreducible representations. This means that it is consistent (and necessary) to throw away all the states except the states for a given irreducible representation. The rest of the states do not couple and therefore do not affect the physics.

The dimension of the representation is the same as the number of fixed points, which is simply D^2. It turns out (see appendix B) that this representation is reducible, and decomposes to D isomorphic irreducible representations of G each of dimension D. For one thing, this proves that D is an integer! Also it implies that we can keep any of the irreducible representations we wish. They all give isomorphic physics and it does not matter which one we keep. This shows that the dimension of the ground state of the twisted sector of an asymmetric orbifold, which was determined by performing modular transformations to be D, is also seen from the viewpoint of forming irreducible representations of vertex operators. This shows a deep connection between modular transformations and the irreducibility of the representations of vertex operators*. We conclude this section by noting that there is a dual asymmetric orbifold, whose momenta are given by $(0, k_2, -k_1, 0)$, which gives rise to a group $\tilde{\mathrm{G}}$ which is isomorphic to G. Under the group $\mathrm{G} \times \tilde{\mathrm{G}}$ the fixed point set forms the representation $(D, \overline{D})$, and so we see that for the symmetric orbifold, the fixed point set forms a single irreducible representation and we should keep all the states, a fact which is again predicted by modular invariance (in the definition of C in (3.8) for $\tilde{\mathrm{G}}$, ω is replaced by $\bar{\omega}$).

In the construction of orbifolds, we will also have to project onto the group invariant subspaces. For cyclic orbifolds, if the order of the group is not prime, then in general the group projection also acts on the fixed point set to give a subspace of it. The group invariant projection can be understood in term of the requirements of modular invariance. For example suppose the order of the cyclic group is $p_1 \cdot p_2$ where p_1 and p_2 are prime. In the g-twisted sector all the states satisfying level matching are physical and so all the fixed point set can be used to form a physical state. But in the g^{p_1} sector the modular transformations $\tau \to \tau + 1$ etc. do not give all the twistings in the τ direction and hence not all the states satisfying level matching are physical. There will be different phases associated with different orbits which are obtained by modular transformation from other twisted sector. The effect of this will be a group projection on the fixed point set for the g^{p_1} sector (which in general is bigger than the fixed point set for the g-twisted sector) to give a subspace of it. This should also result in an irreducible representation of the group invariant vertex operators.

It seems that there must be a general theorem to the effect that whenever there is a modular invariant string theory there is a Hilbert space for string theory, which

* A similar connection is seen in the standard superstring, where starting from the NS sector with periodic boundary condition in time direction and performing a $\tau \to (-1/\tau)$ transformation, one determines the degeneracy of the ground state of the R sector, which is found to be 16, which is also the dimension of the irreducible representation of the Clifford algebra. In that case we could have introduced reducible representations of the Clifford algebra in the R sector, but again, since the vertex operators do not mix different irreducible representations, all but one of the irreducible representations decouple.

forms an irreducible representation of the relevant vertex operators. In fact there is a theorem in the representation theory of affine Lie algebras [12] which asserts that the partition function for representations of the twisted irreducible representations can be obtained by modular transformations on certain characters of the untwisted representation. It seems that there should be a generalization of this and similar theorems to algebras which appear in string theory (not just the affine Lie algebras).

The construction of interactions on asymmetric orbifolds can be done along the lines discussed in ref. [10]. One computes interactions as if they come from symmetric orbifolds (via the imbedding discussed above) and at the end takes the relevant square root. The oscillator contributions to the twist operator correlations are easily obtained for the asymmetric orbifolds, because they naturally factorize into holomorphic times antiholomorphic contributions and so the square root can therefore be easily taken. The only new novelty is in the instanton contribution to the interactions. In this case the contribution does not obviously factorize. However, in the instanton contributions one has to remember that the irreducible representations of the fixed points involve non-trivial phase choices. Doing the instanton sum, with proper care taken for these non-trivial phases, should allow one to take the appropriate square root. It would be nice to work out the details of this.

4. Examples

In this section we will discuss some examples of asymmetric orbifolds. The examples are to be regarded only as illustrative of the new features of asymmetric orbifolds and not of a direct phenomenological relevance. However, similar constructions for arbitrary asymmetric orbifolds are easily carried out along the lines described for these examples, and may lead to interesting physical models.

We will start by noting that when modding the left and right asymmetrically it sometimes is the case that the "left lattice" and the "right lattice" are associated to a Lie group (by a misuse of the term "lattice"). In such cases, interesting additional structures appear. We shall mention three aspects. First, we tend to get enhanced gauge symmetries after compactification (as is discussed in ref. [2]). Second, when modding out, the fixed point set in the twisted sector may itself form a non-trivial representation of the Lie group in question. The reason for this is that the gauge group states (which come from the untwisted sector) include winding states, and the fixed point set forms a non-trivial representation under the action of the winding states, as discussed in sect. 3 and appendix B. For instance, we recall the example discussed in ref. [3], where one considers modding one of the E_8's by a Z_2 automorphism which sends every vector to its minus. In the untwisted sector this projection corresponds to the breaking $E_8 \to O(16)$. In the twisted sector, there are 2^8 fixed points. The irreducible representation of these fixed points, under the action of vertex operators, is, as discussed in the last section, $\sqrt{2^8} = 16$-dimensional, and it corresponds to the vector representation of O(16).

The third interesting phenomenon which occurs when the left and right lattices are group lattices is that if we mod out by an automorphism of the lattice which is in the Weyl group, the modding out is equivalent to a shift in the lattice. This is discussed in the maths literature [13], and is also discussed in the context of some examples in ref. [3] (and in the recent work of Kawai et al. [4]). For instance, in the above case, the Z_2 twisting is equivalent to a shift given by $(1,0^7)$ in the standard basis of E_8. The reason for this is that the twisting of a string, described by a group element, should be isomorphic to twisting by a conjugate group element. To illustrate this, suppose we consider the SU(2) group, represented by the U(1) circle generated by J_3. Then if we take the twisting group element to be $e^{\pi i J_3}$, it corresponds to a shift on the circle. If we consider twisting by the conjugate element $e^{i\pi J_2}$ it corresponds to the Z_2 reflection on the circle $x \to -x (e^{i\pi J_2} J_3 e^{-i\pi J_2} = -J_3)$. However, since the affine Lie algebra representation should only depend on the group and not a particular representation of the group via the circle generated by J_3, the two operations (shifting and reflection) in this case should be equivalent. Indeed they do give identical partition functions, and isomorphic vertex operators (see ref. [13]). The shifting of a lattice is equivalent to compactification on a torus. So what we are seeing is that string propagation on one space which has the topology of a torus could be isomorphic to string propagation on an orbifold! This example emphasizes, once again, that our naive field theoretic intuition which suggests that tori and orbifolds are totally different spaces, may be misleading when applied to string theories. In such cases, where the automorphism could be represented by a shift, the rank of the resulting gauge group is maximal and equal to $16+d$ when compactified on a d-dimensional space.

Note that it is not true that every automorphism of the group lattices is in the Weyl group. For instance, some automorphisms of the lattice may be outer automorphisms of the group. In such cases, there is no reason that the automorphism could be represented by a shift, and in general it is not possible to do so. In these cases in general the rank of the gauge group is reduced (and is less than $16+d$).

We will begin by a compactification which leaves four uncompactified space-time dimensions, corresponding to an asymmetric orbifold compactification of the heterotic string. The case of interest for leaving four uncompactified dimensions corresponds to considering a lattice $\Gamma^{16+6,6}$ and an automorphism (which lies in an $SO(16+6) \times SO(6)$). Generally speaking this lattice will have no automorphisms (except for an overall reflection), but in some cases there will be extra automorphisms. A symmetric orbifold that was discussed in ref. [3] is the compactification of the heterotic string on the Z orbifold. The Z orbifold corresponds to compactifying on a $T^2 \times T^2 \times T^2$, where T^2 is defined by a lattice which has the shape of the root lattice of SU(3), and modding by a simultaneous $2\pi/3$ rotation on each torus. This automorphism could be accompanied by a Z_3 action on the gauge degrees of freedom. The action is restricted by the requirement of modular invariance. This Z_3

action can be represented by shifts on the gauge lattice. A shift consistent with modular invariance is given by $a=(\frac{2}{3},-\frac{1}{3},-\frac{1}{3},0,0,0,0,0)$ (this is motivated by imbedding the spin connection in the gauge connection, although this is not the only shift consistent with modular invariance). In the construction of the symmetric Z orbifold there is no requirement on the overall radii of each of the three tori. For the asymmetric orbifolds, however, all or some of the compactified radii are generally fixed, as has been noted in the fermionic formulation in the second reference in ref. [4].

Now we will discuss an asymmetric orbifold, which for the right-movers looks like a Z orbifold, and for the left-movers looks like a toroidal compactification. Let R and W denote the root and weight lattices of SU(3) respectively. Consider the lattice $\Gamma^{2,2}$ defined by

$$(p_1,p_2) \qquad p_1,p_2\in W, \qquad p_1-p_2\in R. \tag{4.1}$$

We take for our underlying momentum space of the asymmetric orbifold

$$L=\Gamma^8+\Gamma^8+3\Gamma^{2,2},$$

where Γ^8 denotes the root lattice of E_8. L is an even self-dual lorentzian lattice with signature $(16+6,6)$. A vector in L will be denoted by $(v_1,v_2,p_1,\tilde{p}_1,p_2,\tilde{p}_2,p_3,\tilde{p}_3)$, where $v_i\in\Gamma^8$ and $(p_i,\tilde{p}_i)\in\Gamma^{2,2}$. These are the momentum (winding) states obtained by compactifying the $E_8\times E_8$ heterotic string on the product of three tori obtained by (half) the root lattice of SU(3) with the appropriate B field turned on. In particular, it can be used as a starting point for a symmetric Z orbifold with the B turned on (the B field shares the Z_3 symmetry).

Now we consider an "asymmetric" Z_3 action defined by

$$|(v_1,v_2,p_1,\tilde{p}_1,p_2,\tilde{p}_2,p_3,\tilde{p}_3)\rangle\to e^{2\pi i v_2\cdot a}|(v_1,v_2,p_1,\theta\tilde{p}_1,p_2,\theta\tilde{p}_2,p_3,\theta\tilde{p}_3)\rangle, \tag{4.2}$$

where θ denotes a rotation by $2\pi/3$. It is easy to see that this is a symmetry of the lattice ($(p,\theta\tilde{p})\in\Gamma^{2,2}$ if $(p,\tilde{p})\in\Gamma^{2,2}$)). As mentioned in sect. 2, the action on NSR fermions is determined to be (θ,θ,θ) by the requirement of preserving world sheet supersymmetry. Since the rotation on the six right-movers sits in an SU(3) subgroup of SO(6) (the center element of SU(3)), the resulting four-dimensional theory has an $N=1$ supersymmetry. There are three sectors: The untwisted (identity) sector, the twisted sector, and the inverse twisted sector. It is easy to describe the whole partition function for all the sectors, but we shall limit ourselves to the massless states.

The massless states coming from the identity sector are easy to obtain. They simply correspond to taking Z_3 invariant states of the states in the toroidal compactification. They are the four-dimensional graviton, antisymmetric tensor field, the dilaton, the gauge fields of $E_8 \times E_6 \times SU(3)^4$ (one SU(3) comes from breaking one E_8 to $E_6 \times SU(3)$ and the other three come from our choice of the (6,6) lattice), 3 scalar fields in the representation (27,3,1,1,1) (dropping the E_8 representation which is a singlet), and their supersymmetric partners.

The interesting feature of the asymmetric orbifold arises in the twisted sector. The invariant lattice $I = \Gamma^8 + \Gamma^8 + (R,0)^3$ and the normal lattice is $N = (0, R)^3$ (with the obvious notation). It is easy to see that $(1-\theta)L = N$, and so the number of fixed points computed using (3.4) is equal to one (this is to be compared with the symmetric case where there were 27 fixed points). The right-movers give rise to matter fields, and the left-movers have, aside from the oscillator excitations, momenta which belong to

$$I^* + a^* = \Gamma^8 + (\Gamma^8 + a) + (W,0)^3 .$$

Using this we find that the massless particles from the twisted sector give a single matter field in the representation

$$(\overline{27},3,1,1,1) + [(27,1,3,1,1) + (27,1,\overline{3},1,1)]$$
$$+ [(1,\overline{3},3,3,1) + (1,\overline{3},\overline{3},3,1) + +(1,\overline{3},\overline{3},\overline{3},1)] + X ,$$

where X are the representations which make the above representation symmetric with respect to the last three SU(3)'s. The inverse twisted sector simply provides the antiparticles to the states in the twisted sector. It is easy to see that the resulting four-dimensional theory has no anomalies (in fact this follows from the fact that the theory is modular invariant by an application of the generalization of the work [14])*.

The Z_3 action can be represented purely in terms of shifts, because the automorphism in question is an inner automorphism of the SU(3) lattice. In fact we may as well bosonize the NSR fermions. The resulting lattice (to be more precise, shifted lattice) is odd, self-dual and has signature $(16+d, d+4)$ (in the light cone gauge) and in this example, $d=6$ (self-duality is most easily seen in the covariant approach [16]). Moreover, the automorphism acts on this lattice by shifting, which means that

* There is a possibility that the asymmetric Z orbifold example we have given could be interpreted as a symmetric orbifold in the following way. The shift in E_8 can be represented as a Z_3 rotation in three SU(3)'s of E_8 after a change of Cartan basis. Now we can consider these three SU(3)'s as forming the space part of the left string and then this would become a symmetric Z orbifold in the presence of background fields (symmetric orbifolds in the presence of background fields have been considered in detail in ref. [15]). However in the above example if we did not shift the gauge lattice at all we would obtain an asymmetric orbifold which has on obvious relation to any symmetric orbifold.

similar to ref. [2] we can perform a Lorentz rotation to go from the untwisted theory to the twisted one (Lorentz rotations on such spaces have been considered in ref. [16]). In fact, an arbitrary $SO(16+d, d+4)$ rotation would preserve the modular invariance, but only a discrete subgroup of $SO(16+d, d+4)$ will preserve the world sheet supersymmetry. In the standard toroidal compactifications Lorentz rotations were interpreted as switching on background fields of constant metric, antisymmetric tensor, and Wilson lines. In the present case, since the momenta corresponding to bosonized fermions interact with the momenta of left-moving bosons, a possible background field interpretation could be constant $F_{\mu\nu}$'s and $H_{\mu\nu\rho}$ on a six-dimensional torus. The sigma model action in the presence of these fields, however, becomes non-linear and it is not clear if one can solve the model exactly. Work in this direction is in progress [17]. Finally one can relate the asymmetric orbifold construction with the work of ref. [4]. These authors fermionize all compact directions and consider modding out the fermionic theory, compatible with modular invariance, spin-statistics and world sheet supersymmetry, a property which is manifestly preserved in the bosonic formulation (the lattice rotation interpretation above was also independently noted in the first paper in ref. [4]).

To give a flavor of the possibilities available for asymmetric orbifolds we give two more examples, without going into much detail. Since we are simply considering examples, we will take a compactification to 6 space-time dimensions. The lorentzian lattice has signature (20, 4) in this case. We take the left "lattice" to be $E_8 \times E_8 \times O(8)$ and the right "lattice" to be O(8) (the weights of the left and right O(8)'s are correlated, so that they are in the same conjugacy class). We do a Z_2 modification on the right SO(8) by taking each weight to its minus (this is a symmetry of the lattice), accompanied by the Z_2 automorphism of the E_8 lattice described earlier. Level matching conditions are satisfied. The gauge symmetry in this case is $O(8) \times O(16) \times E_8$. The chiral fermions in the untwisted sector are two copies in $(1, 128, 1)$. In the twisted sector we get two copies $(1, 128, 1) + (8_v, 16, 1) + (8_s, 16, 1) + (8_c, 16, 1)$. The reason we get two copies is that by formula (3.4) the degeneracy in the twisted sector is $\sqrt{\frac{16}{4}} = 2$. Note that the fixed point set in this case does not correspond to any gauge representation (even though it corresponds to non-trivial representation of the vertex operators), because they are related to the right SO(8) which has no gauge symmetry significance (the gauge quantum numbers in the heterotic string comes from the left-movers). This example can also be represented by a lorentzian rotation of the totally bosonized theory, because the Z_2 corresponds to an inner automorphism.

As a final example, we give an example in which the gauge symmetry involves a non-simply laced group. Consider the lorentzian lattice corresponding to left $E_8 \times E_6 \times E_6$, and right $SU(3) \times SU(3)$ (the weights of the E_6 lefts and SU(3) rights are correlated via triality). We again consider a Z_2 orbifold by reflection in all SU(3)'s and E_6's, accompanied with a shift $(\frac{1}{2}, \frac{1}{2}, 0^6)$ in the E_8. The Z_2 in this case is not an inner automorphism and so the resulting theory cannot be written by a lorentzian rotation on a totally bosonized theory. The resulting gauge symmetry is $Sp(8) \times Sp(8) \times E_7 \times SU(2)$. The chiral fermions in the untwisted sector are in two copies of $(42,1,1,1)+(1,42,1,1)+(1,1,56,2)$. In the twisted sector the degeneracy coming from the fixed points is $\sqrt{2^2 \cdot 2^2 \cdot 2^6 \cdot 2^6} = 4 \cdot 8 \cdot 8$. We get for the chiral matter field four copies of spinors in the representation $(8,8,1,2)$ (it is conceivable that this last example may be interpretable in terms of a symmetric orbifold, even so this description viewed from an asymmetric point of view is far simpler).

Finally we would like to mention a point which is not particular to the asymmetry of the orbifold, and can also be discussed in the context of symmetric orbifolds, and that is the question of "invariants" for orbifolds. This is the question of whether one orbifold can tunnel into another. If there are invariants which would prevent that, the string theory would seem to lose part of its predictive power, as two worlds which start on orbifolds with different invariants cannot tunnel into one another. There seem to be invariants of the supersymmetric non-linear sigma models which would be precisely saying that. For instance the Euler characteristic of an orbifold can be phrased as the computation of the supersymmetric sigma model with (P, P) boundary conditions on left-right fermions. For instance, if we compactify the superstring on the four-dimensional lattice corresponding to SO(8)-left SO(8)-right, and mod out by a Z_2 reflection of all coordinates on both left and right, we get a symmetric orbifold which corresponds to a compactification on a singular limit of K_3. This has Euler characteristic 24. So one would be inclined to say that this space cannot tunnel to the four-dimensional torus which has zero Euler characteristic (the question of other dimensions is not relevant to the point we wish to make). However, we know that since the Z_2 in this case corresponds to an inner automorphism, if we bosonize all the degree of freedom, we can achieve this modding out by a Lorentz rotation on an odd self-dual lorentzian lattice. So it seems that we can continuously go from a torus to the K_3 theory and thus continuously change the Euler characteristic from 24 to 0. This seems to be in contradiction to what we said about the invariance of the Euler characteristic in supersymmetric sigma models. In fact there is no contradiction, because the one-parameter family breaks the world sheet supersymmetry and the Euler characteristic is invariant as long as we are describing a supersymmetric theory. The requirement to preserve the world sheet supersymmetry seems to be a requirement for solving the string equations. It is similar to conformal invariance which is needed for satisfying string equations. Similar to describing a one-parameter family of string models which connects two conformally invariant theories by theories which are not themselves conformal, we see that we can connect strings on the K_3 theory to strings on a torus via models which are not solutions to string equations. It seems to us, therefore, that it would be misleading to think of the Euler characteristic as an invariant of string theory, even if it is an invariant of the supersymmetric sigma model. Similar arguments apply to the recently discovered invariant of the supersymmetric sigma model called the "elliptic genus" [18].

5. Conclusion

We have seen that it is possible to describe the strings which have left- and right-moving degrees of freedom on different orbifolds. This means that it is possible to retain heteroticity even after certain interesting compactifications. This suggests that the six compact dimensions required to go from 10 to 4 may not be "real" dimensions. On the other hand it may be that string propagation on asymmetric orbifolds could be interpreted as strings propagating on symmetric orbifolds, with background fields turned on (for example $F_{\mu\nu}$ and $H_{\mu\nu\rho} \neq 0$) corresponding to some non-trivial fixed points of the beta function [17].

The study of asymmetric orbifolds, even though it includes the symmetric one as a special case, seems to simplify the general outlook somewhat. Namely, by combining the gauge degrees of freedom with the space degrees of freedom, the orbifold group is completely specified by defining an isometry group of the lattice $\Gamma^{16+d,d}$ (which sits in an $SO(16+d) \times SO(d)$). This means that we do not have to describe the action on the gauge degrees of freedom separately from the space-time coordinates, and in this sense the modding out operation in the construction of orbifolds is unified. It would be of great interest to classify all the compactifications of the heterotic strings on toroidal orbifolds. This is simply a question of studying the symmetries of the lattices $\Gamma^{16+d,d}$ (which do not mix the left- and right-movers).

Also it may be interesting to try to enlarge the class of asymmetric spaces from the ones which have tori as their underlying spaces to more general spaces.

The fact that asymmetric orbifolds are consistent raises a basic question: The standard phenomenology suggests that both the left- and right-moving degrees of freedom of the four uncompactified dimensions propagate on four-dimensional Minkowski space. In view of the possibility of separation of left- and right-moving degrees of freedom, we should ask what mechanism forces this "identification" of four of the left-movers with four of the right-movers.

We would like to thank J. Conway for describing to us some properties of decomposition of lattices used in appendix A, and J. Lepowsky for many patient discussions of his work. We would also like to thank J. Distler and P. Ginsparg for discussions, and P. Ginsparg also for a careful reading of the manuscript. The research of C.V. was supported in part by NSF contract PHY-82-15249, and by a fellowship from the Harvard Society of Fellows.

Note added

It seems that asymmetric orbifolds in general cannot be viewed as symmetric orbifolds even in the presence of background fields. We will present the argument for the superstring type II asymmetric orbifolds which has been recently considered in ref. [19]. In superstrings we do not start with a gauge group. However from a field theory viewpoint it is clear that if the internal space has a continuous symmetry group it can give rise to gauge symmetries in lower dimensions. This is also true in superstrings, even though sometimes there may be enhanced symmetries due to winding states. Can one obtain chiral fermionic representations in lower dimensions in this way?

The argument in the second reference of [18] implies that for any non-linear supersymmetric sigma model (which includes symmetric orbifolds, possibly with background fields turned on) one cannot obtain chiral fermions in lower dimensions by a Kaluza-Klein type mechanism. Making the very plausible assumption that the Cartan generators of any gauge symmetry in string theory come from a Kaluza-Klein symmetry, this implies that one cannot obtain chiral fermions using symmetric orbifolds in type II strings. However it has been found in ref. [19] that using asymmetric orbifold compactification of type II superstrings it is possible to obtain chiral four-dimensional theories. This means that any asymmetric orbifold may not in general be viewed as a symmetric orbifold even in the presence of background fields.

What we just described in mathematical terms means that the U(1) character valued index of operators such as the Rarita-Schwinger are *not* constant for asymmetric orbifolds, even though they are constant for symmetric orbifolds by the arguments in the second reference in [18]. This suggests that asymmetric orbifolds are interesting new spaces which may be worthwhile to investigate for purely mathematical reasons.

Appendix A

In this appendix we will discuss some of the relevant facts about decomposition of a self-dual even lattice to two orthogonal subspaces. The arguments given below apply equally well to euclidean and lorentzian self-dual lattices (with a decomposition defined by an automorphism), and we shall not need to specify the signature of the lattice.

Consider a self-dual even lattice Γ, and two orthogonal subspaces in it I and N (the orthogonality means that each vector in N is orthogonal to each vector in I). Moreover we assume that each vector orthogonal to all vectors in I is in N and each vector orthogonal to all vectors in N is in I. There may be vectors in the lattice which have components in both I and N directions, without being possible to write it as a sum of a vector in I and a vector in N. This means that generally Γ may not be equal to $I+N$. We can choose the coordinates so that the vectors in I look like $(*,0)$ and those of N look like $(0,*)$, in a basis where the metric is diagonal (this is obvious for euclidean lattices, and it is true for the non-euclidean case when N and I are defined by an automorphism of the lattice which does not mix the left-movers with the right-movers which is the case in construction of asymmetric orbifolds). Consider the projection of all vectors in the lattice on I, which form a lattice $\tilde{I}$. This means that for each vector $(k_1, k_2) \in \Gamma$, we consider the vector $(k_1, 0)$. It is clear, because of the integrality of Γ, that $\tilde{I} \subset I^*$, that is, the scalar product of $(k_1, 0)$ with all the vectors in I gives an integer. We will now show that

$$\tilde{I} = I^* . \tag{A.1}$$

We show this by showing that the lattice dual to $\tilde{I}$ is I. Let $(p, 0)$ represent an arbitrary vector in the lattice dual to $\tilde{I}$. Then it follows that $(p, 0)$ has integral dot product with all the vectors in Γ, because each vector in Γ looks like $(v, *)$ with $(v, 0) \in \tilde{I}$, and $(p, 0) \cdot (v, *) = (p, 0) \cdot (v, 0) =$ integer. But Γ is self-dual, and so it follows that $(p, 0) \in \Gamma$, and this implies that $(p, 0) \in I$. So this shows that $(\tilde{I})^* = I$, and therefore eq. (A.1) follows. Of course similar statement holds for N.

We also note that $I \subset I^*$ and $N \subset N^*$ (by the integrality of Γ). We now wish to show that

$$I^*/I = N^*/N . \tag{A.2}$$

We show that for every vector in I^*, we can associate a vector in N^*, modulo N, and vice-versa. Let $v = (k_1, 0) \in I^*$. By what we showed above, this implies that $(k_1, k_2) \in \Gamma$ for some k_2. We associate to v the vector $w = (0, k_2) \in N^*$. However, we may have chosen a different vector $(k_1, k_2') \in \Gamma$. Then we would have associated $(0, k_2')$ to v. But the difference of these two vectors is

$$(0, k_2 - k_2') = (k_1, k_2) - (k_1, k_2') \in \Gamma$$

and this implies that $(0, k_2 - k_2') \in N$, and shows that the choice of w is well defined modulo the addition of a vector in N. Similarly, by symmetry, the same argument applies in the reverse order, and establishes (A.2).

We can consider the following simple example to illustrate the above results. Consider the root lattice of E_8, which is self-dual. Consider two orthogonal directions, corresponding to the decomposition of E_8 into $E_6 \times SU(3)$. The lattice I is isomorphic to the root lattice of E_6 and N is isomorphic to the root lattice of SU(3). The other roots of E_8 correspond to vectors of the form (w_1, w_2), where w_1 is a weight of E_6 and w_2 is a weight of SU(3), such that they belong to identical trialities of E_6 and SU(3) respectively. I^* is isomorphic to the weight lattice of E_6 and is of the form $(w_1, 0)$ and N^* is isomorphic to the weight lattice of SU(3) and is of the form $(0, w_2)$. We see that I^*/I is isomorphic to N^*/N, and are both isomorphic to Z_3.

Appendix B

In this appendix we discuss some properties of the group G described in sect. 3, which are useful in constructing vertex operators for the emission of untwisted states from the twisted sector. We adapt the arguments in the first reference in [5] to the case of interest to us, as well as note some simplifications of the construction due to self-duality of $\Gamma^{p,q}$.

We recall that the group G is defined by the commutation rules

$$T_{p_1}T_{p_2} = C(p_1, p_2)T_{p_2}T_{p_1}, \tag{B.1}$$

where

$$C(p_1, p_2) = \prod_{k=1}^{n} (\omega^k)^{(\theta^k p_1 \cdot p_2)}$$

and $\omega^n = 1$, and $p_i \in N/(1-\theta)\Gamma^{p,q}$. We wish to think of G as an abstract group, and therefore do not identify ω with an nth root of unity, but as another element of G with period n. We also demand that ω commute with all the elements in G, which implies that it is in the center of G. If D^2 denotes the dimension of $N/(1-\theta)\Gamma^{p,q}$, the commutation relations of elements in G implies that the order of the group is $n \cdot D^2$, i.e. each element of G corresponds to $\omega^k T_p$ (with $1 \leqslant k \leqslant n$).

We first show that the set of all vectors $p \in N$ such that $C(p, *) = 1$ for all $*$, is equal to $(1-\theta)\Gamma^{p,q}$ (this means that the "radical" of C, in the terminology of ref. [5] is equal to $(1-\theta)\Gamma^{p,q}$ for our case). This is the main simplification which is particular to our case and is not shared by the more general case of any even integral lattice considered in ref. [5]. It is easy to see that each vector p in $(1-\theta)\Gamma^{p,q}$ satisfies $C(p, *) = 1$. Now we shall use the self-duality of $\Gamma^{p,q}$ and the properties established in appendix A to show the converse. Suppose p is a vector in N with $C(p, *) = 1$. This implies that

$$\sum_{k=1}^{n-1} k\theta^k p \cdot * = 0 \qquad \mathrm{mod}\ n \tag{B.2}$$

for all $*$ in N. This implies that $\Sigma k\theta^k p \in nN^*$, so that it can be written as

$$\sum_{k=1}^{n-1} k\theta^k p = n(0, k_2), \tag{B.3}$$

where $(k_1, k_2) \in \Gamma^{p,q}$ for some k_1. We note the identity

$$(1-\theta)\sum_{k=1}^{n-1} k\theta^k = -n,$$

and using it we see that

$$(1-\theta)(k_1, k_2) = (1-\theta)(0, k_2) = \frac{1}{n}(1-\theta)\sum_{k=1}^{n-1} k\theta^k p = -p \tag{B.4}$$

and so it follows that $p \in (1-\theta)\Gamma^{p,q}$, which is the desired result.

This result, together with the commutation relations (B.1) implies that the center of the group G contains only ω and its powers. Let us for simplicity consider the case where n is prime. In that case, it is clear that there are D^2+n-1 conjugacy classes in G: n of them are given by powers of ω, and D^2-1 of them are of the form $(T_p, \omega T_p, \dots, \omega^{n-1}T_p)$ with $p \neq 0$ (this is due to the fact that n is prime). So G has D^2+n-1 inequivalent irreducible representations. D^2 of the representations are one-dimensional and are easy to construct. They correspond to representations where ω acts trivially as 1, and are representations of the abelian group $G/(1, \omega, \dots, \omega^{n-1})$ which has dimension D^2 (we recall that the irreducible representations of an abelian group are all one-dimensional, and equal in number to the order of the group). So there must be $n-1$ more representations where ω acts as an nth root of unity not equal to 1. In fact if we have any one of those representation, the others can be obtained by replacing ω by any other non-trivial nth root of unity. The dimension of this representation can be deduced from the fact that the sum of the squares of the dimensions of irreducible representations of a group is equal to the order of the group. So if we denote the order of the irreducible representation where ω acts non-trivially by X, we have

$$D^2(1) + (n-1)X^2 = n(D)^2, \tag{B.5}$$

which implies that $X = D$. This finally proves that D is an integer (if n is not prime similar arguments show that there is a unique irreducible representation where ω acts as $e^{2\pi i/n}$, and has dimension D, this could be done along the same lines above or as shown in ref. [20]).

Now we shall give an explicit description of the representation of G which we have encountered in the imbedding of the asymmetric orbifold in a symmetric orbifold. Since ω acts as $e^{2\pi i/n}$ in that case, and the dimension of the representation is the number of inequivalent fixed points, which is D^2, it must be that this D^2 dimensional representation, decomposes into D equivalent D-dimensional irreducible representations, each of which could be used to describe the Hilbert space of the ground state of the twisted sector of the asymmetric orbifold.

To describe the matrices T_p, it suffices to define the action of them on the fixed point set including non-trivial phases. Each fixed point is labeled by $k \in N/(1-\theta)\Gamma^{p,q}$, and we must define $T_p|k\rangle$. Since $N/(1-\theta)\Gamma^{p,q}$ is a finite abelian group, it is isomorphic to a finite product of cyclic groups [21] $Z_{m_1} \times \dots Z_{m_r}$ (in particular this means that $D^2 = m_1 \dots m_r$). Let $p_1, \dots, p_r$ be elements in $N/(1-\theta)\Gamma^{p,q}$ which correspond respectively to generators of the above cyclic groups. This implies that T_{p_i} together with ω generate G.

We can describe each fixed point by operating the matrices T_{p_i} on a given state, which we call $|0\rangle$ (to be thought as the "vacuum"), this mean that (up to phase)

$$|k\rangle = T_{p_1}^{n_1} \dots T_{p_r}^{n_r}|0\rangle, \tag{B.6}$$

with $0 \leqslant n_i \leqslant m_i - 1$. We can fix the phase of the state $|k\rangle$ by demanding that (B.6) be an equality with the T's ordered the way they have been written. So we can denote each fixed point state $|k\rangle$ by $|n_1, n_2, \dots, n_r\rangle$. Now we can easily determine the action of T_p on such a state. Since T_{p_i} generate the T's, it suffices to define it for them. Using the commutation relations (B.1) we see that

$$T_{p_i}|n_1, n_2, \dots, n_i, \dots, n_r\rangle = \prod_{k=1}^{i-1} C(p_i, p_k)^{n_k}|n_1, n_2, \dots, n_i + 1, \dots, n_r\rangle . \quad \text{(B.7)}$$

This gives us an explicit representation of the group G.

So far as the symmetric orbifold is concerned, the group G is associated with the emission of the states of the form $(k_1, 0; 0, -k_2)$ with $(k_1, k_2) \in N$. We can also consider the emission of the states $(0, k_2; -k_1, 0)$, which gives rise to a group $\tilde{G}$ isomorphic to G. The emission vertices corresponding to these states act on the same fixed points described above, and they commute with the matrices representing G. This implies, by Schur's lemma that the representation of $G \times \tilde{G}$ is of the form $(D, \bar{D})$ (note that the phase for $\tilde{G}$ is ω^{-1}). So for the symmetric orbifold the Hilbert space of fixed points forms a single irreducible representation for the vertex operators.

References

[1] D.J. Gross, J.A. Harvey, E. Martinec and R. Rohm, Phys. Rev. Lett. 54 (1985) 502; Nucl. Phys. B256 (1985) 253, B267 (1986) 75

[2] K.S. Narain, Phys. Lett. 169B (1986) 41;
K.S. Narain, M.H. Sarmadi and E. Witten, Nucl. Phys. B279 (1987) 369

[3] L. Dixon, J. Harvey, C. Vafa and E. Witten, Nucl. Phys. B261 (1985) 678, B274 (1986) 285

[4] H. Kawai, D.C. Lewellen and S.-H.H. Tye, Phys. Rev. Lett. 57 (1986) 1832, Nucl. Phys. B269 (1986) 1;
M. Meuller and E. Witten, Princeton preprint (1986);
I. Antoniadis, C. Bachas and C. Kounnas, Nucl. Phys. B, to be published

[5] J. Lepowsky, Proc. Nat. Acad. of Sci. 82 (1985) 8295;
V.G. Kac and D.H. Peterson, *in* Anomalies, geometry and topology, ed. W. Bardeen and A. White (World Scientific, 1985)

[6] C. Vafa, Nucl. Phys. B273 (1986) 592

[7] D. Freed and C. Vafa, Harvard preprint HUTP-86/A090

[8] J. Bagger, D. Nemeschansky, N. Seiberg and S. Yankielowicz, Nucl. Phys. B, to be published;
H. Neuberger, A. Niemi and G. Semenoff, LBL preprint (1986);
H. Kawai et al., ref. [4]

[9] P. Ginsparg, Harvard preprint HUTP-86/A053

[10] S. Hamidi and C. Vafa, Nucl. Phys. B279 (1987) 465;
L. Dixon, D. Friedan, E. Martinec and S. Shenker, Nucl. Phys. B282 (1987) 13

[11] I. Frenkel and V. Kac, Inv. Math. 62 (1982) 23;
G. Segal, Comm. Math. Phys. 80 (1982) 301;
P. Goddard and D. Olive, Proc. Conf. on Vertex operators in mathematics and physics, ed. J. Lepowsky (Springer, 1984)

[12] V.G. Kac and D.H. Peterson, Adv. Math. 53 (1984) 125

[13] I. Frenkel, J. Lepowsky and A. Meurmen, Proc. Conf. on Vertex operators in mathematics and physics, ed. J. Lepowsky (Springer 1984) and references therein

[14] A. Schellekens and N. Warner, Nucl. Phys. B287 (1987) 317
[15] L.E. Ibáñez, H.P. Nilles and F. Quenedo, CERN-TH 4611/86
[16] W. Lerche, D. Lüst and A. Schellekens, Phys. Lett. 181B (1986) 71; Nucl. Phys. B287 (1987) 477
[17] K.S. Narain, M.H. Sarmadi and C. Vafa, work in progress
[18] P.S. Landweber, D. Ranevel and R. Stong, to appear;
E. Witten, Princeton preprint PUPT-1024;
K. Pilch, A. Schellekens and N. Warner, Nucl. Phys. B287 (1987) 362
[19] L. Dixon, V. Kaplunovsky and C. Vafa, to appear
[20] I. Frenkel, J. Lepowsky and A. Meurmen, to appear
[21] L. Pontrjagin, *in* Topological groups, ch. 1 (Gordon and Breach, 1966)

NEW HETEROTIC STRING THEORIES IN UNCOMPACTIFIED DIMENSIONS < 10

K.S. NARAIN

Rutherford Appleton Laboratory, Chilton, Didcot, OX11 0QX, UK

Received 6 December 1985

It is shown that infinitely many heterotic string theories exist in uncompactified dimensions less than 10, that are one-loop finite (for massless external legs). Tachyons are removed by compactifying into tori $(10-d)$ and $(26-d)$ dimensions of the right-moving superstring and left-moving bosonic string sectors, respectively. The condition for modular invariance is shown to be equivalent to self-duality condition on even lorentzian lattices with $(10-d)$ and $(26-d)$ timelike and spacelike directions, respectively. The construction results in a $(10-d)(26-d)$ parameter family of one-loop finite string theories. The zero mass sector of these theories for $d=4$ and 6 correspond to $N=4$ and 2 supergravity coupled to super Yang–Mills with many possible groups, some of which cannot be obtained by compactifying $d=10$ heterotic string theory.

Superstring theories [1] and in particular the heterotic string theory [2] has attracted a lot of attention recently. The latter has two important features – namely it is finite at the one-loop level (and possibly to all orders) and the zero mass sector describes an anomaly-free chiral field theory [3]. Due to these reasons they have become serious candidates for describing nature, and many attempts have been made to obtain $d = 4$ theories by compactifying the extra dimensions into Calabi–Yau manifolds [4], starting either from the $d = 10$ field theory (i.e. massless sector) or by directly defining the string theory on these compact manifolds. The latter is possible due to some remarkable features of the Calabi–Yau manifolds, such as vanishing of the β function in the corresponding two-dimensional field theory and of the one-loop correction to the central charge in the Virasoro algebra [5]. More recently an attempt has been made to compactify via orbiforlds obtained from a six-torus by the action of some appropriate discrete group, and it appears that most of the features of Calabi–Yau manifolds could be obtained this way [6]. This is, of course, very attractive, as orbifolds are simpler to handle than Calabi–Yau manifolds.

All the approaches so far, however, have involved two steps – first the 26-dimensional left-moving sector of the heterotic string is compactified to ten dimensions and then the ten dimensions of the left- and right-moving sector are compactified to lower dimensions. As the scales of these two steps are of the same order, one can envisage an alternative approach of directly compactifying 26-dimensional and ten-dimensional dimensions of left- and right-moving sectors into tori.

In this letter, I shall show that such a construction yields a $(10-d)(26-d)$ parameter family of tachyon-free, one-loop finite string theories. This, of course, does not preclude the existence of other consistent theories on more complicated manifolds. The massless sector of these theories in $d = 6$ and 4 correspond, respectively, to $N=2$ and 4 supergravity coupled to super Yang–Mills with many possible groups (of rank $26-d$). Many of these groups cannot be obtained by a simple torus compactification of ten-dimensional heterotic string theory, and thus may provide new possibilities for phenomenology. For this, however, one will have to break $N = 4$ (in $d = 4$) down to $N = 1$ by invoking some discrete symmetries, such as in the orbifold approach [6], but I shall not dwell upon this here.

A crucial ingredient in the construction is the generalisation of the condition for modular invariance. Let p and q be the number of compactified dimensions in the left- and right-moving sectors, respectively

0370-2693/86/$ 03.50 © Elsevier Science Publishers B.V.
(North-Holland Physics Publishing Division)

$$X^A = q^A + 2L^A u + \sum_{n\neq 0} \frac{i}{2n} \alpha_n^A \exp(-i2nu) , \quad A = 26 - p + 1, ..., 26 ,$$

$$\tilde{X}^I = \tilde{q}^I - 2\tilde{L}^I v + \sum_{n\neq 0} \frac{i}{2n} \tilde{\alpha}_n^I \exp(-i2nv) , \quad I = 10 - q + 1, ..., 10 , \tag{1}$$

where $u = \tau + \sigma$, $v = \tau - \sigma$, and $(L^A, \tilde{L}^I) \in \Gamma$ – a lattice in $(p+q)$-dimensional vector space. The periodicity condition of the closed string at $\sigma = 0$ and π, implies that $(X^A, \tilde{X}^I)$ describe a $(p+q)$-torus defined by the lattice $2\pi \cdot \Gamma$. Note that we take here a general lattice that mixes the left- and right-moving sectors, but this does not pose any problem. Indeed in the orbifold construction [6], the action of discrete symmetries on the six-torus and that on the left-moving 16-torus defining the gauge group are coupled, similarly in ref. [7], the group constraint necessary for modular invariance mixes the two sectors.

The space of maps $S^1 \to T^{p+q}$ can be thought of as the loop space ΩT^{p+q} and has an abelian group structure under pointwise multiplication of maps. Let us symbolically denote a group element as $\exp[i(X, \tilde{X})]$ with the property

$$\exp[i(X, \tilde{X})] \exp[i(X', \tilde{X}')] = \exp[i(X + X', \tilde{X} + \tilde{X}')] , \quad \exp[2\pi i(L, \tilde{L})] = \exp[i(0, 0)] = 1 . \tag{2}$$

At the first quantized level one is actually interested only in the unitary, positive energy ray representation of eq. (2) on a Hilbert space. This is achieved by making use of the 2-cocycle (see Segal [8] for notation)

$$C((X, \tilde{X}), (X', \tilde{X}')) = \exp\{-i[S(X, X') + S(\tilde{X}, \tilde{X}')]\}(-1)^{\sigma((L,\tilde{L}),(L',\tilde{L}'))} , \tag{3}$$

where

$$S(X, X') = \frac{1}{4\pi} \int_0^\pi (dX^A/du)\, X^{A'} du + (1/4\pi)[X^A(\pi) - X^A(0)]\, X^{A'}(0) .$$

It is clear that C defines a 2-cocycle on Γ provided

$$L^A L^{A'} - \tilde{L}^I \tilde{L}^{I'} = \text{integer} , \quad (L, \tilde{L}), (L', \tilde{L}') \in \Gamma .$$

In other words, Γ, regarded as a "lorentzian" lattice with signature $(\underbrace{++ \dots +}_{p}\underbrace{-- \dots -}_{q})$ is integral. In fact I will restrict Γ to be also even, the justification for this will be given following eq. (6) below. Note that S just gives the usual classical Poisson bracket structure for oscillators. σ is a bilinear form, needed to get the correct commutation relations for charged generators [2,9]

$$\sigma\colon\ \Gamma \times \Gamma \to Z/2 , \quad \sigma((L, \tilde{L}), (L, \tilde{L})) = \tfrac{1}{2}(L^2 - \tilde{L}^2) \bmod 2 . \tag{3'}$$

The first quantized operators then satisfy the relation

$$U(\exp[i(X, \tilde{X})]) \cdot U(\exp[i(X', \tilde{X}')]) = C((X, \tilde{X}), (X', \tilde{X}'))\, U(\exp[i(X + X', \tilde{X} + \tilde{X}')]) . \tag{4}$$

Now one constructs a Fock space $H_{(0,0)}$ on which the oscillators α_n and $\tilde{\alpha}_n$ $(n\neq 0)$ act in the usual way. Next one defines infinite copies of this Fock space by specifying the action of the constant operators $U(\exp[i(q, \tilde{q})])$ as follows:

$$U(\exp[i(q, \tilde{q})])|_{H(k,\tilde{k})} = \exp[i(q \cdot k - \tilde{q} \cdot \tilde{k})] .$$

Clearly this makes sense only if $(k, \tilde{k}) \in \Gamma^*$ where Γ^* is the dual of Γ. By making use of eqs. (3), (4) we see that the winding operators

$$U(\exp[i(2Lu, -2\tilde{L}v)])\, H_{(k,\tilde{k})} \to H_{(k+L,\tilde{k}+\tilde{L})} . \tag{5}$$

Thus it is clear that if Z is of the order of Γ^*/Γ, then there are Z irreducible representations of the algebra of string operators.

The generator of rigid translation of a σ parameter must vanish, if we require that there be no preferred point $\sigma = 0$ on the closed string. Since a rigid translation in σ(orτ) does not mix the annihilation and creation operators α_n and α_{-n} $(n > 0)$, they can be determined uniquely [‡1], modulo constants. These constants are obtained as usual by requiring the closure of Lorentz algebra in the physical space–time. The resulting constraint is

$$\tfrac{1}{2}(k^2 - \tilde{k}^2) = \tilde{N} - N + 1 = \text{integer} . \tag{6}$$

Here the constant 1 appears for the 26-dimensional left-moving bosonic string, whereas for the ten-dimensional right-moving superstring this constant is zero. N and $\tilde{N}$ are the oscillator numbers, and hence non-negative integers, in the compactified as well as noncompactified and fermionic degrees of freedom. Eq. (6) implies that only the even subset of Γ^* is physical. In fact from eq. (5) it is clear that only winding operators corresponding to the even subset of Γ interpolate between physical states, thus we do not loose any generality by restricting Γ to be even.

The generator for τ translation gives the mass formula

$$\tfrac{1}{4}m^2 = \tfrac{1}{2}k^2 + \tfrac{1}{2}\tilde{k}^2 + N + \tilde{N} - 1 . \tag{7}$$

Eqs. (6) and (7) imply that $m^2 \geqslant 0$, signifying the absence of tachyons.

Finally we come to the question of interactions. As mentioned above there are Z inequivalent representations of the algebra of string operators and at the level of free strings it is consistent to pick any one of them, that is even, as the state space. Now I will show that modular invariance and hence one-loop finiteness requires that $Z = 1$ (i.e. $\Gamma^* = \Gamma$). Interaction vertices describing the emission of massless particles are defined as usual. From eqs. (6) and (7), it is clear that massless states require

$$(\tilde{k}^2 = \tilde{N} = 0) \quad \text{and either } (k^2 = 2, N = 0) \quad \text{or} \quad (k^2 = 0, N = 1) . \tag{8}$$

Consider, for example, a one-loop diagram with 4 external charged massless states (i.e. with $k^2 = 2, \tilde{k} = 0, N = \tilde{N} = 0$) and $(k, \tilde{k}) \in \Gamma$). One can go through the analysis of ref. [2], where this diagram has been calculated for $d = 10$. In the present case, as $d < 10$, the difference lies in the fact that $(10-d)$ of the additional left-moving (besides the 16 in ref. [2]) as well as $(10-d)$ of the right-moving bosonic momenta are discrete, instead of being continuous. This means that we can obtain our results by dividing by a factor $(-2/\mathrm{Im}\, t)^{(10-d)/2}$ and summing over L and $\tilde{L}$. Since the external momenta $\tilde{K}_i^A = 0$, this is particularly simple. The result is (see ref. [2] for notation)

$$A(1,2,3,4) = \bar{\epsilon}K \int \prod_{i=1}^{3} \mathrm{d}^2\nu_i \, \mathrm{d}^2\tau (2/\mathrm{Im}\,\tau)^5 \prod_{1\leqslant i<j\leqslant 4} \chi(\exp(2\pi \mathrm{i}\nu_{ji}), \exp(2\pi \mathrm{i}\tau))^{p_i \cdot p_j/2}$$

$$\times \exp(2\pi \mathrm{i}\bar{\tau}) f(\exp(-\mathrm{i}2\pi\bar{\tau}))^{-24} \prod_{1\leqslant i<j\leqslant 4} [2\pi \mathrm{i} \exp(-\mathrm{i}\pi\bar{\nu}_{ji}^2/\bar{\tau})\, \Theta_1(\bar{\nu}_{ji}|\bar{\tau})/\Theta_1'(0|\bar{\tau})]^{K_i \cdot K_j}$$

$$\times \left\{ (-\tfrac{1}{2}\mathrm{Im}\,\tau)^{(10-d)/2} \sum_{(L,\tilde{L})\in\Gamma} \exp\left[-\pi \mathrm{i}\bar{\tau}\left(L - \sum_{i=1}^{4} K_i \bar{\nu}_i/\bar{\tau} \right)^2 + \mathrm{i}\pi\tau\tilde{L}^2 \right] \right\} , \tag{9}$$

where $\nu_{ji} = \nu_j - \nu_i$, $\nu_4 = \tau$, f and χ are as defined in ref. [2] and p_i and K_i are the momenta in the space–time and compactified directions of the ith external particle. The curly bracket in the above expression contains the new terms – the factor $(-\tfrac{1}{2}\mathrm{Im}\,\tau)^{(10-d)/2}$ and $\exp(\mathrm{i}\pi\tau\tilde{L}^2)$. Note that the dependence on ν_1, ν_2 and ν_3 is unchanged, hence the periodicity condition

‡1 Even though C is not invariant under the diffeomorphism group, the latter acts on the associated central extension of the maps $S^1 \to T^{p+q}$ (see ref. [8]).

$$\nu_i \to \nu_i + 1\,, \quad \nu_i \to \nu_i + \tau\,, \quad i = 1, 2, 3$$

remains unchanged. Under the transformation $\tau \to \tau + 1$, using the fact that

$$\sum_{j>i} \bar{\nu}_{ji}^2 K_i \cdot K_j = -\left(\sum_i K_i \bar{\nu}_i\right)^2 ,$$

which follows from the conservation law $\Sigma K_i = 0$, and the fact that the Jacobi functions $\Theta_1(\bar{\nu}|\bar{\tau})$ are invariant we see that eq. (9) gets multiplied by a factor $\exp[-i\pi(L^2 - \tilde{L}^2)]$. As Γ is even this factor = 1. Finally to check the invariance of (9) under $\tau \to -1/\tau$, we use the following properties (see eq. (B.7) and (B.4) of ref. [2]):

$$\exp(2\pi i\bar{\tau}) f(\exp(-i2\pi\bar{\tau}))^{-24} \to \exp(-2\pi i/\bar{\tau}) f(\exp(i2\pi/\bar{\tau}))^{-24}(-\bar{\tau})^{+12} ,$$

$$\Theta_1(\bar{\nu}_{ji}|\bar{\tau})/\Theta_1'(0|\bar{\tau}) = [\Theta_1(-\bar{\nu}_{ji}/\bar{\tau}|-1/\bar{\tau})/\Theta_1'(0|-1/\bar{\tau})](-\bar{\tau}) \exp(i\pi\bar{\nu}_{ji}^2/\bar{\tau}) ,$$

and the Poisson summation formula

$$\sum_{(L,\tilde{L})\in\Gamma} \exp\left[-i\pi\tau\left(L - \sum K_i\bar{\nu}_i/\bar{\tau}\right)^2 + i\pi\tau\tilde{L}^2\right]$$

$$= \frac{1}{\sqrt{Z}} (-i\bar{\tau})^{-(26-d)/2} (i\tau)^{-(10-d)/2} \sum_{(L,\tilde{L})\in\Gamma^*} \exp\left[\frac{i\pi}{\bar{\tau}}\left(L - \sum K_i\bar{\nu}_i\right)^2 - \frac{i\pi}{\tau}\tilde{L}^2\right] \exp\left[-i\pi\left(\sum K_i\bar{\nu}_i\right)^2/\bar{\tau}\right] .$$

Substituting these identities in eq. (9) and using the fact that the first line of (9) is invariant by itself, as it represents the (d = 10) superstring II amplitude, we conclude that (9) is invariant under the transformation

$$\tau \to -1/\tau\,, \quad \nu_i \to -\nu_i/\tau\,, \quad i = 1, 2, 3 \quad \text{if and only if } \Gamma = \Gamma^* .$$

This of course means that $Z = 1$. To sum up, we see that the one-loop amplitudes are modular invariant provided the lattice $\Gamma = \{(L, \tilde{L})\}$ is an even, self-dual "lorentzian" lattice with signature $(\underbrace{++\,...\,+}_{26-d}\underbrace{--\,...\,-}_{10-d})$. Modular invariance allows one to restrict the domain of τ integration to $-\frac{1}{2} \leqslant \mathrm{Re}\,\tau \leqslant \frac{1}{2}$, $\mathrm{Im}\,\tau > 0$ and $|\tau| \geqslant 1$. The only divergence that can appear now is of the form $(\nu_{ji})^{-2}$ and $(\nu_{ji})^{-1}$ representing tachyon and gauge boson intermediate states, but they vanish as argued in ref. [2], due to the integration of the phases of ν_i which ensure the constraint eq. (6). Thus the theory is one-loop finite.

What is known about the existence of such even, self-dual lattices $\Gamma_{(p,q)}$ (here p is the number of positive signature and q that of negative signature). There is a general theorem [10] which states that they exist if and only if $p - q = 8n$ for some integer n. This is as well, as in our case $p - q = 16$. Moreover all such lattices for a given q ($q > 0$) and n ($n \geqslant 0$) are isomorphic to the lattice $n\Gamma_8 \oplus q\mathrm{P}_2$, where Γ_8 is the root lattice of E_8 and P_2 is a two-dimensional lattice with signature (+−) and defined by the metric $\left(\begin{smallmatrix}0 & 1\\ 1 & 0\end{smallmatrix}\right)$ on the simple basis of P_2. In other words, any even, self-dual lattice $\Gamma_{(q+8n,q)}$ can be obtained from $n\Gamma_8 \oplus q\mathrm{P}_2$ by means of an $\mathrm{SO}(8n + q, q)$ transformation. Since the mass formula (7) is only invariant under the subgroups $\mathrm{SO}(8n + q)$ and $\mathrm{SO}(q)$, the distinct string theories are characterised by the points of the coset manifold

$$\mathrm{M}_d = \mathrm{SO}(26-d, 10-d)/\mathrm{SO}(26-d) \times \mathrm{SO}(10-d)$$

(recall that for us $q = 10-d$ and $n = 2$) which has $(26-d)(10-d)$ parameters.

Before proceeding further, I shall give some examples of such lattices.

(i) Setting $d = 10$, we get back the euclidean self-duality condition of ten-dimensional heterotic string theory which gives rise to $\mathrm{E}_8 \times \mathrm{E}_8$ or spin $(32)/\mathrm{Z}_2$. These are the unique choices.

(ii) Setting $p = q$ (i.e. $n = 0$, this would not correspond to heterotic strings), $k, \tilde{k} \in$ weight lattice of a simply laced group G with the constraint $k - \tilde{k} \in$ root lattice of G, we again get a self-dual lattice $\Gamma_{q,q}$. This was the construction given in ref. [7].

(iii) A lattice generated by the following vectors:
$(k, 0)$ where $k \in$ root lattice of SO$(2q + 32)$.
$(0, \tilde{k})$ where $\tilde{k} \in$ root lattice of SO$(2q)$.
$(W_s, \tilde{W}_s)$ W_s and $\tilde{W}_s$ are weights of spinor representations of the two groups.
$(W_v, \tilde{W}_v)$ W_v and $\tilde{W}_v$ are the weights of the vector representations.

(iv) $((k_1, k_2), 0)$ where $k_1 \in$ root lattice of E_6, $k_2 \in$ root lattice of E_7.
$((0, 0), \tilde{k})$ where $\tilde{k} \in$ root lattice of SU(6).
$((W_{27}, 0), \tilde{W}_{1\bar{5}})$ and $((0, W_{56}), \tilde{W}_{20})$.

(v) $((k_1, k_2), (\tilde{k}_1, \tilde{k}_2))$ where $k_1, k_2 \in$ root lattice of E_6, $\tilde{k}_1, \tilde{k}_2 \in$ root lattice of SU(3).
$((W_{27}, 0), (\tilde{W}_3 \tilde{W}_3))$, and $((0, W_{27}), (\tilde{W}_3, \tilde{W}_{\bar{3}}))$.

To the above one can add any number of E_8 as well as P_2 lattices. Note that P_2 lattice corresponds to the example (ii) above with G = SU(2).

All of the above examples (for $q > 0$) [‡2] are isomorphic to $n\Gamma_8 \oplus qP_2$ as a "lorentzian" lattice, but they give a different mass formula (7), resulting in different string theories. In fact as stated above there is a $(8n + q)q$ ($n = 2$ for our case) parameter family of such theories. If, however, one were to concentrate on the zero mass sector of the theory (i.e. field theory limit in $(10-q)$ space–time dimensions), then from the condition (8) we get the following massless states:

$$\alpha^i_{-1,\mathrm{L}}, |0_\mathrm{L}, j_\mathrm{R}\rangle, \ \alpha^i_{-1,\mathrm{L}}|0_\mathrm{L}, a_\mathrm{R}\rangle, \ \alpha^i_{-1,\mathrm{L}}|0_\mathrm{L}, I_\mathrm{R}\rangle, \quad i,j = 3, 4, ..., d,$$

$$\alpha^\mathrm{A}_{-1,\mathrm{L}}|0_\mathrm{L}, j_\mathrm{R}\rangle, \ \alpha^\mathrm{A}_{-1,\mathrm{L}}|0_\mathrm{L}, a_\mathrm{R}\rangle, \ \alpha^\mathrm{A}_{-1,\mathrm{L}}|0_\mathrm{L}, I_\mathrm{R}\rangle, \ |0_\mathrm{L}, j_\mathrm{R}\rangle_{(k,0)}, \ |0_\mathrm{L}, a_\mathrm{R}\rangle_{(k,0)}, \ |0_\mathrm{L}, I_\mathrm{R}\rangle_{(k,0)}$$

where $(k, 0) \in \Gamma$ and $k^2 = 2$.

It is clear that for $d = 4$ and 6, for example, the zero mass sector corresponds to, respectively, $N = 4$ and 2 supergravity coupled to super Yang–Mills theory. The Yang–Mills group G is a product of simply laced groups of total rank $26-d$ (as there are precisely $(26-d)$ U(1) gauge fields, $\alpha^\mathrm{A}_{-1,\mathrm{L}}|0_\mathrm{L}, i_\mathrm{R}\rangle$), but the precise non-abelian structure depends on the set of points $(k, \tilde{k}) \in \Gamma$ such that $k^2 = 2$ and $\tilde{k} = 0$. Unfortunately I do not know of a classification of all possible Yang–Mills groups that can be obtained this way, but clearly there is a large number of such possibilities. The generic point in the coset manifold M_d would correspond to no points with $k^2 = 2$ and $\tilde{k} = 0$, this would result in a $\mathrm{U}(1)^{26-d}$ gauge field theory. On the other hand combining the examples given above, we can also have the following possibilities:

(i) $E_8 \times E_8 \times G_{10-d}$,
where G_{10-d} is a simply laced group of rank $(10-d)$.
(ii) For $d = 6$, $E_8 \times E_6 \times E_6$.
(iii) For $d = 5$, $E_8 \times E_7 \times E_6$ or $E_8 \times E_6^2 \times \mathrm{SU}(2)$.
(iv) For $d = 4$, $E_8 \times E_6 \times E_7 \times \mathrm{SU}(2)$, $E_8 \times E_6^2 \times \mathrm{G}$,
where G = SO(4) or SU(3).
(v) SO$(52-2d)$.

Or by rotating some of the axes in the lattice, one may break many of these groups down to some simply laced subgroup and product of several U(1)'s, but always maintaining the total rank = $26-d$. Note that these groups (for example SO$(52-2d)$) cannot be obtained by compactifying ten-dimensional heterotic string theory or the corresponding $N = 1$, $E_8 \times E_8$ or spin $(32)/Z_2$ field theories. The examples given above are actually isolated points on M_d, as any infinitesimal transformation by an element of M_d would destroy the zero mass condition (8). In general, if the resulting group is a product of non-abelian groups and $\mathrm{U}(1)^n$, then the resulting massless sector is unchanged by infinitesimal transformations of SO$(n, 10-d)$/SO$(n) \times$ SO$(10-d)$ and therefore defines locally a $(10-d)n$-dimensional surface in M_d. From the low energy field theory point of view all the points in this $(10-d)n$-dimensional surface would describe the same physics, but the masses of "regulator fields" i.e. the massive sectors

‡2 For $q = 0$, there are other self-dual lattices Γ_{8n} besides $n\Gamma_8$.

needed to render the theory finite (assuming that the results reported here continue to hold for higher loops) form a $(10-d)n$-dimensional space.

To sum up, we see that there exists an infinite number of tachyon-free and one-loop finite string theories in space–time dimensions $d < 10$ and they lie on a $(26-d)(10-d)$-dimensional coset manifold M_d. If one is interested in only the zero mass sector, this manifold M_d splits into a finite number (unknown at present) of submanifolds of $(10-d)n$ dimensions where n is the number of U(1) groups in the zero mass-sector. Classification of all allowed Yang–Mills groups is an important open question. If it turns out that higher loops are also finite (or this criterion may select a subset of M_d) then one would conclude that each point of M_d (or of the selected subset of M_d) provides a stable vacuum. The question would then be – which point (if any of the above) does nature select and why.

Direct torus compactification to four dimensions, as given here, leads unfortunately to $N = 4$ theories, which are, perhaps, not very appealing phenomenologically. To break supersymmetry completely or to $N = 1$, one can perhaps follow the methods of ref. [6], by invoking some discrete symmetry on the right sector and couple it to the one in the left sector. However, this may interfere with the condition of modular invariance and needs a separate investigation. Besides one has to make sure also that the cocycle C in eq. (3) is well defined on the quotient of Γ by this discrete symmetry. Another possible scenario would be to start from one of the above $N = 2$ theories in $d = 6$, and compactify down to $M_4 \times S^2$ via monopoles [11]. If the latter compactification scale can be pushed much below the string scale, the zero mass sector could be a good approximation for this.

I wish to thank M. Daniel, W. Nahm, M.A. Namazie, S.G. Rajeev and M.H. Sarmadi for helpful discussions.

References

[1] M.B. Green and J.H. Schwarz, Nucl. Phys. B181 (1981) 502; B198 (1982) 252; B198 (1982) 441; M.B. Green, J.H. Schwarz and L. Brink, Nucl. Phys. B198 (1982) 474.

[2] D.J. Gross, J.A. Harvey, E. Martinec and R. Rohm, Nucl. Phys. B256 (1985) 253; Heterotic string theory, II – the interacting heterotic string, Princeton preprint (June 1985).

[3] M.B. Green and J.H. Schwarz, Phys. Lett. 149B (1984) 117.

[4] P. Candelas, G.T. Horowitz, A. Strominger and E. Witten, Nucl. Phys. B258 (1985) 46; A. Strominger and E. Witten, New manifolds for superstring compactification, Commun. Math. Phys., to be published.

[5] C.G. Callan Jr., D. Friedan, E. Martinec and M. Perry, Princeton preprint (1985), and references therein.

[6] L. Dixon, J.A. Harvey, C. Vafa and E. Witten, Strings on orbifolds, Princeton preprint (July 1985).

[7] A. Casher, F. Englert, H. Nicolai and A. Taormina, Consistent superstrings as solutions of the $D = 26$ bosonic string theory, CERN preprint CERN-TH 4220/85.

[8] G. Segal, Commun. Math. Phys. 80 (1982) 301.

[9] P. Goddard and D. Olive, Algebras, lattices and strings, DAMTP preprint 83/22.

[10] J.P. Serrey, A course in arithmetic (Springer, Berlin, 1973) Ch. V.

[11] S. Randjbar-Daemi, A. Salam, E. Sezgin and J. Strathdee, Phys. Lett. 151B (1985) 351.

Nuclear Physics B287 (1987) 477–507
North-Holland, Amsterdam

CHIRAL FOUR-DIMENSIONAL HETEROTIC STRINGS FROM SELF-DUAL LATTICES

W. LERCHE

CERN, Geneva, Switzerland

D. LÜST

Max-Planck-Institut für Physik und Astrophysik, Föhringer Ring 6, 8000 München 40, FRG

A.N. SCHELLEKENS

CERN, Geneva, Switzerland

Received 24 November 1986

It is shown how our previous work on lattice constructions of ten-dimensional heterotic strings can be applied to four dimensions. The construction is based on an extension of Narain's lattices by including the bosonized world-sheet fermions and ghosts, and uses conformal field theory as its starting point. A natural embedding of all these theories in the bosonic string is automatically provided. Large numbers of chiral string theories with and without $N = 1$ supersymmetry can be constructed. Many features of their spectra have a simple interpretation in terms of properties of even self-dual lattices. In particular we find an intriguing relation between extended supersymmetry and exceptional groups.

1. Introduction

In the early days of string theory it was considered a major embarrassment when it turned out that string theories could only be formulated consistently in 26 or 10 dimensions. The revival of interest in the subject was for a small, but not unimportant part due to a change in attitude towards extra dimensions, namely the acceptance of the idea that they can be compactified. This idea dates back to the first half of this century, but received serious attention only during the last ten years, after the end of the first string era. When strings were reconsidered one initially attempted to compactify their field theory limits, with the help of the technology developed during the past decade. More recently the attention has slowly shifted towards more "stringy" compactifications.

Actually the concept of a critical dimension for string theory is a misconception. This is clear once one realizes what this concept is based upon. All string theories which might be relevant for four-dimensional physics (and this certainly includes

0550-3213/87/$03.50

the ones considered in this paper) are based on just two conformally invariant two-dimensional field theories. In conventional terms, one consists of 26 bosons whose conformal anomaly cancels that of the reparametrization ghost; the other is a supersymmetric model with ten bosons and ten fermions, cancelling the anomaly of a reparametrization ghost and a superconformal ghost. One may use either one as the left-moving or right-moving sector of a closed string theory. The space-time dimension is not simply the number of bosons, but is determined by how many of the bosons play the role of space-time coordinates. Of course this is not just a matter of choice, and one has to respect the consistency conditions of the interacting theory.

The compactified bosonic string may be regarded as a string theory formulated in less than 26 dimensions, or as a string theory in 26 dimensions, of which some have been compactified. The latter point of view is already less appropriate for the bosonic sector of the heterotic string [1] or a Narain-type [2] "compactification" of the bosonic string. The latter differs from ordinary compactification of n dimensions because left-moving and right-moving coordinates are treated as independent. It is difficult to see such theories as 26-dimensional strings in compactified space-time, since one cannot compactify n space-time dimensions on a (n, n) lorentzian lattice without the notion of strings.

The theories we will consider are a further step away from compactification. Although there may be some way to regard them as compactifications of ten-dimensional theories, they are far more easily described as four-dimensional string theories. The starting point of the construction is a formulation of two-dimensional field theories in which all fields are bosonic. Bosonization of the Neveu-Schwarz-Ramond fermions has been studied recently for two entirely different purposes. The authors of [3,4] were interested in the resulting simplifications of superconformal field theory. The authors of [5,6] used bosonization as a first step towards a construction of fermionic strings from the bosonic string. These two ideas play a crucial role in our construction, and all theories we will obtain can be regarded from both points of view.

Cohn et al. [3] observed that the bosonized fermions plus ghosts of the superstring can be described by an odd self-dual lorentzian lattice $\Gamma_{5,1}$ with metric $(+++++-)$. This was extended in [7] to heterotic strings, where also the relation of the self-duality of the lattice and modular invariance was understood. It was shown that the "old" $E_8 \times E_8$ and SO(32) strings, as well as some of the new ones, such as the $O(16) \times O(16)$ theory can be described by odd lorentzian self-dual lattices $\Gamma_{16;5,1}$ with metric $(-)^{16}(+)^5(-)$ (the semicolon separates left and right-movers). Such lorentzian lattices are unique up to Lorentz transformations, but not every Lorentz transformation gives a sensible theory. The lattices that do give sensible 10-dimensional theories where classified in [8] by mapping the odd self-dual lattices to even self-dual lorentzian lattices $\Gamma_{16;8}$, which in turn could be mapped to the Niemeier lattices Γ_{24}. We will give a brief summary of these ideas in sect. 2.

It was also suggested in [7] and [8] how this construction could be applied in $10-d$ dimensions by extending the lattices to incorporate d "compactified" bosons. In that way one is led to consider odd lorentzian self-dual lattices $\Gamma_{16+d;5+d,1}$ which can be mapped onto even lorentzian self-dual ones $\Gamma_{16+d;8+d}$. An interesting fact is that a priori there does not seem to be a reason why one could not get chiral theories. In Narain's construction the world-sheet fermions are compactified trivially on a torus, and this prevents the appearance of chiral fermions. In the theories we consider the world-sheet fermions are entangled non-trivially with the bosons, and the argument is no longer valid. The purpose of this paper is to explore these lower-dimensional theories.

Similar ideas have been discussed by Kawai et al. [9]. Their starting point is exactly opposite to ours, and is based on a fermionization of all fields except the space-time coordinates. These authors have already demonstrated the existence of four-dimensional chiral theories in [9]. In a recent paper [10] they have bosonized their construction and have also obtained a lattice formulation. Their lattices are different from ours, and we believe that our lattice formulation (and in particular the even self-dual one) has considerable advantages. The difference is however not a fundamental one: In sect. 2.3 we will show that the two approaches lead to the same theories.

The only new ingredient (in comparison with [7] and [8]) needed in the construction of theories below ten dimensions is a mechanism to maintain world-sheet supersymmetry of the right-moving sector. In a completely bosonic formulation one needs a bosonic realization of this supersymmetry, and one cannot expect this to emerge automatically. Indeed, in [9] it was found that an additional constraint is needed, and an elegant formulation of a sufficient condition (the "triplet constraint") was given. In sect. 3 we will show how one can arrive at the same condition by requiring that only massless chiral states appear in the spectrum, and no massive ones. Thus all massive excitations of these chiral ground states must cancel or pair off, no small feat in a theory with an infinite number of excitations.

A consequence of this is that the *chiral* partition function is a holomorphic modular function, which is subject to the same classification theorems which apply in ten dimensions. In general, partitions functions of strings below ten dimensions are not holomorphic, and such functions cannot be classified.

In sect. 4 we discuss the spectrum of the theories we can obtain. Here the use of even lorentzian lattices turns out to be extremely helpful, since many aspects of the spectrum can be directly related to length two vectors (roots) of the lattice. For example, the presence of $N=1$, 2 or 4 supersymmetry in four dimensions can be related to the presence of E_6, E_7 or E_8 sublattices, in which the space-time lattice is embedded.

In sect. 5 we show how large numbers of chiral four-dimensional (and other) theories can be constructed from the Niemeier lattices. It is also made clear that, unlike the ten-dimensional case we obtain only a very small fraction of a finite, but presumably extremely large set of theories. Some examples are presented in sect. 6. In sect. 7 we formulate some conclusions.

2. Bosonic constructions

2.1. THE COVARIANT LATTICE APPROACH

To discuss string spectra, the covariant lattice approach [3,7,11] is particularly useful. By "covariant" we mean that the charges of the bosonized fermions of the NSR-model are combined with those of the bosonized superconformal ghost system into a (self-dual) lattice. In order to review this approach, we focus on the case of the ten-dimensional spinning string and superstring.

Physical states are generated by covariant vertex operators $V(\bar{z})$ at zero momentum, acting on the SL(2, C) non-invariant vacuum $|0\rangle = c(0)|0\rangle_{\mathrm{SL}(2,\mathbb{C})}$, c representing the reparametrization ghost. These operators have the form (c_w denotes a cocycle-generating Klein factor):

$$V_w(\bar{z}) = (\text{derivatives})\, \mathrm{e}^{\lambda \cdot H} \mathrm{e}^{q\phi}(\bar{z})\, c_w, \qquad w = (\lambda, q). \tag{2.1}$$

Here, ϕ is part of the superconformal ghost system β, γ with $\beta\gamma = \partial\phi$, and H denotes a five-vector of bosonized fermion coordinates, $i\psi^{2j}\psi^{2j-1} = \partial H^j$. In (2.1), λ denotes a vector of the O(10) weight lattice D_5. Since the vacuum $|0\rangle$ has level -1 while $\exp(\lambda \cdot H)$ and $\exp(q\phi)$ have conformal weights $\frac{1}{2}\lambda^2$ and $-\frac{1}{2}q(q+2)$*, respectively, we have for the mass of a state generated by V_w

$$\tfrac{1}{8}m^2 = \tfrac{1}{2}\lambda^2 - \tfrac{1}{2}q^2 - q + N - 1, \tag{2.2}$$

where N denotes the total oscillator number (number of derivatives in (2.1)). In the following discussion, we ignore the effect of oscillators.

Let us now discuss the spectrum of the covariant spinning string [11]. The purpose of the factor $\exp(q\phi)$ is to provide the correct "vacuum" state for $\exp(\lambda \cdot H)$ to act on; it is well-known [4] that the vacuum states of the spinning string are characterized by non-zero Bose ghost sea charges $|q\rangle = \exp(q\phi)|0\rangle$. For instance, the tachyonic ground state of the spinning string is given by $|-1\rangle = \exp(-\phi)|0\rangle$, while the massless vector state is given by $|\mu, -1\rangle = \exp(\mu \cdot H)\exp(-\phi)|0\rangle$, where μ is an O(10) vector weight. More generally, all states in the Neveu-Schwarz sector have ghost charge $q = -1$, and are built on the vacuum $|-1\rangle$ with mass $-\frac{1}{2}q(q+2) - 1 = -\frac{1}{2}$. On the other hand, in the Ramond sector the states are built on $\exp(-\frac{1}{2}\phi)|0\rangle$ with level $\frac{3}{8} - 1 = -\frac{5}{8}$. This is precisely the value needed for the physical ground state $|\alpha, -\frac{1}{2}\rangle = \exp(\alpha \cdot H)\exp(-\frac{1}{2}\phi)|0\rangle$ of the Ramond sector to be massless, as the O(10) spinor weight obeys $\frac{1}{2}\alpha^2 = \frac{5}{8}$.

* The linear term in q arises due to the holomorphic anomaly in the ghost number current [4].

In general, all states in the Neveu-Schwarz or Ramond sectors obey $|\lambda, q\rangle = |\text{tensor}, -1\rangle$ or $|\text{spinor}, -\frac{1}{2}\rangle$, respectively [4]. However, $q = -1$ or $q = -\frac{1}{2}$ are not the only sectors in the theory. Since in our operator products ($\langle w, w'\rangle = \lambda \cdot \lambda' - qq'$)*

$$V_v(\bar{z}) \cdot V_w(0) \sim \bar{z}^{\langle v, w\rangle} V_{v+w} \varepsilon(v, w) \tag{2.3}$$

the vectors w and v add, intermediate states factorize in general on different than the canonical ghost sea levels q. Thus the Hilbert space must be extended to include all ghost charge sectors. However, as discussed in [4], sectors differing by integer units of q are physically equivalent and related by the "picture changing" operation. In other words, any physical state has a representative in each sector q mod 1, so that Neveu-Schwarz and Ramond states are characterized by $q \in \mathrm{Z}$ and $q \in \mathrm{Z} + \frac{1}{2}$, respectively.

Allowing thus for arbitrary ghost charges q, we find that the vector $w = (\lambda, q)$ belongs to a lattice extending D_5, on which all vector additions corresponding to operator products are allowed. From (2.3) it is clear that the natural inner product on this lattice is characterized by a lorentzian metric $(+ + + + + -)$; therefore, let us denote it by $\bar{\Gamma}_{5,1}$.

Since all components of $w \in \bar{\Gamma}_{5,1}$ are either integer or half-integer, we can define conjugacy classes (0), (v), (s) and (c) for $\bar{\Gamma}_{5,1}$ [8, 11], analogous to the $\mathrm{O}(2n)$ weight lattices D_n. In particular, (0) contains the massless vector $|\mu, -1\rangle$ of the Neveu-Schwarz sector, (v) contains the scalar tachyon $|-1\rangle$ as well as state $|\mu, 0\rangle = \psi^\mu|0\rangle$ while (s) contains the massless spinor $|\alpha, -\frac{1}{2}\rangle$ of the Ramond sector; (c) contains spinors of opposite chirality. In general (0) with odd ghost charge and (v) with even ghost charge contain states of the Neveu-Schwarz sector with the usual GSO projection, while (0) with even and (v) with odd ghost charge contain the opposite GSO projection. The sectors (s) and (c) yield Ramond states with opposite chirality if their ghost charges are the same.

In operator products (2.3), scalar products $\langle\, , \rangle$ between different conjugacy classes occur. For $\bar{\Gamma}_{5,1}$ one has the multiplication rules [11] ($(i) = (0)$, (v), (s) or (c)):

$$\begin{aligned} &\langle (i), (i)\rangle \in \mathrm{Z}, \\ &\langle (0), (i)\rangle \in \mathrm{Z}, \\ &\langle (v), (s)\rangle \in \mathrm{Z} + \tfrac{1}{2}, \\ &\langle (v), (c)\rangle \in \mathrm{Z} + \tfrac{1}{2}, \\ &\langle (s), (c)\rangle \in \mathrm{Z} + \tfrac{1}{2}. \end{aligned} \tag{2.4}$$

* The minus sign arises due to the opposite statistics of the ghost; $\varepsilon(v, w)$ below is a cocycle phase.

This means the spinning string theory based on $\bar{\Gamma}_{5,1}$ is non-local [4]. The GSO projection, which generates the (local) superstring, reduces $\bar{\Gamma}_{5,1}$ to an integral sublattice $\Gamma_{5,1}$, which contains only the classes (0) and (s) of $\bar{\Gamma}_{5,1}$. In particular, all even rank tensor fields belonging to (v) like the tachyon are projected out.

The covariant superstring lattice $\Gamma_{5,1}$ is not only integral, but odd self-dual, as was first noted in [3]. Actually one-loop modular invariance of the partition function requires invariance (up to phases and weight factors) of the lattice function ($e = (000001)$)*

$$\theta_{5,1}(e|\bar{\tau}) = \sum_{w \in \Gamma_{5,1}} \mathrm{e}^{-i\pi\bar{\tau}\langle w+e,\, w+e\rangle} \mathrm{e}^{2\pi i\langle w,\, e\rangle} \tag{2.5}$$

under $\bar{\tau} \to -1/\bar{\tau}$. This implies the lattice to be self-dual [7]. Invariance under $\bar{\tau} \to \bar{\tau} + 1$ requires the lattice to be odd,

$$\tfrac{1}{2}\langle w, w\rangle = \langle w, e\rangle \bmod 1 = q \bmod 1\,, \tag{2.6}$$

so that Neveu-Schwarz-(Ramond-)states belonging to (0) ((s)) are associated with even (odd) points on the lattice. This in turn ensures also proper (anti-)commutation rules of the vertex operators.

In the canonical ghost sectors, all states have $q = -1$ or $q = -\frac{1}{2}$. Physical light cone states are characterized by fixed two last entries, $x_0 = (0, -1)$ in the Neveu-Schwarz- and $x_0 = (-\frac{1}{2}, -\frac{1}{2})$ in the Ramond sector. These values provide the correct normal ordering constants for both sectors: if $u \in \mathrm{D}_4$ denotes the first four entries, i.e. the light cone part of $w = (u, x_0)$, then the mass of a light cone state can be written as

$$\tfrac{1}{2}m^2 = \tfrac{1}{2}u^2 - \tfrac{1}{2}\,. \tag{2.7}$$

Regarding the fixed two last entries x_0 of w as belonging to a sublattice $\Gamma_{1,1}$, we can identify the physical states of the superstring by decomposing $\Gamma_{5,1}$ to $\mathrm{D}_4 \times \Gamma_{1,1}$:

$$\begin{aligned} (0) &\to ((v),(v))\,, \\ (s) &\to ((s),(s))\,, \end{aligned} \tag{2.8}$$

and fixing the $\Gamma_{1,1}$ components to the entries x_0 defined above. From this the advantage of our approach compared to the conventional light cone treatment considering only D_4 becomes evident: the light cone states $u \in \mathrm{D}_4$ themselves do not form a lattice, since only the classes (v) and (s) of D_4 occur but not the neutral class (0). Only when u is non-trivially combined with vectors $x_0 \in \Gamma_{1,1}$ (or equivalently

* Whenever both string sectors are discussed simultaneously we will use τ to refer to left-movers and $\bar{\tau}$ for right-movers.

D_4, as will be explained below), does one obtain a sensible lattice. This is quite convenient since, for example, investigation of modular invariance requires only a discussion of the general structure of the lattice, and not the analysis of complicated expressions like sums over spin structures.

The above concept generalizes easily to heterotic theories in any dimension $D \leqslant 10$ [7]. Heterotic string states are generated by vertex operators

$$V_w(\bar{z}, z) = e^{w_L \cdot F}(z) e^{\lambda_R \cdot H} e^{q\phi}(\bar{z}), \qquad w = (w_L; w_R = (\lambda_R, q)). \tag{2.9}$$

The vectors w generate a lattice $\Gamma_{26-D;15-D,1}$ with natural metric $(-)^{(26-D)}(+)^{(15-D)}(-)$. The condition for modular invariance becomes that this lattice has again to be odd self-dual [7], obeying (2.6). In general all four conjugacy classes of the right part of the lattice may now appear, but the physical state selection rule (i.e that the last two entries should be $(0, -1)$ or $(-\frac{1}{2}, -\frac{1}{2})$) remains the same.

As is well known, all self-dual lorentzian lattices of given rank are isomorphic to each other in the sense that they are related by Lorentz-type rotations. However, for a lattice to represent a physically sensible string theory, certain physical constraints like correct statistical behavior and Lorentz transformation properties, as well as superconformal invariance have to be imposed.

For the ten-dimensional case, all lattices $\Gamma_{16;5,1}$ satisfying the additional physical constraints have been classified in [8]. They correspond precisely to the theories found in [12, 13], constructed by different methods. To these belong of course the well-known supersymmetric SO(32) and $E_8 \times E_8$ theories, which are based on the direct product lattices $\Gamma_{16;5,1} = (\Gamma_{16})_L \times (\Gamma_{5,1})_R$.

In the following, we will consider theories in less than ten dimensions: $D = 10 - 2n$. To these belong the Narain-type [2] compactifications of the supersymmetric theories associated with odd self-dual lattices

$$\Gamma_{16+2n;5+2n,1} = (\Gamma_{16} \times \Gamma_{2n;2n}) \times (\Gamma_n \times \Gamma_{5-n,1}). \tag{2.10}$$

Here, $\Gamma_{2n;2n}$ corresponds to the compactified left- and right-moving space-time bosons, $\Gamma_n = D_n$ to the compactified fermions and $\Gamma_{5-n,1} = D_{4-n} \times \Gamma_{1,1}$ to the covariant space-time fermions plus ghosts. Lorentz-type rotations acting only inside of the first bracket in (2.10) are not restricted at all, and produce a continuous infinity of compactified theories [2]. However, these are all vectorlike, since the second bracket corresponds to a trivial torus compactification of the superstring.

On the other hand, Lorentz-type rotations acting on the whole lattice do not correspond, in general, to torus compactifications* of ten-dimensional theories. Since in general for these theories the non-trivial relation between conjugacy classes

* Some might be related to orbifold-type [14] compactifications.

of Γ_n and $\Gamma_{5-n,1}$ in (2.10) is destroyed, there arise strong constraints due to superconformal invariance. This will be discussed in sect. 3, where it is shown that if one demands Lorentz invariance for chiral string theories only a finite, discrete set of lattices remains physical. In the next section we show how the lattice $\Gamma_{1,1}$ appended to the light cone "lattice" D_{4-n} can naturally be re-interpreted in the context of embedding NSR strings into the bosonic string.

2.2. THE BOSONIC STRING APPROACH

The idea that fermionic strings can be obtained from the bosonic string was first suggested by [15]. A more concrete realization was proposed in [5] and was worked out in detail, and presented in the most suitable form for our purpose in [6]. We use only the weak version of this idea, namely the fact that the light cone formulation of fermionic strings can be embedded in the bosonic string, provided that one makes certain truncations. This can be demonstrated explicitly. The more interesting strong version, namely that fermionic strings are actually some sort of "ground state" of the bosonic string is of course much harder to prove (see [16] for a discussion of some of the issues involved). The following will only be a brief summary of the construction, with less emphasis on 10 dimensions than was previously the case.

One starts with the 26-dimensional bosonic string compactified on an even lorentzian self-dual lattice $\Gamma_{16+2n;16+2n}$ to $10-2n$ dimensions. This lattice has a Frenkel-Kac symmetry $G_L \times G_R$. To obtain heterotic strings one considers an $SO(8-2n)$ regular subgroup of G_R, and adds its generators to those of the transverse Lorentz group. In this way $SO(8-2n)$ gauge symmetries get a new interpretation as internal symmetries, and in particular if $SO(8-2n)$ spinor representations are present one will get states which transform as space-time fermions. To describe the newly generated (bosonized) world-sheet fermions covariantly one needs to introduce new ghosts, as well as two additional fermionic components to extend $SO(8-2n)$ to $SO(10-2n)$. Altogether these new fields give a contribution of 12 to the conformal anomaly (in units where a boson contributes 1), and conformal invariance can only be maintained if 12 world-sheet bosons are somehow converted to these new fields. This means that their momenta and excitations cannot be present in the light cone spectrum. The fermionic string light cone spectrum can thus only be obtained from the compactified bosonic string spectrum by truncating 12 bosonic string degrees of freedom.

This truncation should respect modular invariance. Before truncation, the lattice partition function of the theory can be written as follows

$$P_\Gamma(\tau,\bar{\tau}) = \sum_i \theta_i^{4-n}(0|\bar{\tau}) P_i(\tau,\bar{\tau}). \tag{2.11}$$

Here we have made use of the fact that the lattice should at least contain a factor

D_{4-n}, which is represented by the theta function. The sum is over all conjugacy classes of D_{4-n} (or more precisely, it is over the sums and differences of conjugacy classes; for example θ_3 represents the sum of the root lattice and the vector weight lattice). The function P_i represents all remaining components of the lattice vectors. The full partition function is given by (2.11) multiplied by $\eta(\tau)^{-24}\eta(\bar{\tau})^{-24}$. To perform the truncation we have to remove 12 factors $\eta(\bar{\tau})^{-1}$ and a 12-dimensional part of the right lattice. In ref. [16] it was shown that one can do this in a modular invariant way by defining

$$\hat{P}_i(\tau,\bar{\tau}) = P_i(\tau,\bar{\tau})/S_i(\bar{\tau}) \tag{2.12}$$

and replacing P_i by $\hat{P}_i$ in (2.11). The functions S_i must transform into each other in the same way as the θ-functions but without phase or weight factors. They must thus transform as follows

$$\begin{aligned} &S_3(\tau+1) = S_4(\tau), \qquad S_4(\tau+1) = S_3(\tau), \qquad S_2(\tau+1) = S_2(\tau), \\ &S_4(-1/\tau) = S_2(\tau), \qquad S_2(-1/\tau) = S_4(\tau), \qquad S_3(-1/\tau) = S_3(\tau). \end{aligned} \tag{2.13}$$

If we assume that S_i is constructed out of factors θ/η (i.e that the right lattice has a decomposition in terms of D_n factors), then a nontrivial solution requires at least 12 such factors, which is interestingly precisely the number required by conformal invariance. One of the solutions is

$$S_i(\tau) = \eta(\tau)^{-12}(-1)^{i+1}\theta_i^4(0|\tau)\sum_j \theta_j^8(0|\tau). \tag{2.14}$$

Dividing by it must have the effect of removing the contribution of 12 bosons. The numerator of (2.14) tells us then that the original lattice must have had a factor $D_{8-n}\times E_8$, where D_{4-n} is embedded in the first factor. In other words, the even lorentzian self-dual lattice on which the bosonic string is compactified has to have the direct product form

$$\Gamma_{16+2n;16+2n} = \Gamma_{16+2n;8+2n}\times E_8\,. \tag{2.15}$$

There is a second independent solution corresponding to an extension of D_{4-n} to D_{16-n}, but then division by the S_i factors cannot be interpreted as a truncation of the bosonic string spectrum [16].

Division by the functions (2.14) (with argument $\bar{\tau}$) is easily seen to correspond to the following truncation of the spectrum:

(i) removal of all states with non-vanishing E_8 lattice momenta and excitations;

(ii) removal of all states with D_4 oscillator excitations, and all D_4 lattice momenta except one fixed vector weight and one fixed spinor weight, both of length 1: for example, $x_e = (1,0,0,0)$ and $x_e = (\frac{1}{2},\frac{1}{2},\frac{1}{2},\frac{1}{2})$;

(iii) adding a − sign to the partition function for those states with a D_4 spinor weight, to compensate for the one in (2.14).
The removal of a D_4 factor from D_{8-n} reduces it to the transverse Lorentz algebra. The sign in (iii) produces the correct spin-statistics relation for the fermions.

These rules where first written like this as a rather ad hoc way of obtaining the light cone spectrum of the superstring from $E_8 \times E_8$ [6]. As we have shown, they follow also from the requirement of modular invariance in the truncation, and have clearly a more general validity.

A third way of getting them is to use the conjugacy class map introduced in [8] on the "covariant" lattices $\Gamma_{5-n,1} = D_{4-n} \times \Gamma_{1,1}$ discussed in sect. 2.1. One simply maps these odd lattices to even lattices $\Gamma_{8-n} = D_{4-n} \times D_4$ by mapping* $\Gamma_{1,1}$ to D_4, i.e., the four conjugacy classes of $\Gamma_{1,1}$ on the same classes of D_4. It is easy to see that all dot-products change in precisely the right way to turn odd self-dual lattices $\Gamma_{16+2n;5+2n,1}$ into even self-dual lattices $\Gamma_{16+2n;8+2n}$. These are precisely the lattices indicated in (2.15). The physical state selection rule for $\Gamma_{1,1}$ described in sect. 2.1 translates exactly to the truncation rules for D_4 given above, by $x_0 \to x_e$.

We find thus that from the bosonic string point of view all theories are on the same footing, and that the truncation is the same in all cases.

2.3. THE LIGHT-CONE APPROACH

In this section, we explain the relation of our approach to the one of Kawai et al.** [10]. For simplicity, we focus on the ten-dimensional superstring case; extension to the heterotic case and to lower dimensions is trivial.

As explained in sect. 2.1, the bosonized light-cone states correspond to the conjugacy classes (v) and (s) of D_4; therefore, the vectors do not form a lattice. An odd (even) self-dual lattice is achieved by appending a covariantizing factor $\Gamma_{1,1}$ (D_4) to D_4. The observation made in [10] is that one can also obtain a lattice by shifting the charge vectors $u \in D_4$ of the light cone states by a constant vector $S = (1000)$. This maps the conjugacy classes (v) and (s) to (0) and (c), respectively, which form an odd self-dual lattice. Modular invariance furthermore requires that $(w = u + S)$

$$\tfrac{1}{2} w^2 = S \cdot w \bmod 1 , \tag{2.16}$$

so that tensors (spinors) correspond to even (odd) points on the lattice. In fact, (2.16) is equivalent to our equation (2.6). S plays the same role as the vector e used in eq. (2.5), which also represents a sum over a shifted lattice. In the covariant approach, shift by e is equivalent to shift by S, because the vectors differ only by a

* We use this mapping here only as a formal operation which establishes the relation between the two approaches, although it clearly suggests a more fundamental connection, which remains to be understood properly.

** Meant here is the bosonic construction, not the fermionic one of [9].

"root" vector. The only difference is thus the lattice factor $\Gamma_{1,1}$; this is irrelevant, however, since, as was shown in [7], all additional states due to $\Gamma_{1,1}$ factor out in the partition function. Hence the two approaches are equivalent, although the odd self-dual lattices which are used are different.

However, we think that our approach has several advantages. First, it is physically motivated by conformal field theory: the shift vector e originates in the ghost number current anomaly. Second, we are able to relate our odd self-dual lattices to even ones by mapping $\Gamma_{1,1} \to D_4$. This allows for an elegant and direct method of constructing such lattices, as will be detailed further below.

3. Chirality and Lorentz invariance

In this section, we investigate properties of lattices associated with chiral theories. The most convenient approach is the one of sect. 2.2: Instead of odd self-dual ones, we consider equivalent even self-dual lattices $\Gamma_{16+2n;8+2n}$. In general, there exists a continuous variety of such lattices. We show below that if one demands chiral theories, this variety is reduced to a discrete, finite one.

The presence of chiral fermions in the spectrum is easily seen to impose a rather strong condition on a string theory. Because of Lorentz invariance, chiral states can only be massless. However, in string theory there will be an infinite number of massive excitations of such a chiral state. Somehow these will all have to be non-chiral.

It is well-known how this works in ten dimensions. The chiral space-time states are due to world-sheet fermions with periodic boundary conditions (PP) along both non-contractable loops on the world-sheet torus. Their one-loop partition function is thus given by

$$P_1(\bar{\tau}) = \eta^{-4}(\bar{\tau})\theta_1^4(0|\bar{\tau}) = 0. \tag{3.1}$$

This vanishes, precisely because this sector of the theory has a chiral ground state. The two chiralities contribute with opposite sign to the partition function, and therefore cancel each other.

What we are really interested in is not (3.1), but the factor of (3.1) that describes the excitations of the chiral ground state. This is easily obtained by factoring out the zero mode that causes (3.1) to vanish. This can be achieved by twisting the boundary conditions, or coupling the theory to a weak gravitational background field, which does see the difference between the two chiralities (see [17] for a more detailed discussion). After factorizing the zero mode we get

$$P_1'(\bar{\tau}) = \eta^{-4}(\bar{\tau})\theta_1'^4(0|\bar{\tau}) \neq 0, \tag{3.2}$$

where the prime on the θ-function indicates differentiation with respect to the first argument.

In addition to fermionic oscillator excitations, the chiral vacuum is excited by bosonic oscillators α_{-n}. They contribute an additional factor $\eta(\bar{\tau})^{-8}$ to (3.2). By using the identity

$$\theta_1'(0|\tau) = 2\eta^3(\tau), \tag{3.3}$$

we find then that the result is a constant. This means that with the exception of the ground state there is an equal number of left-handed and right-handed states at every level, so that they can combine into massive states. This cancellation is a consequence of world-sheet supersymmetry which ensures an exact matching between the bosonic excitations and those of periodic fermions. It is important that not only the multiplicities of chiral pairs are the same, but that they also belong to the same representations of any relevant symmetry, i.e. any symmetry which is gauged. In ten dimensions this gauge symmetry is gravity, and the requirement is that not only $P_1'(\bar{\tau})\eta^{-8}(\bar{\tau})$ is a constant, but that also its character-valued generalization is a constant. The character-valued partition function can be thought of as the integrand of the one-loop string diagram in a background field [18]. It can be shown that it is indeed a constant [17].

Below ten dimensions a more elaborate mechanism will be needed. The cancellation between world-sheet fermions and bosons with space-time indices still occurs in exactly the same way. But in addition we have now the oscillator excitations of the remaining bosons, and the contribution from their soliton sector. If we consider heterotic strings in $D = 10 - 2n$ dimensions, there are $(2+1)n$ additional bosons (cf. eq. (2.10)). The fermionic part of the related even self-dual lattice $\Gamma_{16+2n;8+2n} = \Gamma_{16+2n;3n} \times \mathrm{D}_{8-n}$ has the following structure

$$(\Delta_{\mathrm{L}}^s; \Delta_{\mathrm{R}}^s, (s)) + (\Delta_{\mathrm{L}}^c; \Delta_{\mathrm{R}}^c, (c)), \tag{3.4}$$

where Δ_{L} is a set of $16+2n$ dimensional vectors, Δ_{R} a set of $3n$ dimensional vectors and $(s),(c)$ denote conjugacy classes of the space-time lattice D_{8-n}. The symmetries associated with the left lattice are gauged, so that a chiral pair must have the same lattice vector Δ_{L} in order to become massive. Consider one fixed such vector, $\boldsymbol{w}_{\mathrm{L}}$. We are interested in all pairs of states of the form

$$(\boldsymbol{w}_{\mathrm{L}}; \Delta_{\mathrm{R}}^s, (s)) + (\boldsymbol{w}_{\mathrm{L}}; \Delta_{\mathrm{R}}^c, (c)). \tag{3.5}$$

After cancelling the space-time factor using (3.3), we are left with the following factor from the bosonic excitations and the right lattice Γ_{3n}:

$$\eta(\bar{\tau})^{-3n}\left[\Gamma_{\mathrm{R}}^s(\bar{\tau}) - \Gamma_{\mathrm{R}}^c(\bar{\tau})\right], \tag{3.6}$$

where Γ_{R}^s and Γ_{R}^c are the partition functions of Δ_{R}^s and Δ_{R}^c, respectively. According to our previous arguments, (3.6) should be a constant, which should be non-zero, since otherwise we would not obtain a massless chiral fermion. For arbitrary lattices (3.6) may be difficult to solve in general, but it becomes easy if we assume that the right lattice consists of products of D_m lattices*. Then we know that the partition function must be a sum of products of $3n$ θ-functions, and we are looking for a combination of $3n$ factors θ/η which is equal to one. In particular this combination should be modular invariant with weight zero. In [17] all such modular invariant combinations have been constructed, and there turns out to be just one way to get a modular invariant function out of 3 factors θ/η. Fortunately this combination is indeed equal to a constant:

$$\eta(\tau)^{-3}\theta_2(0|\tau)\theta_3(0|\tau)\theta_4(0|\tau)=2. \tag{3.7}$$

It is this fact that allows us to construct chiral theories. To give a lattice interpretation to (3.7) we write it in terms of D_1 conjugacy classes**, of which the θ-functions are the partition functions. Define

$$\begin{aligned}\theta_0(0|\tau)&=\tfrac{1}{2}[\theta_3(0|\tau)+\theta_4(0|\tau)],\\ \theta_v(0|\tau)&=\tfrac{1}{2}[\theta_3(0|\tau)-\theta_4(0|\tau)],\\ \theta_s(0|\tau)&=\theta_c(0|\tau)=\tfrac{1}{2}\theta_2(0|\tau).\end{aligned} \tag{3.8}$$

Then the identity given above may also be written as

$$\eta^{-3}(\tau)\theta_s(0|\tau)\left[\theta_0^2(0|\tau)-\theta_v^2(0|\tau)\right]=1. \tag{3.9}$$

Here θ_0 is the partition function of the root lattice of D_1, θ_v the partition function of the vector weight lattice, etc. In $D=10-2n$ dimensions we need n factors of the form (3.9). We can now give this result a lattice interpretation by associating the positive terms with Δ_R^s and the negative ones with Δ_R^c.

The identity (3.7) does not hold if the θ-functions have a non-zero argument. This means there should be no gauge symmetries associated with the right lattice, because this would lead to a character valued partition function which is not a constant.

We specialize now to $n=3$, i.e., to 4 dimensions, corresponding to the even lattice $\Gamma_{22;14}=\Gamma_{22;9}\times D_5$ (other even dimensions are not essentially different). The right lattice can be decomposed to $(D_1)^9\times D_5$, where $D_5=D_1\times D_4$ is the space-time lattice, consisting of the light-cone factor D_1 and the appended factor D_4. It may be possible to combine the $(D_1)^9$ factor into larger D_n's, but this is irrelevant. Taking

* The conjugacy classes of these D_m's may be correlated with each other and with the left lattice.

** The root lattice (0) of D_1 is the set of even integers; the (v), (s) and (c) conjugacy classes are the even integers shifted by 1, $\frac{1}{2}$ and $-\frac{1}{2}$, respectively.

the third power of (3.9), we find that the lattice must have the following conjugacy classes of $D_5 \times (D_1)^9$:

$$\begin{aligned}
&((s),(F)^3(0)^6),\\
&3((c),(F)^3(v)^2(0)^4),\\
&3((s),(F)^3(v)^4(0)^2),\\
&((c),(F)^3(v)^6).
\end{aligned} \tag{3.10}$$

Here F stands for c or s. All these conjugacy classes should be associated with the same left vector w_L. Because the D_1's are not gauged, only the total number of states one gets from $(D_1)^9$ is relevant. It is therefore not a priori clear how to distribute the conjugacy classes over the nine D_1's.

If we subtract the first line in (3.10) from the seven others we get a set of lattice vectors with vanishing left components. These vectors must have even length, and generate a certain sublattice. If we write the first $D_5 \times (D_1)^9$ conjugacy class as

$$((s),(s,s,s,0,0,0,0,0,0)), \tag{3.11}$$

then one can show that to reproduce (3.10) the sublattice must contain a set of vectors of the form

$$\begin{aligned}
&((v),(x,v,v,0,0,0,0)),\\
&((v),(y,0,0,v,v,0,0)),\\
&((v),(z,0,0,0,0,v,v)),
\end{aligned} \tag{3.12}$$

where x, y and z are 3-vectors of (v)'s and (0)'s with odd length. Without loss of generality one may assume that the $(F)^3$ terms in (3.10) belong always to the same three D_1 factors. If one chooses them differently, one either generates a sublattice with half-integer length vectors, or a sublattice bigger than (3.12), which contains a set of vectors of the form (3.12). The shortest vector in the conjugacy class $((s),(s^3,0^6))$ has length 2, and will lead to chiral fermions if $w_L^2 = 2$ and if there is no vector $((c),(F^3,0^6))$ associated with w_L. The existence of a set of vectors (3.12) is then a necessary condition for the absence of massive chiral excitations of the chiral ground state. However, there are many other spinors on the lattice, for which either another set (3.12) must exist, or which must come in chiral pairs. The existence of other space-time fermions is restricted by (3.12) and the self-duality of the lattice, because they must have integer dot products with (3.12).

Modulo permutations, there are seven ways to choose the vectors x, y and z. One of these choices has the property that any fermion allowed by (3.12) automatically has a chiral partner if it is massive. This choice is

$$((v),(v,0,0,v,v,0,0,0,0)),$$
$$((v),(0,v,0,0,0,v,v,0,0)),$$
$$((v),(0,0,v,0,0,0,0,v,v)). \tag{3.13}$$

For example, a chiral partner for the vector $((s),(s,s,s,s,s,0,0,0,0))$ associated with any left vector, is obtained by adding the first vector of (3.13) to it; if one has a vector $((s),(0,0,0,s,0,s,0,s,0))$, the complete sublattice generates chiral partners for its massive excitations.

For any other choice of x, y and z one can find allowed spinors that do not automatically have chiral partners. If such spinors are present one has to add chiral partners, which leads to an enlargement of the sublattice (3.12). This leads to further constraints, etc. It is possible that one always ends up with a set of vectors (3.13), or with a vector which destroys the chirality of (3.11). It would be interesting to know if this is true, but it is not essential, because we will see that the constraint imposed by (3.13) already leads to large numbers of chiral theories.

The presence of the vectors (3.13) can also be rephrased as a condition on the spinor conjugacy classes. Because the inner product between (s) and (v) is $\frac{1}{2}$ mod 1, we see that the number of (s) (or (c)) conjugacy classes overlapping with the (v)'s in (3.13) must be even. This implies that the D_1 factors can be grouped into triplets, so that for each vector on the lattice the sum of each of the triplet components is equal, mod 1, to the D_5 components of that vector. Conversely, if the latter condition is satisfied, the self-duality of the lattice implies the existence of vectors (3.13).

The same constraint has been obtained in [9, 10] from a different point of view. These authors observe that the world-sheet supersymmetry of the right-movers should not just be realized on the right-moving bosons and fermions with space-time indices, but also on the n right-moving internal bosons. Although it is not clear what the most general way of doing this is, a sufficient condition is this "triplet constraint". For non-chiral theories it is not a necessary condition for world-sheet supersymmetry; for example, Narain's compactifications [2] do not satisfy it in general.

Our condition of absence of massive chiral states imposes constraints on chiral theories only and is a necessary condition, which comes close to reproducing the sufficient condition of [10]. We cannot rigorously rule out other possibilities, such as $(D_1)^9$ lattices with a more general constraint (3.12), or entirely different lattices which do not admit a D_1 decomposition at all.

In the rest of this paper we will restrict ourselves to the triplet constraint, which ensures both world-sheet supersymmetry and space-time Lorentz invariance. This implies that we only have to consider lattices which admit a D_1 decomposition; this destroys the continuous infinity of lorentzian self-dual lattices.

This can also be understood from a different point of view. As we have seen, the chiral partition function (i.e. the partition function multiplying the chiral ground state) should only depend on τ, and not on $\bar{\tau}$. It should in fact be a modular function of weight $n-4$ (in $10-2n$ dimensions), and be holomorphic apart from possible poles at $\tau = i\infty$ (or $q = e^{i\pi\tau} = 0$) [17]. Functions of this kind are determined almost completely by their modular weight and pole structure, a fact employed a decade ago by Nahm [19] to obtain what we now call the partition function of the left moving sector of the heterotic string.

These arguments do not apply for partition functions which depend on q as well as $\bar{q}$, a situation which typically arises in string theories below ten dimensions. Such non-holomorphic partition functions are not classifiable, a fact most convincingly demonstrated by Narain's construction of continuous families of non-chiral lower-dimensional theories.

For chiral theories we are in a better position, because we can at least say something about the chiral partition function. Unfortunately this function is not very useful in four dimensions: because it has odd weight (namely -1), it vanishes. It is easy to see explicitly why it vanishes: for every lattice vector $(\boldsymbol{w}_L; \boldsymbol{u}_R, (s))$ (in $\Gamma_{22} \times (D_1)^9 \times D_5$ notation) there is a vector $(-\boldsymbol{w}_L; -\boldsymbol{u}_R, (c))$, so that the two precisely cancel against each other in the chiral partition function (they are in fact CPT conjugates).

This is only a minor problem, which is easily circumvented by considering a more interesting object, the character valued partition function. (For a detailed discussion see [17].) It does not just contain information about the zeroth order traces (i.e. the dimensions) of the representations at all levels, but also about all higher order traces. It is easy to see that the cancellation described above will not occur for odd traces (unless only real representations appear, in which case we would not have a chiral theory). The coefficient function $C_k(q)$ of any odd trace of order k, except those proportional to $\operatorname{Tr} F^2 - \operatorname{Tr} R^2$, can now be shown to be a holomorphic modular function of weight $k-1$ (in four dimensions). In fact they cannot even have poles at $q=0$, because the tachyonic ground state of the left moving bosonic sector is in a singlet representation of the gauge group. It is known that any such function can be expressed completely in terms of the two Eisenstein functions G_4 and G_6. For example,

$$
\begin{aligned}
C_1 &= a_1 , \\
C_3 &= 0 , \\
C_5 &= a_5 G_4 , \\
C_7 &= a_7 G_6 , \\
C_9 &= a_9 (G_4)^2 , \\
C_{11} &= a_{11} G_4 G_6 , \\
C_{13} &= a_{13} (G_4)^3 + b_{13} (G_6)^2 \text{ etc.}
\end{aligned}
$$

The fact that $C_3 = 0$ implies that all anomalies cancel* [17]. The coefficients a_n and b_n are not fixed by modular invariance and have to be determined from the lowest level(s). The traces of all representations at higher levels (and hence to a large extent the representations themselves) are thus practically determined by the massless level. This does not tell us very much about the massless level itself, but it probably does restrict it, because it will not in general be possible to find representations at higher levels which produce precisely the right traces. It might be possible to prove general properties of the massless states in this way, but for the time being we will let the lattices take care of these subtle relations.

4. The spectrum

The theories we are considering can be constructed from even or odd lorentzian self-dual lattices. Although the two constructions are completely equivalent, the even lattices have several practical advantages, so that in the rest of this paper we will only consider even lattices. One advantage is that the mass formula is symmetric in left and right sectors; it is simply the one for the bosonic string,

$$\tfrac{1}{8}m_{\mathrm{L}}^2 = \tfrac{1}{2}\boldsymbol{w}_{\mathrm{L}}^2 = N_{\mathrm{L}} - 1\,, \tag{4.1}$$

$$\tfrac{1}{8}m_{\mathrm{R}}^2 = \tfrac{1}{2}\boldsymbol{w}_{\mathrm{R}}^2 + N_{\mathrm{R}} - 1\,, \tag{4.2}$$

$$m_{\mathrm{L}}^2 = m_{\mathrm{R}}^2\,, \tag{4.3}$$

$$m^2 = \tfrac{1}{2}\left(m_{\mathrm{L}}^2 + m_{\mathrm{R}}^2\right), \tag{4.4}$$

where $(\boldsymbol{w}_{\mathrm{L}}, \boldsymbol{w}_{\mathrm{R}})$ is a lorentzian lattice vector, and N_{L} and N_{R} are the bosonic oscillator contributions. From now on, the word "mass" will be used to refer to the value of $\frac{1}{8}m^2$. For simplicity, we will concentrate on the case $D = 10 - 2n = 4$; $D = 6$, 8 and 10 are completely analogous. In four dimensions, the relevant even lorentzian lattice is $\Gamma_{22;14}$; the right moving part should contain a D_5 sublattice; the remainder should be a $(\mathrm{D}_1)^9$ lattice which can be split into 3 triplets, so that $\boldsymbol{w}_{\mathrm{R}}$ decomposes as

$$\boldsymbol{w}_{\mathrm{R}} = (\boldsymbol{x}, \boldsymbol{y}, \boldsymbol{z}, \boldsymbol{v}_{\mathrm{R}}) \equiv (\boldsymbol{u}_{\mathrm{R}}, \boldsymbol{v}_{\mathrm{R}})\,, \tag{4.5}$$

where $\boldsymbol{v}_{\mathrm{R}}$ is a weight of D_5 and $\boldsymbol{x}$, $\boldsymbol{y}$ and $\boldsymbol{z}$ are 3-vectors. Because of the triplet constraint, the sum of the 3 entries of each of the three vectors $\boldsymbol{x}$, $\boldsymbol{y}$ and $\boldsymbol{z}$ should be equal the entries of $\boldsymbol{v}_{\mathrm{R}}$ modulo 1.

The physical states are obtained by decomposing D_5 to $\mathrm{D}_1 \times \mathrm{D}_4$ and keeping only states without D_4 oscillator excitations and with a fixed D_4 vector or spinor weight

* Notice that in this case (but not in general) the anomaly cancellation occurs even at higher levels, among left-mover excitations which do not have right-moving partners.

(of length 1), for example $x_e = (1,0,0,0)$ and $x_e = (\frac{1}{2},\frac{1}{2},\frac{1}{2},\frac{1}{2})$. Hence, the massless states are characterized by $\boldsymbol{w}_L^2 = 0,2$ and $\boldsymbol{u}_R^2 \leqslant 1$ (see table 1). The lattice momenta plus Cartan subalgebra excitations produce states in representations of $SO(2) \times G_L \times G_R$, where G_L is a rank-22 group corresponding to the left lattice and G_R a rank-9 group realized on the $(D_1)^9$ part of the right lattice. The transverse Lorentz group representation of a state is obtained by multiplying the lattice SO(2) representation (from the D_1 subalgebra of D_5) with SO(2) representations generated by the left and right moving uncompactified bosonic oscillators.

This construction is illustrated in table 1, where we summarize all states with masses less than or equal to zero that might occur in the four D_5 conjugacy classes. Right moving oscillators can be ignored, because they only contribute to massive states. The left movers can be divided in space-time oscillators (N_L^S) and internal ones (N_L^I), which generate the Cartan subalgebra of G_L. The mass difference between the ground state and the first excited state can be as small as $\frac{1}{4}$, because one can add two spinor entries to the same triplet of $\boldsymbol{w}_R$ without violating the triplet constraint. Tachyons can only appear in the D_5 conjugacy class (v). The absence of tachyonic fermionic spinors is guaranteed by the triplet constraint.

In four dimensions (and also in eight) a change in the sign of a spinor weight changes the conjugacy class from (s) to (c) and vice versa. The lattice has a vector $(-\boldsymbol{w}_L; -\boldsymbol{w}_R)$ for every vector $(\boldsymbol{w}_L; \boldsymbol{w}_R)$. These two are each other's CPT conjugates, but each such pair should be counted as only one Weyl or Majorana spinor. This is most easily seen by counting gravitinos or gauginos in supersymmetric theories.

Another advantage of the even lorentzian lattices is that all gauge invariances of the field theory can be traced back to roots of the lattice, i.e. vectors with either

TABLE 1
Massless particles and tachyons

Conjugacy classes $D_5 = D_1 \times D_4$			Excitations			Mass	
D_5	D_1	D_4	$\frac{1}{2}\boldsymbol{v}_R^2$	$\frac{1}{2}\boldsymbol{w}_L^2 + N_L^I$	N_L^S	$\frac{1}{8}m^2$	Particle type
(0)	(0)	(0) not physical					
	(v)	(v)	0	1	0	0	G_L gauge boson
	(v)	(v)	0	0	1	0	graviton, $B_{\mu\nu}$, dilaton
(v)	(v)	(0) not physical					
	(0)	(v)	0	1/2	0	−1/2	scalar tachyon
	(0)	(v)	1/4	3/4	0	−1/4	scalar tachyon
	(0)	(v)	1/2	1	0	0	massless scalar
	(0)	(v)	1/2	0	1	0	G_R gauge boson
(s)	(c)	(c) not physical					
	(s)	(s)	3/8	1	0	0	massless fermion
	(s)	(s)	3/8	0	1	0	gravitino
(c)	same as (s), but with opposite chirality.						

$w_L^2 = 0$, $w_R^2 = 2$ or $w_L^2 = 2$, $w_R^2 = 0$. In some cases we may conclude from the presence of such vectors that D_5 is actually embedded in a regular way in a larger simple Lie algebra. The following gauge particles may appear:

gravitons, $B_{\mu\nu}$, *dilaton*. They owe their existence to the root lattice of D_5 (with $w_L = u_R = 0$) and are therefore always present.

G_L *gauge bosons*. Their presence is due to the $G_L \times D_5$ root lattice (with $u_R = 0$) and they are therefore also inevitable. There are at least 22 U(1) gauge bosons, plus one for every vector with $w_L^2 = 2$.

G_R *gauge bosons*. Their existence requires a vector w_R of length 2, with $w_L = 0$. Furthermore the D_5 component of w_R should be a vector. The presence of such a vector indicates that D_5 is a part of a larger algebra, which must be a D_n algebra ($n > 5$). But then the D_n spinors would decompose into D_5 spinors with opposite chirality, and we would get a non-chiral theory*. Thus chiral theories do not have G_R gauge bosons. (This was a crucial assumption in sect. 3, which has now been justified).

Finally, there may appear

Gravitinos. To obtain gravitinos we need length 2 vectors w_R ($w_L = 0$) with spinor components in D_5.

The only regular embedding of D_5 which yields such roots is in E_6, E_7 or E_8. It is easy to see that these choices lead to $N = 1$, 2 and 4 supergravity respectively. Analogous results for other even dimensions are shown in table 2. (In two dimensions there are interesting additional possibilities because one can get spinors from triality rotated embeddings of D_4 in D_n.) This result also explains why only two of the ten-dimensional heterotic strings constructed from Niemeier lattices [8] have space-time supersymmetry. Some expected consequences of supersymmetry follow immediately, such as the presence of gauginos (the presence of $w_L = 0$ for some w_R implies the presence of the entire G_L root lattice combined with the same w_R). Others, such as supersymmetry at higher levels, are less obvious, and depend on rather intriguing properties of exceptional groups.

5. Construction of lattices

Any even self-dual lattice constructed entirely out of D_n factors can be mapped to another even self-dual lattice by changing the dimension of any D_n factor by multiplies of eight, keeping all conjugacy classes the same. Such a transformation

* This is true in the absence of G_R gauge symmetries. One might wonder whether the fermions could be in chiral representations of G_R. This is impossible because the gauged part of G_R has no such representations: The G_R gauge bosons originate from length 1 roots, so that the gauge group can only be a product of SO(3)'s [20]. In particular, this applies to models generalizing type II superstrings, which can straightforwardly be obtained by our construction.

TABLE 2
Conditions for extended supergravity and chirality

Dimension	Embedding of space-time lattice	Supersymmetry	Chiral
$D=2$	$D_4 \subset D_4$	None	possible
	$D_4 \subset D_n$, $5 \leqslant n \leqslant 16$	None	
	$D_4 \subset D_n$, $5 \leqslant n \leqslant 16(*)$	$(2n-8, 0)$	yes
	$D_4 \subset E_6$	$(2,2)$	
	$D_4 \subset E_7$	$(4,4)$	
	$D_4 \subset E_8$	$(8,8)$	
$D=4$	$D_5 \subset D_5$	none	possible
	$D_5 \subset D_n$, $6 \leqslant n \leqslant 14$	none	
	$D_5 \subset E_6$	$N=1$	possible
	$D_5 \subset E_7$	$N=2$	
	$D_5 \subset E_8$	$N=4$	
$D=6$	$D_6 \subset D_6$	none	possible
	$D_6 \subset D_n$, $7 \leqslant n \leqslant 12$	none	
	$D_6 \subset E_7$	$N=1$	yes
	$D_6 \subset E_8$	$N=2$	
$D=8$	$D_7 \subset D_7$	none	possible
	$D_7 \subset D_n$, $8 \leqslant n \leqslant 10$	none	
	$D_7 \subset E_8$	$N=1$	
$D=10$	$D_8 \subset D_8$	none	possible
	$D_8 \subset E_8$	$N=1$	yes

The second column gives the maximal extension of the space-time lattice $D_{8-n} = D_{4-n} \times D_4$ within the right-moving part of the lattice $\Gamma_{16+2n;8+2n}(D=10-2n)$. The entry "possible" means that the conditions for chirality we display are only necessary, but not sufficient; $(*)$ indicates triality rotated embeddings. In type II theories one gets similar structures (with the same or opposite chirality) from the left-moving sector.

leaves the lengths of all vectors unchanged mod 2, and all mutual dot-products mod 1, so that it does not affect self-duality. One may even subtract multiples of eight to make the dimension of a D_n factor negative. This can be interpreted as a change of metric. For example, $(D_5)_R$ can be changed to $(D_3)_L$, where the subscripts refer to the left (negative metric) and right (positive metric) part of the lattice. We will make extensive use of such transformations.

In this way we can associated with any even self-dual *lorentzian* lattice consisting entirely of D_n's a left *euclidean* lattice (obtained by mapping the right lattice to the left) or a right *euclidean* lattice. Because of the triplet constraint, the lattices of chiral theories we consider consist only of D_n factors on the right (as discussed in section 3), so that in all cases we can associate left euclidean lattices with them.

Euclidean lattices have the advantage of having useful classifications. Lorentzian lattices do not: their classification is up to Lorentz transformations, which unfortunately do not leave the physics invariant. This euclidean lattice classification is very successful in ten dimensions, where only 24-dimensional lattices have to be considered [8].

In four dimensions things are far more complicated. In the worst possible case we have a lattice $\Gamma_{22;14} = (\Gamma_{22})_{\mathrm{L}} \times (\mathrm{D}_5 \times (\mathrm{D}_1)^9)_{\mathrm{R}}$, which can be mapped to $(\Gamma_{22} \times \mathrm{D}_3 \times (\mathrm{D}_7)^9)_{\mathrm{L}}$, a euclidean lattice of dimension 88. A lower limit on the total number of such lattices is provided by the Siegel mass formula [21] [22]

$$\sum_{\Lambda} g(\Lambda)^{-1} = (8k)^{-1} B_{4k} \prod_{j=1}^{4k-1} (4j)^{-1} B_{2j}, \tag{5.1}$$

where the sum is over all even self-dual lattices of dimension $8k$, and $g(\Lambda)$ is the order of the automorphism group of Λ. Because $g(\Lambda) \geqslant 1$ the right hand side is a lower limit of the number of lattices (B_{2j} are the Bernoulli numbers). For $k = 11$ this number is of order 10^{1500}! The requirement that Λ should contain $\mathrm{D}_3 \times (\mathrm{D}_7)^9$ with a triplet constraint will reduce the number considerably, but clearly this is not a viable approach towards classification. It only tells us that the number of chiral theories is finite, but most likely extremely large*.

However, it is easy to construct many examples already from the 24-dimensional Niemeier lattices; a list of those may be found in [23]. To do so, one decomposes such a lattice in D_n factors by using one of the following regular embeddings

$$\begin{aligned}
&\mathrm{E}_8 \supset \mathrm{D}_8, \\
&\mathrm{E}_7 \supset \mathrm{D}_6 \times \mathrm{A}_1, \\
&\mathrm{E}_6 \supset \mathrm{D}_5 \times \mathrm{U}(1), \\
&\mathrm{A}_n \supset [\mathrm{D}_3]^k \times \mathrm{U}(1)^{k-1}, \qquad (k = \tfrac{1}{4}(n+1)) \\
&\mathrm{A}_2 \supset \mathrm{A}_1 \times \mathrm{U}(1).
\end{aligned} \tag{5.2}$$

After these decompositions one is left with factors A_1 and several D_1's. An even number of A_1's can be combined to D_2's because $\mathrm{D}_2 = \mathrm{A}_1 \times \mathrm{A}_1$. In all cases we have considered all m remaining factors can be rotated to a $(\mathrm{D}_1)^m$ lattice, and it is not unlikely that this is always possible, even for the Leech lattice. Note that we are not doing anything to the lattices; we simply regard them as root-lattices plus weights of subalgebras of the original algebras. Some of the weights may have length two and

* A more reasonable but less rigorous estimate can be made by observing that the 88-dimensional lattice has (at most) 32 factors, so that combinatorically their classification should be similar to the classification of even self-dual lattices of dimension 32 with D_1 lattices as building blocks. On the basis of such an estimate one would still expect a very large number of solutions.

this indicates the possibility of enlarging the algebras. One should always keep in mind that the euclidean lattices are only used for classification. Their roots are not very relevant; the relevant roots are those of the associated lorentzian lattices.

The lorentzian lattices can be constructed as follows. A euclidean lattice with p factors D_n can be completely specified by a list of all conjugacy classes that appear. For self-dual lattices there are 2^p classes out of a total of 4^p on this list. The list can be condensed to a set of generators, from which all others can be obtained by addition. For definiteness we will construct four-dimensional theories. Then the lattice must contain a D_5 factor, which becomes part of the right lattice. All other D_n's can be decomposed, and put partly on the left and partly on the right. The space-time D_5 cannot be embedded in such a left-right split D_n, because the spectrum would then contain tachyons of mass $-\frac{1}{2}$ (see table 1. The ten-dimensional theories provide a nice illustration: the tachyon-free ones are precisely those with a separate D_8 factor). Furthermore, as discussed in sect. 4, D_5 cannot be embedded in a larger D_n which stays entirely on the right, or else the theory will not be chiral. Thus we should look for a separate D_5 (it may be part of an E_6, but that is irrelevant at present).

The left-right decomposition of the remainder of the lattice should satisfy the triplet constraint. This is easy to achieve. One considers all possible ways of selecting 3 of the remaining D_n factors. (Each factor may be selected more than once in one triplet.) Then one makes a list of basic triplets, satisfying the constraint. Of course only the generators of the conjugacy classes have to be inspected. Finally one considers all possible ways of choosing 3 triplets from the list, where again each triplet may be chosen more than once. The triplet assignment may now be forgotten; the only relevant information is how many times a D_n factor has been selected. If it has been selected m times, one can replace $(D_n)_R$ by $(D_m)_R \times (D_{m-n+8k})_L$, where k is adjusted so that the dimension is positive. The left dimension is now automatically 22 mod 8, and may or may not be adjustable to 22 by changing the values of k. If it cannot be adjusted, the lattice is of no interest.

Because many triplet assignments are possible, one generates many lorentzian lattices from a given euclidean one. Furthermore there are several simple tricks to construct even more euclidean lattices. Of course one may add E_8 factors. One can "blow up" D_n factors, by changing their dimension by multiples of 8. One can decompose D_{n+m} to $D_n \times D_m$. However, there is no advantage in decomposing a D_n lattice in smaller D_n pieces. A more interesting change is splitting off a D_4, making a triality rotation on it and then blowing it up. This process destroys the length two vectors that hint at the fact that D_4 was once part of a larger algebra. Similarly one may recombine A_1 factors from several D_2's and blow up the latter. One can always construct a mirror lattice by changing all D_n's to negative dimensions (using mod 8 reductions), and changing the overall sign of the metric. Using combinations of these tricks one can change any D_n with odd n to a D_5. Several of the Niemeier lattices can be obtained from the dimension 8 and 16 ones in this way.

6. Examples

We now illustrate the above-explained construction of lorentzian lattices $\Gamma_{22;14}$ satisfying the triplet constraint, by means of some examples. We start with a euclidean Niemeier lattice with an E_6 factor, so that we can expect to find some string theories with $N=1$ supersymmetry. There are two such lattices, namely $(E_6)^4$ and $A_{11}\times D_7\times E_6$; in the following, we focus on the latter. The conjugacy classes which are present in addition to the roots can be found in [22]. They are generated by a weight of the representation (12, 64, 27) of $SU(12)\times SO(14)\times E_6$. This is a vector of length-squared four on the lattice. To decompose the lattice into D_n factors one simply considers, for example, the regular embeddings

$$SU(12)\supset SU(4)\times SU(4)\times SU(4)\times U(1)\times U(1) \tag{6.1}$$

and

$$E_6\supset SO(10)\times U(1).$$

The SU(4) factors give D_3 lattices because $A_3=D_3$. The three remaining U(1) factors can be rotated to a $(D_1)^3$ lattice. To see explicitly how that works, consider the decompositions of the (12) of SU(12) and the (27) of E_6 with respect to (6.1):

$$\begin{aligned}(12) &= \left(4,1,1,\tfrac{1}{6}\sqrt{3},\tfrac{1}{6}\sqrt{3}\right)\\ &\quad+\left(1,4,1,\tfrac{1}{4}-\tfrac{1}{12}\sqrt{3},-\tfrac{1}{4}-\tfrac{1}{12}\sqrt{3}\right)\\ &\quad+\left(1,1,4,-\tfrac{1}{4}-\tfrac{1}{12}\sqrt{3},\tfrac{1}{4}-\tfrac{1}{12}\sqrt{3}\right),\\ (27) &= \left(16,\tfrac{1}{6}\sqrt{3}\right)+\left(10,-\tfrac{1}{3}\sqrt{3}\right)+\left(1,\tfrac{2}{3}\sqrt{3}\right).\end{aligned} \tag{6.2}$$

Here the charges have been normalized so that they give the correct length-squared of the SU(12) and E_6 weights (namely $\frac{11}{12}$ and $\frac{4}{3}$ respectively). We can now define $(D_1)^3$ spinor weights as the following vectors on the $U(1)^3$ lattice:

$$\begin{aligned}s_1 &= \left(\tfrac{1}{6}\sqrt{3},\tfrac{1}{6}\sqrt{3},\tfrac{1}{6}\sqrt{3}\right),\\ s_2 &= \left(\tfrac{1}{4}-\tfrac{1}{12}\sqrt{3},-\tfrac{1}{4}-\tfrac{1}{12}\sqrt{3},\tfrac{1}{6}\sqrt{3}\right),\\ s_3 &= \left(-\tfrac{1}{4}-\tfrac{1}{12}\sqrt{3},\tfrac{1}{4}-\tfrac{1}{12}\sqrt{3},\tfrac{1}{6}\sqrt{3}\right).\end{aligned} \tag{6.3}$$

These vectors are mutually orthogonal and have length-squared $\frac{1}{4}$. We can now express all other vectors on the $A_{11}\times D_7\times E_6$ lattice as weights of

$$(D_3)^3\times(D_1)^3\times D_7\times D_5, \tag{6.4}$$

by projecting on these three vectors. By decomposing the representation (12,64,27) completely one gets then a set of generators for the conjugacy classes of the lattice, which is in fact slightly redundant. A minimal set is

$$(s,0,0,s,0,0,c,c),$$

$$(0,s,0,0,s,0,c,c),$$

$$(0,0,s,0,0,s,c,c),$$

$$(0,0,0,s,s,s,0,s). \tag{6.5}$$

By adding these vectors in all possible ways one generates 256 conjugacy classes (the addition rules for D_n, n odd are $(s)+(s)=(v)$, $(s)+(c)=(0)$, $(v)+(s)=(c)$, $(v)+(c)=(s)$, $(v)+(v)=(0)$). Notice that the result will be symmetric under simultaneous permutations of the first three entries (the D_3's) and the second three (the D_1's).

The last entry in the vectors (6.5) indicates the conjugacy class of the space-time factor D_5. We now have to form triplets of the remaining entries that add up to the last one, mod 1. The solutions can be specified by giving the vectors (3.13) which enforce the constraint. One has the following seven possibilities

$$t_{1\ldots3}=(v,0,0,v,0,0,v,v)(+\text{perm}),$$

$$t_{4\ldots6}=(0,v,v,v,0,0,0,v)(+\text{perm}),$$

$$t_7=(0,0,0,v,v,v,0,v). \tag{6.6}$$

Here "perm" indicates the simultaneous permutations of the first and second three entries. These seven vectors have integral dot-products with all the generators, and by self-duality they must therefore be vectors on the euclidean lattice. Now one has to select three triplets out of these seven. Valid choices are, for example, $3\times t_1$, $2\times t_1+t_3$, $t_4+t_6+t_7$ etc.

This determines the right lattice. For example, to obtain the triplet constraint corresponding to $3t_1$ one maps the euclidean lattice to the following lorentzian lattice

$$(D_{3R})\times(D_{5L})\times(D_{5L})\times(D_{2L}\times D_{3R})\times(D_{7L})\times(D_{7L})\times(D_{4L}\times D_{3R})\times(D_{5R}).$$

The factors between parentheses indicate how the eight factors of the original lattice are distributed between left and right. By decomposing the vector t_1 one obtains (among others) a set of three vectors of the form (3.13). The factors on the left lattice are determined mod 8, and we have chosen the minimal dimensions. In this case the total dimension of the left lattice is 30, which is too large. By considering the choice $3 \times \boldsymbol{t}_7$ we get

$$(D_{5L})^3 \times (D_{2L} \times D_{3R})^3 \times (D_{1L}) \times (D_{5R}).$$

This example yields a rank 22 gauge group $(SO(10) \times SO(4))^3 \times U(1)$. In general one simply calculates the sum of the selected vectors $\boldsymbol{t}_i$, replacing the entries (v) by 1. The resulting vector specifies for each D_n factor of the euclidean lattice which part of it is put on the right lattice. The remainder (which may have positive or negative dimension) is moved to the left lattice; it can always be given a positive dimension by adding multiples of 8. Finally one has to check that the left lattice has dimension 22.

By going systematically through all possibilities and removing trivial permutations, we have obtained 21 different string theories from the Niemeier lattice $A_7 \times D_{11} \times E_6$. Not all of these have supersymmetry. To get a supersymmetric theory one has to make sure that the E_6 factor of the euclidean lattice remains intact. Because the U(1), which enlarges D_5 to E_6, is a linear combination of the three D_1's, these must all be selected precisely once, since otherwise the length of the E_6 roots would be destroyed. There are 4 solutions (again up to permutations), namely

$$\begin{aligned} &\boldsymbol{t}_1 + \boldsymbol{t}_2 + \boldsymbol{t}_3, \\ &\boldsymbol{t}_1 + \boldsymbol{t}_2 + \boldsymbol{t}_6, \\ &\boldsymbol{t}_4 + \boldsymbol{t}_5 + \boldsymbol{t}_6, \\ &\boldsymbol{t}_1 + \boldsymbol{t}_5 + \boldsymbol{t}_6. \end{aligned} \tag{6.7}$$

The gauge groups associated with the first three solutions are respectively $SO(12)^3 \times SO(8)$, $SO(14)^2 \times SO(10) \times SO(6)$ and $SO(14)^3 \times U(1)$. The "minimal" left lattice for the last solution is $(D_6)^2 \times D_2$, which has dimension 14. We can enlarge the rank to 22 by extending any D_n factor by eight, or by simply adding an E_8 factor. In any case the result will not be chiral, because none of the groups has complex representations. Of the first three supersymmetric models, two are at least potentially chiral, and we will see further below that they do indeed have chiral fermions.

To obtain the complete spectrum of any of these theories one has to consider all 256 conjugacy classes, and decompose all D_n factors in left and right parts. The decomposition rules for the conjugacy classes are

$$\begin{aligned} (0) &\to (0,0) + (v,v), \\ (v) &\to (v,0) + (0,v), \\ (s) &\to (s,s) + (c,c), \\ (c) &\to (s,c) + (c,s). \end{aligned} \tag{6.8}$$

It does not matter whether one uses these rules for a "euclidean" decomposition $D_{m+n} \supset D_n \times D_m$, $m, n > 0$, or for a "lorentzian" one where n or m can be negative. Obviously many sectors are generated, and a complete description of the theory would be rather lengthy. But if one is only interested in the massless sector (as one usually is) one can eliminate most of the conjugacy classes, because they have a minimal weight length larger than 2 on the left or the right lattice. As an example we consider the non-supersymmetric $(SO(10) \times SO(4))^3 \times U(1)$ model described above. The lattice conjugacy classes that give rise to chiral spinors are (in the order specified in (6.4))

$$
\begin{aligned}
&(s,0,0,s,0,0,c,c) + \text{c.c.},\\
&(0,s,0,0,s,0,c,c) + \text{c.c.},\\
&(0,0,s,0,0,s,c,c) + \text{c.c.},
\end{aligned} \tag{6.9}
$$

where c.c. denotes the same set of vectors, but with opposite sign (so that (s) and (c) are interchanged). A vector and its opposite describe together just one physical state. One can immediately read off the representations of the chiral fermions (all taken to be lefthanded) for the group $SO(10) \times SO(10) \times SO(10) \times SO(4) \times SO(4) \times SO(4) \times SO(2)$:

$$
\begin{aligned}
&4\left(16,1,1,2,1,1,\tfrac{1}{2}\right) + 4\left(1,16,1,1,2,1,\tfrac{1}{2}\right) + 4\left(1,1,16,1,1,2,\tfrac{1}{2}\right)\\
&\quad + 4\left(16,1,1,2',1,1,\tfrac{1}{2}\right) + 4\left(1,16,1,1,2',1,\tfrac{1}{2}\right) + 4\left(1,1,16,1,1,2',\tfrac{1}{2}\right),
\end{aligned} \tag{6.10}
$$

where (2) and (2′) denote the inequivalent spinor representations of SO(4). The multiplicity of 4 is simply the dimension of the spinor representation of a factor $(D_3)_R$ which is not gauged. The spectrum is obviously chiral, and has 16 generations of (16)'s for each of the 3 SO(10) groups. The last entry gives the U(1) (= SO(2)) charge. There is no sign mistake: this U(1) factor is indeed not traceless, and it has anomalies with each of the SO(10) and SO(4) groups. This is in agreement with the theorem proved in [17], which states that because of modular invariance the anomaly is factorizable. For four dimensions the six-form from which the anomaly can be derived is predicted to be proportional to

$$
(\operatorname{Tr} F^2 - \operatorname{Tr} R^2)\operatorname{Tr} F. \tag{6.11}
$$

The factorization itself is a triviality for the theory under consideration, because it has only a U(1) anomaly. But the prediction goes further than that: if F and R are both in the vector representations of $SO(N)$ groups, then $\operatorname{Tr} R^2$ and $\operatorname{Tr} F^2$ should appear with relative factor -1. Furthermore this should be true for each simple factor of the gauge group.

One can calculate the anomaly explicitly by using the index theorem in six dimensions. By the usual arguments [24] the anomaly of a Weyl spinor is generated by

$$(\text{constant}) \times \left(\tfrac{1}{2}\mathrm{Tr}(F_{\mathrm{r}})^2 - \tfrac{1}{48}\mathrm{Tr}\,R^2\right)\mathrm{Tr}\,F_{\mathrm{U}(1)}\,, \tag{6.12}$$

where $\mathrm{Tr}\,F_{\mathrm{r}}$ denotes the trace over the gauge group representation r of the Weyl spinor. ($\mathrm{Tr}\,F$, used above, refers to the vector representation.) Using the fact that for spinor representations of $\mathrm{SO}(N)$ groups

$$\mathrm{Tr}(F_{\mathrm{s}})^2 = \tfrac{1}{8}\dim(s)\mathrm{Tr}\,F^2\,, \tag{6.13}$$

one can easily show that the anomaly of this model does indeed have the predicted form. The triplication of the SO(10) and SO(4) groups is essential to get the correct ratio between gauge and gravitational anomalies.

The appearance of such anomalous U(1)'s is not a new phenomenon. It has been observed before in [25] and [26] (the U(1)'s discussed in these two papers have different origins: those of the first paper come from isotropy groups of the compactifying spaces, whereas those of [26] are U(1) factors left over from the breaking of $\mathrm{E}_8 \times \mathrm{E}_8$). We expect that they are cancelled by the four-dimensional analogue of the Green-Schwarz mechanism.

Returning now to our example, we find that the massless scalars are in the following representations

$$\begin{gathered} 6(10,1,1,1,1,1,\pm 1) + \text{perm}\,, \\ 6(1,1,1,4,4,1,0) + \text{perm}\,, \\ 6(10,10,1,1,1,1,0) + \text{perm}\,, \end{gathered} \tag{6.14}$$

i.e., one has to include the permutations of the $\mathrm{SO}(10) \times \mathrm{SO}(4)$ factors and in addition to that a multiplicity factor of 6. Evidently the model is not supersymmetric, and it does indeed not have gravitinos.

The $N = 1$ supersymmetric $\mathrm{SO}(14)^3 \times \mathrm{U}(1)$ theory discussed above has the following massless fermions in addition to the gravitino

$$\begin{gathered} 2(64,1,1,\tfrac{1}{2}) + \text{perm}\,, \\ 2(64^*,1,1,\tfrac{1}{2}) + \text{perm}\,, \\ (14,14,1,0) + \text{perm}\,, \\ (14,1,1,\pm 1) + \text{perm}\,. \end{gathered} \tag{6.15}$$

This theory is chiral, but only because of the anomalous SO(2). Again the anomalies factorize properly because of the triplication. The other potentially chiral $N=1$ model turns out to be precisely the one discussed in [10].

The 21 models obtained from the euclidean lattice $A_{11} \times D_7 \times E_6$ with the specific embedding (6.1) are distributed as follows over the various possibilities:

2	chiral,	$N=1$ supersymmetric
4	non-chiral,	$N=1$ supersymmetric
11	chiral,	not supersymmetric, tachyon free
3	non-chiral,	not supersymmetric, tachyon free
1	chiral,	not supersymmetric, with tachyons of mass* $-\frac{1}{4}$

Similarly, from the lattice $(E_6)^4$ we obtained seven non-chiral and two chiral $N=1$ supersymmetric models, as well as 27 chiral non-supersymmetric, tachyon free ones.

The spectra of these models show some intriguing features. We find that the number of (16)'s of SO(10) (minus the number of (16*)'s) is always a multiple of 16. (Similarly, there are always 4k chiral (64)'s of SO(14).) Should this be a general phenomenon, then the prospects for these models are not very good. However, we do not know of any reason why this should be true in general.

It might be more attractive to look for models with a smaller gauge group realized on the massless fermions, such as $SU(3) \times SU(2) \times U(1)$. If such groups arise on D_n lattices, they will not be easy to find. They can only appear when several D_2 or D_1 factors turn out to form a sublattice of the lattice of a larger algebra. To find such cases one has to look for additional length-2 vectors on the left lattice. It is not easy to look for such examples systematically, but they might very well exist.

An obvious but important observation is that the only representations of any simply laced group one can encounter in the massless fermion or scalar spectrum are those with weights of length two. Chiral fermions with simple groups can therefore only owe their chirality to the spinor representations of SO(14), SO(10) and SO(2), at least as long as one only considers SO(n) symmetries and not their possible enlargements (SO(6) spinors are anomalous; the anomalies of the SO(2) spinors can be cancelled by the Green-Schwarz mechanism). Furthermore an SO(10) spinor must appear in combination with a spinor representation of another orthogonal group in order to have the correct weight length. Despite the presumably gigantic number of models that may exist**, the possibilities are thus still severely limited in comparison with field theory in four dimensions.

* We remind the reader that by construction tachyons of mass $-\frac{1}{2}$ (the mass of the Neveu-Schwarz tachyon) cannot appear in these models.

** With the help of a computer, it is easy to generate spectra of hundreds of theories within of a few moments.

7. Conclusions

As anticipated in [7, 8], we have found that the covariant lattice construction, which provided a simple and elegant way of classifying all ten-dimensional string theories, has similar advantages for constructing chiral string theories below ten dimensions. We expect that all theories discussed in this paper have the same degree of consistency as the well-known ten-dimensional theories. Although the general class of theories we find has already been constructed fermionically in [9], the lattice approach gives far more immediate insight in their structure. Although the number of chiral theories of this type is finite, our results suggest that there exist very many of them, so that a complete enumeration appears impossible. Perhaps some interesting subclass can be classified completely.

It seems that not much is left of the once celebrated uniqueness of string theory. Of course string theory never really was unique even in ten dimensions, and it is already known for some time that the situation is much worse in four dimensions. Up to now, one may have taken comfort from the fact that four-dimensional theories are just compactifications of the ten-dimensional ones, at least if one believes that it is better to have one string theory with many vacua than many string theories. If this kind of uniqueness is what is desired, one would be better off if all fermionic strings could be shown to originate from the bosonic string, which seems the best candidate for a really unique theory. Our construction puts the ten- and lower-dimensional theories on equal footing in this respect.

Even if all that string theory could achieve would be a completely finite theory of all interactions including gravity, but with no further restrictions on the gauge groups and the representations, it would be a considerable success. But the situation is better than that; although gauge groups are not very much restricted except that in chiral models their rank cannot exceed 22, the representations are. The fact that weights of length larger than 2 cannot appear in the massless sector selects low-dimensional representations; therefore, it is impossible to obtain many models that have been considered in the past, such as those with large Higgs representations or color exotics. Furthermore, one is not free to select fermion and scalar representations in an arbitrary way, and couple them with arbitrary coupling constants.

In our models, there is a built-in mechanism for naturally producing several generations. The multiplicity occurring in the spectra (6.10) and (6.15) of our examples is a quite general phenomenon; it arises due to the possibility of assembling length-squared two vectors in the right-moving sector in different ways.

A rank-22 gauge group may seem excessively large in comparison with the standard model, but this problem can be dealt with in the same way as the "second" E_8 from the ten-dimensional heterotic string. There should be many cases where a large part of the gauge group does not act on the massless chiral fermions or where several parts of the gauge group act on several sets of fields separately. In fact, there is a slight tendency in favor to such a situation because of the limited weight length of massless states; we have indeed found examples where that is the case. Furthermore, it may be possible to reduce the gauge symmetry by dividing out discrete groups along the lines of [14, 27, 28]. It is obvious that there will exist even more string theories in four dimensions than the ones we can construct from self-dual lattices, or which have been constructed in [9, 10].

Of course we do not exclude the possibility that the kind of theories we construct here can in some way be regarded as compactifications of some or all ten-dimensional theories, but this will almost certainly be a rather epicyclical description compared to the one we have given here. In particular, there is no need anymore to study ten-dimensional field theories. If everything works as expected, one can directly calculate all relevant quantities in four dimensions using conformal field theory. After our recent excursion into higher dimensions, it may be difficult to get used to the fact that perhaps the world is four-dimensional after all.

We would like to thank B. Nilsson for discussions. D.L. expresses his gratitude to the theory division of CERN for hospitality; A.N.S. thanks the Max-Planck-Institut für Physik und Astrophysik for hospitality.

References

[1] D. Gross, J. Harvey, E. Martinec and R. Rohm, Nucl. Phys. B256 (1985) 253
[2] K. Narain, Phys. Lett. 169B (1986) 41
[3] J. Cohn, D. Friedan, Z. Qiu and S. Shenker, Nucl. Phys. B278 (1986) 577
[4] D. Friedan, E. Martinec and S. Shenker, Nucl. Phys. B271 (1986) 93
[5] A. Casher, F. Englert, H. Nicolai and A. Taormina, Phys. Lett. 162B (1985) 121
[6] F. Englert, H. Nicolai and A. Schellekens, Nucl. Phys. B274 (1986) 315
[7] W. Lerche and D. Lüst, Phys. Lett. B, to be published
[8] W. Lerche, D. Lüst and A.N. Schellekens, Phys. Lett B 181 (1986) 71
[9] H. Kawai, D. Lewellen and S. Tye, Nucl. Phys. B, to be published; Phys. Rev. Lett. 57 (1986) 1832
[10] H. Kawai, D. Lewellen and S. Tye, Nucl. Phys. B, to be published
[11] V. Kostelecký, O. Lechtenfeld, W. Lerche, S. Samuel and S. Watamura, Nucl. Phys. B, to be published
[12] L. Alvarez-Gaumé, P. Ginsparg, G. Moore and C. Vafa, Phys. Lett. B171 (1986) 155
[13] L. Dixon and J. Harvey, Nucl. Phys. B274 (1986) 93
[14] L. Dixon, J. Harvey, C. Vafa and E. Witten, Nucl. Phys. B261 (1985) 651; Nucl. Phys. B274 (1986) 285
[15] P. Freund, Phys. Lett. 151B (1985) 387
[16] H. Nicolai and A.N. Schellekens, Topics in string theory, CERN-TH.4585/86
[17] A.N. Schellekens and N.P. Warner, Nucl. Phys. B287 (1987) 317
A.N. Schellekens and N.P. Warner, Phys. Lett. B177 (1986) 317; B181 (1986) 339
[18] K. Pilch, A.N. Schellekens and N.P. Warner, Nucl. Phys. B287 (1987) 362
[19] W. Nahm, Nucl. Phys. 114B (1976) 174; 120B (1976) 125; 124B (1977) 121
[20] L. Castellani, R. d'Auria, F. Gliozzi and G. Sciuto, Phys. Lett. 168B (1986) 47
[21] C. Siegel, Ann. of Math. 36 (1935) 527; J.-P. Serre, A course in arithmetic, (Springer, New York/Berlin, 1971)
[22] J. Conway and M. Sloane, J. of Number Theory 15 (1982) 83

[23] H. Niemeier, J. of Number Theory 5 (1973) 142
[24] L. Alvarez-Gaumé and P. Ginsparg, Ann. of Phys. 161 (1985) 423
[25] E. Witten, Phys. Lett. 149B (1984) 637
[26] K. Pilch and A.N. Schellekens, Nucl. Phys. B259 (1985) 637
[27] D. Benett, M. Brene, L. Mizrachi and H. Nielsen, Phys. Lett. 178B (1986) 1832
[28] Da-Xi Li, McGill University preprint

Nuclear Physics B288 (1987) 1–76
North-Holland, Amsterdam

CONSTRUCTION OF FERMIONIC STRING MODELS IN FOUR DIMENSIONS

Hikaru KAWAI, David C. LEWELLEN and S.-H. Henry TYE

Newman Laboratory of Nuclear Studies, Cornell University, Ithaca, New York 14853, USA

Received 3 November 1986

The construction of four-dimensional closed fermionic string models is discussed. The approach is based on a fermionic formulation of all internal (i.e. toroidally compactified) coordinates. Modular invariance, world sheet supersymmetry, (super)conformal invariance and proper space-time spin-statistics impose stringent constraints on the model building. Using these constraints on the boundary conditions (spin structure) of the world sheet fermions, we obtain a simple set of rules for constructing ultraviolet-finite closed fermionic string models. For a large subclass of these models, this "spin structure" construction can be related to bosonic constructions via the fermionic charge lattice. These charge lattices are odd lorentzian self-dual lattices shifted by a fixed vector and form a nontrivial generalization of the lorentzian self-dual even-integer lattices considered by Narain. In particular, four-dimensional models with $N=4$, $N=2$, and $N=1$ supersymmetry as well as non-supersymmetric tachyon-free chiral models can easily be constructed. Some models may be interpreted as charge lattices moded by discrete symmetries – in particular Z_2 type orbifolds. String interactions and other related issues are also discussed.

1. Introduction

Following the work of Green and Schwarz [1], the possibility that the ultimate theory of nature may emerge from superstring theories has attracted considerable attention in the past two years. In particular, the heterotic string theory proposed by Gross, Harvey, Martinec, and Rohm [2], which is likely to be ultra-violet finite, represents a significant step towards making contact between string theories and the real world. Since heterotic strings were originally constructed in ten dimensions, we must compactify six dimensions to obtain a four-dimensional string model. The easiest approach is compactification of the six bosonic space coordinates on a flat torus. Unfortunately, this simple compactification yields models that are phenomenologically unappealing. In general, more realistic compactifications will involve highly non-trivial mathematics. A notable exception is the orbifold construction proposed by Dixon, Harvey, Vafa and Witten [3–5]. Recently, we found that phenomenologically interesting compactifications can be achieved on the flat torus if the bosonic coordinates are fermionized [6]. This follows from the fact that the four real (two left- and two right-moving) fermions originating from a single bosonic

0550-3213/87/$03.50

coordinate can have different boundary conditions. The purpose of this paper is to elaborate on this fermionic compactification approach and relate this to bosonic formulations.

The basic idea of our construction of four-dimensional (or, more generally, $D \leqslant 10$ dimensional) fermionic string models is quite simple. To construct a fermionic string model, it is convenient to consider an analysis of the string one-loop partition function and impose consistency requirements. First, all internal degrees of freedom (i.e., compactified coordinates) are fermionized so that they can be treated as world sheet fermions. These world sheet fermions are specified by their boundary conditions (spin structures) and by their interactions on the world sheet. Here, the only world sheet interactions we consider are (abelian) Thirring interactions, which maintain the (super)conformal invariance.

In general each fermion can have independent boundary conditions on the world sheet torus,

$$\psi_l(t+2\pi m, \sigma+2\pi n) = e^{2\pi i(mu_l+nv_l)}\psi_l(t,\sigma).$$

For each set of boundary conditions, we have a one-loop partition function $Z^{\{v_l\}}_{\{u_l\}}$, and the string one-loop partition function is given as a linear combination of these ($Z^{\{v_l\}}_{\{u_l\}} = Z^v_u$),

$$Z_{\text{fermion}} = \sum_{\{v,u\}} C^v_u Z^v_u. \tag{1.1}$$

In four dimensions in the light-cone gauge, we have, for the heterotic string, 44(22) left-moving and 20(10) right-moving Majorana (complex) fermions. For complex fermions, $0 \leqslant v_l, u_l < 1$ while for Majorana fermions, $u_l, v_l = 0, \frac{1}{2}$. The number of terms in (1.1) can a priori be very large. We can, however, impose consistency requirements to limit the possibilities and in fact derive a simple set of rules for model-building. To be specific, we require the following conditions:

(i) Reparametrization invariance and world sheet supersymmetry (if present); they insure positivity and, in the light cone gauge which we adopt, space-time Lorentz invariance.

(ii) The anomaly free condition; specifically, the partition function must be (super)conformal invariant and free from local two-dimensional gravitational anomalies. Although (super)conformal invariance may not be necessary, it certainly simplifies the analysis considerably.

(iii) Modular invariance, or invariance under discrete reparametrizations; it is this symmetry that ensures the ultra-violet finiteness of the string theory. This requires a proper summation over different spin structures.

(iv) Physically sensible projection; positivity is assured by reparametrization invariance and two-dimensional world sheet supersymmetry (this supersymmetry is maintained by the fermionization procedure). From the space of world sheet states, we perform a sensible projection onto the subspace of physical states with proper space-time spin-statistics. In terms of the partition function, this means the contribution of a physical fermion to the partition function has a negative sign while the contribution of a physical boson has a positive sign. This projection is built into the sum over different spin structures. As we will show, one particular consequence of these requirements is that the graviton is always present in the physical spectrum.

These constraints on the partition function can be translated into a simple set of rules for constructing ultraviolet finite closed fermionic string models. This approach is actually a generalization of the projection of Gliozzi, Scherk, and Olive (GSO) [7] and in 10 dimensions has been analyzed by many groups. [3, 4, 8, 9–11]. We shall refer to this method as the spin structure construction. Models are specified by the choice of a set of basis vectors $\boldsymbol{W}_i$ (whose lth components specify the boundary condition of the lth fermion) and structure constants k_{ij} which determine in part the desired projections. It seems that the number of four-dimensional chiral models given by this construction is quite large. One might hope that multiloop modular invariance would impose further restrictions on the set of consistent models; however, simple arguments suggest that multiloop modular invariance is probably satisfied automatically by the models constructed with our rules and so provides no further constraints. Hence the selection of the string model that describes nature is likely to come from confrontation with experiments and/or from an analysis of string field dynamics. It is clear that the spin structure construction is applicable to any string model in any dimension less than or equal to its critical dimension; however, we will emphasize heterotic string models in 10 and 4 space-time dimensions. A detailed analysis of four-dimensional type II superstrings will be presented elsewhere. Ref. [6] gives a summary of the spin structure construction and its application.

To obtain a better understanding of the set of rules that we obtain, we consider the subclass of models that permit bosonization, which we shall refer to as rebosonization, since the resulting bosons need not have a one-to-one correspondence with the original bosonic coordinates. In these cases we can write all real fermions in terms of complex fermions and construct the fermionic charge (i.e., fermion number) lattice of states. The charge lattice $\boldsymbol{Q}$ is an odd self-dual lorentzian lattice shifted by a constant vector $\boldsymbol{S}$, which takes care of the space-time spin-statistics, that is, in even D-dimensional space-time,

$$\boldsymbol{Q} = \boldsymbol{L} + \boldsymbol{S}, \qquad \boldsymbol{L} \in \Gamma_S = \Gamma_S^* , \tag{1.2}$$

where Γ_S is an odd self-dual $(26-D, 14-D)$ dimensional lorentzian lattice and

$$\tfrac{1}{2}\boldsymbol{Q}^2 = \tfrac{1}{2}\boldsymbol{S}^2 \quad (\text{mod } 1) \tag{1.3}$$

and $\boldsymbol{S}$ is a fixed vector; e.g., for the heterotic string, $\boldsymbol{S} = (1, 0, 0, 0 \dots 0)$. We will refer to $\boldsymbol{S}$ as the spin-statistics vector, since its role is to ensure proper space-time spin-statistics. As a check, it is easy to show that the charge lattice satisfies modular invariance. Also we note that the reflection symmetry of the charge lattice (i.e. $\boldsymbol{Q} \to -\boldsymbol{Q}$) implies the CPT invariance in the effective quantum field theory.

As mentioned earlier, the spin structure construction is more general than the charge lattice construction since the former includes the latter. To be more specific, the spin structure construction allows the construction of models in any odd space-time dimensions, thus including Narain's construction [12] in odd space-time dimensions. In heterotic type string models, the gauge groups (from the left-moving sector) obtainable from the charge lattice always have rank $26-D$ and are always simply laced; on the other hand, the spin structure construction allows lower rank as well as non-simply laced gauge groups. For 4-dimensional $N=4$ supersymmetric string models resulting from the toroidal compactification of a heterotic string theory, the charge lattice reduces to the lorentzian self-dual even-integer lattice considered by Narain [12,13]. The charge lattice, however, is clearly much more general, including lattices that correspond to nonsupersymmetric tachyon-free chiral models as well as $N=2$ and $N=1$ supersymmetric models.

The relation of our work to the interesting work of strings on orbifolds is also briefly discussed. The models represented by some charge lattices may be interpreted as the incorporation of multiple left-right independent Z_2 orbifolds into Narain's construction. Outer automorphisms [4] (or projection) as well as twisting can be incorporated into the construction of the charge lattice. We conjecture that the generalized twisted and projected charge lattice is equivalent to the orbifold construction if the latter construction is suitably generalized to include anti-symmetric $B_{\mu\nu}$ tensor background fields plus outer automorphisms. Further analysis is needed to clarify the connection of the orbifold and the spin structure constructions as well as generalizations of these methods.

Although the derivation of the rules for the spin structure construction and the charge lattice are somewhat involved, the rules themselves are quite simple and straightforward to use. Therefore, we have organized the presentation in a way so that, for first reading, the reader may skip the details of the derivations without losing too much continuity.

The paper is organized as follows. Sect. 2 contains a brief review of the fermionic formulation of strings to clarify our notations and to provide some results necessary for later use. For readers who are not familiar with the fermionic formulation of strings, a simple and well-known example, namely a single left-moving complex fermion, is presented. This model is a generalization of the Ramond-Neveu-Schwarz (RNS) [14] fermion with periodic and anti-periodic boundary conditions. The spectrum (or partition function) is calculated using both the bosonic and fermionic formulations, demonstrating the equivalence of the two-dimensional fermionic and bosonic pictures. This simple toy model illustrates the close relation between the momentum lattice in the bosonic formulation and the fermionic charge lattice in the fermionic formulation and is the building block for the rest of the discussion. Also it provides the starting point for a discussion of the connection to the Z_2 orbifold [15] and the S^1 torus (i.e. circle) and Z_2 orbifold with continuous radius.

The set of rules for the construction of closed fermionic string models is derived in sect. 3 and sect. 5 using the basis vectors $\boldsymbol{W}_i$. In sect. 3, the derivation is quite general but the choice of $\boldsymbol{W}_i$ or spin structure concentrates on ten-dimensional models. In sect. 4, the application of the rules is demonstrated with ten-dimensional examples. Ten-dimensional models have been considered in the literature [2, 4, 8, 9, 11] and there are no new results in sect. 4; however, it is our belief that our rules can be most clearly illustrated by first applying them to models which are relatively simple and which the reader is familiar with, so we give extensive details on reconstructing known ten-dimensional models. The rules for constructing fermionic string models in fewer than 10 dimensions are derived in sect. 5 and represent a simple extension of the rules for constructing 10-dimensional models. Since the constraints on the basis vectors $\boldsymbol{W}_i$ necessary for modular invariance are given in sect. 3, only the constraints from world sheet supersymmetry need to be reexamined. Here we also introduce Thirring interactions and demonstrate the consistency of the fermionization of the compactified space coordinates with world sheet supersymmetry. This results in the introduction of the triplet constraints on the basis vectors $\boldsymbol{W}_i$ In sect. 6, we illustrate the construction of four-dimensional fermionic models with some explicit examples. Models with $N=4$, $N=2$, and $N=1$ supersymmetry as well as non-supersymmetric tachyon-free chiral models can be constructed easily. The models constructed in this section are chosen to illustrate the use of the rules and little attempt is made to systematically search for realistic models.

In sect. 7 the spin structure construction for models involving only complex fermions is rewritten in terms of a fermionic charge lattice (which replaces the set of quantities $\{\boldsymbol{W}_i, k_{ij}\}$), and this is used to prove some formal results (e.g. that only simply-laced Lie groups are generated in heterotic string models with only complex fermions). The charge lattice is an odd self-dual lorentzian lattice shifted by a fixed vector $\boldsymbol{S}$. In sect. 8 we recheck directly that the partition function represented by the fermionic charge lattice is modular invariant, and discuss additional properties of the charge lattice. This section may be read independently, with little reference to the spin-structure construction discussed in sects. 3–6. In sect. 9, connection to Z_2 orbifolds and lorentzian self-dual lattices is illustrated in detail in the simple context of a toy model with a single bosonic coordinate. In addition we use the charge lattice to relate the spin structure construction to Narain's construction in 4 dimensions, i.e., a demonstration that the lorentzian, self-dual, even-integer lattices considered by Narain are special cases of the charge lattices. Finally we give some general discussions and remarks in sect. 10.

2. Preliminaries

First quantized string theories are conveniently represented by two-dimensional fields residing on the world sheet swept out by the string in space-time. The requirements we demand for a physically acceptable string theory, i.e. unitarity, space-time Lorentz invariance and ultraviolet finiteness, then translate into conditions on the two-dimensional theory describing the string. For the sake of concreteness (most of the results we collect here are actually quite generic for closed fermionic strings) consider a ten-dimensional heterotic type superstring theory. The covariant first-quantized action [2,16], which is the action for a two-dimensional $N=\frac{1}{2}$ supergravity theory on the string world sheet, is,

$$S(e,\psi,\lambda,X,\chi)=\int d^2\xi\, e\left[\tfrac{1}{2}g^{mn}\partial_m X^\mu \partial_n X^\mu+\tfrac{1}{2}i\lambda_R^\mu \gamma^m \partial_m \lambda_R^\mu \right.$$
$$\left.+\tfrac{1}{2}i\psi_n\gamma^m\gamma^n\lambda_R^\mu\partial_m X^\mu+\tfrac{1}{2}i\chi_L^J\gamma^m\partial_m\chi_L^J\right]. \quad (2.1)$$

$X^\mu(\xi_1,\xi_2)$ is the string coordinate parametrically defining the string world sheet in space-time; note that the Lorentz index μ appears in a fashion guaranteeing manifest space-time Lorentz invariance. Here, and throughout the present work, the Ramond-Neveu-Schwarz (RNS) formalism is employed for the space-time fermions, that is the right-moving field λ_R^μ describes space-time fermionic (bosonic) degrees of freedom if it has periodic (antiperiodic) boundary conditions around the closed string. The internal symmetry degrees of freedom are represented by left-moving two-dimensional fermions χ_L^J. The zweibein e_m^a and the metric $g_{mn}=e_m^a e_{an}$ appear in the action in the above form to guarantee general coordinate invariance of the two dimensional theory (that is reparametrization invariance of the string) which permits one to eliminate the time-like modes of the string coordinate X^μ and so allows the string theory to satisfy unitarity. Similarly the local two-dimensional supersymmetry of (2.1) (which necessitates the gravitino ψ_m) allows one to eliminate the time-like fermionic modes.

In principle, any term in string perturbation theory can be found by calculating the path integral of (2.1) on the appropriate world sheet topology, e.g. spherical for tree amplitudes or toroidal for one-loop. In practice one must carefully factor out the infinite volume from the path integral due to the superdiffeomorphism invariance of (2.1) (local supersymmetry and reparametrization invariance). In addition, to completely determine the theory, one must specify for a given world sheet topology the boundary conditions of the fields around noncontractible loops on the world sheet.

Our main focus in the present work will be on the simplest topology with noncontractible loops, the torus. If $\mu=0,1,\ldots 9$ and $J=1,2,\ldots 32$ in (2.1) then the superconformal invariance of (2.1) is not broken by anomalies and for the torus topology we can choose the light cone gauge in which the action describes two-dimensional free fields,

$$S_{l.c.}=\int d^2\xi\left[\tfrac{1}{2}(\partial_a X^j)^2+\tfrac{1}{2}i\lambda_R^j\gamma^-\partial_-\lambda_R^j+\tfrac{1}{2}i\chi_L^J\gamma^+\partial_+\chi_L^J\right],\qquad j=1,2,\ldots 8. \quad (2.2)$$

The action (2.2) is integrated over a torus. It is convenient to parametrize the points on the torus by the complex quantity $\sigma_1 + \tau\sigma_2$ with σ_1 and σ_2 periodic (with period 2π). τ is the complex modular parameter which labels conformally inequivalent tori; all that remains of the integration over metrics in the one-loop path integral is an integration over τ. The last remnant of reparametrization invariance is invariance under modular transformations of τ, i.e. transforming τ by

$$\tau \to \frac{a\tau + b}{c\tau + d}, \qquad a, b, c, d \in Z, \qquad ad - bc = 1 \tag{2.3}$$

gives a different parametrization of the *same* torus and so should not change the contribution to the path integral. The modular group is generated by the two transformations,

$$\tau \to -1/\tau \tag{2.4a}$$

$$\tau \to \tau + 1, \tag{2.4b}$$

which correspond respectively to interchanging the σ_1 and σ_2 directions and to cutting the torus on a constant σ_2 slice, twisting by 2π and rejoining. To avoid an infinite multiple counting we must restrict the integration over τ to a fundamental domain of the modular group; this is the key ingredient in the ultraviolet finiteness of string theories.

Finally to completely specify the one-loop amplitude we need the boundary conditions of the fields around the σ_1 and σ_2 loops. Since the string coordinates are space-time coordinates, they must be periodic,

$$X^j(\sigma_1 + 2\pi, \sigma_2) = X^j(\sigma_1, \sigma_2 + 2\pi) = X^j(\sigma_1, \sigma_2). \tag{2.5}$$

The space-time RNS fermions, λ^j_R, can be periodic or antiperiodic in the two directions (giving four possible spin structures),

$$\lambda^j_R(\sigma_1 + 2\pi, \sigma_2) = e^{-2\pi i v}\lambda^j_R(\sigma_1, \sigma_2),$$

$$\lambda^j_R(\sigma_1, \sigma_2 + 2\pi) = e^{-2\pi i u}\lambda^j_R(\sigma_1, \sigma_2), \qquad u, v \in \{0, \tfrac{1}{2}\}. \tag{2.6}$$

The spin structures must be identified for all j. This follows from the action (2.1); the form of the interaction necessitates that all of the λ^j_R have the same spin structure as the gravitino and hence as each other. The left moving fermions χ^J_L are internal rather than space-time degrees of freedom and so are not restricted to having only periodic or antiperiodic boundary conditions. The form in which the χ^J_L appear in (2.1) allows the boundary conditions to include O(32) rotations as well as periodic and antiperiodic spin structure. For the boundary conditions on the torus to be compatible, those for the σ_1 and σ_2 loops must commute and hence can be simultaneously diagonalized. In a complex diagonal basis the boundary conditions corresponding to a general SO(32) rotation are of the form

$$\Psi^l(\sigma_1 + 2\pi, \sigma_2) = e^{-2\pi i W^l}\Psi^l(\sigma_1, \sigma_2),$$

$$\Psi^l(\sigma_1, \sigma_2 + 2\pi) = e^{-2\pi i W'^l}\Psi^l(\sigma_1, \sigma_2), \tag{2.7}$$

where Ψ^l are complex fermions (which have been analyzed by Vafa [17]). For clarity we restrict ourselves to boundary conditions of this form for now. To diagonalize an arbitrary combination of periodic and antiperiodic boundary conditions in conjunction with an SO(32) rotation in general requires some of the basis fermions to be real; our results are extended to include these cases in the last part of sect. 3. Notice that the eight real-right moving fermions may, since they share the same spin structure, be grouped into four complex fermions with boundary conditions of the form given in (2.7). Thus we can specify the σ_1 and σ_2 boundary conditions for all of the two-dimensional fermions according to (2.7) with two vectors $\boldsymbol{W}$ and $\boldsymbol{W}'$, each of the form $\boldsymbol{W} = (W^1, W^2, W^3, W^4 | W^5, \dots W^{20})$ with $W^l = 0$ or $\frac{1}{2}$ for $l = 1, 2, 3, 4$, labeling right movers, $0 \leqslant W^l < 1$, $l = 5, \dots 20$ labeling left movers, and the vertical bar for convenience separating right from left. We will often employ an obvious shorthand for specifying such vectors, for example, writing $\boldsymbol{W} = \left(\left(\frac{1}{2}\right)^4 | (0)^6 \left(\frac{1}{3}\right)^{10} \right)$ for the vector with $W^l = \frac{1}{2}$ for $l = 1, \dots 4$; 0 for $l = 5, \dots 10$; and $\frac{1}{3}$ for $l = 11, \dots 20$.

We have emphasized modular invariance and the boundary conditions for the two-dimensional fields on the torus because of the connection between the two. In general a modular transformation (2.3) will change the boundary conditions of the two-dimensional fermionic fields. Thus to ensure the modular invariance of the one-loop string amplitude we must include some sum over path integral contributions with different boundary conditions for the fermionic fields. While this might seem odd from a two-dimensional field theory point of view, in the string theory it is natural to expect a sum over boundary conditions just as we have a sum over different world sheet topologies.

Under modular transformations the boundary conditions (2.5) on the bosonic coordinates do not change and the contribution to the path integral from these fields is by itself modular invariant; we thus focus our attention on the fermionic degrees of freedom. For simplicity, let us consider a single free left-moving complex fermion with boundary condition ($t = -i\sigma_2$)

$$\lambda(\sigma_1 + 2\pi, t) = e^{-2\pi i v}\lambda(\sigma_1, t).$$

Then $\lambda(\sigma, t)$ has the following normal-mode expansion:

$$\lambda(\sigma, t) = \sum_{n=1}^{\infty} \left\{ b_{n+v-1} e^{-i(n+v-1)(\sigma+t)} + d^{\dagger}_{n-v} e^{i(n-v)(\sigma+t)} \right\}. \tag{2.8}$$

The quantization conditions are given by the anti-commutation relations

$$\{b_a^\dagger, b_b\} = \{d_a^\dagger, d_b\} = \delta_{ab}; \qquad \text{others vanish.} \tag{2.9}$$

The hamiltonian $\hat{H}_v$ and the fermion number operator $\hat{N}_v$ are given by

$$\hat{H}_v = \sum_{n=1}^{\infty} \left[(n+v-1) b_{n+v-1}^\dagger b_{n+v-1} + (n-v) d_{n-v}^\dagger d_{n-v} \right] + \tfrac{1}{2}\left(v^2 - v + \tfrac{1}{6}\right), \tag{2.10}$$

$$\hat{N}_v = \sum_{n=1}^{\infty} \left[b_{n+v-1}^\dagger b_{n+v-1} - d_{n-v}^\dagger d_{n-v} \right]. \tag{2.11}$$

For $v = 0$ and $\frac{1}{2}$, (2.10) gives the familiar Ramond and Neveu-Schwarz hamiltonians (for complex fermions); the objects we are considering are just generalizations of the RNS fermions. The last term in (2.10) is the vacuum energy constant arising from normal ordering the creation and annihilation operators. This can be computed using the Riemann ζ-function regularization (which, since it introduces no scale, maintains the conformal invariance of the theory). In addition we can calculate the vacuum charge by the same method. The vacuum charge arises from filling the negative energy states (defined to have unit charge),

$$Q_v = \sum_{n=0}^{\infty} (1+n-v)^0 = \zeta(0, 1-v) = -B_1(1-v) = v - \tfrac{1}{2}. \tag{2.12}$$

Here B_1 is the first Bernoulli polynomial. Similarly the normal ordering constant (vacuum energy) is related to the second Bernoulli polynomial

$$E_v = -\sum_{n=0}^{\infty} (1-v+n) = -\zeta(-1, 1-v) = \tfrac{1}{2} B_2(1-v) = \tfrac{1}{2}\left(v^2 - v + \tfrac{1}{6}\right). \tag{2.13}$$

Alternatively we can calculate differences of vacuum charges or energies by including in the sums a factor of $e^{-\lambda(1+n-v)}$. Computing the sums and taking the limit $\lambda \to 0$ gives results agreeing with (2.12) and (2.13). Notice that the value of the vacuum charge Q_v in (2.12) agrees with the usual definition in terms of the η-invariant [18] $Q_v = -\frac{1}{2}\eta_H$. The fermion charge operator is given by

$$\hat{Q} = \hat{N}_F + Q_v = \hat{N}_F + v - \tfrac{1}{2}. \tag{2.14}$$

The set of discrete eigenvalues Q of the charge operator $\hat{Q}$ constitute what we will refer to as the fermionic "charge lattice", although strictly speaking for general v this set is not a lattice in that it is not generated by a set of basis vectors (cf. sects. 7 and 8).

Now let us consider the contribution to the one-loop path integral of a single complex left-moving fermion with σ_1 and σ_2 boundary conditions specified by v and u (as in (2.7) with $W^l = v$; $W'^l = u$). This essentially gives the propagation of a string degree of freedom through a "time" $2\pi\tau$ and so, can be computed (up to a phase) in a hamiltonian framework as the partition function ($q = e^{2\pi i\tau}$),

$$Z_u^v = \mathrm{Tr}\left[q^{\hat{H}_v} e^{2\pi i(\frac{1}{2}-u)\hat{N}_v}\right]$$

$$= q^{\frac{1}{2}(v^2 - v + \frac{1}{6})} \prod_{n=1}^{\infty} \left(1 - q^{n+v-1} e^{-2\pi i u}\right)\left(1 - q^{n-v} e^{2\pi i u}\right). \tag{2.15}$$

Here the hamiltonian $\hat{H}_v$ and the fermion number operator $\hat{N}_v$ depend on the σ_1 boundary condition as in (2.10) and (2.11). The contribution of a right moving complex fermion to the one-loop path integral is just the complex conjugate of (2.15). All partition functions discussed in this work can be built up from this single partition function Z_u^v. When $u, v = 0, \frac{1}{2}$, we can in addition consider the square root of Z_u^v, that is, we can consider single real fermion partition functions; these will be employed in the last part of sect. 3.

To determine how Z_u^v transforms under modular transformations we will make use of the well known equivalence of two-dimensional bosonic and fermionic systems. The bosonic version of the fermionic model above turns out to be given by a single chiral boson field $\phi(t, \sigma)$

$$\phi(t, \sigma) = \tfrac{1}{2}\hat{x} - (\sigma + t)\tfrac{1}{2}\hat{p} + \sum_{n=1}^{\infty} \frac{1}{\sqrt{2n}}\left[a_n e^{-in(\sigma+t)} + a_n^\dagger e^{+in(\sigma+t)}\right], \tag{2.16}$$

whose canonical quantization with the commutator,

$$[\phi(t, \sigma), \dot{\phi}(t, \sigma')] = 2\pi i \delta(\sigma - \sigma'), \tag{2.17}$$

implies the commutation relations,

$$[a_n, a_{n'}^\dagger] = \delta_{nn'}. \tag{2.18}$$

The bosonic hamiltonian is

$$H_B = \tfrac{1}{2}\hat{p}^2 + \sum_{n=1}^{\infty} n a_n^\dagger a_n - \tfrac{1}{24}, \tag{2.19}$$

where the last term is the vacuum energy of a single left-moving boson. The eigenvalues of the momentum operator $\hat{p}$ are,

$$P = k + v - \tfrac{1}{2}, \qquad k \in \mathrm{Z}, \tag{2.20}$$

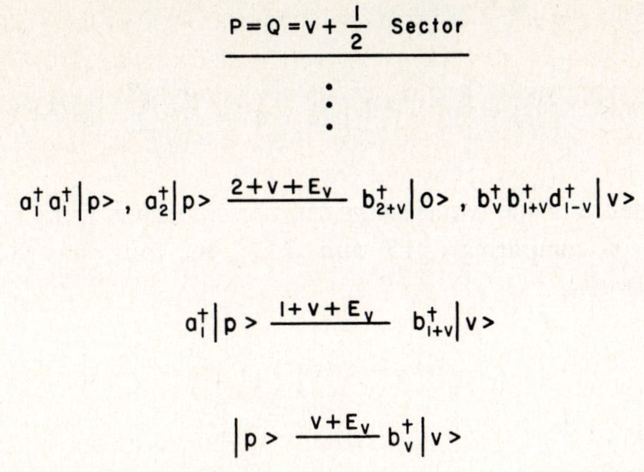

Fig. 1. The lowest few states in the fermionic charge $Q = v + \frac{1}{2}$ sector of a single complex fermion. They correspond precisely to the lowest few states in the momentum $P = v + \frac{1}{2}$ sector of a single chiral boson.

and the Fock space is spanned by

$$a^{\dagger}_{n_1} a^{\dagger}_{n_2} \dots a^{\dagger}_{n_l} |p\rangle \,. \tag{2.21}$$

It is easy to check that the states in the Q sector of the fermionic model coincide precisely with the states in the $P = Q$ sector of the bosonic model. As an example, the lowest few states for the $P = Q = v + \frac{1}{2}$ sector of the two models are shown in fig. 1. This means the set of eigenvalues $\{P\}$ of the momentum operator $\hat{p}$, i.e. the "momentum lattice", is the same as the set of eigenvalues $\{Q\}$ of the fermionic charge operator $\hat{Q}$, i.e. the charge lattice:

$$\text{momentum lattice} = \text{charge lattice} \,. \tag{2.22}$$

This boson-fermion equivalence can easily be demonstrated by expressing the fermion field λ in terms of the boson field ϕ and vice versa:

$$\lambda(t, \sigma) = \mathrm{e}^{i\phi(t, \sigma)} , \tag{2.23}$$

$$\bar{\lambda}(t, \sigma) = \mathrm{e}^{-i\phi(t, \sigma)} , \tag{2.24}$$

$$\frac{\partial \phi}{\partial \sigma} = j(t, \sigma) = \bar{\lambda}(t, \sigma) \lambda(t, \sigma) , \tag{2.25}$$

where normal orderings of the a_n, b_{n+v-1} and d_{n-v} oscillators are implied. The partition function on the torus for the bosonic model corresponding to (2.15) is given by ($q = e^{2\pi i\tau}$, $t = e^{2\pi i(\frac{1}{2} - u)}$)

$$\mathrm{Tr}\left(q^{\hat{H}_{\mathrm{B}}} t^{(\hat{p} - v + \frac{1}{2})}\right) = \left[q^{-1/24} \prod_{n=1}^{\infty} (1 - q^n)^{-1}\right] \sum_{k=-\infty}^{\infty} q^{\frac{1}{2}(k + v - \frac{1}{2})^2} t^k. \quad (2.26)$$

Since the equivalence of the two systems can be established using (2.23), (2.24), and (2.25), we have, by comparing (2.15) and (2.26), the following well-known Jacobi triple-product identity:

$$\sum_k q^{\frac{1}{2}(k + v - \frac{1}{2})^2} t^k = \eta(\tau) q^{\frac{1}{2}(v^2 - v + 1/6)} \prod_{n=1}^{\infty} (1 + q^{n+v-1} t)(1 + q^{n-v} t^{-1}), \quad (2.27)$$

where $\eta(\tau)$ is the Dedekind eta function:

$$\eta(\tau) = q^{1/24} \prod_{n=1}^{\infty} (1 - q^n).$$

Now we want to see how $Z_u^v(\tau)$ transforms under the modular transformation $\tau \to -1/\tau$. We shall calculate this in two steps. First, let us derive the transformation property of $\eta(\tau)$. Choosing $t = 1 + \varepsilon$ (for infinitesimal ε) and $v = 0$ and equating the leading order terms in ε in (2.27), we obtain

$$[\eta(\tau)]^3 = -\sum_k (-1)^k k q^{\frac{1}{2}(k - \frac{1}{2})^2}. \quad (2.28)$$

Introducing the δ function

$$\sum_k \delta(x - k + \tfrac{1}{2}) = \sum_n e^{2\pi i n (x + \frac{1}{2})}, \quad (2.29)$$

the right-hand side of (2.28) becomes

$$-\sum_n \int dx \, (x + \tfrac{1}{2}) e^{2\pi i (n + \frac{1}{2})(x + \frac{1}{2})} e^{\pi i \tau x^2} = \frac{1}{(-i\tau)^{3/2}} \sum_l (-1)^l l \, e^{(-\pi i/\tau)(l - \frac{1}{2})^2}, \quad (2.30)$$

where the integrals are well defined since $\mathrm{Im}\,\tau$ is restricted to the upper half-plane. This is simply the Poisson summation formula. Taking the cube root of (2.28), we obtain ($\omega^3 = 1$)

$$\eta\left(-\frac{1}{\tau}\right) = \omega\sqrt{-i\tau}\, \eta(\tau). \quad (2.31)$$

For pure imaginary τ with $\mathrm{Im}\,\tau > 0$, the left-hand side is positive definite. Positive definiteness of the right-hand side implies $\omega = 1$. Now $Z_u^v(-1/\tau)$ can be evaluated by using (2.31) and applying the Poisson summation formula on the left-hand side of (2.27). One can thus check that under the generators of modular transformations (2.4) the partition functions defined in (2.15) transform as (the transformation under $\tau \to \tau + 1$ is trivially obtained),

$$\tau \to -\frac{1}{\tau}: \qquad Z_u^v \to e^{2\pi i(u-\frac{1}{2})(v-\frac{1}{2})} Z_{-v}^u, \tag{2.32a}$$

$$\tau \to \tau + 1: \qquad Z_u^v \to e^{2\pi i \frac{1}{2}(v^2 - v + 1/6)} Z_{u-v}^v. \tag{2.32b}$$

The partition function of a single chiral fermion has anomalies and is not modular invariant. It turns out modular invariant anomaly-free partition functions can be obtained by considering a set of fermions and a sum over different boundary conditions for the fermions. The path integral for a set of left- and right-moving fermions with the boundary conditions (2.7) is a product of partition functions,

$$Z_{W'}^{W} = \mathrm{Tr}\left[q^{\hat{H}_W^{\mathrm{left}}} \bar{q}^{\hat{H}_W^{\mathrm{right}}} e^{-2\pi i (W_0 + W')\cdot \hat{N}_W} \right]$$

$$= \prod_{l=1}^{4} \bar{Z}_{W'^l}^{W^l} \prod_{l=5}^{20} Z_{W'^l}^{W^l} \tag{2.33a}$$

where

$$\hat{H}_W^{\mathrm{left}} = \sum_{l=5}^{20} \hat{H}_{W^l},$$

$$\hat{H}_W^{\mathrm{right}} = \sum_{l=1}^{4} \hat{H}_{W^l},$$

$$\hat{N}_W^l = \hat{N}_{W^l}. \tag{2.33b}$$

W_0 is defined to be the vector with all components equal to $\frac{1}{2}$ and the dot product between vectors in (2.33) includes a factor of (-1) for right moving components, e.g., $W \cdot \hat{N} = \Sigma_{l=5}^{20} W^l \hat{N}^l - \Sigma_{l=1}^{4} W^l \hat{N}^l$. In section 3 when we consider sums of partition functions it will prove convenient to have all partition functions defined with boundary conditions characterized by u and v restricted to the interval $[0,1)$. For this purpose let us introduce an overbar notation; for any function $f(u,v)$ we define, $\overline{f(u,v)} = f(u,v) - \Delta$ with $0 \leqslant \bar{f} < 1$ and $\Delta \in \mathrm{Z}$. Note that from (2.15) we have $Z_{f(u,v)}^v = Z_{\overline{f(u,v)}}^v$ while in general $Z_u^{f(u,v)} \neq Z_u^{\overline{f(u,v)}}$. With this notation (2.33) transforms under modular transformations as,

$$\tau \to -\frac{1}{\tau}: \qquad Z_{W'}^{W} \to e^{2\pi i (W\cdot W' - W_0\cdot(W + W'))} Z_{\overline{-W}}^{W'} \tag{2.34a}$$

$$\tau \to \tau + 1: \qquad Z_{W'}^{W} \to e^{2\pi i (\frac{1}{2} W\cdot W - W_0 \cdot W)} Z_{\overline{W' - W}}^{W}. \tag{2.34b}$$

Again the dot product between vectors is defined to include a factor (-1) for the right-moving components. The constant (i.e. u, v, τ independent) phase present in (2.32) drops out because the number of left movers (16) minus the number of right movers (4) is always a multiple of 12.

In the following section we will construct sums of partition functions of the form (2.33) over different sets of boundary conditions with coefficients chosen such that the sum is one-loop modular invariant. We will make one other significant demand on this sum of partition functions. For any given set of σ_1 boundary conditions the sum of partition functions over different sets of σ_2 boundary conditions naturally forms a projection onto a subspace of string states so that the total partition function can again be interpreted as a sum over states weighted by $e^{2\pi i\tau H}$. To give a simplified example, for four complex left-moving spacetime RNS fermions the combination,

$$\mathrm{Tr}\left[\exp\left(2\pi i\tau\sum_{i=1}^{4}\hat{H}^{i}_{1/2}\right)\frac{1-(-1)^{\sum_{i=1}^{4}\hat{N}^{i}_{1/2}}}{2}\right]-\mathrm{Tr}\left[\exp\left(2\pi i\tau\sum_{i=1}^{4}\hat{H}^{i}_{0}\right)\frac{1+(-1)^{\sum_{i=1}^{4}\hat{N}^{i}_{0}}}{2}\right] \tag{2.35}$$

is modular invariant (up to an overall phase), and naturally forms a projection onto states with odd total fermion number in the Neveu-Schwarz (NS) sector $(\hat{H}_{1/2})$ and even total fermion number in the Ramond (R) sector $(\hat{H}_0)$. These are just the GSO projections [7]. Note in addition the overall signs of the traces in (2.35). These guarantee that the allowed physical states (those included in the projection) appear with the correct space-time statistics. The space-time bosons (NS sector) contribute positively to the one-loop amplitude while the space-time fermions (R sector) contribute with an overall $(-)$ sign. In the constructions of sect. 3 we will demand that these properties are retained in the general case; the sum over partition functions should form a proper projection (i.e. with eigenvalues 0 and 1) onto a subspace of states and the states should be included with the correct space-time statistics. Notice that with the inclusion of this assumption we have met, by construction, all the requirements for proper physical states. Lorentz invariance is guaranteed by the form of (2.1); positivity is ensured by our choice of the light cone gauge which we could consistently impose because of the reparametrization invariance and local world sheet supersymmetry of (2.1); and now the correct space-time statistics for physical states has been imposed by hand.

For the sake of completeness, let us present explicitly the full one-loop partition function including the contribution from the string coordinates,

$$Z=c\int_{F}\frac{d^2\tau}{\tau_2^2}\left[\eta(\tau)\eta^*(\tau)\sqrt{\tau_2}\right]^{-(D-2)}Z_{\text{fermion}}(\tau), \tag{2.36}$$

where $\tau = \tau_1 + i\tau_2$ and the integration region is over the fundamental domain F: $-\frac{1}{2} \leqslant \tau_1 \leqslant \frac{1}{2}$, $\tau_2 > 0$ and $|\tau| \geqslant 1$. The constant c depends on the particular string model under consideration and its exact value is not relevant for our discussion. $Z_{\text{fermion}}(\tau)$ will be discussed in sect. 3.

3. Construction of modular invariant string models

The most general possible contribution from fermion fields to the string one-loop path integral is a sum, with arbitrary coefficients, over partition functions of the form (2.33) with different sets of boundary conditions,

$$Z_{\text{fermion}} = \sum_{i,j} C^{W_i}_{W_j} Z^{W_i}_{W_j}, \tag{3.1}$$

where $C^{W_i}_{W_j}$ are the coefficients. The analysis in this section is a simple extension of that in ref. [11]. As in the previous section we will for concreteness keep in mind the specific example of a ten dimensional heterotic type theory with boundary conditions of the form (2.7) even though our results will prove to be of considerably wider validity (as we will later demonstrate). If $Z^{W_a}_{W_b}$ appears in the sum in (3.1) then $\boldsymbol{W}_a$ and $\boldsymbol{W}_b$ (or more precisely the boundary conditions that these vectors represent) must commute for the boundary conditions on the torus to be compatible. Furthermore if, for example, $Z^{W_a}_{W_b}$ and $Z^{W_a}_{W_c}$, appear, then $\boldsymbol{W}_b$ and $\boldsymbol{W}_c$ must commute or (3.1) cannot form a good projection onto a subspace of states in the sector of the theory with hamiltonian H_{W_a}. If (3.1) breaks into two separate sums over mutually disjoint sets of boundary conditions then it in fact describes two separate string models which do not interact with each other. Thus all of the boundary conditions present in (3.1) should commute with each other and we can choose a diagonal basis for the fermions as in (2.7).

Our goal is to determine for which sets of boundary conditions the sum in (3.1) forms a proper projection and is modular invariant. For this purpose it will prove advantageous to introduce some notation to effectively organize the sets of boundary conditions which might appear in (3.1). For an arbitrary set of boundary conditions we can choose an independent set of "basis" vectors $\{\boldsymbol{W}_i\}$ which generates them; that is, all of the vectors characterizing boundary conditions can be written in the form $\overline{\alpha W} \equiv \sum_i \alpha_i \boldsymbol{W}_i$ where α_i are integers taking values from 0 to $m_i - 1$ and the $\boldsymbol{W}_i$ are independent in the sense that $\overline{\alpha W} = \mathbf{0}$ (all components zero) only if $\alpha_i = 0$ for all i. Here m_i is the order of the vector $\boldsymbol{W}_i$ defined to be the smallest positive integer such that all of the components of $m_i \boldsymbol{W}_i$ (i not summed) are integers. Here we restrict ourselves to $\boldsymbol{W}_i$ with rational number components, i.e., m_i finite; some comments on the irrational case are included in sect. 9. The overbar notation is defined as in sect. 2 ($\overline{\alpha W} = \alpha W - \Delta(\alpha)$ where $0 \leqslant \overline{\alpha W}^l < 1$ and $\Delta^l(\alpha) \in \mathrm{Z}$). For convenience we will always take the set $\{\boldsymbol{W}_i\}$ to include the vector $\boldsymbol{W}_0$ with all components equal to $\frac{1}{2}$ (we will later show that the boundary condition represented by W_0 *must* always be present). Without loss of generality we may rewrite (3.1) in the form,

$$Z_{\text{fermion}} = \sum_{\{\alpha,\beta\}} C^{\overline{\alpha W}}_{\overline{\beta W}} Z^{\overline{\alpha W}}_{\overline{\beta W}} . \tag{3.2}$$

Of course some of the coefficients in (3.2) may be zero, but by assumption at least some of the coefficients involving W_i are nonzero for any i except perhaps $i=0$ (since W_0 was included in the generating set $\{W_i\}$ by hand). Note that while the set $\{W_i\}$ behaves something like a basis spanning a vector space, in general this is not quite correct since unless m_i is prime the coefficients $\alpha_i = 0, \dots m_i - 1$ do not form a field. In general we cannot choose the W_i such that all of the m_i are prime; we could not construct the vector $((0)^4|(\frac{1}{4})^{16})$, for example, in this case. Because we do not have a vector space, even the number of vectors in $\{W_i\}$ is not fixed. For example $\{W_0, ((0)^4|(\frac{1}{3})^{16}), ((0)^4|(\frac{1}{5})^{16})\}$ and $\{W_0, ((0)^4|(\frac{1}{15})^{16})\}$ are both acceptable sets generating the same collection of vectors. On the other hand $\{W_0, ((0)^4|(\frac{1}{3})^{16})\}$ and $\{(\frac{1}{2})^4|(\frac{1}{6})^{16}\}$ generate the same set of vectors but only the former is in the desired form with the generators including W_0.

We have taken the trouble to rewrite (3.1) in the form (3.2) because, as we will now show, in order for (3.2) to be modular invariant and form a proper projection, all of the coefficients must be nonzero. First consider modular transformations. For (3.2) to be invariant under (2.4a) we must have from (2.34a),

$$\tau \to -1/\tau: \qquad C^{\overline{\alpha W}}_{\overline{\beta W}} Z^{\overline{\alpha W}}_{\overline{\beta W}} \to C^{\overline{\beta W}}_{\overline{-\alpha W}} Z^{\overline{\beta W}}_{\overline{-\alpha W}} ,$$

i.e.,

$$C^{\overline{\alpha W}}_{\overline{\beta W}} \, e^{2\pi i(\overline{\alpha W}\cdot\overline{\beta W} - W_0\cdot(\overline{\alpha W}+\overline{\beta W}))} = C^{\overline{\beta W}}_{\overline{-\alpha W}} . \tag{3.3}$$

Similarly for invariance under (2.4b) eq. (2.34b) gives,

$$\tau \to \tau + 1: \qquad C^{\overline{\alpha W}}_{\overline{\beta W}} Z^{\overline{\alpha W}}_{\overline{\beta W}} \to C^{\overline{\alpha W}}_{\overline{(\beta-\alpha) W}} Z^{\overline{\alpha W}}_{\overline{(\beta-\alpha) W}} ,$$

i.e.,

$$C^{\overline{\alpha W}}_{\overline{\beta W}} \, e^{2\pi i\left(\frac{1}{2}\overline{\alpha W}\cdot\overline{\alpha W} - W_0\cdot\overline{\alpha W}\right)} = C^{\overline{\alpha W}}_{\overline{(\beta-\alpha) W}} . \tag{3.4}$$

In addition to (3.3) and (3.4) we require that for any physical sector characterized by the σ_1 boundary condition $\overline{\alpha W}$ the sum over σ_2 boundary conditions in (3.2) form a proper projection with eigenvalues 0 or 1. Specifically this means (c.f. (2.33)),

$$\sum_{\{\beta\}} C^{\overline{\alpha W}}_{\overline{\beta W}} \, e^{-2\pi i \Sigma_i(\beta_i+\delta_{0i})\phi_i} = e^{2\pi i(\alpha s + \frac{1}{2})} \eta\left(\overline{\alpha W}, \phi\right), \tag{3.5}$$

where $\phi_i(\alpha) \equiv W_i \cdot N_{\overline{\alpha W}}$ takes values l/m_i, $l = 0, 1, \dots m_i - 1$; $\alpha s \equiv \Sigma_i \alpha_i s_i \equiv \Sigma_i \alpha_i W_i^1$, and $\eta(\overline{\alpha W}, \phi)$ takes values 0 or 1 depending on the values of the α_i's and ϕ_j's. Note that when we use, for example, α or ϕ (with no subscript) we mean a set of values of the variables $\alpha_0, \alpha_1, \ldots$ or $\phi_0, \phi_1, \ldots$. Also we in general never use an implied sum on indices, instead defining for example, $\Sigma\phi_i\alpha_i = \phi\alpha$.

Eq. (3.5) must hold for all choices of α and all choices of ϕ, in other words for any combination of fermion numbers defining a state, either that state is included in Z_{fermions} (i.e., $\eta = 1$) or not (i.e., $\eta = 0$). The coefficient of η in (3.5) guarantees that physical states appear in the partition function with an overall sign indicating the correct statistics; $\alpha s = 0, \frac{1}{2}$ designating spacetime fermions (Ramond sector) or bosons (Neveu-Schwarz sector) respectively. Eq. (3.5) is what we mean by a physically sensible projection.

Multiplying (3.5) by $\exp\{2\pi i\Sigma_i(\beta_i' + \delta_{0i})\phi_i\}$ and summing over all combinations of values of the ϕ_j's (to invert the Fourier decomposition in (3.5)) gives a formal solution for the C coefficients,

$$C^{\overline{\alpha W}}_{\overline{\beta W}} = \prod_j m_j^{-1} \sum_{\{\phi\}} \mathrm{e}^{2\pi i\{\Sigma_i(\beta_i+\delta_{0i})\phi_i + \alpha s + \frac{1}{2}\}} \eta(\overline{\alpha W}, \phi), \tag{3.6}$$

where the primes on β_i have been dropped. In particular this gives,

$$C^{W_0}_{\overline{\alpha W}} = \left(\prod m_j^{-1}\right) \sum_{\{\phi\}} \mathrm{e}^{2\pi i\Sigma_i(\alpha_i+\delta_{0i})\phi_i} \eta(W_0, \phi), \tag{3.7}$$

$$C^{\overline{\alpha W}}_{W_0} = \left(\prod m_j^{-1}\right) \mathrm{e}^{2\pi i(\alpha s + \frac{1}{2})} \sum_{\{\phi\}} \eta(\overline{\alpha W}, \phi). \tag{3.8}$$

Now from (3.3) modular invariance requires $C^{W_0}_{\overline{\alpha W}} = C^{\overline{\alpha W}}_{W_0}$ or, using (3.7) and (3.8),

$$\sum_{\{\phi\}} \mathrm{e}^{2\pi i\Sigma_i(\alpha_i+\delta_{0i})(\phi_i - s_i)} \eta(W_0, \phi) = \sum_{\{\phi\}} \eta(\overline{\alpha W}, \phi). \tag{3.9}$$

The right-hand side of (3.9) is always a nonnegative integer and this must hold for all possible values of α. Notice that unless all of the η's are zero (in which case we have no theory) some of the $\eta(W_0, \phi)$'s must be nonzero, i.e., W_0 *must* appear in the set of boundary conditions. Then, by assumption, the entire set $\{W_i\}$ is needed to generate the boundary conditions for which the C coefficients are non-vanishing and so for each W_i some of the $\eta(W_i, \phi)$'s must be nonzero. The general solution to these conditions is $\eta(W_0, s) = 1$ (i.e., $\phi_i = s_i$ for all i) with all other $\eta(W_0, \phi)$'s vanishing (we know of no exceptional solutions meeting our requirements). In this case, for any set of α values, the left-hand side of (3.9) is unity so that $\eta(\overline{\alpha W}, \phi)$ must be nonzero for one and only one choice of ϕ depending on α. Eq. (3.6) becomes,

$$C^{\overline{\alpha W}}_{\overline{\beta W}} = \left(\prod m_i\right)^{-1} e^{2\pi i(\phi(\alpha)\beta + D(\alpha))}, \tag{3.10}$$

with,

$$D(\alpha) = \alpha s + \tfrac{1}{2} + \phi_0(\alpha) \qquad (\mathrm{mod}\ 1), \tag{3.11}$$

$$m_i \phi_i(\alpha) = 0 \qquad (\mathrm{mod}\ 1). \tag{3.12}$$

The form given in (3.10) ensures that the sum (3.2) provides a proper projection of states with correct space-time statistics. It remains to determine completely the constraints that modular invariance places on (3.10). In these terms (3.3) becomes the requirement that,

$$\phi(\alpha)\beta + D(\alpha) + \overline{\alpha W}\cdot\overline{\beta W} - W_0\cdot(\overline{\alpha W} + \overline{\beta W}) = -\phi(\beta)\alpha + D(\beta) \qquad (\mathrm{mod}\ 1). \tag{3.13}$$

This must hold for all choices of α and β. In particular, successively choosing $\alpha = 0$ and $\beta = 0$ gives respectively,

$$D(\beta) = -W_0\cdot\overline{\beta W} + \phi(0)\beta + D(0) \qquad (\mathrm{mod}\ 1), \tag{3.14}$$

$$D(\alpha) = W_0\cdot\overline{\alpha W} - \phi(0)\alpha + D(0) \qquad (\mathrm{mod}\ 1). \tag{3.15}$$

Substituting these expression for D back into (3.13) we have,

$$\phi(\alpha)\beta - \phi(0)\alpha + \overline{\alpha W}\cdot\overline{\beta W} = -\phi(\beta)\alpha + \phi(0)\beta \qquad (\mathrm{mod}\ 1). \tag{3.16}$$

For $\beta_j = \delta_{ij}$ this reduces to,

$$\phi_i(\alpha) = \sum_j k_{ij}\alpha_j + \phi_i(0) - W_i\cdot\overline{\alpha W} \qquad (\mathrm{mod}\ 1), \tag{3.17}$$

where,

$$k_{ij} \equiv \phi_j(0) - \phi_j(\beta_l = \delta_{li}).$$

Again we use this to rewrite (3.16),

$$\sum_{i,j}(k_{ij} + k_{ji})\alpha_i\beta_j - \beta W\cdot\overline{\alpha W} - \alpha W\cdot\overline{\beta W} + \overline{\alpha W}\cdot\overline{\beta W} = 0 \qquad (\mathrm{mod}\ 1). \tag{3.18}$$

Finally choosing $\beta_l = \delta_{li}$ and $\alpha_l = \delta_{lj}$ in (3.18) implies that,

$$k_{ij} + k_{ji} = W_i\cdot W_j \qquad (\mathrm{mod}\ 1) \tag{3.19}$$

and one can check that this is in fact sufficient to solve (3.18) for all α and β.

Turning now to invariance under $\tau \to \tau + 1$ we require (cf. (3.4) and (3.10)),

$$\phi(\alpha)\alpha + \tfrac{1}{2}\overline{\alpha W} \cdot \overline{\alpha W} - W_0 \cdot \overline{\alpha W} = 0 \qquad (\mathrm{mod}\ 1). \tag{3.20}$$

Using (3.17) we rewrite this as,

$$\sum_{i,j} k_{ij}\alpha_i\alpha_j + \alpha\phi(0) - \alpha W \cdot \overline{\alpha W} + \tfrac{1}{2}\overline{\alpha W} \cdot \overline{\alpha W} - W_0 \cdot \overline{\alpha W} = 0 \qquad (\mathrm{mod}\ 1). \tag{3.21}$$

With (3.19) we can reexpress the first term,

$$\begin{aligned}\sum_{i,j} k_{ij}\alpha_i\alpha_j &= \sum_i k_{ii}\alpha_i^2 + \sum_{i>j} \alpha_i\alpha_j(k_{ij} + k_{ji}) \\ &= \sum_i (k_{ii} - \tfrac{1}{2}W_i \cdot W_i)\alpha_i + \tfrac{1}{2}\alpha W \cdot \alpha W \qquad (\mathrm{mod}\ 1),\end{aligned} \tag{3.22}$$

where we have used

$$2(k_{ii} - \tfrac{1}{2}W_i \cdot W_i) = 0 \qquad (\mathrm{mod}\ 1),$$

$$\tfrac{1}{2}\alpha_i = \tfrac{1}{2}\alpha_i^2 = -\tfrac{1}{2}\alpha_i \qquad (\mathrm{mod}\ 1).$$

Note that equalities of this type are frequently used in the following analysis. Also we can eliminate the overbars from the last half of (3.21),

$$\begin{aligned}&\tfrac{1}{2}\overline{\alpha W} \cdot \overline{\alpha W} - W_0 \cdot \overline{\alpha W} - \alpha W \cdot \overline{\alpha W} \\ &\quad = \tfrac{1}{2}\alpha W \cdot \alpha W - \Delta(\alpha) \cdot \alpha W - \tfrac{1}{2}\Delta(\alpha) \cdot \Delta(\alpha) - W_0 \cdot \alpha W \\ &\qquad + W_0 \cdot \Delta(\alpha) - \alpha W \cdot \alpha W + \alpha W \cdot \Delta(\alpha) \qquad (\mathrm{mod}\ 1) \\ &\quad = -\tfrac{1}{2}\alpha W \cdot \alpha W - W_0 \cdot \alpha W \qquad (\mathrm{mod}\ 1).\end{aligned} \tag{3.23}$$

With (3.22) and (3.23), eq. (3.21) becomes,

$$\sum_i \alpha_i\{\phi_i(0) + k_{ii} - \tfrac{1}{2}W_i \cdot W_i - W_0 \cdot W_i\} = 0, \tag{3.24}$$

which must hold for all α, i.e. we have,

$$\phi_i(0) = \tfrac{1}{2}W_i \cdot W_i + W_0 \cdot W_i - k_{ii}. \tag{3.25}$$

Finally we impose eqs. (3.11) and (3.12). The latter together with (3.17) implies, for all α

$$m_i\sum_j k_{ij}\alpha_j + m_i\phi_i(0) - m_iW_i \cdot \overline{\alpha W} = 0 \qquad (\mathrm{mod}\ 1). \tag{3.26}$$

In other words,

$$m_i(k_{ij} - W_i \cdot W_j) = m_i k_{ji} = 0 \qquad (\text{mod } 1), \tag{3.27}$$

$$m_i \phi_i(0) = m_i(\tfrac{1}{2} W_i \cdot W_i + W_0 \cdot W_i) = 0 \qquad (\text{mod } 1). \tag{3.28}$$

Eq. (3.28) is equivalent to the level matching condition of Vafa [17].

Equating the two expressions for $D(\alpha)$, (3.11) and (3.15), and using (3.17) and (3.25) to rewrite the result we have,

$$\sum \alpha_i \{ s_i - k_{0i} + \tfrac{1}{2} W_i \cdot W_i + W_0 \cdot W_i - k_{ii} \} = 0 \qquad (\text{mod } 1) \qquad \text{for all } \alpha \tag{3.29}$$

or

$$\tfrac{1}{2} W_i \cdot W_i + s_i + k_{i0} - k_{ii} = 0 \qquad (\text{mod } 1). \tag{3.30}$$

Note that (3.30) together with (3.27) and (3.19) are sufficient to ensure (3.28). Also one can check that the compatibility of (3.14) and (3.15) follows from (3.19) and (3.25). Finally we can use the above results to express the exponent giving the C coefficient in (3.10),

$$\begin{aligned} &\phi(\alpha)\beta + D(\alpha) \\ &\quad = \sum_i (\beta_i + \delta_{i0}) \phi_i(\alpha) + \alpha s + \tfrac{1}{2} \qquad (\text{mod } 1) \\ &\quad = \sum_i (\beta_i + \delta_{i0}) (k_{ij}\alpha_j + \tfrac{1}{2} W_i \cdot W_i + W_0 \cdot W_i - k_{ii} - W_i \cdot \overline{\alpha W}) + \alpha s + \tfrac{1}{2} \quad (\text{mod } 1) \\ &\quad = \sum_i (\beta_i + \delta_{i0}) \Big(\sum_j k_{ij} \cdot \alpha_j + s_i + k_{0i} - W_i \cdot \overline{\alpha W} \Big) + \alpha s + \tfrac{1}{2} \qquad (\text{mod } 1). \end{aligned} \tag{3.31}$$

Collecting and summarizing the results of this section we have that the sum of partition functions (3.2) is modular invariant and forms a proper projection with correct spacetime statistics if the coefficients are chosen according to (3.10) and (3.31) and the W_i and structure constants k_{ij} satisfy the constraints (3.19), (3.27), and (3.30). Explicitly,

$$\begin{aligned} Z_{\text{fermion}} &= \left(\prod m_i\right)^{-1} \\ &\quad \times \sum_{\{\alpha, \beta\}} \exp\left[2\pi i \left\{ \sum_i (\beta_i + \delta_{i0}) \Big(\sum_j k_{ij}\alpha_j - W_i \cdot \overline{\alpha W} + k_{0i} + s_i \Big) + \alpha s - \tfrac{1}{2} \right\} \right] \\ &\quad \times \mathrm{Tr}\left[\exp\left(2\pi i \left\{ \tau \hat{H}^{\text{left}}_{\alpha W} - \bar{\tau} \hat{H}^{\text{right}}_{\alpha W} - \sum_j (\beta_j + \delta_{j0}) W_j \cdot \hat{N}_{\alpha W} \right\} \right) \right], \end{aligned} \tag{3.32}$$

With the constraints,

$$k_{ij} + k_{ji} = \boldsymbol{W}_i \cdot \boldsymbol{W}_j \qquad (\mathrm{mod}\ 1)\,, \tag{3.33}$$

$$m_j k_{ij} = 0 \qquad (\mathrm{mod}\ 1)\,, \tag{3.34}$$

$$k_{ii} + k_{i0} + s_i - \tfrac{1}{2}\boldsymbol{W}_i \cdot \boldsymbol{W}_i = 0 \qquad (\mathrm{mod}\ 1)\,. \tag{3.35}$$

All states are projected out of the sum in (3.32) except those with fermion numbers satisfying,

$$\boldsymbol{W}_i \cdot \boldsymbol{N}_{\alpha W} = \phi_i(\alpha) = \sum_j k_{ij}\alpha_j + s_i + k_{0i} - \boldsymbol{W}_i \cdot \overline{\alpha W} \qquad (\mathrm{mod}\ 1)\,. \tag{3.36}$$

This is the spectrum generating formula. The states given by (3.36) appear with weight one and phase $\exp\{2\pi i(\alpha s - \frac{1}{2})\} = +1$ or -1 for space-time bosons or fermions respectively. Notice also that the individual terms in the sum in (3.32) (which are products of chiral Dirac determinants) have $\boldsymbol{W}$ dependent phase such that they reproduce the correct anomalies, [13, 17] i.e., under the shifts in boundary conditions $\overline{\alpha W^l} \to \overline{\alpha W^l} + 1$ or $\overline{\beta W^l} \to \overline{\beta W^l} + 1$ they acquire phases $-\exp(i\pi\overline{\beta W^l})$ or $-\exp(-i\pi\overline{\alpha W^l})$ respectively. The constraints (3.36) assume a particularly simple form in the $\boldsymbol{W}_0$ sector,

$$\boldsymbol{W}_i \cdot \boldsymbol{N}_{W_0} = s_i \qquad (\mathrm{mod}\ 1)\,.$$

As a consequence one can easily see that the graviton always appears in the physical spectrum.

We have chosen the vectors generating boundary conditions $\{\boldsymbol{W}_i\}$ and the parameters determining statistics, s_i, of a particular form so that equations (3.32)–(3.36) are appropriate for examining ten-dimensional heterotic type string theories with world sheet fermionic boundary conditions of the form (2.7). Our derivation of these equations, however, was largely independent of the details of these choices so as one might expect the final results are of much wider validity; to construct other types of closed string models we need only the appropriate forms for $\boldsymbol{W}_i$ and s_i and perhaps some minor changes in our equations.

Consider ten-dimensional type II and purely bosonic string models. For the moment let us retain the general form (2.7) for the boundary conditions, i.e. all of the fermions in a complex diagonal basis. For the type II string the right movers are treated as in the heterotic case and the left movers are treated in the same fashion. Thus the vectors specifying boundary conditions should be of the form,

$$\textit{type II}: \quad \boldsymbol{W} = \left((a)^4 | (b)^4\right) \qquad a, b \in \{0, \tfrac{1}{2}\}\,. \tag{3.37}$$

Similarly the left and right movers of the closed, ten-dimensional, purely bosonic string are treated like the heterotic left movers and so we need $\boldsymbol{W}$ of the form,

$$\textit{bosonic}: \quad \boldsymbol{W} = \left(W^1, W^2, \dots W^{16} | W^{17}, \dots W^{32}\right), \qquad 0 \leqslant W^l < 1\,. \tag{3.38}$$

The definition of s_i in these two cases must be changed to insure correct space-time statistics, i.e. $\alpha s = 0(\frac{1}{2})$ should correspond to space-time fermions (bosons) (c.f. (3.5)). For type II models we have spacetime bosons in the $\boldsymbol{W}$ sector if $a+b=0$ in (3.37); space-time fermions if $a+b=\frac{1}{2}$. Thus we can define,

$$\textit{type II}: \qquad s_i = W_i^{\mathrm{R}} + W_i^{\mathrm{L}} \qquad (\mathrm{mod}\ 1)\,, \tag{3.39}$$

where R and L label any right or left spacetime component. All particles in purely bosonic models are spacetime bosons. This is most smoothly incorporated in our formalism by defining,

$$\textit{bosonic}: \qquad s_i = 0 \tag{3.40}$$

and flipping the overall sign of (3.32). With these definitions eqs. (3.32)–(3.36) can be used for constructing type II and purely bosonic string models.

Let us now consider what changes are necessary to incorporate more general boundary conditions into our formalism (we postpone until sect. 5 the extension to strings in fewer than ten dimensions). The formulation given above with all complex fermions is sufficient for any set of commuting boundary conditions which includes only pure rotations. Adding arbitrary reflections we must allow for some of the fermions in a diagonal basis to be real. In our formalism this means that the set of $\{\boldsymbol{W}_i\}$ generating boundary conditions may have components $W_i^l \in \{0, \frac{1}{2}\}$ where l labels a *real* fermion for all i. Notice that if a component of the $\boldsymbol{W}_i$'s representing a complex fermion is always either 0 or $\frac{1}{2}$ then we can split this component into two, both representing real fermions with the same spin structure (i.e., we can diagonalize these boundary conditions with either complex or real fermions); any values other than 0 or $\frac{1}{2}$ require complex fermions while collections of real fermions with boundary conditions not always identified in pairs cannot be represented by complex ones. These latter two classes of boundary conditions do not commute, so in particular we cannot choose $\{\boldsymbol{W}_i\}$ with some group of components labeling boundary conditions requiring complex fermions for some $\boldsymbol{W}_i$ and real fermions for some other $\boldsymbol{W}_j$.

Real fermions with antiperiodic boundary conditions present no particular difficulties in our formalism [11]. The same treatment employed for the complex case can be used, with the single real fermion hamiltonian and number operators given by,

$$\hat{H}^{\mathrm{NS}} = \hat{H}_{\frac{1}{2}\mathrm{r}} = \sum_{n=0}^{\infty} \left(n+\tfrac{1}{2}\right) b_{n+\frac{1}{2}}^{\dagger} b_{n+\frac{1}{2}} - \tfrac{1}{48}\,, \tag{3.41}$$

$$\hat{N}_{\frac{1}{2}\mathrm{r}} = \sum_{n=0}^{\infty} b_{n+\frac{1}{2}}^{\dagger} b_{n+\frac{1}{2}}^{\dagger}\,,$$

$$\left\{ b_a^{\dagger}, b_b \right\} = \delta_{ab}\,. \tag{3.42}$$

(The few times in this paper where we explicitly treat real fermions we will employ the standard notation of b and d for Neveu-Schwarz and Ramond annihilation operators even though we already use these same symbols for complex fermion and anti-fermion annihilation operators; it should be clear from the context which we mean.) The periodic boundary condition requires a bit more care due to the presence of a zero mode. Consider first a single complex fermion (for which the treatment of the zero mode is clear). From (2.9)–(2.11) we have,

$$\hat{H}_0 = \sum_{n=1}^{\infty} n\left(b_n^\dagger b_n + d_n^\dagger d_n\right) + \tfrac{1}{12}, \tag{3.43}$$

$$\hat{N}_0 = \sum_{n=1}^{\infty} \left(b_n^\dagger b_n - d_n^\dagger d_n\right) + b_0^\dagger b_0, \tag{3.44}$$

$$\left\{d_m^\dagger, d_n\right\} = \left\{b_m^\dagger, b_n\right\} = \delta_{mn} \qquad \text{(others vanish)}. \tag{3.45}$$

There are two ground states, $|0\rangle$ (satisfying $b_n|0\rangle = d_n|0\rangle = 0$ for all n) and $b_0^\dagger|0\rangle$, with fermion numbers $N_0 = 0$ and 1 respectively. The quantity needed in our analysis is not $\hat{N}_0$ but $(-1)^{\hat{N}_0}$, which appears in the partition function with periodic boundary conditions in both σ_1 and σ_2 directions (cf. (2.15)). Thus we can equivalently take eq. (3.44) with the antifermions contributing with a positive sign. Then writing this complex fermion in terms of two real ones,

$$b_n = \sqrt{\tfrac{1}{2}}\left(d_n^1 + i d_n^2\right), \qquad d_n = \sqrt{\tfrac{1}{2}}\left(d_n^1 - i d_n^2\right), \tag{3.46}$$

$$\left\{d_m^{i\dagger}, d_n^j\right\} = \delta_{mn}\delta_{ij}, \tag{3.47}$$

we have,

$$\hat{H}^{\text{Ramond}} = \hat{H}_0 = \sum_{n=1}^{\infty} n\left(d_n^{1\dagger} d_n^1 + d_n^{2\dagger} d_n^2\right) + \tfrac{1}{12}, \tag{3.48}$$

$$\hat{N}_0 = \sum_{n=1}^{\infty} \left(d_n^{1\dagger} d_n^1 + d_n^{2\dagger} d_n^2\right) + \tfrac{1}{2}\left(1 + 2i d_0^1 d_0^2\right). \tag{3.49}$$

This can be written as,

$$(-1)^{\hat{N}_0} = -2i d_0^1 d_0^2 (-1)^{\sum_{n=1}^{\infty}\left(d_n^{1\dagger} d_n^1 + d_n^{2\dagger} d_n^2\right)}. \tag{3.50}$$

The two ground states are now distinguished by their $2i d_0^1 d_0^2$ eigenvalues (± 1),

$$2i d_0^1 d_0^2 |0\rangle = -|0\rangle,$$

$$\left(2i d_0^1 d_0^2\right) b_0^\dagger|0\rangle = b_0^\dagger|0\rangle. \tag{3.51}$$

The purpose of this little exercise is two-fold. First we see a familiar result from the Ramond model: the operators $\sqrt{2}\, d_0^i$ satisfy the commutation relations of γ matrices so that in particular the projections formed give states of definite chirality as well as fermion number, e.g. the Ramond sector projection in (2.35) becomes,

$$\tfrac{1}{2}\left(1 + i\gamma^5(-1)^{\Sigma d_n^{i\dagger} d_n^i}\right). \tag{3.52}$$

It is this algebraic structure of the d_0^i operators which allows space-time spinor representations to be formed from operators with vector indices. The second point to observe, which is our immediate concern here, is the fact that we can only define a sensible number operator for an *even* number of real fermionic fields with periodic boundary conditions. For a single real Ramond fermion we in a sense have only the "square root" of the zero mode, which is unacceptable by itself. This means that to get a sensible projection of states from a sum of partition functions, each particular term in the sum must be such that it *could* be written in the form (2.33) with complex fermions. In terms of the generating vectors this is the condition,

$$2 \sum_{\{l:\,\text{real}\}} W_i^l W_j^l \in \mathrm{Z} \qquad \text{for all } i, j \tag{3.53a}$$

where the sum runs over components representing real fermions. In addition, to extend the analysis leading to (3.32) to this case, we must insure in any sector $\overline{\alpha W}$ that the fermion number operators $\boldsymbol{W}_i \cdot \boldsymbol{N}_{\overline{\alpha W}}$ and $\boldsymbol{W}_j \cdot \boldsymbol{N}_{\overline{\alpha W}}$ commute. In fact a careful analysis shows that modular invariance and a physically sensible projection require the commutation of such operators even between different sectors. This condition is equivalent to choosing basis vectors such that given any four, the number of simultaneously nonzero real fermion components is even,

$$\sum_{\{l:\,\text{real}\}} 8 W_i^l W_j^l W_k^l W_n^l \in \mathrm{Z} \qquad \text{for any } i, j, k, n. \tag{3.53b}$$

This does not in general mean that the entire set of boundary conditions can be diagonalized with complex fermions. For example, (using a subscript r to denote the real components) the vectors,

$$\begin{aligned}
\boldsymbol{W}_1 &= \left((0)^4 \middle| \left(\tfrac{1}{2}\right)_\mathrm{r}^{16} (0)_\mathrm{r}^{16}\right),\\
\boldsymbol{W}_2 &= \left((0)^4 \middle| \left(\tfrac{1}{2}\right)_\mathrm{r}^{8} (0)_\mathrm{r}^{8} \left(\tfrac{1}{2}\right)_\mathrm{r}^{8} (0)_\mathrm{r}^{8}\right),\\
\boldsymbol{W}_3 &= \left((0)^4 \middle| \left(\tfrac{1}{2}\right)_\mathrm{r}^{4} (0)_\mathrm{r}^{4} \left(\tfrac{1}{2}\right)_\mathrm{r}^{4} \cdots \left(\tfrac{1}{2}\right)_\mathrm{r}^{4} (0)_\mathrm{r}^{4}\right),\\
\boldsymbol{W}_4 &= \left((0)^4 \middle| \left(\tfrac{1}{2}\right)_\mathrm{r}^{2} (0)_\mathrm{r}^{2} \cdots \left(\tfrac{1}{2}\right)_\mathrm{r}^{2} (0)_\mathrm{r}^{2}\right),\\
\boldsymbol{W}_5 &= \left((0)^4 \middle| \left(\tfrac{1}{2}\right)_\mathrm{r} (0)_\mathrm{r} \left(\tfrac{1}{2}\right)_\mathrm{r} \cdots \left(\tfrac{1}{2}\right)_\mathrm{r} (0)_\mathrm{r}\right),
\end{aligned} \tag{3.54}$$

which will be used to construct a ten-dimensional heterotic type model in sect. 4, satisfy the above conditions but generate a set of boundary conditions which cannot simultaneously be diagonalized using complex fermions. In the case of models which can be formulated with either complex or real fermions we often use both the complex and real notations interchangeably as is convenient.

Given the constraint (3.53b) it is easy to see that (3.3) and (3.4) are still correct expressions for the condition of modular invariance if the inner product of $\boldsymbol{W}$ vectors is defined to include a factor of $\frac{1}{2}$ for the components representing real fermions, e.g. with the vectors defined in (3.54) we have $\boldsymbol{W}_1 \cdot \boldsymbol{W}_2 = 1$. Also (3.53) guarantees that the operators $\boldsymbol{W}_i \cdot \hat{\boldsymbol{N}}_{\overline{\alpha W}}$ (where the inner product is taken without $\frac{1}{2}$ factors) with different values of i commute with each other (as they did in the complex case) so that, in fact, the entire previous analysis leading to (3.32)–(3.36) carries through in this more general case.

To summarize the conditions for constructing ten-dimensional closed string models we have eqs. (3.32)–(3.36) with (in full generality):

$$\boldsymbol{W}_i \cdot \boldsymbol{W}_j \equiv \sum_{\{l:\ \text{complex}\}} \varepsilon(l) W_i^l W_j^l + \tfrac{1}{2} \sum_{\{l:\ \text{real}\}} \varepsilon(l) W_i^l W_j^l,$$

$$\varepsilon(l) = +(-) \qquad \text{for } l\text{: left(right)}.$$

$$\text{heterotic: } \boldsymbol{W}_i = \left((a_i)_r^8 | U_i^1, U_i^2, \dots U_i^s, r_i^1, r_i^2, \dots r_i^{32-2s} \right),$$

$$a_i, r_i^l \in \{0, \tfrac{1}{2}\}; \qquad 0 \leqslant U_i^l < 1,$$

$$s_i = a_i,$$

$$8 \sum_l r_i^l r_j^l r_k^l r_n^l \in \mathrm{Z} \quad \text{for all } i, j, k, n, \tag{3.55}$$

$$\text{type II: } W_i = \left((a_i)_r^8 | (b_i)_r^8 \right),$$

$$a_i, b_i \in \{0, \tfrac{1}{2}\},$$

$$s_i = a_i + b_i; \tag{3.56}$$

$$\text{bosonic: } \boldsymbol{W}_i = \left(U_i^1, U_i^2, \dots U_i^x, r_i^1, \dots r_i^{32-2x} | \tilde{U}_i^1, \dots \tilde{U}_i^y, \tilde{r}_i^1, \dots \tilde{r}_i^{32-2y} \right),$$

$$0 \leqslant U_i^l, \tilde{U}_i^l < 1; \qquad r_i^l, \tilde{r}_i^l \in \{0, \tfrac{1}{2}\},$$

$$s_i = 0 \qquad \text{with overall } (-) \text{ added},$$

$$8\left(\sum_l r_i^l r_j^l r_k^l r_n^l + \tilde{r}_i^l \tilde{r}_j^l \tilde{r}_k^l \tilde{r}_n^l \right) \in \mathrm{Z} \qquad \text{for all } i, j, k, n. \tag{3.57}$$

4. Ten-dimensional models

$Z_{\text{fermion}}(\tau)$ in (3.32) represents the most general fermionic contribution to the one-loop path integral which is modular invariant and includes a sensible projection onto a subspace of states which have correct space-time statistics. Any choice of basis vectors $\boldsymbol{W}_i$ and structure constants k_{ij} consistent with the constraints (3.33)–(3.35) and (3.53) defines a closed fermionic string model. In this section we will show how to extract the physical spectra of these models, in particular the massless particle content and gauge group, from the information contained in (3.32)–(3.36). For now we restrict ourselves to constructing ten-dimensional models. Experience in analyzing these relatively simple examples will prove quite useful when we consider, in sect. 6, string models in fewer than ten dimensions. All of the ten-dimensional models which we will consider here were constructed using essentially the same methods in ref. [11]; readers familiar with that work should bear in mind that in order to present the more general formalism given here we have employed somewhat different notations from those used in ref. [11].

The complete one-loop amplitude is obtained by multiplying (3.32) by the corresponding contribution from the two-dimensional bosonic fields (string coordinates in light cone gauge) and integrating the result over τ in the fundamental domain of the modular group ($\tau = \tau_1 + i\tau_2$: $-\frac{1}{2} \leqslant \tau_1 \leqslant \frac{1}{2}$; $\tau_2 \geqslant 0$; $|\tau| \geqslant 1$). The hamiltonian term in the exponent within the trace of this total expression is of the form,

$$2\pi i\tau_1\left(\hat{H}^{\text{left}}_{\overline{\alpha W}} + \hat{H}^{\text{left}}_{\text{bosonic}} - \hat{H}^{\text{right}}_{\overline{\alpha W}} - \hat{H}^{\text{right}}_{\text{bosonic}}\right)$$

$$-2\pi\tau_2\left(\hat{H}^{\text{left}}_{\overline{\alpha W}} + \hat{H}^{\text{left}}_{\text{bosonic}} + \hat{H}^{\text{right}}_{\overline{\alpha W}} + \hat{H}^{\text{right}}_{\text{bosonic}}\right), \tag{4.1}$$

where

$$\hat{H}^{\text{right}}_{\text{bosonic}} = p^2 + \sum_{i=1}^{8}\left\{\sum_{n=1}^{\infty} n a^{i\dagger}_n a^i_n - \tfrac{1}{24}\right\},$$

$$\hat{H}^{\text{left}}_{\text{bosonic}} = p^2 + \sum_{i=1}^{8}\left\{\sum_{n=1}^{\infty} n \tilde{a}^{i\dagger}_n \tilde{a}^i_n - \tfrac{1}{24}\right\}, \tag{4.2}$$

$[\tilde{a}^{i\dagger}_n, \tilde{a}^j_m] = [a^{j\dagger}_n, a^j_m] = \delta_{mn}\delta^{ij}$, others vanish, and p^2 is the space-time momentum squared. From (4.1), the contribution of any state to the one-loop amplitude will vanish when we integrate over τ_1 unless when acting on that state $\hat{H}^{\text{left}} = \hat{H}^{\text{right}}$. Physically this guarantees that the theory is invariant under shifts of the world sheet parameter σ_1 by any constant σ_c since the operator which performs such shifts is $\exp[i\sigma_c(\hat{H}^{\text{left}} - \hat{H}^{\text{right}})]$. From (4.1) the (mass)^2 of a state is the eigenvalue of $\frac{1}{2}(\hat{H}^{\text{left}} + \hat{H}^{\text{right}} - 2p^2)$ or, given the above result, $m^2 = m^2(\text{left}) = m^2(\text{right}) = H^{\text{left}} - p^2 = H^{\text{right}} - p^2$. Explicitly in terms of occupation numbers for the various excitations we have in a physical sector characterized by $\overline{\alpha W}$,

$$m^2 = m^2(\text{left}, n, \bar{n}, \tilde{M}) = m^2(\text{right}, n, \bar{n}, M), \tag{4.3}$$

$$m^2(\text{left}, n, \bar{n}, \tilde{M}) = \sum_{\{l:\ \text{left}\}}\left\{E_{\overline{\alpha W}^l} + \sum_{q=1}^{\infty}\left[\left(q - \overline{\alpha W}^l\right)\bar{n}^l_q + \left(q + \overline{\alpha W}^l - 1\right)n^l_q\right]\right\}$$

$$-\tfrac{1}{24}(D-2) + \sum_{i=1}^{D-2}\sum_{q=1}^{\infty} q\tilde{M}^i_q \tag{4.4}$$

and similarly for $m^2(\text{right}, n, \bar{n}, M)$ except $\tilde{M}$ is replaced by M and the sum over l runs over those values denoting right movers. $E_{\overline{\alpha W}^l}$ is the vacuum energy from the fermions as given in (2.13); for l labeling real fermions this is replaced by $\frac{1}{2}E_{\overline{\alpha W}^l}$ and the $\bar{n}^l_q$'s are taken to be identically zero. D is the space-time dimension and the $(D-2)$ term gives the contribution of the string coordinates to the vacuum energy ($-\frac{1}{3}$ in the present case). The fermion occupation numbers (which can take values 0 or 1) are related to the fermion number defined in (2.11) by,

$$N^l_{\overline{\alpha W}} = \sum_q n^l_q - \bar{n}^l_q . \tag{4.5}$$

Given a set of W_i and k_{ij} satisfying (3.33)–(3.35) it is straightforward to determine the mass spectrum of the physical states of the model. For a given sector characterized by $\overline{\alpha W}$ the allowed occupation numbers are given by (4.5) subject to (3.36), and the (mass)^2 for this choice of occupation numbers is then given by (4.4) subject to the constraint (4.3). Finally to generate the complete spectrum we must consider the sector $\overline{\alpha W}$ in turn for each value of α (that is all sets of values of $\{\alpha_i\}$ with $\alpha_i \in (0, 1, \dots m_i - 1)$). For a large number of generating vectors, W_i, this can be a long process. In practice, however, we can often make use of symmetries of the set $\{W_i\}$ to speed up the process, for example by showing that several sectors contribute the same spectrum of particles. In any case we are usually only interested in obtaining a small portion of the particle spectrum, for example the tachyons, chiral fermions, or massless gauge bosons, in which case the majority of sectors $\overline{\alpha W}$ will often not contribute.

Let us begin with the simplest possible example, choosing only the single vector W_0 to specify the boundary conditions. Since this vector must be present in any set of boundary conditions this represents the only possible single vector case. We may employ either the complex or real formulation described above, i.e., take either $W_0 = ((\frac{1}{2})^4 | (\frac{1}{2})^{16})$ or $W_0 = ((\frac{1}{2})^8_r | (\frac{1}{2})^{32}_r)$ or in principle some mixture of the two. In general the purely real formulation is the most convenient, when we can use it, because it provides for a more symmetric construction of the physical states. For example in the present case the right-moving excitations form representations of SO(8) (the little group in ten dimensions for massless particles) and the left-moving excitations in the real formulation naturally fall into representations of SO(32), whereas in the purely complex formulation the states fall into representations of the U(16) subgroup. The resulting models (and indeed the application of eqs. (3.33)–(3.35)) is, of course, independent of the formulation chosen. For convenience we begin with the purely real formulation.

To determine the particle content of this simple theory we first need to find the values of k_{00} consistent with (3.33)–(3.35). We have $W_0 \cdot W_0 = \frac{1}{2}(32 \cdot \frac{1}{4} - 8 \cdot \frac{1}{4}) = 3$, $m_0 = 2$ and $s_0 = \frac{1}{2}$ so (3.33)–(3.35) are satisfied for either $k_{00} = 0$ or $\frac{1}{2}$. The k_{ij} only appear in (3.33)–(3.36) and thus need only be defined mod 1. There are two physical sectors in this theory given by $\alpha_0 = 0$ or 1. Consider first the $\boldsymbol{W}_0$ sector which (being a Neveu-Schwarz sector for the space-time right movers) contains the bosonic particles of the model. The total vacuum energies for right and left movers is seen to be (cf. (4.4) and (2.13)) $-\frac{1}{2}$ and -1 respectively or, defining a shorthand notation, $[-\frac{1}{2}, -1]$. To these right and left vacuum levels we may build up the (mass)^2 of a physical state by adding integer quanta (nonzero M_q or $\tilde{M}_q$ in (4.4)) or half integer quanta (nonzero n_q^l), subject to the constraints (4.3) and (3.36). The latter becomes, in the $\boldsymbol{W}_0$ sector with purely real formalism,

$$\frac{1}{2}\sum_{l=1}^{40}\sum_{q=1}^{\infty} n_q^l = \frac{1}{2} \qquad (\text{mod } 1)\,. \tag{4.6}$$

The signs appearing in (4.6) are of no significance since $-\frac{1}{2} = \frac{1}{2}$ (mod 1).

The lowest-lying physical states are tachyons with $m^2 = -\frac{1}{2}$ which are obtained by combining the right-moving vacuum state with a left-moving state with a single $\frac{1}{2}$ quanta, i.e., one of the n_1^l, $l = 9, \dots 40$, equal to one. These 32 states, $|0\rangle_R \otimes b_{1/2}^{l\dagger}|0\rangle_L$, clearly form an SO(32) vector representation, and are space-time scalars. The next lowest-lying physical states, for which there are two distinct possibilities, are massless. For the right movers we require one of the n_1^l ($l = 1, \dots 8$) nonzero; for the left movers we need either one of the $\tilde{M}_1^i = 1$, ($i = 1, \dots 8$) or two of the n_1^l ($l = 9, \dots 40$) nonzero. In the former case, $b_{1/2}^{l\dagger}|0\rangle_R \otimes \tilde{a}_1^{i\dagger}|0\rangle_L$, we have spacetime tensor states (the graviton, antisymmetric tensor and dilaton); in the latter case, $b_{1/2}^{l\dagger}|0\rangle_R \otimes b_{1/2}^{[i\dagger}b_{1/2}^{j]\dagger}|0\rangle_L$ ($i, j = 9, \dots 40$ and antisymmetrized), we have massless space-time vectors in the adjoint representation of SO(32), i.e., gauge bosons.

Adding ever more excitations we could proceed as above to construct an infinite tower of massive bosonic states; instead, let us briefly consider the fermionic states from the $\boldsymbol{0}(\alpha_0 = 0)$ sector. The vacuum energies are $[0, 1]$ so all of the physical fermionic states are massive. Eq. (3.36) becomes,

$$\frac{1}{2}\sum_{l=1}^{40}\sum_{q=1}^{\infty} n_q^l = \frac{1}{2} + k_{00} \qquad (\text{mod } 1)\,. \tag{4.7}$$

The states in this sector are both space-time and SO(32) spinors (or taking nonzero M_q^i or $\tilde{M}_q^i$ we obtain space-time spinoreal vectors). As described in the previous section the spinor structure arises from the Ramond zero modes which form a Clifford algebra. The two possible values for k_{00} (0 or $\frac{1}{2}$) just distinguish the two eigenvalues of the product of all the zero mode operators (the analog of γ^5) and thus do not represent physically distinct models, but just fix our chirality conventions. Recalling that in the $\boldsymbol{W}_0$ sector the k_{00} dependence drops out entirely (cf. (4.6)) we see that the two choices for k_{00} give the same physical model. This represents the simplest example of a very general feature of our method of construction of string models. Many different choices for k_{ij} and W_i (that is many different sets of boundary conditions) produce the same physical string models. This is simply a manifestation of the fact that we can choose many different fermionic representations for a single gauge group.

To construct other string models we must add additional vectors W_i to the minimal case with only W_0. In general the addition of a generating vector has two somewhat compensating effects on the model under construction. On the one hand the fermion occupation numbers must, from (3.36), satisfy an additional constraint so that some states originally present may now be projected out. On the other hand we have increased the number of physical sectors from which we may obtain new physical states. To illustrate these effects consider the vector $W_1 = ((0)^8_r|(\frac{1}{2})^{32}_r)$ along with W_0. We have $W_1 \cdot W_1 = W_1 \cdot W_0 = 4$, $m_1 = 2$, and $s_1 = 0$. To satisfy (3.33)–(3.35) we need $k_{10} = k_{01} = k_{11} = 0$ or $\frac{1}{2}$. As before we have the W_0 sector with the constraint (4.6) but now in addition we must satisfy (from (3.36)),

$$W_1 \cdot N_{W_0} = \tfrac{1}{2} \sum_{l=9}^{40} \sum_{q=1}^{\infty} n^l_q = 0 \qquad (\text{mod } 1)\,. \tag{4.8}$$

In particular this serves to project out the tachyon states which were present leaving the model tachyon free. Eq. (4.8) does not alter the massless boson spectrum, in particular the gauge group is still SO(32). The **0** sector contains massive fermions, as before. We now have two additional physical sectors, W_1 and $W_1 + W_0$. The latter provides some additional massive bosons; the former gives massless (as well as massive) fermions. Specifically in the W_1 sector the vacuum energy is $[0, -1]$ and the constraints from (3.36) are,

$$\tfrac{1}{2} \sum_{l=1}^{40} \sum_{q} n^l_q = \tfrac{1}{2} + k_{01} + k_{00} \qquad (\text{mod } 1)\,,$$

$$\tfrac{1}{2} \sum_{l=9}^{40} \sum_{q} n^l_q = 0 \qquad (\text{mod } 1)\,. \tag{4.9}$$

Thus we have in particular 496 massless Majorana-Weyl fermions in the adjoint representation of SO(32). All sectors taken together give the SO(32) heterotic string model (for any choice of k_{00} and k_{01}). The quickest signal from which to discern the supersymmetry of this (or any such) model is the presence of a massless spin-$\frac{3}{2}$ gravitino. In this case the gravitino appears in the W_1 sector, the right-moving Ramond ground state providing a space-time spinor index while a single $\tilde{a}^{i\dagger}_1$ operating on the left-moving ground state gives the space-time vector index.

We can obtain different gauge groups by adding additional vectors. For example take W_0 and W_1 from above and in addition $W_2 = ((0)^8_r|(\frac{1}{2})^{16}_r(0)^{16}_r)$. The k_{ij}'s must be 0 or $\frac{1}{2}$ and from (3.33)–(3.35) constrained to be of the form,

$$k_{ij} = \begin{bmatrix} k_{00} & k_{10} & k_{20} \\ k_{10} & k_{10} & k_{21} \\ k_{20} & k_{21} & k_{20} \end{bmatrix} \quad (\text{mod } 1)\,. \tag{4.10}$$

We restrict ourselves to considering the gauge bosons. These arise from three sectors W_0, $W_0 + W_2$ and $W_0 + W_1 + W_2$. The associated vacuum energies are $[-\frac{1}{2}, -1]$, $[-\frac{1}{2}, 0]$ and $[-\frac{1}{2}, 0]$ respectively. In each sector the physical states are arranged in representations of SO(8) ⊗ SO(16) ⊗ SO(16). For a gauge boson we must have one of the n_1^l $(l = 1, \dots 8)$ nonzero (all others vanishing) to give an SO(8) vector. Given this the constraints on the left-moving fermion numbers (which will determine the SO(16) ⊗ SO(16) representations) are, in the W_0 sector,

$$\frac{1}{2} \sum_{l=9}^{40} n_1^l = 0 \quad (\text{mod } 1)\,,$$

$$\frac{1}{2} \sum_{l=9}^{24} n_1^l = 0 \quad (\text{mod } 1)\,. \tag{4.11}$$

To satisfy (4.11) and obtain a massless state we can take either two of the first 16 left moving n_1^l nonzero or two of the last 16 nonzero but not one of each. Thus regardless of the values of k_{ij} in (4.10) the W_0 sector contributes gauge bosons in the adjoint representation of SO(16) ⊗ SO(16).

In the $W_0 + W_2$ sector (again taking one of the n_1^l $(l = 1, \dots 8)$ nonzero) we must have $n_1^l = 0$ for $l = 25, \dots 40$ to produce a massless state in which case the constraints become,

$$\frac{1}{2} \sum_{l=9}^{24} n_1^l = k_{20} \quad (\text{mod } 1)\,,$$

$$\frac{1}{2} \sum_{l=9}^{24} n_1^l = k_{21} \quad (\text{mod } 1)\,. \tag{4.12}$$

If $k_{21} \neq k_{20}$ then the two constraints are incompatible in which case the $W_0 + W_2$ sector (and by symmetry the $W_0 + W_1 + W_2$ sector) do not contribute any gauge bosons and we have in fact the SO(16) × SO(16) string model. If $k_{20} = k_{21}$ then the first 16 left movers (which have periodic σ_1 boundary condition and thus form a Ramond sector) give the 128-dimensional spinor representation of SO(16). By symmetry then, we have from the three sectors gauge bosons, in the SO(16) ⊗ SO(16) representations, $(120, 1) \oplus (1, 120) \oplus (128, 1) \oplus (1, 128)$. Taken together these form the 496-dimensional adjoint representation of $E_8 \otimes E_8$. If one went on to examine the fermionic sectors one would find that this construction gives the familiar $E_8 \otimes E_8$ heterotic string model.

In general we can determine the gauge symmetry of a given model using the methods sketched above. Checking each sector giving gauge bosons we obtain a direct sum of products of SO($2n$) representations (or U(n) representations in the complex case) which combine to give the adjoint representation of some larger group. This fact – that the massless vector bosons always fall into the adjoint representation of some group (see sect. 7 for the proof in the complex case) – along with a knowledge of the general structure of the $\boldsymbol{W}_i$'s and the total number of gauge bosons is often sufficient to identify the gauge group. There is, however, a more systematic and unambiguous (though lengthier) method of determining the gauge group as well as the representations of the other particles in the spectrum. In the complex formulation for the left movers (when we can employ it) we can in general label the massless vector states by their associated U(1) charges, i.e., their $[U(1)]^{16}$ representations. This corresponds precisely to specifying the coordinates of the root vectors of the gauge group in an orthonormal basis. For example consider the $E_8 \otimes E_8$ model examined above, in particular the contribution of the first 8 complex left-moving fermions (since the first and last 8 clearly factor). The $\boldsymbol{W}_0$ sector contributes the gauge bosons

$$b^{l\dagger}_{1/2}|0\rangle_{\mathrm{R}} \otimes \left(\tilde{b}^{i\dagger}_{1/2}\tilde{d}^{j\dagger}_{1/2}|0\rangle_{\mathrm{L}} \oplus \tilde{b}^{[i\dagger}_{1/2}\tilde{b}^{j]\dagger}_{1/2}|0\rangle_{\mathrm{L}} \oplus \tilde{d}^{[i\dagger}_{1/2}\tilde{d}^{j]\dagger}_{1/2}|0\rangle_{\mathrm{L}}\right)$$

with $i, j = 1, \dots 8$. Denoting the basis vectors by e_i ($i = 1, \dots 8$) and noting that in this sector the vacuum U(1) charges are all zero (cf. (2.12)) we have the root vectors $\pm e_i \pm e_j$ for $i > j$ and eight zero roots (representing the Cartan subalgebra) for $i = j$ to give the 120 root vectors of the group SO(16). The $\boldsymbol{W}_0 + \boldsymbol{W}_2$ sector gives gauge bosons of the form $b^{l\dagger}_{1/2}|0\rangle_{\mathrm{R}} \otimes b^{i\dagger}_0 b^{j\dagger}_0 \dots b^{k\dagger}_0|0\rangle_{\mathrm{L}}$ where the number of zero mode creation operators ($i \neq j \neq k$ etc.) is either always even or always odd depending on whether $k_{20} = k_{21} = 0$ or $\frac{1}{2}$. The vacuum charge for each of the eight U(1)'s is $-\frac{1}{2}$ so depending on whether $b^{i\dagger}_0$ is present or not the ith U(1) charge is either $+\frac{1}{2}$ or $-\frac{1}{2}$. Thus from this sector we have the root vectors $\frac{1}{2}(\pm e_1 \pm e_2 \dots \pm e_8)$ (with the number of + signs appearing either always even if $k_{20} = 0$ or always odd if $k_{20} = \frac{1}{2}$). Combined with the SO(16) root vectors from the $\boldsymbol{W}_0$ sector this gives the 248 root vectors of the group E_8. If necessary we could in similar fashion explicitly construct the weight vectors for any set of physical states to determine the group representation.

So far we have considered sets of boundary conditions which could be diagonalized with either complex or real fermions. Now let us consider some cases where we do not have this freedom. Take the set of $\boldsymbol{W}_i$ given in (3.54) along with $\boldsymbol{W}_0$ (as always). To diagonalize the boundary conditions generated by this set requires real fermions. We note in passing that eliminating one or more of the vectors in (3.54) leaves a set which could be considered in the complex formulation and yields tachyonic models with gauge groups: $SO(16)\times E_8$ if only one of the vectors in (3.54) is included; $SO(8)\otimes SO(24)$ if we employ two vectors from (3.54); $(E_7\times SU_2)^2$ for three vectors; and U(16) if we use all but one of the vectors in (3.54). Including now the entire set, the constraints (3.33)–(3.35) are easy to impose given the symmetry of the vectors in (3.54). The matrix $\boldsymbol{W}_i\cdot\boldsymbol{W}_j$ is equivalent to $\delta_{i0}\delta_{j0}$ (since for (3.33)–(3.36) we need only define the diagonal elements mod 2 and the off diagonal elements mod 1). The k_{ij} matrix is then symmetric with $k_{ii}=k_{i0}$, all elements equal to 0 or $\frac{1}{2}$ and otherwise unrestricted. To determine the gauge boson spectrum consider first the $\boldsymbol{W}_0$ sector. The constraints (3.36) become,

$$\boldsymbol{W}_0\cdot\boldsymbol{N}_{\boldsymbol{W}_0}=\tfrac{1}{2}\qquad(\mathrm{mod}\ 1),$$

$$\boldsymbol{W}_i\cdot\boldsymbol{N}_{\boldsymbol{W}_0}=0,\qquad i=1,2,\ldots 5\qquad(\mathrm{mod}\ 1).\tag{4.13}$$

Taking one and only one of the right moving n_1^l equal to one ($l=1,\ldots 8$) to obtain a spacetime vector state we see that (4.13) implies that all of the left moving fermion occupation numbers must be zero, making it impossible to construct a massless vector state in this sector. Notice that for any $\{\boldsymbol{W}_i\}$ the constraints in the $\boldsymbol{W}_0$ sector always take the form above, that is $\boldsymbol{W}_i\cdot\boldsymbol{N}_{\boldsymbol{W}_0}=s_i$ (mod 1), so that the spectrum in this sector is always easily obtained by inspection.

The remaining possibilities for gauge bosons are the sectors $\overline{\alpha \boldsymbol{W}}$ with $\alpha_0=1$ and at least some of the other α_i's nonvanishing. There are 31 such sectors each with vacuum energy $[-\frac{1}{2},0]$ and, by symmetry, each contributing the same number of gauge bosons. We need only consider one of these sectors explicitly, for example $\boldsymbol{W}^*=\overline{\boldsymbol{W}_0+\boldsymbol{W}_1}=((\frac{1}{2})_{\mathrm{r}}^{8}|(0)_{\mathrm{r}}^{16}(\frac{1}{2})_{\mathrm{r}}^{16})$. In this sector to obtain massless vector states we again need to excite a single right-moving $\frac{1}{2}$ quantum and in addition must take the last 16 real left-moving fermions in their (Neveu-Schwarz) ground state (to obtain a massless state). The remaining constraints from (3.36) on the first 16 left-moving fermions which are in Ramond ground states are,

$$\boldsymbol{W}_1\cdot\boldsymbol{N}_{\boldsymbol{W}^*}=\tfrac{1}{4}(1+\gamma^1\gamma^2\ldots\gamma^{16})=k_{11},$$

$$\boldsymbol{W}_2\cdot\boldsymbol{N}_{\boldsymbol{W}^*}=\tfrac{1}{4}(1+\gamma^1\gamma^2\ldots\gamma^{8})=k_{21},$$

$$\boldsymbol{W}_3\cdot\boldsymbol{N}_{\boldsymbol{W}^*}=\tfrac{1}{4}(1+\gamma^1\gamma^2\gamma^3\gamma^4\gamma^9\gamma^{10}\gamma^{11}\gamma^{12})=k_{31},$$

$$\boldsymbol{W}_4\cdot\boldsymbol{N}_{\boldsymbol{W}^*}=\tfrac{1}{4}(1+\gamma^1\gamma^2\gamma^5\gamma^6\gamma^9\gamma^{10}\gamma^{13}\gamma^{14})=k_{41},$$

$$\boldsymbol{W}_5\cdot\boldsymbol{N}_{\boldsymbol{W}^*}=\tfrac{1}{4}(1+\gamma^1\gamma^3\gamma^5\gamma^7\gamma^9\gamma^{11}\gamma^{13}\gamma^{15})=k_{51},\tag{4.14}$$

where the γ^i are proportional to the Ramond zero modes $\gamma^i = \sqrt{2}\, d_0^i$ (as in sect. 3) and the constraints should be interpreted as acting on states, e.g., projecting out those Ramond states with $\gamma^1\gamma^2 \dots \gamma^{16}$ eigenvalue of $+1$ or -1 depending on the value of k_{11}. The Ramond ground states satisfying the first constraint are in one of the two 128-dimensional SO(16) spinor representations (again depending on the value of k_{11}). The operators for the different constraints in (4.14) are all independent and mutually commute; each projects out half of the spinor states, leaving 8 states which satisfy (4.14) (regardless of the particular choice for k_{11}). Each of the other 30 sectors with this same structure also contribute 8 massless vector states giving in total the 248 gauge bosons of a single E_8.

Unlike all of the other ten-dimensional models we have constructed, the gauge group of this single E_8 model has rank 8 rather than 16. This reduction of rank was made possible by our choice of boundary conditions which necessitated the use of single real fermions. In general if the boundary conditions can be diagonalized with complex fermions (16 left movers in ten dimensions) then the $\boldsymbol{W}_0$ sector will produce 16 U(1) gauge bosons and the rank of the gauge group (including gauge bosons from all sectors) will be 16. The converse is not true; boundary conditions which cannot be diagonalized in a complex basis do not guarantee a model with gauge group rank less than 16. For example including the vector $((0)_\mathrm{r}^8|(\frac{1}{2})_\mathrm{r}^{32})$ with the set in (3.54) gives rise to the SO(32) heterotic or SO(16) ⊗ SO(16) string model (depending on the choice of k_{ij}'s).

To complete this section let us consider sets of boundary conditions requiring the complex formulation. Take for instance $\boldsymbol{W}_0$ and $((0)_\mathrm{r}^8|(\frac{1}{3})^{12}(0)_\mathrm{r}^8)$; the left movers are grouped into twelve complex fermions and eight real ones. The right moving space time degrees of freedom are treated as before. The physical states will naturally fall into representations of U(12) ⊗ SO(8). The matrix $\boldsymbol{W}_i \cdot \boldsymbol{W}_j$ is (mod 2),

$$\boldsymbol{W}_i \cdot \boldsymbol{W}_j = \begin{bmatrix} 1 & 0 \\ 0 & \frac{4}{3} \end{bmatrix}. \tag{4.15}$$

Eqs. (3.33)–(3.35) requires the k_{ij} matrix to be (note $m_1 = 3$),

$$k_{ij} = \begin{bmatrix} k_{00} & 0 \\ 0 & \frac{2}{3} \end{bmatrix}, \qquad k_{00} = 0 \text{ or } \tfrac{1}{2}. \tag{4.16}$$

There are six sectors in this model; we will consider the three which give space-time bosons, i.e. $\overline{\boldsymbol{W}_0 + \alpha_1 \boldsymbol{W}_1}$ with $\alpha_1 = 0, 1, 2$. The vacuum energies for these three sectors are respectively $[-\frac{1}{2}, -1]$, $[-\frac{1}{2}, -\frac{1}{3}]$, and $[-\frac{1}{2}, -\frac{1}{3}]$. In the $\boldsymbol{W}_0$ sector we have the constraints,

$$\tfrac{1}{2}\left(\sum_{1}^{8} n_1^l + \sum_{1}^{20} \tilde{N}^l\right) = \tfrac{1}{2}(\text{mod } 1)\,,$$

$$\tfrac{1}{3}\sum_{1}^{12} \tilde{N}^l = \tfrac{1}{3}\sum_{l}^{12}\sum_{q}\left(\tilde{n}_q^l - \tilde{\tilde{n}}_q^l\right) = 0 \qquad (\text{mod } 1)\,. \tag{4.17}$$

These have tachyonic solutions (all occupation numbers zero except one of the $\tilde{n}_1^l$ for $l = 13, \ldots 20$) in the $(1, 8_v)$ representation of U(12) ⊗ SO(8). These are easily seen to be the only tachyons in the theory. For gauge bosons we take on the $n_1^l = 1$ and either two of the $\tilde{n}_1^l$ ($l = 13, \ldots 20$) equal to 1 (giving (1, 28)) or one of the $\tilde{n}_q^l$ and one of the $\tilde{\tilde{n}}_1^l$ ($l = 1, \ldots 12$) equal to one giving the (144, 1) of U(12) ⊗ SO(8). No other combinations in this sector give massless vector states satisfying (4.17). Turning to the gauge bosons in the $\overline{W_0 + W_1} = ((\tfrac{1}{2})^8_r|(\tfrac{5}{6})^{12}(\tfrac{1}{2})^8_r)$ sector we must satisfy the constraints (having already taken one of the $n_1^l = 1$),

$$\tfrac{1}{2}\left(\sum_{1}^{12} \tilde{n}_1^l - \tilde{\tilde{n}}_1^l + \sum_{13}^{20} \tilde{n}_1^l\right) = 0 \qquad (\text{mod } 1)\,,$$

$$\tfrac{1}{3}\sum_{1}^{12}\left(\tilde{n}_1^l - \tilde{\tilde{n}}_1^l\right) = \tfrac{1}{3} \qquad (\text{mod } 1)\,. \tag{4.18}$$

In this sector the fermions and antifermions with occupation numbers $\tilde{n}_1^l$ and $\tilde{\tilde{n}}_1^l$ ($l = 1, \ldots 12$) respectively contribute quanta of $\tfrac{5}{6}$ and $\tfrac{1}{6}$ while the remaining real fermions ($\tilde{n}_1^l$, $l = 13, \ldots 20$) carry a quantum of $\tfrac{1}{2}$. To achieve a massless state for the left movers (for which the vacuum energy is $-\tfrac{1}{3}$ as noted previously) the only possibility is to add two $\tfrac{1}{6}$ quanta, i.e. take two of the $\tilde{\tilde{n}}_1^l = 1$. These 66 states do in fact satisfy (4.18) and so contribute gauge bosons in the $(\overline{66}, 1)$ representation of U(12) ⊗ SO(8). Similarly one can readily verify that the $\overline{W_0 + 2W_1}$ sector contributes gauge bosons in the representation (66, 1). Collecting together the massless vector states of the three sectors one finds the gauge bosons of the group SO(24) × SO(8).

This model we have just constructed is in fact just the SO(8) ⊗ SO(24) model previously constructed using only periodic and antiperiodic boundary conditions. This is in fact a general feature of ten-dimensional model construction; introducing W_i with components other than 0 or $\tfrac{1}{2}$ does not appear to give any new heterotic type models. This should not be particularly surprising. In a ten-dimensional heterotic type model the spectrum of right movers is constrained to be integral or half-integral. For physical states the spectrum from left and right movers must coincide and thus any states with $(\text{mass})^2$ of say $\tfrac{1}{3}$ or $\tfrac{2}{7}$ cannot contribute to the physical spectrum. This is not the case for purely bosonic models (in less than 26 dimensions) for for type II or heterotic type models in fewer than ten dimensions. These latter cases will be considered in sect. 6.

5. Construction of fermionic string models in less than ten dimensions

Here we will generalize the spin structure construction to fermionic string models in less than 10 dimensions. Our discussion again concentrates on the heterotic string. In general, compactification of space coordinates gives rise to internal symmetries. In particular, under toroidal compactification of six space dimensions, the $N=1$ supersymmetric heterotic string models become four-dimensional $N=4$ supersymmetric string models. This is phenomenologically unappealing, since $N=4$ supersymmetric models are non-chiral. To obtain 4-dimensional chiral models, the compactification must give models with $N=1$ supersymmetry or non-supersymmetric models. Cutting down the space-time supersymmetry resembles reducing the rank of the internal gauge symmetry. In analyzing ten-dimensional models, we learned that the rank of the gauge group can be reduced (i.e. in the single E_8 model) by replacing the internal coordinates by fermions with different spin structures. Therefore, we are led by similar considerations to fermionize the compactified space coordinates. For the left-moving coordinates in the heterotic string, the fermionization of the compactified coordinates is exactly the same as the fermionization of the 16 left-moving bosonic coordinates of the 10-dimensional model. For the right movers, the fermionization must be carried out in a way such that the world-sheet supersymmetry is maintained. This fermionization turns out to be a well-known procedure. [19]

The derivation of the constraints on the basis vectors W_i and the rules in sect. 3 is applicable in any dimension. The only new point to consider is the choice of W_i. We expect, fermionizing in an obvious fashion, that each compactified string coordinate will give rise to two real left-moving fermions and three real right-moving ones (the original RNS fermion along with two real fermions from fermionizing the bosonic coordinate). Thus in D dimensions in a real formulation the W_i are $2(14-D)+2(26-D)$ dimensional vectors. Now recall that in 10 dimensions the eight right-moving fermions were constrained to all share the same boundary conditions – either periodic or antiperiodic – as a consequence of the world sheet supersymmetry of the action (2.1). In the pertinent interaction terms in (2.1) the right-moving fermions coupled to the string coordinates and to the gravitino. Thus the product of a fermion and a string coordinate, $\lambda^{\mu}\,\partial X^{\mu}$, must have the same boundary conditions as the gravitino for each value of μ; this follows from the requirement that the lagrangian density must be single-valued on the torus. Compactifying and fermionizing some of the string coordinates we then expect the following constraints from world sheet supersymmetry: the right-moving fermions representing space-time degrees of freedom should again all have the same boundary conditions (periodic or antiperiodic) and the product of the triplet of fermions from a compactified dimension should share these same boundary conditions as well. In the fermionic formulation, the super current is given by ($d=10-D$),

$$\tilde{T}_{\mathrm{F}}=i\sum_{j=1}^{D-2}\psi^{j}\,\partial X^{j}+i\sum_{l=1}^{d}\psi_1^l\psi_2^l\psi_3^l\,, \tag{5.1}$$

where $\tilde{T}_{\mathrm{F}}$ must be periodic or antiperiodic on the world sheet torus. Since the space-time X^j are periodic, ψ^j must be periodic (Ramond) or antiperiodic (NS). This imposes a strong constraint on the internal ψ_i^l. This is the general form of the "triplet" constraint. It is clear that the internal fermions ψ_i^l have an $[SO(3)]^d \times S_d$ (permutation) symmetry. For many interesting models, we will impose a more restrictive triplet constraint, which simplifies the analysis considerably; however, it is important to keep in mind the general (and far less restrictive) form of the triplet constraint.

Let us consider the following special cases where the W_i vectors in D dimensions have the form ($d = 10 - D$),

$$W = \left((S)_r^{D-2}(A_1B_1C_1)_r(A_2B_2C_2)_r \ldots (A_dB_dC_d)_r | U_1U_2 \ldots U_x r_1 \ldots r_{52-2x-2D}\right),$$

$$0 \leqslant U_i < 1 ; \qquad r_i = 0, \tfrac{1}{2} . \quad (5.2)$$

Then we have the "triplet constraint",

$$A_l + B_l + C_l = S \qquad (\mathrm{mod}\ 1) . \tag{5.3a}$$

Furthermore, for most of the interesting models in even space-time dimensions we can group pairs of triplets of real fermions represented by $(A_jB_jC_j)$ and $(A_{j+1}B_{j+1}C_{j+1})$ together to form $\frac{1}{2}(10 - D)$ triplets of complex fermions represented by $(a_lb_lc_l)$. In this complex formulation we again have the triplet constraint,

$$a_l + b_l + c_l = S \qquad (\mathrm{mod}\ 1) ,$$

$$a_l, b_l, c_l, s \in Z/2 . \tag{5.3b}$$

Note that there are other ways of pairing up real fermions to form complex ones; a nontrivial example will be given in the following section and a simpler one in sect. 9. The form of the W_i for type II or purely bosonic theories is again found by taking both right and left moving portions to have the same form as the right moving portion or the left moving portion respectively of the W_i for the heterotic models.

The purpose of the remainder of this section is to demonstrate that the fermionization procedure represented by (5.1) and (5.3) is in fact correct. In particular we will show that it preserves the usual supersymmetry algebra and properties of the interaction vertices. In the process we will also discuss Thirring interactions [20] between the world sheet fermions, which are the only interactions which explicitly preserve conformal invariance.

To begin let us review the supercurrent algebra and the super-Virasoro algebra. First, recall the bosonic case [21,22]. A bosonic coordinate $X^\mu(t, \sigma)$, $\mu = 1, 2 \ldots d$ can be decomposed into left- and right-moving parts

$$X^\mu(t, \sigma) = X^\mu(z) + X^\mu(\bar{z}),$$

where, on the euclidean two-dimensional world sheet, $z = \sigma + it$, $\bar{z} = \sigma - it$. The correlation function is given by

$$\langle X^{\mu}(z_1)X^{\nu}(z_2)\rangle = -\delta^{\mu\nu}\ln(z_1 - z_2)\,. \tag{5.4}$$

The current commutation relation, or the current algebra of the current $j^{\mu}_{\mathrm{L}}(z) = \partial_z X^{\mu}(z)$ can be written in operator product form as,

$$j^{\mu}_{\mathrm{L}}(z)\,j^{\nu}_{\mathrm{L}}(w) \sim \frac{-\delta^{\mu\nu}k}{2(z-w)^2} + \cdots\,, \tag{5.5}$$

where the central extension $k = 2$, and regular terms are neglected. The only nonvanishing component of the energy-momentum tensor is given by (normal ordering is implied and $\mu, \nu = 1, 2, \ldots d$),

$$T_{\mathrm{B}}(z) = -\tfrac{1}{2}\delta_{\mu\nu}\,j^{\mu}_{\mathrm{L}}(z)\,j^{\nu}_{\mathrm{L}}(z) \tag{5.6}$$

and the Virasoro algebra is given by

$$T_{\mathrm{B}}(z)T_{\mathrm{B}}(\omega) \sim \frac{d}{2(z-\omega)^4} + \frac{2T_{\mathrm{B}}(\omega)}{(z-\omega)^2} + \frac{1}{z-\omega}\partial_{\omega}T_{\mathrm{B}}(\omega) + \cdots\,, \tag{5.7}$$

where the central charge d is the dimension (i.e. the central charge of a scalar field is one). Fermionization is easily carried out by replacing each bosonic field X^{μ} by two real (or one complex) fermions ψ^{μ}_1 and ψ^{μ}_2:

$$j^{\mu}_{\mathrm{L}} = i\psi^{\mu}_1\psi^{\mu}_2 \qquad (\mu \text{ not summed})\,. \tag{5.8}$$

With the correlation function for ψ^{μ}_i given by

$$\langle \psi^{\mu}_i(z)\psi^{\nu}_j(w)\rangle = \frac{\delta^{\mu\nu}\delta_{ij}}{z-w}\,, \tag{5.9}$$

it is easy to show that

$$j^{\mu}_{\mathrm{L}}(z)\psi^{\nu}_i(\omega) \sim \frac{-\delta^{\mu\nu}}{z-\omega}\varepsilon_{ij}\psi^{\nu}_j(\omega) + \cdots \tag{5.10}$$

$$T_{\mathrm{B}}(z)\psi^{\mu}_i(\omega) \sim \Delta\frac{\psi^{\mu}_i}{(z-w)^2} + \frac{\partial_w\psi^{\mu}_i(w)}{z-w} + \cdots \tag{5.11}$$

where the conformal weight $\Delta = \frac{1}{2}$ for free fermions so that (5.5) and (5.7) remain unchanged. Of course, we can introduce another form for the energy-momentum tensor, given by

$$T_{\mathrm{B}}(z)=\tfrac{1}{2}(\partial\psi_i^{\mu})\psi_i^{\mu}, \tag{5.12}$$

which also satisfies (5.7) and (5.11). For free field theories, T_{B} in (5.6) is equivalent to T_{B} in (5.12), due to normal ordering. However, as we shall see, T_{B} in (5.6) is more suited for generalization to conformal interacting field theories.

Next, recall the supersymmetric case, where

$$T_{\mathrm{B}}(z)=-\tfrac{1}{2}j_L^{\mu}j_L^{\mu}+\tfrac{1}{2}(\partial\lambda^{\mu})\lambda^{\mu}, \tag{5.13}$$

with λ^{μ} a real fermion. The supersymmetric partner of T_{B} is given by

$$T_{\mathrm{F}}(z)=i\lambda^{\mu}(z)\,\partial X^{\mu}(z) \tag{5.14}$$

so that the Ramond-Neveu-Schwarz algebra, or the super-Virasoro algebra, can be obtained from the following operator products expansions:

$$T_{\mathrm{B}}(z)T_{\mathrm{B}}(w)\sim\frac{\frac{3}{2}d}{2(z-w)^4}+\frac{2T_{\mathrm{B}}(w)}{(z-w)^2}+\frac{\partial_w T_{\mathrm{B}}(w)}{z-w}+\cdots, \tag{5.15}$$

$$T_{\mathrm{F}}(z)T_{\mathrm{F}}(w)\sim\frac{d}{(z-w)^3}+\frac{2T_{\mathrm{B}}(w)}{z-w}+\cdots, \tag{5.16}$$

$$T_{\mathrm{B}}(z)T_{\mathrm{F}}(w)\sim\frac{3T_{\mathrm{F}}(w)}{2(z-w)^2}+\frac{\partial_w T_{\mathrm{F}}(w)}{z-w}+\cdots. \tag{5.17}$$

In the fermionized form, we have ($\psi_3^{\mu}=\lambda^{\mu}$)

$$\tilde{T}_{\mathrm{B}}(z)=\tfrac{1}{2}\sum_{\mu,i=1}^{3}(\partial\psi_i^{\mu}(z))\psi_i^{\mu}(z), \tag{5.18}$$

$$\tilde{T}_{\mathrm{F}}(z)=i\sum_{\mu}\psi_1^{\mu}\psi_2^{\mu}\psi_3^{\mu}. \tag{5.19}$$

The operator product expansions given above remain valid. Notice that (5.19) has a manifest $[\mathrm{SO}(3)]^d$ symmetry. It is interesting to point out that there are two ways of constructing an SO(3) Kac-Moody algebra [19]. For the left-moving sector, an SO(3) gauge group (if present) is constructed from a single scalar field, or equivalently, a single complex fermionic field. The algebra should be thought of as being an A_1 algebra. For the right-moving sector, on the other hand, the SO(3) algebra should be thought of as being a B_1 algebra, where three Majorana-Weyl fermions are needed. In this case, the massless vector bosons are in the vector representation of SO(3). In the four-dimensional case, we have $[\mathrm{SO}(3)]^6$ from the right-moving sector. Clearly, $\tilde{T}_{\mathrm{B}}(z)$ and $\tilde{T}_{\mathrm{F}}(z)$ in (5.18) and (5.19) and the super-Virasoro algebra are invariant under the permutation of these six sets of fermion triplets. Hence we have an $[\mathrm{SO}(3)]^6 \times \mathrm{S}_6$ symmetry. The couplings of these massless vector bosons can be quite different from the usual massless gauge bosons obtained from the A_1 algebra. As we shall see, the six massless vector bosons in the $N=4$ supersymmetric gravity multiplet come from the $[\mathrm{U}(1)]^6$ subalgebra of $[\mathrm{SO}(3)]^6$. Their couplings are known to be different form the usual abelian gauge fields [23]. The SO(3) symmetry can be broken in two ways, namely by the introduction of independent boundary conditions (spin structure) for the fermions or by the introduction of interaction terms such as Thirring interactions.

Let us turn our attention to the emission vertices. We shall restrict ourselves to checking the consistency of the fermionization of the bosonic string vertex operators with world sheet supersymmetry. For the RNS string, the emission vertices V_2 can be constructed from super-primary fields V_1 with conformal dimension $\frac{1}{2}$, that is,

$$T_{\mathrm{F}}(z)V_1(w) = \frac{1}{z-w}V_2(w) + \cdots, \tag{5.20}$$

$$T_{\mathrm{B}}(z)V_1(w) = \frac{1}{2(z-w)^2}V_1(w) + \frac{1}{z-w}\partial_w V_1(w) + \cdots. \tag{5.21}$$

The V_2 vertices constructed in this way satisfy the requirement for the superconformal invariance:

$$T_{\mathrm{B}}(z)V_2(w) = \partial_w\left[\frac{1}{z-w}V_2(w)\right] + \cdots, \tag{5.22}$$

$$T_{\mathrm{F}}(z)V_2(w) = \partial_w\left[\frac{1}{z-w}V_1(w)\right] + \cdots. \tag{5.23}$$

For example, the choice $V_1^{\text{tachyon}}(z) = \mathrm{e}^{ik\cdot X(z)}$ $(k^2=1)$ gives the tachyon emission vertex

$$V_2^{\text{tachyon}}(z) = k^\mu \lambda^\mu(z)\,\mathrm{e}^{ik\cdot X(z)}, \tag{5.24}$$

while

$$V_1^{\text{vector}}(z) = \varepsilon^\mu \lambda^\mu(z)\mathrm{e}^{ik\cdot X(z)} \qquad (\varepsilon\cdot k = k^2 = 0) \tag{5.25}$$

gives the vector emission vertex

$$V_2^{\text{vector}}(z) = \varepsilon^\mu(i\partial_z X^\mu + (k\cdot\lambda)\lambda^\mu)\mathrm{e}^{ik\cdot X(z)}. \tag{5.26}$$

In general these two kinds of vertices V_1 and V_2 correspond to the F_1 and the F_2 formalisms [24] respectively.

It is straightforward to generalize the above construction to the fermionized formulation. For example, the "vector" emission vertices are given by choosing $V_1(z)=\psi_3^\mu(z)$, which gives

$$V_2(z)=i\psi_1^\mu(z)\psi_2^\mu(z), \tag{5.27}$$

or the vertices resulting from permutations of $(\psi_1^\mu, \psi_2^\mu, \psi_3^\mu)$. Of course this construction makes sense only if the resulting V_2 vertices are single valued operators on the world sheet; fortunately we can always describe the emissions of the physically allowed states in the $\boldsymbol{W}_0$ sector. This situation is analogous to the standard RNS model, where the emission vertices for particles in the NS sector are single valued, well defined fields on the world sheet in both Ramond and NS sectors. More precisely, in our case any physically allowed state in the $\boldsymbol{W}_0$ sector has quantum numbers $\boldsymbol{N}_{\boldsymbol{W}_0}$ which can be interpreted as the quantum numbers of the corresponding V_1 vertex satisfying

$$\boldsymbol{W}_i\cdot\boldsymbol{N}_{\boldsymbol{W}_0}=s_i \qquad \text{for all } i\,. \tag{5.28}$$

Since the V_2 vertex is obtained by taking a commutator of V_1 and the world-sheet supercharge, its quantum numbers $\boldsymbol{M}$ satisfy

$$\boldsymbol{W}_i\cdot\boldsymbol{M}=\boldsymbol{W}_i\cdot\left(\boldsymbol{N}_{\boldsymbol{W}_0}+\boldsymbol{C}\right)=s_i+\boldsymbol{W}_i\cdot\boldsymbol{C}, \tag{5.29}$$

where $\boldsymbol{C}$ is the vector of quantum numbers of the supercurrent. Here in principle we have several possibilities for $\boldsymbol{C}$, because the world-sheet supercurrent is a sum of contributions from the space-time coordinates and the right-moving triplets. However, in any case thanks to the triplet constraint we have

$$\boldsymbol{W}_i\cdot\boldsymbol{M}=s_i+\boldsymbol{W}_i\cdot\boldsymbol{C}=0 \qquad (\text{mod } 1)\,, \tag{5.30}$$

which means the V_2 vertex is a single-valued quantity on the world sheet for any sector $\alpha\boldsymbol{W}$, because $2\pi\Sigma_i\alpha_i\boldsymbol{W}_i\cdot\boldsymbol{M}$ is precisely the phase change of the vertex operator when going from $\sigma=0$ to $\sigma=2\pi$.

Comparing the $T_\mathrm{B}(z)$ in (5.13) and $\tilde{T}_\mathrm{B}(z)$ in (5.18) in the fermionic formulation, we note that $\tilde{T}_\mathrm{B}(z)$ is quadratic in the fermionic fields while $T_\mathrm{B}(z)$ is quartic in ψ_1^μ, ψ_2^μ and quadratic in $\psi_3^\mu=\lambda^\mu$. For free fermions, they are equivalent, due to normal ordering. However, if we now introduce Thirring interactions [20] which are conformally invariant, $\tilde{T}_\mathrm{B}(z)$ must be appropriately modified while $T_\mathrm{B}(z)$ retains its present form. The dimensions of the fermion fields involved in the Thirring interactions change accordingly, while that of the currents remain canonical. The introduction of non-abelian Thirring interactions in general breaks conformal invariance. Therefore, let us concentrate on abelian Thirring interactions. For heterotic strings in D dimensions, the left-moving fermions can be grouped in

$16+(10-D)$ complex fermions forming $16+(10-D)$ currents $j_{\rm L}^i$ while one right-moving current $j_{\rm R}^i$ can be formed by a pair of real fermions from each of the $(10-D)$ triplets of fermions. The Thirring interactions take the form

$$\sum_{i,j} h_{ij} j_{\rm L}^i j_{\rm R}^j , \tag{5.31}$$

where $i=1,2,\dots,(26-D)$ and $j=1,2,\dots,(10-D)$ so that there are $(26-D)(10-D)$ independent couplings. We note that the Thirring interactions (5.31) must be single-valued on the world sheet. If a particular Thirring coupling h_{ij} in (5.31) is non-zero, then the boundary conditions of the fermions that appear in the corresponding $j_{\rm L}^i j_{\rm R}^j$ must take values such that $j_{\rm L}^i j_{\rm R}^j$ is periodic on the torus. Returning to the bosonic language, these interaction terms may be written as [13]

$$\sum_{i,j} h_{ij} \partial_z X^i \partial_{\bar z} X^j , \tag{5.32}$$

where the couplings h_{ij} are constant background fields of the massless gravitational, antisymmetric tensor, and gauge fields. Recall the operator product for $j_{\rm L}^i$ in (5.5) (the operator product for $j_{\rm R}^j$ is similar while the operator product for $j_{\rm L}^i j_{\rm R}^j$ is regular); since currents remain canonical in the presence of Thirring interactions (5.5) remains valid while (5.8) is no longer true. To solve the Thirring interactions, let us follow ref. [20] and postulate the following operator product expansions (normal ordering on the right-hand side is implicit in the rest of this section):

$$j_{\rm L}^i(z)\psi_{\rm L}^l(w,\bar w) \sim \frac{A_{il}}{z-w}\psi_{\rm L}^l(w,\bar w)+\cdots , \tag{5.33a}$$

$$j_{\rm R}^j(\bar z)\psi_{\rm L}^l(w,\bar w) \sim \frac{h_{lj}}{\bar z-\bar w}\psi_{\rm L}^l(w,\bar w)+\cdots , \tag{5.33b}$$

$$j_{\rm L}^i(z)\psi_{\rm R}^k(w,\bar w) \sim \frac{h_{ik}}{z-w}\psi_{\rm R}^k(w,\bar w)+\cdots , \tag{5.33c}$$

$$j_{\rm R}^j(\bar z)\psi_{\rm R}^k(w,\bar w) \sim \frac{B_{jk}}{\bar z-\bar w}\psi_{\rm R}^k(w,\bar w)+\cdots , \tag{5.33d}$$

where $\psi_{\rm L}^l$ and $\psi_{\rm R}^k$ are complex fermions and l,k are not summed. Here A is a constant $d=10-D$ dimensional square matrix and B is a constant $(16+d)$ dimensional square matrix. In the limit $h_{ij}=0$, A and B reduce to the unit matrices, consistent with (5.10). Using (5.33) and (5.13), (where $j\psi=j_+\psi+\psi j_-$

with j decomposed into creation and annihilation components) we have,

$$T_{\rm B}(z)\psi^l_{\rm L}(w,\bar w)\sim\frac{\Delta^{li}_{\rm L}\psi^i_{\rm L}(w,\bar w)}{(z-w)^2}+\frac{A_{il}j^i_{\rm L}(w)\psi^l_L(w,\bar w)}{z-w}+\cdots,$$

$$\bar T_{\rm B}(\bar z)\psi^l_{\rm L}(w,\bar w)\sim\frac{\bar\Delta^{li}_{\rm L}\psi^i_{\rm L}(w,\bar w)}{(\bar z-\bar w)^2}+\frac{h_{lj}j^j_{\rm R}(\bar w)\psi^l_L(w,\bar w)}{\bar z-\bar w}+\cdots,$$

$$T_{\rm B}(z)\psi^k_{\rm R}(w,\bar w)\sim\frac{\Delta^{kj}_{\rm R}\psi^j_{\rm R}(w,\bar w)}{(z-w)^2}+\frac{h_{ik}j^i_{\rm L}(w)\psi^k_{\rm R}(w,\bar w)}{z-w}+\cdots,$$

$$\bar T_{\rm B}(\bar z)\psi^k_{\rm R}(w,\bar w)\sim\frac{\bar\Delta^{kk'}_{\rm R}\psi^{k'}_{\rm R}(w,\bar w)}{(\bar z-\bar w)^2}+\frac{B_{jk}j^j_{\rm R}(\bar w)\psi^k_{\rm R}(w,\bar w)}{\bar z-\bar w}+\cdots, \tag{5.34}$$

where

$$\Delta^{li}_{\rm L}=\sum_m A_{ml}A_{mi},$$

$$\bar\Delta^{li}_{\rm L}=\sum_m h_{lm}h_{im},$$

$$\Delta^{kj}_{\rm R}=\sum_m h_{mk}h_{mj},$$

$$\bar\Delta^{kj}_{\rm R}=\sum_m B_{mk}B_{mj}. \tag{5.35}$$

Comparing (5.34) with (5.11) and the corresponding operator product for $\bar T_{\rm B}(\bar z)$, we have the equations of motion

$$i\partial_{\bar w}\psi^i_{\rm L}(w,\bar w)=h_{ij}j^j_{\rm R}(\bar w)\psi^i_{\rm L}(w,\bar w),$$

$$i\partial_w\psi^j_{\rm R}(w,\bar w)=h_{ij}j^i_{\rm L}(w)\psi^j_{\rm R}(w,\bar w), \tag{5.36}$$

as well as two additional equations:

$$i\partial_w\psi^i_{\rm L}(w,\bar w)=A_{il}j^l_{\rm L}(w)\psi^i_{\rm L}(w,\bar w),$$

$$i\partial_{\bar w}\psi^k_{\rm R}(w,\bar w)=B_{jk}j^j_{\rm R}(\bar w)\psi^k_{\rm R}(w,\bar w). \tag{5.37}$$

These equations replace the need for the method of point separation in (5.36). Again, normal ordering on the right-hand side is implied. Also, recalling the relation between conformal weights and the dimensions and spins of fields, we have, for spin-half particles, the following constraints on the matrices A and B,

$$2S_{iL}=1 \;\Rightarrow\; \Delta_{L}^{ij}-\bar{\Delta}_{L}^{ij}=\delta_{ij},$$

$$2S_{jR}=1 \;\Rightarrow\; \Delta_{R}^{jk}-\bar{\Delta}_{R}^{jk}=\delta_{jk}. \tag{5.38}$$

Rewriting these entirely in terms of ψ_L^i and ψ_R^j, we have

$$\psi_L^i(z,\bar{z})=\exp\{-iA^{il}(X_L^l(z)-X_L^l(z_0))-ih^{ij}(X_R^j(\bar{z})-X_R^j(\bar{z}_0))\}\psi_L^i(z_0,\bar{z}_0), \tag{5.39}$$

$$\psi_R^j(z,\bar{z})=\exp\{-iB^{jk}(X_R^k(\bar{z})-X_R^k(\bar{z}_0))-ih^{ij}(X_L^i(z)-X_L^i(z_0))\}\psi_R^j(z_0,\bar{z}_0), \tag{5.40}$$

where normal ordering on the right-hand side is implied. It is clear that, in the absence of Thirring couplings h^{ij}, $\psi_L^i(z,\bar{z})=\psi_L^i(z)$ and $\psi_R^j(z,\bar{z})=\psi_R^j(\bar{z})$ as required. Eqs. (5.39) and (5.40) relate the boundary conditions of ψ_L^i and ψ_R^j to the compactification radii of the X variables i.e.,

$$X_L^l(\sigma+t+2\pi(m+n))=X_L^l(\sigma+t)+L_L^l,$$

$$X_R^k(\sigma-t+2\pi(n-m))=X_R^k(\sigma-t)+L_R^k, \tag{5.41}$$

where L_L^l and L_R^k determine some lattice specifying the compactification X variables. In this sense, Thirring interactions are absorbed into the compactification radii of the bosonized variables, as should be expected from (5.32). If, however, the compactification radii of these bosonic variables are all rational numbers then we can express the system in terms of free fermions with non-trivial boundary conditions, as we will illustrate in sect. 9.

Finally we should note that in principle one could consider other interactions among the world sheet fermions which are (super) conformal invariant only at some critical points, or compactifications on non-toroidal manifolds.

6. Four-dimensional models

Using $\boldsymbol{W}_i$ of the form given in (5.2), (5.3) we can employ the rules of sect. 3 to construct consistent string models in fewer than ten space-time dimensions. The constructions follow completely along the lines of those in sect. 4; however, the possible varieties of models in fewer than ten dimensions become more numerous and the constructions at times more intricate than was the case for ten dimensional models because of the increased number of internal degrees of freedom. In this section we will explore some of these new possibilities in the physically interesting context of four-dimensional models. In particular we will stress the construction of models with traits of some phenomenological importance: the presence of $N=1$ supersymmetry (which is probably necessary to avoid a Planck scale cosmological constant), chiral fermions, and interesting gauge groups.

Consider the set of generating vectors,

$$\boldsymbol{W}_0=\left(\left(\tfrac{1}{2}\right)_{\mathrm{r}}^{2}\left(\tfrac{1}{2}\tfrac{1}{2}\tfrac{1}{2}\right)_{\mathrm{r}}^{6}\Big|\left(\tfrac{1}{2}\right)_{\mathrm{r}}^{44}\right),$$

$$\boldsymbol{W}_1=\left((0)_{\mathrm{r}}^{2}\left(0\tfrac{1}{2}\tfrac{1}{2}\right)_{\mathrm{r}}^{6}\Big|\left(\tfrac{1}{2}\right)_{\mathrm{r}}^{44}\right),$$

$$\boldsymbol{W}_2=\left((0)_{\mathrm{r}}^{2}\left(0\tfrac{1}{2}\tfrac{1}{2}\right)_{\mathrm{r}}^{2}\left(\tfrac{1}{2}0\tfrac{1}{2}\right)_{\mathrm{r}}^{4}\Big|\left(\tfrac{1}{2}\right)_{\mathrm{r}}^{28}(0)_{\mathrm{r}}^{16}\right),$$

$$\boldsymbol{W}_3=\left((0)_{r}^{2}\left(\tfrac{1}{2}0\tfrac{1}{2}\right)_{\mathrm{r}}^{2}\left(0\tfrac{1}{2}\tfrac{1}{2}\right)_{\mathrm{r}}^{2}\left(\tfrac{1}{2}\tfrac{1}{2}0\right)_{\mathrm{r}}^{2}\Big|\left(\tfrac{1}{2}\right)_{\mathrm{r}}^{14}(0)_{\mathrm{r}}^{14}\left(\tfrac{1}{2}\right)_{\mathrm{r}}^{6}(0)_{r}^{10}\right). \tag{6.1}$$

Label the models generated by the sets $\{\boldsymbol{W}_0\}$, $\{\boldsymbol{W}_0, \boldsymbol{W}_1\}$, $\{\boldsymbol{W}_0, \boldsymbol{W}_1, \boldsymbol{W}_2\}$ and $\{\boldsymbol{W}_0, \boldsymbol{W}_1, \boldsymbol{W}_2, \boldsymbol{W}_3\}$ by M0, M1, M2 and M3 respectively. As we saw in sect. 4 the addition of a vector to a generating set $\{\boldsymbol{W}_i\}$ produces new physical sectors as well as new constraints on the spectra of old sectors, but does not alter those constraints already present. Thus with little additional work we may simultaneously construct M0, M1, M2, and M3 by successively adding the appropriate constraints and physical sectors. Here we will restrict ourselves to determining the tachyonic and massless particle spectra. The symmetric matrix $\boldsymbol{W}_i \cdot \boldsymbol{W}_j$ (with diagonal elements specified mod 2 and off diagonal elements mod 1) is,

$$\boldsymbol{W}_i \cdot \boldsymbol{W}_j = \begin{bmatrix} 1 & 0 & 0 & 0 \\ 0 & 0 & \frac{1}{2} & \frac{1}{2} \\ 0 & \frac{1}{2} & 0 & 0 \\ 0 & \frac{1}{2} & 0 & 1 \end{bmatrix}. \tag{6.2}$$

To satisfy eqs. (3.33)–(3.35) the k_{ij} must be of the form,

$$k_{ij} = \begin{bmatrix} k_{00} & k_{10} & k_{20} & k_{30} \\ k_{10} & k_{10} & k_{21}+\frac{1}{2} & k_{31}+\frac{1}{2} \\ k_{20} & k_{21} & k_{20} & k_{32} \\ k_{30} & k_{31} & k_{32} & k_{30}+\frac{1}{2} \end{bmatrix} \quad (\mathrm{mod}\ 1), \tag{6.3}$$

where all of the unspecified k_{ij}'s take values of 0 or $\frac{1}{2}$.

All of the models contain the sectors **0** and $\boldsymbol{W}_0$; these are the only sectors in M0. The **0** sector provides massive fermionic states while the $\boldsymbol{W}_0$ sector (with vacuum energy $[-\frac{1}{2}, -1]$) can give rise to tachyons and massless bosons, as well as massive states. The constraints (3.36) on the fermion numbers in the $\boldsymbol{W}_0$ sector are,

$$\boldsymbol{N}_{\boldsymbol{W}_0}\cdot \boldsymbol{W}_0 = \tfrac{1}{2}(N^1 + \cdots N^{20} + \tilde{N}^1 + \cdots + \tilde{N}^{44}) = \tfrac{1}{2} \quad (\mathrm{mod}\ 1), \tag{6.4a}$$

$$\begin{aligned}\boldsymbol{N}_{\boldsymbol{W}_0}\cdot \boldsymbol{W}_1 = \tfrac{1}{2}(&N^4 + N^5 + N^7 + N^8 + N^{10} + N^{11} + N^{13} + N^{14} + N^{16}\\ &+ N^{17} + N^{19} + N^{20} + \tilde{N}^1 + \cdots + \tilde{N}^{44}) = 0 \quad (\mathrm{mod}\ 1),\end{aligned} \tag{6.4b}$$

$$\begin{aligned}\boldsymbol{N}_{\boldsymbol{W}_0}\cdot \boldsymbol{W}_2 = \tfrac{1}{2}(&N^4 + N^5 + N^7 + N^8 + N^9 + N^{11} + N^{12} + N^{14}\\ &+ N^{15} + N^{17} + N^{18} + N^{20} + \tilde{N}^1 + \cdots + \tilde{N}^{28}) = 0 \quad (\mathrm{mod}\ 1),\end{aligned} \tag{6.4c}$$

$$\begin{aligned}\boldsymbol{N}_{\boldsymbol{W}_0}\cdot \boldsymbol{W}_3 = \tfrac{1}{2}(&N^3 + N^5 + N^6 + N^8 + N^{10} + N^{11} + N^{13} + N^{14} + N^{15} + N^{16}\\ &+ N^{18} + N^{19} + \tilde{N}^1 + \cdots + \tilde{N}^{14} + \tilde{N}^{29} + \cdots + \tilde{N}^{34}) = 0 \quad (\mathrm{mod}\ 1).\end{aligned} \tag{6.4d}$$

Considering first M0 we find that (6.4a) allows an SO(44) vector multiplet of tachyons (taking $N^1 = \cdots = N^{20} = 0$ and $\tilde{N}^i = \delta_{ij}$ for some j), and gauge bosons in the adjoint representation of SO(44) ($N^i = \delta_{i1}$ or δ_{i2}; $\tilde{N}^i = \delta_{ij} + \delta_{ik}$ for any $j \neq k$). In contrast to the ten-dimensional models of sect. 4 we now also have the possibility of internal symmetries arising from right-moving degrees of freedom. Exciting a left-moving bosonic degree of freedom (giving a spacetime vector index and a massless state) and taking any one of the $N^3, N^4, \ldots N^{20}$ equal to 1 satisfies (6.4a) and gives a set of gauge bosons in the adjoint representation of $[\mathrm{SO}(3)]^6$ (recall that for SO(3) the vector representation is the adjoint). Thus the full gauge group of M0 is $[\mathrm{SO}(3)]^6 \otimes \mathrm{SO}(44)$. In addition to tachyons and vector bosons the spectrum of M0 includes massless scalar bosons in the representations $(3,1,1,1,1,1,946) \oplus \cdots \oplus (1,1,1,1,1,3,946)$ obtained from (6.4a) by taking one of $N^3, N^4, \ldots N^{20}$ equal to one and $\tilde{N}^i = \delta_{ij} + \delta_{ik}$ for some $j \neq k$, and also a graviton, scalar dilaton and antisymmetric tensor (which is a pseudo scalar in four dimensions) obtained from exciting left- and right-moving space-time degrees of freedom, $b^{i\dagger}_{1/2}|0\rangle_{\mathrm{R}} \otimes \tilde{a}^{j\dagger}_1|0\rangle_{\mathrm{L}}$, $i, j = 1, 2$.

Turning now to M1 in the $\boldsymbol{W}_0$ sector we have the additional constraint (6.4b). This serves to project out all of the tachyonic states found in M0 above as well as all but six of the $[\mathrm{SO}(3)]^6$ gauge bosons (those with $N^l = 1$, $l = 3, 6, \ldots, 18$) and all but six families of massless scalars in the adjoint representation of SO(44). The other $\boldsymbol{W}_0$ states present in M0 remain in the spectrum of M1. To complete the $\boldsymbol{W}_0$ sector we have for M2 the subspace of the M1 states satisfying (6.4c), that is, gauge bosons in the adjoint representation of $\mathrm{SO}(28) \otimes \mathrm{SO}(16)$, two additional vector bosons remaining from the original $[SO(3)]^6$ set (N^3 or $N^6=1$), two families of massless scalars (N^3 or $N^6=1$) in the adjoint of $SO(28)\otimes SO(16)$, four families of scalars (N^9, N^{12}, N^{15} or $N^{18}=1$) in the (28,16) representation, and the graviton, dilaton and antisymmetric tensor as before. Finally the massless $\boldsymbol{W}_0$ sector states contributing to M3 are those M2 states satisfying (6.4d): $SO(14)\otimes SO(14)\otimes SO(6)\otimes SO(10)$ gauge bosons, two families each of scalars in the representations (14,14,1,1), (1,1,6,10) (from N^3 or $N^6=1$), (14,1,1,10), (1,14,6,1) (from N^{15} or $N^{18}=1$) and (14,1,6,1), (1,14,1,10) (from N^9 or $N^{12}=1$) and again the graviton, dilation and antisymmetric tensor states.

To complete the spectrum of massless states for M1, M2, and M3 we must consider those sectors other than $\boldsymbol{W}_0$ which can contribute. Of those sectors providing space-time bosons (that is, $\boldsymbol{W}_0$ plus some combination of $\boldsymbol{W}_1$, $\boldsymbol{W}_2$ and $\boldsymbol{W}_3$) only $\overline{\boldsymbol{W}_0+\boldsymbol{W}_1+\boldsymbol{W}_2}=\left((\tfrac{1}{2})_r^2(\tfrac{1}{2}\tfrac{1}{2}\tfrac{1}{2})_r^2(00\tfrac{1}{2})_r^4\,|\,(\tfrac{1}{2})_r^{28}(0)_r^{16}\right)$ has vacuum energies for both right and left movers which are non-positive (in this case [0,0]) and thus can contribute massless states. The massless states in this sector (which is present only in M2 and M3) consists solely of right- and left-moving Ramond ground states (any excitations give a massive state) for which the constraints (3.36) are,

$$\tfrac{1}{4}(1+\gamma^9\gamma^{10}\gamma^{12}\gamma^{13}\gamma^{15}\gamma^{16}\gamma^{18}\gamma^{19}\tilde{\gamma}^{29}\tilde{\gamma}^{30}\dots\tilde{\gamma}^{44})=k_{10}+k_{20}+\tfrac{1}{2}\quad(\text{mod }1)\qquad(6.5\text{a})$$

$$\tfrac{1}{4}(1+\gamma^{10}\gamma^{13}\gamma^{16}\gamma^{19}\tilde{\gamma}^{29}\tilde{\gamma}^{30}\dots\tilde{\gamma}^{44})=k_{10}+k_{21}\quad(\text{mod }1)\qquad(6.5\text{b})$$

$$\tfrac{1}{4}(1+\gamma^9\gamma^{12}\gamma^{15}\gamma^{18})=k_{21}+k_{20}+\tfrac{1}{2}\quad(\text{mod }1)\qquad(6.5\text{c})$$

$$\tfrac{1}{4}(1+\gamma^{10}\gamma^{13}\gamma^{15}\gamma^{16}\gamma^{18}\gamma^{19}\tilde{\gamma}^{29}\tilde{\gamma}^{30}\dots\tilde{\gamma}^{34})=k_{31}+k_{32}\quad(\text{mod }1).\qquad(6.5\text{d})$$

As in (4.14) these equations should be interpreted as acting on states and the γ^i and $\tilde{\gamma}^j$ are proportional to right- and left-moving Ramond zero-mode operators, respectively. Eqs. (6.5a)–(6.5c) are not independent but are compatible (i.e., (6.5a) and (6.5b) imply (6.5c)) for any choices of k_{ij}. For M2 these are the only constraints to be satisfied and this sector provides four sets of massless scalar bosons in the $SO(28)\times SO(16)$ representation (1,128) (that is ground states $|a\rangle_R\otimes|b\rangle_L$ where

$$\gamma^9\gamma^{12}\gamma^{15}\gamma^{18}|a\rangle_R=(-1)^{2k_{21}+2k_{20}}|a\rangle_R,$$

$$\gamma^{10}\gamma^{13}\gamma^{16}\gamma^{19}|a\rangle_R=(-1)^{2k_{10}+2k_{21}+1}|a\rangle_R,$$

$$\tilde{\gamma}^{29}\dots\tilde{\gamma}^{44}|b\rangle_L=|b\rangle_L)$$

and four sets of scalars in the representation (1,128′), (i.e. states,

$$|a'\rangle_R\otimes|b'\rangle_L$$

with

$$\gamma^9\gamma^{12}\gamma^{15}\gamma^{18}|a'\rangle_R = (-1)^{2k_{21}+2k_{20}}|a'\rangle_R,$$

$$\gamma^{10}\gamma^{13}\gamma^{16}\gamma^{19}|a'\rangle_R = (-1)^{2k_{10}+2k_{21}}|a'\rangle_R,$$

$$\tilde{\gamma}^{29}\ldots\tilde{\gamma}^{44}|b'\rangle_L = -|b'\rangle_L).$$

For M3 we have the subspace of these states satisfying (6.5d) which is two sets each of scalars in the SO(14) ⊗ SO(14) ⊗ SO(6) ⊗ SO(10) representations $(1,1,4,16)$, $(1,1,\bar{4},16)$, $(1,1,4,\overline{16})$ and $(1,1,\bar{4},\overline{16})$.

Consider now the sectors of these models which can contribute massless space-time fermions. The only possibilities with small enough vacuum energies are $\boldsymbol{W}_1$ with $[0,-1]$ and $\boldsymbol{W}_2$ with $[0,0]$. Massless states in the former sector must satisfy the constraints,

$$\tfrac{1}{4}(1+\gamma^1\gamma^2\gamma^3\gamma^6\gamma^9\gamma^{12}\gamma^{15}\gamma^{18}) + \tfrac{1}{2}(\tilde{N}^1 + \cdots \tilde{N}^{44}) = k_{00} + k_{10} + \tfrac{1}{2} \quad (\text{mod } 1), \tag{6.6a}$$

$$\tfrac{1}{2}(\tilde{N}^1 + \cdots \tilde{N}^{44}) = 0 \quad (\text{mod } 1), \tag{6.6b}$$

$$\tfrac{1}{4}(1+\gamma^9\gamma^{12}\gamma^{15}\gamma^{18}) + \tfrac{1}{2}(\tilde{N}^1 + \cdots \tilde{N}^{28}) = k_{21} + k_{20} + \tfrac{1}{2} \quad (\text{mod } 1), \tag{6.6c}$$

$$\tfrac{1}{4}(1+\gamma^3\gamma^6\gamma^{15}\gamma^{18}) + \tfrac{1}{2}(\tilde{N}^1 + \cdots \tilde{N}^{14} + \tilde{N}^{29} + \cdots \tilde{N}^{34}) = k_{31} + k_{30} + \tfrac{1}{2} \quad (\text{mod } 1). \tag{6.6d}$$

From the first two we find that the model M1 contains eight sets of massless fermion states in the adjoint representation of SO(44), four each of positive and negative space-time helicity (depending on the eigenvalue of $\gamma^1\gamma^2$). Specifically these are given by $|a\rangle_R \otimes \tilde{b}^{i\dagger}_{1/2}\tilde{b}^{j\dagger}_{1/2}|0\rangle_L$ with $i \neq j \in \{1,2,\ldots 44\}$ and $|a\rangle_R$ representing the eight right-moving ground states satisfying $\gamma^1\gamma^2\gamma^3\gamma^6\ldots\gamma^{18}|a\rangle_R = (-1)^{2k_{00}+2k_{10}}|a\rangle_R$. In addition there are four sets of spinoreal-vector states, $|a\rangle_R \otimes \tilde{a}^{i\dagger}_1|0\rangle_L$, each with four polarizations which can be recombined into two polarizations of a spin-$\frac{1}{2}$ state and two polarizations of a spin-$\frac{3}{2}$ gravitino. The presence of four gravitinos suggests that the model M1 has an $N=4$ supersymmetry; checking the rest of the massless spectrum given above one can verify that this is in fact the case. The six vector bosons from the right-moving internal symmetry degrees of freedom are just the spin-1 states of the $N=4$ supergravity multiplet.

In similar fashion taking into account (6.6c) one can determine the massless fermion content of M2 in the $\boldsymbol{W}_1$ sector: four sets of fermionic states (two each of positive and negative space-time helicity) in the representations (378,1), (1,120), and (28, 16) of SO(28) ⊗ SO(16), and two gravitinos and spin-$\frac{1}{2}$ fermions each with two polarizations. Satisfying in addition (6.6d) gives the following states of M3: massless fermions of both positive and negative space-time helicity in the SO(14) ⊗ SO(14) ⊗ SO(6) ⊗ SO(10) representations (91, 1, 1, 1), (1, 91, 1, 1), (1, 1, 15, 1), (1, 1, 1, 45), (14, 14, 1, 1), (14, 1, 6, 1), (1, 14, 6, 1), (14, 1, 1, 10), (1, 14, 1, 10), and (1, 1, 6, 10) and a single gravitino and spin-$\frac{1}{2}$ fermion (again each with both helicities).

To finish up the massless spectra of M2 and M3 consider the W_2 sector. Since the vacuum energies are [0, 0] only the ground states contribute to the massless spectrum and we have the constraints,

$$\tfrac{1}{4}(1+\gamma^1\gamma^2\gamma^3\gamma^6\gamma^{10}\gamma^{13}\gamma^{16}\gamma^{19}\tilde{\gamma}^{29}\tilde{\gamma}^{30}\ldots\tilde{\gamma}^{44}) = k_{20}+k_{00}+\tfrac{1}{2} \quad (\text{mod } 1), \tag{6.7a}$$

$$\tfrac{1}{4}(1+\gamma^{10}\gamma^{13}\gamma^{16}\gamma^{19}\tilde{\gamma}^{29}\tilde{\gamma}^{30}\ldots\tilde{\gamma}^{44}) = k_{10}+k_{21} \quad (\text{mod } 1), \tag{6.7b}$$

$$\tfrac{1}{4}(1+\gamma^{3}\gamma^{6}\gamma^{10}\gamma^{13}\gamma^{16}\gamma^{19}\tilde{\gamma}^{29}\ldots\tilde{\gamma}^{34}) = k_{30}+k_{32} \quad (\text{mod } 1). \tag{6.7c}$$

There are only three constraints because that for $W_2 \cdot N_{W_2}$ is satisfied identically for massless states. Eqs. (6.7a) and (6.7b) permit four sets of massless fermionic states of M2 (two sets with each space-time helicity) in the SO(28) ⊗ SO(16) representations (1, 128) and (1, 128′). This completes the massless spectrum of M2, and, as expected by the presence of two gravitinos, one finds that these states are consistent with an $N = 2$ supersymmetry.

Imposing all of the constraints in (6.7) gives the physical states of M3 in the W_2 sector. Here the constraints serve to correlate the space-time helicity with the gauge group representation of the states. To see this assume (just for clarity) that the k_{ij} are chosen such that the right-hand side of each equation in (6.7) is $\frac{1}{2}$; then the states in the physical spectrum are those with positive eigenvalues of the strings of γ's appearing in (6.7). The space-time helicity is determined by the $\gamma^1\gamma^2$ eigenvalues; the group representation by the eigenvalues of $\tilde{\gamma}^{29}\ldots\tilde{\gamma}^{34}$ and $\tilde{\gamma}^{35}\ldots\tilde{\gamma}^{44}$. The states consistent with (6.7) comprise two sets of fermions (distinguished, for example, by their $\gamma^{10}\gamma^{13}$ eigenvalues) in SO(14) ⊗ SO(14) ⊗ SO(6) ⊗ SO(10) representations (1, 1, 4, 16) and (1, 1, $\overline{4}$ 16) with positive space-time helicity and (1, 1, 4, $\overline{16}$) and (1, 1, $\overline{4}$, $\overline{16}$) with negative helicity. Notice that these are chiral fermions in representations consistent with CPT invariance (16 families of SO(10) if we interpret the SO(6) as a horizontal symmetry group). As one can check from the massless states given above, M3 possesses an $N = 1$ supersymmetry; again this is as expected from the presence of a single gravitino. Note that the supergravity sector contains a scalar multiplet in addition to the gravity multiplet.

To summarize, the set of boundary condition generating vectors in (6.1) gives rise to the following four-dimensional, closed string models: a tachyonic $[\mathrm{SO}(3)]^6$ ⊗ SO(44) model with no massless fermions (M0); an $N = 4$ supergravity, super-Yang-

Mills model with gauge group SO(44) (M1); an $N=2$ supergravity, super-Yang-Mills $SO(28)\otimes SO(16)$ model (M2); and an $N=1$ supergravity (with scalar multiplet), super-Yang-Mills $SO(14)\otimes SO(14)\otimes SO(6)\otimes SO(10)$ model with chiral fermions.

The gauge groups of M2 and M3 arose, of course, from the particular structure of the boundary conditions for left movers given in (6.1) (specifically W_2 and W_3); by changing these we could construct many different gauge groups. The general features of M0–M3, on the other hand, are largely fixed by the structure of the boundary conditions for right movers given in (6.1). In particular W_1 is in general necessary to obtain supersymmetry (providing the fermonic partners of the W_0 sector bosons) and the structure for right movers in W_2 and W_3 serves in general to cut the $N=4$ supersymmetry down to $N=1$ (though this is not always the case, e.g. W_0, W_1, W_2 from (6.1) with the left-moving portion of W_2, replaced by $(\frac{1}{2})_r^{44}$ reproduces the SO(44) model with $N=4$, not $N=2$). In addition the structure of the left-mover portions of W_2 and W_3, which allowed the appearance of $SO(4n+2)$ representations (and hence chiral fermions in M3 above) was possible because the right mover portions of W_2 and W_3, $(0(0\frac{1}{2}\frac{1}{2})(\frac{1}{2}0\frac{1}{2})^2)$ and $(0(\frac{1}{2}0\frac{1}{2})(0\frac{1}{2}\frac{1}{2})(\frac{1}{2}\frac{1}{2}0))$ respectively, share an odd number (in the complex representation) of nonzero components. Six other choices for right movers (neglecting linear combinations) also have these properties, (that is an "odd overlap" and a form which can cut $N=4$ supersymmetry to $N=1$ given W_1 as in (6.1)) and so can easily be employed to generate $N=1$ models with chiral fermions. For example two of them are,

$$\left(0(0\tfrac{1}{2}\tfrac{1}{2})(000)^2\right),\left(0(000)(\tfrac{1}{2}0\tfrac{1}{2})^2\right),\left(0(\tfrac{1}{2}0\tfrac{1}{2})(000)(\tfrac{1}{2}0\tfrac{1}{2})\right)$$

and these three together with,

$$\left(0(000)(0\tfrac{1}{2}\tfrac{1}{2})(000)\right).$$

The model M3 given by (6.1) has 16 chiral families of SO(10). In general, we can add more basis vectors to decrease the number of chiral families. Also, we can add basis vectors to obtain non-supersymmetric tachyon-free chiral models. To illustrate this, let us consider a model with the basis vectors of the M3 model in (6.1) plus one more vector:

$$W_4=\left(0(0\tfrac{1}{2}\tfrac{1}{2})(\tfrac{1}{2}\tfrac{1}{2}0)(\tfrac{1}{2}\tfrac{1}{2}0)\middle|(0)^7(0)^7(\tfrac{1}{2})^2(0)(0)^5\right). \tag{6.8}$$

One can easily show that the gauge group in this "M4" model is $SO(14)\otimes SO(14)\otimes SU(2)\otimes SU(2)\otimes U(1)\otimes SO(10)$. If we consider $SU(2)\otimes SU(2)\otimes U(1)$ as some horizontal symmetry, the model has 8 chiral families of SO(10). More specifically, the chiral fermion content (coming from the W_2 sector) is $2(1,1,2,1,16)(1)+2(1,1,1,2,16)(1)$ where the U(1) charge is shown in the last bracket. To see if the model is supersymmetric or not, all we need to do is count the number of gravitinos.

This depends on the structure constants k_{ij}. A simple examination reveals that the model has one gravitino (i.e. $N = 1$ supersymmetry) if

$$k_{21} + k_{02} = k_{41} + k_{04} \qquad (\text{mod } 1)\,. \tag{6.9}$$

If this condition is not satisfied, then the M4 model is not supersymmetric. Notice that the introduction of $\boldsymbol{W}_1$ serves two purposes: it introduces $N = 4$ supersymmetry and removes the tachyonic modes from the $\boldsymbol{W}_0$ sector. The supersymmetry can be removed by introducing other vectors and appropriate structure constants k_{ij} making possible non-supersymmetric tachyon free models.

Let us consider some other interesting features of the spin structure construction by introducing yet another vector to the model above,

$$\boldsymbol{W}_5 = \left(0(\tfrac{1}{2}\tfrac{1}{2}0)(\tfrac{1}{2}0\tfrac{1}{2})(0\tfrac{1}{2}\tfrac{1}{2})\middle|(\tfrac{1}{2})^2(0)^5(0)^7(\tfrac{1}{2})(0)(0)(0)^2(\tfrac{1}{2})^3\right). \tag{6.10}$$

The gauge group is,

$$\mathrm{SO}(4) \otimes \mathrm{SO}(10) \otimes \mathrm{SO}(14) \otimes \mathrm{U}(1)^3 \otimes \mathrm{SU}(2) \otimes \mathrm{SU}(2) \otimes \mathrm{SU}(4)\,. \tag{6.11}$$

From the $\boldsymbol{W}_2$ sector, we find the chiral fermions to be singlets under the $\mathrm{SO}(4) \otimes \mathrm{SO}(10) \otimes \mathrm{SO}(14)$ gauge group; if $\mathrm{U}(1)^3$ is treated as some horizontal symmetry (three extra Z bosons), we end up with four families of $(2,1,4) + (1,2,\bar{4})$ of the $\mathrm{SU}(2) \otimes \mathrm{SU}(2) \otimes \mathrm{SU}(4)$ group. In standard grand unification schemes, this can be obtained from 4 chiral families of SO(10) broken to $\mathrm{SU}(2) \otimes \mathrm{SU}(2) \otimes \mathrm{SU}(4)$ [25]. In this "M5" model, even though the SO(10) grand unified gauge group never appears by construction, the gauge couplings of the gauge groups must have the same value at the tree level and baryon number violating processes can be mediated by massive string states. So this M5 model appears to resemble SO(10) broken to the left-right symmetric $\mathrm{SU}(2)_\mathrm{L} \otimes \mathrm{SU}(2)_\mathrm{R} \otimes \mathrm{SU}(4)$ model. However, it turns out that the left-right symmetry is actually explicitly broken by particles from another sector. Let us concentrate on fermions in the spinor representations of $\mathrm{SU}(2)_\mathrm{L} \otimes \mathrm{SU}(2)_\mathrm{R} \otimes \mathrm{SU}(4)$. In the $\overline{\boldsymbol{W}_2 + \boldsymbol{W}_3 + \boldsymbol{W}_4 + \boldsymbol{W}_5}$ sector, namely,

$$\overline{\boldsymbol{W}_2 + \boldsymbol{W}_3 + \boldsymbol{W}_4 + \boldsymbol{W}_5} = \left(0(0\tfrac{1}{2}\tfrac{1}{2})(\tfrac{1}{2}0\tfrac{1}{2})(\tfrac{1}{2}\tfrac{1}{2}0)\middle|(\tfrac{1}{2})^2(0)^5(\tfrac{1}{2})^7(\tfrac{1}{2})(0)(\tfrac{1}{2})(0)^2(\tfrac{1}{2})^3\right),$$

we have fermions with only $\mathrm{SU}(2)_\mathrm{R}$ (or only $\mathrm{SU}(2)_\mathrm{L}$, depending on the choice of structure constants) doublets. The fermions are in the $(1,16,1,1,2,1)$ and $(1,\overline{16},1,1,2,1)$ representations of the $\mathrm{SO}(4) \otimes \mathrm{SO}(10) \otimes \mathrm{SO}(14) \otimes \mathrm{SU}(2)_\mathrm{L} \otimes \mathrm{SU}(2)_\mathrm{R} \otimes \mathrm{SU}(4)$ group. For the $N = 1$ supersymmetric M5 model, that is where

$$k_{21} + k_{02} = k_{41} + k_{04} = k_{31} + k_{03} + k_{51} + k_{05} + \tfrac{1}{2}\,, \tag{6.12}$$

we have extra $SU(2)_R$ Higgs doublets which may make $SU(2)_R$ gauge bosons massive via the Higgs mechanism. Such left-right asymmetry may provide a mechanism to make the $SU(2)_R$ gauge bosons substantially heavier than the $SU(2)_L$ gauge bosons. Alternatively, one may consider the SO(10) group in (6.11) as some technicolor group so that $SU(2)_R$ techni-pions can make the $SU(2)_R$ gauge bosons heavy.

The possible gauge group structure in four dimensions arising from the present method of construction is extremely rich, although some groups, especially $SO(2n)$, are considerably easier to include than others. As a non-trivial example let us determine the gauge group of the model specified by the generating vectors,

$$\boldsymbol{W}_0 = \left(\tfrac{1}{2}\left(\tfrac{1}{2}\tfrac{1}{2}\tfrac{1}{2}\right)^3 \middle| \left(\tfrac{1}{2}\right)^{22}\right),$$

$$\boldsymbol{W}_1 = \left(0\left(0\tfrac{1}{2}\tfrac{1}{2}\right)^3 \middle| \left(\tfrac{1}{2}\right)^{22}\right),$$

$$\boldsymbol{W}_2 = \left(0\left(0\tfrac{1}{2}\tfrac{1}{2}\right)\left(\tfrac{1}{2}0\tfrac{1}{2}\right)^2 \middle| (0)^{16}\left(\tfrac{1}{2}\right)^{6}\right),$$

$$\boldsymbol{W}_3 = \left(0\left(\tfrac{1}{2}\tfrac{1}{2}0\right)\left(\tfrac{1}{2}\tfrac{1}{2}0\right)\left(0\tfrac{1}{2}\tfrac{1}{2}\right) \middle| \left(\tfrac{1}{2}\right)^2(0)^3\left(\tfrac{1}{4}\right)^8\left(\tfrac{1}{3}\right)^9\right). \tag{6.13}$$

There are many sectors contributing to the massless spectrum, but only $\overline{\alpha W}$ with a right mover portion of the form $(\frac{1}{2}(\frac{1}{2}\frac{1}{2}\frac{1}{2})^3)$ can possibly contribute massless vector bosons (since other sectors have vacuum energy greater than $-\frac{1}{2}$). This leaves four sectors to consider, $\boldsymbol{W}_0$, $\boldsymbol{W}_0 + 6\boldsymbol{W}_3$, $\boldsymbol{W}_0 + 4\boldsymbol{W}_3$, and $\boldsymbol{W}_0 + 8\boldsymbol{W}_3$. We will not consider these in detail (the calculations are completely analogous to those considered previously) but just sketch the determination of the gauge group. The $\boldsymbol{W}_0$ sector contributes (almost by inspection) gauge bosons in the adjoint representation of $SO(4) \otimes SO(6) \otimes U(8) \otimes U(3) \otimes U(6)$. In the $\boldsymbol{W}_0 + 6\boldsymbol{W}_3$ sector the constraints turn out to eliminate all possible massless vector bosons. The two remaining sectors, $\boldsymbol{W}_0 + 4\boldsymbol{W}_3$ and $\boldsymbol{W}_0 + 8\boldsymbol{W}_3$, both with vacuum energies $[-\frac{1}{2}, -\frac{1}{2}]$, are in some sense conjugates, contributing vector bosons in the $SO(4) \otimes SO(6) \otimes U(8) \otimes U(3) \otimes U(6)$ representations $(1,6,1,\{\frac{1}{3},\frac{1}{3},\frac{1}{3}\}, \{\frac{1}{3},\ldots\frac{1}{3}\})$, $(1,1,1,\{-\frac{2}{3},-\frac{2}{3},-\frac{2}{3},\}, \{\frac{1}{3},\ldots\frac{1}{3}\})$, $(1,1,1,\overline{3},\overline{15})$ and $(1,6,1,\{-\frac{1}{3},-\frac{1}{3},-\frac{1}{3}\}, \{-\frac{1}{3},\ldots-\frac{1}{3}\})$, $(1,1,1,\{\frac{2}{3},\frac{2}{3},\frac{2}{3}\}, \{-\frac{1}{3},\ldots-\frac{1}{3}\}$, $(1,1,3,15)$ respectively, where we have included the U(1) charges with respect to the Cartan subalgebras of U(3) and U(6) in brackets. The only subtle point in calculating these is remembering to include the U(1) charges of the ground states as well as the excitations, i.e.,

$$\boldsymbol{Q} = \overline{\alpha W} - \boldsymbol{W}_0 + \boldsymbol{N}_{\overline{\alpha W}} \tag{6.14}$$

in the $\overline{\alpha W}$ sector.

What is the gauge group of this model? From the gauge bosons given above the group clearly contains a direct product factor $SO(4) \otimes U(8)$. To determine the rest of the gauge group requires some recombination of the representations listed. Separating U(1) factors from both U(3) and U(6) we have gauge bosons in the $SU(3) \otimes SU(6)$ representations, (8,1), (1,35), (3,15), and $(\bar{3},\overline{15})$. These add up to form precisely the 133-dimensional adjoint representation of E_7. In similar fashion the two (1,6,1,1,1)'s together with the (1,15,1,1,1) and one of the U(1)'s combine to form the 28 of SO(8) and the other U(1) along with $(1,1,1,\{-\frac{2}{3},-\frac{2}{3},-\frac{2}{3}\},\{\frac{1}{3},\dots\frac{1}{3}\})$ and its conjugate form an SU(2). Thus the complete gauge group is found to be $SO(4) \otimes U(8) \otimes E_7 \otimes SO(8) \otimes SU(2)$. If the decomposition of the subgroup representations had not been clear we could have alternatively taken the root vectors, $\boldsymbol{Q} = \overline{\alpha \boldsymbol{W}} - \boldsymbol{W}_0 + \boldsymbol{N}_{\overline{\alpha W}}$, corresponding to each gauge boson, chosen a subset of simple roots, and directly calculated the Dynkin diagram to determine the gauge group.

To illustrate how the rank of the gauge group can be reduced, let us incorporate the structure (3.54) used in constructing the 10-dimensional single E_8 model into the construction of a four-dimensional model with rank $22 - 8 = 14$. Consider the following set of vectors:

$$
\begin{aligned}
&\boldsymbol{W}_0, \boldsymbol{W}_1,\\
&\quad \boldsymbol{W}_2 = \left((0)^{10} \middle| (0)^6 (\tfrac{1}{2})_{\mathrm{r}}^{16} (0)_{\mathrm{r}}^{16}\right),\\
&\quad \boldsymbol{W}_3 = \left((0)^{10} \middle| (0)^6 (\tfrac{1}{2})_{\mathrm{r}}^{8} (0)_{\mathrm{r}}^{8} (\tfrac{1}{2})_{\mathrm{r}}^{8} (0)_{\mathrm{r}}^{8}\right),\\
&\quad \boldsymbol{W}_4 = \left((0)^{10} \middle| (0)^6 (\tfrac{1}{2})_{\mathrm{r}}^{4} (0)_{\mathrm{r}}^{4} (\tfrac{1}{2})_{\mathrm{r}}^{4} (0)_{\mathrm{r}}^{4} (\tfrac{1}{2})_{\mathrm{r}}^{4} (0)_{\mathrm{r}}^{4} (\tfrac{1}{2})_{\mathrm{r}}^{4} (0)_{\mathrm{r}}^{4}\right),\\
&\quad \boldsymbol{W}_5 = \left((0)^{10} \middle| (0)^6 (\tfrac{1}{2})_{\mathrm{r}}^{2} (0)_{\mathrm{r}}^{2} \dots (\tfrac{1}{2})_{\mathrm{r}}^{2} (0)_{\mathrm{r}}^{2}\right),\\
&\quad \boldsymbol{W}_6 = \left((0)^{10} \middle| (0)^6 (\tfrac{1}{2})_{\mathrm{r}} (0)_{\mathrm{r}} \dots (\tfrac{1}{2})_{\mathrm{r}} (0)_{\mathrm{r}}\right).
\end{aligned}
\tag{6.15}
$$

From the $\boldsymbol{W}_0$ sector, we obtain only SO(12) gauge bosons. As in the 10-dimensional model with a specific choice of k_{ij}, the following 31 sectors:

$$
\boldsymbol{W}_0 + \sum_{i=2}^{6} \alpha_i \boldsymbol{W}_i \qquad (\text{not all } \alpha_i = 0) \tag{6.16}
$$

each contributes 8 gauge bosons combining into the 248 gauge bosons of E_8. Hence this model is an $N = 4$ supersymmetric string model with $E_8 \otimes SO(12)$ gauge group. A simpler example which demonstrates in addition the construction of a non-simply laced gauge group is given by, $\boldsymbol{W}_0$,

$$\boldsymbol{W}_1 = \left((0)_{\mathrm{r}}^2 (0\tfrac{1}{2}\tfrac{1}{2})_{\mathrm{r}} (000)_{\mathrm{r}}^5 \middle| (\tfrac{1}{2})_{\mathrm{r}} (\tfrac{1}{2})_{\mathrm{r}} (0)_{\mathrm{r}} (0)_{\mathrm{r}} (0)_{\mathrm{r}}^{40}\right),$$

$$\boldsymbol{W}_2 = \left((0)_{\mathrm{r}}^2 (\tfrac{1}{2}0\tfrac{1}{2})_{\mathrm{r}} (000)_{\mathrm{r}}^5 \middle| (\tfrac{1}{2})_{\mathrm{r}} (0)_{\mathrm{r}} (\tfrac{1}{2})_{\mathrm{r}} (0)_{\mathrm{r}} (0)_{\mathrm{r}}^{40}\right). \tag{6.17}$$

It is easily seen that all of the gauge bosons appear in the $\boldsymbol{W}_0$ sector where any pair of the last 41 left-moving fermions are excited giving rise to the gauge group SO(41).

Recall that the set of internal fermions for the right-moving sectors has an $[\mathrm{SO}(3)]^d \otimes \mathrm{S}_d$ symmetry where d is the number of compactified dimensions and S_d is the permutation symmetry. So far we have considered simple cases in which the six triplets of real fermions could be combined to form three complex triplets with boundary conditions satisfying (5.3b). To illustrate the above symmetry and constraint in a non-trivial context, consider the subgroup S_3 of the permutation group S_6. We can complexify the real fermions to obtain 3 complex triplets $(\psi_1^{(i)}\psi_2^{(i)}\psi_3^{(i)})$ $i = 1, 2, 3$ and combine the transformations under S_3 with sign changes ($\sigma_j = \pm 1$), to specify non-trivial transformations of the fermions around a non-contractible loop on the world sheet,

$$\begin{aligned} \psi_j^{(1)} &\to \psi_j^{(2)}\sigma_j, \\ \psi_j^{(2)} &\to \psi_j^{(3)}\sigma_j, \\ \psi_j^{(3)} &\to \psi_j^{(1)}\sigma_j. \end{aligned} \tag{6.18}$$

The eigenstates are

$$\begin{aligned} \alpha_j &= \psi_j^{(1)} + \psi_j^{(2)} + \psi_j^{(3)}, \\ \beta_j &= \psi_j^{(1)} + \omega\psi_j^{(2)} + \omega^2\psi_j^{(3)}, \\ \gamma_j &= \psi_j^{(1)} + \omega^2\psi_j^{(2)} + \omega\psi_j^{(3)}, \end{aligned} \tag{6.19}$$

where $\omega = \exp(\frac{2}{3}\pi i)$ so that $\omega^3 = 1$ and $1 + \omega + \omega^2 = 0$. If we choose $\sigma_1 = 1$, $\sigma_2 = \sigma_3 = -1$, then we obtain the boundary conditions

$$(000)(\tfrac{1}{3}\tfrac{1}{3}\tfrac{1}{3})(\tfrac{2}{3}\tfrac{2}{3}\tfrac{2}{3}) + (0\tfrac{1}{2}\tfrac{1}{2})(0\tfrac{1}{2}\tfrac{1}{2})(0\tfrac{1}{2}\tfrac{1}{2}) = (0\tfrac{1}{2}\tfrac{1}{2})(\tfrac{1}{3}\tfrac{5}{6}\tfrac{5}{6})(\tfrac{2}{3}\tfrac{1}{6}\tfrac{1}{6}). \tag{6.20}$$

Let us consider the model generated by,

$$\boldsymbol{W}_0 \quad \text{and} \quad \boldsymbol{W}_1 = \left(0(0\tfrac{1}{2}\tfrac{1}{2})(\tfrac{1}{3}\tfrac{5}{6}\tfrac{5}{6})(\tfrac{2}{3}\tfrac{1}{6}\tfrac{1}{6}) \middle| (\tfrac{1}{2})^{22}\right). \tag{6.21}$$

The $\boldsymbol{W}_0$ sector gives the SO(44) gauge group while the $\boldsymbol{W}_1$, $3\boldsymbol{W}_1$ and $5\boldsymbol{W}_1$ sectors, each with vacuum energy [0, −1], give one, two, and one gravitinos respectively, yielding an $N=4$ supersymmetric SO(44) model (cf. the $N=4$ supersymmetric SO(44) model M1).

Finally for completeness let us consider a four-dimensional type II model. In ten dimensions the possibilities for type II models are severely limited (there are only four [8]) because all of the fermions represent space-time degrees of freedom. In four dimensions there is considerably more freedom for constructing different models; by considering generalized forms of the supercurrent, some of the gauge groups produced can be of phenomenological interest. To give a simple (but not phenomenologically interesting) example the vectors,

$$W_0 = \left(\tfrac{1}{2}\left(\tfrac{1}{2}\tfrac{1}{2}\tfrac{1}{2}\right)^3 \middle| \tfrac{1}{2}\left(\tfrac{1}{2}\tfrac{1}{2}\tfrac{1}{2}\right)^3\right),$$

$$W_1 = \left((0)\left(0\tfrac{1}{2}\tfrac{1}{2}\right)^3 \middle| \tfrac{1}{2}\left(\tfrac{1}{2}\tfrac{1}{2}\tfrac{1}{2}\right)^3\right),$$

$$W_2 = \left((0)\left(0\tfrac{1}{2}\tfrac{1}{2}\right)\left(\tfrac{1}{2}0\tfrac{1}{2}\right)^2 \middle| \tfrac{1}{2}\left(\tfrac{1}{2}00\right)^2\left(\tfrac{1}{2}\tfrac{1}{2}\tfrac{1}{2}\right)\right),$$

$$W_3 = \left((0)\left(\tfrac{1}{2}0\tfrac{1}{2}\right)\left(0\tfrac{1}{2}\tfrac{1}{2}\right)\left(\tfrac{1}{2}0\tfrac{1}{2}\right) \middle| \tfrac{1}{2}\left(00\tfrac{1}{2}\right)^2\left(\tfrac{1}{2}\tfrac{1}{2}\tfrac{1}{2}\right)\right),$$

generate an $N=1$ supersymmetric model with gauge symmetry $SO(3)^2$ which arises from the last three left movers. A detailed discussion of type II models in four dimensions is under preparation.

7. Derivation of fermionic charge lattice construction

The rules given in sect. 3 and 5 and employed in sect. 4 and 6 are convenient for constructing many interesting closed string models, but are somewhat awkward for the purposes of clarifying general properties of models and comparing with alternative methods of model construction. For these purposes it is convenient to reformulate the rules of sect. 3 in terms of quantities of some direct physical importance, namely the fermionic charge vectors $\boldsymbol{Q}$ of physical states. If the boundary conditions appearing in a model can be diagonalized with complex fermions then the charge vectors for states in the sector $\overline{\alpha W}$ (given in the complex formulation) is,

$$\boldsymbol{Q} = \overline{\alpha W} - W_0 + N_{\overline{\alpha W}} . \tag{7.1}$$

The first two terms give the charge vector of the ground state of the $\overline{\alpha W}$ sector, while the last gives the charges from the fermionic excitations. If the boundary conditions are such that some real fermions must be employed, there is no simple way to define a charge; in this section we restrict our attention to the formulation with complex fermions.

Taking the dot product of (7.1) with a generating vector $\boldsymbol{W}_i$, using (3.36) for the last term and (3.33) to simplify, we find that the charge vectors for physical states satisfy,

$$\boldsymbol{W}_i \cdot \boldsymbol{Q} = s_i + \sum_j k_{ij}(\alpha_j - \delta_{0j}) \qquad (\text{mod } 1). \tag{7.2}$$

We will formally solve (7.2) and use this to investigate some of the properties of the "lattice" formed by the set of charge vectors. First some definitions: let G denote the set of vectors with integer components whose inner products with all of the $\boldsymbol{W}_i$ vanish (mod 1), i.e.,

$$\mathrm{G} = \{\boldsymbol{g}\colon \boldsymbol{W}_i \cdot \boldsymbol{g} = 0 \ (\text{mod } 1)\ \forall i;\ g^l \in \mathrm{Z}\}. \tag{7.3}$$

Define another constant vector, $\boldsymbol{S}$, and a set of vectors $\boldsymbol{\Delta}_i$, all with integer components, such that,

$$\boldsymbol{W}_i \cdot \boldsymbol{S} = s_i, \qquad S^l \in \mathrm{Z}, \tag{7.4}$$

$$\boldsymbol{\Delta}_i \cdot \boldsymbol{W}_j = k_{ij} \qquad (\text{mod } 1), \qquad \Delta_i^l \in \mathrm{Z}. \tag{7.5}$$

For heterotic-type models we can simply take $S^l = \delta_{1l}$; S will be referred to as the spin-statistics vector since its purpose is to keep track of the space-time spin-statistics; the existence of $\boldsymbol{\Delta}_i$ satisfying (7.5) will be demonstrated in a moment. Finally defining,

$$\boldsymbol{U}_i = \boldsymbol{W}_i - \boldsymbol{\Delta}_i, \tag{7.6}$$

we have that (making use of (3.33)),

$$\boldsymbol{W}_i \cdot \boldsymbol{U}_j = k_{ij} \qquad (\text{mod } 1). \tag{7.7}$$

We can thus write a formal general solution to (7.2),

$$\boldsymbol{Q} = \boldsymbol{g} + \boldsymbol{S} + \sum_j \boldsymbol{U}_j(\alpha_j - \delta_{0j}); \qquad \boldsymbol{g} \in \mathrm{G}. \tag{7.8}$$

The restriction of G to vectors with integer components is necessary to preserve the form of $\boldsymbol{Q}$ in (7.1). To complete this formulation we should express the constraints (3.33)–(3.35) on $\boldsymbol{W}_i$, k_{ij} and s_i in terms of the new quantities $\boldsymbol{U}_i$ and $\boldsymbol{S}$. Eq. (3.34) is automatically satisfied given (7.5) (i.e., (3.34) is necessary for the existence of $\boldsymbol{\Delta}_i$). Inserting (7.6) into (3.33) and (3.35) we easily obtain the corresponding new constraints,

$$\boldsymbol{U}_i \cdot \boldsymbol{U}_j = 0 \qquad (\text{mod } 1)\,, \tag{7.9}$$

$$\tfrac{1}{2}\boldsymbol{U}_i \cdot \boldsymbol{U}_i = \boldsymbol{U}_i \cdot \boldsymbol{S} \qquad (\text{mod } 1)\,. \tag{7.10}$$

Before investigating properties of the charge lattice formed by the points given in (7.8), let us sketch the proof of the existence of $\boldsymbol{\Delta}_i$ satisfying (7.5); some of the intermediate steps in the proof will prove useful later on. If the vectors $\boldsymbol{W}_j$ have L components and j runs from 0 to $n-1$, then (7.5) represents a set of n linear equations for the L variables which are the components of $\boldsymbol{\Delta}_i$. We wish to show that given the constraints placed on $\boldsymbol{W}_j$ and k_{ij} there are always integer solutions to these equations. To do this we will first define a set of invertible operations which produce equivalent sets of equations and then use these operations to bring the equations into a standard form for which the existence of integer solutions is clear. The operations, all defined mod 1 on the augmented matrix formed by appending the column of solutions $(k_i)_j$ to the matrix $\boldsymbol{W}_j^l$, are: (i) replacing any row by that row plus some other row; (ii) the same operation for columns; (iii) interchanging rows; and (iv) interchanging columns. These operations are invertible and preserve an important property of the original eq. (7.5), namely that the "order" of the left-hand side of the jth equation (that is m_j the smallest positive integer such that $m_j\boldsymbol{W}_j$ has integer components) is a multiple of the order of the right-hand side (i.e., $m_j k_{ij} \in \mathrm{Z}$). That this is true for the last three operations is obvious; however, for (i) to preserve this property requires the assumption of independence included in our definition of the set $\{\boldsymbol{W}_i\}$. Recall our demand that the linear combination αW have all integer components if and only if $\alpha_1 = 0$ (mod m_1), $\alpha_2 = 0$ (mod m_2),$\ldots$ $\alpha_i = 0$ (mod m_i),$\ldots$ etc. Now repeatedly employing the operation (i) forms some linear combination of the $\boldsymbol{W}_i$, e.g. αW; if this produces a row of the augmented matrix for which the order of the left-hand side (say m) is not a multiple of the order of the right hand side, then the linear combination $m\alpha W$ has all integer components but does not satisfy $m\alpha_i = 0$ (mod m_i) for all i. In other words in this case the $\boldsymbol{W}_i$ were not properly chosen to be independent.

Now repeatedly using the operations (i) and (ii) we can produce the smallest possible positive element of the matrix (call it r_1) and then place this in the upper left hand corner using (iii) and (iv). Note that this element divides the other elements in the first row and first column, otherwise using (i) and (ii) we could produce a smaller positive element but, by construction r_1 is the smallest. Given this we can add an integer multiple of the first row to subsequent rows such that the only remaining non-vanishing element in the first column is r_1. Similarly via the column operation (ii) we can eliminate the other elements of the first row. Now we may repeat the above operations on the submatrix which does not include the first row and column. Continuing in this fashion we are left with the equivalent set of equations (assuming $n < L$),

$$\begin{bmatrix} r_1 & 0 & \cdots & 0 & \cdots & 0 \\ 0 & r_2 & \cdots & 0 & \cdots & 0 \\ \vdots & \vdots & \ddots & \vdots & \ddots & \vdots \\ 0 & 0 & \cdots & r_n & \cdots & 0 \end{bmatrix} \begin{bmatrix} d_1 \\ \vdots \\ d_n \\ \vdots \\ d_L \end{bmatrix} = \begin{bmatrix} \kappa_1 \\ \vdots \\ \kappa_n \end{bmatrix} \pmod 1, \qquad (7.11)$$

where r_i and κ_i are positive rational numbers and r_{i+1} is a multiple of r_i. If $n > L$ then the form is as in (7.11) except that the last $n - L$ rows of the reduced matrix R (rather than the last $L - n$ columns) vanish as do $\kappa_{L+1}, \kappa_{L+2}, \ldots \kappa_n$. Eq. (7.11) always has integer solutions for the d_i ($r_i d_i = \kappa_i$ (mod 1)) since by construction the order of r_i is a multiple of the order of κ_i. The operations leading to (7.11) may all be inverted, in particular some linear combinations of the d_j's give integer solutions for the components of $\boldsymbol{\Delta}_i$ in (7.5). To recapitulate, we have used the elementary row and column operations (i)–(iv) to bring (7.5) to the form (7.11), in effect implicitly constructing invertible integer valued $n \times n$ and $L \times L$ matrices A and B (which implement the row and column operations) such that,

$$A_{ij} W_j^l B_{lk} = r_i \delta_{ik} . \qquad (7.12)$$

With the formal results (7.8) and (7.12) we can establish some general properties of the charge lattice of a consistent string model, $\Gamma = \{\boldsymbol{Q}\colon \boldsymbol{Q} = \boldsymbol{g} + \boldsymbol{S} + \Sigma \boldsymbol{U}_j(\alpha_j - \delta_{0j}),\ \boldsymbol{g} \in \mathrm{G}\}$. First given the definition of G and $\boldsymbol{S}$ and the constraint (7.9) on the set of $\{\boldsymbol{U}_i\}$, it is clear that Γ is a lorentzian (signature $(- - \cdots - | + \cdots +)$), half-integer "lattice", i.e. given any $\boldsymbol{Q}_1$ and $\boldsymbol{Q}_2 \in \Gamma$, we have $2\boldsymbol{Q}_1 \cdot \boldsymbol{Q}_2 \in \mathrm{Z}$. Strictly speaking, the charge lattice Γ is not a lattice. For reasons that will be clear soon, we call Γ a shifted lattice. Furthermore we can show that it is even (odd) if $\boldsymbol{S}^2$ is even (odd) i.e. for any $\boldsymbol{Q} \in \Gamma$,

$$\begin{aligned} \tfrac{1}{2}\boldsymbol{Q}^2 &= \tfrac{1}{2}\left(\boldsymbol{g}^2 + \boldsymbol{S}^2 + 2s_j(\alpha_j - \delta_{0j}) + (\alpha_j - \delta_{0j})^2 \boldsymbol{U}_j^2\right) \pmod 1 \\ &= \tfrac{1}{2}\boldsymbol{S}^2 \pmod 1 . \end{aligned} \qquad (7.13)$$

Here we used (7.10) to cancel the last two terms (mod 1), and the fact that $\frac{1}{2}\boldsymbol{g}^2 = \boldsymbol{U}_0 \cdot \boldsymbol{g} = 0$ (mod 1), to eliminate the first term. Thus Γ is even for purely bosonic and type II models ($\boldsymbol{S}^2 = 0$) and odd for heterotic type models ($\boldsymbol{S}^2 = -1$). Finally we will show that the lorentzian lattice $\Gamma_S = \{\boldsymbol{Q} - \boldsymbol{S}\} = \{\boldsymbol{g} + \Sigma_i \boldsymbol{U}_i \cdot \alpha_j;\ \alpha_i \in \mathrm{Z},\ \boldsymbol{g} \in \mathrm{G}\}$ is self dual, i.e. if $\boldsymbol{x} \cdot \boldsymbol{q} = 0$ (mod 1) for all $\boldsymbol{q} \in \Gamma_S$ then $\boldsymbol{x} \in \Gamma_S$ (note that Γ_S is an integer lattice but in general not even). To show this consider first a vector $\boldsymbol{x}$ for which $\boldsymbol{x} \cdot \boldsymbol{g} = 0$ (mod 1) for all $\boldsymbol{g} \in \mathrm{G}$. Using (7.12) we can obtain a formal general expression for $\boldsymbol{g}$ and use this to determine the general form of $\boldsymbol{x}$. By definition $\boldsymbol{g} \in \mathrm{G}$ are solutions of the equations $\boldsymbol{g} \cdot \boldsymbol{W}_i = 0$ (mod 1) for all i (or equivalently $\boldsymbol{g} \cdot \boldsymbol{U}_i = 0$ (mod 1)). Using (7.12) we can rewrite this in matrix form as,

$$AUBB^{-1}\mathbf{g} = RB^{-1}\mathbf{g} = 0 \qquad (\mathrm{mod}\ 1), \tag{7.14}$$

from which the general solution for $\boldsymbol{g}$ is seen to be,

$$\boldsymbol{g} = B\begin{bmatrix} m_1 k_1 \\ \vdots \\ m_l k_l \end{bmatrix}, \qquad k_i \in \mathrm{Z}, \tag{7.15}$$

where m_i is the order of the element r_i of R. Now if $\boldsymbol{x} \cdot \boldsymbol{g} = 0$ (mod 1) for all $\boldsymbol{g} \in \mathrm{G}$ (i.e., all k_i in (7.15)) it must have the form,

$$\boldsymbol{x} = (a_1, \dots a_l)RB^{-1} = (a_1, \dots a_l)AU \qquad (\mathrm{mod}\ 1). \tag{7.16}$$

In other words $\boldsymbol{x}$ is a linear combination of the $\boldsymbol{U}_i$ plus some integer vector. If we demand in addition that $\boldsymbol{x} \cdot \boldsymbol{U}_i = 0$ (mod 1) for all $\boldsymbol{U}_i$ then this integer vector must in fact be in G. This is the desired result: if $\boldsymbol{x} \cdot \boldsymbol{q} = 0$ (mod 1) for all $\boldsymbol{q} \in \Gamma_S$ then $\boldsymbol{x}$ must be of the form $\boldsymbol{g} + \Sigma_i \boldsymbol{U}_i \alpha_i$ and hence be in Γ_S.

We have shown that Γ is a shifted (by the spin-statistic vector $\boldsymbol{S}$) lorentzian lattice, even or odd depending on $\boldsymbol{S}^2$, and that Γ_S is a lorentzian self-dual integer lattice. In sect. 8 we will show that these properties are sufficient to give a modular invariant string theory. Not surprisingly then, the converse of these results is also true: if Γ_S is an integer self-dual lattice and Γ is even or odd (depending on $\boldsymbol{S}^2$) then Γ can be represented in the form (7.8) with the constraints (7.9) and (7.10), and treated in the formalism of sect. 3. The only somewhat subtle step involved in showing this is proving that $\boldsymbol{U}_0$ is in Γ_S. The constraint on Γ, that is $\frac{1}{2}\boldsymbol{Q}^2 = \frac{1}{2}\boldsymbol{S}^2$ (mod 1) for all $\boldsymbol{Q} \in \Gamma$, necessarily requires that $\frac{1}{2}\boldsymbol{g}^2 = 0$ (mod 1) for all $\boldsymbol{g} \in \mathrm{G}$. This is equivalent, however, to having $\boldsymbol{W}_0 \cdot \boldsymbol{g} = 0$ (mod 1) for all $\boldsymbol{g} \in \mathrm{G}$ and by (7.16) we then have that $\boldsymbol{W}_0$ plus some integer vector (this sum is precisely of the form of $\boldsymbol{U}_0$) must be in Γ_S.

Finally let us prove the result alluded to in sects. 4 and 6, that the massless vector bosons of any heterotic string model constructed by the rules of sects. 3 and 5 (with only complex fermions) transform as the adjoint representation of a simply laced Lie group. Here a right-moving excitation provides the spacetime vector structure and the internal symmetry arises from the left movers. Let us define Γ_{L} to be the charge lattice Γ projected onto the sub-space pertaining to left movers (i.e. with elements consisting of just the left mover components of $\boldsymbol{Q}$). We will show that the subset of elements of Γ_{L} which represents massless vector bosons (call this set Φ with elements φ) consists of the root vectors of a simply laced Lie algebra. The right mover portion of $\boldsymbol{Q}$ for massless vector bosons has one nonzero component of magnitude 1. Thus since we have previously found that Γ is an odd ($\boldsymbol{S}^2 = -1$ for heterotic type models) "lattice" we have that Φ is a finite subset of an even integer lattice. Moreover for massless bosons we must have,

$$\varphi^2 = 0 \text{ or } 2\,. \tag{7.17}$$

This follows because of the close relation between φ^2 and the (mass)^2 from the left movers, specifically,

$$\begin{aligned} \tfrac{1}{2}\varphi^2 &= \tfrac{1}{2} \sum_{l:\,\text{left}} \left(N^l_{\overline{\alpha W}} + \overline{\alpha W}^{\,l} - \tfrac{1}{2}\right)^2, \\ &= E^{\text{left}}_{\text{vacuum}} + 1 + \sum_{L:\,\text{left}} \tfrac{1}{2} N^l_{\alpha W}\left(N^l_{\alpha W} - 1\right) + N^l_{\alpha W}\overline{\alpha W}^{\,l}, \\ &= m^2 + 1 + \sum \left\{ \tfrac{1}{2} N^l_{\alpha W}\left(N^l_{\alpha W} - 1\right) - \bar{n}^l \right\}. \end{aligned} \tag{7.18}$$

As one can readily verify, for massless states the last term in (7.18) is -1 or 0 giving (7.17). In addition it is clear that any $\varphi \in \Gamma_{\text{L}}$ satisfying (7.17) represents some massless vector boson and so is in Φ.

Eq. (7.17) guarantees that all nonzero elements of Φ have the same length and that if φ is in Φ then the only multiples of φ which are in Φ are $\pm\varphi$ and the zero vector. All that remains to demonstrate that Φ is a system of root vectors of a simply laced Lie algebra is to show that if φ_1 and φ_2 are nonzero elements of Φ with $\varphi_1 \cdot \varphi_2 > 0$ then $\varphi_1 - \varphi_2$ is also in Φ. This is obviously satisfied since $\varphi_1 - \varphi_2 \in \Gamma_{\text{L}}$ and $(\varphi_1 - \varphi_2)^2 = 4 - 2\varphi_1 \cdot \varphi_2 = 2$, so by our previous observation $\varphi_1 - \varphi_2 \in \Phi$.

8. Properties of the fermionic charge lattice

In the previous section we reexpressed the bulk of the spin structure construction of sects. 3–6 in terms of the fermionic charge lattice. As pointed out in sect. 2 and to be reiterated in sect. 9 this charge lattice is equivalent to the "momentum lattice" (i.e., the set of allowed momentum eigenvalues) of the equivalent two-dimensional bosonic system. Not surprisingly then we can re-examine in a purely bosonic formulation some properties of the charge lattice proved in sect. 7 using the spin structure construction. In particular we will demonstrate sufficient conditions on the charge lattice for modular invariance and correct space-time spin-statistics using a bosonic construction (little knowledge of the spin structure construction is necessary) which is essentially a non-trivial generalization of the lorentzian even self-dual lattices considered by Narain. We close this section with some comments on string interactions in the charge lattice formulation.

Once again let us consider for concreteness a heterotic type string model (in light cone gauge) in an even number, D, of space-time dimensions. We can now bosonize all of the two-dimensional degrees of freedom, the compactified string coordinates and, since D is even, the space-time RNS fermions and the "compactified" fermions as well. This leaves us with $26-D$ left-moving and $14-D$ right-moving "compactified" bosonic "coordinates" as well as the $D-2$ light cone string coordinates. We want to specify the compactification of the bosonic coordinates by giving the corresponding momentum "lattice", and are specifically concerned with choosing the momentum lattice so that the partition function is modular invariant. This is essentially the problem considered in string compactification but now with the added wrinkle of treating pairs of space-time RNS fermions as additional compactified bosonic "coordinates". These degrees of freedom are on exactly the same footing as the compactified string coordinates except in two respects: (i) they must satisfy certain symmetry requirements amongst themselves to insure unitarity and Lorentz invariance of the physical string theory (e.g. the space-time degrees of freedom must be treated symmetrically) and (ii) the space-time degrees of freedom carry information about space-time statistics, i.e., determining whether a given state is a space-time boson or fermion. For the details of handling (i) above we refer the reader to sects. 5 and 7 since these considerations do not affect the discussion of modular invariance given here; (ii) is properly handled in the construction we now describe.

The momentum "lattice", Γ, (which is just the charge lattice given in sect. 7) is $\{(14-D)+(26-D)\}$ dimensional with a lorentzian signature, in other words an element $\boldsymbol{Q}\in\Gamma$ can be written $\boldsymbol{Q}=(\boldsymbol{Q}_{\mathrm{R}},\boldsymbol{Q}_{\mathrm{L}})$ where $\boldsymbol{Q}_{\mathrm{R}}$ and $\boldsymbol{Q}_{\mathrm{L}}$ are $(14-D)$ and $(26-D)$ dimensional vectors respectively and the inner product is given by $\boldsymbol{Q}_1\cdot\boldsymbol{Q}_2 = \boldsymbol{Q}_{\mathrm{L}1}\cdot\boldsymbol{Q}_{\mathrm{L}2}-\boldsymbol{Q}_{\mathrm{R}1}\cdot\boldsymbol{Q}_{\mathrm{R}2}$. Γ is taken to have the following properties:

(i) half-integer inner products, i.e. if $\boldsymbol{X},\boldsymbol{Y}\in\Gamma$ then $2\boldsymbol{X}\cdot\boldsymbol{Y}\in Z$;

(ii) odd in the sense that for all $\boldsymbol{X}\in\Gamma$, $\boldsymbol{X}\cdot\boldsymbol{X}=$ odd integer;

(iii) symmetrical in its first $\frac{1}{2}D-1$ components, i.e., permuting these components does not alter the set Γ; these components must take integer (i.e. boson) or half-integer (i.e. fermion) values; the rest of the right-moving coordinates come in triplets and each triplet satisfies the triplet constraint, i.e.

$$Q^1 = Q_i^1 + Q_i^2 + Q_i^3 \qquad (\mathrm{mod}\ 1)\,, \tag{8.1}$$

where

$$\boldsymbol{Q}_{\mathrm{R}} = (Q^1,(Q_1^1,Q_1^2,Q_1^3),(Q_2^1,Q_2^2,Q_2^3),(Q_3^1,Q_3^2,Q_3^3))\,;$$

(iv) notice that strictly speaking Γ is not a lattice at all because it cannot be generated by a set of basis vectors (because of property ii); however, it is such that $\Gamma_S=\{\boldsymbol{L}\colon \boldsymbol{L}=\boldsymbol{Q}-\boldsymbol{S};\ \boldsymbol{Q}\in\Gamma,\ \boldsymbol{S}=(1,0,\dots 0)\}$ is an integer lattice and,

(v) the lattice Γ_S is an odd self-dual lattice, i.e., $\Gamma_S^* = \Gamma_S$.

The purpose of the spin-statistics vector $\boldsymbol{S}$ is to sort out the space-time bosons and fermions in the sense that $\boldsymbol{L} \in \Gamma_S$ represents a space-time boson or fermion depending on whether $\boldsymbol{L} \cdot \boldsymbol{S}$ is integral or half integral, i.e. Γ_S is an odd self-dual lorentzian lattice with $\boldsymbol{L} \in \Gamma_S$ satisfying the condition

$$\tfrac{1}{2}\boldsymbol{L}^2 = \boldsymbol{L} \cdot \boldsymbol{S} \qquad (\text{mod } 1). \tag{8.2}$$

Now the partition function of the system of bosons with momenta taking values in Γ is (cf. (2.26)),

$$Z(\tau, \bar{\tau}) = [\eta(\tau)]^{D-26}[\eta^*(\tau)]^{D-14} h(\tau, \bar{\tau}), \tag{8.3a}$$

where

$$h(\tau, \bar{\tau}) = \sum_{\boldsymbol{Q} \in \Gamma} \exp\{i\pi\tau \boldsymbol{Q}_{\mathrm{L}}^2 - i\pi\bar{\tau}\boldsymbol{Q}_{\mathrm{R}}^2 + 2\pi i \boldsymbol{Q} \cdot \boldsymbol{S}\}, \tag{8.3b}$$

and $\eta(\tau)$ is the Dedekind eta function. As mentioned the $\boldsymbol{S}$-dependent term in (8.3b) guarantees the correct statistics for states contributing to (8.3). We will now show how the properties we have bestowed upon Γ guarantee that (8.3) is modular invariant. Under the transformation $\tau \to \tau + 1$ it is easy to see that,

$$\begin{aligned} Z(\tau + 1, \bar{\tau} + 1) &= \exp\{2\pi i(\tfrac{1}{2}\boldsymbol{Q}^2 - \tfrac{1}{24}((26 - D) - (14 - D)))\} Z(\tau, \bar{\tau}), \\ &= Z(\tau, \bar{\tau}), \end{aligned}$$

where we have used property (ii) of Γ. To consider the transformation $\tau \to -1/\tau$ we first rewrite $h(-1/\tau, -1/\bar{\tau})$ in terms of Γ_S,

$$h\left(-\frac{1}{\tau}, -\frac{1}{\bar{\tau}}\right) = \sum_{\boldsymbol{L} \in \Gamma_S} \exp\left\{\frac{-i\pi}{\tau}\boldsymbol{L}_{\mathrm{L}}^2 + \frac{i\pi}{\tau}(\boldsymbol{L}_{\mathrm{R}}^2 - 2\boldsymbol{L} \cdot \boldsymbol{S} + 1) + 2\pi i \boldsymbol{L} \cdot \boldsymbol{S}\right\}. \tag{8.4}$$

Now inserting a δ-function and using Poisson's summation formula (given that Γ_S is self-dual),

$$\sum_{\boldsymbol{L} \in \Gamma_S} \delta(\boldsymbol{V} - \boldsymbol{L}) = \sum_{\boldsymbol{L} \in \Gamma_S^* = \Gamma_S} \mathrm{e}^{2\pi i \boldsymbol{V} \cdot \boldsymbol{L}}, \tag{8.5}$$

we can write ($\boldsymbol{V} = (\boldsymbol{v}, \tilde{\boldsymbol{v}})$),

$$\begin{aligned} h\left(-\frac{1}{\tau}, \frac{-1}{\bar{\tau}}\right) = \sum_{\boldsymbol{L} \in \Gamma_S} \int \mathrm{d}^{14-D} v \, \mathrm{d}^{26-D} \tilde{v}, \\ \times \exp\left\{\frac{-i\pi}{\tau}\tilde{\boldsymbol{v}}^2 + 2\pi i \tilde{\boldsymbol{v}} \cdot \boldsymbol{L}_{\mathrm{L}} + \frac{i\pi}{\tau}(\boldsymbol{v}^2 - 2\boldsymbol{V} \cdot \boldsymbol{S} + 1) \right. \\ \left. + 2\pi i \boldsymbol{V} \cdot \boldsymbol{S} - 2\pi i \boldsymbol{v} \cdot \boldsymbol{L}_{\mathrm{R}}\right\}. \end{aligned} \tag{8.6}$$

Performing the gaussian integrations we are left with,

$$h\left(-\frac{1}{\tau},-\frac{1}{\bar{\tau}}\right)=(-i\tau)^{(26-D)/2}$$

$$\times(i\bar{\tau})^{(14-D)/2}\sum_{L\in\Gamma_S}\exp\left\{i\pi\tau L_{\mathrm{L}}^2-i\pi\bar{\tau}\left(L_{\mathrm{R}}^2-2L\cdot S+1\right)+2\pi iL\cdot S\right\}$$

$$=(-i\tau)^{(26-D)/2}(i\bar{\tau})^{(14-D)/2}h(\tau,\bar{\tau}). \tag{8.7}$$

Combining this result with the transformation of $\eta(\tau)$ given in (2.31) we obtain finally the desired result,

$$Z\left(-\frac{1}{\tau},-\frac{1}{\bar{\tau}}\right)=Z(\tau,\bar{\tau}). \tag{8.8}$$

As an illustration we give the lattice Γ explicitly for two simple models, the $E_8\otimes E_8$ heterotic string in 10 dimensions and the $N=4$ supersymmetric SO(44) model in 4 dimensions. In the former case we have,

$$\Gamma=\left\{Q=L+S\colon L=(L_{\mathrm{R}},L_{\mathrm{L}})\right\}, \tag{8.9}$$

where $L_{\mathrm{R}}\in$ weight lattice of SO(8) and $L_{\mathrm{L}}\in\Gamma_8\otimes\Gamma_8=$ root lattice of $E_8\otimes E_8$. The 16-dimensional sublattice of Γ which does not describe the space-time RNS fermions is simply the even integer self-dual lattice from which the $E_8\otimes E_8$ heterotic string is conventionally obtained. Turning now to the $N=4$ SO(44) model we can take $\Gamma=\{Q=L+S\colon L\in\Gamma_S\}$ where Γ_S is generated by $(k_1,0,0),(0,k_2,0),(0,0,\tilde{k})$ where k_1, k_2 and $\tilde{k}$ are root vectors of SO(8), SO(12), and SO(44) respectively and $[022]=(0,W_v^2,\tilde{W}_v)$, $[100]=(W_s^1,0,0)$ and $[011]=(0,W_s^2,\tilde{W}_s)$ with $W^1_{v(s)}$, $W^2_{v(s)}$ and $\tilde{W}_{v(s)}$ weights of the vector (spinor) representation of SO(8), SO(12), and SO(44) respectively. The lattice obtained from Γ by removing the first four components (the SO(8) weights) is a 6 + 22 dimensional, even-integer, self-dual, lorentzian lattice such as those considered in ref. [12]. In general, charge lattices can easily be obtained from the spin structure construction as described in sect. 7. Both the names "charge lattices" and "momentum lattices" are derived from their two-dimensional field theoretic interpretation. In terms of space-time, it is probably more appropriate to call them "gauge lattices".

Since properties of self-dual lattices have been extensively studied, there exists a large body of known results and methods which may be used to facilitate the study of charge lattices. Let us illustrate this approach by reproducing the list of ten-dimensional string models with rank-16 gauge group constructed in ref. [4, 11]. These are the models where the fermionic formulation can be written in terms of the charge lattice. In 10 dimensions, we have ($\boldsymbol{L}^2 = \boldsymbol{L}_{\rm L}^2 - \boldsymbol{L}_{\rm R}^2$)

$$\boldsymbol{L} = (\boldsymbol{L}_{\rm R}, \boldsymbol{L}_{\rm L}),$$

$$\boldsymbol{L}_{\rm R} = (l_1, l_2, l_3, l_4), \tag{8.10}$$

where the right-moving sector corresponds to the SO(8) group. Here l_i are all half-integers or all integers, so that $\mathrm{L}_{\rm R}^2 = \Sigma l_i^2$ is always an integer. Therefore, instead of the lorentzian lattice, we can consider the euclidean lattice

$$\boldsymbol{L}_{\rm E}^2 = \boldsymbol{L}_{\rm L}^2 + \boldsymbol{L}_{\rm R}^2 = \boldsymbol{L}^2 + 2\boldsymbol{L}_{\rm R}^2, \tag{8.11}$$

so that

$$\tfrac{1}{2}\boldsymbol{L}_{\rm E}^2 = \boldsymbol{L}_{\rm E} \cdot \boldsymbol{S}_{\rm E} \ (\mathrm{mod}\ 1) \tag{8.12}$$

where $\boldsymbol{S}_{\rm E} = \boldsymbol{S}$. Thus we need to consider 20-dimensional self-dual integer lattices $\mathrm{L}_{\rm E}$ that satisfy the constraint (8.12). Since the complete classification of 20-dimensional self-dual integer lattices are known, all we need to do is select the subset that satisfies (8.12) and contains the root lattice of $\mathrm{D}_4 = \mathrm{SO}(8)$. The latter is required for space-time Lorentz properties.

A d-dimensional integral lattice Λ^d containing vectors of length 1 can be decomposed in the form,

$$\Lambda^d = \mathrm{Z}^m \oplus \mathrm{K}^{d-m}, \tag{8.13}$$

where Z^m is the m-dimensional cubic lattice and K^{d-m} is a $(d-m)$-dimensional lattice whose minimum length squared is at least 2. The only indecomposable lattices which are generated by vectors of length squared 2 are A_n ($n \geqslant 1$), D_n ($n \geqslant 4$), E_6, E_7 and E_8. So if Λ is a lattice with minimum length squared 2, the lattice Λ is generated by a sublattice $\Lambda_{\rm s}$ where $\Lambda_{\rm s}$ is a direct sum of the simply-laced lattices $\Lambda_i = \mathrm{A}_n, \mathrm{D}_n, \mathrm{E}_n$ and a set $\{y\}$ of weights:

$$\Lambda = \langle \Lambda_s = \Lambda_1 \oplus \Lambda_2 \oplus \cdots \oplus \Lambda_k \quad \text{and a set of } y = [y_1, y_2, \dots y_k] \rangle, \tag{8.14}$$

where y_i is a weight lattice of Λ_i.

In general, a 20-dimensional lattice that contains a factor of Z^m cannot satisfy the constraint (8.12) unless Z^m overlaps with the first component, since otherwise the product $\boldsymbol{L}_{\rm E} \cdot \boldsymbol{S}_{\rm E}$ is independent of Z^m while $\frac{1}{2}\boldsymbol{L}_{\rm E}^2$ depends on these components. Consider then the case where Z^m overlaps with the first component. Recall that the D_4 root lattice must be included in the first four components as a sublattice because of the space-time SO(8) symmetry. To contain a Z-lattice, it is necessary and sufficient for this D_4 to contain a spinor weight $[s] = ((\frac{1}{2})^4)$, which has length squared 1. The resulting lattice corresponds to $\mathrm{Spin}(8)/\mathrm{Z}_2$ which is self-dual. Since the Spin(8)/Z_2 lattice contains the four orthogonal points $(\frac{1}{2}, \frac{1}{2}, \frac{1}{2}, \frac{1}{2})$, $(\frac{1}{2}, \frac{1}{2}, -\frac{1}{2}, -\frac{1}{2})$, $(\frac{1}{2}, -\frac{1}{2}, \frac{1}{2}, -\frac{1}{2})$ and $(\frac{1}{2}, -\frac{1}{2}, -\frac{1}{2}, \frac{1}{2})$ which have length squared 1, this lattice is isomorphic to the Z^4 lattice. Hence we have a lattice of the type

$$\Lambda = (\mathrm{Spin}(8)/Z_2) \oplus Z^{m-4} \oplus \Lambda_{\mathrm{Rest}}\,.$$

It is clear that, for $\boldsymbol{L}_{\mathrm{R}} \in \mathrm{Spin}(8)/Z_2$ lattice,

$$\tfrac{1}{2}\boldsymbol{L}_{\mathrm{R}}^2 + \boldsymbol{L}\cdot\boldsymbol{S} = 0 \qquad (\mathrm{mod}\ 1)\,. \tag{8.15}$$

This implies that the left-moving and right-moving sectors factorize and from (8.12),

$$\tfrac{1}{2}\boldsymbol{L}_{\mathrm{L}}^2 = 0 \qquad (\mathrm{mod}\ 1)\,, \tag{8.16}$$

which means $Z^m = Z^4$ and all 16-dimensional even self-dual euclidean lattices are acceptable, i.e.

$$\Lambda = \{E_8 \oplus E_8;\ \text{no weight lattice}\}\,,$$

or

$$\Lambda = \{D_{16}; [s]\}\,, \tag{8.17}$$

where $[s] = ((\frac{1}{2})^{16})$ is the weight lattice of the spinor representation of D_{16}. These choices give the $E_8 \otimes E_8$ and Spin(32)/Z_2 heterotic string models; the left-right factorization provides the space-time supersymmetry.

Now all that are left to consider are 20-dimensional self-dual lattices with no Z^m factors. These lattices have been classified by Conway and Sloane [27]. There are 12 such lattices, all of type I, i.e., odd lattices. Out of this set of lattices, three of them do not contain D_4 as a regular embedding (to be explained in a moment) and may be ignored for our purposes. The remaining 9 lattices are shown in table 1. To explain table 1, let us consider the first 2 cases. The "D_{20}" lattice is generated by the root lattice of the D_{20} algebra plus its spinor weight lattice $[1] = ((\frac{1}{2})^{20})$. For any root lattice, $\boldsymbol{L}_{\mathrm{E}}\cdot\boldsymbol{S}_{\mathrm{E}} = 0$ (mod 1) and $\frac{1}{2}\boldsymbol{L}_{\mathrm{E}}^2 = 0$ (mod 1). Let $D_{20} \supset D_4 \oplus D_{16}$. For the [1] class, $\frac{1}{2}\boldsymbol{L}_{\mathrm{E}}^2 = \frac{5}{2}$ and $\boldsymbol{L}_{\mathrm{E}}\cdot\boldsymbol{S}_{\mathrm{E}} = \frac{1}{2}$. Hence D_{16} for the left-moving sector satisfies the constraint (8.12). Next, consider the $A_{15}D_5$ case. In A_n,

$$[i] = \left(\left(\frac{i}{n+1}\right)^j \left(\frac{-j}{n+1}\right)^i\right),$$

where $i + j = n + 1$. This vector has length $ij/(n+1)$. Eq. (8.12) is always satisfied by the root lattice, hence all we have to worry about are the weight lattices. In this case, the weight lattice is $y = ((\frac{2}{16})^{14}(-\frac{14}{16})^2(\frac{1}{2})^5)$. If we embed D_4 in A_{15}, $\boldsymbol{L}_{\mathrm{E}}\cdot\boldsymbol{S}_{\mathrm{E}} = \frac{1}{8}$ or $\frac{7}{8}$, clearly unacceptable (since this implies parastatistics for space-time particles).

TABLE 1
Lattices for ten-dimensional non-supersymmetric heterotic string models

Sublattice Λ_s	Weight lattice y	Left-moving sector Λ_{left}	Constraint satisfied?
D_{20}	[1]	D_{16}	yes
$A_{15}D_5$	[21]	$A_{15}D_1$	yes
$D_{12}D_8$	[12],[21]	$D_{12}D_4$	yes
		D_8D_8	no
$D_{12}E_8$	[10]	D_8E_8	yes
		$D_{12}D_4$	no
$D_8D_8D_4$	[101],[013],[330]	D_8D_8	yes
		$D_8D_4D_4$	no
		$D_4D_8D_4$	no
$E_7E_7D_6$	[101],[013]	$E_7E_7D_2$	yes
$A_7A_7D_5[4]$	[1112],[1721]	$A_7A_7D_1[4]$	no
$D_6D_6D_6A_1A_1$	[(123)00],[11110][33301]	$D_6D_6D_2A_1A_1$	no
$D_4D_4D_4D_4D_4$	[11111],[(02332)]	$D_4D_4D_4D_4$	no

All 20-dimensional odd self-dual euclidean lattices that contain D_4 as a regular embedding are listed. This list is given in ref. [27]. We adopt their notations. Each of the nine lattices is given by a sublattice Λ_s (column 1) plus a set of weight lattices (column 2). For D_n, [0], [1], [2], and [3] stands for the weight lattice [zero], [spinor], [vector], and [conjugate spinor] respectively; for A_n, $[i]$ stands for the class containing the tensor of rank i; for E_8, [0] stands for the zero vector; for E_7, the weight $[1]=((\frac{1}{4})^6,(-\frac{3}{4})^2)$ corresponding to the $\underline{56}$ representation. For weights in curly bracket, cyclic permutation is implied. Column 3 is the sub-lattice obtained by removing the right-moving sector (i.e. D_4). Column 4 indicates whether the particular lattice satisfies the spin-statistic condition (8.12).

For the same reason, the condition that $\boldsymbol{L}\cdot\boldsymbol{S}$ must be an integer or half-integer rules out the embedding of the right-moving D_4 in A_n, E_6, and E_7. Hence we must embed D_4 in D_5 of $A_{15}D_5$. In this case $\frac{1}{2}\boldsymbol{L}_E^2=\frac{1}{2}y^2=\frac{3}{2}$ and $L_E\cdot S_E=\frac{1}{2}$ so that constraint (8.12) is satisfied. Using the constraint (8.12), we generate the list of acceptable 10-dimensional models. This analysis is quite similar to that of ref. [28] where 24-dimensional self-dual even lattices were examined.

This method can be extended to analyze four-dimensional models. For example, for a subclass of models where all inner products of charge vectors of the right-moving sector are integers or half-integers, we can consider 32-dimensional euclidean self-dual integer lattices that satisfy the triplet constraint (8.1) and the constraint

$$\tfrac{1}{2}\boldsymbol{L}_E^2=\boldsymbol{L}_R^2+\boldsymbol{L}\cdot\boldsymbol{S}\qquad(\mathrm{mod}\ 1),\tag{8.18}$$

where $\boldsymbol{L}_R^2$ is an integer or half-integer. In general, however, for phenomenologically interesting models we need to examine odd self-dual lorentzian lattices. This analysis is complicated by the large size of the lattice as well as the presence of Thirring interactions (or background fields). Such an analysis is non-trivial but tractable. Progress on this approach will be discussed elsewhere.

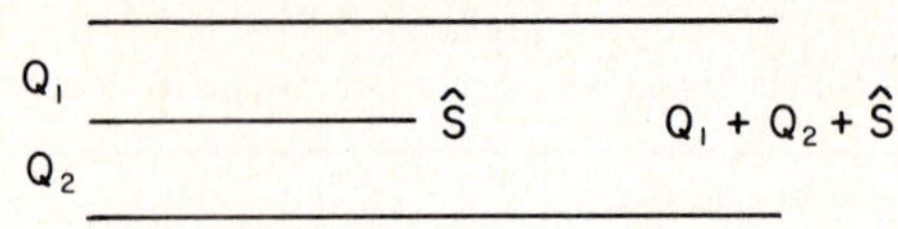

Fig. 2. The three-point vertex of a closed string theory in the light-cone picture. A string in the Q_1 state joins with a string in the Q_2 state to form a string in the $Q_1 + Q_2 + \hat{S}$ state.

In principle, the treatment of string interactions is straightforward. The bosonic vertex operators in the fermionic formulation are discussed in sect. 5. It is quite easy to reinterpret them for the charge lattice formulation. Although the fermionic vertex operators are more complicated, the standard construction of refs. [21, 29] can be adopted. Here, we shall simply concern ourselves with a simple but important feature of string interactions in the charge lattice picture.

As pointed out earlier, the charge lattice $\Gamma = \{\boldsymbol{Q}\}$ is strictly speaking not a lattice. Let $\boldsymbol{Q}_1, \boldsymbol{Q}_2 \in \Gamma$ so that $\boldsymbol{Q}_1 = \boldsymbol{L}_1 + \boldsymbol{S}$, $\boldsymbol{Q}_2 = \boldsymbol{L}_2 + \boldsymbol{S}$ with $\boldsymbol{L}_1, \boldsymbol{L}_2 \in \Gamma_S$ where Γ_S is a lattice. Then $\boldsymbol{Q}_1 + \boldsymbol{Q}_2 = \boldsymbol{L}_1 + \boldsymbol{L}_2$ is in Γ_S but not in Γ, i.e.

$$\boldsymbol{Q}_1 + \boldsymbol{Q}_2 \notin \Gamma \qquad \text{for } \boldsymbol{Q}_1, \boldsymbol{Q}_2 \in \Gamma . \tag{8.19}$$

If a string in the $\boldsymbol{Q}_1$ state joins with a string in the $\boldsymbol{Q}_2$ state, then the resulting string is apparently in the $\boldsymbol{Q}_1 + \boldsymbol{Q}_2$ state, which is unacceptable. This problem can easily be remedied if we associate with each vertex operator (in the light-cone gauge) a fermion charge $\hat{\boldsymbol{S}}$ so that

$$\boldsymbol{Q}_1 + \boldsymbol{Q}_2 + \hat{\boldsymbol{S}} \in \Gamma \qquad \text{for } \boldsymbol{Q}_1, \boldsymbol{Q}_2 \in \Gamma . \tag{8.20}$$

We can usually take simply $\hat{\boldsymbol{S}} = \boldsymbol{S}$. This is illustrated in fig. 2. Let us explain this in more detail by considering a specific example, the 10-dimensional $\mathrm{E}_8 \otimes \mathrm{E}_8$ heterotic string model. The charge lattice points of the graviton are given by

$$\boldsymbol{Q}_{\mathrm{g}} = \left(\pm 1, 0, 0, 0 \,|\, (0)^{16} \right),$$

to which we can add root vectors of SO(8) to the right-moving sector such that $\boldsymbol{Q}_{\mathrm{g}}^2 = -1$. There are eight such lattice points, each state of the form $b^{\dagger}_{1/2\mu} \tilde{\alpha}^{\dagger}_{\nu} |0\rangle$ which also includes the dilaton and $B_{\mu\nu}$ fields. The gravitino sector is given by

$$\boldsymbol{Q}_{\tilde{\mathrm{g}}} = \left(\tfrac{1}{2} \tfrac{1}{2} \tfrac{1}{2} \tfrac{1}{2} \,|\, (0)^{16} \right)$$

to which we can add root lattice vectors of SO(8) such that $\boldsymbol{Q}_{\tilde{\mathrm{g}}}^2 = -1$. There are eight such points. Now $\hat{\boldsymbol{S}}$ is given by $\hat{\boldsymbol{S}} = \boldsymbol{S}$, to which we can add root vectors of SO(8). Now we have, with appropriate choices of root lattice points of SO(8)

(including the zero lattice point)

$$Q_{\mathrm{g}} + Q_{\mathrm{g}} + \hat{S} = Q_{\mathrm{g}} \,,$$

$$Q_{\tilde{\mathrm{g}}} + Q_{\mathrm{g}} + \hat{S} = Q_{\tilde{\mathrm{g}}} \,,$$

$$Q_{\tilde{\mathrm{g}}} + Q_{\tilde{\mathrm{g}}} + \hat{S} = Q_{\mathrm{g}} \,,$$

where the various Q_{g} ($Q_{\tilde{\mathrm{g}}}$) in general differ by some root vectors. Here $\hat{S}$ has unit fermion charge and plays the role of the supercurrent charge. In fact, the necessity and the consistency of the inclusion of such a factor to each vertex in the light-cone gauge was pointed out a long time ago by Mandelstam [29]. This is also clear from the construction of emission vertices where we obtain F_2 vertices from F_1 vertices by taking commutators with the world-sheet supercharge as discussed in sect. 5.

9. Relation to other works

The preceding two sections established the formal connection between charge lattice constructions of string models and the spin-structure construction developed in sects. 3–6. Here we clarify this connection, first in the context of a toy model (a single compactified boson), and then with some more general comments. In particular we will demonstrate how a change in the bosonic radius of compactification manifests itself in the spin structure construction and how a simple Z_2 orbifold is handled. The relation of these simple bosonic and fermionic models has been discussed elsewhere [15] so we only stress here those features which clarify the spin-structure construction. Finally, we also explain, via the charge lattice, how Narain's construction can be obtained from the spin structure construction.

A one-dimensional compactified closed string can be described by a free compactified two dimensional boson (neglecting the non-vanishing conformal anomaly which will not affect the present discussion). In terms of normal mode operators the hamiltonian and momentum operators are,

$$\hat{H} = \frac{\hat{k}^2}{a^2} + \frac{a^2\hat{r}^2}{4} + \sum_{n=1}^{\infty} \left\{ n a_n^\dagger a_n + n \tilde{a}_n^\dagger \tilde{a}_n \right\} - \tfrac{1}{12} \,, \tag{9.1}$$

$$\hat{P} = \hat{k}\hat{r} + \sum \left\{ n a_n^\dagger a_n - n \tilde{a}_n^\dagger \tilde{a}_n \right\} , \tag{9.2}$$

where a is the compactification radius, $[a_n, a_{n'}^\dagger] = [\tilde{a}_n, \tilde{a}_{n'}^\dagger] = \delta_{nn'}$, and the eigenvalues k and r of the internal momentum and winding number operators are necessarily integers. The partition function, on the torus is given by,

$$Z(\tau_1, \tau_2) = \mathrm{Tr}[\mathrm{e}^{-2\pi\tau_2 \hat{H}} \mathrm{e}^{2\pi i \tau_1 \hat{P}}] \,, \tag{9.3}$$

where there is no space-time spin statistics factor. To compare with the spin structure construction we rewrite this in terms of right- and left-moving chiral bosons ($\tau = \tau_1 + i\tau_2$),

$$Z(\tau, \bar{\tau}) = \mathrm{Tr}\left[e^{2\pi i \tau \hat{H}_L} e^{-2\pi i \bar{\tau} \hat{H}_R}\right], \tag{9.4}$$

where

$$\hat{H}_L = \frac{\hat{H} + \hat{P}}{2} = \frac{1}{2}\left(\frac{k}{a} + \frac{ar}{2}\right)^2 + \sum n a_n^\dagger a_n - \frac{1}{24}, \tag{9.5a}$$

$$\hat{H}_R = \frac{\hat{H} - \hat{P}}{2} = \frac{1}{2}\left(\frac{k}{a} - \frac{ar}{2}\right)^2 + \sum n \tilde{a}_n^\dagger \tilde{a}_n - \frac{1}{24}. \tag{9.5b}$$

With this system we associate the two-dimensional "momentum lattice",

$$\{P\} = \left\{\left(\left(\frac{k}{a} - \frac{ar}{2}\right), \left(\frac{k}{a} + \frac{ar}{2}\right)\right), \quad k, r \in Z\right\}. \tag{9.6}$$

We noted in sect. 2 the equivalence of the fermionic and bosonic systems when the fermionic charge lattice $\{Q\}$ and chiral boson momentum lattice $\{P\}$ coincide. Let us see which fermionic models in the spin-structure construction (extended in obvious fashion to this one dimensional case) possess charge lattices of the form (9.6) and so are equivalent to single bosons compactified with some particular radii. Limiting ourselves to free fermions we can at best generate rational compactification radii, $a = m/n$. For convenience assume that the integers m and n are both odd. The set $\{P\}$ in (9.6) can be put in the form of eq. (7.8) (S is zero in this purely bosonic toy model), by choosing $\{U_j\} = \{U_0 = (\frac{n}{2} | \frac{n}{2})$; $U_1 = (\frac{m}{n} | \frac{m}{n})$; $U_2 = (\frac{n}{m} | -\frac{n}{m})\}$. In the spin-structure construction this amounts to building a model with the generating vectors $W_0 = (\frac{1}{2} | \frac{1}{2})$, $W_1 = (\frac{1}{n} | \frac{1}{n})$ and $W_2 = (\frac{1}{m} | \frac{m-1}{m})$. The k_{ij} are all fixed by the constraining equations except for k_{00} and k_{10} both of which (from (7.5) and (7.6)) take the value zero. The simplest model, generated by W_0 alone (and $k_{00} = 0$) is thus equivalent to the single boson with $a = 1$. One can easily show that choosing $k_{00} = \frac{1}{2}$ generates the momentum lattice for the dual compactification radius, $2/a = 2$. The two systems (compactification radii a and $2/a$) give identical partition functions because of the symmetry under $\tau \to -1/\tau$ which effectively interchanges k and r in (9.5).

We cannot achieve irrational compactification radii with the free fermion formulation of sect. 3. It might seem that we could choose W_i with irrational components but in general this limit ($m_i \to \infty$) is ill defined. The exception is a set of $\{W_i\}$ for which we can choose $k_{ij} = 0$ if either W_i or W_j contains irrational components. This case turns out to be rather uninteresting in that an infinitely dense set of massive physical states appear in the spectrum; essentially a would-be-compactified dimension with irrational radius has "decompactified". Of course, we can choose m and n such that n/m is arbitrarily close to any given radius. In any case we can obtain an arbitrary compactification radius by choosing the model generated by say W_0 and adding the appropriate Thirring interactions among the fermions to change the effective bosonic radius (see sect. 5).

So far we have illustrated the spin-structure construction with complex fermions. What about the one simple example which requires real fermions:

$$W_0 = \left((\tfrac{1}{2})_r(\tfrac{1}{2})_r|(\tfrac{1}{2})_r(\tfrac{1}{2})_r\right),$$

$$W_1 = \left((\tfrac{1}{2})_r(0)_r|(\tfrac{1}{2})_r(0)_r\right). \tag{9.7}$$

For simplicity take all $k_{ij} = 0$. The W_0 and $\mathbf{0}$ sectors contribute to the spectrum those states in the model generated by W_0 alone which satisfy the additional constraint,

$$\tfrac{1}{2}\left(N^1_{W_0} + \tilde{N}^1_{W_0}\right) = 0 \qquad (\mathrm{mod}\ 1) \tag{9.8a}$$

in the W_0 sector and,

$$\tfrac{1}{4}(1 + \gamma^1\tilde{\gamma}^1) = 0 \qquad (\mathrm{mod}\ 1) \tag{9.8b}$$

in the $\mathbf{0}$ sector. One can check that in the bosonic formulation this is equivalent to that portion of the spectrum of the single boson with $a = 1$ which is invariant under the transformation $\phi \to -\phi$. Thus the W_0 and $\mathbf{0}$ sectors of the model (9.7) are equivalent to the untwisted sector of the "one-dimensional string" on the Z_2 orbifold obtained by identifying ϕ with $-\phi$. Not surprisingly one finds that the two twisted sectors corresponding to the fixed points $\phi = 0$ and $\phi = \pi$ contribute the same spectrum of states as the two sectors W_1 and $W_1 + W_0$ of the model (9.7). Thus the model generated via the spin structure construction from (9.7) which could not be expressed in terms of a charge lattice is equivalent to the single boson with Z_2 orbifold target space for which we correspondingly cannot define a momentum lattice.

Note that we cannot obtain the string model on the Z_2 orbifold with compactification radius $a \neq 1$ by adding generating vectors such as $(\frac{1}{n}|\frac{1}{n})$ and $(\frac{1}{m}|\frac{m-1}{m})$ since these boundary conditions do not commute with those represented by (9.7). In this case we must add Thirring interactions as discussed in sect. 5. It should be pointed out that the Thirring couplings are in one-to-one correspondence with the constant background fields discussed in ref. [13]. In D dimensions there are $(10-D)(26-D)$ Thirring couplings. They correspond precisely to the $\frac{1}{2}(10-D)(11-D)$ gravitational background fields $\langle g_{\mu\nu}\rangle$, $\frac{1}{2}(10-D)(9-D)$ antisymmetric tensor background fields $\langle B_{\mu\nu}\rangle$ and $16(10-D)$ gauge background fields $\langle A_\mu\rangle$ which give in total $(10-D)(26-D)$ parameters. In 4-dimensional string models, there are 132 parameters; for some models, some of these parameters can be absorbed into the boundary conditions of the world sheet fermions as we have seen. For some models, some of the background fields must be zero, as discussed in sect. 5.

Let us now clarify the relation between our construction and Narain's work on heterotic strings. In sects. 7 and 8 we discussed charge lattices with elements $\boldsymbol{Q} = \boldsymbol{L} + \boldsymbol{S}$, where the first component l_1 of $\boldsymbol{L}$ (in $\boldsymbol{L}_{\mathrm{R}}$) corresponds to a space-time degree of freedom and so must take integer or half-integer values.

Consider the subclass of such charge lattices with all elements satisfying,

$$l_2, l_3, l_4 \in \mathrm{Z} \qquad \text{if } l_1 \in \mathrm{Z},$$

$$l_2, l_3, l_4 \in \mathrm{Z} + \tfrac{1}{2} \qquad \text{if } l_1 \in \mathrm{Z} + \tfrac{1}{2}, \tag{9.9}$$

where l_2, l_3, l_4 are three of the remaining $13 - D \geqslant 4$ right-moving components. Then with $\boldsymbol{L}_{\mathrm{R}} = (l_1 l_2 l_3 l_4, \hat{\boldsymbol{L}}_{\mathrm{R}})$ and $\frac{1}{2}\boldsymbol{Q}^2 \in \mathrm{Z} + \frac{1}{2}$, we have

$$\tfrac{1}{2}\left(\boldsymbol{L}_{\mathrm{L}}^2 - \hat{\boldsymbol{L}}_{\mathrm{R}}^2\right) = \sum_{i=1}^{4} \tfrac{1}{2} l_i^2 + \boldsymbol{L} \cdot \boldsymbol{S} = 0 \qquad (\mathrm{mod}\ 1), \tag{9.10}$$

where $\hat{\boldsymbol{L}}_{\mathrm{R}}$ has $(10 - D)$ components and we have used

$$\tfrac{1}{2} l_i^2 = \tfrac{1}{8} \qquad (\mathrm{mod}\ 1) \qquad \text{for } l_i \in \mathrm{Z} + \tfrac{1}{2},$$

$$\tfrac{1}{2} l_i^2 = 0, \tfrac{1}{2} \qquad (\mathrm{mod}\ 1) \qquad \text{for } l_i \in \mathrm{Z}.$$

Therefore, in this special class of charge lattices, the constraint on the $(14 - D, 26 - D)$ lorentzian lattice is reduced to the requirement that the $(10 - D, 26 - D)$ lorentzian lattice $(\hat{\boldsymbol{L}}_{\mathrm{R}}, \boldsymbol{L}_{\mathrm{L}})$ must be an even integer self-dual lattice. These are the lattices considered by Narain [12]. We still have to check the triplet constraint. This can be trivially satisfied if the fermion charges are identified in a different way. Let us explain this in the $D = 4$ case, where as in eq. (5.1) the 20 right-moving real fermions are labeled ψ^j $(j = 1,2)$ and ψ_i^l $(i = 1,2,3;\ l = 1,2,\dots 6)$. To obtain complex fermions from pairs of these real fermions, we have taken the combinations,

$$\lambda = \psi^1 + i\psi^2,$$

$$\lambda_j^l = \psi_j^{2l-1} + i\psi_j^{2l}, \qquad j, l = 1,2,3 \tag{9.11}$$

with boundary conditions specified by s, a_j, b_j and c_j for λ, λ_j^1, λ_j^2 and λ_j^3 respectively satisfying (5.3b). We can, however, form complex fermions from an alternative combination,

$$\lambda_0 = \psi^1 + i\psi^2 ,$$

$$\lambda_j = \psi_1^{2j-1} + i\psi_1^{2j} , \qquad j = 1, 2, 3 ,$$

$$\hat{\lambda}_j = \psi_2^j + i\psi_3^j , \qquad j = 1, 2, \ldots 6 , \tag{9.12}$$

and demand boundary conditions l_j for λ_j and $\hat{l}_j$ for $\hat{\lambda}_j$. For l_i satisfying the special condition (9.9) (with the index j shifted by 1), the triplet constraint is always trivially satisfied for any $\hat{l}_i$, because the super current (5.1) can be rewritten as,

$$\tilde{T}_F = i \sum_{j=1}^{2} \psi^j \partial X^j + \tfrac{1}{2} \sum_{j=1}^{6} \psi_1^j \hat{\lambda}_j^* \hat{\lambda}_j . \tag{9.13}$$

Now it is clear that the triplet constraint is trivial, i.e.

$$l_j + \hat{l}_j - \hat{l}_j = l_j = l_0 . \tag{9.14}$$

It should be clear from previous examples that the spin structure construction often involves multiple left-right independent Z_2 orbifolds which are correlated to guarantee modular invariance and a physically sensible projection. Such left-right independent Z_2 twists have also been considered in ref. [30]. Let us turn our attention to the connection of the charge lattice to orbifolds. Starting from the charge lattice, we can introduce outer automorphisms as suggested by Dixon and Harvey [4]. This generalization of the charge lattice should be equivalent to the spin structure construction. Furthermore, we may introduce twists on the charge lattice so that the orbifold construction can be incorporated. On the other hand, if we start from the orbifold, we should introduce outer automorphisms as well as antisymmetric tensor background fields $\langle B_{\mu\nu} \rangle$. This generalized orbifold may then be equivalent to the twisted charge lattices. Further analysis is needed to clarify these possibilities.

It is intriguing to observe that the list of ten-dimensional models generated by the orbifold approach [4] coincides precisely with the list generated by the spin structure construction [11] while the two lists of $N = 1$ supersymmetric 4-dimensional models constructed as yet have no overlap. This peculiar feature may be understood in the following way. In general, orbifolds involve twists and so have fixed points. We may classify orbifolds into two types: orbifolds with curvatures (at the fixed points) and orbifolds without curvatures. The orbifolds used in constructing 10-dimensional models do not involve fixed points with curvatures. Also, since the Z_2 orbifolds only involve one-dimensional circles, their fixed points have no curvatures. Therefore, it is not surprising that models generated by orbifolds with no curvature can also be generated by the toroidal compactification in the fermionic formulation. On the other hand, the Z_n orbifold in general involves fixed points with curvatures. For example, the Z_3 orbifold constructed from a two dimensional torus [3] is topologically a sphere and each of the three fixed points has a curvature (i.e. deficit angle), of $\frac{4}{3}\pi$. It is clear that toroidal compactification with free fermions cannot generate such a sphere.

A complete classification of all four-dimensional string models is essentially reduced to a complete classification of modular invariant representations of the Kac-Moody algebra (or Virasoro algebra) which have the correct dimension and proper central charge, and satisfy the space-time spin-statistics constraint. We believe this problem is non-trivial but tractable. It will be interesting to see within the context of phenomenologically interesting models whether or not the generalized orbifold or twisted lattice give a complete classification.

10. Discussions and remarks

A popular view among string enthusiasts is that the world we live in is actually a 10-dimensional space-time described by the $E_8 \otimes E_8$ heterotic string, where 6 of the spatial dimensions are dynamically compactified so that the real world looks 4-dimensional; that is, the fermionic string model which describes our universe originates from the $E_8 \otimes E_8$ string model. It has been realized by many, however, that it may not be meaningful to identify a string model in less than 10 dimensions with a particular 10-dimensional string model. This becomes clear when it is pointed out that, [12,13] in 9 dimensions, for example, there is a unique lorentzian self-dual even-integer lattice, $U \otimes 2\Gamma_8$ (where U is a two-dimensional lorentzian lattice). One cannot uniquely identify any lattice in 10 dimensions as the origin of this lattice from compactification. From the rules of the spin structure or charge lattice construction discussed in this paper, it is clear that all models in $D \leqslant 10$ dimensions obtainable from these rules should be treated, in the first quantized version, on an equal footing, even though the 10-dimensional $E_8 \otimes E_8$ and $\mathrm{Spin}(32)/Z_2$ heterotic strings are particularly easy to construct.

It is also clear that in contrast to the 10-dimensional case the number of 4-dimensional chiral models is very large. As yet, a complete classification of all consistent string models is unavailable. In this work, we have given a complete treatment of fermionic string models in $D \leqslant 10$ dimensions obtainable from toroidal compactification in the fermionic formulation. This subclass of models is already quite large. In the first quantized formalism, all consistent string models should be treated on an equal footing. It is plausible that string dynamics may select a subset of the first quantized string models (i.e. second quantized vacuum states) as locally stable (e.g., by considering solutions of the (as yet unknown) closed string field theory). However, even if string dynamics eventually selects a unique ground state, it does not necessarily imply that this is the state representing our universe. Since the tunneling from some locally stable state to this unique ground state typically involves the Planck scale and hence may take a time many orders of magnitude longer than the age of the universe, our real world may be sitting at a locally stable point (but not the unique ground state) in the string field space. From this argument, we believe that the ultimate choice of the string ground state or the string model should be selected by direct confrontations with experiments. It is obvious that a very efficient approach is to construct four-dimensional models directly and compare their properties with experimental observations.

For most low-energy physics, the string model reduces to an effective gauge field theory with or without supergravity. Since many of the Thirring couplings can be absorbed into the boundary conditions or are unobservable at low energies each model has relatively few parameters and has completely specified low energy particle content. Properties of gauge theories are relatively well understood, so we see no insurmountable difficulties in comparing 4-dimensional string models with experiments. In practice, however, such a test is not necessarily easy due to the large size of the gauge group and non-perturbative effects in the gauge sector.

As pointed out in sect. 6 by a specific example, namely the chiral $SU(4) \otimes SU(2) \otimes SU(2)$ string model, the massless particle content of a string model need not reduce to an effective grand unified theory and string models can have properties, i.e. left-right asymmetry, unexpected from the point of view of grand unified theories.

The compactification to $D < 10$ dimensions in the fermionic formulation should be viewed as a mathematical compactification by construction, akin to the compactification of the sixteen left-moving coordinates in the heterotic string, rather than as a dynamical compactification. This is because, in the fermionic formulation of the compactified bosonic coordinates, the center-of-mass coordinate x^{μ} of the compactified dimension is absent. This is in contrast to the compactification in the bosonic formulation, where x^{μ} is present. Of course, one can question if this difference is actually physically real or not. The conversion of space-time to internal symmetry, or color, is well studied in Kaluza-Klein theories. In string theories, the reverse is also possible. For example, in any of the supersymmetric string models in $D < 10$ dimensions, there are massless scalar fields in some color or gauge group. The gauge symmetry is spontaneously broken if this scalar field develops (dynamically) a vacuum expectation value $\langle \phi_i \rangle$. This is the Goldstone mechanism in usual field theory. If this vacuum expectation value $\langle \phi_i \rangle$ is identified as the center of mass, x_i, of a coordinate X_i then, for large radius, color is converted to space-time. This almost equivalence of space-time and color is a very intriguing feature in string theories. Of course, any possible conversion of space-time $\leftarrow \rightarrow$ color depends on the details of the dynamics of the string theory. We should point out that the equivalence of space and color was known in Buddhism many centuries ago.

The ultraviolet finiteness of the $E_8 \times E_8$ or $Spin(32)/Z_2$ heterotic string can be analyzed by considering the holomorphic properties of the partition function since the left- and right-moving sectors factorize. For interesting four-dimensional models and non-supersymmetric 10-dimensional models, such factorization is absent and hence the issue of finiteness looks more involved. The work of Seiberg and Witten [8] and others suggest that physically sensible projection and one-loop modular invariance are probably sufficient to ensure multiloop modular invariance. Note that the collection of boundary conditions for the multiloop topologies is fixed by the requirement of factorization of amplitudes, once we have chosen the one-loop boundary conditions. For example near a string state pole the two-loop amplitude should factor into a product of a propagator and two one-loop amplitudes; in turn the sum of terms with different boundary conditions around the four non-contractible loops of the two handled manifold should agree with the product of the sums of boundary conditions around the two loops of the two tori giving the one-loop amplitude. In fact, it was demonstrated rather explicitly by Parkes [31] for the RNS case that one-loop modular invariance and physically sensible projections are enough to ensure two-loop modular invariance. Generalization of his argument for the general spin structure construction is straightforward. The multi-loop modular invariance of the models constructed by our approach can be placed on the same footnoting as that of the original 10-dimensional heterotic string.

In the context of fermionic string models, it is not clear that space-time supersymmetry is either necessary or desirable. In the usual grand unification schemes, supersymmetry provides a mechanism for resolving the gauge hierarchy, or equivalently the fine tuning, problem. In four-dimensional string models, since Higgs and Yukawa couplings are fixed in terms of the gauge coupling (and possibly the values of the background fields if they are non-zero), it is not clear a priori that there is a gauge hierarchy problem. A more compelling argument for supersymmetry may be the vanishing of the cosmological constant. However, a four-dimensional $N=1$ supersymmetric string model, which has a zero cosmological constant to start with, will in general develop a non-zero cosmological constant when supersymmetry is broken (probably spontaneously at a mass scale between the Planck scale and the electroweak scale). For this cosmological constant to vanish, the string model must have other parameters, such as vacuum expectation values of some scalar fields and/or radii of the compactified dimensions, which adjust themselves to minimize the vacuum energy to zero. However, if this mechanism occurs in supersymmetric string models, it probably occurs in non-supersymmetric chiral string models as well [32]. In fact, for non-supersymmetric string models, it is possible (probable) to have a cosmological constant in the very early universe to provide the desired inflation. Further investigation is clearly needed to re-examine the desirability of space-time supersymmetry. Of course, world sheet supersymmetry is an entirely separate issue. It is absolutely necessary to describe fermionic strings in terms of supersymmetric world-sheet fermions.

The success of string theory will be recognized if a four-dimensional chiral string model is found which has properties that are in agreement with observations.

Recalling that the standard model involves some of the simplest among all possible gauge groups, namely SU(3) ⊗ SU(2) ⊗ U(1), we may hope that nature is again kind in choosing a string model that is relatively simple to construct. A systematic approach to test the string theory would be to completely classify all consistent four-dimensional chiral string models and then examine them one by one. We believe such a complete classification is a tractable problem and that the relevance of string theory to nature can be tested.

We thank Lance Dixon, Paul Ginsparg, and Jeffrey Harvey and our colleagues at Cornell, in particular Joel Schwartz, for valuable discussions. This work is supported in part by the National Science Foundation.

Note added in proof

Thirring interactions in string theory have also been studied in two recent works, refs. [33] and [34].

References

[1] M.B. Green and J.H. Schwarz, Nucl. Phys. B255 (1985), 92;
J.H. Schwarz, Phys. Reports 89 (1982) 223;
M.B. Green, Surveys in High Energy Phys. 3 (1983) 127

[2] D.J. Gross, J.A. Harvey, E. Martinec and R. Rohm, Phys. Rev. Lett. 54 (1985) 502; Nucl. Phys. B256 (1985) 253 and B267 (1986) 75

[3] L. Dixon, J. Harvey, C. Vafa, and E. Witten, Nucl. Phys. B261 (1985) 651; B274 (1986) 285

[4] L.J. Dixon and J.A. Harvey, Nucl. Phys. B274 (1986) 93

[5] L.J. Dixon, D. Friedan, E. Martinec and S. Shenker, Nucl. Phys. B282 (1987) 13;
S. Hamidi and C. Vafa, Nucl. Phys. B279 (1987) 465

[6] H. Kawai, D.C. Lewellen and S.-H.H. Tye, Phys. Rev. Lett. 57 (1986) 1832; errata, to be published

[7] F. Gliozzi, J. Scherk and D. Olive, Nucl. Phys. B122 (1977) 253

[8] N. Seiberg and E. Witten, Nucl. Phys. B276 (1986) 272;
E. Witten, in Proc. of the Argonne Symp. on Anomalies, geometry and topology, ed. W.A. Bardeen and A.R. White (World Scientific, 1985)

[9] L. Alvarez-Gaumé, P. Ginsparg, G. Moore and C. Vafa, Phys. Lett. 171B (1986) 155

[10] S. Chaudhuri, H. Kawai, and S.-H.H. Tye, Cornell preprint CLNS-86/723 (1986);
W. Nahm, Bonn-HE-86-01;
R. Bluhm and L. Dolan, Phys. Lett. 169B (1986) 347;
E. D'Hoker and D.H. Phong, Nucl. Phys. B278 (1986) 225

[11] H. Kawai, D.C. Lewellen, and S.-H.H. Tye, Phys. Rev. D34 (1986) 3794

[12] K.S. Narain, Phys. Lett. 169B (1986) 41

[13] K.S. Narain, M.H. Sarmadi and E. Witten, Nucl. Phys. B279 (1987) 369
P. Ginsparg, Harvard preprint, HUTP-86/A053 (1986)

[14] P. Ramond, Phys. Rev. D3 (1971) 2415;
A. Neveu and J.H. Schwarz, Nucl. Phys. B31 (1971) 86

[15] S. Elitzur, E. Gross, E. Rabinovici and N. Seiberg, Nucl. Phys. B283 (1987) 413

[16] L. Brink, P. Di Vecchia and P. Howe, Phys. Lett. B65 (1976), 471;
S. Deser and B. Zumino, Phys. Lett. B65 (1976) 369

[17] C. Vafa, Nucl. Phys. B273 (1986) 592;
L. Alvarez-Gaumé, G. Moore and C. Vafa, Comm. Math. Phys. 106 (1986) 1
[18] M.F. Atiyah, V.K. Patodi, and I.M. Singer, Proc. Cambridge Philos. Soc. 77 (1975) 78, 405; 79 (1976) 71
[19] E. Witten, Nucl. Phys. B142 (1978) 285;
P. Di Vecchia, V.G. Knizhnik, J.L. Petersen, and P. Rossi, Nucl. Phys. B252 (1985) 701;
I. Antoniadis, C. Bachas, C. Kounnas and P. Windey, Phys. Lett. 171B (1986) 51;
P. Windey, Comm. Math. Phys. 105 (1986) 511;
P. Goddard, W. Nahm and D. Olive, Phys. Lett. 160B (1985) 111;
O. Alvarez, P. Windey, and M. Mangano, Berkeley preprint UCB-PTH-85/51 (1985)
[20] B. Klaiber, *in* Lectures in theoretical physics, ed. A.O. Barut and W.E. Britten (Gordon and Breach, New York, 1968);
G.F. Dell'Antonio, Y. Frishman, and D. Zwanziger, Phys. Rev. D6 (1972) 988;
S.-H.H. Tye, Phys. Rev. D9 (1974) 2821
[21] D. Friedan, E. Martinec and S. Shenker, Nucl. Phys. B271 (1986) 93
[22] A.A. Belavin, A.M. Polyakov, and A.B. Zamolodchikov, Nucl. Phys. B241 (1984) 333;
V. Knizhnik and A.B. Zamolodchikov, Nucl. Phys. B247 (1985) 83
[23] P. van Nieuwenhuizen, Phys. Reports 68 (1981) 189
[24] A. Neveu, J.H. Schwarz, and C.B. Thorn, Phys. Lett. 35B (1971) 529;
A. Neveu and J.H. Schwarz, Phys. Rev. D4 (1971) 1109;
J. Scherk, Rev. Mod. Phys. 47 (1975) 123
[25] J.C. Pati and A. Salam, Phys. Rev. D10 (1974) 275
[26] J.-P. Serre, A course in arithmetic (Springer, New York, 1973);
P. Goddard and D. Olive, *in* Vertex operators in mathematics and physics, ed. J. Lepowsky, S. Mandelstam and I.M. Singer (Springer, New York, 1985)
[27] J.H. Conway and N.J.A. Sloane, J. Number Theory 15 (1982) 83
[28] W. Lerche, D. Luest, and A.N. Schellekens, CERN-TH 4540/86 (1986)
[29] S. Mandelstam, Phys. Reports 13C (1974) 259
[30] M. Mueller and E. Witten, Princeton preprint (1986)
[31] A. Parkes, CERN-TH 4525/86 (1986);
G. Moore, Phys. Lett. 176B (1986) 369
[32] V.P. Nair, A. Shapere, A. Strominger and F. Wilczek, Nucl Phys. B287 (1987) 402;
N. Sakai and I. Senda, Vacuum energies of string compactified on torus, Tokyo Inst. of Tech. TIT/HEP-88 (1985)
[33] J. Bagger, D. Nemeschansky, N. Seiberg and S. Yankielowicz, Nucl. Phys. B, to be published; P. Ginsparg, private communication
[34] H. Neuberger, A. Niemi and G. Semenoff, LBL preprint

Nuclear Physics B289 (1987) 87–108
North-Holland, Amsterdam

FOUR-DIMENSIONAL SUPERSTRINGS

I. ANTONIADIS*

CERN, Genève, Switzerland

C.P. BACHAS

Centre de Physique Théorique, Ecole Polytechnique, 91128 Palaiseau Cedex, France

C. KOUNNAS**

Lawrence Berkeley Laboratory, Berkeley, California 94720, USA

Received 30 December 1986

We solve completely the constraints of factorization and multiloop modular invariance for closed string theories in which all internal quantum numbers of the string are carried by free periodic and antiperiodic world-sheet fermions. We derive a simple set of necessary and sufficient rules, and illustrate how they can be used to find the spectrum, one-loop amplitudes and low-energy lagrangian of many realistic four-dimensional chiral models. We prove that modular invariance and factorization ensure the presence of a massless graviton and the correct connection between spin and statistics. We also prove that the existence of a massless spin-$\frac{3}{2}$ state ensures the absence of tachyons and the vanishing of the one-loop cosmological constant.

1. Introduction

It is presently believed that in order to realize the program of string unification [1] of all particle interactions, one must eventually arrive at a theory in four flat space-time dimensions, with $N=1$ supersymmetry and chiral matter fields. This would presumably be the first step in any effort to affront the real world, and see whether string theories may provide the answers to such long-standing questions in particle physics as the vanishing of the cosmological constant, the gauge hierarchy problem, the explanation of the observed spectrum of fermion masses etc.

A first approach to carrying out this program, was to compactify the known ten-dimensional superstrings [1,2] on a Calabi-Yau manifold [3] or an orbifold [4]. A much simpler proposal [5,6,11] is to construct string theories directly in four dimensions with nothing fancier than the tools used for constructing the consistent ten-dimensional superstrings [2,7–10]: all of the string's internal quantum numbers

* On leave from Ecole Polytechnique, 91128 Palaiseau, France.
** On leave from Ecole Normale Supérieure, 75231 Paris, France.

0550-3213/87/$03.50 © Elsevier Science Publishers B.V.
(North-Holland Physics Publishing Division)

are carried, either by extra flat dimensions compactified on tori, or by free periodic or antiperiodic fermions among which world-sheet supersymmetry is non-linearly realized [11]. Restricting ourselves to this latter case, we solve explicitly in this paper the constraints of factorization and modular invariance of multiloop string amplitudes [8, 12] which ensure the existence of a consistent perturbation theory. Using the simple set of rules that generate all solutions, we deduce several general results, namely that factorization and modular invariance imply the presence of a massless graviton and the correct connection between spin and statistics [13], and that the presence of a massless spin-$\frac{3}{2}$ particle in the spectrum ensures the absence of a tachyon and the vanishing of the one-loop corrections to the cosmological constant. We also explain the basic steps that should be followed for constructing realistic four-dimensional models with $N=1$ supersymmetry and chiral matter fields in complex representations of the gauge group.

The question that naturally arises is whether these theories are particular compactifications of the ten-dimensional superstrings, possibly in the presence of non-trivial background gauge fields [14]. Although this might well be the case, they still stand out by virtue of their simplicity: as will be clear in this paper, their complete spectrum, vertex operators, one-loop amplitudes, effective low-energy lagrangian etc. are readily calculable since one uses little more than the old Neveu-Schwarz-Ramond formalism. If nothing else, they thus provide a playground in which the low-energy properties of string theories can be tested. Besides, both these four-dimensional and the ten-dimensional models presumably correspond to perturbatively stable vacua of a second-quantized string theory, and should in this sense be treated on an equal footing; which of those vacua will be dynamically preferred, is a non-perturbative and highly non-trivial question for which we do not have at present any hint.

This paper is organized as follows: in sect. 2 we introduce our notation and review the constraints of invariance under all super-reparametrizations of the world-sheet. In sect. 3 we solve completely these constraints, and find a simple set of rules that are both necessary and sufficient for the consistency of string perturbation theory to all orders. In sect. 4 we show how these rules can be interpreted in terms of generalized GSO [17] projections in the Hilbert space of the string theory, and prove that they imply a "sensible" particle interpretation of the one-loop amplitude, including the correct spin-statistics connection; this was the starting point in the work of Kawai, Lewellen and Tye [5, 10]. We also prove by means of elementary set algebra that a massless spin-$\frac{3}{2}$ state ensures the absence of a tachyon, the existence of equal numbers of bosons and fermions at all mass levels of the string and the vanishing of the one-loop cosmological constant. Finally in sect. 5 we outline the basic steps that one should follow to construct realistic models: we show how the gauge group, number of supersymmetries and number of chiral families can be modified, and explain a simple trick for deriving the tree-level lagrangian in the low energy limit.

2. Stringy constraints

Our notation is as follows: the non-supersymmetric sector of right-moving string excitations contains in addition to the four space-time coordinates $\partial_{\bar{z}} X^{\mu}$ an extra 44 real free fermions $\bar{\phi}^{A}$ $(A=1,\ldots,44)$. The supersymmetric left-moving sector has in addition to $\partial_z X^{\mu}$ and their fermionic partners ψ^{μ}, another 18 real free fermions χ^{I}, y^{I} and ω^{I} $(I=1,\ldots,6)$ among which supersymmetry is non-linearly realized. For the world-sheet supercurrent we take

$$G(z)=\sum_{\mu}\psi^{\mu}\partial_z X_{\mu}+\sum_{I}\chi^{I}y^{I}\omega^{I}, \tag{2.1}$$

since it can be shown that all other possibilities will lead to a string theory with only massive fermions and broken space-time supersymmetry [11]. It can be easily verified that with the above choices the (super)conformal anomaly cancels in each sector separately, so that the theory is invariant under infinitesimal (super)reparametrizations of the world-sheet. For topologically non-trivial world-sheets we must still specify the spin-structures, i.e. the properties of spinor fields under parallel transport around non-contractible loops of the surface, and verify that our choice is invariant under large diffeomorphisms not continuously connected to the identity [15], also called modular transformations. Different choices of spin-structures correspond to different string-theories.

To be more precise we consider first a world-sheet with the topology of a torus. After a reparametrization and a Weyl transformation the torus can be represented by a flat parallelogram in the complex plane with sides $\mathbf{1}$ and τ corresponding to its two non-contractible loops. The one-loop vacuum to vacuum string amplitude can be written as:

$$Z=\int\left[\frac{\mathrm{d}\tau\,\mathrm{d}\bar{\tau}}{(\operatorname{Im}\tau)^{2}}\right]Z_{\mathrm{B}}^{2}\sum_{\text{spin str}}C\begin{bmatrix}\boldsymbol{a}\\ \boldsymbol{b}\end{bmatrix}Z_{\text{long},3/2}\begin{bmatrix}a_{\psi}\\ b_{\psi}\end{bmatrix}\prod_{f=1}^{64}Z_{\mathrm{F}}\begin{bmatrix}a_{\mathrm{f}}\\ b_{\mathrm{f}}\end{bmatrix}. \tag{2.2}$$

We now explain this formula: the measure in square brackets is invariant under the transformations

$$\tau\to\tau+\mathbf{1} \tag{2.3a}$$

and

$$\tau\to-\frac{1}{\tau} \tag{2.3b}$$

which generate the modular group of the torus; the integration is over the upper complex plane moded out by the modular group; Z_{B} is the bosonic contribution (the determinant of the laplacian):

$$Z_{\mathrm{B}}=|\operatorname{Im}\tau|^{-1/2}|\eta(\tau)|^{-2},$$

where $\eta(\tau)=q^{1/12}\Pi_n(1-q^{2n})$ is the Dedekind eta function with $q=e^{i\pi\tau}$. $Z_F\left[\begin{smallmatrix}a_f\\b_f\end{smallmatrix}\right]$ is the contribution of the fermion f (the determinant of the chiral Dirac operator) which depends on its spin structure $(a_f, b_f)\in Z_2\times Z_2$; following standard conventions $a_f=1$ or 0 denotes periodic or antiperiodic respectively boundary conditions in the direction **1**, and similarly for b_f in the direction τ. Explicitly, considering τ as a complex temperature, and using standard path-integral techniques we have:

$$Z_F\begin{bmatrix}0\\0\end{bmatrix}=\mathrm{Tr}[e^{i\tau H_{NS}}]=\frac{\Theta_3^{1/2}(\tau)}{\eta^{1/2}(\tau)}, \tag{2.4a}$$

$$Z_F\begin{bmatrix}0\\1\end{bmatrix}=\mathrm{Tr}\left[(-)^F e^{i\tau H_{NS}}\right]=\frac{\Theta_4^{1/2}(\tau)}{\eta^{1/2}(\tau)}, \tag{2.4b}$$

$$Z_F\begin{bmatrix}1\\0\end{bmatrix}=\mathrm{Tr}[e^{i\tau H_R}]=\frac{\Theta_2^{1/2}(\tau)}{\eta^{1/2}(\tau)}, \tag{2.4c}$$

$$Z_F\begin{bmatrix}1\\1\end{bmatrix}=\mathrm{Tr}\left[(-)^F e^{i\tau H_R}\right]=\frac{\Theta_1^{1/2}(\tau)}{\eta^{1/2}(\tau)}, \tag{2.4d}$$

where H_{NS} and H_R stand for the hamiltonians in the Neveu-Schwarz and Ramond sectors respectively, $(-)^F$ is the fermion number operator*, and Θ_i are the well known theta functions. These formulae hold for the left-movers and should be complex conjugated for the right-movers.

To finish with the explanations concerning eq. (2.2), note that in this formula the summation runs over all possible spin-structure assignements that are consistent with two-dimensional supersymmetry (i.e. such that $G(z)$ has well defined parallel transport properties), with coefficients $C\left[\begin{smallmatrix}a\\b\end{smallmatrix}\right]=C\left[\begin{smallmatrix}a_1,\dots,a_{64}\\b_1,\dots,b_{64}\end{smallmatrix}\right]$ that are yet to be specified. Note also that two of the components of X^μ were used to cancel the contribution of the graviton ghosts. Likewise $Z_{\mathrm{long},3/2}\left[\begin{smallmatrix}a_\psi\\b_\psi\end{smallmatrix}\right]$ is the contribution of two of the components of ψ^μ and of the world-sheet gravitino ghosts; it equals 1 in the case of the torus and can thus be dropped, but we keep it for later reference. Finally note that the fermions ψ^μ, the supercurrent $G(z)$ and the two-dimensional gravitino must all have the same spin structure, that we denoted by $\left[\begin{smallmatrix}a_\psi\\b_\psi\end{smallmatrix}\right]$.

The modular transformations generated by (2.3a, b) correspond to cutting the torus along its non-contractible loops, then gluing it back together after an integral number of relative ("Dehn") twists by 2π. This has the effect of mixing the spin

* The origin of this operator in eqs. (2.4b, d) can be easily understood if one notes that a path integral over periodic fermions in the τ direction can be recast into an integral over antiperiodic fermions by a change of variables: $\psi=e^{i\pi\,\mathrm{Im}(z)/\mathrm{Im}(\tau)}\hat{\psi}$.

structures $\left[\begin{smallmatrix}1\\0\end{smallmatrix}\right]$, $\left[\begin{smallmatrix}0\\0\end{smallmatrix}\right]$ *and* $\left[\begin{smallmatrix}0\\1\end{smallmatrix}\right]$, which are called even because they have a even (in fact zero) number of Dirac zero-eigenmodes. Explicitly under $\tau \to \tau + 1$

$$\eta \to e^{i\pi/12}\eta\,, \qquad \Theta_1 \to e^{i\pi/4}\Theta_1\,, \qquad \Theta_2 \to e^{i\pi/4}\Theta_2\,, \qquad \Theta_3 \leftrightarrow \Theta_4\,, \tag{2.5a}$$

while under $\tau \to -1/\tau$

$$\eta \to (-i\tau)^{1/2}\eta\,, \qquad \frac{\Theta_1}{\eta} \to e^{-i\pi/2}\frac{\Theta_1}{\eta}\,, \qquad \frac{\Theta_2}{\eta} \leftrightarrow \frac{\Theta_4}{\eta}\,, \qquad \frac{\Theta_3}{\eta} \to \frac{\Theta_3}{\eta}\,. \tag{2.5b}$$

To ensure modular invariance we must therefore impose the following conditions on the coefficients C:

$$C\begin{bmatrix}\boldsymbol{a}\\ \boldsymbol{b}\end{bmatrix} = -\exp\left(\tfrac{1}{8}i\pi \sum a_{\mathrm{f}}\right) C\begin{bmatrix}\boldsymbol{a}\\ \boldsymbol{a}+\boldsymbol{b}+\boldsymbol{1}\end{bmatrix}, \tag{2.6a}$$

$$C\begin{bmatrix}\boldsymbol{a}\\ \boldsymbol{b}\end{bmatrix} = \exp\left(\tfrac{1}{4}i\pi \sum a_{\mathrm{f}} b_{\mathrm{f}}\right) C\begin{bmatrix}\boldsymbol{b}\\ \boldsymbol{a}\end{bmatrix}, \tag{2.6b}$$

where Σ here stands for $\Sigma_{\text{left-movers}} - \Sigma_{\text{right-movers}}$, and $\boldsymbol{a}$ and $\boldsymbol{b}$ are of course 64-component vectors. In order to establish these conditions we have actually:

(i) Kept formally track of the transformation properties of the completely periodic, odd spin structure $\left[\begin{smallmatrix}1\\1\end{smallmatrix}\right]$ even though its contribution to (2.2) vanishes due to the existence of a Dirac zero eigenmode (the constant field).

(ii) Treated the contributions of different fermions as distinct, even when they had the same spin-structure.

Strictly speaking this is not necessary for establishing the modular invariance of the one-loop vacuum amplitude, eq. (2.2), but it is necessary for one-loop amplitudes with external momentum-dependent legs, and also as we will now see for multiloop vacuum amplitudes.

A multiloop vacuum diagram is a linear chain of donuts. A spin structure is specified by giving the spin structure of every fermion on each donut separately. The coefficient with which it contributes must be a product of one-loop coefficients:

$$C\begin{bmatrix}\boldsymbol{a}^{(1)}\boldsymbol{a}^{(2)}\ldots\boldsymbol{a}^{(l)}\\ \boldsymbol{b}^{(1)}\boldsymbol{b}^{(2)}\ldots\boldsymbol{b}^{(l)}\end{bmatrix} = C\begin{bmatrix}\boldsymbol{a}^{(1)}\\ \boldsymbol{b}^{(1)}\end{bmatrix} C\begin{bmatrix}\boldsymbol{a}^{(2)}\\ \boldsymbol{b}^{(2)}\end{bmatrix}\cdots C\begin{bmatrix}\boldsymbol{a}^{(l)}\\ \boldsymbol{b}^{(l)}\end{bmatrix}$$

since by cluster decomposition the amplitude should factorize [8,12] when the donuts are pulled infinitely far apart (put differentily, the vacuum should dominate in the sum over intermediate states that propagate infinitely far).

Now the modular transformation of the multiloop surface can be generated by [12]:

(i) Elementary Dehn twists of each donut separately; these leave the amplitude invariant thanks to factorization and one-loop modular invariance*.

(ii) An additional Dehn twist for each pair of neighboring donuts that mixes their spin-structures; thanks to factorization, we need only consider such a twist in the case of the double torus. After conformal gauge fixing the double torus is described by a 2×2 symmetric period matrix Ω; a modular transformation sends Ω to $(A\Omega+B)(C\Omega+D)^{-1}$ where $\begin{pmatrix} A & B \\ C & D\end{pmatrix}$ is a 4×4 integer matrix with unit determinant. The choice $A=D=\begin{pmatrix}1&0\\0&1\end{pmatrix}$, $B=-\begin{pmatrix}0&1\\1&0\end{pmatrix}$ and $C=\begin{pmatrix}0&0\\0&0\end{pmatrix}$ corresponds to a modular transformation that mixes the spin-structures on the two donuts:

$$\begin{bmatrix} \boldsymbol{a} & \boldsymbol{b}' \\ \boldsymbol{b} & \boldsymbol{b}' \end{bmatrix} \to \begin{bmatrix} \boldsymbol{a} & \boldsymbol{a}' \\ \boldsymbol{b}+\boldsymbol{a}' & \boldsymbol{b}'+\boldsymbol{a} \end{bmatrix}$$

and thus leads to the required additional constraint among the one-loop coefficients that reads:

$$(-)^{a_\psi + a'_\psi} \exp\left(\tfrac{1}{4} i\pi \sum a_f a'_f\right) C\begin{bmatrix} \boldsymbol{a} \\ \boldsymbol{b} \end{bmatrix} C\begin{bmatrix} \boldsymbol{a}' \\ \boldsymbol{b}' \end{bmatrix} = C\begin{bmatrix} \boldsymbol{a} \\ \boldsymbol{b}+\boldsymbol{a}' \end{bmatrix} C\begin{bmatrix} \boldsymbol{a}' \\ \boldsymbol{b}'+\boldsymbol{a} \end{bmatrix}. \tag{2.7}$$

The second phase on the left hand-side of the above equation comes from the known transformation properties of Θ-functions on multiloop surfaces [16]**. The first phase comes from the transformation properties of

$$Z_{\text{long},3/2}\begin{bmatrix} a_\psi^{(1)} a_\psi^{(2)} \dots a_\psi^{(l)} \\ b_\psi^{(1)} b_\psi^{(2)} \dots b_\psi^{(l)} \end{bmatrix}.$$

Since the world-sheet gravitino determinant on a multiloop surface is not known explicitly, the modular transformation properties of this factor are only known up to a phase. We can nevertheless fix this unknown phase by demanding that the consistent ten-dimensional superstrings [15]*** satisfy the condition (2.7). Note that $(-)^{a_\psi}$ is -1 or $+1$ according to whether the fermions ψ^μ, and hence also the supercurrent $G(z)$ and the world-sheet gravitino, are periodic or antiperiodic respectively in the direction **1** of the torus.

Eqs. (2.6) and (2.7) suffice to ensure the absence of global reparametrization anomalies on an arbitrary genus surface. It is another one of the miracles of string

* It is here that the full conditions (2.6a, b) become necessary, since fermions with identical or with odd spin structures on one of the l donuts, need not make an identical or a vanishing, respectively, contribution to the l-loop amplitude.

** Our definition of the Θ-functions differs from [16] by an extra phase $\exp\{-\frac{1}{2} i\pi(a^{(1)}b^{(1)} + \dots + a^{(l)}b^{(l)})\}$.

*** We may restrict ourselves to the 8 left-movers of the 10d superstrings, assuming holomorphic factorization of the amplitude. Then the condition (2.7) must be satisfied by the coefficients $C\begin{bmatrix}0\\0\end{bmatrix} = C\begin{bmatrix}1\\1\end{bmatrix} = -C\begin{bmatrix}1\\0\end{bmatrix} = -C\begin{bmatrix}0\\1\end{bmatrix}$.

theory that they also suffice, as we will soon see, to ensure the presence of a massless graviton, and the correct spin-statistics of string excitations. We first turn our attention, however, to deriving the general solution of these consistency conditions.

3. The general solution

In order to solve the consistency conditions, we will need to rewrite them in a more convenient form by introducing some new notation. First we will denote by F the set of all 64 fermionic coordinates, and by $\mathbf{2}^F$ the set of all subsets of F. Now we can specify a general spin structure assignement on the torus in either of the following two ways:

(i) As an ordered tetraplet $[A_1; A_2; A_3; A_4]$ where A_1, A_2, A_3, and A_4 are the sets of fermions with spin structure $\begin{bmatrix}1\\1\end{bmatrix}$, $\begin{bmatrix}1\\0\end{bmatrix}$, $\begin{bmatrix}0\\0\end{bmatrix}$ and $\begin{bmatrix}0\\1\end{bmatrix}$ respectively; clearly A_1, A_2, A_3, and A_4 form a partition of F.

(ii) As an ordered pair $(\alpha|\beta)$ where α and β are the sets of fermions that are periodic in the directions **1** and τ, respectively, of the torus; α and β are arbitrary subsets of F.

The bilingual dictionary allowing us to pass from one notation to the other is:

$$[A_1; A_2; A_3; A_3] = (A_1 \cup A_2 | A_1 \cup A_4), \tag{3.1a}$$

$$(\alpha|\beta) = [(\alpha \cap \beta); (\alpha - \alpha \cap \beta); (F - (\alpha \cup \beta)); (\beta - \alpha \cap \beta)]. \tag{3.1b}$$

Notation (ii) will prove more convenient for our purposes, but some equations may be more transparent when translated back to language (i).

It will be useful to associate to every set of fermions X the phases:

$$\varepsilon_X = e^{i\pi n(X)/8}$$

and

$$\delta_X = \begin{cases} -1, & \text{if } \psi^\mu \in X \\ +1, & \text{otherwise}, \end{cases}$$

as well as a parity operator that counts the fermions in X modulo two:

$$(-)^X f = \begin{cases} -f(-)^X & \text{if } f \in X \\ f(-)^X, & \text{otherwise}. \end{cases}$$

We also need to define $n_L(X)$ and $n_R(X)$ which are the numbers of left-moving and right-moving, respectively, fermions in the group X, as well as the net number of fermions in the group X: $n(X) = n_L(X) - n_R(X)$. Finally, we define the product of two fermion sets to be their symmetric difference:

$$\alpha\beta = \alpha \cup \beta - \alpha \cap \beta .$$

With this product rule, $\mathbf{2}^F$ becomes a commutative group with the empty set $\varnothing$ as the identity, and with all elements equal to their own inverse; note also that $F\alpha$ is the compliment of α.

Consider now an arbitrary spin-structure assignement $(\alpha|\beta)$. In order that it be consistent with world-sheet supersymmetry, which is required for eliminating negative metric states, we should make sure that the supercurrent $G(z)$, eq. (2.1), is either periodic or antiperiodic under parallel transport around the non-contractible loops of the torus. This means that the sets $F-\alpha$ and $F-\beta$ of fermions that are antiperiodic in the directions $\mathbf{1}$ and τ, respectively, of the torus should either

(a) contain an odd, 1 or 3, number of fermions from every additive term of the supercurrent (and hence also contain the ψ^μ), or

(b) contain an even, 0 or 2, number of these fermions (and hence not the ψ^μ).

A little thought will convince the reader that these conditions are conveniently summarized by demanding that

$$\left[G(z),(-)^{\alpha}\right]_{-\delta_\alpha} = \left[G(z),(-)^{\beta}\right]_{-\delta_\beta} = 0 ,$$

where $[\ ,\ ]_+$ here stands for the anticommutator and $[\ ,\ ]_-$ for the commutator. Let $\Upsilon \subset \mathbf{2}^F$ be the collection of all fermion sets that satisfy the relation $(-)^X G = \delta_X G(-)^X$; since $(-)^{XY} = (-)^X(-)^Y$ and $\delta_{XY} = \delta_X \delta_Y$, it follows easily that Υ is closed under symmetric differences, and is thus a subgroup of $\mathbf{2}^F$. Spin structure assignments that are consistent with world-sheet supersymmetry correspond to elements in $\Upsilon \times \Upsilon$.

Let us return now to the one-loop amplitude. We will use the same symbol that specifies a spin structure assignment, to also denote its contribution to this amplitude; thus

$$[A_1; A_2; A_3; A_4] = \int_\tau \Theta_1^{A_1} \Theta_2^{A_2} \Theta_3^{A_3} \Theta_4^{A_4} , \tag{3.2}$$

where we have here omitted the integration measure and the bosonic contribution, and we used the shorthand Θ_i^X for $(\Theta_i/\eta)^{n_L(X)/2}(\bar{\Theta}_i/\bar{\eta})^{n_R(X)/2}$. The full one-loop amplitude is of course a linear combination

$$Z = \sum_{\alpha,\beta \in \Upsilon} C_{(\alpha|\beta)} (\alpha|\beta) .$$

The coefficients $C_{(\alpha|\beta)}$ are constrained by the string consistency conditions, eqs. (2.6a, b) and (2.7), which ensured the absence of anomalies under global reparametrizations of the string world-sheet (for any number of loops). Translated in our new language these conditions take the elegant form:

$$C_{(\alpha|\beta)} = -\varepsilon_\alpha C_{(\alpha|F\alpha\beta)}, \tag{3.3a}$$

$$C_{(\alpha|\beta)} = +\varepsilon^2_{\overline{\alpha\cap\beta}} C_{(\beta|\alpha)}, \tag{3.3b}$$

and

$$C_{(\alpha_1|\beta_1)} C_{(\alpha_2|\beta_2)} = \delta_{\alpha_1}\delta_{\alpha_2}\varepsilon^2_{\alpha_1\cap\alpha_2} C_{(\alpha_1|\alpha_2\beta_1)} C_{(\alpha_2|\alpha_1\beta_2)}. \tag{3.4}$$

These are the basic equations which we will now solve completely. Our method will be to first derive some properties that must necessarily characterize any solution, then in the end show that they are also sufficient.

Let us begin by writing down some special (and actually redundant) cases of eq. (3.4) which can be easily derived with the help also of the transposition property (3.3b):

$$C_{(\varnothing|\varnothing)} C_{(\alpha|\varnothing)} = \delta_\alpha C^2_{(\alpha|\varnothing)}, \tag{3.5a}$$

$$C_{(\alpha|\varnothing)} C_{(\beta|\varnothing)} = \delta_\alpha C_{(\alpha|\varnothing)} C_{(\alpha\beta|\varnothing)}, \tag{3.5b}$$

$$C^2_{(\alpha|\beta)} = \delta_\alpha\delta_\beta C_{(\alpha|\varnothing)} C_{(\beta|\varnothing)}, \tag{3.5c}$$

$$C_{(\alpha|\beta)} C_{(\alpha|\gamma)} = \delta_\alpha\delta_\gamma C_{(\alpha|\beta\gamma)} C_{(\gamma|\varnothing)}. \tag{3.5d}$$

The first relation shows that for every fermion set α either $C_{(\alpha|\varnothing)} = 0$, or else $C_{(\alpha|\varnothing)} = \delta_\alpha C_{(\varnothing|\varnothing)}$. We may for convenience set $C_{(\varnothing|\varnothing)} = 1$ in what follows*, since an overall rescaling of the one-loop amplitude amounts to a redefinition of the string theory coupling constant. Let us denote by

$$\Xi = \{\alpha_1, \alpha_2, \ldots\}$$

the collection of fermion sets, for which $C_{(\alpha_i|\varnothing)} = \delta_{\alpha_i}$. Clearly $\varnothing \in \Xi$, and setting $\alpha = \beta = \varnothing$ in eq. (3.3a) we see that also $F \in \Xi$ always. Next, eq. (3.5b) shows that Ξ is closed under symmetric differences. It is therefore a subgroup of Υ, generated by some basis $\{\mathrm{b}_0 = \mathrm{F}, \mathrm{b}_1, \ldots, \mathrm{b}_N\}$; its cardinality is 2^{N+1}, and its generic element is of the form $\mathrm{b}_0^{m_0}\mathrm{b}_1^{m_1}\ldots\mathrm{b}_N^{m_N}$, with $m_i = 0$ or 1. Finally eq. (3.5c) proves that a general spin structure $(\alpha|\beta)$ contributes to the amplitude if and only if both $\alpha \in \Xi$ and

* The case $C_{(\varnothing|\varnothing)} = 0$ can be easily seen to lead to $C_{(\alpha|\beta)} = 0$ for all α and β, and is therefore trivial.

$\beta \in \Xi$, and in this case its coefficient is a sign: $C^2_{(\alpha|\beta)} = 1$. We summarize these conclusions in a

Lemma 1. The one-loop amplitude of a consistent string theory, in which all fermions are either periodic or antiperiodic, is always of the form

$$Z = \sum_{\alpha, \beta \in \Xi} C_{(\alpha|\beta)} (\alpha|\beta), \tag{3.6}$$

with Ξ a subgroup (under symmetric differences) of Υ containing F, and with $C^2_{(\alpha|\beta)} = 1$ for all α and β in Ξ.

We will refer to Ξ as the group of parities of the string theory, anticipating its natural interpretation in terms of generalized GSO projections (see the following section). Not all subgroups of Υ are acceptable groups of parities however, since an assignment of signs $C_{(\alpha|\beta)} \in Z_2$, consistent with all constraints (3.3a, b) and (3.4) is not in general possible. Indeed from eq. (3.3a) it follows immediately that for all $\alpha \in \Xi$ we must have $\varepsilon^2_\alpha = 1$, so that the net number of fermions should be a multiple of eight

$$n(\alpha) = 0 \pmod 8 \qquad \forall \alpha \in \Xi .$$

It is not sufficient to impose this condition on the basis elements of Ξ, because ε_α is not a group homomorphism. Instead it satisfies the following identity that is easy to derive by induction, keeping in mind also that $\varepsilon^{16}_{\mathrm{X}} = 1$ always:

$$\varepsilon_{\alpha_1 \alpha_2 \ldots \alpha_M} = \Big(\prod_i \varepsilon_{\alpha_i}\Big)\Big(\prod_{i>j} \varepsilon^{-2}_{\alpha_i \cap \alpha_j}\Big)\Big(\prod_{i>j>k} \varepsilon^{4}_{\alpha_i \cap \alpha_j \cap \alpha_k}\Big)\Big(\prod_{i>j>k>l} \varepsilon^{-8}_{\alpha_i \cap \alpha_j \cap \alpha_k \cap \alpha_l}\Big).$$

By raising this equation to the square, we can see that the condition $n(\alpha) = 0 \pmod 8$ for all $\alpha \in \Xi$ is equivalent to the following conditions among the basis elements b_i:

$$n(\mathrm{b}_i) = 0 \pmod 8, \tag{3.7a}$$

$$n(\mathrm{b}_i \cap \mathrm{b}_j) = 0 \pmod 4, \tag{3.7b}$$

$$n(\mathrm{b}_i \cap \mathrm{b}_j \cap \mathrm{b}_k) = 0 \pmod 2. \tag{3.7c}$$

There is one final consistency condition that requires the net number of fermions in the intersection of any quadruplet of basis elements to be even

$$n(\mathrm{b}_i \cap \mathrm{b}_j \cap \mathrm{b}_k \cap \mathrm{b}_l) = 0 \pmod 2. \tag{3.8}$$

This is actually stronger than (3.7c), since the basis elements include F. In order to derive this condition we must now do some work. To start with, from eq. (3.5d) we find that for all α, β, and γ in Ξ

$$C_{(\alpha|\beta)}C_{(\alpha|\gamma)} = \delta_\alpha C_{(\alpha|\beta\gamma)}, \tag{3.9}$$

which says that for all α in Ξ, the mapping

$$\delta_\alpha C_{(\alpha|x)}: \Xi \ni x \to Z_2$$

is a group homomorphism. Clearly

$$C_{(\alpha|x_1x_2\ldots x_K)} = \delta_\alpha^{K-1}\prod_{i=1}^{K} C_{(\alpha|x_i)}.$$

This, together with the transposition property (3.3b) allows us to express any $C_{(\alpha|\beta)}$ in terms of the coefficients $C_{ij} = C_{(b_i|b_j)}$ of pairs of basis elements. Explicitly, if $\alpha = b_{\alpha_1}b_{\alpha_2}\ldots b_{\alpha_A}$ and $\beta = b_{\beta_1}b_{\beta_2}\ldots b_{\beta_B}$, we have

$$C_{(\alpha|\beta)} = \left(\prod_{i=1}^{A}\delta_{b_{\alpha_i}}^{B-1}\right)\left(\prod_{j=1}^{B}\delta_{b_{\beta_j}}^{A-1}\varepsilon^2_{b_{\beta_j}\cap(b_{\alpha_1}\ldots b_{\alpha_A})}\right)\left(\prod_{i=1}^{A}\prod_{j=1}^{B} C_{\beta_j\alpha_i}\right). \tag{3.10}$$

The C_{ij} themselves are not all independent; they must satisfy the symmetry relation (3.3b), as well as the condition

$$C_{ii} = -\varepsilon_{b_i}C_{i0},$$

which follows from (3.3a) and the factorization property (3.9). This leaves us with $\frac{1}{2}N(N+1)+1$ independent signs, for instance C_{00} and C_{ij} for all $i>j=0,1\ldots, N$, which we can choose at will. The claim now is that to any such choice there corresponds a consistent string theory, if and only if the basis elements of Ξ obey the condition (3.7a, b) and (3.8). We prove this by brute force (the dum way), by showing that the coefficients $C_{(\alpha|\beta)}$ explicitly constructed in (3.10), obey the general stringy constraints, if and only if these conditions hold.

Since the proof is straightforward, we will only sketch it here. Consider for example the transposition constraint (3.3b) which is satisfied by assumption by the coefficients C_{ij} for pairs of basis elements. Simple book-keeping shows that it is also satisfied by the generic coefficients (3.10), if and only if the following holds:

$$\prod_{j=1}^{B}\varepsilon^2_{b_{\beta_j}\cap(b_{\alpha_1}\ldots b_{\alpha_A})} = \varepsilon^2_{(b_{\beta_1}\ldots b_{\beta_B})\cap(b_{\alpha_1}\ldots b_{\alpha_A})}\left(\prod_{i=1}^{A}\varepsilon^2_{b_{\alpha_i}\cap(b_{\beta_1}\ldots b_{\beta_B})}\right)\left(\prod_{i=1}^{A}\prod_{j=1}^{B}\varepsilon^2_{b_{\alpha_i}\cap b_{\beta_j}}\right).$$

This equation is not as hopeless as it looks. Using the identity

$$\varepsilon_{\alpha\cap\beta\gamma} = \varepsilon_{\alpha\cap\beta}\varepsilon_{\alpha\cap\gamma}\varepsilon^{-2}_{\alpha\cap\beta\cap\gamma}$$

and the fact that $\varepsilon_X^{16}=1$, we can express both sides as products of phases whose arguments are either double, or triple or quadruple intersections: $\varepsilon^2_{b_i\cap b_j}$, $\varepsilon^4_{b_i\cap b_j\cap b_k}$ and $\varepsilon^8_{b_i\cap b_j\cap b_k\cap b_l}$. The first two types of phases always occur in pairs, and therefore cancel thanks to the conditions (3.7). This is not the case for the phases of quadruple intersections however, which only disappear if they are explicitly set equal to the identity for all basis elements. This means that unless the condition (3.8) is also imposed, there is an obstruction in trying to extend the assignment of signs from the basis elements to all of Ξ, because the operations of transposition and factorization do not commute. In a similar manner we show that the stringy constraints (3.3a) and (3.4) are automatically satisfied by the coefficients (3.10), provided (3.7) and (3.8) hold. This means that there are no further obstructions, and completes the demonstration of the following theorem that summarizes the results of this section:

Theorem. Choose any subgroup Ξ of fermion sets, generated by a basis $\{b_0 = F, b_1, \dots, b_N\}$ with the following properties:

$$n(b_i) = 2n(b_i \cap b_j) = 4n(b_i\cap b_j \cap b_k \cap b_l) = 0 \bmod 8\,, \tag{3.11}$$

$$(-)^{b_i}G(z) = \delta_{b_i}G(z)(-)^{b_i}\,. \tag{3.12}$$

For any such Ξ there exist $2^{N(N+1)/2+1}$ consistent string theories, corresponding to all possible choices of signs ($+1$ or -1) for the coefficients $C_{(F|F)}$ and $C_{(b_i|b_j)}$ for $i>j=0,1,\dots,N$. Any given choice can be uniquely extended to a mapping $C_{(\alpha|\beta)}$: $\Xi\times\Xi\to Z_2$, with the properties:

$$C_{(\alpha|\beta)} = \varepsilon^2_{\alpha\cap\beta}C_{(\beta|\alpha)}\,, \tag{3.13}$$

$$C_{(\alpha|\alpha)} = -\varepsilon_\alpha C_{(\alpha|F)}\,, \tag{3.14}$$

and

$$C_{(\alpha|\beta)}C_{(\alpha|\gamma)} = \delta_\alpha C_{(\alpha|\beta\gamma)}\,. \tag{3.15}$$

The full one-loop amplitude is given by

$$Z = \sum_{\alpha,\beta\in\Xi} C_{(\alpha|\beta)}(\alpha|\beta)\,.$$

Furthermore this construction exhausts all consistent theories for which the internal quantum numbers of the string are carried by free periodic or antiperiodic fermions on the world sheet.

4. The spectrum and other properties

We will now give a simple interpretation to Ξ and C, in terms of generalized GSO projections [17] in the Hilbert space of the string theory. The notation introduced in the previous section will prove particularly handy, since it will allow us to derive several general results in a few lines. We will in particular prove that all consistent string theories have a massless graviton, that modular invariance ensures the spin-statistics connection, and that the presence of a massless spin-$\frac{3}{2}$ particle ensures the absence of tachyons as well as the vanishing of the one-loop cosmological constant.

The Hilbert space of a string theory is a direct sum of different sectors of type $R^{\alpha}N^{\mathrm{F}\alpha}$, in which all fermions in the set α have Ramond or periodic boundary conditions, while all remaining fermions have Neveu-Schwarz or antiperiodic boundary conditions. In a self-explanatory, though slightly abusive notation, we may write the contribution of the spin structure $(\alpha|\beta)$ as

$$(\alpha|\beta)=(-)^{\beta}R^{\alpha}N^{\mathrm{F}\alpha},$$

meaning that one sums the contributions of all states in the sector $R^{\alpha}N^{\mathrm{F}\alpha}$, with a sign given by the value of $(-)^{\beta}$ on this state; this follows easily from eqs. (3.2), (3.1a) and (2.4). Changing once more the overall normalization of the full amplitude, that can be reabsorbed in the string theory coupling constant, we may write

$$Z=\frac{1}{2^{N+1}}\sum_{\alpha\in\Xi}\left[\sum_{\beta\in\Xi}C_{(\alpha|\beta)}(-)^{\beta}\right]R^{\alpha}N^{\mathrm{F}\alpha}.$$

Using the factorization property (3.15), and the decomposition of a generic element of Ξ as $\beta=\mathrm{b}_0^{m_0}\mathrm{b}_1^{m_1}\ldots\mathrm{b}_N^{m_N}$ where $m_i=0$ or 1, we can easily deduce that the term inside the square brackets is a product of projectors:

$$Z=\sum_{\alpha\in\Xi}\delta_{\alpha}\left[\prod_{i=0}^{N}\tfrac{1}{2}\left(1+\delta_{\alpha}C_{(\alpha|\mathrm{b}_i)}(-)^{b_i}\right)\right]R^{\alpha}N^{\mathrm{F}\alpha}, \tag{4.1a}$$

or equivalently, since the extra projectors are redundant

$$Z=\sum_{\alpha\in\Xi}\delta_{\alpha}\left[\prod_{\beta\in\Xi}\tfrac{1}{2}\left(1+\delta_{\alpha}C_{(\alpha|\beta)}(-)^{\beta}\right)\right]R^{\alpha}N^{\mathrm{F}\alpha}. \tag{4.1b}$$

The meaning of the subgroup Ξ and the coefficients $C_{(\alpha|\beta)}$ should be now clear; to each element $\beta\in\Xi$, there corresponds (i) a sector $R^{\beta}N^{F\beta}$ of the Hilbert space, and (ii) a generalized GSO projection that leaves only those states in $R^{\alpha}N^{\mathrm{F}\alpha}$, whose

β-parity is $(-)^{\beta} = C_{(\alpha|\beta)}\delta_{\alpha}$. The factorization property (3.15) ensures that these projections are mutually compatible.

Let us recall some standard facts: the sector $R^{\alpha}N^{\mathrm{F}\alpha}$ of the Hilbert space consists of states formed by acting on the vacuum $|0\rangle_{\alpha}$ with the creation operators of left- and right-moving string excitations, X^{μ}_{m} and $\bar{X}^{\mu}_{m}$ respectively, as well as with the creation operators of fermionic excitations, f_m. For $f \in \mathrm{F}\alpha$ the oscillators have half-integer frequencies m; all other frequencies are integer. The vacuum $|0\rangle_{\alpha}$ must represent the Clifford algebra of zero-modes of all fermions in α. There are two cases:

(i) If $\psi^{\mu} \in \alpha$, it is a space-time spinor as well as a spinor of an internal $\mathrm{SO}(n_{\mathrm{L}}(\alpha) + n_{\mathrm{R}}(\alpha) - 2)$; furthermore all states in $R_{\alpha}N_{\mathrm{F}\alpha}$ have half-integer spin.

(ii) If $\psi^{\mu} \in \mathrm{F}\alpha$, it is a space-time scalar, in a spinor representation of an internal $\mathrm{SO}(n_{\mathrm{L}}(\alpha) + n_{\mathrm{R}}(\alpha))$; furthermore all states built on it have integer spin.

From eq. (4.1) we see that the contribution of a state to the vacuum amplitude is negative in the former case and positive in the latter; this is in accordance with the fact that integer and half-integer spin excitations must be quantized with Bose and Fermi statistics, respectively. Let us point out here that the requirement that the one-loop amplitude has such a sensible particle interpretation, was the starting point in the work of Kawai, Lewellen and Tye [5, 10]; the fact that this is equivalent to factorization and multiloop modular invariance was argued for by Seiberg and Witten [8], and has been demonstrated for ten-dimensional superstrings by Parkes [13].

The effect of making a projection $(-)^{\beta} = \rho$ in the sector $R_{\alpha}N_{\mathrm{F}\alpha}$ depends on whether α and β are disjoint or not. If they are we must only keep those states built on all components of the vacuum $|0\rangle_{\alpha}$ with an even (when $\rho = 1$) or odd (when $\rho = -1$) number of β-oscillators. If on the other hand $\alpha \cap \beta$ is not empty, then we define the generalized chirality operator

$$\Gamma_{\alpha\cap\beta} = \prod_{f\in\alpha\cap\beta} f_0 , \tag{4.2}$$

where if $\psi^{\mu} \in \alpha \cap \beta$ the product should, by Lorentz invariance, include all Dirac matrices transverse to the momentum of the state. Now admissible states are those built with an even number of β-oscillators on the components of the vacuum that have generalized chirality ρ, as well as those built with an odd number of β-oscillators on the vacuum of opposite chirality.

Consider in particular the value of the parity $(-)^{\mathrm{F}\alpha} = \delta_{\alpha}C_{(\alpha|\mathrm{F}\alpha)}$ in the sector $R_{\alpha}N_{\mathrm{F}\alpha}$; from (3.13) and (3.15) it follows easily that this is equal to $-\varepsilon_{\alpha}$ so that states with an even (if $\varepsilon_{\alpha} = +1$) or an odd (if $\varepsilon_{\alpha} = -1$) number of half-integer oscillators are always projected out. This is consistent with the fact that these states could never satisfy the mass-shell conditions (that come from the zeroth-moment

Virasoro conditions):

$$M^2 = -\tfrac{1}{2} + \tfrac{1}{16} n_{\mathrm{L}}(\alpha) + \sum_{\text{left-movers}} (\text{frequencies})$$

$$= -1 + \tfrac{1}{16} n_{\mathrm{R}}(\alpha) + \sum_{\text{right-movers}} (\text{frequencies}) . \qquad (4.3)$$

There is one other set of projections, whose values have been completely fixed by the string consistency conditions (3.13)–(3.15), and which are therefore not free parameters. These are the values of all β-parities in the pure Neveu-Schwarz sector N^{F}: $(-)^{\beta} = C_{(\varnothing|\beta)} = \delta_{\beta}$. This is particularly important; it means that the massless state $\psi^{\mu}_{1/2} \bar{X}^{\nu}_{1}|0\rangle_{\varnothing}$, which contains the graviton, the dilaton and a two-index anti-symmetric tensor, can never be projected out of the Hilbert space of a consistent string theory. Let us here pause and summarize these conclusions in a

Corollary. The Hilbert space of a consistent string theory is

$$\boldsymbol{H} = \bigoplus_{\alpha \in \Xi} \left(\prod_{\beta \in \Xi} \tfrac{1}{2}\left(1 + (-)^{\beta} \delta_{\alpha} C_{(\alpha|\beta)}\right)\right) R^{\alpha} N^{\mathrm{F}\alpha} .$$

Modular invariance ensures the presence of a massless graviton, and the correct connection between spin and statistics.

Let us consider next under what conditions the spectrum contains a massless spin-$\frac{3}{2}$ particle. It follows easily from the mass-shell conditions eqs. (4.3), that this must be a state $\bar{X}^{\mu}_{1}|0\rangle_{\mathrm{S}}$ in $R^{\mathrm{S}} N^{\mathrm{FS}}$, with S a set of exactly 8 left-movers, including in particular the fermions ψ^{μ}. Without loss of generality we may take

$$\mathrm{S} = \{\psi^{\mu}, \chi^{1}, \ldots, \chi^{6}\} . \qquad (4.4)$$

Since $\bar{X}^{\mu}_{1}|0\rangle_{\mathrm{S}}$ has an internal SO(6) spinor index, it actually contains as many components as 8 Weyl spin-$\frac{3}{2}$ particles as well as 8 Weyl spin-$\frac{1}{2}$ particles. Half of them are always eliminated by the projection $(-)^{\mathrm{S}} = \prod_{f \in \mathrm{S}} f_0 = \delta_{\mathrm{S}} C_{(\mathrm{S}|\mathrm{S})}$. In order to analyze the effect of the remaining, if any, β-parity projections on this state, consider first the mapping

$$\Xi \ni \beta \to \beta \cap \mathrm{S} \in \mathbf{2}^{S} ,$$

which is a group homomorphism from Ξ into the group of all subsets of S. Because of the conditions (3.11) all elements in the image $\mathrm{Im}(\Xi)$ must satisfy: $n(\mathrm{X}) = 0 \pmod 4$ and $n(\mathrm{X} \cap \mathrm{X}' \cap \mathrm{X}'') = 0 \pmod 2$. It can be easily verified that a subgroup of $\mathbf{2}^{S}$ containing S and satisfying these conditions, can have at most three generators that may, without loss of generality, be taken to be S, $S_1 = \{\psi^\mu, \chi^1, \chi^2\}$ and $S_2 = \{\psi^\mu, \chi^3, \chi^4\}$. The claim now is that, if {S}, or $\{S, S_1\}$, or $\{S, S_1, S_2\}$ is a basis for Im(Ξ), then we can always choose some corresponding basis $\{F, S, \zeta_1 \dots \zeta_K\}$ or $\{F, S, b_1, \zeta_1 \dots \zeta_K\}$, or $\{F, S, b_1, b_2, \zeta_1 \dots \zeta_K\}$ of Ξ, such that:

$$b_1 \cap S = S_1 = \{\psi^\mu, \chi^1, \chi^2\}, \tag{4.5a}$$

$$b_2 \cap S = S_2 = \{\psi^\mu, \chi^3, \chi^4\}, \tag{4.5b}$$

and for all $i = 1, \dots, K$

$$\zeta_i \cap S = \emptyset . \tag{4.5c}$$

We will refer to such a basis as a canonical basis for Ξ; the reason that we can always find a canonical basis, is that if $\zeta_i \cap S = S^m S_1^{m_1} S_2^{m_2}$ is not empty for some basis element ζ_i, then it suffices to redefine our basis replacing ζ_i by $\zeta_i' = \zeta_i S^m b_1^{m_1} b_2^{m_2}$ which satisfies (4.5c). We will also choose by convention the sets b_1 and b_2 to be the sets of minimal cardinality that satisfy (4.5a, b); because of the condition (3.12), each of these sets contains at least 8 left-movers.

Let us go back now to the massless spin-$\frac{3}{2}$ states $\bar{X}_1^\mu|0\rangle_S$. Their ζ_i-parity is $+1$, and hence in order to survive the $(-)^{\zeta_i} = \delta_S C_{(S|\zeta_i)}$ projections, we must necessarily impose $C_{(S|\zeta_i)} = -1$ for all i. This implies that $C_{(S|\beta)} = -1$ for all $\beta \in \Xi$ disjoint from S as can be easily verifies with the help of the consistency conditions (3.13)–(3.15). We still have at most two other independent projections, that correspond to the basis elements b_1 and b_2, if they exist, and which set $\Gamma_{S_i} = -C_{(S|b_i)}$ on these states. Since each of these generalized chirality projections truncates half of the components of $\bar{X}_1^\mu|0\rangle_S$, it follows that the cases of one, two or three generators for Im(Ξ), generally correspond to theories with $N = 4$, $N = 2$ and $N = 1$ respectively massless Weyl gravitinos. Actually there is an exception to this rule: if b_1 is a set of exactly 8 left-movers and no right-movers, then new massless spin-$\frac{3}{2}$ states $\bar{X}_1^\mu|0\rangle_{b_1}$ may appear doubling the number of gravitinos; similarly if both b_1 and b_2 are sets of precisely 8 left-movers, the number of gravitinos can be quadrupled giving us back an $N = 4$ theory. We will return to this comment in the following section.

For the time being let us point out that the necessary and sufficient condition for the existence of at least one massless gravitino, is that there exist at least one set S containing precisely 8 left-moving fermions, including the ψ^μ, and such that for all $\beta \in \Xi$ disjoint from S, $C_{(S|\beta)} = -1$*. We are now ready to prove a

Lemma 2. The presence of at least one massless spin-$\frac{3}{2}$ state, ensures the existence of an equal number of fermions and bosons at each mass-level of the

* If S did not contain the ψ^μ, this condition would be inconsistent with the fact that $C_{(S|FS)} = -\delta_S \varepsilon_S = +1$.

spectrum, the vanishing of the one-loop cosmological constant, and the absence of tachyons.

Proof. The one-loop vacuum-to-vacuum amplitude, eq. (3.6), can be rewritten trivially as

$$Z = \sum_{\alpha,\beta\in\Xi} \tfrac{1}{4}\left[C_{(\alpha|\beta)}(\alpha|\beta) + C_{(S\alpha|\beta)}(S\alpha|\beta) + C_{(\alpha|S\beta)}(\alpha|S\beta) + C_{(S\alpha|S\beta)}(S\alpha|S\beta)\right],$$

since when X runs over all elements of Ξ, so does SX. We will now show that the sum of the four terms inside the square brackets, which we will refer to for short as SUM, is identically zero for all $\alpha, \beta \in \Xi$. Indeed, note first that since the Θ_1 function vanishes identically, $(X|Y) = 0$ unless $X \cap Y = \varnothing$. Hence we need only concern ourselves with the case in which at least one of the four pairs of elements in square brackets are disjoint; without loss of generality we take $\alpha \cap \beta = \varnothing$. The intersection of S with α on the other hand can contain 0, 4 or 8 elements, and similarly with β. We consider these cases separately:

(i) If $S \cap \alpha = S \cap \beta = \varnothing$, then $(S\alpha|S\beta) = 0$, and from the condition of existence of a gravitino: $C_{(S|\alpha)} = C_{(S|\beta)} = -1$. Using the factorization and symmetry properties of the coefficients C we easily deduce:

$$\begin{aligned}\mathrm{SUM} &= C_{(\alpha|\beta)}\left[(\alpha|\beta) - (S\alpha|\beta) - (\alpha|S\beta)\right] \\ &= C_{(\alpha|\beta)}\int_{\mathcal{T}} \Theta_2^{\alpha}\Theta_3^{\mathrm{FS}\alpha\beta}\Theta_4^{\beta}\left[\Theta_3^{\mathrm{S}} - \Theta_2^{\mathrm{S}} - \Theta_4^{\mathrm{S}}\right] = 0,\end{aligned}$$

where the last step follows from the fact that S contains exactly 8 left-movers, and from the well known identity among Θ-functions.

(ii) If $S \subset \alpha$ or $S \subset \beta$, the same argument goes through is one replaces α by $S\alpha$, or β by $S\beta$.

(iii) If $S \cap \alpha = \varnothing$ and $S \cap \beta$ contains precisely 4 left-movers, then $(S\alpha|\beta) = (S\alpha|S\beta) = 0$, and $C_{(S|\alpha)} = -1$. We therefore find easily

$$\mathrm{SUM} = C_{(\alpha|\beta)}\int_{\mathcal{T}}\left[\Theta_2^{\alpha}\Theta_3^{\mathrm{F}\alpha\beta}\Theta_4^{\beta} - \Theta_2^{\alpha}\Theta_3^{\mathrm{FS}\alpha\beta}\Theta_4^{\mathrm{S}\beta}\right] = 0,$$

since β and $S\beta$ contain exactly the same number of both left- and right-movers. The same argument of course goes through if we interchange the role of α and β.

(iv) If $S \cap \alpha$ and $S \cap \beta$ contain both precisely 4 left-moving fermions, then we have $(\alpha|S\beta) = (S\alpha|\beta) = 0$, and also $C_{(S|S\alpha\beta)} = -1$ since $S\alpha\beta$ and S are necessarily disjoint (draw a little picture to believe it). But then using repeatedly the symmetry and factorization properties of the coefficients C we find $C_{(S\alpha|S\beta)} = -C_{(\alpha|\beta)}$ and thus

$$\mathrm{SUM} = C_{(\alpha|\beta)} \int_{\tau} \left[\Theta_2^{\alpha} \Theta_3^{\mathrm{F}\alpha\beta} \Theta_4^{\beta} - \Theta_2^{\mathrm{S}\alpha} \Theta_3^{\mathrm{F}\alpha\beta} \Theta_4^{\mathrm{S}\beta} \right] = 0$$

as before.

What we have actually shown here is that the one-loop vacuum diagram vanishes even before one integrates over the modular parameter τ; this means that the partition functions of space-time bosons and fermions are identical at all temperatures, and hence there are as many bosons as fermions at each mass level of the spectrum. Finally we prove the absence of a tachyon*: if the tachyon were a fermion it would belong to some sector $R^{\beta}N^{\mathrm{F}\beta}$ with $\psi^{\mu} \in \beta$; but then condition (3.12) requires β to have at least 8 left-movers, and the mass-shell condition, eq. (4.3), shows that no states in this sector can have a negative mass. Thus there can be no tachyons that are space-time fermions, and since fermions and bosons occur in pairs at all mass levels of the theory, there can be no tachyons, period.

A last remark: everything we have done until now can be taken over with trivial modifications to any dimension of space-time $D \leqslant 10$. The set F of fermion fields contains in this case $D-2$ "transverse" ψ^{μ}, $3(10-D)$ additional left-movers χ^{I}, y^{I} and ω^{I}, and $32+2(10-D)$ right-movers $\bar{\phi}^{a}$, and the supercharge is appropriately modified. The counting of massless gravitinos of course changes also. We may also easily extend our analysis to the non-heterotic superstring: in four dimensions the set F contains the transverse left- and right-moving fermions ψ^{μ} and $\bar{\psi}^{\mu}$, as well as 18 extra fermionic coordinates on each side. The theorem, corollary and two lemmas of sects. 3 and 4 all hold if only the phase δ_{X} is redefined as -1 if either ψ^{μ} or $\bar{\psi}^{\mu}$ but not both are in X and as $+1$ otherwise. It is amusing to observe that the left-right asymmetric theories can be constructed in this case if one chooses projectors appropriately.

5. Realistic models

The number of consistent four-dimensional string theories is so huge that classifying them all would be both impractical and not very illuminating. Instead we will limit ourselves here to a few examples of models with realistic low-energy spectra; these will illustrate the simple set of rules given in the theorem of sect. 3, and the way in which the gauge group and massless matter fields are obtained. Armed with this simple set of rules, and a good knowledge of group theory, the interested model-builder may then try to construct the theory of his choice.

We will restrict ourselves here to theories that are space-time supersymmetric at the tree-level; this both ensures the vanishing of the one-loop corrections to the cosmological constant, and may help explain the gauge-hierarchy problem. Thus the

* We thank V. Rivasseau for a discussion on this point.

basis of Ξ, which we will denote by ξ for short, contains at least two elements: F and S. Consider first the simplest choice $\xi_1 = \{F, S\}$. As we explained in the previous section this is a theory* with $N=4$ space-time supersymmetries. The graviton multiplet contains: the states $\psi^{\mu}_{1/2}\bar{X}^{\mu}_{1}|0\rangle_{\varnothing}$ which include the graviton, a scalar and an antisymmetric tensor field, the states $\bar{X}^{\mu}_{1}|0\rangle_{S}$ which include four gravitinos and four spin-$\frac{1}{2}$ particles in the representation (**4**) of the internal symmetry group SU(**4**) $\simeq$ SO(6), and finally the vectors $\chi^{I}_{1/2}\bar{X}^{\mu}_{1}|0\rangle_{\varnothing}$ in the real representation (**6**) of SU(**4**). Besides these states the massless spectrum also includes a vector multiplet in the adjoint representation of the gauge group SO(**44**). Explicitly, this multiplet contains, for each adjoint index, a vector, four spinors and six scalars:

$$\psi^{\mu}_{1/2}\bar{\phi}^{a}_{1/2}\bar{\phi}^{b}_{1/2}|0\rangle_{\varnothing}, \qquad \bar{\phi}^{a}_{1/2}\bar{\phi}^{b}_{1/2}|0\rangle_{S}, \qquad \chi^{I}_{1/2}\bar{\phi}^{a}_{1/2}\bar{\phi}^{b}_{1/2}|0\rangle_{\varnothing}. \tag{5.1}$$

Suppose we add now to the basis ξ_1 a new element, ζ_1, that doesn't break any supersymmetries, that is $\zeta_1 \cap S = \varnothing$ and $C_{(S|\zeta_1)} = -1$. It can be easily checked that $(-)^{\zeta_1} = +1$ in both the $R(\varnothing)N(F)$ and the $R(S)N(FS)$ sector, consistent with the fact that the entire $N=4$ graviton multiplet survives these projections. The vector multiplet (5.1) on the other hand does not, since the $\bar{\phi}^{a}$ that belong to ζ_1 must now occur in pairs. As a result the gauge group SO(**44**) is broken down to SO($n_R(\zeta_1)$) $\otimes$ SO(**44** $- n_R(\zeta_1)$). In fact there is an exception: if ζ_1 is a set of precisely 8 or 16 right-movers, then new vector multiplets coming from the sector $R(\zeta_1)N(F\zeta_1)$ will appear, enlarging the gauge group to SO(**44**) or $E_8 \times$ SO(28) respectively. More generally for any supersymmetric theory let $(R_1 R_2 \ldots R_{M_\Xi})$ be the partition of the **44** right-movers induced by the basis of Ξ in the following sense: every R_i is an intersection of some number of basis elements, and no intersection of basis elements is a proper subset of R_i. Then the gauge group is at least $G = \otimes_i$ SO($n(R_i)$), and may be enlarged if Ξ contains elements with precisely 8 or 16 right-movers. Note that in order to reduce the rank of the gauge group below 22 some of the $n(R_i)$ need be odd. In general a large variety of gauge groups can be obtained in four dimensions, but we will limit ourselves here to orthogonal groups, since these suffice to yield realistic models.

A realistic model must of course have chiral matter fields in a complex representation of the gauge group, which could eventually describe the observed low-energy matter particles. A little thought and the mass-shell conditions (4.3) will convince the reader that such states can only be one of the following two kinds:

(i) $|0\rangle_{\beta}$ with β a set of 8 left-movers that include ψ^{μ}, and of 16 right-movers, or

(ii) $\bar{\phi}^{a}_{1/2}|0\rangle_{\beta}$ with β a set of 8 left-movers that include the ψ^{μ}, and of 8 right-movers.

In both these cases the set β_R of right-movers of β should not itself be an element of Ξ, since if it were the above states would have supersymmetric vector partners

* There are actually four theories, corresponding to all possible choices of signs for $C_{(S|S)}$ and $C_{(F|F)}$, but they only differ in the chiralities of the spinor representations of the internal symmetry groups.

($\psi^{\mu}_{1/2}|0\rangle_{\beta_R}$ or $\psi^{\mu}_{1/2}\bar{\phi}^{a}_{1/2}|0\rangle_{\beta_R}$ respectively), and would therefore belong to real representations of the gauge group. As we discussed in sect. 4, a β with the above properties will necessarily eliminate half of the gravitinos; this is of course due to the well known fact that there are no scalar matter multiplets of $N=4$ supersymmetry.

Let us further pursue case (i). The basis of our theory must thus include at least $\xi_2 = \{F, S, b_1\}$, where without loss of generality

$$b_1 = \{\psi^{\mu}, \chi^1, \chi^2, y^3, \ldots, y^6, \bar{\phi}^1, \ldots \bar{\phi}^{16}\}.$$

The theory described by ξ_2 has $N=2$ supersymmetry. Its massless spectrum includes the graviton multiplet, a vector multiplet in the adjoint of the gauge group SO(16) ⊗ SO(28), and a scalar multiplet in the real (16, 28) representation of the gauge group; these are the states of the $N=4$ graviton and SO(**44**) vector multiplets that survived the b_1 parity projections. In addition there is the new scalar matter multiplet $|0\rangle_{\beta_1}$ in the (spinor, 1) representation of the gauge group; this is a real representation, since out of all orthogonal groups only the SO($2+4n$) have complex representations. In order to obtain a chiral theory we must thus break SO(16) down to factors that include at least one of the groups SO(6), SO(10) or SO(14).

Suppose we opt for SO(10) grand unification; let $h = \prod_{\text{transverse}} \psi^{\mu}_0$ be the helicity operator, and $\Gamma_{10} = \bar{\phi}^1_0\bar{\phi}^2_0 \ldots \bar{\phi}^{10}_0$ the chirality operator that distinguishes the two inequivalent complex conjugate spinor representations of SO(10). For chiral asymmetry we must demand $h\Gamma_{10}$ to have a well defined value*; taking conditions (3.11), (3.12) and (4.5b) into account, it can be shown that there is an almost unique set that will impose such a projection, namely

$$b_2 = \{\psi^{\mu}, \chi^3, \chi^4, \omega^3, \ldots, \omega^6, \bar{\phi}^1, \ldots \bar{\phi}^{10}, \bar{\phi}^{17}, \ldots \bar{\phi}^{k}\},$$

with $k = 22$, 30 or 38. For $k=30$ for instance the theory described by $\xi_3 = \{F, S, b_1, b_2\}$ has $N=1$ supersymmetry, an SO(10) ⊗ SO(6) ⊗ SO(**14**) ⊗ SO(**14**) gauge symmetry, and four chiral matter multiplets, two in the (16, **4**, **1**, **1**) and two in the (16, $\bar{\mathbf{4}}$, **1**, **1**) representations of the gauge group. This makes a total of 16 chiral families of SO(10).

The number of chiral families can in fact be easily modified. Adding for instance the basis element

$$\zeta_1 = \{y_3, \ldots, y_6, \omega_3, \ldots, \omega_6, \bar{\phi}_{17}, \bar{\phi}_{18}, \bar{\phi}_{31}, \ldots \bar{\phi}_{44},\}$$

* Recall that a Weyl spinor contains positive helicity particles in some representation r of the gauge group, and negative helicity antiparticles in the complex conjugate representation r^*. If r were real, a Majorana mass would not be forbidden by gauge invariance and will in general be generated by quantum corrections.

eliminates half of the above chiral families. Further reductions by factors of two are possible provided one breaks the "horizontal" SO(6) and/or the SO(10) gauge symmetries. We can also obtain a number of families that is not a power of 2; choosing for instance b_2 with $k=22$, and adding a basis element $\zeta_1' = \{\bar{\phi}_{11}, \dots \bar{\phi}_{14}, \bar{\phi}_{23}, \dots \bar{\phi}_{26}\}$ leads to a theory with 3×16 families of SO(10); these can then be reduced by means of further projections. We stop here since a classification of all interesting models is beyond the scope of this paper.

We would like to conclude with an important observation that illustrates the main advantage of these theories, namely their calculability. Let (Ξ, C) be some arbitrary theory; to every subgroup $\Xi' \subset \Xi$ there corresponds a naturally induced theory obtained by restricting the coefficients C to Ξ'. From eq. (4.1) it follows easily that part of the Hilbert space of the theory (Ξ, C) is an exact truncation of the Hilbert space of (Ξ', C). To see how this remark can be useful, let $\xi = \{F, S, b_1, b_2, \zeta_1, \dots, \zeta_K\}$ be a canonical basis for a $N=1$ supersymmetric theory; then the subbasis $\xi' = \{F, S, \zeta_1, \dots, \zeta_K\}$ corresponds to an induced $N=4$ theory. Now the tree-level lagrangian of the massless modes of an $N=4$ theory is uniquely specified [18], up to the unknown coupling constants of the semi-simple components of the gauge group. Thus we can immediately obtain by truncation part of the low-energy lagrangian of the original $N=1$ theory, including in particular the couplings of all vector multiplets; this is not trivial because this lagrangian is not polynomial. We can even determine the ratios of gauge coupling constants, by noting that the vector bosons of $\otimes\, SO(n(R_i))$ come from the truncation of the SO(**44**) multiplet (5.1), which has a single arbitrary coupling. The interactions of the additional massless scalar multiplets, other than those left over by the truncation of the $N=4$ vector multiplets, cannot be similarly determined; they are however considerably restricted [19] by the fact that they can be obtained by truncating an $N=2$ supersymmetric theory.

We acknowledge useful conversations with E. Cremmer, J. Iliopoulos, N. Marcus, K.S. Narain and V. Rivasseau.

Noted added

While this work was being completed we received preprints [20,21]; the first contains the details of the work of Kawai, Lewellen and Tye, on which we already commented in the text. The second is based on group lattices of bosonized world-sheet fermions. Finally we should point out that Thirring interactions are studied in ref. [20], and also by Bagger, Nemeschansky, Seiberg and Yankielowicz.

References

[1] J. Schwartz, Phys. Reports 89 (1982) 223;
M.B. Green, Surv. High Energy Phys. 3 (1983) 128, and references therein

[2] D.J. Gross, J.A. Harvey, E. Martinec and R. Rohm, Phys. Rev. Lett. 54 (1985) 502; Nucl. Phys. B256 (1985) 253, B267 (1986) 75
[3] P. Candelas, G. Horowitz, A. Strominger and E. Witten, Nucl. Phys. B258 (1985) 46
[4] L. Dixon, J.A. Harvey, C. Vafa and E. Witten, Nucl. Phys. B261 (1985) 651 and B274 (1986) 286
[5] H. Kawai, D.C. Lewellen and S.H.H. Tye, Phys. Rev. Lett. 57 (1986) 1832
[6] K.S. Narain, Phys. Lett. 169B (1986) 41
[7] L.J. Dixon and J.A. Harvey, Nucl. Phys. B274 (1986) 93
[8] N. Seiberg and E. Witten, Nucl. Phys. B276 (1986) 272
[9] L. Alvarez-Gaumé, P. Ginsparg, G. Moore and C. Vafa, Phys. Lett. 171B (1986) 155
[10] H. Kawai, D.C. Lewellen and S.H.H. Tye, Nucl. Phys. B 269 (1986) 1
[11] I. Antoniadis, C. Bachas, C. Kounnas and P. Windey, Phys. Lett. 171B (1986) 51
[12] L. Alvarez-Gaumé, G. Moore and C. Vafa, Comm. Math. Phys. 106 (1986) 1
[13] A. Parkes, CERN preprint TH.4525/86
[14] K.S. Narain, M.H. Sarmadi and E. Witten, Nucl. Phys. B279 (1986) 369;
P. Ginsparg and C. Vafa, Nucl. Phys. B, to be published
[15] E. Witten, *in* Proc. Argonne Symp. on Anomalies, geometry and topology, eds. W. Bardeen and A.R. White (World Scientific, 1985)
[16] H.E. Rauch and H.M. Farkas, Theta functions with applications to Riemann surfaces (Williams and Wilkins, 1974)
[17] F. Gliozzi, J. Scherk and D. Olive, Nucl. Phys. B122 (1977) 253
[18] M. de Roo, Nucl. Phys. B255 (1985) 515;
E. Bergshoeff, I.G. Koh and E. Sezgin, Phys. Lett. 155B (1985) 71;
M. de Roo and P. Wagemans, Nucl. Phys. B262 (1985) 644
[19] J.A. Bagger, Lecture in Bonn NATO Advanced Institute on Supersymmetry (1984)
[20] H. Kawai, D.C. Lewellen and S.H.H. Tye, Nucl. Phys. B288 (1987) 1
[21] W. Lerche, D. Lust and A.N. Schellekens, Nucl. Phys. B287 (1987) 477

Nuclear Physics B296 (1988) 757–778
North-Holland, Amsterdam

SPACE-TIME SUPERSYMMETRY IN COMPACTIFIED STRING THEORY AND SUPERCONFORMAL MODELS

Doron GEPNER†

Joseph Henry Laboratories, Princeton University, Princeton, New Jersey 08544, USA

Received 29 April 1987

Space-time supersymmetric compactified string theories which obey spin and statistics, tachyon free, and modular invariant, are constructed by compactifying on arbitrary sums of $N = 2$ minimal superconformal models. Heterotic-type compactifications with realistic phenomenology are described. The theories offer a considerably simpler structure than standard supersymmetric compactifications, gauge groups which contain E_6 with chiral fermions in realistic grand unification representations, and no free parameters. Also given are the partition functions and the classification of modular invariants for the $N = 2$ minimal superconformal models.

1. Introduction

Space-time supersymmetric string compactifications [1–3] have attracted considerable attention for several reasons. First, in such theories the problem of nonvanishing cosmological constant is absent, as long as supersymmetry remains unbroken. Second, supersymmetry was seen to dramatically restrict the otherwise dazzling array of possibilities for compactification. Third, supersymmetry seems to simplify the analysis of compactified string theory.

The purpose of this paper is to introduce new supersymmetric compactifications of both the closed superstring [4] and the heterotic string [5] (see ref. [6] for a review). These theories offer a surprising degree of simplicity. For example, some of them are merely toroidal compactifications. This is in contrast with the far greater complexity of the conventional Calabi-Yau and orbifold compactifications. In addition to supersymmetry, these theories also obey modular invariance, the right

† Research supported in part by a Chaim Weizmann fellowship and by NSF grant PHY-80-19754.

0550-3213/88/$03.50 © Elsevier Science Publishers B.V.
(North-Holland Physics Publishing Division)

connection between spin and statistics, and the absence of tachyons. Thus, they are fully viable candidates for realistic phenomenology.

Our basic strategy is to take for the internal space any solvable conformal field theory with $N=2$ world-sheet supersymmetry [7]. Solvable conformal field theories are models in which all correlation functions and partition functions may be computed exactly, by the existence of an infinite dimensional algebra of symmetries (for a review see ref. [9]). As was first demonstrated in ref. [10], in the context of string propagation on group manifolds, the general structure of the underlying conformal field theory guarantees the full consistency of the string theory, including vertex operators and modular invariance. Actually, in order for modular invariance to work one needs to take a conformal field theory where the left and right structures are identical. We shall thus enforce this restriction. One can think about this as a compactificationon some sigma model (possibly a fictitious one).

For compactifying the superstring* we need also an $N=1$ world sheet supersymmetry, in order to satisfy the conformal gauge conditions. Thus, any $N=1$ solvable conformal field theory may be used in a left-right symmetric superstring or heterotic string compactifications, as long as the total trace anomaly has the correct value.

The reason we actually wish to consider theories with $N=2$ world-sheet supersymmetry is to be able to achieve space-time supersymmetry. The $N=2$ superconformal algebra contains a U(1) Kac-Moody sub-algebra, which, as will be seen here, plays a central role in supersymmetry. In the infinite radius limit of Calabi-Yau compactifications an $N=2$ structure was seen to arise [11].

The novelty in this supersymmetry projection is that for the first time one can get SUSY in exact string theory which is not just a free fermion system. In contrast, other solvable compactification schemes such as ref. [25, 28], are based on free fermions or bosons, and one uses the usual GSO projection to get space-time SUSY.

The $N=2$ conformal theories that will be considered here are the simplest, yet non-trivial ones. Namely, the so called minimal $N=2$ superconformal models which form a discrete unitary series with a trace anomaly $c<3$ [12],

$$c=\frac{3k}{k+2}, \qquad k=1,2,\ldots,\infty .$$

We shall take an arbitrary combination of such models where the trace anomaly sums up to the right value.

* We use here the term superstring and heterotic string compactification, somewhat freely, to denote any string theory in low dimension with an $N=1$, or mixed $N=1$ and $N=0$, gauge conditions. This does not imply that these string theories are derived of, or related to, any 10-dimensional theory.

Our conundrum is how to make the requirement of charge integrality compatible with modular invariance. Usually, modular transformations mix all possible representations, or at least a large portion of them. For this purpose we study in sect. 2 the problem of modular invariance for the $N=2$ minimal conformal models. We use the relation found in ref. [13] to the Z_k parafermionic field theories [14,15] to write down character formulae for the $N=2$ superconformal models, which involve Hecke indefinite modular forms and SU(2) classical theta functions. These characters form a unitary representation of the modular group. The transformation matrices factorize into three parts which are an affine $A_1^{(1)}$ part, a theta function at level $k+2$, and a theta function at level 2. This also solves the problem of classifying modular invariant partition functions for a single $N=2$ theory, since any such invariant is composed out of an $A_1^{(1)}$ invariant and two theta function invariants. Moreover, all $A_1^{(1)}$ invariants are conjectured to be known [16,17], and all theta invariants are known from a theorem proved in ref. [15].

Superstring compactifications using the $N=2$ models are studied in sect. 3. We find a basic mechanism by which charge integrality and modular invariance are made compatible. For any combination of the minimal models with the correct value of the trace anomaly we find supersymmetric theories with integral U(1) charges. The theories are also modular invariant and obey spin and statistics.

Heterotic string compactifications are addressed in sect. 4. We first describe a completely general procedure via which any consistent superstring compactification may be converted to a heterotic one. This procedure may be thought of as the stringy formulation of the embedding of the spin connection in the gauge group. We then apply this procedure to convert the superstring compactifications of sect. 3 into heterotic ones. In four dimensions we find the gauge groups $E_8 \times E_6$ or SO(26). An example of a compactification to four dimensions on 9 copies of the $k=1$ theory is worked out in detail. The gauge group in this case is $E_8 \times E_6 \times U(1)^8$. The massless spectrum is then seen to contain in addition to the gravity and gauge multiplets, 84 generations of chiral fermions in the 27 of E_6 along with their scalar superpartners, and 252 E6 singlets.

2. The minimal superconformal model

2.1. THE $N=2$ SUPERCONFORMAL ALGEBRA

We shall start by giving the partition functions for the $N=2$ minimal superconformal models, and by obtaining their transformation properties under modular transformations. We shall use the relation found in ref. [13] between the minimal superconformal models and the Z_k parafermionic field theories.

The $N=2$ superconformal algebra is encoded in the operator product expansions,

$$T(z)T(w)=\frac{c/2}{(z-w)^4}+\frac{2T(w)}{(z-w)^2}+\frac{T'(w)}{(z-w)}+\cdots,$$

$$T(z)J(w)=\frac{J(w)}{(z-w)^2}+\frac{J'(w)}{(z-w)}+\cdots,$$

$$T(z)G_{\pm}(w)=\frac{3G_{\pm}(w)}{2(z-w)^2}+\frac{G'_{\pm}(w)}{(z-w)}+\cdots,$$

$$J(z)J(w)=\frac{c/3}{(z-w)^2}+\cdots,$$

$$J(z)G_{\pm}(w)=\pm\frac{G_{\pm}(w)}{(z-w)}+\cdots,$$

$$G_+(z)G_-(w)=\frac{2c/3}{(z-w)^3}+\frac{2J(w)}{(z-w)^2}+\frac{2T(w)+J'(w)}{(z-w)}+\cdots. \tag{2.1}$$

T is the energy momentum tensor; c is the trace anomaly; $G_{\pm}$ and J are the $N=2$ supersymmetry partners of the stress energy tensor. In the Neveu-Schwarz (NS) sector G is periodic and has an expansion in terms of integral oscillators, in the Ramond (R) sector G is anti-periodic and is expanded in half-integral oscillators. There is a standard $N=1$ superconformal sub-algebra of the $N=2$ algebra which is used in the superstring conformal gauge condition. The $N=1$ superpartner of T is given by $G=G_++G_-$.

The $N=2$ minimal superconformal models are labeled by an integer $k=1,2,3,\dots,\infty$. The trace anomaly is given by [12],

$$c=\frac{3k}{k+2}. \tag{2.2}$$

The primary fields are labeled by two integers l and q where l, the principal quantum number takes its value in the range $0,\dots,k$. q labels the U(1) charge of the field, and is defined modulo $2(k+2)$. In addition, the fields are labeled by their sector, either NS or R. In the NS sector we find the dimensions Δ and U(1) charges Q (for the primary fields),

$$\Delta=\frac{l(l+2)}{4(k+2)}-\frac{q^2}{4(k+2)},\qquad Q=\frac{q}{k+2}, \tag{2.3}$$

for $l=0,\dots,k$ and $q=-l,-l+2,\dots,l$. In the R sector the dimensions are

$$\Delta=\frac{l(l+2)}{4(k+2)}-\frac{(q\pm1)^2}{4(k+2)}+\frac{1}{8},\qquad Q=\frac{q\pm1}{k+2}\mp\frac{1}{2}. \tag{2.4}$$

2.2. THE PARAFERMIONIC THEORIES

The parafermionic theories were introduced by Zamolodchikov and Fateev [14] as models for self-dual Z_k symmetric critical points. In particular, there is considerable evidence that the integrable critical points of the Z_k Ising model realize these theories, as well as some antiferromagnetic critical points of the RSOS models [18].

The Z_k theories are the direct generalization of the critical Ising model, where the Majorana fermion is replaced by non-local analytic fields which are called parafermions. The models are labeled by the integer k. The value of the trace anomaly is

$$c=\frac{2(k-1)}{k+2}. \tag{2.5}$$

The different fields that appear in the Z_k theories have the dimensions

$$\Delta_{l,m}=\frac{l(l+2)}{4(k+2)}-\frac{m^2}{4k}+\text{integer}, \tag{2.6}$$

and a similar formula for the right dimension $\bar{\Delta}_{\bar{l},\bar{m}}$. Here, l and m are integers which obey, $0\leqslant l\leqslant k$, $-k+1\leqslant m\leqslant k$, and $l-m=0 \bmod 2$. The fields carry also a Z_{2k} charge whose value is m. The actual symmetry of the theory is Z_k, where the Z_k charge is $\frac{1}{2}(m+\bar{m})$. The lowest dimensional fields for a given value of l and m are denoted by $\Phi^{l,\bar{l}}_{m,\bar{m}}$. For these fields the integer in eq. (6) vanishes if $|m|\leqslant l$. Also, $\Phi^{l,\bar{l}}_{m,\bar{m}}=\Phi^{k-l,k-\bar{l}}_{k+m,k+\bar{m}}$. The parafermions are $\psi_l=\Phi^{0,0}_{2l,0}$, for $l=0,1,\dots,k-1$. They are analytic fields, $\partial_{\bar{z}}\phi(z)=0$, and thus correspond to non-local currents.

The partition functions on the torus for the parafermionic theories were found in ref. [15] to be given by Hecke indefinite modular forms. These are the string functions of the affine algebra $A_1^{(1)}$ at level k [19]. The origin of this relation is that the Hilbert space of an SU(2) WZW theory may be built as a direct product of the Hilbert spaces of the parafermionic theory and the Hilbert space of a free boson [15], where l corresponds to the SU(2) isospin and m corresponds both to $2I_3$ of the current algebra and to the bosonic charge.

Denote by $\mathscr{H}^{l,\bar{l}}_{m,\bar{m}}$ the parafermionic Hilbert space which contains the field $\Phi^{l,\bar{l}}_{m,\bar{m}}$. Then the partition function for this subspace is given by,

$$Z^{l,\bar{l}}_{m,\bar{m}}(\tau)=|\eta(\tau)|^2 c^l_m(\tau)c^{\bar{l}*}_{\bar{m}}(\tau), \tag{2.7}$$

where c^l_m are the string functions of $A_1^{(1)}$ which are Hecke modular forms,

$$c^l_m(\tau)=\eta(\tau)^{-3}\sum_{\substack{-|x|<y\leq|x|\\ (x,y)\text{ or }(\frac{1}{2}-x,\frac{1}{2}+y)\in\left(\frac{l+1}{2(k+2)},\frac{m}{2k}\right)+Z^2}}\text{sign}(x)e^{2\pi i\tau[(k+2)x^2-ky^2]}, \tag{2.8}$$

and $\eta(\tau)$ is the Dedekind η-function, which is a modular form of weight $\frac{1}{2}$.

Under the modular transformation S: $\tau\to-1/\tau$ the string functions transform as,

$$c^l_m(-1/\tau)=k^{-1/2}(k+2)^{-1/2}(-i\tau)^{-1/2}\sum_{l',m'}b(l,m,l',m')c^{l'}_{m'}(\tau), \tag{2.9}$$

where

$$b(l,m,l',m')=e^{i\pi mm'/k}\sin\left(\frac{\pi(l+1)(l'+1)}{k+2}\right). \tag{2.10}$$

Under T: $\tau\to\tau+1$,

$$c^l_m(\tau+1)=e^{2\pi i s^l_m}c^l_m(\tau),$$

where

$$s^l_m=\frac{l(l+2)}{4(k+2)}-\frac{m^2}{4k}-\frac{c}{24}. \tag{2.11}$$

The relation between the $N=2$ superconformal algebra and the Z_k parafermions is expressed by [13],

$$G_+(z)=\sqrt{\frac{2k}{k+2}}\,\psi_1\exp(i\alpha\phi(z)),$$

$$G_-(z)=\sqrt{\frac{2k}{k+2}}\,\psi_1^\dagger(z)\exp(-i\alpha\phi(z)),$$

$$J(z)=\frac{i}{\alpha}\partial_z\phi(z), \tag{2.12}$$

where $\alpha=\sqrt{(k+2)/k}$. A primary field for the superconformal theory will be a product of a parafermionic primary field and a bosonic exponential. For each parafermionic primary field Φ_q^l we have the superconformal primary field

$$\Phi_q^l{:}\mathrm{e}^{i\alpha_q\phi}{:}, \tag{2.13}$$

where

$$\alpha_q = \frac{q - \left(\frac{1}{2}\mathrm{sign}(0) + a\right)k}{\sqrt{k(k+2)}}, \qquad q = \dots, l-2, l,$$

$$a_q = \frac{q - \left(\frac{1}{2}\mathrm{sign}(0) + 1 + a\right)k}{\sqrt{k(k+2)}}, \qquad q = l, l+2, \dots, \tag{2.14}$$

where in the Ramond sector $a = 0$ and $\mathrm{sign}(0) = \pm 1$ and in the NS sector $a = \frac{1}{2}$ and $\mathrm{sign}(0) = -1$. The conformal dimensions and U(1) charges will then agree with eqs. (3)–(4).

2.3. THE CHARACTERS OF $N = 2$

The next step is to compute the characters of the superconformal blocks, using the relation to the parafermionic models. Define the classical theta functions associated with SU(2) at level m by

$$\Theta_{n,m}(\tau, z, u) = \mathrm{e}^{-2\pi i u} \sum_{j \in \mathrm{Z} + n/2m} \mathrm{e}^{2\pi i \tau m j^2 + 2\pi i j z}, \tag{2.15}$$

where n is defined modulo $2m$. The partition functions for the Z_k parafermionic models which are Hecke indefinite modular forms obey the relations $c_n^l = 0$ for $l - n \neq 0 \bmod 2$, $c_n^l = c_{-n}^l = c_{n+k}^{k-l}$, and

$$\sum_{n \bmod 2k} c_n^l \Theta_{n,k} = A_k^l, \tag{2.16}$$

where A_k^l is the affine character at level k,

$$A_k^l = \frac{\Theta_{l+1,k+2} - \Theta_{-l-1,k+2}}{\Theta_{1,2} - \Theta_{-1,2}}. \tag{2.17}$$

Non-primary fields are obtained by acting with $G_\pm$ and J on the primary fields, in the operator product sense. Thus, a superconformal block would contain all fields obtained from some primary field. Actually, for the purpose of studying modular invariance it is necessary to split each conformal block into two parts carrying opposite "fermion" numbers, where $G_\pm$ are defined to carry the fermion number -1. Operating with G on the field $\Phi_q^l \exp(i\alpha_q\phi)$ we get in the leading order the field

$\Phi^l_{q+2}\exp(i(\alpha_q+\sqrt{(k+2)/k}\,)\phi)$. It will be convenient to assume that k is odd, at first. Then applying $2k$ times G, the parafermion ψ_1 cancels and we are left with a net shift in the bosonic momenta. This shows that the bosonic piece is given by a theta function at level $2k(k+2)$. Instead of labeling our sectors by NS and R we have to separate them into four sectors labeled by s which is defined modulo 4. $s=0,2$ compose the NS sector, and $s=1,3$ compose the R sector. Denote the partition function of the s sector by $\chi^{l(s)}_q(\tau,z)$. Here z measures the U(1) charge. Then*,

$$\chi^{l(s)}_q(\tau,z)=\sum_{j \bmod k} c^l_{q+4j-s}(\tau)\Theta_{2q+(4j-s)(k+2),2k(k+2)}(\tau,kz,0), \tag{2.18}$$

where in the R sector we have replaced $q\to q+s$. Note that $\chi^{l(s)}_q$ is invariant under any of the transformations, $s\to s+4$, $q\to q+2(k+2)$, which shows that q is defined modulo $2(k+2)$ and s modulo 4. Also $\chi^{l(s)}_q=0$ if $l+q+s\neq 0 \bmod 2$. In addition, $\chi^{l(s)}_q$ is invariant under the simultaneous interchange: $l\to k-l$, $q\to q+k+2$, and $s\to s+2$. The result (18) is valid also for even k.

Under the modular transformations S: $\tau\to -1/\tau$ and T: $\tau\to\tau+1$, the classical theta functions transform as,

$$\Theta_{n,m}(\tau+1,z,u)=e^{\pi i n^2/(2m)}\Theta_{n,m}(\tau,z,u), \tag{2.19}$$

and

$$\Theta_{n,m}\left(-\frac{1}{\tau},\frac{z}{\tau},u+\frac{z^2}{2m^2\tau}\right)=\frac{1}{\sqrt{2m}}(-i\tau)^{1/2}\sum_{l \bmod 2m} e^{-\pi i l n/m}\Theta_{l,m}(\tau,z,u). \tag{2.20}$$

Using these transformation formulae and eqs. (9)–(11), we find that the characters $\chi^{l(s)}_q$ transform under modular transformations by

$$\chi^{l(s)}_q(\tau+1,0)=e^{2\pi i\gamma^{l(s)}_q}\chi^{l(s)}_q(\tau,0), \tag{2.21}$$

where

$$\gamma^{l(s)}_q=\frac{l(l+2)}{4(k+2)}-\frac{q^2}{4(k+2)}+\frac{s^2}{8}, \tag{2.22}$$

* Character formulae for the $N=2$ minimal models as an infinite series of rational functions were found in ref. [20] by generalizing the method of Feigin and Fuchs [21] and Rocha-Caridi [22]. These characters seem to be too complicated for further analysis.

and

$$\chi_q^{l(s)}\left(-\frac{1}{\tau},0\right)=C\sum_{\substack{l',s',q'\\ l+s+q=0 \bmod 2}}\sin\left(\pi\frac{(l+1)(l'+1)}{k+2}\right)e^{\pi i q q'/(k+2)}e^{-\pi i s s'/2}\chi_{q'}^{l'(s')}, \tag{2.23}$$

and C is a constant: $C^{-1}=\sqrt{2}(k+2)$. The above sum extends over all $0 \leqslant l \leqslant k$, $q \bmod 2(k+2)$ and $s \bmod 4$ such that $l+s+q=0 \bmod 2$. These transformation formulae can be summarized by the following identity,

$$\sum_{q \bmod 2(k+2)}\chi_q^{l(s)}(\tau,z)\Theta_{q,k+2}(\tau,-z,0)=A^l(\tau,0,0)\Theta_{s,4}(\tau,z,0), \tag{2.24}$$

which can be proved by noting that both sides are holomorphic modular forms which transform similarly under modular transformations. Since also they behave similarly at the cusps (the points at infinity in moduli space) they are equal up to a constant, which can be seen to be 1 from the leading behavior. The characters are invariant under the finite index subgroup of the modular group: $\Gamma(4(k+2)) \cap \Gamma(8)$.

We see that there is a factorization of the transformation properties of the superconformal characters into three natural pieces, reflected by eq. (24). The transformation of the character $\chi_q^{l(s)}$ contains an l piece which is just an affine $A_1^{(1)}$ transformation, a q piece which is a level $k+2$ theta function system, and an s piece which is a level-2 theta function.

In view of this factorization we can build a modular invariant solution for the $N=2$ superconformal system out of any $A_1^{(1)}$ invariant, and two invariants for the level $k+2$ and 2 theta systems. So assume that $Z_A=\sum_{l,\bar{l}}N_{l,\bar{l}}A^l A^{\bar{l}*}$, $Z_{k+2}=|\eta|^{-2}\sum_{n,\bar{n}}L_{n,\bar{n}}\Theta_{n,k+2}\Theta^*_{\bar{n},k+2}$, and $Z_2=|\eta|^{-2}\sum_{s,\bar{s}}S_{s,\bar{s}}\Theta_{s,2}\Theta^*_{\bar{s},2}$ are modular invariant. Then a modular invariant solution for the $N=2$ system will be given by*,

$$W=\tfrac{1}{2}\sum_{\substack{l,q,s\\ \bar{l},\bar{q},\bar{s}}}M_{l,q,s,\bar{l},\bar{q},\bar{s}}\chi_q^{l(s)}\chi_{\bar{q}}^{\bar{l}(\bar{s})*}, \qquad \text{where } M_{l,q,s,\bar{l},\bar{q},\bar{s}}=N_{l,\bar{l}}L_{q,\bar{q}}S_{s,\bar{s}}. \tag{2.25}$$

The complete system of invariants for the theta system was found in ref. [15]. They consist of the left-right symmetric solution, and orbifolds of it by any subgroup of the full Z_m symmetry, where m is the level. For $A_1^{(1)}$ the list of known physical solutions consists of the SU(2) and SO(3) WZW theories [10] and three sporadic

* Actually, since $\Theta_{n,m}(\tau,0,0)=\Theta_{-n,m}(\tau,0,0)$ and $c^l_m=c^l_{-m}=c^{k-l}_{k+m}$, there is a certain ambiguity in reading off the spectrum from the partition functions. This ambiguity can be removed by imposing the closure of the operator product algebra, or equivalently demanding locality. Alternatively, we can switch on twisted boundary conditions amounting to $z\neq 0$. Thus we get the correct spectrum by formally assuming that all characters are different.

solutions at levels $k=10,16$ [16] and $k=28$ [17]. There are reasons to believe that these are all the solutions. Thus, the most general solution to modular invariance for a single theory is obtained by taking one of the affine solutions for the $A_1^{(1)}$ part, and the left-right symmetric solutions for the two theta systems, and then making an orbifold in some subgroup of the $Z_{k+2}\times Z_2$ symmetry group. Note, however that some of the affine solutions at level k are also invariant under a discrete symmetry group, Z_2 for the even k SU(2) solutions and for the sporadic $k=10$ solution, Z_3 for the $k=4$ SO(3) solution etc. One can obtain more solutions by twisting with a discrete symmetry of the combined system. It is natural to conjecture that these are all the solutions.

3. String compactifications

Let us consider now the problem of constructing space time supersymmetric superstring compactifications out of sums of $N=2$ minimal conformal models. Suppose that the dimension of Minkowski space is $D=d+2$, where d is the number of transverse dimensions. The first condition is that the total trace anomaly will be $c=12$, in the light-cone gauge, in both the left and right sectors of the theory. This means that the internal space will have a trace anomaly $c_0=12-\frac{3}{2}d$, which can be made as a sum of the minimal models,

$$c_0=12-\tfrac{3}{2}d=\sum_{i=1}^{l}\frac{3k_i}{k_i+2}\,. \tag{3.1}$$

The easiest way to achieve modular invariance is to assume that the left and right compactifying spaces are identical. This is also in harmony with the idea that one should be able to think about it as a compactification on some manifold.

Actually, we will restrict d to be even. In that case the full theory in the light-cone gauge has an $N=2$ superconformal structure which will play an important role. To get this structure we need to group together pairs of the d fermions and bosons into complex fermions and bosons. The energy-momentum $N=2$ superfield is then

$$\mathcal{T}_i=\bar\psi_i\psi_i-\sqrt2\,\theta\psi_i\partial_z\phi_i-\sqrt2\,\bar\theta\bar\psi_i\partial_z\bar\phi_i+\theta\bar\theta\left(\tfrac12\bar\psi_i\partial_z\psi_i-\tfrac12\partial_z\bar\phi_i\partial_z\phi_i\right), \tag{3.2}$$

where $i=1,\dots,\frac12 d$. Thus the total theory will have $N=2$ world sheet SUSY generated by

$$\mathcal{T}_i=\sum_{i=1}^{d/2}\mathcal{T}_i^{\mathrm{M}}+\sum_{i=1}^{l}\mathcal{T}_i^{\mathrm{int}}\,. \tag{3.3}$$

The $N=2$ superconformal structure opens the way for a rather simple mechanism for achieving space-time supersymmetry, as will be seen in the following. The lowest component of the total energy-momentum multiplet, given in eq. (3.3), is a $U(1)$ current, $J(z)$, which generates a $U(1)$ current algebra. The operator product expansion (2.1) implies that J is essentially a free boson, $J(z)=\sqrt{c/3}\,\partial_z\phi(z)$.

There is a natural candidate for the space-time supersymmetry charge, in the present framework of an aribtrary $N=2$ superconformal field theory. This is the operator $Q=e^{i\phi}$. The operator Q contains a piece which is a D-dimensional space-time spinor; it is the bosonization of the $SO(d)$ spin field. In the special case of 10-dimensional flat space superstrings, this is precisely the form of the light-cone supersymmetry charge in the GSO projection [29], and consequently also in any compactification based on free fermions such as toroidal or orbifold [30] (discrete quotients of tori) compactifications, or the compactifications of ref. [28].

In the GSO projection supersymmetry is a consequence of the projection: the elimination of all states with odd fermion number. Under the modular transformation $\tau \to -1/\tau$ a state with odd fermion number goes to a supersymmetric state, and vice versa. Thus, modular invariance essentially dictates the GSO projection (e.g. see ref. [2]).

Here we want to generalize the structure of the GSO projection to the context of any conformal field theory, not necessarily one which can be entirely made out of free fermions (or their discrete quotients). In the infinite radius limit of Calabi-Yau sigma models, the $N=2$ algebra was seen to play an important role in finding a supersymmetric regularization [11]. Our purpose is to build such structure in the exact string theory. The rule for this "general projection" is simply the following. The total $U(1)$ charge must be an odd integer. The $U(1)$ charge may be thought of as the "fermion number" even though there are no fermions, and this charge is not necessarily an integer. As for the GSO projection, we shall see that the transformation $\tau \to -1/\tau$ takes a state with odd integral $U(1)$ charge to a supersymmetric state, and vice versa.

Assuming the general projection, the massless states are the ones with the total dimension $\Delta = \pm\frac{1}{2}$ and the $U(1)$ charges ± 1 (larger charges correspond to higher dimensional operators). The operator $Q(Q^\dagger)$ takes a massless state with the charge $-1(1)$ to its supersymmetric partner massless state with the charge $1(-1)$. Thus, imposing this projection, consistently with modular invariance, gives rise to explicit space-time supersymmetry.

The problem is to make the condition for charge integrality compatible with modular invariance. This is quite tricky since modular transformations tend to mix all the possible fields in a theory with one another, and we need to eliminate non-integral representations separately in the left and right sectors. Consider then the modular transformations of the fields in this collection of minimal superconformal models and d-dimensional superstring. As was seen in sect. 2, the modular transformations of the minimal models factorize into three pieces, an affine $A_1^{(1)}$ piece, a level $k+2$ theta function, and a level-2 theta function. The affine $A_1^{(1)}$ part does not carry any charge (eq. (2.24)) and thus any modular invariant combination for this piece will do. We may use any of the $A_1^{(1)}$ solutions mentioned in sect. 2. Next, we need to worry about the theta functions. We have l level-2 theta functions coming from the l minimal models. The partition functions of the superstring part are also given by level-2 theta functions (when grouped into $d/2$ complex bosons and fermions)

$$Z_s = \frac{\Theta_{s,2}(\tau, z, 0)}{\eta^3}, \tag{3.5}$$

where s which is defined modulo 4 denotes the four possible helicities, $\frac{1}{2}s$, coming from the fermions. Again z measures the U(1) charges which are integral in the Neveu-Schwarz sector ($s=0$ or 2) and half-integral in the Ramond sector ($s=1$ or 3). (Note that this is the $k\to\infty$ limit of eq. (2.24).) So, collectively we have $l+\frac{1}{2}d$ level-2 theta functions, which is the same as the partition function obtained from a theory of $2l+d$ Majorana fermions.

Let us introduce an important general method for making modular invariance work with a restricted set of representations. Notably, these have certain charges. Consider a system which is composed out of j SU(2) classical theta functions at levels $n_1, n_2, \dots, n_j$. A partition function for this system will be,

$$Z_{r_1, r_2, \dots, r_j} = \Theta_{r_1, n_1}\Theta_{r_2, n_2} \cdots \Theta_{r_j, n_j}, \tag{3.6}$$

where r_i is defined modulo $2n_i$. It will be convenient to group $r_1, r_2, \dots, r_j$ into a vector $\boldsymbol{r} = (r_1, r_2, \dots, r_j)$, and to define a scalar product for two such vectors by

$$\boldsymbol{rm} = \frac{r_1 m_1}{2n_1} + \frac{r_2 m_2}{2n_2} + \cdots + \frac{r_j m_j}{2n_j}. \tag{3.7}$$

Denote also $\Theta_{\boldsymbol{r}} = Z_{r_1, r_2, \dots, r_j}$.

Now, suppose that we have a vector $\boldsymbol{\beta}$ which obeys $\boldsymbol{\beta\beta} =$ integer. We shall then define the following theta functions,

$$\Theta_{\boldsymbol{\mu}}^{(s,t)} = \sum_n (-1)^{\beta^2 ns}\Theta_{\boldsymbol{\mu}+n\boldsymbol{\beta}}, \tag{3.8}$$

where $\boldsymbol{\mu}$ obeys $\boldsymbol{\beta\mu} + \frac{1}{2}t\beta^2 =$ integer. s and t are integers which are defined modulo 2 if $\boldsymbol{\beta}^2$ is odd, and modulo 1 if $\boldsymbol{\beta}^2$ is even (in which case the indices s and t are superfluous), and n runs on integers modulo α, where α is the least integer for which $\beta_i\alpha/(2n_i) =$ integer. The important point is that under modular transformations $\Theta_{\boldsymbol{\mu}}^{(s,t)}$ transform among themselves, i.e. they form a unitary representation of the modular group. It is straightforward to compute that under S: $\tau\to -1/\tau$ and T: $\tau\to\tau+1$ these theta functions transform as,

$$\Theta_{\boldsymbol{\mu}}^{(s,t)}\Big|_S = C(-i\tau)^{j/2}\sum_{\nu} e^{-2\pi i\boldsymbol{\mu\nu}}\Theta_{\boldsymbol{\nu}}^{(t,s)}, \tag{3.9}$$

$$\Theta_{\boldsymbol{\mu}}^{(s,t)}\Big|_T = e^{\pi i\boldsymbol{\mu}^2}\Theta_{\boldsymbol{\mu}}^{(t+s+1,t)}. \tag{3.10}$$

Both transformations are unitary (this fixes also the positive constant C).

Now, the condition on the theta functions Θ_μ which appear in this representation of the modular group is that $\beta\mu$ will be an integer if β^2 is even, and half integer if β^2 is odd. This looks very much like the condition for the integrality of a certain charge, namely, the charge defined by $Q_\beta(\mu)=\delta\beta\mu$ ($\delta=1$ for even β^2 and $\delta=2$ for odd β^2). Thus we will be able to guarantee that the U(1) charge will be an integer if it can be identified with Q_β for some β which obeys $\beta^2=$ integer. This is the fundamental mechanism which will serve to guarantee the compatibility of space-time SUSY and modular invariance in the minimal superconformal models. In the following it will be referred to as the β-method.

Of course, we can also use a second β_2, which obeys $\beta_2^2=$ integer and $\beta_1\beta_2=$ integer, to further group the theta functions. This will additionally restrict the μ's by $Q_{\beta_2}\mu=$ integer. In general, we may use an arbitrary set $\beta_1, \beta_2, \dots, \beta_n$ where $\beta_i^2=$ integer, and $\beta_i\beta_j=$ integer for $i\neq j$.

Once we have eliminated non-integral charges from our representation of the modular group, forming a modular invariant combination of the theta functions is easy. This is because of the unitarity of the representation (3.9)–(3.10) which entails modular invariance for the left-right symmetric combination

$$Z_{\mathrm{B}}=|\eta|^{-2j}\sum_{\mu,s,t}\Theta_\mu^{(s,t)}\Theta_\mu^{(s,t)*}, \tag{3.11}$$

where the sum extends over all μ obeying $\beta\mu+\frac{1}{2}\beta^2t=$ integer, and all s, t modulo 2. Z_{B} is a bosonic-type partition function, i.e. all the multiplicities are non-negative integers. For odd β^2, we can also form the fermionic modular invariant partition function,

$$Z_{\mathrm{F}}=|\eta|^{-2j}\sum_{\mu}\Theta_\mu^{(1,1)}\Theta_\mu^{(1,1)*}, \tag{3.12}$$

in which all the multiplicities are either positive or negative integers. As will be seen, the negative multiplicities correspond to space-time fermions, as required by the connection between spin and statistics. These partition functions will be referred to as the standard bosonic and fermionic invariants. The standard fermionic invariant may be thought of as a projection on odd fermion numbers of the theory which is described by the standard bosonic invariant. Thus also, the bosonic invariant forms a closed operator product algebra, whereas the fermionic one does not.

As we have seen, eq. (2.24) means that there is a factorization of the character transformations into affine $A_1^{(1)}$ pieces, and various theta functions. Since the level k_i+2 theta functions appear in the l.h.s. of eq. (2.24), their modular transformations are complex conjugates of the usual ones (2.19)–(2.20). We can take this into account in the β-method by allowing the corresponding n_i's in the definition of the scalar product, eq. (3.7), to be negative. With this change the scalar product will become lorentzian but the β-method will continue to work, mutatis mutandis.

Now we are ready to describe the partition function for the superstring compactification. The U(1) charges will be given by the Q_β for β which is one for all the theta functions, the l level-$n_i = -(k_i+2)$ theta functions, and the $l+\frac{1}{2}d$ level-2 theta functions. We can compute,

$$\beta^2 = \frac{d}{8} + \sum_{i=1}^{l}\left(-\frac{1}{2(k_i+2)} + \frac{1}{4}\right) = \frac{c}{12} = 1, \tag{3.13}$$

showing that β is valid. Thus the β-method will give us integral U(1) charges. The SUSY generator is the state with all $l_i = 0$ and β for the theta functions. It is in the Ramond sector of all theories and has the correct dimension of $\frac{1}{2}$. Thus, we can have a superstring compactification by taking the standard fermionic combination. β corresponds, in fact, precisely to the state obtained by acting with the supersymmetry charge, Q, on the vacuum. In other words, it is the space-time fermion vertex operator. Note, that the β-method precisely guarantees the factorization of the NS and R sectors in space-time between the left and right sectors of the theory. This is indeed required for space-time supersymmetry. In bosonic-type partition functions, such as the partition functions described in sect. 2 for a single $N=2$ minimal model, the R (NS) of the left part is coupled only to the R (NS) of the right part.

There is one additional constraint we need to implement. In order to conserve the total $N=1$ superconformal sub-algebra, which is required for the superconformal gauge condition, all the fields in the theory should be either in the Ramond sector of all the sub-theories, or in the Neveu-Schwarz sector of all the sub-theories, thus giving rise to states that are either in the total NS sector or the total R sector [25]. This condition can be implemented in our partition function by using the additional vectors β_r, where $r = 1, 2, \ldots, l+\frac{1}{2}d-1$. β_r is zero in all the level k_i+2 theta functions and all the level-2 theta functions, except at the rth and the last level-2 theta functions where it is equal to 2. Since $\beta_r^2 = 2$, using this vector in the β-method will imply that $\beta_r \mu =$ integer for all the fields in the theory. Denoting by s_i the value of μ in the ith level-2 theta function, this condition reads $s_r = s_{l+d/2} \bmod 2$. Since s_i is even when the field is in the Neveu-Schwarz sector, and odd if it is in the Ramond sector, the set of conditions from β_r will precisely guarantee that R fields will be coupled only to R fields and NS to NS.

The usage of the vectors β_r which correspond to the level-2 theta functions of the space-time part of the theory, will automatically guarantee also Lorentz invariance when compactifying to more that four space-time dimensions. These vectors span the root lattice of the little group SO(d) and thus ensure that the spectrum respects this symmetry.

Thus we have found a superstring partition function for a compactification on an arbitrary sum of minimal conformal models, with space-time supersymmetry. In addition we can easily see that the theories have the correct connection between spin and statistics. Adding β to some vector μ which is in the Ramond sector of space-time will take it to the Neveu-Schwarz, and vice versa. Now, the states in the partition function $\Theta_\mu \Theta_\mu^*$, which are space-time bosons, will appear in the partition function Z_F with a positive integral coefficient. Adding β to μ in either the left or right sectors will flip the sign of this term in Z_F (eq. (8)) but would also take a space-time boson into a space-time fermion and vice versa. We conclude that spin-statistics is obeyed.

The fact that we have space-time supersymmetry will guarantee also the absence of tachyons. The Ramond states in each of the sub-theories have a dimension which is at least $\frac{1}{24}c$. Thus also the total Ramond states have dimensions which are at least $\frac{1}{24}c_{\text{total}} = \frac{1}{2}$, and are thus not tachyonic. The Neveu-Schwarz fields are also non-tachyonic since they have their superpartners in the Ramond sector.

So far we have been discussing superstring compactifications which are not of much phenomenological significance because of their typically small gauge groups. In the next section we will see that the same compactifications work with only little modifications for the heterotic string, producing theories with semi-realistic phenomenology.

4. Heterotic string compactifications

In this section we will consider compactifications of the heterotic string on arbitrary sums of $N = 2$ minimal superconformal models. In addition to the required consistency conditions of modular invariance and spin-statistics, the heterotic string compactifications have $N = 1$ space-time supersymmetry and sufficiently large gauge groups with chiral matter multiplets to accommodate realistic physics.

In the heterotic case, the left sector of the theory is superstring-like, with an $N = 1$ gauge condition, whereas the right sector is bosonic-like with an $N = 0$ gauge condition. The light-cone critical trace anomaly for the right sector is 24, and for the left sector 12. Thus we can compactify the left sector to four space-time dimensions using a $c = 9$ theory. We shall use precisely the same theory in the right sector, in order to achieve modular invariance at more ease. In the right sector we will have 2 transverse bosons with $c = 2$, and the $c = 9$ compactifying theory. Thus we are left with $c = 13$ unmatched trace anomaly in the right sector which will be saturated by free bosons moving on the maximal torus of a rank-13 Lie group. As we will see, a consistent theory will be formed by taking either SO(26) or $E_8 \times SO(10)$ for this Lie-group. More generally, we will be able to compactify to $d + 2$ space-time dimensions by taking an $N = 2$ minimal theory with $c = 12 - \frac{3}{2}d$, and in the right sector a rank-$12 + \frac{1}{2}d$ Lie group. A consistent theory will be formed by taking either $SO(24 + d)$ or $E_8 \times SO(8 + d)$ for the gauge group.

Let us start by introducing a very general mechanism via which every superstring compactification may be converted to an heterotic one. We shall then use this mechanism to convert the superstring compactifications of the previous section into heterotic ones. Suppose then that we have a superstring compactified to $d+2$ space-time dimensions. We shall assume that the Minkowski part of the theory has the usual structure in terms of free bosons and fermions. In this case, the light-cone gauge partition function of the full theory will be composed out of products of the partition functions of the d transverse fermions and bosons, times the internal partition functions. Since the transverse partition functions should respect the SO(d) internal symmetry, they are given by SO(d) classical theta functions at level one, which are the basically specialized characters of affine SO(d) (for a discussion of the characters see the appendix of ref. [10]).

The level k classical theta functions associated with a Lie algebra G are defined by [26],

$$\Theta_\lambda(\tau, z, u) = \mathrm{e}^{-2\pi iku} \sum_{\gamma \in \mathrm{M}+\lambda/k} \mathrm{e}^{\pi ik\tau\gamma^2 - 2\pi ik\gamma z}, \tag{4.1}$$

where λ is any weight vector of the algebra G (an element of the weight lattice), and M is the lattice spanned by the long roots of the algebra. Under modular transformations these theta functions transform by,

$$\Theta_\lambda(\tau+1, z, u) = \mathrm{e}^{\pi i\lambda^2/k}\Theta_\lambda(\tau, z, u),$$

$$\Theta_\lambda\left(-\frac{1}{\tau}, -\frac{z}{\tau}, u+\frac{z^2}{2\tau}\right) = \left|\frac{M^*}{kM}\right|^{1/2} (-i\tau)^{l/2} \sum_{\mu \bmod kM} \mathrm{e}^{-2\pi i\mu\lambda/k}\Theta_\mu(\tau, z, u), \tag{4.2}$$

where l is the rank of G and the summation extends over all weights modulo k times the long root lattice. The level-k theta functions mix among themselves under modular transformations. The basic specialization amounts to taking $z=u=0$. In the case where G is a simply laced Lie algebra the characters are given by

$$B_\lambda = \frac{\Theta_\lambda(\tau, z, u)}{\eta(\tau)^l}, \tag{4.3}$$

where the theta functions are at level $k=1$. The B_λ's form a unitary representation of the modular group [10].

In the case of SO($2n$) the relevant weights are these obeying the integrability condition at level one (correlation functions containing primary fields in other representations vanish identically as a consequence of the Ward identities [10]), which are the singlet (0), vector (v), spinor (s) and antispinor ($\bar{\mathrm{s}}$). Let us arrange the characters corresponding to these weights into a vector, $\boldsymbol{B} = (B_0, B_\mathrm{v}, B_\mathrm{s}, B_{\bar{\mathrm{s}}})$. Then $\boldsymbol{B}$ transforms under the modular transformation S: $\tau \to -1/\tau$ as

$$\boldsymbol{B}|_S = S_{2n}\boldsymbol{B}, \quad \text{where } S_{2n} = \frac{1}{2}\begin{pmatrix} 1 & 1 & 1 & 1 \\ 1 & 1 & -1 & -1 \\ 1 & -1 & i^{-n} & -i^{-n} \\ 1 & -1 & -i^{-n} & i^{-n} \end{pmatrix}, \tag{4.4}$$

and under T: $\tau \to \tau + 1$ as,

$$\boldsymbol{B}|_T = T_{2n}\boldsymbol{B}, \qquad \text{where } T_{2n} = \mathrm{e}^{-\pi i n/12}\mathrm{diag}(1, -1, \mathrm{e}^{\pi i n/4}, \mathrm{e}^{\pi i n/4}). \tag{4.5}$$

Now, suppose that we have a modular invariant superstring compactification in $d+2$ space-time dimensions. The partition function for this theory is then,

$$Z(\tau, \bar{\tau}) = \left|(\mathrm{Im}\,\tau)^{-d/2}\eta(\tau)^{-2d}\right| \sum_{\lambda, \bar{\lambda}} B_\lambda(\tau) B_{\bar{\lambda}}(\bar{\tau})^* Z_{\lambda,\bar{\lambda}}(\tau, \bar{\tau}), \tag{4.6}$$

where $\lambda, \bar{\lambda}$ range over the level one representations of the SO(d) algebra (which comes from the transverse fermions), (0), (v), (s), and ($\bar{\mathrm{s}}$). Here $Z_{\lambda,\bar{\lambda}}$ correspond to the partition functions of the internal part of the theory. From our assumption the total partition function Z is modular invariant. In order to make this theory heterotic, we shall replace the B_λ in the right sector by another set of characters which transform in precisely the same representation of the modular group as B_λ does. This will guarantee that the resulting heterotic theory will continue to be modular invariant.

There are two possibilities to achieve this corresponding to the two gauge groups $\mathrm{E}_8 \times \mathrm{SO}(8+d)$ and $\mathrm{SO}(24+d)$. These algebras can be realized by free bosons moving on the maximal torus of the group. The total trace anomaly in the right sector will then be $c = 24$ as required, composed of the d transverse bosons, the $12 + \frac{1}{2}d$ internal bosons, and the $c = 12 - \frac{3}{2}d$ internal theory.

Consider the $\mathrm{SO}(24+d)$ case first. Now, the point is that the representation of the modular group for the $\mathrm{SO}(24+d)$ theta functions is precisely the same as the representation of SO(d) if we interchange the vector representation with the singlet one $B_\mathrm{v} \to B_0$, $B_0 \to B_\mathrm{v}$, and flip the signs of the spinor partition functions, $B_\mathrm{s} \to -B_\mathrm{s}$, and $B_{\bar{\mathrm{s}}} \to -B_{\bar{\mathrm{s}}}$. The matrix which implements this change of basis is,

$$M = \begin{pmatrix} 0 & 1 & 0 & 0 \\ 1 & 0 & 0 & 0 \\ 0 & 0 & -1 & 0 \\ 0 & 0 & 0 & -1 \end{pmatrix}. \tag{4.7}$$

Explicitly, we find the relations,

$$T_d = M T_{24+d} M,$$

$$S_d = M S_{24+d} M. \tag{4.8}$$

Similar relations hold in the case of $\mathrm{E}_8 \times \mathrm{SO}(8+d)$, where the relevant representations are the singlet, vector and two spinors of $\mathrm{SO}(8+d)$, multiplied by the singlet of E_8 (the partition function of E_8, B_0, is invariant under S, and under T it is multiplied by the phase $\mathrm{e}^{-2\pi i/3}$).

We conclude that the total partition function Z will continue to be modular invariant if we replace the singlet of SO(d) by the vector of SO($24+d$), the vector of SO(d) by the singlet of SO($24+d$), and flip the signs of the two spinors. The partition function for the heterotic compactification would then read,

$$Z_{\rm het} = \left|(\mathrm{Im}\,\tau)^{-d/2}\eta(\tau)^{-2d}\right| \sum_{\lambda,\bar\lambda} B_\lambda^{\mathrm{SO}(d)}(\tau)(MB^{\mathrm{SO}(24+d)})^*_{\bar\lambda}(\bar\tau)Z_{\lambda,\bar\lambda}(\tau,\bar\tau). \quad (4.9)$$

The total partition function is then modular invariant. In addition, the condition of spin statistics will continue to be obeyed, if it was obeyed by the superstring partition function, since the signs of the bosonic representations (singlet and vector) remain unchanged, while the fermionic ones (the spinors) change sign. We conclude that $Z_{\rm het}$ indeed represents a consistent heterotic compactification. This procedure to get a heterotic compactification out of a superstring one may be thought of as the stringy formulation of the embedding of the spin connection in the gauge connection in Calabi-Yau compactifications [1].

We can now apply this procedure to convert the superstring compactifications of the previous section into heterotic ones. Since, the left (superstring) sector remains unchanged, space-time supersymmetry will continue to hold. The resulting theories have $N=1$ space-time supersymmetry, with chiral fermions.

Suppose then that we have a collection of the $N=2$ minimal conformal models at levels $k_1, k_2, \ldots, k_r$, such that

$$c = 12 - \frac{3d}{2} = \sum_{i=1}^{r} \frac{3k_i}{k_i+2}. \quad (4.10)$$

In order to describe explicitly the partition function of the theory we need to introduce some notation. As in the previous section we group into the vector $\boldsymbol{v}$ the various theta functions indices. The first element of the vector $\boldsymbol{v}$, v_1, stands for any weight of the algebra SO(d) in the left sector, or SO($d+24$) or $\mathrm{E}_8\times\mathrm{SO}(d+8)$ in the right sector. The other elements of $\boldsymbol{v}$ correspond to the indices of the r level-2 theta functions, and the r level k_i+2 theta functions. In addition, we group into the vector $\boldsymbol{l}$ the affine indices $l_1, l_2, \ldots, l_r$. Denote also the partition functions,

$$Z_{\boldsymbol{v}}^{\boldsymbol{l}} = B_{v_1}^{G} \prod_{i=1}^{r} \chi_{v_{i+1}}^{l_i(v_{i+r+1})}, \quad (4.11)$$

where the χ's stand for the $N=2$ minimal conformal characters (eq. (2.18)), and G stands for SO(d) in the left sector, and SO($24+d$) or $\mathrm{E}_8\times\mathrm{SO}(8+d)$ in the right sector. The vectors we are using in the β-method in the left sector are the same ones described in sect. 3. The first β is the one which guarantees charge integrality,

$$\boldsymbol{\beta} = (\bar{s},1,1,\ldots,1). \quad (4.12)$$

In addition, in order to ensure that (R) will be coupled to (R) and (NS) to (NS) we need to use the additional vectors

$$\beta_i = (2 \text{ on } i,\ v \text{ on } 1,\ 0 \text{ elsewhere}), \qquad \text{for } i = 2,\dots,r+1, \tag{4.13}$$

where v stands for the highest weight of the vector representation. We shall denote the lattice spanned by these vectors over the integers by Q. The vectors $\boldsymbol{v}$ are composed out of the three vectors, $\boldsymbol{v}_1$ which is an SO($2n$) index, the vector $\boldsymbol{s} = (\boldsymbol{v}_2,\dots,\boldsymbol{v}_{r+1})$ which specifies the sector in the internal sub-theories, and $\boldsymbol{q} = (\boldsymbol{v}_{r+2},\dots,\boldsymbol{v}_{2r+1})$ which corresponds to the U(1) charges. The partition function of the theory (without the contribution from the transverse bosons) is given by

$$W = \sum_{\substack{\boldsymbol{v},\,\boldsymbol{\bar v},\,\boldsymbol{l},\,\boldsymbol{\bar l}\\ \boldsymbol{v}\to\boldsymbol{\bar v}\in Q\\ \boldsymbol{v}\beta \text{ and } \boldsymbol{\bar v}\beta\in\frac{1}{2}+Z,\ \boldsymbol{v}\beta_i \text{ and } \boldsymbol{\bar v}\beta_i\in Z}} \pm A_{l,\bar l} Z^{l}_{\boldsymbol{v}} Z^{\bar l *}_{\mu+\boldsymbol{\bar v}}, \tag{4.14}$$

where μ is the vector (v on 1, 0 elsewhere). $A_{l,\bar l}$ stands for a product of the multiplicities of any of the $A_1^{(1)}$ invariants,

$$A_{l,\bar l} = \prod_{i=1}^{r} A^{(k_i)}_{l_i,\bar l_i}, \tag{4.15}$$

where $A^{(k_i)}$ is an affine invariant at level k_i. The signs are determined by spin-statistics. The known affine invariants consist of the left-right symmetric one, $A_{l,\bar l} = \delta_{l,\bar l}$, which appears at any level (this is the partition function of the SU(2) WZW theory at level k), the SO(3) WZW theory which is a Z_2 orbifold of this solution which appears at any even level [10], and three sporadic solutions at levels $k = 10, 16$ [16] and $k = 28$ [17]. There is strong evidence that these are all the physical solutions (there are more solutions [16] which do not form a closed operator algebra, and are thus unsuitable for our purposes).

The spectrum of this theory is actually quite simple. The massless states are obtained by combining massless states in the left (superstring sector) with massless states in the right (bosonic sector). The states that appear in the spectrum are precisely the ones obeying the following conditions:

(i) The left and right states separately have odd integral U(1) charges (the U(1) charge in the right sector is computed according to the replacement by M).

(ii) The left and right states separately are either all in the (R) sector or all in the (NS) sector.

(iii) The vectors of the left and right states differ by an element of Q. The l's and $\bar l$'s are tied up according to one of the affine solutions.

This theory has $N = 1$ space-time supersymmetry, as well as chiral fermions. It is modular invariant and obeys spin-statistics.

It would be useful to work out an example in some detail. Consider a compactification to four space-time dimensions, on 9 copies of the $k=1$ theory. Since the $k=1$ theory is equivalent to a theory of a free boson moving on a circle [27], it is particularly easy to deal with. Also, the low energy predictions are not harder to extract in this theory, than they are for flat bosonic strings. The gauge group of this theory is $E_8 \times E_6 \times U(1)^8$. The only matter fields are 84 left-handed fermion generations which are singlets of E_8 and in the 27 of E_6, along with their antiparticles and scalar superpartners in the 27.

For $k=1$ the characters reduce to level-6 theta functions from eq. (2.18) (divided by an η function which is the only string function at level 1, $c_0^0 = c_1^1 = \eta^{-1}$, the l index is irrelevant in this case). So we will work directly with these theta functions. We shall use a ten-component vector $\boldsymbol{v}$, whose first component is an element of the SO($2n$) weight lattice, and the other indices are theta indices at level 6. The vector $\boldsymbol{\beta}$ which guarantees integral U(1) charges is then given by $\boldsymbol{\beta} = (\bar{s}, 1, 1, \dots, 1)$, where $\bar{s}$ is the highest weight of the antispinor representation. The other β's which serve to guarantee the coupling of NS to NS and R to R take the form, $\beta_i = (v \text{ on } 1, 6 \text{ on } i, 0 \text{ elsewhere})$, for $i = 2, \dots, 10$. The allowed fields in the spectrum are those which have an odd integral U(1) charge and are either entirely R or entirely NS. It is easy to see that there are only three types of cosets modulo Q which contribute to the massless spectrum. These are: $(s, 1^9)$, $(\bar{s}, 1^6, -1^3)$, and $(s, -1^6, 1^3)$. The first coset gives the graviton and gauge multiplets, the second and third give the matter and antimatter multiplets. The order in the last two cosets is arbitrary, giving rise to the multiplicity, $9!/(3!6!) = 84$.

Let us consider the first coset. In the left sector we have contributions from the three vectors: $(v, 0^9)$, $(s, 1^9)$, and $(\bar{s}, -1^9)$. Massless states in the right sector receive contribution from the three vectors: $(s, 1^9)$, $(0, 0^9)$, $(\bar{s}, -1^9)$. Taking in the right sector the vector $(0, 0^9)$ with any of the two transverse bosons excited, we get the usual gravity multiplet, graviton, antisymmetric tensor field and dilaton, along with their superpartners. The gauge group is $E_8 \times E_6 \times U(1)^8$. The E_6 gauge bosons are obtained by taking the vector in the left sector, and by combining in the right sector the adjoint (45) of SO(10) from $(0, 0^9)$ with the spinor (16) and antispinor ($\overline{16}$) of SO(10) (coming from the $(s, 1^9)$ and $(\bar{s}, -1^9)$ vectors) and the U(1) internal boson in the right sector (which is the superconformal U(1)). In total we have $45 + 16 + 16 + 1 = 78$, which is the dimension of E_6. The $E_8 \times U(1)^8$ gauge bosons are obtained by taking the vector in the left sector along with the E_8 adjoint in the right sector, or any of the eight internal bosons in the right sector, other than the boson corresponding to the superconformal U(1). Taking the spinor or antispinor in the left sector we get the superpartners of the gauge bosons which are fermions in the adjoint.

Let us consider now the coset of the type $(\bar{s}, 1^6, -1^3)$, which is responsible for the matter multiplets. In the left sector we get contributions from the vectors: $(\bar{s}, 1^6, -1^3)$, and $(0, 0^6, -2^3)$. In the right sector we find the vectors, $(0, 2^6, 0^3)$, $(v, 0^6, -2^3)$, $(\bar{s}, 1^6, -1^3)$, $(0|0^6, -2^2, 4)$, $(0|0^6, 4, -2^2)$, and $(0|0^6, -2\ 4 -2)$. The right vectors combine to give us the 27 of E_6 according to: $1+10+16=27$. There are 84 arrangements of this coset. Combining left and right states we find 84 left handed fermion generations in the 27 of E_6, which are singlets of E_8, along with their scalar superpartners in the same representation. These fields have different charges under the $U(1)^8$ gauge symmetry, which are given by the components $v_2, \ldots, v_{10}$, of the right-handed vectors they correspond to (modulo the superconformal U(1) which became part of E_6). The coset $(s, -1^6, 1^3)$ will give the antiparticles of this multiplet. In addition there are 252 fermions which are E_6 singlets and their superpartners. There are no other massless fields in this theory. We see that this is an 84 generations model in the 27 of E_6, which is a viable grand unification model.

Presumably other models can be found in this scheme, which have a realistic number of generations. Indeed, the number of consistent models is very large, perhaps in the thousands. The relatively large number of generations we found in this example is entirely due to the large symmetry of the compactifying theory. Other compactifications have much less symmetry.

Note added in proof

Recently, I have shown [31] that the models described in this paper all correspond to exactly solvable Calabi-Yau compactifications. For example, the theory made out of five copies of the $k=3$ minimal model corresponds to an exact solution for string propagation on the quintic hypersurface in CP^4. Moreover, it is demonstrated in [31] that any $N=2$ superconformal field theory with the correct trace anomaly describes (after the general projection) string propagation on some Calabi-Yau manifold, and that the structure described in this paper is the exact and general structure of Calabi-Yau compactifications.

I wish to thank J. Harvey, V. Kaplunovski, and E. Witten for much helpful conversation. I am particularly grateful to S. Shenker for a crucial discussion.

References

[1] P. Candelas, G. Horowitz, A. Strominger and E. Witten, Nucl. Phys. B258 (1985) 46

[2] L. Dixon, J. Harvey, C. Vafa and E. Witten, Nucl. Phys. B261 (1985) 678; B274 (1986) 285

[3] E. Witten, Nucl. Phys. B268 (1986) 79

[4] M.B. Green and J.H. Schwarz, Nucl. Phys. B255 (1985) 92

[5] D.J. Gross, J. Harvey, E.J. Martinec and R. Rohm, Phys. Rev. Lett. 54 (1985) 502; Nucl. Phys. B256 (1985) 253

[6] M. Green, J. Schwarz, and E. Witten, Superstring theory, vols. I and II (Cambridge University Press, 1987)

[7] M. Ademollo et al., Phys. Lett. 62B (1976) 105

[8] A.A. Belavin, A.M. Polyakov and A.B. Zamolodchikov, Nucl. Phys. B241 (1984) 333;
D. Friedan, Z. Qiu and S. Shenker, Phys. Rev. Lett. 52 (1984) 1575

[9] D. Gepner, Nucl. Phys. B290 [FS20] (1987) 10; talk given at the Santa Barbara Conference on Superstrings (December 1986), and at the Irvine Conference on Nonperturbative methods in field theory (January 1987), Nucl. Phys. B (Proc. Suppl.) 1A (1987) 1

[10] D. Gepner and E. Witten, Nucl. Phys. B278 (1986) 493
[11] A. Sen, Nucl. Phys. B278 (1986) 289; B284 (1987) 423;
D. Friedan et al., unpublished
[12] W. Boucher, D. Friedan and A. Kent, Phys. Lett. B172, 316 (1986);
J. Petersen, Niels Bohr Institute preprint NBI-HE-85-31 (1985);
P. di Vecchia, J.L. Petersen and M. Yu, Niels Bohr Inst. preprint NBI-HE-86-04 (1986);
P. di Vecchia, J.L. Petersen and H.B. Zheng, Phys. Lett. 174 (1986) 280;
S. Nam, Phys. Lett. B172 (1986) 323
[13] Z. Qiu, IAS preprint (October 1986)
[14] A.B. Zamolodchikov and V.A. Fattev, Sov. Phys. JETP 62 (1985) 215
[15] D. Gepner and Z. Qiu, Nucl. Phys. B285 [FS19] (1987) 423
[16] D. Gepner, Nucl. Phys. B287 (1987) 111
[17] A. Cappelli, C. Itzykson, and J.B. Zuber, Nucl. Phys. B280 [FS18] (1987) 445
[18] D.A. Huse, Phys. Rev. B30 (1984) 3908;
G.F. Andrews, R.J. Baxter and P.J. Forrester, J. Stat. Phys. 35 (1984) 193
[19] V.G. Kac and D. Peterson, Adv. Math. 53 (1984) 125
[20] E.B. Kiritsis, Caltech preprint CALT-68-1347 (September 1986);
D. Kastor and A. Kent (unpublished);
V.K. Dobrev, ICTP Trieste preprints IC/86/123, IC/86/241 (June, August 1986)
[21] B.L. Feigin and D.B. Fuchs, Funct. Anal. Appl. 17 (1983) 241
[22] A. Rocha-Caridi, *in* Vertex operators in mathematics and physics, eds. J. Lepouski, S. Mandelstam, I.M. Singer, (Springer, New York, 1984)
[23] A. Schwimmer and N. Seiberg, Weizmann Inst. preprint (1986)
[24] S. Helgason, Differential geometry and symmetric spaces (Academic Press, New York, 1978)
[25] M. Mueller and E. Witten, Phys. Lett. 182B (1986) 28
[26] V.G. Kac, Infinite dimensional Lie algebras (Birkhauser, Boston, 1983)
[27] G. Waterson, Phys. Lett. B171 (1986) 77
[28] K.S. Narain, M.H. Sarmadi and C. Vafa, Nucl. Phys. B288 (1987) 551;
H. Kawai, D. Lewellen and H. Tye, Phys. Rev. Lett. 57 (1986) 1832; Nucl. Phys. B288 (1987) 1;
R. Bluhm, L. Dolan and P. Goddard, Nucl. Phys. B289 (1987) 364;
W. Lerche, D. Lust and A. Schellekens, Nucl. Phys. B287 (1987) 477;
L. Antoniadis, C. Bachas and C. Kounnas, Nucl. Phys. B289 (1987) 87
[29] F. Gliozzi, J. Scherk and D.I. Olive, Nucl. Phys. B122 (1977) 253
[30] N. Seiberg and E. Witten, Nucl. Phys. B276 (1986) 272
[31] D. Gepner, Princeton Univ. preprint PUPT-1066

EXACTLY SOLVABLE STRING COMPACTIFICATIONS ON MANIFOLDS OF SU(N) HOLONOMY

Doron GEPNER [1]

Joseph Henry Laboratories, Princeton University, Princeton, NJ 08544, USA

Received 23 September 1987

A variety of heterotic string compactifications on the K3 surface, manifolds of SU(3) holonomy, and higher holonomy manifolds, are solved exactly. An example of the quintic hypersurface in CP_4 is worked out in detail. It is conjectured, and demonstrated in part, that any supersymmetric compactification of the heterotic string with an $N=2$ superconformal theory is equivalent to a compactification on a manifold of SU(N) holonomy, and in particular an arbitrary gluing of the discrete models with $c=9$ gives a solvable heterotic string compactification on some Calabi–Yau manifold. Calabi–Yau compactifications are seen to be exact vacua of string theory, retaining their topological and geometrical characteristics. Previously unknown enhanced gauge symmetries are found to arise for certain backgrounds.

The discovery of a consistent theory of quantum gravity described by the superstring [1] and the heterotic string [2] was all the more exciting because of the uniqueness of these theories. Unfortunately, in the process of compactification down to four dimensions this uniqueness is replaced by a vast number of possibilities, each leading to different physical predictions. Moreover, the geometrical interpretation of strings was lost due to the constructions of apparently non-geometrical compactifications based on free fermions or bosons [3]. Our purpose here is to restore to a degree both the uniqueness, and the geometrical interpretation of the (2,2) string compactifications. This is done by establishing that any supersymmetric heterotic compactification with an arbitrary conformal field theory with $N=2$ superconformal symmetry is equivalent to string propagation on a Calabi–Yau manifold [4]. Moreover, by constructing such theories out of tensor products of $N=2$ minimal models, which are exactly solvable as two-dimensional quantum field theories, we find a vast number of solvable Calabi–Yau compactifications.

The constructions of heterotic string compactifications out of tensor products of $c<3$ discrete minimal models were described by this author in ref. [5] as a two stage process. First it was shown that any tensor product of $c<3$ models with a total trace anomaly $c=3k$ leads to a space–time supersymmetric superstring compactification which is fully consistent as a string theory. This was done by the elimination of all states with U(1) charges that are not odd integers (the generalized GSO projection), which is a consequence of having in the theory all states which are related by exp $(i\phi)$ where ϕ is the appropriately normalized analytic free boson of the total $N=2$ algebra (including space–time), and modular invariance. The operator $Q=\exp(i\phi)$ contains a piece which is a space–time spinor, and is thus the space–time supersymmetry charge. In the large-radius limit of Calabi–Yau compactifications a similar structure for the SUSY generator is known to arise [6]. This leads to a modular invariant, space–time supersymmetric string theory, as was shown in ref. [5]. The second stage is to convert the superstring compactification to a heterotic one by interchanging the singlet and vector of SO(d), for the left movers (where d is the number of transverse dimensions of space–time), with the vector and singlet of SO$(8+d)\times E_8$ or SO$(24+d)$ which correspond to the heterotic internal bosons, and replacing SO(d) spinors with SO$(8+d)\times E_8$ or SO$(24+d)$

[1] Research supported in part by NSF grant PHY-80-19754.

spinors (taking the singlet representation for E_8). For any consistent superstring compactification, this leads to a consistent heterotic compactification, leaving the theory modular invariant, and with consistent spin-statistics.

This conversion gives rise to heterotic compactifications with gauge groups that are $E_8 \times E_6$ in four dimensions, $E_8 \times E_7$ in six dimensions, and $E_8 \times E_8$ in eight dimensions, times a possible extra gauge symmetry. This shows that this procedure is the exact string theory analog of the embedding of the spin connection in the gauge group, described in ref. [4] for the leading term in $1/r$. Indeed, in four dimensions the spin connection is in SU(3) and breaks E_8 to E_6, in six dimensions the SU(2) holonomy breaks E_8 to E_7, and in eight dimension the SU(1) holonomy leaves E_8 unbroken.

The reason we actually use the discrete $N=2$ series as building blocks of our string theories is that they are exactly solvable conformal field theories, in which all the physical predictions can be computed exactly, most without too much difficulty. Thus, these string theories are solvable space–time supersymmetric string compactifications. As will be seen in the sequel they correspond to exactly solvable points in the moduli space of manifolds of SU(N) holonomy.

The discrete $N=2$ models are labeled by a positive integer, k. The trace anomaly of the kth theory is [7]

$$c=\frac{3k}{k+2}\,. \tag{1}$$

The primary fields of the $N=2$ algebra are labeled by three integers, the principal quantum number l, which obeys $0\leqslant l\leqslant k$; the U(1) charge quantum number q which is defined modulo $2(k+2)$, and the sector quantum number, s which is defined modulo 4, in addition $l+q+s=0 \bmod 2$. The fields are labeled by two sets of such integers, one for the left movers, and one for the right movers, $\Phi^{l,\bar{l}}_{q,\bar{q},s,\bar{s}}$, with the identification $\Phi^{l,\bar{l}}_{q,\bar{q},s,\bar{s}}=\Phi^{k-l,k-\bar{l}}_{q+k+2,\bar{q}+k+2,s+2,\bar{s}+2}$. The dimensions and U(1) charges are given by

$$\Delta=\frac{l(l+2)-q^2}{4(k+2)}+\frac{s^2}{8}+\text{integer}\,,$$

$$Q=-\frac{q}{k+2}+\frac{s}{2}+2(\text{integer})\,. \tag{2}$$

For $|q|<l$ and $s=0$ these integers vanish. Each such primary field corresponds to a conformal block. The multiplicities of states in a conformal block are generated by the characters $\chi^l_{q,s}$. These characters were computed in ref. [5] in terms of modular forms, and they obey the relation

$$\sum_{q \bmod 2(k+2)} \chi^l_{q,s}(\tau,z)\Theta_{q,k+2}(\tau,-z,0)$$
$$=A^l(\tau,0,0)\Theta_{s,4}(\tau,z,0)\,, \tag{3}$$

where τ is the modulus of the torus on which we compute the characters, and z measures the U(1) charge (see ref. [5] for more detail). We denote by $\Theta_{n,m}$ the classical theta functions of level k, defined by

$$\Theta_{n,m}(\tau,z,u)=\exp(-2\pi i u)$$
$$\times\sum_{j\in Z+n/2m}\exp(2\pi i\tau m j^2+2\pi i j z)\,, \tag{4}$$

where n is defined modulo $2m$, and by A_l is the character of the affine A^{1}_{1} at level k,

$$A^l_k=\frac{\Theta_{l+1,k+2}-\Theta_{-l-1,k+2}}{\Theta_{1,2}-\Theta_{-1,2}}\,. \tag{5}$$

Under modular transformations the three indices of $\chi^l_{q,s}$ transform independently, where l transforms like an affine character, q like a level-$(-(k+2))$ theta function (that is, a complex conjugate of a level-$(k+2)$ theta function), and s like a level-2 theta function. This is encapsulated in the identity (3). Thus the general modular invariant partition function of a single theory is given by a tensoring of three such invariants for the affine A^{1}_{1} system and for the levels $(k+2)$ and 2 theta functions, which have been classified in refs. [8,9].

The fact that q transforms like a level-$(k+2)$ theta function is responsible for the somewhat surprising fact that these theories are invariant under a Z_{k+2} discrete symmetry. This can be seen by writing the partition function on the torus with boundary conditions in the world-sheet space and time directions given by some arbitrary elements of Z_{k+2}. Denote by $Z(r,s)$ the partition function with (r,s) boundary conditions, where $x,y\in Z_{k+2}$,

$$Z(x,y)=\tfrac{1}{2}\exp[2\pi i xy/(k+2)]$$
$$\times\sum_{l,q,s}\exp[2\pi i xq/(k+2)]\chi^l_{q+2y,s}\chi^{l*}_{q,s}\,. \tag{6}$$

Under the modular transformation $\tau\to(a\tau+b)/(c\tau+d)$ these partition functions transform as $Z(x,y)\to Z(ax+by,cx+dy)$, as they should. Since $Z(x,y)$ is well defined for $x,y\in Z_{k+2}$, there are no anomalies in the modular invariance and thus this discrete symmetry is seen to be realized in the theory.

In order to compactify the heterotic string from 10 to $D=d+2$ space–time dimensions we tensor a collection of the discrete models with different k so as to make the correct total trace anomaly

$$c=\sum_{i=1}^{l}\frac{3k_i}{k_i+2}=\tfrac{3}{2}(10-D)\,. \tag{7}$$

The partition functions of the theory are not just a product of the separate partition functions. The states that appear in the spectrum are precisely the ones obeying the following conditions:

(C1) The left and right states separately have odd integral U(1) charges. Note that this is the total U(1) charge of both the internal and the space–time part of the theory. In the right (bosonic) sector the charges are computed according to the replacement of the vector by the singlet described above. This is the generalization of the GSO projection.

(C2) The left and right states both for the left movers and the right movers separately are either all in the Ramond sector or all in the Neveu–Schwarz sector. This ensures the proper coupling of the world-sheet gravitino.

(C3) In each of the discrete theories the l and $\bar{l}$ indices are correlated according to any of the affine A_1^1 invariants at level k which were classified in ref. [8]. These include the standard invariant at any level [10] which is given by $\delta_{l,\bar{l}}$ summed over all the l. We will use here only the standard invariant. The q and s indices are equal up to some vectors which are generated by β_0 which is 1 in all theta functions and β_i which is a space–time vector, is 2 in the ith s index, and 0 elsewhere, for any i, $1<i<l$.

For any combination of the discrete models this prescription leads to a fully consistent string theory which is modular invariant, obeys the correct connection between spin and statistics, and is space–time supersymmetric. More detail can be found in ref. [5].

The symmetries of the discrete $N=2$ sub-theories lead to such descrete symmetries for the product theories. Consider the theory made by gluing the k_1 theory, the k_2 theory, etc. The discrete symmetry of such a product theory is $G=Z_{k_1+2}\times Z_{k_2+2}\times\ldots\times Z_{k_l+2}$. In addition there are various permutation symmetries involving identical sub-theories. Denote by $\{s_1,s_2,\ldots,s_l\}$ an element of G. Assume that the space–time dimension is $D=4n$. Now, the summation over the vector β_0 in condition (C3) means that elements of G related by $4\beta_0$ should be considered the same. In other words the symmetry of the product theory is the quotient group $\tilde{G}=G/H$ where H is the subgroup of G generated by the element $\{2,2,\ldots,2\}$. (Along with the permutation symmetries which are not affected by β_0). In $D=4n+2$ and l even, H is the bigger subgroup generated by $\{1,1,\ldots,1\}$.

Actually, not all the elements of $\tilde{G}$ commute with supersymmetry. Such elements, which are good symmetries of the theory, but do not respect supersymmetry are called R symmetries. From (6) the conditions for the element $\{s_1,s_2,\ldots,s_l\}$ to commute with supersymmetry is

$$\sum_{i=1}^{l}\frac{s_i}{k_i+2}=\text{integer}\,. \tag{8}$$

Further discrete symmetries result from the affine part. For example, when k is even the standard solution has a Z_2 symmetry which is the center of SU(2). For odd k, this symmetry is broken by anomalies [10].

Consider compactifications to four dimensions, which is the physically interesting case. For this purpose we need an internal theory with $c=9$, which can be made by combining the discrete models in about 140 different ways (using the standard modular invariant for A_1^1). Here we will concentrate on one example.

The quintic hyper-surface in CP^4, denoted by Q, is the simplest Calabi–Yau manifold (see ref. [11] for a nice discussion). It is described by the equation

$$Z_1^5+Z_2^5+Z_3^5+Z_4^5+Z_5^5=0\,. \tag{9}$$

The Hodge numbers for this manifold are $h^{1,0}=h^{0,1}=0, h^{0,0}=h^{3,0}=1$ (this is true for any manifold whose holonomy is precisely SU(3)), $h^{2,1}=101, h^{1,1}=1$. Thus, the leading result in the large-radius expansion predicts that a compactification of the heterotic string on Q should have 101 generations (spinors in the 27 of E_6) and one anti-generation (spinor in the $\overline{27}$). The Hodge number

$h^{2,1}$ counts the number of deformations of the complex structure, i.e., there is a 101-parameter family of manifolds which are topologically equivalent to Q but have different complex structures. They can all be obtained through polynomial perturbations of the defining equation (9). The relevant polynomials are the ones which are at most cubic in any of the Z_i.

The hyper-surface Q enjoys the large global symmetry group $G=(S_5 \ltimes Z_5^5)/Z_5$ where the nth Z_5 is generated by $Z_n \to \exp(\frac{2}{5}\pi i)\, Z_n$ and the S_5 permutes the different variables. Since an overall phase is irrelevant in CP^4 there is an identification of $\{s_i\}$ and $\{s_i+1\}$, where $\{s_i\} \in Z_5^5$, accounting for the quotient by Z_5. Note that any perturbation of Q would destroy part of this symmetry. The group G has 75 000 elements. Some of the elements of G are R symmetries, which do not commute with SUSY. These are given by $\{s_i\}$ where $\sum s_i \neq 0 \bmod 5$.
There is one theory made out of the discrete model which has precisely the same symmetry group as the quintic hyper-surface Q. This is the theory made out of 5 copies of the $k=3$ theory, denoted by 3^5. As explained earlier the symmetry group of this theory is $(S_5 \ltimes Z_5^5)/Z_5$. From (8) the R symmetries are the elements $\{s_i\}$ for which $\sum s_i \neq 0 \bmod 5$. Thus there is an exact correspondence between the symmetries.

It is straightforward to compute the massless spectrum of this theory, using the conditions (C1)–(C3). The spectrum contains the gauge bosons of the group $E_8 \times E_6 \times U(1)^4$. Apart from the standard $N=1$ supergravity and gauge multiplets, one finds matter in the following representations. All matter fields are singlets of E_8. There are 101 left-handed fermions in the 27 of E_6 (generations) along with their scalar superpartners and *CPT* conjugates which are right-handed fermions in the $\overline{27}$ of E_6. There is one anti-generation which is a right-handed fermion in the $\overline{27}$ along with its *CPT* conjugate and one complex scalar superpartner. In addition there are 330 spinors which are E_6 singlets along with their anti-spinor *CPT* conjugates and 330 complex scalar superpartners which are E_6 singlets.

We see that the theory 3^5 has precisely the right number of generations and anti-generations as string theory on Q should have. Moreover, this theory has precisely the same discrete symmetry group G as the hyper-surface Q has. We are thus led to suspect that this theory describes heterotic string propagation on Q.

In order to further substantiate this correspondence, we note that the 101 generations are related to the 101 polynomial deformations of the complex structure. Under the global group G the 101 generations should transform in the same way as these polynomials do. Thus also the generations in the exact spectrum we found should form a representation of G which is isomorphic to the representation of these polynomials. The following is an enumeration of the quantum numbers of the 101 anti-spinors in the 27 which are found in the theory 3^5 along with their isomorphic polynomial deformations

$$\begin{array}{lll} (20) & (0,0,0)\,(1,9,0)\,(3,7,0)^3 & Z_1^3 Z_2^2, \\ (30) & (0,0,0)\,(2,8,0)^2\,(3,7,0)^2 & Z_1^3 Z_2 Z_3, \\ (30) & (1,9,0)^2\,(2,8,0)\,(3,7,0)^2 & Z_1^2 Z_2^2 Z_3, \\ (20) & (1,9,0)\,(2,8,0)^3\,(3,7,0) & Z_1^2 Z_2 Z_3 Z_4, \\ (1) & (2,8,0)^5 & Z_1 Z_2 Z_3 Z_4 Z_5, \end{array}$$

where we denote by the triplet (l,q,s) the three quantum numbers of a field in any of the sub-theories. (More precisely, the 27 is constructed by combining the 1, 10 and 16 of SO(10), which are related by the vector β_0; the numbers above correspond to the 1 of SO(10).) The Z_5 charges of each of the fields, which are given by $q \bmod 5$, are the same as those of the polynomials up to an overall Z_5 constant which is irrelevant. Also the fields permute in precisely the same way as the polynomials. Since the global symmetry structure is very rich, this is a highly unambiguous identification.

The one anti-generation corresponds to the Kähler class which is the only harmonic (1,1) form on Q. Thus the anti-generation is a singlet of G. In the solvable theory it is represented by the field $(2,8,0)^5$, which is indeed a singlet of G.

Consider now the E_6 singlets, in the leading order in $1/r$. (We will refer to the large-radius index theorems as "topological predictions".) The scalar E_6 singlets in Calabi–Yau compactifications divide into three types. First, scalars associated with the deformations of the complex structure. Their number is given by the Hodge number $h^{2,1}$. Second, scalars associated with the change of radius (blowing up), their number is $h^{1,1}$. Third, scalars which are related by supersymmetry to massless fermions predicted by an index theorem with values in the adjoint representation of SU(3). These are counted by (0,1) forms with values in the endomorphisms of the tangent bundle, H^1 (End T) [12]. The quintic hyper-surface has 101 deformations of the complex structure, one perturbation of the Kähler class (radius), and 224 deformations of the tangent bundle. All together the topology predicts 326 massless complex scalars.

In the theory 3^5 we find 330 massless complex scalars which are E_6 singlets. Thus the numbers match quite nicely, and there are only four scalars more than topology predicts. These four scalars are the Higgs particles needed to give masses to the $U(1)^4$ gauge bosons when moving away slightly from this special point in the moduli space of Q. Under the global group G the 101 massless scalars which are related to deformations of the complex structure should transform like the polynomials enumerated above. The one scalar related to the Kähler form should be invariant under G. In addition, the 224 deformations of the tangent bundle of Q are in 1-1 correspondence with the tensors, P_{abcde}, which are symmetric in the last four indices and vanish when symmetrized in all five indices, where the indices are in the range 1-5. Thus also the 224 scalars should transform under G in the same way that the tensors do. The following is a list of the scalars found in the solvable theory along with their isomorphic polynomials and tensors

(20) $(0,8,0)(0,0,0)^3(2,2,0)$ $Z_1^3Z_5^2$

(30) $(0,8,0)(0,0,0)^2(1,1,0)^2$ $Z_1^3Z_4Z_5$

(20) $(0,0,0)^3(1,7,0)(3,3,0)$ P_{44555}

(30×2) $(0,0,0)^2(1,1,0)(1,7,0)(2,2,0)$

$P_{43455}P_{53445}$

(30×2) $(0,0,0)^2(1,1,0)(2,6,0)(3,3,0)$

$P_{34555}P_{43555}$

(30) $(0,0,0)^2(2,6,0)(2,2,0)^2$ $Z_3Z_4^2Z_5^2$

(20) $(0,0,0)(1,1,0)^3(1,7,0)$ $Z_2Z_3Z_4Z_5^2$

(20×3) $(0,0,0)(1,1,0)^2(2,6,0)(2,2,0)$

$P_{23455}P_{32455}P_{42355}$

(20) $(1,9,0)^3(2,0,0)(3,3,0)$ P_{45555}

(5) $(1,1,0)^4(2,6,0)$

$Z_1Z_2Z_3Z_4Z_5P_{12345}P_{21345}P_{31245}P_{41235}$

(5) $(1,9,0)^4(2,4,0)$ Kähler+4 extra

(Some small freedom in this identification remains.)

A $Z_5 \times Z_5$, subgroup of G, H, which acts freely on Q, is generated by the two transformations, $A(Z_n) = \exp(\frac{2}{5}\pi i n) Z_n$, and $B(Z_n) = Z_{n+1}$. The quotient manifold Q/H is a Calabi–Yau manifold with $h^{2,1} = 5$, $h^{1,1} = 1$, leading to a four-generation model. For simplicity, consider string propagation on Q/(A), which is a Calabi–Yau manifold with $h^{2,1} = 21$ and $h^{1,1} = 1$.

The partition function of an heterotic string propagating on Q/(A) can be computed exactly by implementing eq. (6) in the solvable theory. Namely, the propagation of a closed string on Q/H is equivalent to the propagation of an open string in Q whose end points differ by elements of H. In addition, all the states are projected onto the H invariant subspace. Thus, if we denote by $Z(x,y)$ the partition function of Q on the torus with boundary conditions (x,y), the partition function of Q/H will be given by the sum $\sum_{x,y\in H} Z(x,y)$. This amounts to a simple modification of conditions (C1) and (C3) in the prescription for computing the spectrum. Using this method, the exact spectrum of an heterotic string propagating on Q/(A) was computed, and we find 21 generations, one anti-generation, and 210 massless scalars, 70 of which come from the untwisted sector. The generations and anti-generations all come from the untwisted sector as predicted by the topology. Thus the geometrical information that (A) acts freely on Q is recovered. This completes the identification of the theory 3^5 as an exact solution for string propagation on Q.

The leading $1/r$ results service in the exact theory to a surprisingly good degree. It should have been expected that quantum effects would, at least, break some of the global symmetries, especially the R symmetries. However, all global symmetries survive. Furthermore, instanton calculations [13] suggest that the modes associated with H^1(End T) could receive non-perturbative masses. Again, as seen here, this is not the case.

Consider now compactifications down to eight dimensions. For this purpose we need a $c=3$ theory which can be made in three different ways. These are, $c=1+1+1$, $c=\frac{2}{2}+\frac{3}{2}$, and $c=1+2$. It can be easily seen that the first theory corresponds to string propagation on the SU(3) torus [#1]. This of course gives rise to an extra SU(3) gauge symmetry by the Frenkel–Kac mechanism, and the gauge group is $E_8\times E_8\times SU(3)$. Similarly, the second and third theories describe string propagation on the $SU(2)^2$ torus. Tori are indeed the only manifolds of SU(1) (trivial) holonomy.

A number of important features are already apparent in these simple examples. The tori described by these theories have extra "accidental" gauge symmetries which are not predicted by topology, but are stringly in nature. In addition, these tori have discrete symmetries. The SU(3) torus has a Z_3 symmetry, and the $SU(2)^2$ torus has a Z_2 symmetry. These are in fact the only possible holomorphic global symmetries allowed for tori, and the SU(3) and $SU(2)^2$ tori are in fact the only ones with any nontrivial such symmetry at all. The SU(3) torus may be described by one cubic equation in CP_2,

$$Z_1^3+Z_2^3+Z_3^3=0 . \tag{10}$$

This equation nicely demonstrates its symmetry: $(S_3\ltimes Z_3^3)/Z_3$. Two of the Z_3 are embedded in the $U(1)^2$ continuous symmetry, and only the third Z_3 is special to this torus. Since the $k=1$ theory has a Z_3 symmetry, the theory made out of three copies of it has the same symmetry.

Let us turn now to six-dimensional compactifications. In this case we need $c=6$ for the internal theory, and here are 17 combinations of the discrete series which give this value. We denote by $k_1^{m_1}k_2^{m_2}...k_l^{m_l}$ the theory composed out of m_1 copies of the k_1th discrete theory, m_2 copies of k_2th, etc. The theory $1^3 2^2$ can be easily seen to correspond to string propagation on the $SU(3)\times SU(2)^2$ torus. The spectra of the other 16 theories are listed in table 1. From this table it is apparent that these spectra are identical, and differ only in the enhanced gauge symmetries and a number of E_7 singlets equal to twice the dimension of this group.

The explanation for this is as follows. Topologically, there are precisely two manifolds of SU(2) holonomy in two complex dimensions. These are the complex torus T^2 and the K3 surface. Indeed, the 16 theories listed in table 1 all correspond to string propagation on the K3 surface, with an assortment of complex structures. (None though corresponds to the standard orbifold limits of K_3 [15].) On this surface the topology predicts 20 spinors in the 56 of E_7 [#2], and no anti-spinors in the 56. The spinors come from the $h^{1,1}=20$, deformations of the complex structure. In addition, we should find 45 spinors which are E_6 singlets from $H^1(\mathrm{End}\,T)$, another 45 from their *CPT* conjugates, 20 singlets from the deformation of the complex structure corresponding to harmonic (1,1) forms, and another 20 singlets from the 20 radii of K_3 also corresponding to (1,1) forms. In total 130 spinors which are E_6 singlets. The scalars associated with the radii and the scalars associated to the complex structure form an $N=1$ SUSY multiplet along with their corresponding spinors.

As two-dimensional quantum field theories these models are very different from one another. For example, their discrete symmetries are completely different. Thus they describe string propagation on different K3 surfaces. The most symmetric shape for the K3 surface is described by the quartic equation in CP_3,

$$Z_1^4+Z_2^4+Z_3^4+Z_4^4=0 . \tag{11}$$

The discrete symmetry of this surface is $G=(S_4\ltimes Z_4)/Z_4$. The theory 2^4 has the same discrete symmetry. As in the case of the quintic hypersurface, it is straightforward to verify that the 20 spinors in the 56 of E_7, and the 130 singlets found in the theory 2^4 transform under G in the same way that their corresponding deformations, and tensors do. Thus, this theory describes string propagation on the surface (11).

The only significant property the sixteen models listed in table 1 share in common is the $N=2$ superconformal invariance, which leads to space–time

[#1] More precisely the SU(3) WZW theory at level one. This amounts to a constant background anti-symmetric tensor field [14], at $\theta=\pi$. Similar instanton terms with $\theta=\pi$ (because of parity in the sigma model) can appear also in the other compactifications, by the addition of the complex partners of the fields that change the radius.

[#2] The 20 spinors are 10 left-handed fermions and another 10 *CPT* conjugates which are also spinors. Similarly, the 130 E_6 singlets are 65 left-handed fermions.

Table 1

Theory	s in 56	$\bar{s}$ in 56	E_6 singlets	Gauge
1^6	20	0	140	$U(1)^5$
$1^4 4^1$	20	0	140	$U(1)^5$
$1^2 2^1 10^1$	20	0	136	$U(1)^3$
$1^2 4^2$	20	0	136	$U(1)^3$
$1^1 2^2 4^1$	20	0	146	$SU(2)^2 \times U(1)^2$
$1^1 5^1 40^1$	20	0	134	$U(1)^2$
$1^1 6^1 22^1$	20	0	134	$U(1)^2$
$1^1 7^1 16^1$	20	0	134	$U(1)^2$
$1^1 8^1 13^1$	20	0	134	$U(1)^2$
$1^1 10^2$	20	0	134	$U(1)^2$
2^4	20	0	136	$U(1)^3$
$2^1 3^1 18^1$	20	0	134	$U(1)^2$
$2^1 4^1 10^1$	20	0	134	$U(1)^2$
$2^1 6^2$	20	0	134	$U(1)^2$
$3^2 8^1$	20	0	134	$U(1)^2$
4^3	20	0	134	$U(1)^2$

supersymmetric string compactifications as explained in ref. [5]. This procedure is, in fact, general for any $N=2$ superconformal field theory. Since the models we have been considering are by no means special, table 1 demonstrates that any $c=6$ $N=2$ superconformal theory would lead to either the T^2 or K3, which are the only surfaces of SU(2) holomomy.

In another direction, since six and eight dimensions should also not be special these results demonstrate that compactifying the heterotic string down to four dimensions on an arbitrary $N=2$ superconformal field theory with $c=9$ is equivalent to a Calabi–Yau compactification, and in particular, the compactifications we have been considering using tensoring of $c<3$ models correspond to solvable points in the moduli space of Calabi–Yau manifolds. We have studied the spectrum of the 140 or so four-dimensional models obtained by taking the standard modular invariant, and, indeed, they all fall into the general profile of Calabi–Yau compactifications. Several of the models have the same number of generation and anti-generations as known Calabi–Yau manifolds predict. Most, however, are new ones.

Why should an arbitrary $N=2$ superconformal field theory with $c=9$ be equivalent to a Calabi–Yau compactification? As shown in ref. [5] using the $N=2$ models we get space–time supersymmetric heterotic string compactifications. In ref. [4] it was shown that the field equations of ten-dimensional superstring in the leading order have supersymmetric solutions for an internal manifold only if this manifold is a complex one with SU(3) holonomy. Thus also a sigma model describing heterotic string propagation would give rise to space–time supersymmetry only if it corresponds to a Calabi–Yau space. On the other hand, experience with solvable conformal field theory, in general, suggests that an action of some sort should always exist. For example, any conformal field theory invariant under current algebra, in the Sugawara form, can be described by a WZW action. Thus, what singled out Calabi–Yau manifolds to begin with, space–time supersymmetry, tells us also that any (2,2) compactification is equivalent to a Calabi–Yau one.

We can add to the above correspondence by contemplating its continuation to a higher number of dimensions. For example, consider the theory made out of n copies of the $k=n-2$ discrete theory. This theory has the trace anomaly $c=3(n-2)$, and symmetry group $(S_n \ltimes Z_n^n)/Z_n$. This is precisely the symmetry of the hyper-surface

$$Z_1^n + Z_2^n + \ldots + Z_n^n = 0 , \tag{12}$$

which is a hyper-surface of $n-2$ complex dimensions, with vanishing first Chern class (as can be seen by writing the globally defined holomorphic $(0,n)$ form). Indeed, for $n=3,4,5$ we have seen in detail that the theory $(n-2)^n$ describes string propagation on this hyper-surface. Thus, it is natural to extrapolate that for any n this theory corresponds to a sigma model on the manifold (12). Remarkably, the entire discrete series can then be interpreted as a piece of the appropriate complex manifold.

In view of the foregoing analysis it is conjectured by this author that any heterotic string theory made by using a $c=3k, N=2$ superconformal theory, and the generalization of the embedding of the spin connection in the gauge group described in ref. [5] is equivalent to some compactification on a manifold of SU(k) holonomy. Using solvable building blocks we then obtain exactly solvable Calabi–Yau compactifications. By tensoring the discrete series, we can get at least several hundred such solvable points. Furthermore, we can twist these theories by any subgroup of the discrete symmetry group which does not contain R symmetries, as was discussed for the quintic hyper-surface. The resulting theories are perfectly acceptable even when these subgroups do not act freely on the manifold, in which case they correspond to string propagation on Calabi–Yau manifolds with conical singularities. This would lead to something about 10^6 such theories. Still more, we can contemplate modifying these product theories in a variety of other ways, which do not break supersymmetry. Thus, the number of solvable points is huge, perhaps dense in the moduli space of all Calabi–Yau manifolds. (Solvability can be defined by demanding the partition functions to be a sum of products of holomorphic and anti-holomorphic modular forms of some finite index subgroup of the modular group). A classification of these models will appear elsewhere [16].

A number of previously unknown features of Calabi–Yau compactifications are apparent from the exact solutions. The solvable models have an enhanced gauge symmetry which is at least $U(1)^{l-1}$, where l is the number of discrete models, as well as large global symmetry groups. Thus we see that in the moduli space of Calabi–Yau manifolds there are many points of enhanced gauge symmetry, similar to those encountered in toroidal and orbifold compactifications. Moreover, the selection rules on Yukawa couplings can be considerably stronger than topology predicts. For instance, in the Q manifold example each of the Z_5 charges is actually embedded in a U(1), and many Yukawa couplings allowed by the conservation of the Z_5 charges, are forbidden by U(1) conservation. Moreover, the solvable models naturally incorporate in them, previously unsuspected, global symmetries. Such selection rules and global symmetries are crucial in model building, enabling the reduction of the number of generations, Wilson line breaking of E_6, and the suppression of proton decay.

Perhaps most importantly, the results presented here demonstrate that Calabi–Yau compactifications indeed exist as conformal field theories and are exact solutions of the string equations of motion, despite the breakdown of conformal invariance at the four-loop level [17], and the possibility that non-perturbative effects on the world sheet would destabilize the vacuum. The fact that we find in the spectrum all the modes corresponding to the deformations of the complex structure and the radii that for any complex structure, and radii, the models are conformally invariant, space–time supersymmetric, and fully retain their topological and geometrical properties.

Finally, a degree of uniqueness and geometrical interpretation is restored to four-dimensional string theories, by demonstrating that any compactification of the heterotic string of the generalized type of "spin connection embedded in the gauge group" is equivalent to some Calabi–Yau compactification.

It is a pleasure for me to acknowledge interesting discussions with L. Dixon, A.B. Zamolodchikov, and in particular E. Witten.

References

[1] M.B. Green and J.H. Schwartz, Nucl. Phys. B 255 (1985) 92.

[2] D.J. Gross, J. Harvey, E.J. Martinec and R. Rohm, Nucl. Phys. B 256 (1985) 253.

[3] K.S. Narain, M.H. Sarmadi and C. Vafa, Nucl. Phys. B 279 (1987) 369;
H. Kawai, D. Lewellen and H. Tye, Phys. Rev. Lett. 57 (1986) 1832; and Cornell preprint CLNS 86/751;
R. Bluhm, L. Dolan and P. Goddard, preprint RU/B/187 (1986);
M. Mueller and E. Witten, Phys. Lett. B 182 (1986) 28;
I. Antoniadis, C. Bachas and C. Kounnas, preprint LBL-22709.

[4] P. Candelas, G. Horowitz, A. Strominger and E. Witten, Nucl. Phys. B 258 (1985) 46.

[5] D. Gepner, Princeton University preprint PUPT-1056, April (1987), to be published in Nucl. Phys. B.

[6] A. Sen, Nucl. Phys. B 278 (1986) 289; B 284 (1987) 423; D. Friedan et al., unpublished.
[7] A.B. Zamolodchikov and V.A. Fateev, Zh.Eksp. Theor. Fiz. 90 (1986) 1553;
W. Boucher, D. Friedan and A. Kent, Phys. Lett. B 172 (1986) 316;
P. Di-Vecchia, J.L. Petersen and H.B. Zheng, Phys. Lett. B 174 (1986) 280;
S. Nam, Phys. Lett. B 172 (1986) 323.
[8] D. Gepner, Nucl. Phys. B 287 (1987) 111;
A. Cappelli, C. Itzykson and J.B. Zuber, Nucl. Phys. B 280 [FS16] (1987) 455.
[9] D. Gepner and Z. Qiu, Nucl. Phys. B 285 [FS19] (1987) 423.
[10] D. Gepner and E. Witten, Nucl. Phys. B 278 (1986) 493.
[11] M. Green, J. Schwartz, and E. Witten, Superstring theory, Vol II (Cambridge U.P., Cambridge, 1987).
[12] E. Witten, Nucl. Phys. B 268 (1986) 79.
[13] M. Dine, N. Seiberg, X.G. Wen and E. Witten, Nucl. Phys. B 278 (1986) 769; and IAS preprint (1986).
[14] K.S. Narain, M.H. Sarmadi and E. Witten, Nucl. Phys. B 279 (1987) 369.
[15] L. Dixon, J. Harvey, C. Vafa and E. Witten, Nucl. Phys. B 261 (1985) 678.
[16] D. Gepner, in preparation.
[17] M.T. Grisaru, A. Van de Ven and D. Zanon, Phys. Lett. B 173 (1986) 423.

6. Four-Dimensional Type-II Strings

This chapter is devoted to a rather short-lived revival of type-II strings. Before the birth of heterotic strings, type-II strings were, from a string theorist's point of view, the most elegant of string theories. Unfortunately they were less attractive phenomenologically, since there was no way to obtain a reasonable gauge group. There is no gauge group in ten dimensions, and naive compactification did not appear to give anything but $U(1)^6$, in both sectors. The construction of the heterotic string led to new hope that a less naive compactification might yield something more interesting.

Indeed, it was shown in paper 19 and [139] that also the group $SU(2)^6$ could be obtained. This discovery is easily understandable from our present point of view. In the bosonic string representation, to construct a type-II string in four dimensions one looks for some CFT with a left- and right-moving level-1 D_5 KM algebra, combined with an internal sector with central charges $c_L = c_R = 9$. Obvious solutions to the modular invariance conditions are (among many others) the lattice $(E_8D_6)_L(D_{14})_R$ (with D_5 embedded in the E_8 factor) and $(D_{14})_L(D_{14})_R$, with left–right pairing of conjugacy classes between the D lattices.

World-sheet supersymmetry can be realized as follows. In the E_8D_6 sector one can choose a supercurrent $\psi^a\, \partial X_a$, where ∂X_a is a Cartan sub-algebra generator of D_6, and ψ^a spans the rest of the internal sector within E_8. This sector gives rise to an $N = 4$ space–time supersymmetry. In the internal sector of the D_{14} lattice one can choose a supercurrent of the form $f_{abc}\chi^a\chi^b\chi^b$, where χ^a, $a = 1, \dots, 18$, is a set of fermions spanning the internal sector. As was shown in [121], this is a good supercurrent if and only if f_{abc} are, up to normalization, structure constants of a semi-simple Lie algebra. In four dimensions, there are just three semi-simple Lie algebras $\mathcal{G}$ of dimension 18: $SU(2)^6$, $SU(4) \times SU(2)$ and $SU(3) \times SO(5)$. The monodromy conditions for this supercurrent are obviously satisfied: the NS sector is formed by the identity and vector conjugacy class of D_{14}, and the R sector by the spinor conjugacy class.

As explained in section 2.3.2, the fermions transforming in the adjoint representation of these groups are members of a superfield with $h = \frac{1}{2}$. Their partners are 18 currents J^a with $h = 1$, which form a sub-algebra of the SO(18) KM algebra of the right-moving sector. The embedding is simple: the vector representation of SO(18) contains the adjoint representation of the gauge group. The spectrum of the $(E_8D_6)_L(D_{14})_R$ theory contains the gauge bosons of $\mathcal{G}$ and in addition six U(1) bosons associated with the left-moving sector. Having $N = 4$ supersymmetry, it is certainly not chiral. The $(D_{14})_L(D_{14})_R$ theory has a left- and a right-moving factor $\mathcal{G}$ (the choice among the three possibilities need not be the same for left- and right-movers). However, the latter theory is tachyonic, and does not even have massless spinors. Another obvious possibility is an $(E_8D_6)_L(E_8D_6)_R$ theory with $N = 8$ supersymmetry.

This explains not only the appearance of $SU(2)^6$, but also points to the existence of two other gauge groups. Type-II strings realizing these other possibilities were first obtained in paper 20 which was also the first to properly discuss modular invariance for these theories. The model described in paper 20 is the $N = 4$ string discussed above.

If one wants to insist on having 18 gauge bosons, the sector containing their vertex operators is completely constrained to be D_{14}. This limits the possible solutions to precisely the non-chiral ones listed above (see paper 22).

If one is satisfied with fewer gauge bosons there are more possibilities. The attraction of the groups $SU(3) \times SO(5)$ and $SU(4) \times SO(3)$ is that they contain the standard model gauge group

SU(3) × SU(2) × U(1). The latter has only 12 gauge bosons, so that is would not seem such a bad idea to consider this option. One might furthermore hope to get chiral fermions.

Such generalizations were consider in paper 21 for fermionic constructions and in paper 22 for lattice constructions. No chiral models were found, and in paper 22 it was shown that one cannot get the standard model with lattice constructions. The matter was finally closed by Dixon, Vafa and Kaplunovsky (paper 23) who showed without making any assumption other than the usual string consistency conditions (in particular world-sheet superconformal invariance is important here) that one cannot obtain the standard model from type-II strings. Interestingly, these authors did find chiral models, a remarkable result since one cannot obtain chiral fermions from compactification of pure gravity in field theory.

Time will tell whether the negative result of paper 23 is the final nail in the coffin of type-II strings, or just another no-go theorem that can be evaded because one of the assumptions was too strong in an unforeseen way.

ON THE COMPACTIFICATION OF THE CLOSED SUPERSYMMETRIC STRING

L. CASTELLANI [a], R. D'AURIA [a,b], F. GLIOZZI [a] and S. SCIUTO [a,b]

[a] *Istituto Nazionale di Fisica Nucleare, Sezione di Torino, I-10125 Turin, Italy*

[b] *Dipartimento di Fisica Teorica, Università di Torino, I-10125 Turin, Italy*

Received 6 December 1985

If D dimensions of the closed $N = 1$ supersymmetric string are compactified on a torus, solitonic excitations can enlarge the $[\mathrm{U}(1)]^D$ Kaluza–Klein gauge symmetry to $[\mathrm{SU}(2)]^D$. The "old formalism" is extended to obtain the $N = 2$ supersymmetric string. In its compactification to four dimensions, massless solitons break $N = 8$ to $N = 4$ supersymmetry.

The known superstring theories [1,2] require a ten-dimensional spacetime to be consistent at the quantum level. The ambitious goal of the recent superstring Renaissance is to provide a finite, unified quantum theory of gravity and matter. A necessary step in this direction is a study of possible $10 \to 4$ compactification schemes.

For the limiting ten-dimensional field theories, prospects are not very bright ("ten into four won't go" [3]). Hence, to obtain a consistent four-dimensional theory via compactification, the full superstring must be taken into consideration.

In this letter we concentrate on closed ten-dimensional strings, and discuss a compactification mechanism that produces massless nonabelian vector states in $(10 - D)$ dimensions.

This mechanism was first proposed in the context of the heterotic string [2], where the left-moving (LM) modes of the bosonic string in 26 dimensions are compactified on a 16-torus, and combine with the right-moving (RM) modes of the ten-dimensional superstring. The interplay between the LM tachyon and the solitonic excitations on the 16-torus gives rise to the gauge vectors of a Lie algebra of rank 16 [2] [$E_8 \times E_8$ or $\mathrm{Sp}(32)/Z_2$].

In ref. [4], the same mechanism is applied to a ten-dimensional string, made out of the RM modes of the $N = 1$ superstring [5] and the LM modes of the Neveu–Schwarz [6] spinning string. The presence of a tachyon state in the NS sector allows to compactify the theory on a particular D-torus, so that massless solitonic vectors arise in $(10 - D)$ dimensions. As in the heterotic string, the compactified theory is tachyon-free.

The open $N = 1$ superstring contains two sectors, that we briefly recall by using the "old formalism" [5] in the light-cone gauge:

(i) *The bosonic sector*, described by the oscillators a_n^i ($n \in \mathbf{Z}$, $i = 1, ..., 8$), b_r^i ($r \in \mathbf{Z} + 1/2$), acting on a Fock-space vacuum $|0_{\mathrm{NS}}\rangle$.

The states are constrained by the requirement $G = +1$, where G is the conserved multiplicative quantum number

$$G = (-1)^{\sum_{r=1/2}^{\infty} b_{-r} b_r - 1} . \tag{1}$$

The mass in this sector is given by

$$m_{\mathrm{R}}^2/2 = N - 1/2 , \tag{2}$$

where

$$N = \sum_{m=1}^{\infty} n a_{-n} a_n + \sum_{r=1/2}^{\infty} r b_{-r} b_r . \tag{3}$$

The tachyonic ground state $|0_{\mathrm{NS}}\rangle$ is removed from the spectrum by the restriction $G = +1$.

(ii) *The fermionic sector*, described by the a_n^i and by the Ramond oscillators [7] d_n^i ($n \in \mathbf{Z}$), acting on a Majorana–Weyl ground state $|0_{\mathrm{R}}\rangle^a$ ($a = 1, ..., 8$ is the spinorial index).

The states are chosen to have positive Weyl parity, i.e. $G_{\mathrm{W}} = +1$, where

0370-2693/86/$ 03.50 © Elsevier Science Publishers B.V.
(North-Holland Physics Publishing Division)

$$G_W = \gamma_{11}(-1)^{\Sigma_{n=1}^{\infty} d_{-n} d_n} . \tag{4}$$

The mass formula is

$$m_R^2/2 = R , \tag{5}$$

where

$$R = \sum_{n=1}^{\infty} n a_{-n} a_n + \sum_{n=1}^{\infty} n d_{-n} d_n . \tag{6}$$

The ten-dimensional closed string considered in ref. [4] is obtained by matching together the RM modes of the sectors (i) and (ii) with those of a LM Neveu–Schwarz (bosonic) string, described by the oscillators $\tilde{a}_n{}^i$ and $\tilde{b}_n{}^i$. The LM mass formula reads

$$m_L^2/2 = \tilde{N} - 1/2 . \tag{7}$$

The mass m of the closed string states is related to the RM and LM masses as follows:

$$m^2/2 = m_R^2 + m_L^2 . \tag{8}$$

Moreover, the invariance of the closed string under an arbitrary shift in σ implies the constraint:

$$m_R^2 = m_L^2 . \tag{9}$$

Before compactification, (9) together with eqs. (2), (5), (7) gives $N = \tilde{N}$ in the bosonic sector of the closed string, and $\tilde{N} = R + 1/2$ in the fermionic sector. Notice that $N = \tilde{N}$ forces the LM string to have $\tilde{G} = G = +1$.

The massless modes of the supersymmetric string combine into the $N = 1, d = 10$ supergravity multiplet. The graviton, the antisymmetric tensor and the dilaton correspond to the symmetric traceless, antisymmetric and trace part of the state

$$b^i_{-1/2}\tilde{b}^j_{-1/2}|0_{NS}\rangle\, |\tilde{0}_{NS}\rangle , \tag{10}$$

while the gravitino and the spin-1/2 are contained in

$$\tilde{b}^i_{-1/2}|0_R\rangle^a\, |\tilde{0}_{NS}\rangle . \tag{11}$$

The compactification on a D-torus does not change eqs. (8) and (9), while (2), (5) and (7) are replaced by:

$$m_R^2/2 = N - 1/2 + P^I P^I/2 , \tag{2'}$$

$$m_R^2/2 = R + P^I P^I/2 , \tag{5'}$$

$$m_L^2/2 = \tilde{N} - 1/2 + \tilde{P}^I\tilde{P}^I/2 , \tag{7'}$$

where $P^I = p^I/2 - L^I, \tilde{P}^I = p^I/2 + L^I$ and L^I is the soliton winding number around the Ith circle.

Now the constraint (9) does not imply $\tilde{G} = G$ any-more. In fact it becomes

$$N + P^I P^I/2 = \tilde{N} + \tilde{P}^I\tilde{P}^I/2 \tag{12}$$

in the bosonic sector, and

$$R + P^I P^I/2 = \tilde{N} - 1/2 + \tilde{P}^I\tilde{P}^I/2 \tag{13}$$

in the fermion sector. $\tilde{G} = -G$ is allowed when $P^I P^I - \tilde{P}^I\tilde{P}^I$ is an odd integer, and the LM open string can be the full Neveu–Schwarz string with $\tilde{G} = \pm 1$. In particular, LM $\tilde{G} = -1$ states with $\tilde{N} = 0$ and $\tilde{P}^I\tilde{P}^I = 1$ are massless [see eq. (7′)]. Thus, if the D-torus is chosen such that the momenta $\tilde{P}^I$ lie on a lattice containing a finite sublattice Λ_1 of unit length, the $(10-D)$-dimensional spectrum includes massless solitons.

For $D = 6$ the massless states of the $10 \to 4$ compactified model arrange themselves into the $N = 4$, $d = 4$ supergravity multiplet coupled to $[n(\Lambda_1) + 6]$ $N = 4$ vector multiplets, $n(\Lambda_1)$ being the number of points in Λ_1 [4].

As the generalized momenta $\tilde{P}^I$ lie on a lattice, one is tempted to use the Frenkel–Kac–Goddard–Olive [8,9] construction to build a Lie algebra. The vectors $\tilde{P}^I$ (corresponding to the massless states) have unit length and then the construction of refs. [8,9] would lead to a graded algebra; however an ordinary Lie algebra can be recovered if also the anticommuting Neveu–Schwarz operators b_r^I come into play.

In fact one can write the hybrid vertex operators

$$V^I_{\pm e_J} = \frac{1}{2\pi i} c(e_J) \oint \frac{dz}{z} H^I(z)\, U(\pm e_J, z) , \tag{14}$$

where the e_J are the six unit vectors on the lattice Λ_1,

$$H^I(z) = \sum_{\substack{r=-\infty \\ r\in \mathbf{Z}+1/2}}^{\infty} \tilde{b}^I_r z^{-r} , \tag{15}$$

and

$$U(e, z) = :\exp[ie\tilde{Q}(z)]:$$

$$= \exp\left(ie \sum_{n=1}^{\infty} \tilde{a}_{-n} \frac{z^n}{\sqrt{n}}\right) \exp\left(ie \sum_{n=1}^{\infty} \tilde{a}_n \frac{z^{-n}}{\sqrt{n}}\right)$$

$$\times \exp(ieq)\, z^{e^2/2 + e\cdot\tilde{p}} . \tag{16}$$

As $H^I(z)$ is two-valued, the single valuedness of the vertices (14) is assured by the conditions $e_J^2 = 1, e \cdot \tilde{p} \in \mathbf{Z}$; then Λ_1 must be integral, i.e. hypercubic; finally the coefficients $c(e)$ are the Frenkel–Kac factors satisfying:

$$c(u)\, c(v) = c(v)\, c(u)(-1)^{(u\cdot v)^2 - u^2 v^2} ,$$

$$c(u)\, c(v) = \epsilon(u,v)\, c(u+v) , \quad \epsilon(u,v) = \pm 1 . \tag{17}$$

The 6 × 12 vertices $V^I_{\pm e_J}$ together with the SO(12) generators: $\tilde{p}^I$ and

$$E(u) = c(u) \oint \frac{\mathrm{d}z}{z}\, U(u,z) ,$$

$$u = \pm e_I \pm e_J , \quad I \neq J , \tag{18}$$

and with the SO(6) generators

$$M^{IJ} = -\mathrm{i} \sum_{r=1/2}^{\infty} (\tilde{b}^I_{-r} \tilde{b}^J_r - \tilde{b}^J_{-r} \tilde{b}^I_r) \tag{19}$$

close an SO(18) algebra [‡1]. It is easy to verify that the 18 massless spin-1 states

$$B^{\mu I} = b^\mu_{-1/2} \tilde{b}^I_{-1/2} |0_{\mathrm{NS}}\rangle\, |\tilde{0}_{\mathrm{NS}}\rangle ,$$

$$C^\mu_{\pm e_J} = b^\mu_{-1/2} |0_{\mathrm{NS}}\rangle\, |(e_J)^2 = 1\rangle , \tag{20}$$

of the matter multiplet span the vector representation of this SO(18) algebra.

However this fact should be really significant only if the symmetry were preserved by interaction. Unfortunately this is not the case. The operators V, E, P, M commute with the Virasoro operators L_n (then everything would work in a Bardakci–Halpern [10]-like model) but do not commute with the supergauge operators G_r. The requirements of invariance under the G_r's break the SO(18) algebra into six commuting SU(2) subalgebras, each of them generated by the triplet of operators:

$$\tilde{P}^I , \quad V^{\pm I} = V^I_{\pm e_I} \quad (I \text{ fixed}) ; \tag{21}$$

analogously each triplet of particles (20), for fixed I, transforms according to the adjoint representation of the corresponding SU(2) algebra [and as 3 scalars under the other SU(2)'s].

‡1 In general this technique allows to build the SO$(2N + M)$ algebra (not simply laced if M is odd) by means of N oscillators a_n and M oscillators b_r.

Now it is easy to see that such a $[\mathrm{SU}(2)]^6$ group is actually the only gauge symmetry of our string model. Indeed it is sufficient to investigate the trilinear coupling of the massless vector states, which is proportional to the structure constants of the gauged group.

As every three-particle amplitude of closed strings, it factorizes into the product of two open string vertices, left and right moving respectively. The coupling with an odd number of solitonic vectors (*CCC* or *CBB*) is forbidden by the conservation of quantized momenta; similarly, the non-solitonic vertex *BBB* is zero because it is proportional to the momentum along the compactified directions. The only non-vanishing amplitude involves two *C*'s and one *B* belonging to the same SU(2) representation.

Denoting by $\epsilon_1, \epsilon_2, \epsilon_3$ and k_1, k_2, k_3 the four-dimensional polarizations and momenta of a *CBC* vertex, and by P_1, P_3 the generalized momenta along the compactified dimensions, we have

$$\epsilon_1^\mu \epsilon_2^\nu \epsilon_3^\rho\, W^I_{\mu\nu\rho}(k_i; P_1, P_3)$$

$$= \epsilon_1^\mu \epsilon_2^\nu \epsilon_3^\rho\, V_{\mu\nu\rho}(\tfrac{1}{2} k_i)\, \tilde{V}^I(\tfrac{1}{2} k_i; P_1, P_3) , \tag{22}$$

where

$$\epsilon_1^\mu \epsilon_2^\nu \epsilon_3^\rho\, V_{\mu\nu\rho}(\tfrac{1}{2} k_i) = -(k_1 \cdot \epsilon_2)(\epsilon_1 \cdot \epsilon_3)$$

$$-(k_2 \cdot \epsilon_3)(\epsilon_2 \cdot \epsilon_1) - (k_3 \cdot \epsilon_1)(\epsilon_3 \cdot \epsilon_2) \tag{23}$$

is the well known "three-gluon vertex" of the NS model, and

$$\tilde{V}^I(\tfrac{1}{2} k_i; P_1, P_3) = -\tfrac{1}{2} e_I (P_1 - P_3) \tag{24}$$

is the two-tachyon photon vertex. Note that eq. (23) is odd under the exchange 1 ↔ 3: the full three-particle amplitude (sum of all orderings) vanishes in the ordinary open string model, unless a Chan–Paton factor is introduced. In the present *closed* string model the role of such a factor is played by the LM vertex (24), which gives essentially the structure constants of the SU(2) group defined in eq. (21).

Thus, like in the heterotic string, the solitonic excitations are the crucial ingredients to build up non-abelian symmetries.

Finally, we present the extension of the "old formalism" to the $N = 2$ supersymmetric string. Here one needs all the four sectors (corresponding to the four combinations of boundary conditions on the spinor fields [5]):

$$\left.\begin{array}{l} \mathrm{NS} \oplus \widetilde{\mathrm{NS}} \\ \mathrm{R} \;\;\oplus \widetilde{\mathrm{NS}} \end{array}\right\} N = 1 \text{ oriented}$$

$$\mathrm{NS} \oplus \widetilde{\mathrm{R}}$$

$$\mathrm{R} \;\;\oplus \widetilde{\mathrm{R}} \,. \tag{25}$$

The last two were discarded by the authors of ref. [5] as unnecessary for the $N = 1$ formulation. In doing so, they obtained an *oriented* $N = 1$ superstring.

The massless $d = 10$ spectrum is given by the $N = 2$ supermultiplet:

$$b^{i}_{-1/2}\tilde{b}^{j}_{-1/2}|0_{\mathrm{NS}}\rangle\,|0_{\mathrm{NS}}\rangle \quad \text{graviton, antisymmetric tensor, dilaton}$$

$$\left.\begin{array}{l} \tilde{b}^{i}_{-1/2}|0_{\mathrm{R}}\rangle^{a}\,|0_{\mathrm{NS}}\rangle \\ b^{i}_{-1/2}|0_{\mathrm{NS}}\rangle\,|\tilde{0}_{\mathrm{R}}\rangle^{a} \end{array}\right\} \quad 2 \text{ gravitini, } 2 \text{ spin-}\tfrac{1}{2}$$

$$|0_{\mathrm{R}}\rangle^{a}\,|\tilde{0}_{\mathrm{R}}\rangle^{b} \,. \tag{26}$$

For the $|0_{\mathrm{R}}\rangle^{a}\,|\tilde{0}_{\mathrm{R}}\rangle^{b}$ states, one has to distinguish two cases, corresponding to type IIA and IIB superstrings:

(IIA) $|0_{\mathrm{R}}\rangle^{a}$ and $|\tilde{0}_{\mathrm{R}}\rangle^{b}$ are in inequivalent SO(8) spinor representations. Then their tensor product decomposes into

$$|a\rangle \otimes |\dot{b}\rangle = \square + \begin{array}{|c|}\hline \\ \hline \\ \hline \\ \hline \end{array} \,,$$

i.e. a vector T_i and a three-index antisymmetric tensor T_{ijk};

(IIB) $|0_{\mathrm{R}}\rangle^{a}$ and $|\tilde{0}_{\mathrm{R}}\rangle^{b}$ are in the same SO(8) irrep. Then

$$|a\rangle \otimes |b\rangle = \cdot + \begin{array}{|c|}\hline \\ \hline \\ \hline \end{array} + \begin{array}{|c|}\hline \\ \hline \\ \hline \\ \hline \\ \hline \end{array}_{\,\text{self dual}} \quad :$$

$$\phi,\, T_{ij},\, (T_{ijkl})_{\text{self dual}} \,.$$

In IIB superstrings both right- and left-handed Ramond states have the same Weyl parity $G_{\mathrm{W}} = \tilde{G}_{\mathrm{W}}$, whereas IIA corresponds to $G_{\mathrm{W}} = 1 = -\tilde{G}_{\mathrm{W}}$.

The matching of bosons and fermions at each mass level can be derived from the NS ↔ R matching of ref. [5], and holds separately for the first two and the last two sectors of (25).

The $N = 2$ supersymmetric string can be compactified in $(10 - D)$ dimensions by the same mechanism discussed in the first part of this letter, allowing states with $\tilde{G} = -1$ to be present in the Neveu–Schwarz sector of the LM string. However a problem immediately arises: as the emission of a $\tilde{G} = -1$ state from a Ramond string changes its Weyl parity [11], it seems that the Ramond sector of the LM open string should necessarily contain states of both Weyl parities. However we will show that for our *closed* string this emission process is forbidden, so that the choices IIA and IIB are separately consistent.

To this aim it is convenient to introduce a new quantum number: the "soliton parity" $S = (-1)^{(P^I)^2}$ [and $\tilde{S} = (-1)^{(\tilde{P}^I)^2}$]; as the generalized momenta P^I and $\tilde{P}^I$ lie on an *integral* lattice and are independently conserved, both S and $\tilde{S}$ are multiplicatively conserved. Then in any process involving m closed string states the following selection rules hold

$$\prod_{i=1}^{m} S_i = 1 \,, \quad \prod_{i=1}^{m} \tilde{S}_i = 1 \,, \tag{27,28}$$

$$\prod_{\alpha=1}^{n} G_\alpha \prod_{\beta=1}^{r} (G_{\mathrm{W}})_\beta = \prod_{\beta=1}^{r} (G_{\mathrm{W}})_\beta = 1$$

$$(G = +1 \text{ for all RM NS states}) \,, \tag{29}$$

$$\prod_{\alpha=1}^{\tilde{n}} \tilde{G}_\alpha \prod_{\beta=1}^{\tilde{r}} (\tilde{G}_{\mathrm{W}})_\beta = 1 \,, \tag{30}$$

where n ($\tilde{n}$) and r ($\tilde{r}$) are the number of Neveu–Schwarz and Ramond states in the RM (LM) string ($n + r = \tilde{n} + \tilde{r} = m; r, \tilde{r}$ are even because of fermion number conservation).

The constraint $m_{\mathrm{R}}^2 = m_{\mathrm{L}}^2$ implies a further relationship among the quantum numbers previously defined; it is easy to check that for any state of the compactified closed string

$$T = \tilde{T} \,, \tag{31}$$

where $T = GS$ ($\tilde{T} = \tilde{G}\tilde{S}$) for the Neveu–Schwarz states and $T = S$ ($\tilde{T} = \tilde{S}$) for the Ramond states. Since $G = 1$ for the RM open string, eq. (27) implies $\Pi_{i=1}^{m} T_i = 1$ and from (28), (31) we deduce:

$$1 = \prod_{i=1}^{m} \tilde{T}_i = \prod_{\alpha=1}^{\tilde{n}} \tilde{G}_\alpha \,. \tag{32}$$

Comparison with eq. (30) shows that also $\Pi_{\beta=1}^{\widetilde{r}}(\widetilde{G}_\mathrm{W})_\beta = 1$: Ramond LM states cannot change their $\widetilde{G}_\mathrm{W}$ parity by emission of a $\widetilde{G} = -1$ state. Hence IIA and IIB do not mix even after compactification to $(10 - D)$ dimensions.

The case $D = 6$ yields the following massless spectrum:

$b^i\widetilde{b}^j\,|0_\mathrm{NS}\rangle\,|\widetilde{0}_\mathrm{NS}\rangle$	graviton, antisymmetric tensor (= pseudoscalar), dilaton
$b^i\widetilde{b}^J\,|0_\mathrm{NS}\rangle\,|\widetilde{0}_\mathrm{NS}\rangle$, $b^I\widetilde{b}^j\,|0_\mathrm{NS}\rangle\,|\widetilde{0}_\mathrm{NS}\rangle$	6 + 6 vectors
$b^I\widetilde{b}^J\,|0_\mathrm{NS}\rangle\,|\widetilde{0}_\mathrm{NS}\rangle$	6 × 6 scalars
$\widetilde{b}^i\,|0_\mathrm{R}\rangle^a\,|\widetilde{0}_\mathrm{NS}\rangle$, $b^\mu\,|0_\mathrm{NS}\rangle\,|\widetilde{0}_\mathrm{R}\rangle^a$	8 gravitini, 8 spin-$\frac{1}{2}$
$b^I\,|0_\mathrm{R}\rangle^a\,|\widetilde{0}_\mathrm{NS}\rangle$, $b^I\,|0_\mathrm{NS}\rangle\,|\widetilde{0}_\mathrm{R}\rangle^a$	6 × 8 spin-$\frac{1}{2}$
(IIA) T_i, T_{iIJ}	1 + 15 vectors
T_I, T_{IJK}, T_{ijI}	6 + 20 + 6 scalars
(IIB) T_{iI}, T_{iIJK}	6 + 10 vectors
T_{IJ}, T_{IJKL}, ϕ	15 + 15 + 2 scalars ($T_{ijKL} \sim T_{IJMN}$ by selfduality) . (33)

We see that the massless $P^I = \widetilde{P}^I = 0$ states combine into the $N = 8$, $d = 4$ supergravity multiplet (1 graviton, 8 gravitini, 28 vectors, 56 spin-$\frac{1}{2}$, 70 scalars). The solitonic states, however, give only the $N = 4$ vector multiplets:

$$
\begin{array}{ll}
n(\Lambda_1)\ \text{vectors:} & b^i\,|0_\mathrm{NS}\rangle\,|\widetilde{P}^2 = 1\rangle\,, \\
4 \times n(\Lambda_1)\ \text{spin-}\tfrac{1}{2}\text{:} & |0_\mathrm{R}\rangle^a\,|\widetilde{P}^2 = 1\rangle\,, \\
6 \times n(\Lambda_1)\ \text{scalars:} & b^I\,|0_\mathrm{NS}\rangle\,|\widetilde{P}^2 = 1\rangle\,.
\end{array}
\tag{34}
$$

We know that $N = 4$ supergravity cannot couple to matter, so that (33) together with the solitonic states (34) is a consistent massless limit only if the $N = 8$ of the graviton multiplet is at least broken to $N = 4$.

How this happens appears evident by inspection of the massless spectrum: only half of the 8 gravitini are coupled to the matter sector, namely the four $\widetilde{b}^i\,|0_\mathrm{R}\rangle^a\,|\widetilde{0}_\mathrm{NS}\rangle$. Only these can mediate transitions from spin-1 to spin-$\frac{1}{2}$ in the matter sector, since they must transform RM NS states into RM R states. Another argument is the following. The states added to the $N = 4$ supergravity multiplet to complete the $N = 8$ multiplet come from the NS ⊕ $\widetilde{\mathrm{R}}$ and R ⊕ $\widetilde{\mathrm{R}}$ sectors in (25): they are built by combining the RM superstring with the LM Ramond string. As we have already discussed, they do not couple to $\widetilde{G} = -1$ solitonic states [and in particular to the solitonic matter multiplet in (34)].

Because of the presence of the massless solitons (34), the $N = 8$ supermultiplet in (33) breaks into $N = 4$ supermultiplets:

$$
\begin{array}{l}
\text{spin} \\
2 \\ 3/2 \\ 1 \\ 1/2 \\ 0
\end{array}
\quad
\begin{pmatrix} \\ 1 \\ 8 \\ 28 \\ 56 \\ 70 \end{pmatrix}
\rightarrow
\begin{pmatrix} \\ 1 \\ 4 \\ 6 \\ 4 \\ 1+1 \end{pmatrix}
+ 4 \times
\begin{pmatrix} \\ \\ 1 \\ 4 \\ 7 \\ 8 \end{pmatrix}
+ 6 \times
\begin{pmatrix} \\ \\ \\ 1 \\ 4 \\ 6 \end{pmatrix} .
\tag{35}
$$

Thus the four-dimensional massless limit of our compactified theory, if consistent, describes $N = 4$ supergravity coupled to $n(\Lambda_1)$ soliton vector multiplets, and to 4 spin-$\frac{3}{2}$, 6 spin-1 $N = 4$ multiplets. We conclude by observing that the conservation laws (31) and (32) suggest that in both the compactifications of the $N = 1$ and $N = 2$ strings, the states with $T = \widetilde{T} = -1$ can be consistently removed; since all these states are massive, the massless spectrum is unchanged.

Another compactified closed string can be obtained, with the same massless spectrum of the string of ref. [4], by combining the RM superstring with an LM string made out of a Neveu–Schwarz sector with $\widetilde{G} = \widetilde{S} = \pm 1$ ($\widetilde{T} = 1$) and of a Ramond sector with $\widetilde{S} = \widetilde{T} = -1$.

Finally we remark that all the closed strings discussed in this paper are intrinsically oriented. To retrieve the $N = 1$ *unoriented* version, one just symmetrizes in the RM and LM modes of the $N = 2$ string. Notice, however, that in compactifying the symmetrized string, solitonic interactions would produce $G = -1$ states in the RM modes, re-introducing tachyons and destroying supersymmetry.

To summarize: we have discussed symmetries and possible gaugings of the compactified model of ref. [4]. The $N = 2$ supersymmetric string has been formulated in the "old formalism" of the spinning string. In its compactification on a six-torus, massless solitons break $N = 8$ supersymmetry to $N = 4$. This "soliton supersymmetry breaking" is, to our knowledge, a novel feature in string theories.

References

[1] M.B. Green and J.H. Schwarz, Nucl. Phys. B181 (1981) 502; B198 (1982) 252, 441; Phys. Lett. 109B (1982) 444;
M.B. Green, J.H. Schwarz and L. Brink, Nucl. Phys. B198 (1982) 474;
J.H. Schwarz, Phys. Rep. 89 (1982) 223.

[2] D. Gross, J. Harvey, E. Martinec and R. Rohm, Nucl. Phys. B256 (1985) 253; Princeton preprint (1985).

[3] D.Z. Freedman, G.W. Gibbons and P.C. West, Phys. Lett. 124B (1983) 491.

[4] L. Castellani, Phys. Lett. 166B (1986) 54.

[5] F. Gliozzi, J. Scherk and D. Olive, Phys. Lett. 65B (1976) 282; Nucl. Phys. B122 (1977) 253.

[6] A. Neveu and J.H. Schwarz, Nucl. Phys. B31 (1971) 86; Phys. Rev. D4 (1971) 1109.

[7] P. Ramond, Phys. Rev. D3 (1971) 2415.

[8] I.B. Frenkel and V.G. Kac, Inv. Math. 62 (1980) 23.

[9] P. Goddard and D. Olive, DAMPT preprint (1983).

[10] K. Bardakci and M.B. Halpern, Phys. Rev. D3 (1971) 2493.

[11] E. Cremmer and J. Scherk, Nucl. Phys. B103 (1976) 399.

Nuclear Physics B289 (1987) 364–384
North-Holland, Amsterdam

A NEW METHOD OF INCORPORATING SYMMETRY INTO SUPERSTRING THEORY

R. BLUHM and L. DOLAN[1]

The Rockefeller University, New York, NY 10021, USA

P. GODDARD[2]

Institute for Theoretical Physics, University of California, Santa Barbara, CA 93106 USA and Department of Applied Mathematics and Theoretical Physics, University of Cambridge, Silver Street, Cambridge CB3 9EW, UK

Received 8 January 1987

Type II superstrings are compactified to four dimensions giving rise to a dimension-18 semi-simple Lie group. Two of these groups, SU(3) × SO(5) and SU(2) × SU(4) contain the standard gauge group. The four-dimensional string satisfies perturbative unitarity and is finite at least to one loop.

1. Introduction

The Frenkel-Kac-Segal (FKS) [1] construction provides a natural way of introducing internal symmetry into bosonic string theories [2] which has been successfully exploited in the heterotic string [3] to provide string theories with phenomenologically viable gauge groups. Here the symmetry is incorporated by using the FKS construction for sixteen of the left-moving twenty-six-dimensional bosonic degrees of freedom of a closed string while the right-moving degrees of freedom are those of the ten-dimensional superstring. There remains the puzzle of whether it is possible to incorporate non-abelian gauge symmetry directly into type II superstring theory by a suitable compactification or comparable mechanism. In this paper, we shall describe what might be regarded as being the analogue for the Neveu-Schwarz-Ramond spinning string of the FKS mechanism for the bosonic string and use it to construct a unitary, Lorentz-invariant, tachyon-free, interacting closed string theory in four dimensions, which is finite at least to one loop. This extends and generalizes our treatment of $SU(2)^6$ [4], in which the non-abelian algebraic compactification of interacting ten-dimensional strings to four dimensions on odd lattices was first considered.

[1] Supported in part by the US Department of Energy under contract No. DE-AC 02-81 ER 40033B.
[2] Supported by The National Science Foundation under Grant No. PHY82-17853, supplemented by funds from the National Aeronautics and Space Administration.

0550-3213/87/$03.50 © Elsevier Science Publishers B.V.
(North-Holland Physics Publishing Division)

When the FKS construction is used to incorporate a symmetry algebra g, the states at different mass levels of a single string together form level-one representations of the affine Kac-Moody algebra associated with g (at least before the left-right constraint is used to eliminate the tachyon). The highest weight states of level-one representations cannot transform according to the adjoint representation of g. In the heterotic string, the massless states in the adjoint representation, which provide the gauge vector bosons, are at the first excited level of the representation of the affine algebra, being obtained by the application of generators to the scalar tachyon (which itself decouples). For the spinning string, the massless states must themselves be highest-weight states of the affine algebra because the smaller mass gap means that they cannot be reached from the tachyon by application of the generators. In order to obtain a representation of the affine algebra with adjoint representation highest weight states, we need affine representations of level greater than one. In fact, such representations must have level number greater than or equal to h, where h is the dual Coxeter number of the algebra g, which is n for $SU(n)$ and $n-2$ for $SO(n)$, and so is $(3,3)$ for $SU(3)\times SO(5)$, $(2,4)$ for $SU(2)\times SU(4)$ and $(2,2,2,2,2,2)$ for $SU(2)^6$. For a review of the representation theory of Kac-Moody algebras see, eg., ref. [13].

The construction we describe in this paper realizes a level h representation of the affine algebra. In the FKS construction the number of compactified dimensions determines the *rank* of the gauge group, while here the *dimension* of the gauge group is determined by the number of internal b_s or d_n oscillators used in a fermionic description of the compactified degrees of freedom.

The low energy limit of our four-dimensional string is $N=4$ supergravity coupled to $N=4$ supersymmetric Yang-Mills theory with the gauge group, whose algebra we have denoted by g, given by $SU(3)\times SO(5)$, $SU(2)\times SU(4)$, or $SU(2)^6$. The first case provides an example of a non-simply laced algebra realized in string theory. Further amendments to the model are necessary to reduce to one the number of residual supersymmetries and to obtain chiral fermions, but the model described here shows how to incorporate in superstring theories symmetry groups big enough to have the standard gauge group $SU(3)\times SU(2)\times U(1)$ as a subgroup. Although the massless fermion content is at this stage not yet realistic, this method provides a mechanism by which the more economical type II superstring may be phenomenologically viable.

We are guided to this particular compactification, which involves various Neveu-Schwarz and Ramond sectors with various boundary conditions on the internal bosonic coordinates, by observing that the one-loop amplitude for the scattering of four gauge bosons in the $SU(2)^6$ version of this theory is equal to the amplitude (invariant under a subgroup of the modular group) calculated in the compactified closed oriented string whose right- and left-moving degrees of freedom are the Green-Schwarz superstring and the Neveu-Schwarz bosonic string respectively ($SS\times NS$) [4]. This amplitude can be rewritten as a different, modular invariant, integrand integrated over a fundamental region of the full modular group. It is now straightforward to check perturbative unitarity: this amplitude satisfies unitarity on the combination of (i) the original spectrum of the SS $\times$ NS model compactified with internal momentum eigenvalues of the Z^6 lattice and (ii) a purely massive sector whose right and left movers are, respectively, the Green-Schwarz superstring and the Ramond fermionic string compactified with internal momenta on the shifted $(Z+\frac{1}{2})^6$ "lattice". *These two sectors comprise our four-dimensional supersymmetric string*. An interesting property of this compactification mechanism is that the latter sector, although scaled by the Planck mass, carries spinor representations of the gauge group, and thus correlates the Lorentz spinors of the Ramond model with quark-like representations, a feature of low energy phenomenology. In the case with $SU(2)^6$ symmetry, we can take the ten-dimensional limit of the two sectors which form this theory and obtain the type II superstring. Thus even in ten dimensions, the theory has no chiral anomalies.

In sect. 2, we use the covariant formulation to describe the mechanism by which symmetry is incorporated in the dimensional reduction of type II superstrings. In sect. 3, we discuss the light-cone formulation. In sect. 4, we calculate the one-loop amplitude for four external gauge bosons in our model and show it satisfies perturbative unitarity. In the concluding section, the connection among the different gauge groups and the relation between the hyperbolic Kac-Moody algebra associated with the odd ten-dimensional lorentzian lattice and the superstring is discussed.

2. Compactification and symmetry

To describe the way in which we shall reduce the dimension of the type II superstring in order to incorporate internal symmetry, we begin by considering the left-moving modes of a ten-dimensional closed string described in the covariant Neveu-Schwarz-Ramond formalism by bosonic fields

$$X^{\mu}(z)=\tfrac{1}{2}x^{\mu}+\tfrac{1}{2}\alpha' p^{\mu}(-i\ln z)+\tfrac{1}{2}i\sqrt{2\alpha}\sum_{n\neq 0}\frac{1}{n}\alpha_n^{\mu}z^{-n}, \tag{1a}$$

where the sum is over $n\in Z$, and fermionic fields

$$b^{\mu}(z)=\sum_r b_r^{\mu}z^{-r}, \tag{1b}$$

where the sum is over $r\in Z+\frac{1}{2}$ in the Neveu-Schwarz (NS) sector and over $r\in Z$ in the Ramond sector. The Lorentz index μ runs over ten values and we shall consider compactifying six euclidean dimensions (so $\mu=\{\hat{\mu},I\}$, $\hat{\mu}=0,1,2,3$, $I=1,\dots,6$) to obtain a four-dimensional theory, although we evidently could consider theories in other numbers of space-time dimensions. In the NS sector we take the internal momenta to lie on the Z^6 lattice. Then each of the six bosonic fields $X^I(z)$ is equivalent to a pair of real Neveu-Schwarz fields $b_1^I(z)$ and $b_2^I(z)$. (Since the internal space partition function is

$$\prod_{n=1}^{\infty}(1-w^n)^{-6}\prod_{s=1/2}^{\infty}(1+w^s)^6(\theta_3(0|\tau))^6=\prod_{s=1/2}^{\infty}(1+w^s)^{18}, \tag{2}$$

where $\Pi_{n=1}^{\infty}(1-w^n)^{-1}$ counts A_n's, $\Pi_{s=1/2}^{\infty}(1+w^s)$ count b_s's and $\theta_3(0|\tau)$ counts internal momentum states, the same spectrum is generated by A_n^I, b_s^I, $|p^I\rangle$ on Z^6 as by $b_s^a=\{b_{(1)s}^I, b_{(2)s}^I, b_s^I\}$, $a=1,\ldots,18$.) Here

$$\sqrt{\tfrac{1}{2}}\left\{b_{(1)}^I(z)\pm ib_{(2)}^I(z)\right\}=:\exp\{\pm 2iX^I(z)\}:c_{\pm}^I, \tag{3}$$

or equivalently

$$A_n^I=-i\sum_s b_{(1)n-s}^I b_{(2)s}^I \tag{4}$$

(no sum on I). The cocycle factor $c_{\pm}^{\mu}$ or Klein transformation is introduced in eq. (3) to insure that the b_s^a anticommute for different a. The world sheet fermions $b_{(1)}^I(z)$, $b_{(2)}^I(z)$ together with $b_{(3)}^I(z)=b^I(z)$ form triplets under the six SU(2) groups generated for each fixed I, and $i=1,2,3$ by

$$T_i^I=-\frac{1}{4\pi}\varepsilon_{ijk}\oint\frac{\mathrm{d}z}{z}b_{(j)}^I(z)b_{(k)}^I(z). \tag{5}$$

The spectrum of the NS sector of the free string will respect this $SU(2)^6$ symmetry provided that the physical states are defined in the usual way by the conditions

$$L_0|\psi\rangle=\tfrac{1}{2}|\psi\rangle;\qquad L_n|\psi\rangle=0,\qquad n>0; \tag{6}$$

$$G_r|\psi\rangle=0,\qquad r>0, \tag{7}$$

where the generators of the super-Virasoro algebra are defined by

$$L(z)\equiv\sum_n L_n z^{-n}=\sum_\mu \tfrac{1}{2}:P^\mu(z)P_\mu(z):+\tfrac{1}{2}\overset{\circ}{\circ}z\frac{\mathrm{d}b^\mu(z)}{\mathrm{d}z}b_\mu(z)\overset{\circ}{\circ}, \tag{8}$$

$$G(z)=\sum_r G_r z^{-r}=\sum_\mu P^\mu(z)b_\mu(z), \tag{9}$$

because the L_n's and G_r's so defined commute with the symmetry generators. In eq. (9), $P^\mu(z)=(2iz/\sqrt{2\alpha'})\mathrm{d}X^\mu(z)/\mathrm{d}z$, and the open dots indicate normal ordering with respect to the fermionic oscillators as opposed to the bosonic normal ordering used in eq. (3). To maintain the symmetry it will also be necessary that any projections we do to remove tachyons or to produce a modular invariant loop integrand commute with the generators T_i^I.

It is convenient to split each of L_n and G_r into two parts. One corresponding to the space-time coordinates μ and the other corresponding to the internal compactified coordinates I. We shall denote the latter by $\bar{L}_n$ and $\bar{G}_r$. We can rewrite this part entirely in terms of the NS fields

$$\bar{L}(z) = \tfrac{1}{2}z \,\mathring{\circ}\, \frac{\mathrm{d}b^a(z)}{\mathrm{d}z} b^a(z) \,\mathring{\circ}\,, \tag{10}$$

$$\bar{G}(z) = -\tfrac{1}{6} i f_{abc} b^a(z) b^b(z) b^c(z). \tag{11}$$

The notation $b^a(z)$, denoting all the fermion fields we have formed, puts them on equal footing, as they should be if we desire to maintain an SU(2)^6 symmetry. We shall use the summation convention on the a, b, c indices and f_{abc} denotes the structure constants of SU(2)^6.

Put in this form we can generalize our symmetry from SU(2)^6 to any other dimension-18 semi-simple Lie group. Actually there are just two such possibilities: SU(3) $\times$ SO(5) and SU(2) $\times$ SU(4). If now f_{abc} denotes the totally antisymmetric structure constants of any of these groups and we normalize the structure constants so that $c_\psi = 2$, where c_ψ is the value of the quadratic Casimir in the adjoint representation

$$f_{abc} f_{abd} = c_\psi \delta_{cd}, \tag{12}$$

then it remains true that [5]

$$[\bar{L}_n, \bar{L}_m] = (n-m)\bar{L}_{m+n} + \tfrac{1}{24} d n (n^2 - 1)\, \delta_{n,-m}, \tag{13a}$$

$$[\bar{L}_n, \bar{G}_r] = (\tfrac{1}{2}n - r)\bar{G}_{n+r}, \tag{13b}$$

$$\{\bar{G}_r, \bar{G}_s\} = 2\bar{L}_{r+s} + \tfrac{1}{6} d (r^2 - \tfrac{1}{4})\delta_{r,-s}, \tag{13c}$$

with $d = 18$ the number of free independent fermi fields. We can also introduce the generators of an affine Kac-Moody algebra associated with the symmetry

$$T^a(z) = \sum_n T^a_{-n} z^n = -\tfrac{1}{2} i f_{abc} b^b(z) b^c(z), \tag{14a}$$

$$[T^a_n, T^b_m] = i f_{abc} T^c_{n+m} + n\delta^{ab}\delta_{n,-m}, \tag{14b}$$

and the zero-graded elements of this generate our gauge symmetry. (Note that eq. (14b) defines a level h representation because we have chosen the normalization

$c_\psi = 2$.) In fact, since

$$[L_n, T^a_m] = -mT^a_{n+m}, \qquad [G_r, T^a_n] = -nb^a_{n+r}, \tag{15}$$

we see that T^a_0 maps the physical space, defined by the conditions eqs. (6) and (7) into itself. This generalization of the group corresponds to a redefinition of G_r.

The generalized models considered here are free from ghosts (negative norm states) because the proof of their absence can be formulated [6] entirely in terms of the algebra eq. (13), for the complete super-Virasoro generators L_n, G_r (with $d = 30$ corresponding to ten bosons and ten fermions). Alternatively, we can consider the light-cone formulation of these models rather than the covariant one. Then the ghost problem is translated into that of proving the closure of the Lorentz algebra, and the generalization of the gauge groups is made possible by the modification of the SO(3,1) Lorentz generators, through the redefinition of G_r. This is described in detail in sect. 4. Other authors [7] have considered the incorporation of symmetry using Neveu-Schwarz-Ramond fermions, but they have made assumptions that imply that the form of the generators G_r remains unchanged.

We conclude this section by considering the vertices describing the coupling of the massless states of the theory. These correspond to massless states in the NS sector of the left-moving modes multiplied by the massless right-moving superstring modes. As the second factor is described in the usual way, here we shall concentrate on the first. (For the full vertices in light-cone gauge, see sect. 4.) These massless NS states are the eighteen adjoint representation states

$$e^{ix\cdot k} b^a_{-1/2}|0\rangle, \tag{16}$$

where k is a light-like four-momentum.

The coupling of a physical Neveu-Schwarz state $|A\rangle$ is described by a vertex $V^1_A(z)$ which satisfies the equation

$$[G_r, V^0_A(z)]_+ = z^{r+1/2} V^1_A(z) \tag{17}$$

for some $V^0_A(z)$, and

$$[L_n, V^1_A] = z^{n+1}\frac{d}{dz}V^1_A + z^n(n+1)V^1_A, \tag{18}$$

with

$$\lim_{z\to 0} V^0_A|0\rangle = |A\rangle, \qquad \lim_{z\to 0} V^1_A(z)|0\rangle = G_{-1/2}|A\rangle. \tag{19}$$

For the states (eq. (16)),

$$V^1_a(k,z) = z^{-1}\left\{\frac{\sqrt{2\alpha'}}{2}k\cdot b(z)b^a(z) - \tfrac{1}{2}if_{abc}b^b(z)b^c(z)\right\}e^{ik\cdot \bar{X}(z)}, \tag{20}$$

$$V^0_a(k,z) = z^{-1/2}b^a(z)e^{ik\cdot\bar{X}(z)}. \tag{21}$$

For convenience, the right-moving zero-mode position coordinate has been included in

$$\overline{X}^{\hat{\mu}}(z) = X^{\hat{\mu}} + \tfrac{1}{2}\alpha' p^{\hat{\mu}}(-i\ln z) + \frac{i\sqrt{2\alpha'}}{2}\sum_{n\neq 0}\frac{\alpha_n^{\hat{\mu}}}{n}z^{-n},$$

note $\alpha_0^{\hat{\mu}} \equiv \sqrt{2\alpha'}/2p^{\hat{\mu}} \equiv \tilde{\alpha}_0^{\mu}$.The three-point function for these states is proportional to the structure constants of the symmetry group, and acquires the appropriate kinematic structure for vector mesons when multiplied by the factor corresponding to the massless vector mesons of the superstring, demonstrating that these states correspond to the gauge bosons of the symmetry group. See eq. (58).

3. Light-cone formulation

The SO(3,1) Lorentz generators are given as a subgroup of the ten-dimensional Lorentz algebra for this closed string theory. In the SS × NS sector [4], they are M^{+-}, $M^{\hat{i}\hat{j}}$, $M^{\hat{i}\pm}$ where

$$M^{+-} = l^{+-}, \qquad M^{\hat{i}+} = l^{\hat{i}+} \qquad (l^{\mu\nu} \equiv x^{\mu}p^{\nu} - x^{\nu}p^{\mu}), \tag{22}$$

$$M^{\hat{i}\hat{j}} = l^{\hat{i}\hat{j}} - i\sum_{n=1}^{\infty}\frac{1}{n}\left(\tilde{A}^{\hat{i}}_{-n}\tilde{A}^{\hat{j}}_{n} - \tilde{A}^{\hat{j}}_{-n}\tilde{A}^{\hat{i}}_{n}\right) - i\sum_{s=1/2}^{\infty}\left(b^{\hat{i}}_{-s}b^{\hat{j}}_{s} - b^{\hat{j}}_{-s}b^{\hat{i}}_{s}\right)$$

$$-i\sum_{n=1}^{\infty}\frac{1}{n}\left(A^{\hat{i}}_{-n}A^{\hat{j}}_{n} - A^{\hat{j}}_{-n}A^{\hat{i}}_{n}\right) + \tfrac{1}{8}i\sum_{n=-\infty}^{\infty}\overline{S}_{-n}\gamma^{\hat{i}\hat{j}-}S_n, \tag{23a}$$

$$M^{\hat{i}-} = l^{\hat{i}-} - \frac{i}{\sqrt{2\alpha'}\,p^+}\sum_{n=1}^{\infty}\frac{1}{n}\left(\tilde{A}^{\hat{i}}_{-n}\tilde{A}^{-}_{n} - \tilde{A}^{-}_{-n}\tilde{A}^{\hat{i}}_{n}\right) - \frac{i}{\sqrt{2\alpha'}\,p^+}\sum_{s=1/2}^{\infty}\left(b^{\hat{i}}_{-s}G_s - G_{-s}b^{\hat{i}}_{s}\right)$$

$$-\frac{i}{\sqrt{2\alpha'}\,p^+}\sum_{n=1}^{\infty}\frac{1}{n}\left(A^{\hat{i}}_{-n}A^{-}_{n} - A^{-}_{-n}A^{\hat{i}}_{n}\right)$$

$$+\frac{i}{8}\frac{1}{\sqrt{2\alpha'}\,p^+}\sum_{n=-\infty}^{\infty}\sum_{m=-\infty}^{\infty}\overline{S}_{-n-m}\gamma^{\hat{i}\hat{j}-}S_m A^{\hat{j}}_{n}, \tag{23b}$$

where

$$G_s = \sum_{m=-\infty}^{\infty} b^{i}_{s-m}A^{i}_{m}, \tag{24}$$

for

$$n \neq 0 \qquad L_n = \tfrac{1}{2}\sum_{m=-\infty}^{\infty}\left[A^{i}_{n-m}A^{i}_{m} + \tfrac{1}{2}\left(m - \tfrac{1}{2}n\right)\overline{S}_{n-m}\gamma^{-}S_n\right], \tag{25}$$

$$\tilde{L}_n = \tfrac{1}{2}\sum_{m=-\infty}^{\infty}\tilde{A}^{i}_{n-m}\tilde{A}^{i}_{m} + \tfrac{1}{2}\sum_{s=-\infty}^{\infty}\left(s - \tfrac{1}{2}n\right)b^{i}_{n-s}b^{i}_{s} \tag{26}$$

and

$$p^- = \frac{1}{2\alpha' p} + \left[\alpha'(p^{\hat{i}})^2 + 2\alpha'(\tilde{p}^I)^2 + 2\alpha'(\bar{p}^I)^2 + 2N + 2\tilde{N} - 1\right], \tag{27}$$

where N and $\tilde{N}$ are given in eq. (37), and $A_n^- \equiv 2L_n$, $\tilde{A}_n^- \equiv 2\tilde{L}_n$. These SO(3,1) generators commute with the $(\mathrm{SU}(2))^6$ gauge generators ($n = 0$) of A_n^I and

$$X_n(r) = (c_r/2\pi i)\oint(\mathrm{d}z/z)z^n\left[r^I \sum_{s=-\infty}^{\infty} b_s^I z^{-s}{:}\exp(2ir/\sqrt{2\alpha'})^{I x I(z)}{:}\right]$$

(r^I are the roots of $\mathrm{SU}(2)^6$, $(r^I)^2 = 1$) which mix the states of a given $\mathrm{SU}(2)^6$ multiplet at all mass levels. A_n^I and $X_n(r)$ are the generators of affine $\mathrm{SU}(2)^6$ with level number (2,2,2,2,2,2).

We now change the definition of SO(3,1) and thus enlarge the possible gauge groups to include SU(3) × SO(5) and SU(2) × SU(4). In accordance with eqs. (10) and (11), we find that eq's (26), (27) and (24) become

$$\tilde{L}_n = \tfrac{1}{2}\sum_m \tilde{A}^{\hat{i}}_{n-m}\tilde{A}^{\hat{i}}_m + \tfrac{1}{2}\sum_s (s - \tfrac{1}{2}n) b^{\hat{i}}_{n-s} b^{\hat{i}}_s + \bar{L}_n ,$$

$$\bar{L}_n = \tfrac{1}{2}\sum_s (s - \tfrac{1}{2}n) b^a_{n-s} b^a_s , \tag{28}$$

$$p^- = \frac{1}{2\alpha' p} + \left[\alpha'(p^{\hat{i}})^2 + \alpha' m_4^2\right],$$

$$\tfrac{1}{2}\alpha' m_4^2 = N + \sum_{n=1}^{\infty} \tilde{A}^{\hat{i}}_{-n}\tilde{A}^{\hat{i}}_n + \sum_{s=1/2}^{\infty} s b^{\hat{i}}_{-s} b^{\hat{i}}_s$$

$$+ \sum_{s=1/2}^{\infty} s b^a_{-s} b^a_s - \tfrac{1}{2} + \alpha'(\bar{p}^I)^2, \tag{29}$$

and

$$G_s = \sum A^i_n b^i_{s-n} + \bar{G}_s ,$$

$$\bar{G}_s = -\tfrac{1}{6} i f_{abc} \sum_{n,s'} b^a_{n-s'} b^b_{s'} b^c_{s-n} . \tag{30}$$

f_{abc} are the totally antisymmetric structure constants of $\mathrm{SU}(2)^6$: $f_{1713} = 1$, $f_{1714} = 0$ etc. To generalize the gauge group $\mathrm{SU}(2)^6$, now change the definition of G_s and hence the definition of $M^{\hat{i}-}$ in SO(3,1) by letting f_{abc} be the structure constants of any dimension-18 semi-simple Lie algebra. The fact that $\bar{L}_n$ and $\bar{G}_s$ still satisfy the super-Virasoro algebra, eq. (13), ensures that the Lorentz algebra SO(3,1) still closes.

4. Unitarity

We first consider the compactified SS × NS model [4]. In light-cone gauge the transverse degrees of freedom for $\alpha = 1,\dots,32$; $\mu = 0,1,\dots,9$; $i = 1,\dots,8 = \hat{i}, I$; $\hat{i} = 1,2$; $I = 1,\dots,6$; $n \in \mathrm{Z}$, $s \in \mathrm{Z} + \frac{1}{2}$, are:

$$X^{\hat{i}}(\tau-\sigma) = \tfrac{1}{2}x^{\hat{i}} + \alpha' p^{\hat{i}}(\tau-\sigma) + \tfrac{1}{2}i\sqrt{2\alpha'} \sum_{n\neq 0} \frac{1}{n} A_n^{\hat{i}} \mathrm{e}^{-2in(\tau-\sigma)}, \tag{31a}$$

$$X^{I}(\tau-\sigma) = \bar{x}^{I} + 2\alpha' \bar{p}^{I}(\tau-\sigma) + \tfrac{1}{2}i\sqrt{2\alpha'} \sum_{n\neq 0} \frac{1}{n} A_n^{I} \mathrm{e}^{-2in(\tau-\sigma)}, \tag{31b}$$

$$S^{\alpha}(\tau-\sigma) = \sum_{n=-\infty}^{\infty} S_n^{\alpha} \mathrm{e}^{-2in(\tau-\sigma)}, \tag{31c}$$

$$X^{\hat{i}}(\tau+\sigma) = \tfrac{1}{2}x^{\hat{i}} + \alpha' p^{\hat{i}}(\tau+\sigma) + \tfrac{1}{2}i\sqrt{2\alpha'} \sum_{n\neq 0} \frac{1}{n} \tilde{A}_n^{\hat{i}} \mathrm{e}^{-2in(\tau+\sigma)}, \tag{32a}$$

$$X^{I}(\tau+\sigma) = \tilde{x}^{I} + 2\alpha' \tilde{p}^{I}(\tau+\sigma) + \tfrac{1}{2}i\sqrt{2\alpha'} \sum_{n\neq 0} \frac{1}{n} \tilde{A}_n^{I} \mathrm{e}^{-2in(\tau+\sigma)}, \tag{32b}$$

$$b^{i}(\tau+\sigma) = \sum_{s=-\infty}^{\infty} b_s^{i} \mathrm{e}^{-2is(\tau+\sigma)}. \tag{32c}$$

These physical modes have the following non-zero commutation relations:

$$\begin{aligned} &[A_n^i, A_m^j] = [\tilde{A}_n^i, \tilde{A}_m^j] = n\delta_{n,-m}\delta^{ij}, \qquad \{b_s^i, b_{s'}^j\} = \delta_{s,-s'}\delta^{ij}, \\ &\{S_n^{\alpha}, \bar{S}_m^{\beta}\} = (\gamma^{+}h)^{\alpha\beta}\delta_{m,-n}, \qquad [x^{\hat{i}}, p^{\hat{j}}] = i\delta^{\hat{i}\hat{j}}, \\ &[\bar{x}^I, \bar{p}^J] = [\tilde{x}^I, \tilde{p}^J] = \tfrac{1}{2}i\delta^{IJ}. \end{aligned} \tag{33}$$

In ten dimensions the right-moving fermions are light-cone Majorana-Weyl spinors

$$h^{\alpha\beta}S_n^{\beta} = 0, \qquad (\gamma^{+})^{\alpha\beta}S_n^{\beta} = 0, \qquad h^{\alpha\beta} = \tfrac{1}{2}(1+\gamma_{11})^{\alpha\beta}, \qquad \bar{S}_n^{\alpha} = S_n^{\beta}(\gamma^0)^{\beta\alpha}. \tag{34}$$

$(\gamma^{\mu})^{\alpha\beta}$ are pure imaginary in the Majorana representation, γ^0 is anti-symmetric, γ^i and γ^9 are symmetric, $\gamma_{11} = \gamma^0\gamma' \dots \gamma^9$ is real symmetric; $\{\gamma^{\mu}, \gamma^{\nu}\} = -2\eta^{\mu\nu}$; the metric is $\eta^{\mu\nu} = \mathrm{diag}(-1,1,\dots 1)$.

This mass operator is (sum on I):

$$\tfrac{1}{2}\alpha' m_4^2 = N + \tilde{N} + \alpha'(\bar{p}^I)^2 + \alpha'(\tilde{p}^I)^2 - \tfrac{1}{2}, \tag{35}$$

where the internal momenta are fixed by the boundary condition on the internal coordinates $X^I(0) = X^I(\pi) + \sqrt{2\alpha'}\,\Sigma N^L \delta_L^I \pi$ and are given by discrete eigenvalues labelled by points on the six-dimensional hypercubic lattice $\Lambda^I = Z^6$, with basis vectors $\alpha_L^I = \delta_L^I$

$$\sqrt{2\alpha'}\,\bar{p}^I = \Lambda^I = \sum N^L \alpha_L^I, \tag{36a}$$

$$\sqrt{2\alpha'}\,\tilde{p}^I = \Lambda^I = \sum N^L \alpha_L^I. \tag{36b}$$

Z^6 is the root + vector conjugacy class weight lattice of SO(12), or equivalently the root lattice of $SU(2)^6$ where the roots have length squared one. The number of points p_1^I with length squared one $(2\alpha'(p_1^I)^2 = 1)$ is 12, etc.

The number operators N and $\tilde{N}$ are given by

$$N = \sum_{n=1}^{\infty} A_{-n}^i A_n^i + \sum_{n=1}^{\infty} \tfrac{1}{2} n \bar{S}_{-n} \gamma^- S_n, \tag{37a}$$

$$\tilde{N} = \sum_{n=1}^{\infty} \tilde{A}_{-n}^i \tilde{A}_n^i + \sum_{s=1/2}^{\infty} s b_{-s}^i b_s^i. \tag{37b}$$

The left-right light-cone gauge constraint is

$$N + \alpha'(\bar{p}^I)^2 = \tilde{N} + \alpha'(\tilde{p}^I)^2 - \tfrac{1}{2}. \tag{38}$$

This constraint eliminates the tachyon. The physical massless states are $\{|\hat{i}\rangle_R, |I\rangle_R, |a\rangle_R\} \times b_{-1/2}^{\hat{j}}|0\rangle_L$ which is the $d=4$, $N=4$ supergravity multiplet with spin content $(\pm 2, 4(\pm\tfrac{3}{2}), 6(\pm 1), 4(\pm\tfrac{1}{2}), 2(0))$ and $\{|\hat{i}\rangle_R, |I\rangle_R, |a\rangle_R\} \times \{b_{-1/2}^J|0\rangle_L, |k^I\rangle_L\}$, the $d=4$, $N=4$ supersymmetric Yang-Mills multiplet with spin content $(\pm 1, 4(\pm\tfrac{1}{2}), 6(0))$ in the adjoint of $SU(2)^6$. Expressions for the generalized groups are given in eqs. (54)–(65).

The vertex operator for the massless boson states $|i\rangle_R \times b_{-1/2}^i|0\rangle_L$ is given by

$$V(k^{\hat{i}}, z) = \frac{1}{\pi}\varepsilon^{ij}\int_0^{\pi} \mathrm{d}\sigma\, W^{ij}(k^{\hat{i}}, z, \sigma), \tag{39}$$

$$W^{ij}(k^{\hat{i}}, z, \sigma) = g\left(\frac{1}{\sqrt{2\alpha'}}\sum_n A_n^i \mathrm{e}^{-2in(\tau-\sigma)} + \tfrac{1}{2}k^{\hat{k}} R^{i\hat{k}}\right)\mathrm{e}^{ik^{\hat{i}}x^{\hat{i}}(\sigma,\tau)}$$
$$\times\left(\frac{1}{\sqrt{2\alpha'}}\sum_n A_n^j \mathrm{e}^{-2in(\tau+\sigma)} - \tfrac{1}{2}k^{\hat{l}}\sum_{s,r} b_s^j b_r^{\hat{l}} \mathrm{e}^{-2i(s+r)(\tau+\sigma)}\right). \tag{40a}$$

Here $R^{ik} = \frac{1}{8}\bar{S}^a(\tau-\sigma)(\gamma^{ik-})^{ab}S^b(\tau-\sigma)$ and $X^{\hat{i}}(\sigma,\tau) = X^{\hat{i}}(\tau-\sigma) + X^{\hat{i}}(\tau+\sigma)$. The polarizations and momenta are transverse: $k^{\hat{i}}\varepsilon^{\hat{i}j} = \varepsilon^{i\hat{j}}k^{\hat{j}} = 0 = \varepsilon^{\pm i} = k^{\pm}$ etc.

The vertex operators for the massless $SU(2)^6$ "charged" vector bosons $|\hat{i}\rangle_R \times |\tilde{k}^J\rangle_L$ and scalars $|I\rangle_R \times |\tilde{k}^J\rangle_L$ are

$$V(\tilde{k}^j, z) = \frac{1}{\pi}\varepsilon^i \int_0^{\pi} d\sigma\, W^i(\tilde{k}^j, z, \sigma),$$

where

$$W^i(\tilde{k}^j, z, \sigma) = g\left\{\frac{1}{\sqrt{2\alpha}}\sum_n A_n^i e^{-2in(\tau-\sigma)} + \tfrac{1}{2}k^{\hat{k}}R^{i\hat{k}}\right\} e^{ik^{\hat{i}}X^{\hat{i}}(\sigma,\tau)}$$

$$\times\left\{\tilde{k}^J\sum_s b_s^j e^{-2is(\tau+\sigma)}\right\} :e^{2ik^I X^I(\tau+\sigma)}: c_k. \tag{40b}$$

Here $\tilde{k}^j = (\frac{1}{2}k^{\hat{j}}, k^J)$, $(\sqrt{2\alpha'}k^I)^2 = 1$, and c_k are operator cocycles (see ref. [4] for more notation). The propagator is derived from eq. (35).

$$\Delta_c = \frac{\alpha'}{2\pi}\int_D d^2z\, z^{N+\alpha'(\bar{p}^I)^2}\bar{z}^{\tilde{N}+\alpha'(\tilde{p}^I)^2-1/2}|z|^{\alpha'(p)^2/2-2}. \tag{41}$$

The integration region D will be fixed in the loop amplitudes by the periodicity of the integrand.

The one-loop amplitude [4] for external charged vector mesons from (40b), (41) is

$$A_{\text{loop}}(1234) = \varepsilon_1^{\hat{i}}\varepsilon_2^{\hat{j}}\varepsilon_3^{\hat{k}}\varepsilon_4^{\hat{l}}\int dp\, \mathrm{tr}\,\Delta_c W^{\hat{i}}(1)\Delta_c W^{\hat{j}}(2)\Delta_c W^{\hat{k}}(3)\Delta_c W^{\hat{l}}(4) \tag{42a}$$

$$= \left(\frac{\alpha' g}{2\pi}\right)^4 \cdot \frac{1}{16}K\frac{(2\pi)^8}{(\alpha')^2}\int_{F_{SM}} d^2\tau\,(\mathrm{Im}\,\tau)^{-2}g(\tau), \tag{42b}$$

where K is the superstring kinematic factor and the integrand $g(\tau)$ is given in eq. (44) and is invariant under the theta subgroup Γ_θ of modular transformations generated by $\tau \to \tau + 2$ and $\tau \to -1/\tau$, and F_{SM} is a fundamental region of this subgroup (F_{SM}: $-1 < \mathrm{Re}\,\tau < 1$, $\mathrm{Im}\,\tau > 0$, $|\tau| > 1$.) Since this fundamental region F_{SM} is a finite union of images of F under Γ_θ (where F is a fundamental region of the full modular group Γ, F: $-\frac{1}{2} < \mathrm{Re}\,\tau < \frac{1}{2}$, $\mathrm{Im}\,\tau > 0$, $|\tau| > 1$), we can rewrite eq. (42) using

$$\int_{F_{SM}} d^2\tau\,(\mathrm{Im}\,\tau)^{-2}g(\tau) = \int_F d^2\tau\,(\mathrm{Im}\,\tau)^{-2}\left[g(\tau) + g(\tau+1) + g\left(-\frac{1}{\tau}+1\right)\right]. \tag{43}$$

From (42a), we find

$$\int \mathrm{d}^2\tau\,(\mathrm{Im}\,\tau)^{-2} g(\tau) = \int \mathrm{d}^2\tau \left(\frac{-2\pi}{\ln|w|}\right)^2 \int \prod_{I=1}^{3} \mathrm{d}^2\nu_I \cdot \prod_{I<J} (\chi_{IJ})^{\alpha' k_I^{\hat{i}} k_J^{\hat{i}}} \cdot (f(\bar{w}))^{-8}$$

$$\times \bar{w}^{-1/2} \prod_{s=1/2}^{\infty} (1+\bar{w}^s)^8 \cdot \sum_{\sqrt{2}\,\alpha' p^I \in \mathrm{Z}^6} w^{\alpha'(\bar{p}^I)^2} \cdot \prod_{I<J} (\Psi_{IJ})^{2\alpha' k_I^K \cdot k_J^K}$$

$$\times \sum_{\sqrt{2\alpha'}\,\tilde{p}^I \in \mathrm{Z}^6} \exp\left(\alpha' \ln\bar{w}\left[\tilde{p}^I - \sum_{m=1}^{4} \frac{\ln\bar{z}_m}{\ln\bar{w}} Q_m^I\right]^2\right)$$

$$\times\left[\tilde{k}_1\cdot\tilde{k}_2\tilde{k}_3\cdot\tilde{k}_4 \chi_{43}^{+}\chi_{21}^{+} - \tilde{k}_1\cdot\tilde{k}_3\tilde{k}_2\cdot\tilde{k}_4\chi_{31}^{+}\chi_{42}^{+} + \tilde{k}_1\cdot\tilde{k}_4\tilde{k}_2\cdot\tilde{k}_3\chi_{41}^{+}\chi_{32}^{+}\right]. \tag{44a}$$

Using transformation properties of the Jacobi theta functions, we have from (44a),

$$\int \mathrm{d}^2\tau\,(\mathrm{Im}\,\tau)^{-2} g(\tau+1)$$

$$= \int \mathrm{d}^2\tau \left(\frac{-2\pi}{\ln|w|}\right)^2 \int \prod_{I=1}^{3} \mathrm{d}^2\nu_I \cdot \prod_{I<J} (\chi_{IJ})^{\alpha' k_I^{\hat{i}} k_J^{\hat{i}}} (f(\bar{w}))^{-8}$$

$$\times\left[-\bar{w}^{-1/2}\prod_{s=1/2}^{\infty}(1-\bar{w}^s)^8 \cdot \sum_{\sqrt{2\alpha'}\,\tilde{p}^I\in\mathrm{Z}^6} w^{\alpha'(\bar{p}^I)^2}(-1)^{2\alpha'(\bar{p}^I)^2}\right]\cdot\prod_{I<J}(\Psi_{IJ})^{2\alpha' k_I^K\cdot k_J^K}$$

$$\times \sum_{\sqrt{2\alpha'}\,\tilde{p}^I \in \mathrm{Z}^6} \exp\left(\alpha' \ln\bar{w}\left[\tilde{p}^I - \sum_{m=1}^{4} \frac{\ln\bar{z}_m}{\ln\bar{w}} Q_m^I\right]^2\right)(-1)^{2\alpha'(\tilde{p}^I)^2}\right)$$

$$\times\left[\tilde{k}_1\cdot\tilde{k}_2\tilde{k}_3\cdot\tilde{k}_4 \tilde{\chi}_{43}^{+}\tilde{\chi}_{21}^{+} - \tilde{k}_1\cdot\tilde{k}_3\tilde{k}_2\cdot\tilde{k}_4\tilde{\chi}_{31}^{+}\tilde{\chi}_{42}^{+} + \tilde{k}_1\cdot\tilde{k}_4\tilde{k}_2\cdot\tilde{k}_3\tilde{\chi}_{41}^{+}\tilde{\chi}_{32}^{+}\right], \tag{44b}$$

$$\int \mathrm{d}^2\tau\,(\mathrm{Im}\,\tau)^{-2} g\left(-\frac{1}{\tau}+1\right)$$

$$= \int \mathrm{d}^2\tau \left(\frac{-2\pi}{\ln|w|}\right)^2 \int \prod_{I=1}^{3} \mathrm{d}^2\nu_I \cdot \prod_{I<J} (\chi_{IJ})^{\alpha' k_I^{\hat{i}} k_J^{\hat{i}}} \cdot (f(\bar{w}))^{-8}$$

$$\times\left[-\prod_{n=1}^{\infty}{}^{(1+\bar{w}^n)^8}\cdot\frac{16}{\sqrt{2\alpha'}}\cdot\sum_{p^I\in(\mathrm{Z}+1/2)^6} w^{\alpha'(\bar{p}^I)^2}\prod_{I<J}(\Psi_{IJ})^{2\alpha' k_I^K\cdot k_J^K}\right]$$

$$\times \sum_{\sqrt{2\alpha'}\,\tilde{p}^I \in (\mathrm{Z}+1/2)^6} \exp\left(\alpha' \ln\bar{w}\left[\tilde{p}^I - \sum_{m=1}^{4} \frac{\ln\bar{z}_m}{\ln\bar{w}} Q_m^I\right]^2\right)$$

$$\times\left[\tilde{k}_1\cdot\tilde{k}_2\tilde{k}_3\cdot\tilde{k}_4 \tilde{\tilde{\chi}}_{43}^{+}\tilde{\tilde{\chi}}_{21}^{+} - \tilde{k}_1\cdot\tilde{k}_3\tilde{k}_2\cdot\tilde{k}_4\tilde{\tilde{\chi}}_{31}^{+}\tilde{\tilde{\chi}}_{42}^{+} + \tilde{k}_1\cdot\tilde{k}_4\tilde{k}_2\cdot\tilde{k}_3\tilde{\tilde{\chi}}_{41}^{+}\tilde{\tilde{\chi}}_{32}^{+}\right]. \tag{44c}$$

The functions χ^{+}_{JI}, $\tilde{\chi}^{+}_{JI}$, and $\tilde{\tilde{\chi}}^{+}_{JI}$ are

$$\chi^{+}_{JI}=\sum_{s=1/2}^{\infty}\frac{(\bar{\rho}_J/\bar{\rho}_I)^s+(\bar{w}\bar{\rho}_I/\bar{\rho}_J)^s}{1+\bar{w}^s}=\tfrac{1}{2}i\bar{\theta}_2(0|\tau)\bar{\theta}_4(0|\tau)\frac{\bar{\theta}_3(\nu_{JI}|\tau)}{\bar{\theta}_1(\nu_{JI}|\tau)}, \tag{45a}$$

$$\tilde{\chi}^{+}_{JI}=\sum_{s=1/2}^{\infty}\frac{(\bar{\rho}_J/\bar{\rho}_I)^s-(\bar{w}\bar{\rho}_I/\bar{\rho}_J)^s}{1-\bar{w}^s}=\tfrac{1}{2}i\bar{\theta}_2(0|\tau)\bar{\theta}_3(0|\tau)\frac{\bar{\theta}_4(\nu_{JI}|\tau)}{\bar{\theta}_1(\nu_{JI}|\tau)}, \tag{45b}$$

$$\tilde{\tilde{\chi}}^{+}_{JI}=\tfrac{1}{2}+\sum_{n=1}^{\infty}\frac{(\bar{\rho}_J/\bar{\rho}_I)^n+(\bar{w}\bar{\rho}_I/\bar{\rho}_J)^n}{1+\bar{w}^n}=\tfrac{1}{2}i\bar{\theta}_3(0|\tau)\bar{\theta}_4(0|\tau)\frac{\bar{\theta}_2(\nu_{JI}|\tau)}{\bar{\theta}_1(\nu_{JI}|\tau)}. \tag{45c}$$

We now identify the contribution of eq. (44a, b) as

$$\left(\frac{\alpha' g}{2\pi}\right)^4\frac{1}{16\alpha'^2}K\cdot(2\pi)^8\int_F \mathrm{d}^2\tau\,(\mathrm{Im}\,\tau)^{-2}\left(\frac{g(\tau)+g(\tau+1)}{2}\right)$$
$$=\varepsilon_1^{\hat{i}}\varepsilon_2^{\hat{j}}\varepsilon_3^{\hat{k}}\varepsilon_4^{\hat{l}}\int \mathrm{d}p\,\mathrm{tr}\left(\frac{1+(-1)^P}{2}\right)\Delta_c W^{\hat{i}}(1)\left(\frac{1+(-1)^P}{2}\right)\Delta_c W^{\hat{j}}(2)$$
$$\times\left(\frac{1+(-1)^P}{2}\right)\Delta_c W^{\hat{k}}(3)\left(\frac{1+(-1)^P}{2}\right)\Delta_c W^{\hat{l}}(4), \tag{46}$$

where

$$P=\sum_{s=1/2}^{\infty}b^i_{-s}b^i_s-1+2\alpha'(\bar{p}^I)^2+2\alpha'(\tilde{p}^I)^2.$$

Therefore $g(\tau)+g(\tau+1)$ is a projection onto states with $(-1)^P=1$. This projection in fact does not eliminate any states from the original SS × NS spectrum since $(-1)^P=1$ is automatically satisfied by the left-right constraint eq. (38).

Similarly, eq. (44c) is identified as

$$\left(\frac{\alpha' g}{2\pi}\right)^4\frac{1}{16\alpha'^2}K\cdot(2\pi)^8\int_F \mathrm{d}^2\tau\,(\mathrm{Im}\,\tau)^{-2}\frac{g(-1/\tau+1)}{2}$$
$$=-\varepsilon_1^{\hat{i}}\varepsilon_2^{\hat{j}}\varepsilon_3^{\hat{k}}\varepsilon_4^{\hat{l}}\tfrac{1}{2}\int \mathrm{d}p\,\mathrm{tr}\left(\frac{1+(-1)^{\bar{P}}}{2}\right)\bar{\Delta}_c\bar{W}^{\hat{i}}(1)\left(\frac{1+(-1)^{\bar{P}}}{2}\right)\bar{\Delta}_c\bar{W}^{\hat{j}}(2)$$
$$\times\left(\frac{1+(-1)^{\bar{P}}}{2}\right)\bar{\Delta}_c\bar{W}^{\hat{k}}(3)\left(\frac{1+(-1)^{\bar{P}}}{2}\right)\bar{\Delta}_c\bar{W}^{\hat{l}}(4), \tag{47}$$

where

$$(-1)^{\bar{P}} = 2^4 d_0^1 d_0^2 \dots d_0^8 (-1) \sum_{n=1}^{\infty} d_{-n}^i d_n^i + 2\alpha'(\bar{p}^I)^2 + 2\alpha'(\tilde{p}^I)^2;$$

$\overline{W}^{\hat{i}}(k)$ is the vertex emission operator for the charged vector bosons $|\hat{i}\rangle_R \times |k^J\rangle_L$ *from a fermion line* (see for e.g. ref. [8]); and $\bar{\Delta}_c$ is calculated from the mass operator for a massive SS × Ramond sector compactified on a shifted internal momenta "lattice":

$$\overline{W}^{\hat{i}}(k^j, z, \sigma) = g\left(\frac{1}{\sqrt{2\alpha'}} \sum_n A_n^{\hat{i}} e^{-2in(\tau-\sigma)} + \tfrac{1}{2} k^{\hat{k}} R^{\hat{i}\hat{k}}\right) e^{ik^{\hat{i}} X^{\hat{i}}(\sigma,\tau)}$$

$$\times \tilde{k}^j \left[\gamma^j + i\sqrt{2}\,\gamma_9 \sum_{n\neq 0} d_n^j e^{-2in(\tau+\sigma)}\right] \gamma_9 (-1)^{\sum_{n=1}^{\infty} d_{-n}^i d_n^i} \cdot \frac{1}{i\sqrt{2}}$$

$$\times :e^{2ik^I X^I(\tau+\sigma)}: c_k, \tag{48}$$

$$\bar{\Delta}_c = \frac{\alpha'}{2\pi} \int d^2 z\, z^{\bar{N}+\alpha'(\bar{p}^I)^2} \bar{z}^{\tilde{\bar{N}}+\alpha'(\tilde{p}^I)^2} |z|^{\alpha' p^2/2-2}, \tag{49}$$

$$\bar{N} = \sum_{n=1}^{\infty} A_{-n}^i A_n^i + \tfrac{1}{2} n \bar{S}_{-n} \gamma^- S_n, \tag{50a}$$

$$\tilde{\bar{N}} = \sum_{n=1}^{\infty} \tilde{A}_{-n}^i \tilde{A}_n^i + \sum_{n=1}^{\infty} n d_{-n}^i d_n^i. \tag{50b}$$

From eq. (44c), the allowed internal momenta for this sector cannot be zero and are given by

$$\sqrt{2\alpha'}\,\bar{p}^I = \sum_L (N + \tfrac{1}{2})^L \alpha_L^I,$$

$$\sqrt{2\alpha'}\,\tilde{p}^I = \sum_L (N + \tfrac{1}{2})^L \alpha_L^I. \tag{50c}$$

Therefore, the left- and right-moving degrees of freedom which contribute as intermediate states in eq. (47) are given by another sector also satisfying eqs. (31) and (32), but where now the momenta in eqs. (31b) and (32b) are given by eq. (50c) corresponding to a twisted boundary condition on these internal coordinates i.e. $X^I(0) = X^I(\pi) + \sqrt{2\alpha}\sum N^L \alpha_L^I \pi + \sqrt{2\alpha'} \cdot \frac{1}{2}\pi$, and eq. (32c) is replaced by

$$d^i(\tau+\sigma) = \sum_{n=-\infty}^{\infty} d_n^i e^{-2in(\tau+\sigma)}, \qquad \{d_n^i, d_m^j\} = \delta_{n,-m}\delta^{ij}. \tag{51}$$

This is the Green-Schwarz superstring × the Ramond fermionic string compactified on the $(Z+\frac{1}{2})^6$ "lattice". The lowest mass state in this sector is

$$e^{i\hat{x}^I\bar{p}_1^I}\{|\hat{i}\rangle_R, |I\rangle_R, |a\rangle_R\} \times e^{i\hat{x}^I\tilde{p}_1^I}\{|a\rangle_L\} \qquad \text{with } \alpha' m_4^2 = 3,$$

since in this sector the lowest momentum is $\sqrt{2\alpha'}\,p_l^I = (\frac{1}{2},\dots,\frac{1}{2})$. The left-right constraint is

$$\bar{N} + \alpha'(\bar{p}^I)^2 = \tilde{\bar{N}} + \alpha'(\tilde{p}^I)^2 \tag{52}$$

and the mass operator is

$$\tfrac{1}{2}\alpha' m_4^2 = \bar{N} + \tilde{\bar{N}} + \alpha'(\bar{p}^I)^2 + \alpha(\tilde{p}^I)^2 . \tag{53}$$

The projection $(-1)^{\bar{P}} = 1$ indicates the left-moving Ramond fermions are Majorana-Weyl. In fact, this projection does not contribute to eq. (47) but it is required for the unitarity of one-loop amplitudes for external particles in the SS × Ramond sector.

To check perturbative unitarity, we calculate the imaginary part of eq. (42) and show that it is equal to products of tree amplitudes. Since eq. (42) can be reexpressed as the sum of eqs. (46) and (47), the discontinuity can be calculated in the standard way using the Cutkosky prescription of replacing propagators with delta functions, and we find that the imaginary part of eq. (42) is an infinite sum of theta step functions, multiplied by tree amplitudes, with thresholds given by the free particle spectrum of the compactified SS × NS and SS × Ramond sectors.

For the generalized groups, eq. 32(b, c) become

$$b^a(\tau+\sigma) = \sum_{s=-\infty}^{\infty} b_s^a\, e^{-2is(\tau+\sigma)} . \tag{54}$$

The mass operator replacing eq. (35) is

$$\tfrac{1}{2}\alpha' m_4^2 = N + \sum_{n=1}^{\infty} \tilde{A}^{\hat{i}}_{-n}\tilde{A}^{\hat{i}}_{n} + \sum_{s=1/2}^{\infty} s b^{\hat{i}}_{-s} b^{\hat{i}}_{s} + \sum_{s=1/2}^{\infty} s b^{a}_{-s} b^{a}_{s} - \tfrac{1}{2} + \alpha'(\bar{p}^I)^2 . \tag{55}$$

The left-right constraint is now

$$N + \alpha'(\bar{p}^I)^2 = \sum_{n=1}^{\infty} \tilde{A}^{\hat{i}}_{-n}\tilde{A}^{\hat{i}}_{n} + \sum_{s=1/2}^{\infty} s b^{\hat{i}}_{-s} b^{\hat{i}}_{s} + \sum_{s=1/2}^{\infty} s b^{a}_{-s} b^{a}_{s} - \tfrac{1}{2} . \tag{56}$$

In addition to the $N = 4$ supergravity multiplet, the massless states are $\{|\hat{i}\rangle_R, |I\rangle_R, |a\rangle_R\} \times \{b^a_{-1/2}|0\rangle_L\}$, the $d = 4$, $N = 4$ supersymmetric Yang-Mills multiplet in any dimension-18 semi-simple Lie group.

For these generalized groups, the vertex operators for $|i\rangle_R \times b^a_{-1/2}|0\rangle_L$ which replace eq. (40) are

$$W^{ia}(k^{\hat{i}}, z, \sigma) = g\left(\frac{1}{\sqrt{2\alpha'}} \sum_n A^i_n e^{-2in(\tau-\sigma)} + \tfrac{1}{2} k^{\hat{k}} R^{i\hat{k}}\right) e^{ik^{\hat{i}} X^{\hat{i}}}(\sigma, \tau)$$

$$\times \left[-\frac{i}{2} \frac{1}{\sqrt{2\alpha'}} f_{abc} H^b(z) H^c(z) - H^a(z) \tfrac{1}{2} k^{\hat{i}} H^{\hat{i}}(z)\right], \qquad (57)$$

where $H^a(z) \equiv \sum_{s=-\infty}^{\infty} b^a_s e^{-2is(\tau+\sigma)}$. The three-point gauge coupling is the amplitude involving three states $|\hat{i}\rangle_R \times b^a_{-1/2}|0\rangle_L$. In tree approximation,

$$A^{abc}(k^{\hat{i}}_1, k^{\hat{i}}_2, k^{\hat{i}}_3) = \langle \varepsilon_1, k^{\hat{k}}_1 | b^a_{1/2} \frac{1}{\pi} \varepsilon^{\hat{i}}_2 \int_0^{\pi} d\sigma\, W^{\hat{i}b}(k^{\hat{i}}_2, 1, \sigma) b^c_{-1/2} | \varepsilon_3, k^{\hat{i}}_3\rangle$$

$$= i f_{abc} g \varepsilon^{\mu}_1 \varepsilon^{\nu}_2 \varepsilon^{\rho}_3 \left[(k_2 - k_1)^{\mu} \delta^{\nu\rho} + (k_3 - k_2)^{\nu} \delta^{\mu\rho} + (k_1 - k_3)^{\rho} \delta^{\mu\nu}\right]. \qquad (58)$$

Eq. (58) is the standard Yang-Mills three-point coupling, and proves that the affine algebraic compactification gives rise to standard interacting non-abelian gauge bosons.

The propagator eq. (41) becomes

$$\Delta_c = \frac{\alpha'}{2\pi} \int_D d^2 z\, z^{N + \alpha'(\bar{p}^I)^2 \bar{z}^{\sum_{n=1}^{\infty} \tilde{A}^{\hat{i}}_{-n} \tilde{A}^{\hat{i}}_n + \sum_{s=1/2}^{\infty} s b^{\hat{i}}_{-s} b^{\hat{i}}_s + \sum_{s=1/2}^{\infty} s b^a_{-s} b^a_s - 1/2}}$$

$$\times |z|^{\alpha' p^2/2 - 2}.$$

The one-loop amplitude for external vector mesons is

$$A^{abcd}_{\text{loop}}(1234) = \varepsilon^{\hat{i}}_1 \varepsilon^{\hat{j}}_2 \varepsilon^{\hat{k}}_3 \varepsilon^{\hat{l}}_4 \int dp\, \mathrm{tr}\, \Delta_c W^{\hat{i}a}(1) \Delta_c W^{\hat{j}b}(2) \Delta_c W^{\hat{k}c}(3) \Delta_c W^{\hat{l}d}(4)$$

$$= \left(\frac{\alpha' g}{2\pi}\right)^4 \cdot \frac{1}{16} \cdot K \frac{(2\pi)^8}{(\alpha')^2} \int_{F_{SM}} d^2\tau\, (\mathrm{Im}\,\tau)^{-2} g^{abcd}(\tau), \qquad (59)$$

where the integrand is now given in eq. (61) and is again invariant under the modular subgroup Γ_θ. We rewrite eq. (59) using

$$\int_{F_{SM}} d^2\tau\, (\mathrm{Im}\,\tau)^{-2} g^{abcd}(\tau)$$

$$= \int_F d^2\tau\, (\mathrm{Im}\,\tau)^{-2} \left[g^{abcd}(\tau) + g^{abcd}(\tau + 1) + g^{abcd}\left(-\frac{1}{\tau} + 1\right)\right]. \qquad (60)$$

From eq. (59), we find

$$\int \mathrm{d}^2\tau\,(\operatorname{Im}\tau)^{-2} g^{abcd}(\tau) = \int \mathrm{d}^2\tau \left(\frac{-2\pi}{\ln|w|}\right)^2 \int \prod_{I=1}^{3} \mathrm{d}^2\nu_I \cdot \prod_{I<J} (\chi_{IJ})^{\alpha' k_I^{\hat{i}} k_J^{\hat{i}}} \cdot (f(\bar{w}))^{-2}$$

$$\times \bar{w}^{-1/2} \prod_{s=1/2}^{\infty} (1+\bar{w}^s)^{20} \cdot \sum_{\sqrt{2\alpha'}\, p^I \in Z^6} w^{\alpha'(\bar{p}^I)^2} a^{abcd}(\chi^+), \tag{61a}$$

where $a^{abcd}(\chi)$ is given in eq. (61b):

$$\begin{aligned}
a^{IJKL}(\chi) &= \frac{1}{16}\left(\frac{1}{2\alpha'}\right)^2 \Big[4\cdot\big[g_{IJ}g_{KL}\chi_{43}^2\chi_{21}^2 + g_{IK}g_{JL}\chi_{42}^2\chi_{31}^2 + g_{IL}g_{JK}\chi_{41}^2\chi_{32}^2\big] \\
&\quad + 8\big[(T_{IJKL}+T_{ILKJ})\chi_{43}\chi_{41}\chi_{32}\chi_{21} - (T_{IKJL}+T_{ILJK})\chi_{42}\chi_{41}\chi_{32}\chi_{31} \\
&\qquad - (T_{IJLK}+T_{IKLJ})\chi_{43}\chi_{42}\chi_{31}\chi_{21}\big] \\
&\quad + (2\alpha')\big[4\{(sM_{ILJK} - uM_{IJKL})\chi_{43}\chi_{41}\chi_{32}\chi_{21} \\
&\qquad + (tM_{ILJK} + uM_{IKJL})\chi_{42}\chi_{41}\chi_{32}\chi_{31} \\
&\qquad + (-sM_{IKJL} - tM_{IJKL})\chi_{43}\chi_{42}\chi_{31}\chi_{21}\} \\
&\qquad + s(\delta_{IJ}g_{KL} + g_{IJ}\delta_{KL})\chi_{43}^2\chi_{21}^2 \\
&\qquad + t(\delta_{JL}g_{IK} + g_{JL}\delta_{IK})\chi_{21}^2\chi_{31}^2 \\
&\qquad + u(\delta_{IL}g_{JK} + g_{IL}\delta_{JK})\chi_{41}^2\chi_{32}^2\big] \\
&\quad + (2\alpha')^2\cdot\tfrac{1}{4}\cdot(s^2\chi_{43}\chi_{21} - t^2\chi_{42}\chi_{31} + u^2\chi_{41}\chi_{32}) \\
&\quad \cdot(\delta_{IJ}\delta_{KL}\chi_{43}\chi_{21} - \delta_{IK}\delta_{JL}\chi_{42}\chi_{31} + \delta_{IL}\delta_{JK}\chi_{41}\chi_{32})\Big],
\end{aligned}$$

where

$$\begin{aligned}
g_{IJ} &= f_{IPQ}f_{JPQ}, \\
T_{IJKL} &= f_{IPQ}f_{JQR}f_{KRS}f_{LSP}, \\
M_{IJKL} &= f_{IJR}f_{KLR}.
\end{aligned} \tag{61b}$$

Using transformation properties of the Jacobi theta functions, we have from (44a),

$$\int \mathrm{d}^2\tau\,(\operatorname{Im}\tau)^{-2} g^{abcd}(\tau+1) = \int \mathrm{d}^2\tau \left(\frac{-2\pi}{\ln|w|}\right)^2 \int \prod_{I=1}^{3} \mathrm{d}^2\nu_I \cdot \prod_{I<J} (\chi_{IJ})^{\alpha' k_I^{\hat{i}} k_J^{\hat{i}}}$$
$$\times \Bigg\{ (f(\bar{w}))^{-2} \left[-\bar{w}^{-1/2} \prod_{s=1/2}^{\infty} (1-\bar{w}^s)^{20} \right]$$
$$\times \sum_{\sqrt{2\alpha'}\,\bar{p}^{\prime} \in Z^6} w^{\alpha'(\bar{p}^{\prime})^2} (-1)^{2\alpha'(\bar{p}^{\prime})^2} a^{abcd}(\tilde{\chi}^+) \Bigg\} \tag{61c}$$

and

$$\int \mathrm{d}^2\tau\,(\operatorname{Im}\tau)^{-2} g^{abcd}\left(-\frac{1}{\tau}+1\right) = \int \mathrm{d}^2\tau \left(\frac{-2\pi}{\ln|w|}\right)^2 \int \prod_{I=1}^{3} \mathrm{d}^2\nu_I \cdot \prod_{I<J} (\chi_{IJ})^{\alpha' k_I^{\hat{i}} k_J^{\hat{i}}}$$
$$\times \Bigg\{ (f(\bar{w}))^{-2} 2^{10} \left[-\bar{w}^{3/4} \prod_{n=1}^{\infty} (1+w^n)^{20} \right]$$
$$\times \sum_{\sqrt{2\alpha'}\,\bar{p}^{\prime} \in (Z+1/2)^6} w^{\alpha'(\bar{p}^{\prime})^2} a^{abcd}(\tilde{\tilde{\chi}}^+) \Bigg\}. \tag{61d}$$

As in eq. (44), we identify the contribution of eq. (61a, c) as given by

$$\left(\frac{\alpha' g}{2\pi}\right)^4 \frac{1}{16\alpha'^2} K \cdot (2\pi)^8 \int_{\mathrm{F}} \mathrm{d}^2\tau\,(\operatorname{Im}\tau)^{-2} \left(\frac{g(\tau)+g(\tau+1)}{2}\right)^{abcd}$$
$$= \varepsilon_1^{\hat{i}} \varepsilon_2^{\hat{j}} \varepsilon_3^{\hat{k}} \varepsilon_4^{\hat{l}} \int \mathrm{d}p\,\operatorname{tr}\left(\frac{1+(-1)^P}{2}\right) \Delta_{\mathrm{c}} W^{\hat{i}a}(1) \left(\frac{1+(-1)^P}{2}\right) \Delta_{\mathrm{c}} W^{\hat{j}b}(2)$$
$$\times \left(\frac{1+(-1)^P}{2}\right) \Delta_{\mathrm{c}} W^{\hat{k}c}(3) \left(\frac{1+(-1)^P}{2}\right) \Delta_{\mathrm{c}} W^{\hat{l}d}(4),$$

where

$$P = \sum_{s=1/2}^{\infty} b_{-s}^{a} b_{s}^{a} - 1 + 2\alpha'(\bar{p}^{\prime})^2 + \sum_{s=1/2}^{\infty} b_{-s}^{\hat{i}} b_{s}^{\hat{i}}. \tag{62}$$

Similarly, eq. (61d) is identified as

$$\left(\frac{\alpha' g}{2\pi}\right)^4 \frac{1}{16\alpha'^2} K (2\pi)^8 \int_{\mathrm{F}} \mathrm{d}^2\tau\,(\operatorname{Im}\tau)^{-2} \frac{g^{abcd}\left(-\frac{1}{\tau}+1\right)}{2}$$
$$= -\varepsilon_1^{\hat{i}} \varepsilon_2^{\hat{j}} \varepsilon_3^{\hat{k}} \varepsilon_4^{\hat{l}} \tfrac{1}{2} \int \mathrm{d}p\,\operatorname{tr}\left(\frac{1+(-1)^{\bar{P}}}{2}\right) \bar{\Delta}_{\mathrm{c}} \overline{W}^{\hat{i}a}(1) \left(\frac{1+(-1)^{\bar{P}}}{2}\right) \bar{\Delta}_{\mathrm{c}} \overline{W}^{\hat{j}b}(2)$$
$$\times \left(\frac{1+(-1)^{\bar{P}}}{2}\right) \bar{\Delta}_{\mathrm{c}} \overline{W}^{\hat{k}c}(3) \left(\frac{1+(-1)^{\bar{P}}}{2}\right) \bar{\Delta}_{\mathrm{c}} \overline{W}^{\hat{l}d}(4), \tag{63}$$

where

$$(-1)^{\bar{P}} = 2^{10} d_0^1 d_0^2 \dots d_0^{18} d_0^{\hat{1}} d_0^{\hat{2}} (-1)^{\sum_{n=1}^{\infty} d_{-n}^a d_n^a + 2\alpha'(\bar{p}^I)^2} + \sum_{n=1}^{\infty} d_{-n}^{\hat{1}} d_n^{\hat{1}}$$

$\overline{W}^{\hat{i}a}(k^{\hat{j}})$ is the vertex emission operator for the vector bosons $|\hat{i}\rangle_R \times b^a_{-1/2}|0\rangle_L$ from a *fermion line*, and $\bar{\Delta}_c$ is calculated from the mass operator for a massive SS × Ramond sector where the right movers have been compactified on a shifted $(Z+\frac{1}{2})^6$ "lattice": eq. (48) becomes

$$\overline{W}^{\hat{i}a}(k^{\hat{i}}, z, \sigma) = g\left(\frac{1}{\sqrt{2\alpha'}} \sum_n A_n^{\hat{i}} e^{-2in(\tau-\sigma)} + \tfrac{1}{2} k^{\hat{k}} R^{\hat{i}\hat{k}}\right) e^{ik^{\hat{i}} X^{\hat{i}}(\sigma,\tau)}$$

$$\times\left(-\frac{i}{2}\frac{1}{\sqrt{2\alpha'}} f_{abc} \frac{\Gamma^b(z)}{i\sqrt{2}} \frac{\Gamma^c(z)}{i\sqrt{2}} - \frac{k^{\hat{l}}}{2} \frac{\Gamma^a(z)}{i\sqrt{2}} \frac{\Gamma^{\hat{l}}(z)}{i\sqrt{2}}\right), \tag{64}$$

where

$$\Gamma^a(z) = \gamma^a + i\sqrt{2}\,\gamma_{21} \sum_{n\neq 0} d_n^a e^{-2in(\tau+\sigma)}, \qquad \gamma_{21} \equiv 2^{10} d_0^1 \dots d_0^{18} d_0^{\hat{1}} d_0^{\hat{2}},$$

$$\Gamma^{\hat{i}}(z) = \gamma^{\hat{i}} + i\sqrt{2}\,\gamma_{21} \sum_{n\neq 0} d_n^{\hat{i}} e^{-2in(\tau+\sigma)}, \qquad \Gamma_3 = \gamma_3 (-1)^{\sum_{n=1}^{\infty} d_{-n}^{\hat{i}} d_n^{\hat{i}}},$$

$$\bar{\Delta}_c = \frac{\alpha'}{2\pi} \int_D d^2 z\, z^{\bar{N} + \alpha'(\bar{p}^I)^2} \bar{z}^{\sum_{n=1}^{\infty} A_{-n}^{\hat{i}} A_n^{\hat{i}} + \sum_{n=1}^{\infty} n d_{-n}^{\hat{i}} d_n^{\hat{i}} + \sum_{n=1}^{\infty} n d_{-n}^a d_n^a + 3/4}$$

$$\times |z|^{\alpha' p^2/2 - 2}. \tag{65}$$

The lowest mass state in the SS × Ramond sector is

$$e^{i\hat{x}^I \bar{p}^I}\{|\hat{i}\rangle_R, |I\rangle_R, |a\rangle_R\} \times \{|\bar{a}\rangle_L\}$$

where $|\bar{a}\rangle_L$ is the dimension 2^9 spinor ground state of the internal Ramond modes.

5. Conclusion

We have described a compactification technique which is useful in reducing a string from ten to four dimensions, and which, for theories with no gauge group in ten dimensions, inserts a non-abelian gauge symmetry in four dimensions in the original spirit of Kaluza-Klein. The motivation for this is that other compactifications to four dimensions either start with a gauge group in ten dimensions or pick up the gauge group via compactification from twenty-six dimensions, so that before symmetry breaking these give rise to rank 22 = 26 − 4 gauge groups. The compactification procedure for type II superstrings described here gives rise to gauge groups with rank less than or equal to $6 = 10 - 4$, and thus appears to be a more economical model for low energy phenomenology. Indeed, this mechanism, although it does not explain independently why we live in four dimensions, has, with its new way of introducing symmetry, exactly matched the fact that we live in four dimensions with the size of the symmetry group which governs four-dimensional interactions. An interesting and important question is whether the three models we have presented in this paper, based on the gauge groups $SU(2)^6$, $SU(3) \times SO(5)$, and $SU(2) \times SU(4)$ can be continuously related in a sense similar to that in which Narain's work relates the SO(32) and $E_8 \times E_8$ heterotic string [9].

An associated question is whether, if we compactify all the remaining dimensions, the Lie algebra obtained from the vertex operators describing the physical states depends on the way the compactification has been performed. Since there are seven dimension-30 semi-simple Lie groups, there are presumably that many ways of compactifying. Associated with each way there is a Lie algebra formed by the operators

$$\frac{1}{2\pi i}\oint \mathrm{d}z\, V_A^1(z),$$

where we just consider the left-moving modes of the string in the NS sector. Here the $V_A^1(z)$ is the vertex operator associated with an arbitrary physical state $|A\rangle$. In this algebra, the operators corresponding to null states form an ideal and, if we divide them out, we obtain an algebra which is isomorphic (as a vector space) with the physical states modulo the null states. For the straightforward compactification on a $Z^{9,1}$ lattice, which corresponds to the group $SU(1,1) \times SU(2)^9$, we obtain the Lie algebra described by Borcherds [10]. This is not quite the same as that generated by the tachyon vertex operators which Borcherds stated has a Dynkin diagram given by the whole of the E_8 root lattice. We would suggest that this algebra bears the same relation to the NS string as the "Monster" Lie algebra, whose Dynkin diagram is the Leech lattice, and which is associated with the even self-dual 26-dimensional lorentzian lattice [11], bears to the bosonic string.

It would be of interest to know whether the Lie algebras so obtained by using the seven compactifications are all isomorphic. It is also natural to conjecture that Borcherds' algebra can be turned into a superalgebra by the addition of fermion emission operators and GSO projections [12].

References

[1] I.B. Frenkel and V.G. Kac, Inv. Math. 62 (1980) 23;
G. Segal, Commun. Math. Phys. 80 (1981) 301

[2] P. Goddard and D. Olive, in Vertex operators in mathematics and physics, MSRI Publication #3, ed. J. Lepowsky, S. Mandelstam and I.M. Singer (Springer, Heidelberg, 1985) p. 55

[3] D. Gross, J.A. Harvey, E. Martinec and R. Rohm, Phys. Rev. Lett. 54 (1985) 502; Nucl. Phys. B256 (1985) 253

[4] R. Bluhm and L. Dolan, Phys. Lett. B169 (1986) 347

[5] P. Goddard and D. Olive, Nucl. Phys. B257 (1985) 245
[6] P. Goddard and C. Thorn, Phys. Lett. B40 (1972) 235
[7] I. Antoniadis, C. Bachas, C. Kounnas and P. Windey, Phys. Lett. B171 (1986) 51;
H. Kawai, D.C. Lewellen and S.-H. Tye, Nucl. Phys. B288 (1987) 1
[8] M.B. Green, Phys. Lett. B46 (1973) 392
[9] K.S. Narain, Phys. Lett. B169 (1986) 41
K.S. Narain, M.H. Sarmadi, E. Witten, Nucl. Phys. B279 (1986) 369
[10] R.E. Borcherds, in Proc. US National Academy of Sciences, to be published
[11] R.E. Borcherds, J.H. Conway, L. Queen and N.J.A. Sloan, Adv. Math 53 (1984) 75
[12] F. Gliozzi, J. Scheck, and D. Olive, Nucl. Phys. B122 (1977) 253
[13] P. Goddard and D. Olive, Int. Jour. of Mod. Phys. A1 (1986) 303;
L. Dolan, *Lewes String Theory Workshop*, ed. L. Clavelli and A. Halprin (World Scientific, Singapore, 1985) p. 103

FOUR-DIMENSIONAL TYPE II STRINGS AND THEIR EXTENSIONS: TYPE III STRINGS

Hikaru KAWAI, David C. LEWELLEN and S.-H. Henry TYE
Newman Laboratory of Nuclear Studies, Cornell University, Ithaca, NY 14853, USA

Received 22 January 1987; revised manuscript received 6 March 1987

Four-dimensional type II string models and their generalizations are analyzed. In particular, models referred to as type III strings, with spin structures and world sheet supercurrents which are asymmetric between the left and right moving degrees of freedom, are analyzed. Models with $N=8$, 4, 2, and 1 supersymmetry as well as nonsupersymmetric models can be constructed. Furthermore, many of them have interesting gauge groups. Attempts to construct phenomenologically interesting models are outlined.

Superstring theories offer the possibility of describing all known physical interactions in a single mathematically consistent framework (see, e.g., ref. [1]). In particular fermionic closed string theories automatically include Einstein gravity as an effective low-energy theory and appear to be completely finite and anomaly free [2]. Using the spin structure construction described in ref. [3] it has proved possible to construct many heterotic-type string models (that is, with world sheet supersymmetric right moving excitations and bosonic left movers) in four space–time dimensions which are phenomenologically viable, e.g., are nonsupersymmetric or possess $N=1$ space–time supersymmetry and contain chiral fermions and potentially realistic gauge groups. Unfortunately, the near uniqueness of heterotic string models in ten dimensions [4] is entirely absent in four dimensions. The chief source of the abundance of candidate models is the large number of left moving degrees of freedom unconstrained by any local world sheet supersymmetry.

String theories with both left and right moving sectors possessing world sheet supersymmetry (e.g. the type II strings in ten dimensions [1]) offer an attractive alternative to the multitude of heterotic string models, since they have fewer degrees of freedom constrained by more world sheet symmetries. The question we address in this paper is whether such models in four dimensions retain enough flexibility to include something as rich and complex as the $SU(3)\otimes SU(2)\otimes U(1)$ standard model. We will show in particular that the most interesting of these models share a significant property of heterotic strings which distinguishes them from type II models in ten dimensions and their compactifications: the spin structures and world sheet supercurrents are asymmetric between the left and right movers. To underscore this important distinction we refer to these as type III string models.

To explicitly construct a class of finite and anomaly-free type III models, those which can be represented with free world sheet fermions, we employ the spin structure construction technique described in detail in ref. [3]. Models with $N=8$, 4, 2, and 1 supersymmetry as well as nonsupersymmetric models can easily be constructed; however, the attempts at constructing phenomenologically realistic type III string models given here, while suggestive, are as yet unsuccessful. (Some possible generalizations which may improve this situation are mentioned at the end of this work.) Nonetheless, the construction of type III models illustrates a number of important features of string theories and the spin construction method. To elaborate, we observe: (1) Four-dimensional type III models are prime examples of string models which cannot be obtained from a simple compactification of some ten-dimensional string theory; that is, they are best constructed directly in four dimensions.

0370-2693/87/$ 03.50 © Elsevier Science Publishers B.V.
(North-Holland Physics Publishing Division)

Many of the four-dimensional chiral models of the heterotic type constructed by the spin structure method are similarly not simple compactifications, in contrast to, for example, those four-dimensional models considered by Narain [5] or constructed on Calabi–Yau manifolds [6]. (2) Type III string models are left–right asymmetric and thus do not preclude the possibility of chiral fermions; although, as we shall see, the models constructed solely from free real world sheet fermions are in fact nonchiral. The theorem on the incompatibility of massless fermions and non-abelian gauge fields in type II theories [7] is avoided by choosing right movers which provide gauge quantum numbers along with left movers which allow for massless fermions. (3) The gauge symmetries of type III models arise not from the Frenkel–Kac construction employed in heterotic string models, but from the structure of the world sheet supercurrent. This mechanism allows gauge groups larger than SU(2) if the space–time dimension is five or less. (4) The construction of type III models given here illustrates some of the power of the spin structure construction formalism. In particular, these models cannot be simply formulated in terms of a bosonic momentum lattice (or fermionic charge lattice).

Let us begin with a summary of the spin structure construction formalism with the particular case of type III models in mind. We direct the reader to ref. [3] for the pertinent derivations, notations, and further references. In order to ensure a physical theory which is Lorentz invariant, satisfies positivity, and is finite, we construct a two-dimensional theory on the string world sheet which is reparametrization invariant, world sheet supersymmetric, superconformal invariant and modular invariant. We employ the Ramond–Neveu–Schwarz (RNS) formalism [8] for the space–time fermions and represent all of the string's internal degrees of freedom with world sheet fermions; thus in four dimensions, in the light cone gauge which we employ, there are, besides the two transverse bosonic string coordinates, two left and two right moving RNS fermions representing space–time degrees of freedom and an additional eighteen right moving and eighteen left moving Majorana fermions on the string world sheet. This number of fermionic variables is required for superconformal invariance.

The requirements for world sheet supersymmetry in theories represented by free fermions have been analyzed in refs. [9,10]. Let us review them in some detail since the results are crucial in discovering what models can be constructed. We can consider independent $N=\frac{1}{2}$ supersymmetries for the left and right moving fermions. The general form of the contribution from internal degrees of freedom for the, say, left moving supercurrent (which must have conformal dimensions $\Delta=\frac{3}{2}$ and $\bar{\Delta}=0$) in terms of the left moving world sheet fields is then

$$S(z)=\tfrac{1}{3}\mathrm{i}\sum_{3\leqslant j,k,l\leqslant 20} f^{jkl}\psi^j\psi^k\psi^l\,. \qquad (1)$$

To ensure world sheet supersymmetry the constants f^{jkl} should be chosen so that $S(z)$ satisfies the Ramond–Neveu–Schwarz (i.e. super-Virasoro) algebra. In terms of operator product expansions this is (for four dimensions in the light cone gauge)

$$S(z)S(w)\sim\frac{6}{(z-w)^3}+\frac{2T(w)}{z-w}$$
$$+\text{regular terms}\,, \qquad (2a)$$

$$T(z)S(w)\sim\frac{3S(w)}{2(z-w)^2}+\frac{\partial_w S(w)}{z-w}$$
$$+\text{regular terms}\,, \qquad (2b)$$

$$T(z)T(w)\sim\frac{9}{2(z-w)^4}+\frac{2T(w)}{(z-w)^2}+\frac{\partial_w T(w)}{z-w}$$
$$+\text{regular terms}\,, \qquad (2c)$$

where $T(z)$ is the energy–momentum tensor for the internal degrees of freedom given by

$$T(z)=\frac{1}{2}\sum_{i=3}^{20}:(\partial_z\psi^i)\psi^i:\,. \qquad (3)$$

As one can explicitly verify, the operator products (2) are satisfied if and only if the f^{jkl} in (1) are chosen so that [9,10]

$$f^{ijk}f^{lmk}+f^{ilk}f^{mjk}+f^{imk}f^{jlk}=0\,, \qquad (4a)$$

$$f^{ijk}f^{ljk}=\tfrac{1}{2}\delta^{il}\,. \qquad (4b)$$

These are just the conditions (up to some overall normalization) that the f^{ijk} are the structure constants of a semi-simple Lie group. The only semi-simple Lie algebras of dimension eighteen are $[SU(2)]^6$, $SU(3)\otimes SO(5)$ and $SU(4)\otimes SU(2)$. Note that in D space–time dimensions the f^{ijk} are structure constants of a dimension $3(10-D)$ semi-simple Lie group so that if $D\geqslant 6$ the only possible group structure for both the left and right moving supercurrents is $[SU(2)]^{10-D}$.

In order to fully determine a string model we must specify the boundary conditions (spin structures) of all of the two-dimensional fermions around noncontractible loops on the string world sheet. Modular invariance in fact requires that we take a linear combination of contributions with different sets of boundary conditions, since each piece individually is not invariant under modular transformations. In particular, the fermionic contribution to the one-loop vacuum amplitude is a sum of partition functions, each determined by a set of spin structures (ss);

$$Z_{\text{fermion}}=\sum_{\{ss\}} C(ss)Z(ss)\,. \tag{5}$$

The coefficients $C(ss)$ should be chosen consistent with one-loop modular invariance and also so that Z_{fermion} can be interpreted as a sensible sum of physical states in four-dimensional space–time, i.e., the sum over spin structures for one direction on the torus should naturally form a projection onto a subspace of world sheet states with correct space–time statistics. In the simplest example this is just the GSO projection [11]. Explicit one-loop modular invariance and a physically sensible projection are probably sufficient to guarantee multiloop modular invariance as well [3,12]. Subject to these requirements the general solution to eq. (5) can be written down and its content reduced to a set of rules giving the spectrum of states included in the projection formed in (5). The fermion boundary conditions are conveniently represented by 40-dimensional vectors whose lth component takes the value 0 or $\frac{1}{2}$ depending on whether the lth fermion has periodic or antiperiodic boundary conditions. (We restrict ourselves to these possibilities here for simplicity, though in general we may choose any boundary conditions for which $S(z)$ and $S(\bar{z})$ are periodic or antiperiodic on the string world sheet [3].) The appropriate sum over sets of boundary conditions runs over all boundary condition vectors spanned by some set of basis vectors $\{\boldsymbol{W}_i\}$ which includes, in particular, the vector $\boldsymbol{W}_0$ with all components equal to $\frac{1}{2}$. The f^{jkl} and $\boldsymbol{W}_i$, together with a matrix of constants k_{ij} which determines in part the projection formed in (5), specifies a given string model. To guarantee modular invariance and physically sensible projections these quantities must satisfy the constraints

$$k_{ij}+k_{ji}=\boldsymbol{W}_i\cdot\boldsymbol{W}_j \pmod 1\,, \tag{6}$$

$$2k_{ij}=0 \pmod 1\,, \tag{7}$$

$$k_{ii}+k_{i0}+s_i-\tfrac{1}{2}\boldsymbol{W}_i\cdot\boldsymbol{W}_i=0 \pmod 1\,. \tag{8}$$

The quantity s_i appears to ensure that states contribute to the partition function (5) with the correct space–time statistics. It takes the value 0 or $\frac{1}{2}$ depending on whether the first and the twenty-first components of $\boldsymbol{W}_i$ (which pertain to left and right moving space–time degrees of freedom) agree or not, i.e., $s_i=W_i^1+W_i^{21}$ (mod 1). The inner product between $\boldsymbol{W}$ vectors in (6) and (8) is defined with lorentzian signature (left components minus right components) and, conforming to the notation of ref. [3], implicitly includes an overall factor of $\frac{1}{2}$ since we define our $\boldsymbol{W}_i$ to describe real world sheet fermions rather than complex ones, i.e. $\boldsymbol{W}_i\cdot\boldsymbol{W}_j=\frac{1}{2}(\Sigma_{l=1}^{20}W_i^lW_j^l-\Sigma_{l=21}^{40}W_i^lW_j^l)$.

In addition to (6), (7), and (8), we must choose the set $\{\boldsymbol{W}_i\}$ to be consistent with world sheet supersymmetry. Specifically the world sheet supercurrents for left and right movers must be either periodic or antiperiodic around the world sheet torus. Given the expression (1) for the supercurrent $S(z)$ in terms of the world sheet fermions, this represents a strong constraint on the allowed choices of spin structures. Assume that $S(z)$ is periodic (antiperiodic) around one of the noncontractible loops of the world sheet torus. Then from (1) ψ^1 and ψ^2 must have periodic (antiperiodic) boundary conditions and so the first two components of $\boldsymbol{W}$ are 0 ($\frac{1}{2}$). Group the remaining world sheet fermions into two sets, those with periodic (antiperiodic) boundary conditions, labeled by a, b, c, and those with antiperiodic (periodic) boundary conditions, labeled by i, j, k. $S(z)$ will be periodic (antiperiodic) if the structure constants of the form f^{abi} and f^{ijk} are all zero. This is just the statement that f^{abc} are the structure constants of a symmetric subgroup, H, of the original, G, with all of the f as structure constants (i.e., G/H is a sym-

Table 1
Symmetric subalgebras. g denotes the Lie algebra for groups G appearing in the text. h are the Lie algebras of the subgroups H such that G/H is a symmetric space. (Condensed from ref. [15].)

g	h
SU(2)	U(1)
SU(2)⊗SU(2)	SU(2); SU(2)⊗U(1); U(1)⊗U(1)
SU(3)	SU(2); SU(2)⊗U(1)
SO(5)	SU(2)⊗SU(2); SU(2)⊗U(1)
SU(4)	SU(3)⊗U(1); SO(5); SU(2)⊗SU(2), SU(2)⊗SU(2)⊗U(1)

metric space). The symmetric subgroups of SU(2), SU(2)⊗SU(2), SU(3), SU(4), and SO(5) are given in table 1. If the supercurrent has, for example, the $[SU(2)]^6$ structure, then the first twenty components of the $\boldsymbol{W}_i$ may take forms such as

$$(aaA_1B_1C_1A_2B_2C_2\ldots A_6B_6C_6)$$

$$\text{with } A_i+B_i+C_i=a \quad (\text{mod } 1)\,, \tag{9a}$$

or

$$(aaA_1B_1C_1\ldots A_4B_4C_4(a)^3(a+\tfrac{1}{2})^3)\,, \tag{9b}$$

where in the latter case we have employed the SU(2) symmetric subgroup of SU(2)⊗SU(2) for the last six fermions. On the other hand, a quite different structure such as $(aa(a)^9(\frac{1}{2}+a)^6ABC)$ with $A+B+C=a$ (mod 1) is allowed if the SU(4)⊗SU(2) structure constants appear in the supercurrent. The last twenty components of the $\boldsymbol{W}_i$ are constrained in similar fashion, depending on the structure constants appearing in the construction of the other (right moving) supercurrent.

The spectrum of states of a given model is divided into different sectors characterized by the boundary conditions of the fermions around the string, and so labeled by a linear combination of the basis vectors $\{\boldsymbol{W}_i\}$, i.e., $\overline{\alpha \boldsymbol{W}=\sum \alpha_i \boldsymbol{W}_i}$, where $\alpha_i=0$ or 1 and the overbar signifies that all of the components of $\alpha\boldsymbol{W}$ are shifted by integers to have values of 0 or $\frac{1}{2}$. The projection formed by part of the sum in (5) includes those states in the $\overline{\alpha \boldsymbol{W}}$ sector with fermion numbers (written as a vector with 40 components) which satisfy [3]

$$\boldsymbol{W}_i\cdot\boldsymbol{N}_{\alpha W}=s_i-\boldsymbol{W}_i\cdot\overline{\alpha\boldsymbol{W}}+\sum_j k_{ij}\alpha_j+k_{0i}$$

$$(\text{mod } 1)\,. \tag{10}$$

Those states appearing in the on-shell physical string spectrum must in addition have equal contributions to the $(\text{mass})^2$ from the left and right moving degrees of freedom. For sectors with some Ramond fermions, that is $\overline{\alpha \boldsymbol{W}}$ with some components zero, the left-hand side of (10) is a formal expression which should be interpreted as including chiral projections acting on the Ramond ground states, e.g. factors like $\frac{1}{4}(1+\gamma^i\gamma^j\gamma^k\gamma^l)$ for Ramond components labeled by i, j, k, and l. In addition, the number operators $\boldsymbol{W}_i\cdot\hat{\boldsymbol{N}}_{\alpha W}$ and $\boldsymbol{W}_j\cdot\hat{\boldsymbol{N}}_{\beta W}$ should commute (this is required in the derivation of the rules for modular invariant sums of the form (5) which form proper projections). This is most easily ensured if the set $\{\boldsymbol{W}_i\}$ is such that the 40 real world sheet fermions could be represented by 20 complex ones, i.e., the boundary conditions can always be identified in pairs. This condition is not necessary, however. The necessary and sufficient condition in addition to (6), (7), and (8) is that, given any four (not necessarily distinct) basis vectors, the number of components which are simultaneously nonzero for all four vectors should be even.

Let us now consider the simplest possible models, those with boundary conditions generated solely by the vector $\boldsymbol{W}_0=((\frac{1}{2})^{20}|(\frac{1}{2})^{20})$. The spectrum splits into two sectors characterized by $\boldsymbol{W}_0$ ($\alpha_0=1$, i.e. all antiperiodic fermions) and 0 ($\alpha_0=0$, i.e. all periodic fermions) with left and right ground state energies $(-\frac{1}{2}, -\frac{1}{2})$ and $(\frac{3}{4}, \frac{3}{4})$, respectively. In general, replacing a $(\frac{1}{2})$ by a (0) in $\boldsymbol{W}$ raises the ground state energy by $\frac{1}{16}$. The lowest state in the $\boldsymbol{W}_0$ sector satisfies the spectrum generating condition (10) and so this tachyonic state appears in the physical spectrum. Exciting a single, lowest-energy, NS degree of freedom for both the left and right movers also satisifes (10) and gives rise to a number of massless bosonic states. In particular, taking $N^i_{W_0}$ and $N^j_{W_0}$ equal to one for $i\in\{1, 2\}$, $j\in\{23, 24, \ldots, 40\}$ or $i\in\{3, 4, \ldots, 20\}$, $j\in\{21, 21\}$ produces a set of 36 massless space–time vector bosons (with two polarizations) which carry the internal quantum numbers of the adjoint representation of the dim 18+dim 18 groups whose structure constants appear in the left and right supercurrents. To see that these gauge fields have the appropriate non-abelian couplings in the effective field theory we need only recall that [3,13], in the light cone gauge, the vertex operator includes a factor of the supercurrent $S(z)$, which carries precisely the correct structure constants f^{ijk} for the Yang–Mills coupling. Thus the structure chosen for the world sheet supercurrents directly affects the gauge group of the string model constructed. Possible gauge groups are any dim 18 + dim 18 semi-simple Lie group [i.e. $G_L \otimes G_R$ with G_L and G_R chosen independently from $[SU(2)]^6$, $SU(3)\otimes SO(5)$ and $SU(4)\otimes SU(2)$] for the simple models built from only W_0, or, more generally, some symmetric subgroup of one of these groups.

The models generated by W_0 alone have tachyons and no fermions; our main interest, however, is in models with massless fermions, so let us turn to implementing this condition. A massless fermion requires a fermionic sector with left and right vacuum energies of 0 or $-\frac{1}{2}$; in other words, a generic W of the form $((\frac{1}{2})^{20}|(0)^2(\frac{1}{2})^{12}(0)^6)$ or $((\frac{1}{2})^2(0)^8(\frac{1}{2})^{10}|(0)^2(\frac{1}{2})^{12}(0)^6)$ or one obtained from these by interchanging the left and right movers. Notice that in the first case we can, by exciting one of the first two degrees of freedom, obtain massless spin-$\frac{3}{2}$ states – gravitinos; the spin-$\frac{1}{2}$ states are in the adjoint representation of the gauge group (obtained by taking $N^i_W = 1$ for $2 < i \leqslant 20$) and thus give the gauginos expected in a supersymmetric model. Notice as well that the presence of *any* fermionic sector is sufficient to project out the single tachyon possible in these models (cf. ref. [10]).

The structure of the world sheet supercurrents must, of course, be chosen compatible with the form of W; this represents a very strong constraint. In particular, the structure $|(0)^2(\frac{1}{2})^{12}(0)^6)$, which is required for massless fermions, is *only* compatible with the supercurrent with $[SU(2)]^6$ symmetry [7,10]. Thus more properly the right half of W should be written $|(0)^2(\frac{1}{2}\,\frac{1}{2}\,0)^6)$ to illustrate how each SU(2) is broken by the boundary conditions to its symmetric subgroup U(1). Because of this, left–right symmetric type II models with massless fermions do not include non-abelian gauge groups. Note as well that for heterotic-type models with massless fermions the $[SU(2)]^6$ form for the supercurrent given in (9a) is the only one which we need to consider [3].

Let us consider some examples where both the left and right moving supercurrents are chosen with $[SU(2)]^6$ symmetry. As already noted the model generated by W_0 alone is tachyonic with $[SU(2)]^{12}$ symmetry. Adding $W_1 = ((\frac{1}{2})^2(\frac{1}{2}\,\frac{1}{2}\,\frac{1}{2})^6|(0)^2(\frac{1}{2}\,\frac{1}{2}\,0)^6)$ (the different choices of k_{ij} do not affect the model in this case) breaks the gauge symmetry to $[SU(2)]^6\otimes[U(1)]^6$, but also gives rise to an $N=4$ space–time supersymmetry (properly speaking the $[U(1)]^6$ gauge bosons are part of the supergravity multiplet). The set $\{W_0,\ W_1,\ W_2 = ((0)^2(\frac{1}{2}\,\frac{1}{2}\,0)^6|(\frac{1}{2})^2(\frac{1}{2}\,\frac{1}{2}\,\frac{1}{2})^6)\}$ with the k satisfying $k_{00}+k_{01} = k_{12}+k_{02}$ generates a model whose massless states form a pure $N=8$ space–time supergravity. This includes 28 massless vector bosons, 12 from the W_0 sector and 16 from the $W_0+W_1+W_2$ sector. This left–right symmetric (type II) model can be interpreted as a simple toroidal compactification of the ten-dimensional type II string. The set $\{W_0,\ W_1,\ W_2 = ((\frac{1}{2})^2(\frac{1}{2}\,0\,0)^4(\frac{1}{2}\,\frac{1}{2}\,\frac{1}{2})^2|(0)^2(\frac{1}{2}\,\frac{1}{2}\,0)^2(\frac{1}{2}\,0\,\frac{1}{2})^4)\}$ generates an $N=2$ supersymmetric model with gauge group $[U(1)]^4\otimes[SU(2)]^2$. Adding the vector $W_3 = ((\frac{1}{2})^2(0\,0\,\frac{1}{2})^4(\frac{1}{2}\,\frac{1}{2}\,\frac{1}{2})^2|(0)^2(\frac{1}{2}\,0\,\frac{1}{2})^2(\frac{1}{2}\,\frac{1}{2}\,0)^2(\frac{1}{2}\,0\,\frac{1}{2})^2)$ to this set breaks the supersymmetry to $N=1$ and the gauge group to $[SU(2)]^2$.

Nonsupersymmetric models are also easily obtained; for example, W_0 and $W_1 = ((\frac{1}{2})^2(\frac{1}{2}\,0\,0)^4(\frac{1}{2}\,\frac{1}{2}\,\frac{1}{2})^2|(0)^2(\frac{1}{2}\,\frac{1}{2}\,0)^6)$ together generate a model with $[SU(2)]^2\otimes[U(1)]^{10}$ gauge symmetry and massless fermions (these are, however, singlets under the $[SU(2)]^2$ group). To cite a somewhat more unusual example, W_0 and $W = ((0)^2(\frac{1}{2}\,\frac{1}{2}\,0)^6|(0)^2(\frac{1}{2}\,\frac{1}{2}\,0)^6)$ together generate a model with 44 massless vector bosons (32 from the W sector) – apparently a model with $[U(1)]^{44}$ symmetry.

As the examples above illustrate we can use the spin structure construction to produce a wide range of consistent models including ones with $N=8$, 4, 2 and 1 supersymmetry as well as nonsupersymmetric ones; however, none of the examples given with massless fermions have gauge groups large enough to contain the standard model. For this we must consider models with one supercurrent with $SU(3)\otimes SO(5)$ or $SU(4)\otimes SU(2)$ supersymmetry and the other with $[SU(2)]^6$ symmetry (to allow for massless fermions). For example, W_0 and $W_1 = ((\frac{1}{2})^2(\frac{1}{2})^{18}|(0)^2(\frac{1}{2}\,\frac{1}{2}\,0)^6)$ produce $N=4$ supersymmetric models with gauge groups $SU(3)\otimes SO(5)$

or SU(4)⊗SU(2), depending on the form chosen for the supercurrent. We can reduce the degree of supersymmetry, but in the process we must cut the gauge symmetry as well; e.g., adding $W_2 = ((\frac{1}{2})^2(\frac{1}{2})^9(0)^6(\frac{1}{2}\,0\,0)\,|\,(0)^2(\frac{1}{2}\,\frac{1}{2}\,0)^2(\frac{1}{2}\,0\,\frac{1}{2})^4)$ to the $N=4$ SU(4)⊗SU(2) model above gives an $N=2$ SU(3)⊗U(1)⊗U(1) model. Turning to nonsupersymmetric type III models, W_0 and $W_1 = ((\frac{1}{2})^2(\frac{1}{2})^9(0)^6(\frac{1}{2}\,\frac{1}{2}\,\frac{1}{2})\,|\,(0)^2(\frac{1}{2}\,\frac{1}{2}\,0)^3(0\,0\,0)^3)$ (with one supercurrent with SU(4)⊗SU(2) symmetry) generate a model with gauge group $SU(3)\otimes[SU(2)]^4\otimes[U(1)]^4$, but unfortunately no massless fermions. On the other hand, W_0 and $W_1 = ((\frac{1}{2})^2(\frac{1}{2})^9(0)^6(\frac{1}{2}\,0\,0)\,|\,(0)^2(\frac{1}{2}\,\frac{1}{2}\,0)^6)$ give a model with massless fermions, but gauge group broken to $SU(3)\otimes[U(1)]^8$. Note that the breaking of SU(4) to SU(3)⊗U(1) allows the fermions to transform as 3 and $\bar{3}$ of SU(3) [the SU(4) adjoint breaks to $15 = 18 + 3 + \bar{3} + 1$]. Hence this model has eight quarks and eight leptons [i.e. SU(3) singlets].

The conclusion to be drawn from the examples above is that simple type III models can easily be constructed which contain many of the features of the standard model, though not all of them simultaneously. Unlike four-dimensional heterotic string models, which appear to favor large gauge groups (typically rank 22) and can readily incorporate large numbers of families of quarks and leptons, type III models in four dimensions can only barely contain the standard model gauge group and are almost certainly limited to four families or fewer. (This follows because a sector, such as W_1 in the last model, which gives massless fermions, must have eight Ramond fermions in both the left and the right moving sectors. This gives at most 128 states or, equivalently, 64 chiral fermions. In general, adding another sector with massless fermions will also project out half of these fermions.) In dimensions higher than four, type III models possess even fewer internal degrees of freedom. It is plausible, then, that some generalization of the simple type III models constructed from free world sheet fermions given here might contain the SU(3)⊗SU(2)⊗U(1) standard model with three or four families, and at the same time offer some explanation of why this gauge group, and indeed our four-dimensional space–time, appears to be special. Even if this does not prove to be the case, the type III models discussed here serve as a good testing ground for string ideas; in particular they appear to contradict the widely held belief that all four-dimensional fermionic string models can be interpreted as compactifications of ten-dimensional models on some manifold or (left–right symmetric) orbifold.

Finally, let us mention some possible generalizations of the constructions given here which may prove useful in building more realistic type III models. In general we require some representation of the super-Virasoro algebra which can be constructed consistent with superconformal and modular invariance. In the examples above we restricted ourselves to free, real fermion representations (precluding the possibility of chiral models). We can, however, generalize our construction to fermions with boundary conditions other than periodic and antiperiodic [3] or bosonic representations. For example, it is possible to replace a pair of world sheet fermions by a single bose field which by itself satisfies an $N=2$ world sheet supersymmetry [14]. Then the remaining fermions may be treated as before using (1) and the spin structure construction, but now the gauge symmetries appearing in the supercurrent have dimension 16, 14, 12, etc., depending on the number of bose fields introduced. This allows the appearance of, for example, the group G_2, but is unlikely to lead to more realistic type III models since the gauge symmetries are always reduced.

Taking this idea a step further, we can use bose fields to write a generalized form for (1),

$$S(z) = \sum_{\{k^2=3\}} B(\boldsymbol{k}):\mathrm{e}^{\mathrm{i}\boldsymbol{k}\cdot\boldsymbol{\phi}}: + \sum_{\{l^2=1\}} C^i(\boldsymbol{l}):\partial_z\phi_i \mathrm{e}^{\mathrm{i}\boldsymbol{l}\cdot\boldsymbol{\phi}}:, \tag{11}$$

where $B(\boldsymbol{k})$ and $C^i(\boldsymbol{l})$ are some coefficients, $\boldsymbol{\phi}$ are a set of nine bose fields, $\boldsymbol{k}$ and $\boldsymbol{l}$ are nine-dimensional vectors and the sums run over the set of $\boldsymbol{k}$ and $\boldsymbol{l}$ such that $\boldsymbol{k}^2=3$ and $\boldsymbol{l}^2=1$ (to ensure that S has conformal dimension $\frac{3}{2}$). The bosonized form of (1) is a special case of (11) for which all of the $B(\boldsymbol{k})$ and $C^i(\boldsymbol{l})$ vanish except for some with $\boldsymbol{k}$ and $\boldsymbol{l}$ possessing only integer components. In the general case, demanding that $S(z)$ satisfies the super-Virasoro algebra (2) leads to a system of quadratic constraints for the B and C coefficients. A general solution to these constraints may prove useful for type III string model building.

After the completion of this work, we learned that

similar ideas have been considered by L. Dolan and P. Goddard.

We thank our colleagues at Cornell for discussions. This work is supported in part by the National Science Foundation.

References

[1] J.H. Schwarz, Phys. Rep. 89 (1982) 223;
M.B. Green, Surv. High En. Phys. 3 (1983) 127;
M.B. Green, J.H. Schwarz and E. Witten, Superstring theory (Cambridge U.P., Cambridge, 1986).

[2] M.B. Green and J.H. Schwarz, Nucl. Phys. B 255 (1985) 92.

[3] H. Kawai, D.C. Lewellen and S.-H.H. Tye, Phys. Rev. D 34 (1986) 3794; Phys. Rev. Lett. 57 (1986) 1832; Nucl. Phys. B 288 (1987) 1.

[4] D.J. Gross, J.A. Harvey, E. Martinec and R. Rohm, Phys. Rev. Lett. 54 (1985) 502; Nucl. Phys. B 256 (1985) 253; B 267 (1986) 75.

[5] K.S. Narain, Phys. Lett. B 169 (1986) 41;
K.S. Narain, M.H. Sarmadi and E. Witten, Nucl. Phys. B 279 (1987) 369;
see also P. Ginsparg, Comment on toroidal compactification of heterotic superstrings, Harvard preprint HUTP-86/A053 (1986).

[6] P. Candelas, G. Horowitz, A. Strominger and E. Witten, Nucl. Phys. B 258 (1985) 46;
see also L. Dixon, J. Harvey, C. Vafa and E. Witten, Nucl. Phys. B 261 (1985) 651; B 274 (1986) 285.

[7] D. Friedan, Z. Qiu and S. Shenker, Phys. Lett. B 151 (1985) 37;
see also D. Friedan and S. Shenker, Proc. Santa Fe Meeting of the APS Division of Particles and Fields (1984).

[8] P. Ramond, Phys. Rev. D 3 (1971) 2415;
A. Neveu and J.H. Schwarz, Nucl. Phys. B 31 (1971) 86.

[9] P. Goddard, W. Nahm and D. Olive, Phys. Lett. B 160 (1985) 111;
P. Goddard, A. Kent and D. Olive, Commun. Math. Phys. 103 (1986) 105.

[10] I. Antoniadis, C. Bachas, C. Kounnas and P. Windey, Phys. Lett. B 171 (1986) 51.

[11] F. Gliozzi, J. Scherk and D. Olive, Nucl. Phys. B 122 (1977) 253.

[12] N. Seiberg and E. Witten, Nucl. Phys. B 276 (1986) 272;
A. Parkes, Phys. Lett. B 184 (1987) 19.

[13] S. Mandelstam, Phys. Rep. 13 (1974) 259.

[14] G. Waterson, Phys. Lett. B 171 (1986) 77;
D. Friedan and S. Shenker, Supersymmetric critical phenomena and the two dimensional gaussian model, University of Chicago preprint EFI 86-17 (1986).

[15] S. Helgason, Differential geometry, Lie groups and symmetric spaces (Adademic Press, New York, 1978) p. 518.

Nuclear Physics B294 (1987) 136–156
North-Holland, Amsterdam

COVARIANT LATTICES, SUPERCONFORMAL INVARIANCE AND STRINGS

W. LERCHE, B.E.W. NILSSON and A.N. SCHELLEKENS

CERN, Geneva, Switzerland

Received 2 April 1987

We investigate the implications of various ways of realizing superconformal invariance on the world-sheet, in the framework of the bosonic covariant lattice approach to fermionic strings. We discuss in detail the role of a certain type of gauge interactions, and clarify their group theoretical structure. The results are relevant for lower dimensional heterotic and type II string theories. We show that (in our approach) there exists no four-dimensional type II theory containing the standard model.

1. Introduction

The covariant lattice approach, originating in the work of [1,2], and further developed in [3–6], has proved to be a very convenient tool for constructing various fermionic string theories in their "critical" (or, more properly, maximal) dimension and below. Like all consistent string theories, these theories are conformally invariant, modular invariant and superconformally invariant. The details of this construction are explained in [6]. Here we will remind the reader merely of the main features.

What our method has in common with other approaches [7–14], is that for the left- and right-moving sectors of the closed string one chooses independently one of the following two conformal field theories: either the bosonic one with 26 bosonic matter fields or the $N=1$ superconformal one with 15 bosonic matter fields (or equivalently, 10 bosons and 10 fermions). By assembling these in various ways one can construct the generic bosonic, heterotic [15] or type II [16] string theories. To obtain a theory in D space-time dimensions, one chooses D bosonic left- and right-moving fields as the space-time coordinates $X^{\mu}(z, \bar{z})$. The various methods differ in the way the remaining fields are treated. In the covariant lattice approach the remaining bosons, together with the bosonized spinor ghosts have their internal momenta quantized on an odd self-dual lorentzian lattice. Some of these correspond to D bosonized NSR fermions ψ^{μ}, while the remaining ones play the role of internal degrees of freedom. In general, it is *not* appropriate to regard them as arising from a space-time compactification process.

0550-3213/87/$03.50 © Elsevier Science Publishers B.V.
(North-Holland Physics Publishing Division)

The advantage of our approach becomes evident in discussing modular invariance: it is simply guaranteed by the self-duality of the lattice, provided [4] one assigns appropriate statistics to tensor and spinor representations of $SO(D)_{Lorentz}$. By the arguments of [17] we expect the theory to be not only one-loop, but *all-loop* modular invariant★ [19].

In this paper, we will focus on the third consistency condition, namely superconformal invariance. In [6] chiral heterotic strings in less than ten dimensions were constructed, and a consistency condition was obtained by requiring the absence of massive chiral states. This condition turned out to be equivalent to a condition derived in [11] from superconformal invariance. That condition is clearly not the most general one; for example, it is not satisfied by Narain's compactification [7].

The main purpose of this paper is to investigate generalizations of this condition. The results will be relevant for the right-moving (fermionic) sector of the heterotic string and for both sectors of the type II string. For most of this paper we will consider only one sector, which we will refer to as "right-moving". The lattice has the following structure:

$$\Gamma = \Gamma_L \times \Gamma_R . \tag{1.1}$$

Note that Γ_L and Γ_R are in general not self-dual lattices by themselves. In $D = 10 - 2n$ dimensions one has

$$\Gamma_R = \Gamma_{3n} \times \Gamma_{5-n,1} , \tag{1.2}$$

where $\Gamma_{5-n,1}$ is the lorentzian space-time lattice defined in [6]. In type II theories, Γ_L is of the same form as Γ_R; in heterotic theories, $\Gamma_L = \Gamma_{16+2n}$. The lattice Γ in (1.1) is odd lorentzian self-dual, and can be mapped on an even lorentzian self-dual one by mapping [4] $\Gamma_{5-n,1}$ into D_{8-n}. Both formulations have their specific advantages, and we will freely use either one depending on which is more convenient. To distinguish them we will use superscripts (e) or (o). In either case a vector on Γ_R can be decomposed as follows:

$$w^{(o)} = (u, v^{(o)})$$

or (1.3)

$$w^{(e)} = (u, v^{(e)}) ,$$

where $u \in \Gamma_{3n}$, while $v^{(o)}$ belongs to $\Gamma_{5-n,1}$ and $v^{(e)}$ to D_{8-n}. As explained in [6], in the even lattice formulation massless states correspond to certain length-squared two vectors on Γ_R; on Γ_L they correspond also to length-squared two vectors, and, in the case of heterotic strings, also to bosonic oscillator excitations. We remind the

★ Note also that self-duality of the lattice of a compactified bosonic string is sufficient for all-loop modular invariance [18].

reader that because of the physical state selection rule (explained in [3,5,6]), the conjugacy classes (o), (v), (s) and (c) of $\Gamma_{5-n,1}$ (or D_{8-n}) contain Lorentz vectors, scalars and spinors of both chiralities, respectively. This selection rule applies to the right-moving sector of the heterotic string and to both sectors of the type II string. In particular, massless vector bosons are Lorentz vectors in one sector and Lorentz scalars in the other*. The latter sector determines the gauge group. One has to distinguish two cases. If this sector is the bosonic string, the gauge group is the usual Frenkel-Kac group G_{FK} [20] generated by length-squared two vectors plus oscillators. If the sector in question is the NSR string, we have massless gauge bosons for

$$\boldsymbol{v}^{(e)} \in (\mathrm{v}) \text{ of } D_{8-n}, \qquad \boldsymbol{u}^2 = 1. \tag{1.4}$$

In this case, we will call the gauge group G_{NSR}. Thus, in general, in heterotic strings one can have gauge groups $G^{L}_{FK} \times G^{R}_{NSR}$; in type II strings one can have $G^{L}_{NSR} \times G^{R}_{NSR}$ (where the groups can be different). As we will see below, it turns out that G_{NSR} is not just determined by the structure of the lattice, but also on the way superconformal invariance is realized.

Gauge symmetries of this kind have already been discussed in the literature (in fermionic language) [11,21–23]. The purpose of this paper is to show how they appear in the bosonic lattice construction. Picture changing, which is built into our formulation from the very beginning, will be seen to play a crucial role in determining the gauge interactions. In sects. 2 and 3 we discuss general bosonic realizations of the supercurrent. In the rest of the paper we then turn to a somewhat simpler special case, namely a realization in terms of free fermions** [25,26].

2. Superconformal invariance and picture changing

World-sheet supersymmetry is generated by a supercurrent $T_F(z)$ which has to have the following properties

$$T_F(z)T_F(w) = \frac{M}{(z-w)^3} - \frac{2T(w)}{z-w} + \text{finite},$$

$$T(z)T_F(w) = \frac{\frac{3}{2}}{(z-w)^2}T_F(w) + \frac{1}{z-w}\,\partial_w T_F(w) + \text{finite}, \tag{2.1}$$

where $T(z)$ is the energy-momentum tensor and $M = 10$. In ten dimensions the

* In type II strings, Lorentz vectors can appear also as bispinors where the spinor indices come from different left- and right-moving sectors. We will return to this issue in section 6.

** While writing this paper we received [24] which arrives at conclusions similar to the ones obtained in sects. 4 and 6 below, although with different methods.

operator product algebra is satisfied by

$$T_F(z) = \psi^\mu \partial X_\mu(z), \tag{2.2}$$

where ψ^μ are the NSR fermions which also may be written in bosonized form [1]. Because of Lorentz invariance, in $10-2n$ dimensions the space-time part $T_F^{st}(z)$ of the supercurrent has the same form, but we are free to modify the remaining part $T_F^{int}(z)$. What we are looking for is the most general operator constructed out of $3n$ bosons associated with Γ_{3n}, which satisfies (2.1) with $M=2n$. The second of these equations implies that it must have conformal weight $\frac{3}{2}$. It must then be a linear combination of the form

$$T_F^{int}(z) = \sum_t A(t) e^{it\cdot X}(z) + i\sum_l B(l)\cdot \partial X e^{il\cdot X}(z),$$

$$t^2 = 3, \qquad l^2 = 1, \qquad l\cdot B(l) = 0, \tag{2.3}$$

where X^I denotes the $3n$ bosons, and where multiplication with appropriate cocycle operators is understood. The superconformal algebra (2.1) imposes certain conditions on A and B; however, there is no need here to write these equations down. In general, there are many choices for $A(t)$ and $B(l)$ that satisfy (2.1), and we will discuss some of them below.

The supercurrent plays an important role in determining interactions. The covariant lattice approach is based on conformal field theory, and it provides us with covariant vertex operators for all physical states. In the "canonical" ghost charge vacuum, the emission vertices for the Neveu-Schwarz sector have ghost charge -1 and those for the Ramond sector ghost charge $-\frac{1}{2}$. In order to calculate scattering amplitudes one has to combine the vertices in such a way as to cancel the background ghost charge [1]. To calculate arbitrary amplitudes we need thus also vertex operators in different ghost charge sectors. These can be obtained by applying the picture changing operator ($T_F = T_F^{st} + T_F^{int}$)

$$P_{+1} = e^{i\phi} T_F + \text{irrelevant terms}, \tag{2.4}$$

where ϕ is the bosonized superconformal ghost as defined in* refs. [1,6]. Picture changing is achieved as follows:

$$V_{q+1}(z) = :P_{+1} V_q(z):, \tag{2.5}$$

where V_q is a vertex operator with ghost charge q. For example, to calculate R-R-NS scattering at tree level one can consider the correlation function

* Our convention is to keep explicit factors of i in the exponentials in vertex operators.

$\langle V_{-1/2}V_{-1/2}V_{-1}\rangle$; for $(\mathrm{NS})^3$ couplings one considers $\langle V_{-1}V_{-1}V_0\rangle$. Both expressions have the correct ghost charge -2 for the sphere. (For type II strings these considerations apply of course to both sectors separately.) We will calculate such three-point functions in the next section. Notice that only in the second case does one need the picture changing operator.

One can only have a consistent string theory if both V_q and V_{q+1} are well-defined. In particular the picture changing operator should act on all vertices without introducing branch cuts. Consider a vertex operator $V_q(\boldsymbol{u}) = \exp(i\boldsymbol{u}\cdot\boldsymbol{X})\exp(iq\phi)$ for example with lattice momentum $\boldsymbol{u}$ on Γ_{3n}. The operator product of P_{+1} with V_q will have a branch cut unless

$$q-\boldsymbol{u}\cdot\boldsymbol{t}\in\mathbb{Z},$$

and

$$q-\boldsymbol{u}\cdot\boldsymbol{l}\in\mathbb{Z}, \tag{2.6}$$

where $\boldsymbol{t}$ and $\boldsymbol{l}$ are the vectors appearing in (2.3). Since odd ghost charges are part of the conjugacy class (v) of $\Gamma_{5-n,1}$, this implies that all vectors $w^{(\mathrm{o})} = (\boldsymbol{t},(\mathrm{v}))$ and $w^{(\mathrm{o})} = (\boldsymbol{l},(\mathrm{v}))$ should have integer dot-products with all other lattice vectors. Because the lattice is self-dual, this implies that they must lie on the lattice themselves. The corresponding even lattice must then contain all vectors $w^{(\mathrm{e})} = (\boldsymbol{l},\mathrm{v}))$ and $w^{(\mathrm{e})} = (\boldsymbol{t},(\mathrm{v}))$, for all $\boldsymbol{l}$ and $\boldsymbol{t}$ appearing in the supercurrent. This constrains the choice of the supercurrent with respect to the lattice (or vice versa). The shortest vectors in the two sets have length-squared 2 and 4 respectively. We will refer to such vectors, whose existence is enforced by the supercurrent, as "constraint vectors". From the selection rules [6] it follows that every length-squared 2 vector of the form $(\boldsymbol{l},(\mathrm{v}))$ implies the existence of a $\mathrm{G}_{\mathrm{NSR}}$ gauge boson. This fact will play an important role in our discussion. The space-time part $T_{\mathrm{F}}^{\mathrm{st}}\exp(i\phi)$ of P_{+1} (associated with $w^{(\mathrm{e})} = (\boldsymbol{u},\boldsymbol{r})$, $\boldsymbol{u}=0$, $\boldsymbol{r}^2=2$) belongs to the (0) conjugacy class of D_{8-n}. Thus the length-squared two vectors required by $T_{\mathrm{F}}^{\mathrm{st}}$ are always present*, and imply, in fact, the existence of gravitons.

The simplest way to satisfy (2.1) is to keep the form (2.2) even below ten dimensions. In four dimensions, after bosonization one obtains then an expression of the form (2.3) with $A=0$ and six non-vanishing coefficients $\boldsymbol{B}(\boldsymbol{l})$. The resulting theory will have at least six $\mathrm{G}_{\mathrm{NSR}}$ gauge bosons (of course there may be gauge bosons in addition to those required by the supercurrent). This situation arises in torus compactifications, including the ones discussed in [7] (here, the six gauge bosons belong to the $N=4$ supergravity multiplet; they arise due to the usual Kaluza-Klein mechanism).

* Therefore, the argumentation of [3] discussing modular invariance can always be applied.

3. Three-point interactions

In this section we will calculate three-point couplings in the right-moving sector for three (Lorentz) scalars and investigate their dependence on the supercurrent. (The scalars can of course acquire Lorentz indices from the left-moving sector to become G_{NSR} gauge bosons or space-time fermions.) In addition, we also calculate couplings between left- and right-moving sectors.

Consider first the three-scalar coupling. In the canonical ghost charge vacuum, the emission vertex for a state with lattice vector $(\boldsymbol{w}_L, \boldsymbol{w}_R)$, $\boldsymbol{w}_R = (\boldsymbol{u}, \boldsymbol{v})$, where $\boldsymbol{v}$ is a vector weight of $\Gamma_{5-n,1}$ (or D_{8-n}), is given by*

$$V(\boldsymbol{w}_L, \boldsymbol{w}_R, z, \bar{z}) = V^L(\boldsymbol{w}_L, z) V^R(\boldsymbol{w}_R, \bar{z}) e^{ik\cdot X}(z, \bar{z}),$$

$$V^R(\boldsymbol{w}_R, \bar{z}) = e^{i\boldsymbol{u}\cdot X} e^{-i\phi}(\bar{z}) \equiv V_{-1}(\boldsymbol{u}, \bar{z}). \tag{3.1}$$

The picture changed vertex operator $V_0(\boldsymbol{u}, \bar{z})$ is then obtained by evaluating

$$\lim_{\bar{w}\to\bar{z}} \left((T_F e^{i\phi})(\bar{w}) V(\boldsymbol{w}_L, \boldsymbol{w}_R, z, \bar{z}) - \text{singular terms} \right) \tag{3.2}$$

and extracting the factor that depends on the right lattice. In the special (massless) case $\boldsymbol{u}^2 = 1$ the result is

$$V_0(\boldsymbol{u}) = \sum_{\substack{\boldsymbol{t} \\ (\boldsymbol{t}+\boldsymbol{u})^2 = 2}} A(\boldsymbol{t}) e^{i(\boldsymbol{t}+\boldsymbol{u})\cdot X} + i\boldsymbol{B}(-\boldsymbol{u}) \cdot \partial X$$

$$- \sum_{\substack{\boldsymbol{l} \\ (\boldsymbol{l}+\boldsymbol{u})^2 = 2}} \boldsymbol{B}(\boldsymbol{l}) \cdot \boldsymbol{u}\, e^{i(\boldsymbol{l}+\boldsymbol{u})\cdot X} + \tfrac{1}{2} i (k_\mu \psi^\mu) e^{i\boldsymbol{u}\cdot X}. \tag{3.3}$$

The constraint $(\boldsymbol{l}+\boldsymbol{u})^2 = 2$ in the third term is a consequence of the identity $\boldsymbol{B}(\boldsymbol{l}) \cdot \boldsymbol{l} = 0$ which follows from (2.1). The vertex operator (3.3) can be used to calculate the coupling of a G_{NSR} gauge boson to any other state which is also a Lorentz scalar in the right sector. The correlation function in the right-moving sector to consider is

$$\langle V_{-1}(\boldsymbol{u}_1, \bar{z}_1) V_{-1}(\boldsymbol{u}_2, \bar{z}_2) V_0(\boldsymbol{u}_3, \bar{z}_3) \rangle \tag{3.4}$$

with $\boldsymbol{u}_3^2 = 1$, while $\boldsymbol{u}_1$ and $\boldsymbol{u}_2$ are unrestricted. To get the physical three-point function C_{u_1,u_2,u_3} one has to multiply (3.4) with the contribution of the reparametrization ghosts, the space-time factors $\exp(i\boldsymbol{k}\cdot \boldsymbol{X}(z,\bar{z}))$ and all remaining contributions from the left sector. The reparametrization ghosts yield a factor $|z_{12} z_{23} z_{13}|^2$,

* We use $\bar{z}$ for right-movers, z for left-movers.

where $z_{ij} = z_i - z_j$. The space-time exponentials also contribute a factor $|z_{12}|^{k_1k_2/2}|z_{23}|^{k_2k_3/2}|z_{13}|^{k_1k_3/2}$, where k_i is the space-time momentum corresponding to $V(\boldsymbol{u}_i)$. The rest of the left sector gives z_{ij} dependent factors which determine the Lorentz tensor structure T_{Lorentz} of the result, which is however not of interest to us here.

Putting all $\bar{z}$ factors together we get for the contribution of the first term in (3.3)

$$\sum A(\boldsymbol{t})(\bar{z}_{12})^{k_1\cdot k_2/4+\boldsymbol{u}_1\cdot\boldsymbol{u}_2}(\bar{z}_{13})^{1+k_1\cdot k_3/4+\boldsymbol{u}_1\cdot(\boldsymbol{t}+\boldsymbol{u}_3)}(\bar{z}_{23})^{1+k_2\cdot k_3/4+\boldsymbol{u}_2\cdot(\boldsymbol{t}+\boldsymbol{u}_3)}. \quad (3.5)$$

If the momenta satisfy the on-shell conditions

$$\tfrac{1}{4}k_1^2 = 1 - \boldsymbol{u}_1^2,$$

$$\tfrac{1}{4}k_2^2 = 1 - \boldsymbol{u}_2^2,$$

$$k_3^2 = 0, \quad (3.6)$$

then one can reduce the expression to a constant by using lattice and space-time momentum conservation and the condition $(\boldsymbol{t}+\boldsymbol{u}_3)^2 = 2$. (The complete 3-point function has to be independent of $\bar{z}_i$ (and z_i) because of Möbius invariance.) The result is

$$C_{u_1,u_2,u_3} = A(\boldsymbol{t})T_{\text{Lorentz}}, \quad (3.7)$$

with $\boldsymbol{t} = -\boldsymbol{u}_1 - \boldsymbol{u}_2 - \boldsymbol{u}_3$.

In a similar way we find for the contribution of the second term in (3.3):

$$-(\bar{z}_{12})^{\frac{1}{2}(\boldsymbol{u}_1+\boldsymbol{u}_2)^2-1}\Big(\boldsymbol{u}_1\cdot \boldsymbol{B}(-\boldsymbol{u}_3)(\bar{z}_{13})^{\frac{1}{2}(\boldsymbol{u}_2^2-\boldsymbol{u}_1^2)}(\bar{z}_{23})^{\frac{1}{2}(\boldsymbol{u}_1^2-\boldsymbol{u}_2^2)+1}$$

$$+\boldsymbol{u}_2\cdot \boldsymbol{B}(-\boldsymbol{u}_3)(\bar{z}_{13})^{\frac{1}{2}(\boldsymbol{u}_2^2-\boldsymbol{u}_1^2)+1}(\bar{z}_{23})^{\frac{1}{2}(\boldsymbol{u}_1^2-\boldsymbol{u}_2^2)}\Big). \quad (3.8)$$

This is a constant if and only if $\boldsymbol{u}_1 = -\boldsymbol{u}_2$; the result is then

$$C_{u_1,-u_1,u_3} = \boldsymbol{u}_1\cdot \boldsymbol{B}(-\boldsymbol{u}_3)\cdot T_{\text{Lorentz}}. \quad (3.9)$$

Analogously one determines the contribution of the third term in (3.3):

$$-\sum_{\substack{\boldsymbol{l}\\(\boldsymbol{l}+\boldsymbol{u}_3)^2=2}} \boldsymbol{B}(\boldsymbol{l})\cdot\boldsymbol{u}_3(\bar{z}_{12})^{k_1\cdot k_2/4+\boldsymbol{u}_1\cdot\boldsymbol{u}_2}(\bar{z}_{13})^{1+k_1\cdot k_3/4+\boldsymbol{u}_1\cdot(\boldsymbol{l}+\boldsymbol{u}_3)}(\bar{z}_{23})^{1+k_2\cdot k_3/4+\boldsymbol{u}_2\cdot(\boldsymbol{l}+\boldsymbol{u}_3)}. \quad (3.10)$$

This gives also a constant ($\boldsymbol{l}=-\boldsymbol{u}_1-\boldsymbol{u}_2-\boldsymbol{u}_3$):

$$C_{u_1, u_2, u_3} = -\boldsymbol{B}(\boldsymbol{l})\cdot\boldsymbol{u}_3\cdot T_{\text{Lorentz}}\,. \tag{3.11}$$

Finally, the momentum-dependent last term in (3.3) gives no contribution in this case, where all states are Lorentz scalars in the right-moving sector.

We thus see that certain couplings (in particular, three-point $\mathrm{G}_{\mathrm{NSR}}$ gauge boson couplings) depend on A and $\boldsymbol{B}$, i.e., on the way superconformal invariance is realized. (We will see below how A and $\boldsymbol{B}$ are related to the structure constants of $\mathrm{G}_{\mathrm{NSR}}$.)

From the above it is clear that the group $\mathrm{G}_{\mathrm{NSR}}$ simply acts as a subgroup of the Frenkel-Kac group $\mathrm{G}_{\mathrm{FK}}^{3n}$ of the Γ_{3n} part of the right lattice (1.2). The group $\mathrm{G}_{\mathrm{FK}}^{3n}$ is generated by vertex operators ∂X^I $(I=1,\dots,3n)$ and $\exp(i\boldsymbol{r}\cdot\boldsymbol{X})$, where $\boldsymbol{r}\in\Gamma_{3n}$ with $\boldsymbol{r}^2=2$. Notice that the lattice vector $\boldsymbol{w}_{\mathrm{R}}=(\boldsymbol{u},\boldsymbol{v})$, $\boldsymbol{v}=0$, $\boldsymbol{u}=\boldsymbol{r}$ does not correspond to a physical state because of the projection rules of [6]. In particular, one does not have gauge bosons of the complete group $\mathrm{G}_{\mathrm{FK}}^{3n}$. Rather, the gauge bosons we do get correspond to vectors of the form $\boldsymbol{v}^2=1$, $\boldsymbol{u}^2=1$. After picture changing, their vertex operators are linear combinations of generators of $\mathrm{G}_{\mathrm{FK}}^{3n}$, as one can see from (3.3) (the last term does not contribute here). The correlation function (3.4) can thus be interpreted as the matrix element of a linear combination of $\mathrm{G}_{\mathrm{FK}}^{3n}$ generators between lattice vectors. The gauge group $\mathrm{G}_{\mathrm{NSR}}$, generated by the picture changed vertex operators $V_0(\boldsymbol{u})$, must thus be a subalgebra of the "unphysical" Frenkel-Kac group $\mathrm{G}_{\mathrm{FK}}^{3n}$. The representation content of a given level of the string with respect to $\mathrm{G}_{\mathrm{NSR}}$ is obtained by simply decomposing all representations of G_{FK}^{3n} at that level. The embedding of $\mathrm{G}_{\mathrm{NSR}}$ in $\mathrm{G}_{\mathrm{FK}}^{3n}$ depends of course on the coefficients A and $\boldsymbol{B}$ in (3.3).

The problem of determining the embedding splits into two parts. Because (in the even formulation) gauge bosons originate from vectors $\boldsymbol{v}^2=1$, $\boldsymbol{u}^2=1$, which are roots of Γ_{R}, the space-time lattice D_{8-n} extends to D_{8-n+m} if there are $2m$ gauge bosons [6]. Thus one has the structure

$$\Gamma_{\mathrm{R}}=\Gamma_{3n-m}\times\mathrm{D}_{8-n+m}\,,$$

$$\mathrm{G}_{\mathrm{FK}}^{3n}=\mathrm{G}_{\mathrm{FK}}^{3n-m}\times\mathrm{D}_m\,, \tag{3.12}$$

where $\mathrm{G}_{\mathrm{FK}}^{3n-m}$ is the Frenkel-Kac group of Γ_{3n-m}. It is easy to determine the embedding of $\mathrm{G}_{\mathrm{NSR}}$ in D_m: clearly, the adjoint representation of $\mathrm{G}_{\mathrm{NSR}}$ is embedded in the vector representation of D_m. (Notice that $\mathrm{G}_{\mathrm{NSR}}$ is a conformal subalgebra* of D_m [27,28].) More specific information is needed to discuss the embedding of $\mathrm{G}_{\mathrm{NSR}}$ in $\mathrm{G}_{\mathrm{FK}}^{3n-m}$. Before we examine a special case of supercurrents in the next section, let us briefly consider further three-point couplings.

* This means that $\mathrm{G}_{\mathrm{NSR}}$ has the same Sugawara tensor as D_m.

To what particles the couplings (3.7), (3.9) and (3.11) precisely correspond, depends on the left-moving part which determines the Lorentz structure. For instance, by choosing for two of the three vertex operators a fermion emission vertex as the left factor, one can use the expressions derived above to describe the fermion-fermion-G_{NSR} gauge boson coupling. Hence these couplings have a dependence on the supercurrent, just as those of the G_{NSR} gauge bosons. Massless fermions of this type have length-squared 2 vectors on the right lattice, just as the gauge bosons. But while for the gauge bosons these vectors are roots of D_{8-n+m} (i.e. vectors of D_m and of D_{8-n}), the lattice vectors that describe fermions may also have components in Γ_{3n-m}. This is what might in principle allow complex fermion representations. The roots of D_{8-n+m} lead only to adjoints of G_{NSR}, as explained above.

In general, there exist also states which owe their Lorentz index to the *right*-moving sector. Let us consider, for instance, for type II strings the coupling of a G^R_{NSR} gauge boson to two fermions which are Lorentz scalars in the left-moving sector, and which get their spinor index from the right-moving sector (hence $V^L(z) = V_{-1}(\boldsymbol{u}_i, z)$, $i = 2, 3$ with $\boldsymbol{u}_i^2 = 1$). After application of the picture changing operator to one fermion, the relevant correlation function in the left-moving sector is

$$\left\langle \left(\psi_\mu e^{-i\phi}\right)(z_1) V_{-1}(\boldsymbol{u}_2, z_2) V_0(\boldsymbol{u}_3, z_3)\right\rangle , \tag{3.13}$$

which leads to a coupling proportional to the momentum k_3^μ of that fermion. It arises due to the momentum dependent last term in (3.3), and carries no dependence on the supercurrent. Clearly, this "mixed" coupling between different sectors describes a derivative and hence non-minimal interaction. From a low energy point of view these fermions are thus singlets of G^R_{NSR}. The same result is obtained also for heterotic strings. This strengthens the claim made in [6] that in heterotic strings fermions can only be chiral with respect to the gauge symmetries originating from the left sector. The three-point coupling between G^R_{NSR} and G^L_{NSR} gauge bosons turns out to be zero. This proves the assertion made in sect. 1 that the gauge group is always a direct product of left- and right-moving parts.

4. The supercurrent in terms of free fermions

In this section we discuss special solutions to (2.1) constructed out of free fermions. Here we make use of the fact that the most general form of the supercurrent in terms of free fermions is already known [25, 26, 23]. According to these authors, the most general expression is

$$T_F^{\text{int}}(z) = \tfrac{1}{6} f_{ABC} \psi^A \psi^B \psi^C , \tag{4.1}$$

where the fermion fields satisfy

$$\psi^A(z)\psi^B(w) \sim \frac{1}{z-w}\delta^{AB} \tag{4.2}$$

and $f_{\rm ABC}$ are structure constants of some semi-simple $N = 6n$ dimensional (where $D = 10 - 2n$) group $G_{\rm SC}$, normalized as follows:

$$f_{ABC} f_{DBC} = 2\delta_{AD}. \tag{4.3}$$

Accordingly, there are N free fermions.

Of course (4.1) can always be cast into the bosonic form (2.3) by replacing $6n$ fermions by $3n$ bosons, using

$$\psi_{I1} = \frac{1}{\sqrt{2}} (\mathrm{e}^{ie_I X} + \mathrm{e}^{-ie_I X}),$$

$$\psi_{I2} = \frac{1}{i\sqrt{2}} (\mathrm{e}^{ie_I X} - \mathrm{e}^{-ie_I X}). \tag{4.4}$$

where $\boldsymbol{e}_I$ is a set of $3n$ orthonormal vectors. It is then clear that the role of the structure constants of $G_{\rm SC}$ is played by A and $\boldsymbol{B}$ in (2.3).

Notice that there is still a lot of freedom in this construction. First of all one may change the choice of the basis $\boldsymbol{e}_I$ with respect to the lattice by SO($3n$) rotations. Secondly, one can rotate the ψ's by SO($6n$) rotations respecting (4.2). Because $G_{\rm SC}$ rotations do not affect the structure constants, the true degrees of freedom are parametrized by the coset space SO($6n$)/$G_{\rm SC}$. Both kinds of rotations are restricted by the lattice. By rotating the basis $\boldsymbol{e}_I$, one rotates the constraint vectors. Clearly "small" SO($3n$) rotations will therefore not be allowed, but some discrete ones may be acceptable. The number of constraint vectors is determined by the number of non-vanishing structure constants f_{ABC}. In general, SO($6n$)/$G_{\rm SC}$ rotations change that number.

Despite all this freedom, it is important to realize that one does not obtain the most general bosonic supercurrent in this way. For example

$$T_{\rm F}^{\rm int}(z) = \sqrt{\tfrac{1}{3}} \sum_{I=1}^{3n} (\mathrm{e}^{i\sqrt{3}X^I} + \mathrm{e}^{-i\sqrt{3}X^I}) \tag{4.5}$$

is always a solution to (2.1), but e.g. for $n = 1$ there is no choice of basis for which it is of the form (4.1).

Consider now a lattice that produces exactly $2m$ $G_{\rm NSR}$ gauge bosons. Then we know that the space-time lattie D_{8-n} is embedded in D_{8-n+m}; furthermore D_{8-n+m} is the largest D-type algebra of which D_{8-n} is a subalgebra on the lattice. For the first m orthonormal vectors $\boldsymbol{e}^I$ we choose m orthonormal vector weights of D_m. We split the $6n$ fermion indices $A, B, C, \ldots$ into sets labeled $a, b, c, \ldots$ ($a = 1, \ldots, 2m$) and $i, j, k, \ldots$ ($i = 2m+1, \ldots, 6n$), so that the ψ^a's belong to D_m. Then the supercurrent takes the form

$$T_{\mathrm{F}}^{\mathrm{int}}(z)=\tfrac{1}{6}\left(f_{abc}\psi^a\psi^b\psi^c+3f_{abi}\psi^a\psi^b\psi^i+3f_{aij}\psi^a\psi^i\psi^j+f_{ijk}\psi^i\psi^j\psi^k\right). \tag{4.6}$$

The resulting constraint vectors fall into two classes: terms which are the product of three fermions which bosonize to different X_I's require length-squared 4 constraint vectors; if two fermions in one term bosonize to the same X_I's, the bosonic form of that term is $\partial X_I \exp(i\boldsymbol{l}\cdot \boldsymbol{X})$, which yields a length-squared 2 constraint. Decomposing vectors according to

$$\Gamma_{\mathrm{R}}=\Gamma_{3n-m}\times \mathrm{D}_m\times \mathrm{D}_{8-n},$$

we find then that the four types of terms require the existence of the following kinds of constraint vectors

(a) f_{abc} $(0,(\mathrm{v}),(\mathrm{v}))$,
(b) f_{abi} $(\boldsymbol{l},(0),(\mathrm{v}))$,
(c) f_{aij} $(\boldsymbol{r},(\mathrm{v}),(\mathrm{v}))$,
(d) f_{ijk} $(\boldsymbol{t},(0),(\mathrm{v}))$ or $(\boldsymbol{l},(0),(v))$,

where $\boldsymbol{l}^2=1$, $\boldsymbol{r}^2=2$ and $\boldsymbol{t}^2=3$. Constraints of type (a) are automatically satisfied, because the conjugacy class (v) of the D_m lattice contains all possible length 1 and 3 vectors that can appear. Constraint vectors of type (b) are not allowed, since they would enlarge D_{8-n+m} to a larger algebra. Therefore $f_{abi}=0$, and hence f_{abc} are the structure constants of some subalgebra H of $\mathrm{G_{SC}}$. For the same reason the second constraint in case (d) is not allowed. Suppose now that indices i and j exist so that $f_{ija}\neq 0$ and $f_{ijk}\neq 0$ for some a and k. Then the difference of the constraint vectors implied by these two non-vanishing structure constants is of the form $(\boldsymbol{l},(\mathrm{v}),(0))$, which is also not allowed. Therefore for every i and j either f_{ija} or f_{ijk} (or both) must vanish. Only the following possibilities are consistent with this:

(A) H is trivial; $\mathrm{G_{SC}}/\mathrm{H}$ is a group manifold,
(B) $\mathrm{H}=\mathrm{G_{SC}}$; $\mathrm{G_{SC}}/\mathrm{H}$ is trivial,
(C) H is a factor of $\mathrm{G_{SC}}$; $\mathrm{G_{SC}}/\mathrm{H}$ is a group manifold,
(D) $\mathrm{G_{SC}}/\mathrm{H}$ is a symmetric space,
(E) $\mathrm{G_{SC}}/\mathrm{H}$ is the product of a symmetric space and a group manifold.

Case (A) is relevant for chiral strings, for which at least one sector has to be of this type [6]. Case (B) is the one studied in [21, 23]; the compactified theories à la Narain [7] belong to class (D), with a symmetric space $(\mathrm{SU}(2)/\mathrm{U}(1))^6$.

The group H is precisely the gauge group $\mathrm{G_{NSR}}$ of the previous section. To see this more clearly, write the vertex operators (3.1) as $\mathrm{e}^{-i\phi}\psi^a$. Picture changing with (4.6) produces then the fermionic form of (3.3):

$$V_0(\boldsymbol{e}_a,z)=f_{abc}\psi^b\psi^c+f_{aij}\psi^i\psi^j+\tfrac{1}{2}i\left(k_\mu\psi^\mu\right)\psi^a. \tag{4.7}$$

In this fermionic formulation, the three-point $\mathrm{G_{NSR}}$ gauge boson couplings (3.7),

(3.9) and (3.11) are then proportional to the H structure constants f_{abc}. G_{NSR} gauge boson-fermion-fermion couplings may depend on both f_{abc} and f_{aij}.

The first term in (4.7) generates a subalgebra of SO($2m$) isomorphic to H. The second term acts on the Γ_{3n-m} part of the lattice as a subalgebra of the tangent space group associated to G_{SC}/H generated by $\psi^{[i}\psi^{j]}$. (This tangent space group need not be a Frenkel-Kac symmetry of the lattice; the constraint vectors listed under (c) above only require that H be non-trivially realized.) Notice that the presence of vectors $(\boldsymbol{r},(v),(v))$ implies the presence of vectors $(\boldsymbol{r},0,0)$, roots of Γ_{3n-m}. Here we see once more that H acts on Γ_{3n-m} as a subalgebra of G_{FK}^{3n-m}. Because G_{SC}/H is a symmetric space, the embedding of H in the tangent space group (and hence in G_{FK}^{3n-m}) is conformal*. As observed in sect. 3, $H \subset SO(2m)$ is also conformal. (The diagonal embedding $H \subset SO(2m) \times G^{3n-m}$ however is not conformal.) The third term in (4.7) corresponds to generators which mix the internal with the space-time part of D_{8-n+m}.

In $D = 10 - 2n$ dimensions, the possibilities for G_{SC} are limited to $SU(2)^{2n}$ [21] plus $SU(4) \times SO(3)$ and $SO(5) \times SU(3)$ if $D = 4$ $(n = 3)$ [23] (there are more possibilities if $n > 3$). From the above it is clear that the only possible gauge groups $H \equiv G_{NSR}$ in four dimensions are these groups themselves, or their symmetric subgroups. (In the next section we will show that groups of type (E), i.e. a product of a factor of G_{SC} and a symmetric subgroup of the other factor, are possible only if $G_{SC} = SU(2)^6$.)

For example, in four dimensions $(\Gamma_R = \Gamma_{9-m} \times D_{5+m})$ one can obtain a string theory with gauge group $H = SU(2) \times U(1) \times SO(5)$ as a symmetric subgroup of $G_{SC} = SU(3) \times SO(5)$. There are 14 gauge bosons $(m = 7)$, so that the space-time lattice D_5 must extend to D_{12}. The remaining part Γ_{3n-m} has dimension 2. The SO(5) factor occupies a D_5 subalgebra of D_{12}, and does not act on the tangent space. It does not require further discussion. The $SU(2) \times U(1)$ factor is embedded in the remaining D_2 subalgebra of D_{12}, and acts non-trivially on $\Gamma_{3n-m} = \Gamma_2$. To discuss this example in detail we need the SU(3) structure constants, normalized as in (4.3). Apart from this unusual normalization, we take the standard choice

$$
\begin{aligned}
&f_{123} = \sqrt{\tfrac{2}{3}}\,,\\
&f_{147} = f_{246} = f_{257} = f_{345} = -f_{156} = -f_{367} = 1/\sqrt{6}\,,\\
&f_{458} = f_{678} = 1/\sqrt{2}\,.
\end{aligned}
\tag{4.8}
$$

The $SU(2) \times U(1)$ subalgebra is generated by T^1, T^2, T^3 and T^8. We denote the two coordinates in the D_2 subalgebra of D_{12} as X_1 and X_2, and the two coordinates in

* One can check this explicitly from the tables of [27] and [28]. It also follows from the results of [29] as observed in [28].

Γ_{3n-m} as X_3 and X_4. We can then define the free fermions as follows

$$\psi_1=\sqrt{\tfrac{1}{2}}\,(e^{iX_1}+e^{-iX_1})C_1 \qquad \psi_2=(1/i\sqrt{2})(e^{iX_1}-e^{-iX_1})C_1,$$

$$\psi_3=\sqrt{\tfrac{1}{2}}\,(e^{iX_2}+e^{-iX_2})C_2, \qquad \psi_8=(1/i\sqrt{2})(e^{iX_2}-e^{-iX_2})C_2,$$

$$\psi_4=\sqrt{\tfrac{1}{2}}\,(e^{iX_3}+e^{-iX_3})C_3, \qquad \psi_5=(1/i\sqrt{2})(e^{iX_3}-e^{-iX_3})C_3,$$

$$\psi_6=\sqrt{\tfrac{1}{2}}\,(e^{iX_4}+e^{-iX_4})C_4, \qquad \psi_7=(1/i\sqrt{2})(e^{iX_4}-e^{-iX_4})C_4, \tag{4.9}$$

where C_i is a cocycle factor. Using (4.8) and (4.9) one construct the bosonic form of the supercurrent and the SU(2) × U(1) generators. For the latter we find

$$T^3=-\sqrt{\tfrac{2}{3}}\,\partial X_1-\sqrt{\tfrac{1}{6}}\,\partial X_3+\sqrt{\tfrac{1}{6}}\,\partial X_4,$$

$$T^8=-\sqrt{\tfrac{1}{2}}\,\partial X_4-\sqrt{\tfrac{1}{2}}\,\partial X_3,$$

$$T^1=-\tfrac{1}{2}i\sqrt{\tfrac{2}{3}}\,(e^{i(X_1+X_2)}-e^{-i(X_1+X_2)}+e^{i(X_1-X_2)}-e^{-i(X_1-X_2)})C_1C_2$$
$$+\tfrac{1}{2}i\sqrt{\tfrac{2}{3}}\,(e^{i(X_3-X_4)}-e^{-i(X_3-X_4)})C_3C_4,$$

$$T^2=-\tfrac{1}{2}\sqrt{\tfrac{2}{3}}\,(e^{i(X_1+X_2)}+e^{-i(X_1+X_2)}+e^{i(X_1-X_2)}+e^{-i(X_1-X_2)})C_1C_2$$
$$+\tfrac{1}{2}\sqrt{\tfrac{2}{3}}\,(e^{i(X_3-X_4)}+e^{-i(X_3-X_4)})C_3C_4. \tag{4.10}$$

It should be clear that this is certainly not the most general way to realize SU(2) × U(1), because of the freedom to change bases discussed earlier in this section. It is however the most efficient way, since it requires a Frenkel-Kac group no larger than SU(2) × U(1), generated by ∂X_3 and ∂X_4 and the roots $\boldsymbol{e}_3-\boldsymbol{e}_4$ and $-\boldsymbol{e}_3+\boldsymbol{e}_4$. Notice in particular that the additional SO(4) roots $\pm(\boldsymbol{e}_3+\boldsymbol{e}_4)$ are not required to be present. It is not difficult to construct lattices which allow this structure, and we will indeed find an example in sect. 6.

5. Constraint vectors

In this section we study the constraint vectors due to non-vanishing structure constants f_{ijk} for a fixed number $(2m)$ of gauge bosons. All these vectors must be of the form $(\boldsymbol{t},(\mathrm{v}))$ on the lattice $\Gamma_{\mathrm{R}}=\Gamma_{3n-m}\times \mathrm{D}_{8-n+m}$, with $\boldsymbol{t}^2=3$. Length one vectors $\boldsymbol{t}$ are not allowed (to avoid enlargement of D_{8-n+m}), nor should any be generated by adding an odd number of constraint vectors (by adding an even number, one gets vectors of the form $(\boldsymbol{r},(\mathrm{o}))$, $\boldsymbol{r}^2$ even).

The set of constraint vectors can be obtained by associating the $6n-2m$ generators of G_{SC} which are not in H with the $6n-2m$ free fermions constructed out of the $3n-m$ bosons whose momenta lie on Γ_{3n-m}. One may exploit $SO(6n)/G_{SC}$ rotations to choose a basis for the Lie algebra which minimizes the number of structure constants, and hence the number of constraint vectors. In an arbitrary basis one gets all possible constraint vectors, which can all be satisfied if and only if the entire lattice Γ_R is D_{8+2n}. But then one automatically obtains the maximal number (i.e. $6n$) of gauge bosons.

The $6n-2m$ free fermions can be grouped into $3n-m$ pairs which are related to exponentials of the same Bose field, as in (4.5). Two fermions of the same pair cannot be contracted with the same structure constant as this would give rise to a constraint vector $(\boldsymbol{l},(\mathrm{v}))$, $\boldsymbol{l}^2=1$. Therefore we have to group the generators of G_{SC}/H into commuting pairs, each of which can be represented by a fermion pair. If this is not possible, the constraint cannot be satisfied, and it is impossible to find a covariant lattice theory with the gauge symmetry H. We will call two generators of G_{SC} "associated" if they are represented by one fermion pair.

A term $f_{ijk}\psi_i\psi_j\psi_k$ in the supercurrent produces constraint vectors containing $(\dots \pm 1, \dots \pm 1, \dots \pm 1 \dots)$, with all eight combinations of signs. In some cases there may be a cancellation of two terms in the supercurrent, so that some of these constraint vectors do not appear. This is only possible if there is another structure constant f_{mnk}, with m, n associated to i, j, and $|f_{ijk}| = |f_{mnk}|$ (the example at the end of sect. 4 has this property). If there is no such cancellation, the entries of all lattice vectors in the 3-dimensional subspace of i, j and k must be integer or half-integer in order to have integer dotproducts with all constraint vectors, so that the sublattice is of the form $(D_1)^3$. The relevant part of the constraint vectors is then most conveniently written as $((\mathrm{v}),(\mathrm{v}),(\mathrm{v}))$.

From now on we will specialize to four dimensions. Consider first coset spaces G_{SC}/H of type (C), as defined in sect. 4. We must thus satisfy the constraint vectors of a simple factor H′ of dimension $2k$ of G_{SC} on a k-dimensional lattice. Then we must be able to group all generators of H′ into commuting pairs. This is possible for D_n, n even (the roots $\boldsymbol{e}_i+\boldsymbol{e}_j$ and $\boldsymbol{e}_i-\boldsymbol{e}_j$ may be associated) but not for other simple Lie algebras. In particular it cannot be done for SO(5), SU(3), SU(4) or SO(3). Thus a solution of type (C) is only possible for $G_{SC}=SU(2)^6$, where one can always associate two SU(2)'s with each other. More about this below.

Next we consider cosets of type (E), which are the product of a symmetric space and a group manifold. We find again that this is only possible for $G_{SC}=SU(2)^6$. This can be seen as follows: In all other cases there exists always at least one structure constant f_{ijk} of the group manifold with two unpaired indices i and j. The corresponding group generators have to be associated with two coset generators λ_i and λ_j. Now we make use of the fact that if G/H is a symmetric coset space, then the action of H on the tangent space is real irreducible. In particular this means that there does not exist a basis choice for which the coset generators split into two mutually commuting sets. It is easy to see that therefore at least one of the following three statements must be true:

(i) $[\lambda_i, \lambda_j] \neq 0$,

(ii) there exists a coset generator λ_m so that $[\lambda_m, \lambda_i] \neq 0$ and $[\lambda_m, \lambda_j] \neq 0$,

(iii) there exist coset generators λ_m and λ_n so that $[\lambda_m, \lambda_i] \neq 0$, $[\lambda_n, \lambda_j] \neq 0$ and $[\lambda_m, \lambda_n] \neq 0$.

These commutators lead to constraint vectors of the form given under (c) in sect. 4. Adding all the constraint vectors implied by the non-vanishing commutators listed above to the constraint vector implied by f_{ijk} produces in each of the cases (i), (ii) and (iii) an unwanted additional gauge boson.

Next we turn to case (A), with trivial gauge group. In chiral string theories [6] at least one of the two sectors *has* to be of this kind (because non-trivial H implies enlargement of D_5 to D_{5+m}, which in turn implies that all D_5 fermions appear in non-chiral pairs). Again the only possibility appears to be $G_{SC} = SU(2)^6$, although the argument is not quite as rigorous as the previous ones. In this case one has to find a way to "overlap" the constraint vectors of SU(4) and SO(3) or SO(5) and SU(3). Separately, these four groups need at least 8, 3, 7 and 7 dimensions respectively to avoid roots of Γ_R among the constraint vectors. (e.g. in SU(3) one can only find one commuting pair of generators; the remaining six have to be related to different bosons). By inspection of the simplest choices of structure constants we find that it is not possible to overlap the constraint vectors without

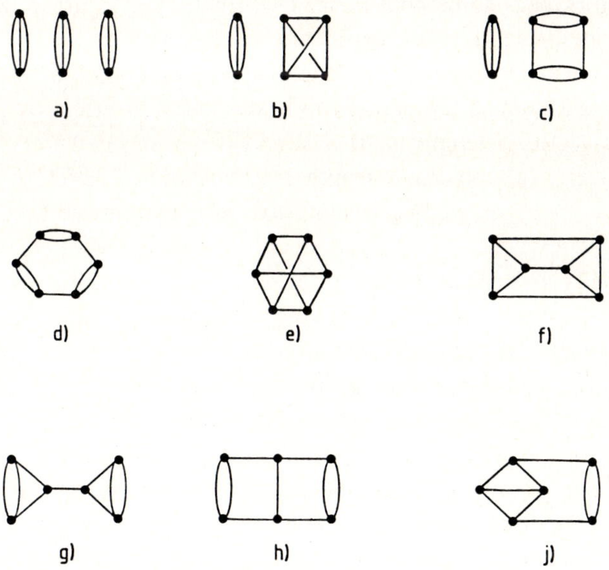

Fig. 1. Graphical representation of the possibilities for length-squared four constraint vectors in the case $G_{SC} = (SU(2))^6$.

producing roots of Γ_R. Choosing more complicated sets of structure constants, leading to more constraint vectors, makes it even harder to overlap the constraint vectors.

If $G_{SC} = SU(2)^6$, the minimal set of structure constants consists of those of the six SU(2)'s, but we still have the freedom to overlap the constraint vectors in different ways. The various possibilities can be represented graphically by associating a 3-point vertex to each of the six structure constants, and connecting them with 9 lines corresponding to the 9 fermion pairs (ψ_{I1}, ψ_{I2}). From these diagrams one can read off which 3 coordinates are linked to each of the six SU(2) factors, and hence which constraint vectors are required. A self-contraction leads to a length 1 vector associated with (v) of D_5, and is therefore not allowed. The graphs without self-contractions are shown in fig. 1. The simplest set of constraint vectors is the one corresponding to (a). It requires the existence of the following vectors $\boldsymbol{t}$ on Γ_9 (associated with (v) of D_5)

$$((\mathrm{v}),(\mathrm{v}),(\mathrm{v}),0,0,0,0,0,0),$$

$$(0,0,0,(\mathrm{v}),(\mathrm{v}),(\mathrm{v}),0,0,0),$$

$$(0,0,0,0,0,0,(\mathrm{v}),(\mathrm{v}),(\mathrm{v})). \tag{5.1}$$

This is equivalent to the "triplet" constraint used in [11] precisely as discussed in [6]. Several other graphs (namely (c), (d), (e) and (h)) lead to the same three constraint vectors, plus some additional ones (these graphs are precisely the ones with three vertices which are not contracted to each other).

One can also read off which vector length-squares can be obtained by closing the basic set under addition. If a diagram can be cut into two pieces by cutting n lines, then there is a vector of length^2 n. If n is odd it is associated to (v) of D_5, if n is even to (0). Clearly (g) contains a length 1 vector, and is therefore ruled out. The remaining ones, (b), (f) and (j) correspond to new constraint vectors, which are not of the type given above.

This possibility was already hinted at in [6]. By requiring that all massive excitations of a *given* massless chiral spinor be non-chiral, it was deduced that certain lattice vectors had to be present. The set given in (5.1) emerged as just one of seven possibilities, an the only one that solved the problem for *all* massive spinors. The cases (b), (f) and (j) turn out to correspond to some of the other six possibilities, but have additional constraint vectors to take care of other spinors. It is in fact possible to prove diagrammatically that in any of the nine cases depicted in fig. 1 massive spinors come in non-chiral pairs. The necessary cancellations are due to the θ function identity $\theta_2\theta_3\theta_4 = 2\eta^3$, as explained in [6] (this applies to all chiral string theories we know).

6. Four-dimensional type II theories

The results we have obtained thus far are of special interest for type II strings in four dimensions. Originally it was thought that type II strings could only give $U(1)^6$ gauge groups. Later it was realized that also the group $SU(2)^6$ could be obtained [21, 22]. From the results of sect. 4 it is clear that one can get other gauge groups as well, some of which are just large enough to contain the standard model. In [23] the cases with 18 gauge bosons and $N=4$ supergravity were discussed. Various models have been discussed in [24]. In this section we will present a few other examples.

As in [6] we will consider even self-dual lorentzian lattices which are decomposable into D_n factors, and for classification purposes we will make use of the possibility to map these lorentzian lattices to even self-dual euclidean ones. In four dimensions the lattices to consider have 14 left and 14 right dimensions. They must contain a space-time lattice D_5 on both sides, and should contain one of the possible sets of constraint vectors on both sides. Of course the constraint vectors on the left and the right lattice need not be the same.

In general the lattices are thus of the form $D_5 \times (D_1)^9$ on both sides (where the factors can combine to larger D_n's). To find out whether it is possible to satisfy a certain constraint, one has to inspect the list of conjugacy classes of the lattice. In some cases there are easier ways to determine potential constraint vectors. Constraint vectors of length-squared 2 can only exist if the D_5 factor is part of a larger D_n. If one wants to have only such constraints, at least 6 length-squared 2-vectors are needed, so that the D_5 factor must extend to D_8. If D_5 cannot be extended at all, the constraint vectors must all have length-squared four. Furthermore (if the constraints are due to a free fermion supercurrent) they must be of the form $((v),(v),(v),(v),0^6)$, where the first entry belongs to D_5, and the other three (v)'s must be vector conjugacy classes of *different* D_n factors. Thus in this case the lattice must have at least three separate D_n factors in addition to D_5.

To illustrate this we consider a class of models which has the maximum number (i.e. 18) of G_R gauge bosons, and which we therefore expect to contain those of [23]. To get 18 gauge bosons, the lattice must be $(\Gamma_{14})_L \times (D_{14})_R$, and can therefore be mapped to a 16-dimensional euclidean lattice $\Gamma_{14} \times D_2$. Thus it is either D_{16} or $E_8 \times E_8$. In the first case the lorentzian lattice is $D_{14} \times D_{14}$, and we get 18 gauge bosons from *each* side. This theory has a tachyon and no massless fermions. (Note that for D_{14} all possible constraints are automatically satisfied. On each side we can therefore have one of the three possible G_{SC} groups $SU(2)^6$, $SU(3) \times SO(5)$ or $SU(4) \times SO(3)$ as gauge group G_{NSR}.) If the lattice is $E_8 \times E_8$ there are two ways to embed D_2, namely $E_8 \times E_8 \supset E_8 \times D_6 \times D_2$ and $E_8 \times E_8 \supset E_7 \times E_7 \times D_2$. In the first case we can embed the space-time part D_5 of the left lattice in either E_8 or D_6. Using E_8 one gets a $N=4$ supergravity theory, which is the one discussed in [23]. The second case, as well as the second embedding of D_2, does not yield anything new, because the constraints on the left sector cannot be satisfied.

Many properties of the spectrum follow directly from simple features of the even lattice, and in particular from its roots. For the heterotic strings this was discussed in detail in [6] and type II strings are virtually identical in most respects. Some of the more interesting aspects are listed below:

Tachyons. As for heterotic strings, in the even lattice formulation there is a simple criterion for the presence of tachyons of minimal mass in type II strings: they are present if and only if the D_5 factors from left and right are both embedded in the same D_n factor of the euclidean lattice. In this way one understands easily why the 36 gauge boson model constructed above is tachyonic. It is also easy to find a non-tachyonic model with many gauge bosons. For example, one can construct a lorentzian lattice $(D_{12} \times D_2)_L \times (D_{12} \times D_2)_R$ from the Niemeier lattice $D_{12} \times D_{12}$; the left and right D_2 factors combine to a self-dual Englert-Neveu type [30] lattice. This model has $14 + 14$ gauge bosons which may, for example, gauge the group $SO(5) \times SU(2) \times U(1)$ on both sides, as in the example discussed in sect. 4. There are no massless fermions.

Massless fermions. One can only get massless fermions if the space-time group D_5 is embedded in a D_n factor, with $n \leqslant 8$ (since otherwise the spinor weights have length-squares larger than 2). Therefore the gauge group in that sector can only have dimension 6 or less. One can only have a symmetric subgroup of G_{SC} which is this small if $G_{SC} = SU(2)^6$. This agrees with and generalizes an observation made in [26]. *Chiral* fermions can occur only if D_5 is not part of a larger D_n group (this need to be the case only in one sector).

Supersymmetry. As in the heterotic case [6], the number of supersymmetries is directly related to embeddings of $(D_5)_L \times (D_5)_R$ in exceptional algebras. If in one sector D_5 is embedded in E_6, E_7 or E_8, one gets respectively one, two or four gravitini from this sector. By combining both sectors one can thus obtain $N = 1, \dots 6$ and $N = 8$ supergravity theories; note that there are two ways to get $N = 2$ and $N = 4$ supersymmetry.

Bispinor gauge bosons. It is possible to get gauge bosons as tensor products of massless left and right spinors, in addition to those of G_{NSR}. Upon inspecting their couplings it turns out that they can gauge only U(1) factors. Their multiplicity can be quite large since it is the product of the multiplicities of the left and right spinors. For example, if D_5 is embedded in D_8 in both sectors, there can be 32 such bosons in addition to the $6 + 6$ G_{NSR} gauge bosons. A model of this kind is discussed in [24].

Anomalies. Before turning to the question which fermion spectra can be obtained in type II strings, we want to explain why they are free of local anomalies. As was shown in [31], modular invariance guarantees Green-Schwarz factorization [32] of the anomaly of the full Frenkel-Kac group G^{3n} in type II theories (the factor is actually only TrF^2, and not $\mathrm{TrF}^2 - \mathrm{TrR}^2$ as in the heterotic string). Normally it would be enough to say that since the actual gauge group G_{NSR} is a subgroup of this group, its anomalies will also cancel. This is certainly correct for the polygon graphs, as well as for the one-loop anomaly cancelling term, calculated in [33]. Here one gets the correct answer by simply replacing group generators by subgroup generators in all traces. But it is not correct for the BAA vertex, which has a fixed normalization independent of the G_{NSR} structure constants f_{abc}. One may however check that the normalization given in (4.3) is precisely the right one to make the cancellation mechanism work. We do not know whether there exist four dimensional type II strings with non-trivial Green-Schwarz cancellation, but it is good to know that string theory has already anticipated that possibility.

The standard model? Finally we turn to the important question whether type II theories can produce the standard model. Within our approach, the answer is negative. There are two symmetric subgroups of G_{SC} which are locally isomorphic to $SU(3) \times SU(2) \times U(1)$, namely $SU(3) \times SO(5) \supset SU(3) \times SO(3) \times SO(2)$ and $SU(4) \times SO(3) \supset SU(3) \times U(1) \times SO(3)$. In these two cases the groups SU(3) and SO(3), respectively, do not act on the tangent space of G_{SC}/H. It follows then that massless fermions can only be in the singlet or adjoint representations of these groups, and hence not in the representations of the standard model.

Although it is of little relevance, one might still ask whether it is possible to get chiral fermions at all. This possibility is severely restricted by anomaly cancellation. As we have seen in sect. 3, the H-representations of the massless fermions are embedded in representations with weights of length 1 of the Frenkel-Kac group. There are usually very few possibilities, so that one is forced to pair them off into real representations in order to cancel anomalies. In many cases, the possibility of getting chiral fermions can be quickly ruled out in this way, but we have not attempted to do this systematically*.

7. Summary

In the covariant lattice approach, we have elucidated the role of world-sheet superconformal invariance in lower dimensional heterotic and type II string theories. In ten dimensions, there is only one Lorentz-invariant choice for the supercurrent. Below ten dimensions, Lorentz invariance is less restrictive, because one is free to modify the "internal" part of the supercurrent. It is in this respect that the covariant lattice construction is different from torus compactifications which leave the ten-dimensional form of the supercurrent unchanged.

We have shown that the requirement of superconformal invariance implies the existence of certain vectors on the self-dual lattice, which (in the even lattice formulation) have to have length-squares two and/or four. These vectors represent the picture changing operator in bosonic language. As is always the case in the even

* After finishing this paper we received [34], in which it is proved that the standard model particle spectrum cannot be obtained in type II strings, and in which some chiral type II theories are constructed.

lattice formulation, the existence of length-squared two-vectors (roots) implies the existence of gauge particles. The length-squared two-vectors arising from the space-time part of the supercurrent imply gravitons; those arising from the internal part imply gauge bosons, which in general form a subset of all G_{NSR} gauge bosons. The picture changing operator maps the vertex operators of these gauge bosons to linear combinations of the Frenkel-Kac generators of the lattice. In this way we can understand why G_{NSR} depends on both the lattice and on the choice of the picture changing operator, ie., on the way superconformal invariance is realized.

The presence of G_{NSR} gauge bosons in one left- or right-moving sector precludes the existence of chiral fermions arising from that sector. Consequently, chiral string theories must have at least one fermionic sector whose superconformal invariance is realized entirely in terms of length-squared four-vectors in the internal part. As already discussed in [6], the existence of such vectors severely restricts the possibility of performing continuous Lorentz rotations acting on the lattice, in contrast to the non-chiral torus-compactified case [7].

G_{NSR} gauge bosons are not of much interest in heterotic string theories, because they render these non-chiral. On the other hand, in type II theories one might get G_{NSR} gauge symmetries from one and chiral fermions from the other sector. However, as we have seen, it turns out that the possible gauge groups and fermion representations are so limited, that the standard model cannot be obtained.

We thank P. Ginsparg for discussions.

References

[1] D. Friedan, E. Martinec and S. Shenker, Nucl. Phys. B271 (1986) 93
[2] J. Cohn, D. Friedan, Z. Qiu and S. Shenker, Nucl. Phys. B278 (1986) 577
[3] W. Lerche and D. Lüst, Phys. Lett. 187B (1987) 45
[4] W. Lerche, D. Lüst and A.N. Schellekens, Phys. Lett. 181B (1986) 71
[5] V. Kostelecký, O. Lechtenfeld, W. Lerche, S. Samuel and S. Watamura, Nucl. Phys. B288 (1987) 173
[6] W. Lerche, D. Lüst and A.N. Schellekens, Nucl. Phys. B287 (1987) 477
[7] K. Narain, Phys. Lett. 169B (1986) 41
[8] Da-Xi Li, McGill University preprint (1986)
[9] L. Dixon, J. Harvey, C. Vafa and E. Witten, Nucl. Phys. B261 (1985) 651; B274 (1986) 285
[10] H. Kawai, D. Lewellen and S. Tye, Phys Rev. D34 (1986) 3794; Phys. Rev. Lett. 57 (1986) 1832
[11] H. Kawai, D. Lewellen and S. Tye, Nucl. Phys. B288 (1987) 1
[12] I. Antoniadis, C. Bachas and C. Kounnas, Nucl. Phys. B289 (1987) 87
[13] L.E. Ibañez, H.P. Nilles and F. Quevedo, Phys. Lett. 187B (1987) 25
[14] K. Narain, M. Sarmadi and C. Vafa, Nucl. Phys. B288 (1987) 551
[15] D. Gross, J. Harvey, E. Martinec and R. Rohm, Nucl. Phys. B256 (1985) 253
[16] J. Schwarz, Phys. Reports 89 (1982) 223
[17] A. Parkes, Phys. Lett. 184B (1987) 19
[18] T. Ohrndorf, Nucl. Phys. B281 (1987) 470
[19] N. Seiberg and E. Witten, Nucl. Phys. B276 (1986) 272
[20] M. Halpern, Phys. Rev. D12 (1975) 1684;
I. Frenkel and V. Kac, Inv. Math. 62 (1980) 23;
G. Segal, Commun. Math. Phys. 80 (1981) 301;

P. Goddard and D.I. Olive, *in* Vertex operators in mathematics and physics, MSRI Publication #3 (Springer, Heidelberg 1985)
[21] L. Castellani, R. d'Auria, F. Gliozzi and G. Sciuto, Phys. Lett. 168B (1986) 47
[22] R. Bluhm and L. Dolan, Phys. Lett. 169B (1986) 347
[23] R. Bluhm, L. Dolan and P. Goddard, Nucl. Phys. B289 (1987) 364
[24] H. Kawai, D. Lewellen and S. Tye, preprint CLNS 87/760
[25] P. Goddard and D.I. Olive, Nucl. Phys. B257 (1985) 245
[26] I. Antoniadis, C. Bachas, C. Kounnas and P. Windey, Phys. Lett. 171B (1986) 51;
P. Windey, preprint LBL-20423
[27] A.N. Schellekens and N.P. Warner, Phys. Rev. D34 (1986) 3092
[28] A. Bais and P. Bouwknegt, Nucl. Phys. B279 (1987) 561
[29] P. Goddard, W. Nahm and D.I. Olive, Phys. Lett. 160B (1985) 111
[30] F. Englert and A. Neveu, Phys. Lett. 163B (1985) 349
[31] A.N. Schellekens and N.P. Warner, Nucl. Phys. B287 (1987) 317; Phys. Lett. B177 (1986) 317; B181 (1986) 339
[32] M.B. Green and J. Schwarz, Phys. Lett. 149B (1984) 117
[33] W. Lerche, B.E.W. Nilsson and A.N. Schellekens, Nucl. Phys. B289 (1987) 609
[34] C. Vafa, L. Dixon and V. Kaplunovsky, SLAC-PUB-4282

Nuclear Physics B294 (1987) 43–82
North-Holland, Amsterdam

ON FOUR-DIMENSIONAL GAUGE THEORIES FROM TYPE II SUPERSTRINGS

Lance J. DIXON*

Stanford Linear Accelerator Center, Stanford University, Stanford, CA 94305, USA

Vadim S. KAPLUNOVSKY†

Physics Department, Stanford University, Stanford, CA 94305, USA

Cumrun VAFA‡

Lyman Laboratory of Physics, Harvard University, Cambridge, MA 02138, USA

Received 4 May 1987

We study the four-dimensional gauge theories which are associated with classical vacua of the type II superstring, i.e. which correspond to a superconformal field theory on the world sheet. Using the fact that gauge symmetry arises from a supersymmetric affine Kac-Moody algebra, and demanding unitarity of the underlying world-sheet field theory, we show that no such vacua can yield the particle spectrum of the standard model. Of the gauge theories which are permitted by unitarity, we find that many can be constructed explicitly as orbifolds which twist the left- and right-moving degrees of freedom of the string asymmetrically; among these are three $N=4$ supersymmetric models – which have previously been constructed in a quite different fashion – and two $N=1$ supersymmetric models with chiral gauge representations for the massless fermions.

1. Introduction

Recent attempts to extract realistic physics from superstring theories [1] have largely focused on the heterotic string [2]. Particular compactification schemes for the $E_8 \otimes E_8$ heterotic string (e.g., on Calabi-Yau manifolds [3] or orbifolds [4]) give rise to a large variety of four-dimensional (4d) models with $N=1$ supersymmetry, chiral fermions and realistic gauge groups. From the phenomenological point of view, the principal problem with this situation is the vast number of quasi-realistic models, and the lack at present of any dynamical reason for preferring one of them

* Work supported by the Department of Energy, contract DE-AC03-76SF00515.
† Research supported in part by the NSF under grant #PHY-86-12280.
‡ Research supported in part by the NSF under grant #PHY-82-15249, and in part by a Fellowship from the Harvard Society of Fellows.

0550-3213/87/$03.50

over the others. Without a better understanding of the dynamics of compactification, finding a single model which correctly describes the four-dimensional world (if such a model exists) is akin to finding a needle in a haystack. Loosely speaking, the abundance of 4d models with realistic gauge groups is related to the existence of a very large Yang-Mills multiplet in the massless spectrum of the ten-dimensional heterotic string. This embarrassment of riches has prompted us to consider in this paper the opposite extreme: the type II superstring theory, which has no Yang-Mills fields in ten dimensions. Under this circumstance, one might expect that any four-dimensional gauge symmetry must arise via a Kaluza-Klein mechanism; unfortunately, it has been shown [5] that symmetries generated by this mechanism act the same way on left-handed and right-handed fermions. In addition, Friedan et al. [39, 19] showed using superconformal field theory techniques that massless fermions are incompatible with non-abelian gauge symmetries in a left-right symmetric background.

For these reasons type II superstrings were not considered to be suitable for realistic model building until the recent construction by Bluhm et al. [6] and by Kawai et al. [7] of several type II superstring models that have four flat space-time dimensions and gauge groups large enough to contain the standard $SU(3) \otimes SU(2) \otimes U(1)$ model. The three models in ref. [6] all have $N = 4$ space-time supersymmetry in four dimensions, so they contain gauginos – massless fermions in the adjoint representation of the gauge group. Additional models constructed in ref. [7] also contain massless fermions in non-trivial gauge representations. The arguments of the previous paragraph do not apply to these models for two reasons: First, the gauge groups do not come entirely from a Kaluza-Klein mechanism, but also utilize inherently stringy states (soliton or winding-type configurations of the string). (Although the models were constructed out of free fermions in [6, 7], they will be given a somewhat more geometric interpretation later in this paper.) Second, the models treat the left- and right-moving degrees of freedom of the string asymmetrically. None of the models in refs. [6, 7] contain chiral (4d) fermions; however, they leave open the questions as to whether other left-right asymmetric constructions can give rise to chiral models and, more importantly, whether models with massless fermions in realistic gauge group representations can be constructed.

The aim of this paper is to investigate further the possible gauge groups and fermion representations for 4d models based on the type II superstring. We will give a complete list of 4d gauge groups that can appear in such models, and will show that most of them (possibly all of them) can be generated by compactifying the type II superstring on an asymmetric orbifold, [8] that is, an orbifold in which the left- and right-moving modes of the bosonic fields coordinatizing the internal space are twisted differently. Included in the list of groups are several containing the standard model gauge group. We will also show that the three $N = 4$ supersymmetric models in ref. [6] (which also appear in ref. [7]), with gauge groups $SU(2)^6$ (this group was first noted by Castellani et al. [9]), SU(4) ⊗ SU(2) and SO(5) ⊗ SU(3) can be constructed as asymmetric orbifolds, although their original construction was quite different. In particular, the models were first constructed entirely from free fermions (replacing the internal bosonic coordinates); the last two of the three models used an unconventional form for the world-sheet supersymmetry generator $T_F(z)$, which is the dimension $\frac{3}{2}$ superpartner of the stress-energy tensor $T_B(z)$. The orbifold construction of the models uses the 'standard' supersymmetry generator for the internal dimensions, $T_F^{\rm int} = -\frac{1}{2}\psi^i \partial X^i$. However, the Neveu-Schwarz-Ramond fermions ψ^i and the bosonic fields ∂X^i appearing in $T_F^{\rm int}$ are twisted. The equivalence between the two types of constructions also demonstrates the relation between the two forms of T_F. The various other models found in [7] may be constructed in turn as asymmetric orbifolds of the three $N=4$ supersymmetric models (although we will not do this explicitly). On the other hand, asymmetric orbifolds also yield models which cannot be described using free fermions alone, for example $N=4$ supersymmetric models with gauge group $SU(3)^2$ or $SO(5) \otimes SU(2)^2$.

The existence of a number of four-dimensional $N=4$ supersymmetric models with gauge groups containing the standard model's suggests that models with chiral fermions (having at most $N=1$ supersymmetry) might also exist, and in fact construction of such models is particularly straightforward from the orbifold point of view. In the $N=4$ models mentioned above all four supersymmetries arise from modes with the same two-dimensional chirality, say right-moving. The gauge symmetries, on the other hand, are due to the presence of left-moving currents on the world-sheet. If one twists an $N=4$ model by a symmetry which breaks all but one of the right-moving supersymmetries, and which simultaneously acts on the left-moving gauge currents, one can build chiral 4d models. However, it is difficult to build models whose fermion content approximates that of the standard model. In fact, the main result of this paper is to show, on very general grounds, that *no models based on the type II superstring can possibly contain the standard model gauge group and its fermion content*! This result is derived using superconformal techniques similar to those used by Friedan et al. [39, 19, 20].

Given the strength of this result, it is important to outline the assumptions leading to it. In particular, we need to define what we mean by a 'model based on the type II superstring.' First of all, we define the type II superstring theory itself as a theory of closed orientable strings that has local (1, 1) world-sheet supersymmetry; this definition involves only the geometrical properties of the world sheet. (We consider 2d supersymmetry to be a geometrical feature since it makes the world sheet a superspace.) Classical space-time backgrounds in which the string might propagate are described in terms of a world-sheet action, i.e. in terms of a two-dimensional quantum field theory; we will refer to different backgrounds as being different string models (based on the same string theory). For example, the flat ten-dimensional space-time background corresponds to the world-sheet action of ref. [10] that describes ten free superfields $\boldsymbol{X}^{\mu}$, $\mu = 0,1,\ldots,9$. One of these superfields, namely $\boldsymbol{X}^0$, has negative metric; however, it can be shown that negative norm states decouple from the physical amplitudes provided that the 2d theory has $(1,1)$ superconformal invariance [11]. In this paper we study models with four-dimensional Minkowski space-time left intact; this (apparently) requires four of the above superfields, $\boldsymbol{X}^{0,1,2,3}$, to be present in the 2d field theory describing a given model. We therefore insist on $(1,1)$ superconformal invariance of the 2d theory in order to decouple negative norm states created by $\boldsymbol{X}^0$.

A better way to justify our insistence on $(1,1)$ superconformal invariance is via the equivalence of this condition to the classical equations of motion of the type II superstring [3,12,13]. Thus classical superstring vacua correspond to superconformally invariant 2d field theories while solutions of the quantum superstring theory that need quantum effects for their stability do not. Hence the arguments of this paper apply to classical superstring models only; however, quantum string vacua have yet to be constructed. On the other hand, we would like to stress that *superconformal invariance* and *existence of 4d Minkowski space-time* are the only requirements we impose on the models; in particular, we do not assume any particular compactification scheme, orbifold or otherwise, space-time supersymmetric or not. In fact the superconformal theory which replaces the six 'missing' spatial dimensions need not have any geometric interpretation at all.

This paper is organized as follows. In sect. 2 we discuss relations between 4d gauge symmetries and properties of the underlying superconformal field theory. The Lie algebra associated with the gauge group extends to a supersymmetric affine Kac-Moody algebra [39,19,41,14], which is generated by purely left-moving (or purely right-moving) supercurrents – world-sheet superfields of dimension $\frac{1}{2}$. Other sources of gauge bosons, such as the Ramond-Ramond sector of the type II superstrings, or both left- and right-moving supercurrents, are excluded. The requirement that the superconformal theory replacing the six internal coordinates is unitary and has super-Virasoro central charge $\hat{c} = 6$ constrains the possible super-Kac-Moody algebras and thus allows us to classify the possible gauge groups for 4d models based on the type II superstring. In sect. 3 we detour from the general approach to construct asymmetric orbifolds which give rise to many of the 4d gauge groups permitted by the results of sect. 2, including the three models of ref. [6]. In addition we describe two chiral $N = 1$ supersymmetric models. In sect. 4 we return to general considerations, focusing now on gauge quantum numbers of massless particles. We find that the representations of the 4d gauge group which can appear in the massless spectrum of a type II superstring model are severely restricted by the requirement that all physical states belong to unitary representation of the entire super-Kac-Moody algebra. In particular, a 4d model whose gauge group contains $SU(3) \otimes SU(2) \otimes U(1)$ can have either massless SU(3) triplets or massless SU(2) doublets, but never both; this rules out the existence of realistic models based on the type II superstring. Our conclusions are presented in sect. 5.

2. Four-dimensional gauge groups

2.1. SUPERCONFORMAL INVARIANCE

We begin this section by reviewing the general properties of a model based on the type II superstring which satisfies the classical string equations of motion. As emphasized in refs. [3,12,13], the latter condition is equivalent to superconformal invariance of the 2d field theory describing the model. Together with superreparametrization invariance, this allows us to describe the 2d theory in a locally flat superspace parametrized by holomorphic and anti-holomorphic coordinates (z, θ) and $(\bar{z}, \bar{\theta})$. Whatever the matter superfields making up the theory, they must give rise to a holomorphic super-stress tensor $T(z,\theta) = T_F(z) + \theta T_B(z)$ and its anti-holomorphic counterpart $\bar{T}(\bar{z},\bar{\theta}) = \bar{T}_F(\bar{z}) + \bar{\theta}\bar{T}_B(\bar{z})$. The operator product expansion (OPE) of $T(z,\theta)$ is given by

$$T(z_1,\theta_1)\cdot T(z_2,\theta_2) = \frac{\frac{1}{4}\hat{c}}{z_{12}^3} + \frac{3\theta_{12}}{2z_{12}^2}T(z_2,\theta_2) + \frac{\frac{1}{2}}{z_{12}}D_2T + \frac{\theta_{12}}{z_{12}}\partial_2 T + \cdots, \quad (2.1)$$

where $z_{12} = z_1 - z_2 - \theta_1\theta_2$ and $\theta_{12} = \theta_1 - \theta_2$. In addition to the matter fields, there are ghosts for the local super-reparametrization invariance giving rise [15] to a stress tensor T^{gh} with central charge $\hat{c}^{\mathrm{gh}} = -10$; hence $\hat{c} = 10$ is required in (2.1) in order for the total superconformal anomaly to cancel ($\hat{c}^{\mathrm{tot}} = 0$).

Expanding $T(z,\theta)$ in a Laurent series, one has

$$T(z,\theta) = \tfrac{1}{2}\sum_{r\in Z+\kappa} G_r z^{-r-3/2} + \theta\sum_{n\in Z} L_n z^{-n-2}, \quad (2.2)$$

where κ is either $\frac{1}{2}$ or 0 according to whether T_F and other world-sheet fermions satisfy anti-periodic or periodic boundary conditions. These two cases give rise respectively to the Neveu-Schwarz (NS) and Ramond (R) versions of the super-Virasoro algebra, which is the algebra obeyed by the Laurent coefficients L_m and G_r as a consequence of the OPE (2.1):

$$[L_m, L_n] = \tfrac{1}{8}\hat{c}(m^3 - m)\delta_{m+n,0} + (m-n)L_{m+n},$$
$$\{G_r, G_s\} = \tfrac{1}{2}\hat{c}(r^2 - \tfrac{1}{4})\delta_{r+s,0} + 2L_{r+s},$$
$$[L_m, G_r] = (\tfrac{1}{2}m - r)G_{m+r}. \quad (2.3)$$

Superconformal theories of physical interest are not completely arbitrary. In particular, four free superfields $X^\mu(z,\theta,\bar{z},\bar{\theta}) = X^\mu(z,\bar{z}) + \theta\psi^\mu(z) + \bar{\theta}\bar{\psi}^\mu(\bar{z})$, $\mu = 0,1,2,3$ are needed to make four-dimensional Minkowski space-time; other, 'internal' fields should commute with X^μ. Thus we have a direct product of two superconformal theories, space-time (4d) and internal, whose respective super-stress tensors $T^{(4d)}(z,\theta) = -\frac{1}{2}\mathrm{D}X^{\mu}\mathrm{D}^2X_{\mu}$ and T^{int} anticommute with each other. It follows from (2.1) that $T(z,\theta) \equiv T^{(4d)}(z,\theta) + T^{\text{int}}(z,\theta)$ gives rise to a super-Virasoro algebra with central charge $\hat{c} = \hat{c}^{(4d)} + \hat{c}^{\text{int}}$; since $\hat{c}$ is normalized to be 1 for a free superfield, $\hat{c}^{(4d)} = 4$ and we need $\hat{c}^{\text{int}} = 6$ in order to cancel the total superconformal anomaly. Note that while local (on the world sheet) properties of the internal and the space-time components of the theory need not be related to each other, the boundary conditions obeyed by $T_{\mathrm{F}}^{\text{int}}$ and by $T_{\mathrm{F}}^{(4d)} = -\frac{1}{2}\psi_{\mu}\,\partial X^{\mu}$ should be the same [16]. (This follows from the fact that the world-sheet gravitino couples to the total T_{F} of all matter fields.) Thus we can refer to sectors of the Hilbert space as being Neveu-Schwarz (NS) or Ramond (R) even when the superconformal theory describing the integral degrees of freedom is highly non-trivial.

2.2. RAMOND-RAMOND VECTOR BOSONS

The type II superstring has both a left- and a right-moving superconformal algebra, so there are four sectors altogether. Clearly states belonging to (R, NS) and (NS, R) sectors are space-time fermions while (NS, NS) and (R, R) sectors contain space-time bosons. (R, R) bosons are peculiar to the type II superstring, so let us take a closer look at them. The first question we would like to ask is whether any *massless* scalar or vector particles could be of this type. In general, the answer to this question is yes; for example, consider a type IIA (non-chiral) ten-dimensional superstring compactified on a Calabi-Yau manifold or even on a torus. In ten dimensions, massless (R, R) bosons form a vector and a three-index anti-symmetric tensor of the (10d) Lorentz group. After this compactification, they give rise to $B_2 + 2$ massless (4d) vectors and $B_3 + 2B_1$ massless scalars. ($B_{1,2,3}$ are the Betti numbers of the manifold.)

Our next question concerns the couplings of massless (R, R) bosons to other massless particles. Consider a vertex operator $V_{\mathrm{s}}(k)$ for an (R, R) scalar. Unlike the (NS, NS) case, this operators must behave like a (4d) spinor with respect to both left- and right-moving superconformal algebras. Moreover, these two spinors should have opposite helicities and thus opposite chiralities since they combine into a scalar rather than into a vector. Thus the vertex operator should look like $V^{\alpha,\dot{\beta}}(k)$ or $V^{\dot{\alpha},\beta}(k)$. Due to the absence of any polarization vector, spinor indices have to be contracted with $k_{\mu}\gamma^{\mu}$: $V_{\mathrm{s}}(k) \equiv k_{\mu}\sigma^{\mu}_{\alpha\dot{\beta}}V^{\alpha,\dot{\beta}}(k)$ or $k_{\mu}\sigma^{\mu}_{\dot{\alpha}\beta}V^{\dot{\alpha},\beta}(k)$. Massless (R, R) vectors also have vertex operators of spinor-spinor type, but now the two spinors have the same chirality. Such bispinors cannot be contracted with a polarization vector ζ^{ν} alone, so again the momentum vector k^{μ} is required: $\mathrm{V}_{\nu}(k,\zeta) \equiv k_{\mu}\zeta_{\nu}(\sigma^{\mu\nu}_{\alpha\beta}V^{\alpha,\beta} + \sigma^{\mu\nu}_{\dot{\alpha}\dot{\beta}}V^{\dot{\alpha},\dot{\beta}})$.

How does an explicit momentum factor in the vertex of an (R, R) boson affect its couplings? In a three-point function involving massless particles only, external momenta cannot be contracted with themselves or with each other ($k_i \cdot k_j = 0$ on shell, $i, j=1,2,3$). Therefore, all cubic couplings of a massless (R, R) field A to other massless fields cannot involve A itself, but only its derivatives. If the other two fields are space-time fermions, then we have a coupling of dimension five (or more, if other derivatives are present), and thus non-renormalizable. If the other two fields are space-time bosons, then one of them must also be of the (R, R) type; this gives us a three-boson coupling with at least two derivatives. Again, the dimension of such a coupling is at least five, so it is also non-renormalizable. Therefore, massless (R, R) bosons do not have any renormalizable cubic couplings, such as gauge couplings, to any massless particles. We immediately conclude that *massless particles are neutral with respect to any gauge groups generated by* (R, R) *gauge bosons*; and, as a corollary, *all* such gauge groups must be *abelian*.

2.3. SUPER-KAC-MOODY ALGEBRAS

Phenomenologically, only gauge symmetries that act non-trivially on massless particles are relevant. As we have just seen, such symmetries must be generated by gauge bosons coming from the (NS, NS) sector, to which we will henceforth confine our attention. In the (0,0) picture for the superconformal ghost system [13] vertex operators for these bosons are superfields of conformal dimension $(\frac{1}{2}, \frac{1}{2})$:

$$V^a(k, \zeta) = \zeta_\mu \bar{D} X^\mu(\bar{z}, \bar{\theta}) J^a(z, \theta) e^{ik\cdot X}. \tag{2.4}$$

The superfield $\bar{D}X^\mu = \bar{\psi}^\mu + \bar{\theta}\bar{\partial}X^\mu$ is required here in order to make a four-dimensional vector; the conformal dimension of $\bar{D}X^\mu$ is $(0, \frac{1}{2})$. (Of course the dimension $(\frac{1}{2}, 0)$ superfield DX^μ would have served equally well; this would give us the complex conjugate of (2.4).) For $k^2=0$ the dimension of $e^{ik\cdot X}$ is (0,0); thus the dimension of J^a must be $(\frac{1}{2}, 0)$. This means that J^a is a holomorphic superfield, i.e. a function of (z, θ) and not $(\bar{z}, \bar{\theta})$, and that the operator product expansion of $J(z, \theta)$ with $T(z, \theta)$ is given by:

$$T(z_1, \theta_1)\cdot J^a(z_2, \theta_2) = \frac{\theta_{12}}{z_{12}^2} J^a(z_2, \theta_2) + \frac{\frac{1}{2}}{z_{12}} D_2 J^a(z_2, \theta_2) + \frac{\theta_{12}}{z_{12}} \partial_2 J^a(z_2, \theta_2) + \cdots . \tag{2.5}$$

In terms of Laurent series

$$\begin{aligned} J^a(z, \theta) &= \mathcal{J}^a(z) + \theta J^a(z) \\ &= \sum_{r\in Z+\kappa} \mathcal{J}_r^a z^{-r-1/2} + \theta \sum_{n\in Z} J_n^a z^{-n-1}, \end{aligned} \tag{2.6}$$

where κ is $\frac{1}{2}$ (0) in the NS (R) sector, eq. (2.5) becomes

$$\begin{aligned}
[J^a_m, L_n] &= mJ^a_{m+n}, \\
[\mathscr{J}^a_r, L_n] &= (r+\tfrac{1}{2}n)J^a_{r+n}, \\
[J^a_m, G_s] &= m\mathscr{J}^a_{m+s}, \\
\{\mathscr{J}^a_r, G_s\} &= J^a_{r+s}.
\end{aligned} \tag{2.7}$$

Let us consider the algebra generated by the supercurrents $\boldsymbol{J}^a$. The product of two supercurrents $\boldsymbol{J}^a(z_1,\theta_1)$ and $\boldsymbol{J}^b(z_2,\theta_2)$ must be single-valued as $z_1 \to z_2$, so only integral powers of $1/z_{12}$ are allowed in the operator product expansion. Since the conformal dimension of $\boldsymbol{J}^a$ is $\frac{1}{2}$, the OPE is limited on dimensional grounds to

$$\boldsymbol{J}^a(z_1,\theta_1)\cdot\boldsymbol{J}^b(z_2,\theta_2) = \frac{1}{2z_{12}}\cdot k^{ab}\mathbf{1} + \frac{\theta_{12}}{z_{12}}\cdot if^{abc}\boldsymbol{J}^c(z_2,\theta_2) + \cdots, \tag{2.8}$$

which means that the Laurent coefficients of the supercurrents – J^a_n and $\mathscr{J}^a_r$ – obey a supersymmetrized affine Kac-Moody algebra [39, 19, 41, 14] (henceforth called an SKM algebra):

$$\begin{aligned}
[J^a_m, J^b_n] &= \tfrac{1}{2}mk^{ab}\delta_{+n,0} + if^{abc}J^c_{m+n}, \\
[J^a_m, J^b_r] &= if^{abc}\mathscr{J}^c_{m+r}, \\
\{\mathscr{J}^a_r, \mathscr{J}^b_s\} &= \tfrac{1}{2}k^{ab}\delta_{r+s,0}.
\end{aligned} \tag{2.9}$$

As with the super-Virasoro algebra, there is a Neveu-Schwarz ($\kappa=\frac{1}{2}$) and a Ramond ($\kappa=0$) version of the SKM algebra, whose respective representations describe states in the NS and R sectors of the superconformal theory. (The boundary conditions on $\mathscr{J}^a(z)$ must match those on $T_F(z)$.) In either sector, zero modes of the bosonic currents, J^a_0, form a closed Lie algebra with structure constants f^{abc}; it is this algebra that generates the 4d gauge symmetry.

Consider the case of a simple gauge group G. Note that although we have written (2.8) with an arbitrary matrix k^{ab} of Kac-Moody central charges, G-invariance requires $k^{ab} = k\cdot\delta^{ab}$. (From now on we assume that supercurrents are normalized such that f^{abc} have their conventional values, which means $f^{abc}f^{dbc} = C_G(\mathrm{adj})\delta^{ad}$, where $C_G(\mathrm{adj})$ is the quadratic Casimir of the adjoint representation of the group: $C_{SU(N)}(\mathrm{adj}) = N$, etc.) Moreover, the existence of unitary representations of the SKM algebra requires the Kac-Moody central charge k to be an integer and

$$\hat{k} \equiv k - C_G(\mathrm{adj}) \tag{2.10}$$

to be non-negative [14]. (We will further discuss the restrictions imposed by unitarity in sect. 4.)

The super-stress tensor T^{SKM} for an SKM algebra can be written as a normal-ordered product of the supercurrents [14, 17]:

$$T^{\mathrm{SKM}}(z,\theta)=\frac{-1}{k}:DJ^a(z,\theta)\cdot J^a(z,\theta):+\frac{2}{3k^2}:f^{abc}J^a(z,\theta)J^b(z,\theta)J^c(z,\theta):. \tag{2.11}$$

(The relation (2.11) is analogous to the construction [18] of the stress tensor $T_{\mathrm{B}}^{\mathrm{KM}}(z)$ for an ordinary, i.e. non-supersymmetric, affine Kac-Moody algebra.) The super-stress tensor T^{SKM} generates a super-Virasoro algebra with central charge $\hat{c}$ given by [14, 17]

$$\hat{c}(\mathrm{G})=\frac{d(\mathrm{G})}{3}+\frac{2d(\mathrm{G})}{3}\frac{k-C_{\mathrm{G}}(\mathrm{adj})}{k}, \tag{2.12}$$

where $d(\mathrm{G})$ is the dimension of G.

In general the 4d gauge group generated by J_0^a is not simple, but rather a direct product of several simple or abelian subgroups. It is easy to see that in this case the SKM algebra is also a direct product of several subalgebras (supercurrents belonging to different subalgebras anticommute with each other); the value of k may differ from subalgebra to subalgebra. For each subalgebra we may use equations (2.11) and (2.12) to compute its share of T and $\hat{c}$; for the combined SKM algebra these shares add up. Note that for any non-abelian group G, eq. (2.12) gives $\frac{1}{3}d(\mathrm{G})\leqslant\hat{c}(\mathrm{G})<d(\mathrm{G})$ since $C_{\mathrm{G}}(\mathrm{adj})\leqslant k<\infty$. (In the limit of infinite k, rescaling the currents in (2.9) and (2.11) shows that one actually has an abelian group $\mathrm{U}(1)^{d(\mathrm{G})}$ instead of G.) On the other hand, for an abelian group $\hat{c}(\mathrm{G})=d(\mathrm{G})$ regardless of k; this follows from the fact that abelian supercurrents are free superfields ($J\sim DX$), whose normalization can absorb any positive k.

The SKM algebra, whether simple or not, does not have to constitute the entire internal superconformal theory. In fact, if $\hat{c}^{\mathrm{SKM}}<6$, there should be a 'left-over' piece, whose super-stress tensor $T^{\mathrm{L}}=T^{\mathrm{int}}-T^{\mathrm{SKM}}$ anticommutes with J^a and T^{SKM} and has central charge $\hat{c}^{\mathrm{L}}=6-\hat{c}^{\mathrm{SKM}}$. This left-over piece is rather arbitrary; however, the representations of its super Virasoro algebra must be unitary. The latter restriction is actually rather useful since for $\hat{c}^{\mathrm{L}}<1$ all unitary representations of the super-Virasoro algebra have been classified [40, 19], and the only allowed values of $\hat{c}^{\mathrm{L}}$ are

$$\hat{c}_m=1-\frac{8}{m(m+2)},\qquad m=2,3,4,\dots. \tag{2.13}$$

Thus, either $\hat{c}^{\mathrm{SKM}}\leqslant 5$, or $\hat{c}^{\mathrm{SKM}}=5+8/m(m+2)$ for some integer $m\geqslant 2$ ($m=2$ corresponds to $\hat{c}^{L}=0$ and $\hat{c}^{\mathrm{SKM}}=6$, i.e. $T^{\mathrm{int}}=T^{\mathrm{SKM}}$). Combining this restriction on $\hat{c}^{\rm SKM}$ with (2.12), we can write down all physically allowed SKM algebras and thus give the complete list of 4d gauge groups allowed in type II superstrings:

$$\begin{array}{ll}
\text{(A)} & \mathrm{SU}(2)^6, \\
\text{(B)} & \mathrm{SU}(4)\otimes\mathrm{SU}(2), \\
\text{(C)} & \mathrm{SO}(5)\otimes\mathrm{SU}(3), \\
\text{(D)} & \mathrm{SO}(5)\otimes\mathrm{SU}(2)\otimes\mathrm{SU}(2), \\
\text{(E)} & \mathrm{SU}(3)\otimes\mathrm{SU}(3), \\
\text{(F)} & \mathrm{G}_2, \\
\text{(G)} & \text{all proper subgroups of (A)–(F).}
\end{array} \tag{2.14}$$

Central charges of the allowed SKM algebras are also restricted: algebras (A), (B), (C), (E) and (F) must have $\hat{k}=0$ for each of their simple factors; however, $\hat{k}>0$ is allowed in cases (D) and (G).

2.4. RIGHT-MOVING SUPERCURRENTS

There are two remarks to be made with respect to the above result. First, we have not shown that all the groups in (2.14) can actually be realized as 4d gauge groups of consistent string models; this will be the subject of the next section. Second, we have concentrated on the holomorphic SKM algebras, but have ignored the anti-holomorphic ones. If both kinds of SKM algebras were present in the same string model, we could have doubled the gauge group. We are now going to show that this never happens in models with massless fermions.

In the type II superstring theory fermions belong to the (R, NS) and (NS, R) sectors of the Hilbert space; for the sake of definiteness, we will demand the presence of massless (R, NS) fermions and concentrate on the consequences for the left-handed currents. The key fact about massless states $|\psi\rangle$ in the Ramond sector is that they satisfy

$$G_0^{\rm int}|\psi\rangle = 0. \tag{2.15}$$

The easiest way to deduce (2.15) is to use the light-cone relation between masses of states and eigenvalues of L_0: $(L_0^{\rm l.c.}-\frac{1}{2})|\psi\rangle = \frac{1}{4}\alpha'{\rm mass}^2|\psi\rangle$*. In the Ramond sector, $L_0^{\rm l.c.}-\frac{1}{2}=(G_0^{\rm l.c.})^2$, so massless Ramond states are annihilated by $(G_0^{\rm l.c.})^2$. Now $G_0^{\rm l.c.}$ is a sum of $G_0^{\rm int}$ and the transverse part of $G_0^{\rm 4d}$, which are both hermitian and

* The definition of $L_0^{\rm l.c.}$ used here was chosen for compatibility with the super Virasoro algebra (2.3); it differs from the usual light-cone convention for the Ramond sector by an additive $\frac{1}{2}$.

anticommute with each other. By squaring this decomposition, we immediately see that any state annihilated by $(G_0^{\text{l.c.}})^2$ is also annihilated by $(G_0^{\text{int}})^2$ – and thus by G_0^{int} itself. This verifies (2.15). Alternatively, we could have proved (2.15) in the covariant Neveu-Schwarz-Ramond formalism with the help of the Dirac-Ramond equation $G_0^{\text{tot}}|\psi\rangle = 0$ and the fact that massless states obey $G_0^{\text{4d}}|\psi\rangle = 0$.

Now consider the SKM algebra for the non-abelian (simple) group G. As shown by Kac and Todorov [14], eigenvalues of L_0^{SKM} in the Ramond sector are given by $h^{\text{SKM}} = \frac{1}{16}d(\text{G}) +$ positive terms. Using $G_0^2 = L_0 - \frac{1}{16}\hat{c}$ and (2.12), we find that for any non-null (physical) state $|\psi\rangle$:

$$\langle\psi|(G_0^{\text{int}})^2|\psi\rangle \geqslant \langle\psi|(G_0^{\text{SKM}})^2|\psi\rangle \geqslant \frac{d(\text{G})C_{\text{G}}(\text{adj})}{24k}\langle\psi|\psi\rangle > 0 \qquad (2.16)$$

and eq. (2.15) cannot be satisfied. Thus, if a string model has any non-abelian factor in the gauge group coming from a holomorphic, i.e. left-moving, SKM algebra, then there are no massless (R, NS) fermions in that model; there are also no massless (R, R) bosons. Similarly, if the source of a non-abelian gauge group is a right-moving (anti-holomorphic) SKM algebra, then there are no massless fermions of the (NS, R) type. Consequently, if both left-moving and right-moving non-abelian SKM algebras are present, then there are no massless fermions at all [19]! In order to avoid such a calamity, only left-moving SKM algebras should be allowed in string models of phenomenological interest; then all massless bosons come from the (NS, NS) sector and all massless fermions come from the (NS, R) sector. (We choose this convention rather than its opposite by analogy with the heterotic string convention.) As a corollary, *all type II superstring models that treat left- and right-moving degrees of freedom symmetrically are excluded.*

The problem with the above argument is that it does not apply to abelian gauge groups. And in fact there do exist string models with massless fermions and with abelian gauge bosons coming from both left- and right-moving internal algebras. For example, the ten-dimensional superstring compactified on a torus has all of these features. However, one can show that massless (NS, R) fermions are always neutral with respect to charges arising from any right-handed currents [20]. (Similarly any massless (R, NS) fermions are neutral with respect to left-handed charges.) The argument is based on (2.15) and on commutation relations (2.7). Let $|\psi\rangle$ be a massless (NS, R) state and let $\bar{\boldsymbol{J}}^a(\bar{z}, \bar{\theta})$ be a right-moving supercurrent. The charge operator is $\bar{J}_0^a$ – the zero mode of the bosonic part of $\bar{\boldsymbol{J}}^a$ – so the charge of $|\psi\rangle$ is given by

$$\langle\psi|\bar{J}_0^a|\psi\rangle = \langle\psi|\{\bar{\mathscr{J}}_0^a, \bar{G}_0^{\text{int}}\}|\psi\rangle\,.$$

But the latter expression vanishes because a massless state $|\psi\rangle$ satisfies $\langle\psi|\bar{G}_0^{\text{int}} = 0 = \bar{G}_0^{\text{int}}|\psi\rangle$ (this is just the right-moving version of (2.15) and its hermitian conjugate).

While decoupling of the abelian gauge groups generated by right-moving supercurrents from massless fermions is already sufficient to render such groups phenomenologically worthless, their very presence also causes phenomenological problems. Specifically, the existence of a free right-moving superfield $\bar{J}^a$ is incompatible with having a chiral 4d gauge theory. The easiest way to see this is to notice that the contribution of the fermionic current $\bar{\mathcal{J}}^a$ to the stress tensor $\bar{T}_B$ and to the GSO projector $(-1)^{\bar{F}}$ is in no way distinguishable from contributions of fermionic partners $\bar{\psi}^\mu$ of space-time dimensions X^μ. So, as far as right-moving fermions are concerned, there are five (or more) flat space-time dimensions, just like in the case of a toroidal compactification. Consequently, Ramond states form spinor representations not just of the ordinary Lorentz group SO(1, 3), but of SO(1, 4) (or an even bigger Lorentz-like group, if there are several supercurrents $\bar{J}^a$). With respect to SO(1, 3), these bigger spinors transform as non-chiral pairs $(\mathbf{2}+\bar{\mathbf{2}})$ of Lorentz spinors. Thus, for every left-handed 4d fermion there is a right-handed one with the same internal quantum numbers and vice versa; in particular, all gauge interaction are vector-like.

Let us now summarize the results of this section. We have considered the most general classical solutions of the type II superstring that incorporate flat four-dimensional Minkowski space-time and thus lead to effective 4d field theories at sufficiently low energies. If the effective 4d theory contains both a non-abelian gauge group and massless fermions that are chiral (these are basic features of any realistic model), then the gauge group is one of the groups listed in (2.14). Moreover, only left-right asymmetric solutions can have these features: The gauge group should be generated by the left-moving SKM algebra, while massless fermions come just from the (NS, R) sector of the superstring; also, only the (NS, NS) sector contains massless bosons.

3. Explicit models via asymmetric orbifolds

3.1. GENERAL CONSIDERATIONS

In this section we show how to construct various 4d type II models which explicitly realize many of the gauge groups permitted by the considerations of the previous section. Those considerations involve only classical (tree-level) string physics; they rely only on local properties of a 2d superconformal theory as formulated on the two-sphere, without regard to the theory's consistent extension to Riemann surfaces of higher genus. We are now interested in constructing models which are also consistent at the quantum level, i.e. for which higher-loop string amplitudes can be defined. Thus we require the 2d theory to be invariant under modular transformations – diffeomorphisms of higher-genus surfaces which are not continuously connected to the identity.

We could try to construct modular-invariant models directly in terms of the representations of the super-Kac-Moody algebra which they contain. At one loop this might be feasible, using the known modular transformation properties of partition functions for representations of Kac-Moody algebras [21]. However at higher loops this approach appears difficult. On the other hand, orbifold techniques provide an efficient way to construct string models which are modular-invariant to all orders in the loop expansion [22]. The general orbifold construction starts with some initial modular-invariant string model with a first-quantized action and Hilbert space which both have some discrete symmetry P; then one twists the initial model by the group P in order to construct a new model. The procedure of twisting a model by P has two steps. One first projects out those states in the original Hilbert space which are not invariant under P. The surviving states form the *untwisted sector* of the new model. Then one adds states in *twisted sectors*, where the boundary conditions on the fields in the original action are modified – the string need only close modulo an element of P. If the group P is non-abelian the twisting procedure is slightly more complicated; for details see reference [4]. We will be using only abelian (in fact cyclic) groups P in this paper.

The orbifolds we consider here start from a generalized toroidal compactification [23, 24] of the type II superstring. The metric of the internal six-dimensional space is that of a torus, but we also allow for constant background values $\langle B^{ij}\rangle$ of the anti-symmetric tensor field. In any compactification of this kind, the eigenvalues $p^i_{\rm L}$ and $p^i_{\rm R}$ of the left- and right-moving zero-modes of the internal coordinates $X^i(z,\bar z)$ $(i=1,\dots,6)$ lie on an even, self-dual lorentzian lattice of signature (6,6), $\Gamma^{6,6}=\{(p^i_{\rm L},p^i_{\rm R})\}$.* These models have $N=8$ supersymmetry in four dimensions, with four supersymmetries each coming from left- and right-moving fermion zero-modes. The gauge group is $\mathrm{U}(1)^6\otimes \mathrm{U}(1)^6$, with 6 U(1)'s each from left- and right-moving supercurrents (the internal free superfields DX^i and $\bar D X^i$).

To construct models with more interesting gauge groups, we will twist these $N=8$ models by rotations of the lattice $\Gamma^{6,6}$ taking lattice points to lattice points, i.e. automorphisms of $\Gamma^{6,6}$. We will accompany these rotations by translations or *shifts* $(v^i_{\rm L},v^i_{\rm R})$ on the lattice. The general transformation we consider therefore takes

$$X^i_{\rm L}(z)\to\omega^{ij}_{\rm L}X^j_{\rm L}(z)+2\pi v^i_{\rm L},$$

$$X^i_{\rm R}(\bar z)\to\omega^{ij}_{\rm R}X^j_{\rm R}(\bar z)+2\pi v^i_{\rm R}, \tag{3.1}$$

where $\omega_{\rm L}$ and $\omega_{\rm R}$ are SO(6) rotations and the combined rotation $(\omega_{\rm L},\omega_{\rm R})$ is an

* Strictly speaking, all lattices $\Gamma^{6,6}$ are equivalent under SO(6,6) rotations; but the generic SO(6,6) rotation does not commute with the left- and right-moving hamiltonians L_0 and $\bar L_0$, so different embeddings of $\Gamma^{6,6}$ with respect to the division into left- and right-moving zero-modes lead to different physics. When we refer to *a* lattice $\Gamma^{6,6}$ we really mean an embedding of *the* lattice $\Gamma^{6,6}$ into the momentum space of left- and right-moving zero modes.

automorphism of $\Gamma^{6,6}$. Eq. (3.1) is a somewhat schematic representation of an asymmetric twist, because there is only one center-of-mass coordinate for the field $X^i(z,\bar{z}) = X^i_{\rm L}(z) + X^i_{\rm R}(\bar{z})$. The proper definition is as an action on the bosonic Hilbert space, which takes $|p_{\rm L}; p_{\rm R}\rangle \to e^{2\pi i(p_{\rm L}\cdot v_{\rm L} - p_{\rm R}\cdot v_{\rm R})}|\omega_{\rm L} p_{\rm L}; \omega_{\rm R} p_{\rm R}\rangle$, etc. [8]. In order to preserve the world-sheet supersymmetry generators $T_{\rm F}(z) = -\frac{1}{2}\psi^i\,\partial X^i$ and $\bar{T}_{\rm F}(\bar{z}) = -\frac{1}{2}\bar{\psi}^i\,\bar{\partial} X^i$ (see the previous section), the fermions ψ^i and $\bar{\psi}^i$ must transform in the same way as ∂X^i and $\bar{\partial} X^i$ respectively:

$$\psi^i(z) \to \omega^{ij}_{\rm L}\psi^j(z),$$
$$\bar{\psi}^i(\bar{z}) \to \omega^{ij}_{\rm R}\bar{\psi}^j(\bar{z}). \tag{3.2}$$

So we can denote an arbitrary element of P by $(\omega_{\rm L}, v_{\rm L}; \omega_{\rm R}, v_{\rm R})$; this completely specifies its action on the fields X^i, ψ^i and $\bar{\psi}^i$.

In the previous section we saw that if there are to be any massless fermions in a 4d model then all gauge bosons that couple to them must arise from, say, left-moving supercurrents. Now we would like to show that in an orbifold model these gauge bosons come only from the sectors of the Hilbert space that are twisted by elements of P of the form $(\omega_{\rm L}, v_{\rm L}; 1, 0)$, that is, from the sectors in which the right-moving fields $X^i_{\rm R}$ and $\bar{\psi}^i$ are completely untwisted. Consider the vector index μ on a gauge boson whose gauge index is given by left-moving supercurrents, as in eq. (2.4). The vector index requires the use of a right-moving Minkowski oscillator such as $\bar{\psi}^{\mu}_{-1/2}$, which raises the $\bar{L}_0$ eigenvalue (the 'right-moving energy') by at least one-half unit. (Note that the Minkowski field $\bar{\psi}^{\mu}$ is necessarily untwisted.) We can ignore the 'left-moving energy' (L_0 eigenvalue) in this discussion. Given the energy cost of the vector index, a massless vector particle requires the vacuum energy to be $-\frac{1}{2}$ or lower. The untwisted NS sector indeed has vacuum energy $-\frac{1}{2}$; in order to complete the proof we have to show that all twisted sectors (with respect to the right-moving fields) have higher vacuum energies. This follows from the fact that a complex free boson with boundary conditions twisted by $e^{2\pi i\beta}$ with $0 \leqslant \beta < 1$ (i.e. $\partial X(e^{2\pi i}z) = e^{2\pi i\beta}\,\partial X(z)$) contributes

$$\tfrac{1}{2}\beta(1-\beta) - \tfrac{1}{12} \tag{3.3}$$

to the vacuum energy [4], while the contribution of a complex fermion with the same boundary conditions has the same absolute value but the opposite sign. The expression (3.3) is minimized for periodic bosons ($\beta = 0$) and anti-periodic fermions ($\beta = \frac{1}{2}$), just the combinations which occur in the untwisted NS sector; any other combination will thus have higher vacuum energy. This argument excludes any non-trivial rotation $\omega_R \neq 1$,* and the case against a non-zero shift v_R is even simpler: Sectors with $v_R \neq 0$ do not contain states with $p_R = 0$, and $p_R \neq 0$ costs $\frac{1}{2}p_R^2 > 0$ energy units.

Consider now the model obtained by twisting the initial toroidal compactification, not by P but by the (normal) subgroup P_L of P consisting of all elements of P of the form $(\omega_L, v_L; 1, 0)$. Then the part of the gauge group of the P model coming from left-handed currents is contained in the gauge group of the P_L model. This is because, as we have just shown, all such P gauge bosons come from sectors twisted by elements of P_L. If they survive the projection onto P-invariant states, then clearly they also survive the projection onto P_L-invariant states, since $P_L \subseteq P$.** So if we are interested only in the maximal gauge symmetries of 4d type II models, we can confine ourselves to studying the $P = P_L$ models, in which the right-handed fields have been left completetly untwisted – thus they are examples of *asymmetric orbifolds* [8].

If we are to generate models with non-abelian gauge symmetry, we know from sect. 2 that there must be no massless (R, NS) states. In particular we must break all four left-moving supersymmetries, by projecting out the four (R, NS) gravitinos of the toroidal compactification. This can be done by choosing rotations ω_L which do not sit in an SU(3) subgroup of SO(6). However, we must also ensure that no massless (R, NS) states appear in any of the twisted sectors. But in the Ramond sector, whether twisted or not, the vacuum energy contributions from X_L^i and from ψ^i always cancel. One way to avoid massless twisted (R, NS) states is by having a nonzero shift v_L in each twisted sector. Then the lowest energy (R, NS) state has nonzero momentum p_L and therefore positive energy, $\frac{1}{2}p_L^2$. Since the P_L models of interest all have $N = 4$ space-time supersymmetry, we will refer to them as $N = 4$ models, and to the generalized toroidal compactifications as $N = 8$ models.

The right-moving $U(1)^6$ gauge symmetry of the $N = 8$ models survives in the $N = 4$ models, but the massless (NS, R) fermions are neutral with respect to this symmetry (see sect. 2), which will in any case be broken if one breaks the space-time supersymmetry to $N = 1$ by further twisting. So we will omit the $U(1)^6$ factor when discussing the gauge group of an $N = 4$ model. Starting from a few different lattices $\Gamma^{6,6}$ and twisting by discrete groups P_L such as Z_2 or Z_3, we can generate a variety of $N = 4$ models, whose gauge groups include all even rank groups in the list (2.14). (Some of our constructions use a succession of cyclic twists.) We would like to give some details of the orbifold construction of these models, but first we need to

* We must be a little more careful in the exceptional case of the twist which rotates one of the right-handed superfields by 2π ($\beta = 1$). This twist does not actually change the boundary conditions on any of the fields, and hence does not change the vacuum energy. However, it does alter the GSO projection (see below), so that the state $\bar{\psi}^{\mu}_{-1/2}$ is projected out by GSO. Hence there are no twisted massless vectors in this case either.

** If P is non-abelian, the projections are not in general onto P-invariant states, but one can show that the same conclusion still holds, using the fact that P_L is a normal subgroup of P.

describe a special class of lattices $\Gamma^{6,6}$ which have automorphisms acting only on the left-moving zero-modes p^i_L.

The generic toroidal compactification does not admit purely left-moving symmetries. However, suppose we choose [23] p_L and p_R to lie on the weight lattice $\Lambda_W(\mathcal{G})$ of a semi-simple, simply-laced Lie algebra $\mathcal{G}$ of rank six, and restrict the difference $p_L - p_R$ to lie on the root lattice $\Lambda_R(\mathcal{G})$. If $\Lambda_R(\mathcal{G})$ is normalized so that all roots α of $\mathcal{G}$ have length $\alpha^2 = 2$, then $\Lambda_R(\mathcal{G})$ is even and is the dual lattice of $\Lambda_W(\mathcal{G})$, and it is easily verified that the lattice

$$\Gamma^{6,6}(\mathcal{G}) \equiv \{(p_L, p_R)\}, \qquad p_L, p_R \in \Lambda_W(\mathcal{G}), \qquad p_L - p_R \in \Lambda_R(\mathcal{G}),$$

is even and self-dual. The Lie algebra $\mathcal{G}$ should not be confused with the gauge group G of a 4d model constructed using $\Gamma^{6,6}(\mathcal{G})$; the two groups bear no direct relation to each other.

Consider now which automorphisms of $\Gamma^{6,6}(\mathcal{G})$ can act only on the p_L. Clearly they must take $(\alpha, 0)$ to $(\alpha', 0)$, where α and α' are roots of $\mathcal{G}$, so they must be elements of the automorphism group of $\Lambda_R(\mathcal{G})$. But the automorphisms must also take a weight λ in $\Lambda_W(\mathcal{G})$ to weights λ' differing from λ by a root vector; this means they must be in the Weyl group of $\mathcal{G}$ – the group of automorphisms generated by reflections in the hyperplanes orthogonal to the simple roots of $\mathcal{G}$. Indeed, a reflection in the hyperplane orthogonal to α takes $\lambda \to \lambda - 2(\alpha \cdot \lambda)\alpha$, which differs from λ by a root vector, since $\alpha \cdot \lambda$ is an integer. The remaining (outer) automorphisms of the root lattice correspond to symmetries of the Dynkin diagram for $\mathcal{G}$; they exchange weights of representations which transform differently under the center of the Lie group for $\mathcal{G}$ and hence do not differ by root vectors – for example $N \leftrightarrow \overline{N}$ for SU(N). Thus the group of automorphisms acting only on the left-moving modes p_L of the lattice $\Gamma^{6,6}(\mathcal{G})$ is precisely the Weyl group $\mathcal{W}(\mathcal{G})$.

As far as the bosons X^i are concerned, any automorphism of the lattice $\Gamma^{6,6}$ is an allowed rotation. However, the fermions ψ^i impose an additional condition, namely that ω_L (and ω_R) should preserve parity, i.e., be an element of SO(6) rather then O(6). Note that this restriction is important for selecting an appropriate subgroup $P_L \subset \mathcal{W}(\mathcal{G})$ since the Weyl group $\mathcal{W}(\mathcal{G})$ does contain parity-violating reflections. The reason parity must be preserved is that a Ramond sector twisted by a parity-violating rotation has an odd number of (real) fermionic zero modes and the GSO projection cannot be defined. On the other hand, given a well-defined GSO projection one can distinguish between two twists that differ by a 2π rotation around some axis and thus act on the bosons in exactly the same way. Therefore, the proper way to describe a rotation ω_L is to treat it as an element of Spin(6) ≡ SU(4) rather then SO(6) or O(6).

The only physical condition on P_L other than $P_L \subset \text{Spin}(6) \cap \mathcal{W}(\mathcal{G})$ is that the orbifold model generated by P_L should be modular invariant. Modular invariance at one loop, for the case of an abelian group P_L, reduces to a simple 'level-matching' condition [25]: In any twisted sector there should exist physical states, which of course must satisfy $L_0 = \bar{L}_0$. In a sector twisted by an element of order l, applying twisted oscillators to the vacuum changes the L_0, $\bar{L}_0$ eigenvalues by multiples of $1/l$. Therefore the left- and right-moving vacuum energies in that sector should differ by multiples of $1/l$. This condition constrains the lengths of shifts v_L which can accompany a given rotation ω_L. One-loop modular invariance for the explicit models we construct below is easily verified just by noting that there are physical states in every twisted sector. It appears that the level-matching conditions are sufficient for an abelian orbifold to be modular invariant to all orders; this has been proved [22] when P acts arbitrarily on any free world-sheet fermions but acts symmetrically on the bosons. The models constructed below involve asymmetric twists of bosons, so the arguments of ref. [22] do not apply directly. However, some of the models can alternatively be described entirely in terms of free fermions; then level-matching does suffice for all-loop modular invariance.

3.2. SU(2)^6 MODEL

The first $N=4$ model we will construct is the SU(2)^6 model of refs. [9, 6, 7], which is also a useful intermediate step in the orbifold construction of the two other models of [6, 7], with gauge groups SU(4) ⊗ SU(2) and SO(5) ⊗ SU(3). The starting point for these three $N=4$ models is the $N=8$ model which uses the lorentzian lattice $\Gamma^{6,6}(D_6)$. The most convenient way to write this lattice is as a union of four conjugacy classes of D_6 weight vectors. That is,

$$\Gamma^{6,6}(D_6) = (A_6; A_6) + (V_6; V_6) + (S_6; S_6) + (C_6; C_6), \tag{3.4}$$

where in an orthonormal basis

$$\begin{aligned}
A_{2l} &= \{(n_1, \dots, n_{2l})\}, \quad && n_i \in Z, \quad \textstyle\sum n_i \in 2Z, \\
V_{2l} &= \{(n_1, \dots, n_{2l})\}, \quad && n_i \in Z, \quad \textstyle\sum n_i \in 2Z+1, \\
S_{2l} &= \{(n_1, \dots, n_{2l})\}, \quad && n_i \in Z+\tfrac{1}{2}, \quad \textstyle\sum n_i \in 2Z, \\
C_{2l} &= \{(n_1, \dots, n_{2l})\}, \quad && n_i \in Z+\tfrac{1}{2}, \quad \textstyle\sum n_i \in 2Z+1,
\end{aligned} \tag{3.5}$$

define the four classes of representations of the Lie algebra D_{2l}: adjoint, vector, spinor and conjugate spinor. Two weight vectors within a given class differ from each other by a root vector.

The $N=8$ model using the lattice $\Gamma^{6,6}(D_6)$ can alternatively be described in terms of free fermions. One replaces the six bosonic coordinates $X^i(z, \bar{z})$ by twelve left-moving and twelve right-moving Majorana-Weyl fermions, denoted by λ^i, μ^i and $\bar{\lambda}^i$, $\bar{\mu}^i$ respectively. Then one sums over the 2^{2g} spin structures for the Riemann surface of genus g, treating all twenty-four fermions as one group. The sum over spin structures at genus one is a sum over periodic and antiperiodic boundary conditions in the σ and τ directions on the torus. States in the sector of the Hilbert space with anti-periodic (periodic) boundary conditions in σ correspond to the entries in (3.4) all being integers (half-integers); the sum over τ boundary conditions is a GSO projection which discards the $(A_6; V_6)$, $(V_6; A_6)$, $(S_6; C_6)$ and $(C_6; S_6)$ classes. The free-fermion interpretation of this $N=8$ model will allow us to show explicitly an equivalence between the three $N=4$ models described in [6, 7] and our orbifold construction of them. (We will use other lattices later in the section which do not have as simple an interpretation in terms of free fermions.)

The $SU(2)^6$ model is obtained from the above $N=8$ model by twisting by the group $P=P_L=Z_2$ generated by

$$\omega_L=(-1)^{F_L}=(e^{2\pi i J_{12}})_L\,,\qquad v_L=(1,0^5)\,,\qquad \omega_R=1\,,\qquad v_R=0\,.$$

Here $(e^{2\pi i J_{12}})_L$ denotes a rotation of a pair of internal left-moving free superfields, say DX^1 and DX^2, by an angle of 2π in the 1-2 plane. The shift v_L is written in the orthonormal basis for $\Gamma^{6,6}(D_6)$ with exponents standing for repeated entries, so v_L is a weight in the vector representation of D_6.* Let us concentrate first on the massless spectrum of this model, using light-cone gauge for simplicity, and denoting the eight left- (right-) moving NSR fermions by $\psi^{\mu,i}$ ($\bar{\psi}^{\mu,i}$), where $\mu=1,2$ is a light-cone Minkowski index and $i=1,2,\ldots,6$ is an internal index. The 256 massless states of the initial $N=8$ model are the tensor product of a left-moving and a right-moving $N=4$ supermultiplet. Of the eight massless left-moving NS states, the pair $\psi^{\mu}_{-1/2}|0\rangle$ has helicity ± 1, and the remaining six states $\psi^{i}_{-1/2}|0\rangle$ have helicity 0. The eight Ramond states, $\psi_0^{i_1}\ldots\psi_0^{i_k}|0\rangle$, form four helicity $\pm\frac{1}{2}$ pairs. The right-movers have the same set of helicities. States in the left-moving Neveu-Schwarz sector are unaffected by the rotation ω_L; however, left-moving Ramond states acquire a minus sign. The shift v_L does not affect the massless states because they carry no momentum on the lattice. So the projection onto P-invariant states simply removes the 128 (R, NS) and (R, R) states from the spectrum, leaving an $N=4$ supergravity multiplet and a $U(1)^6$ $N=4$ Yang-Mills multiplet in the untwisted sector.

In the twisted sector the Fock space for the ψ modes is the same as in the untwisted sector, but the GSO projection is now onto states with an odd number of ψ^{μ}, ψ^{i} excitations (see below). The bosonic zero-modes now lie on the *shifted* lattice

$$\Gamma^{6,6}(D_6)+v_L=(V_6;A_6)+(A_6;V_6)+(C_6;S_6)+(S_6;C_6)\,. \tag{3.6}$$

* Precisely the same group element can be used to obtain the non-supersymmetric SO(32) heterotic string from the supersymmetric $\mathrm{Spin}(32)/Z_2$ heterotic string in ten dimensions [26], if v_L is taken to be a vector weight of D_{16} rather than D_6, and of course $(-1)^{F_L}$ becomes $(-1)^{F_R}$ since the heterotic string NSR fermions are right-moving.

The NS vacuum $|0\rangle$ with energy $-\frac{1}{2}$ can be raised up to the massless level using the 12 momentum states $(\underline{\pm 1, \ldots, 0}; 0, \ldots, 0)$ (the underline denotes permutations of the six left-moving coordinates). These 12 modes (with zero space-time helicity), when tensored with the right-moving $N=4$ supermultiplet, complete the gauge symmetry from $U(1)^6$ to $SU(2)^6$. To verify that the gauge group is $SU(2)^6$ we should compute the charges of the additional 12 vector bosons under the $U(1)^6$ Cartan subalgebra; we will perform this check shortly using an alternative formulation of the same model.

The $N=4$ $SU(2)^6$ model, like the $N=8$ $\Gamma^{6,6}(D_6)$ model, can be described entirely in terms of free fermions. The only difference is that the spin structures for the eight left-moving NSR fermions $\psi^{\mu,i}$ and for the twenty-four fermions $\lambda^i, \mu^i, \bar{\lambda}^i, \bar{\mu}^i$ are now summed over together, whereas in the $N=8$ model they were summed over independently. To see the equivalence between these two descriptions, we first need to express the translation $v_L = (1, 0^5)$ on the lattice $\Gamma^{6,6}(D_6)$ as an action on the fermions. It is convenient to group the fermions into complex pairs, defining

$$\chi^i \equiv \sqrt{\tfrac{1}{2}}\,(\lambda^i + i\mu^i), \qquad \phi^i \equiv \sqrt{\tfrac{1}{2}}\,(\lambda^i - i\mu^i), \qquad \bar{\chi}^i \equiv \sqrt{\tfrac{1}{2}}\,(\bar{\lambda}^i + i\bar{\mu}^i), \qquad \bar{\phi}^i \equiv \sqrt{\tfrac{1}{2}}\,(\bar{\lambda}^i - i\bar{\mu}^i).$$

(Note that the bars refer to complex conjugation on the world-sheet rather than on the space-time torus.) In general, the translation of a boson by β, $X \to X + 2\pi\beta$, is equivalent to the rotations $\chi \to e^{2\pi i\beta}\chi$, $\phi \to e^{-2\pi i\beta}\phi$ acting on the corresponding complex fermions. This is apparent from the bosonization dictionary:*

$$\chi = -i\,e^{iX}, \qquad \phi = -i\,e^{-iX}, \qquad \partial X = i\chi\phi. \tag{3.7}$$

In the case at hand, $\beta_1 = 1$ and $\beta_i = 0$ for $i = 2, \ldots, 6$. For a single fermion pair χ and ϕ with $\beta = 1$, the boundary conditions are unchanged, so it might seem that the fermion twist corresponding to v_L acts trivially. However, if one continuously varies β from 0 to 1, one finds that the fermion number of the vacuum state for $\beta = 1$ has been changed by one unit from that for $\beta = 0$. Therefore the GSO projection onto even fermion states in the twisted sector selects precisely those states which were projected out by GSO in the untwisted sector. (See the first example in the appendix of the second reference in [4].) In order to project onto P-invariant states we also need the eigenvalues of states in the untwisted sector under the twist, remembering that there are two subsectors, with χ and ϕ either both anti-periodic or both periodic. For $\beta = 1$ the modes of χ and ϕ have eigenvalue 1, so all states in the anti-periodic (periodic) subsector of the untwisted sector have the same eigenvalue: $+1$ (-1). Note that the action of v_L on the fermions χ^i and ϕ^i is identical to the action of $\omega_L = (-1)^{F_L}$ on the NSR fermions $\psi^{\mu,i}$, so the same results hold for the ψ system. In the twisted sector the condition $L_0 = \bar{L}_0$ requires the states in the anti-periodic (periodic) subsector with respect to χ and ϕ to be tensored with the

* Our conventions for operator products of free bosons ∂X and free Majorana fermions λ are $\partial_z X \partial_w X \sim -1/(z-w)^2 + \cdots$ and $\lambda(z)\lambda(w) \sim -1/(z-w)$.

NS (R) states of the ψ system. Combining the results for the two systems, one finds that in both the untwisted and twisted NS (R) sectors – i.e., $\psi^{\mu,i}$ anti-periodic (periodic) – χ^i and ϕ^i are also anti-periodic (periodic), and the projection is onto states with even total fermion number $F_\psi + F_{\chi,\phi}$. This verifies that the spin structures for $\psi^{\mu,i}$ and χ^i, ϕ^i are now summed over together (along with that for $\bar{\chi}^i, \bar{\phi}^i$).

Thus the 18 'internal' fermions ψ^i, χ^i, ϕ^i are on an equal footing now, and in fact they are just the fermionic components $\mathscr{J}^a(z)$ of the supercurrents $\boldsymbol{J}^a(z,\theta)$ for the $SU(2)^6$ SKM algebra, up to a normalization factor. To check that the group is $SU(2)^6$, one needs the bosonic currents $J^a(z)$, because the $\mathscr{J}_r^a$ commutation relations are just those of free fermions, for any SKM algebra. The $J^a(z)$ are obtained from the $\mathscr{J}^a(z)$ by a world-sheet supersymmetry transformation, generated by

$$T_F^{\text{int}} = -\tfrac{1}{2}\psi^i\,\partial X^i = -\tfrac{1}{2}i\sum_i \psi^i\chi^i\phi^i = -\tfrac{1}{2}\sum_i \psi^i\lambda^i\mu^i\,. \tag{3.8}$$

Let us consider a supersymmetry generator of the general form

$$T_F(z) = \frac{f_G^{abc}}{3\sqrt{2C_G(\text{adj})}}\eta^a(z)\eta^b(z)\eta^c(z), \tag{3.9}$$

where $\eta^a(z)$ are free Majorana fermions, and f_G^{abc} are the structure constants for a (simple) Lie algebra G. One can verify [14, 42, 27] that $T_F(z)$ in (3.9) has the appropriate super-Virasoro operator product relations. It also generates from the fermions $\eta^a(z)$ the bosonic currents

$$J^a(z) = \tfrac{1}{2}if_G^{abc}\eta^b(z)\eta^c(z),$$

which obey an ordinary Kac-Moody algebra with $k = C_G(\text{adj})$, corresponding to the minimal value $\hat{k}=0$ for the SKM algebra. Note that $\mathscr{J}^a(z) = -i(\tfrac{1}{2}C_G(\text{adj}))^{1/2}\eta^a(z)$ are the correctly normalized fermionic currents, so that we can also write

$$J^a(z) = \frac{-i}{k}f_F^{abc}\mathscr{J}^b(z)\mathscr{J}^c(z) \tag{3.10}$$

for this minimal case.

Obviously the six terms of T_F^{int} in (3.8) are of the form (3.9) with $G = SU(2)$ ($f^{abc} = \varepsilon^{abc}$ and $C_G(\text{adj}) = 2$), so all we need to complete the identification of the gauge group as $G = SU(2)^6$ is to check that all of the supercurrents correspond to physical states, i.e. are not projected out. This is indeed the case here: the tensor products of the 18 left-moving states $\psi^i_{-1/2}|0\rangle, \chi^i_{-1/2}|0\rangle, \phi^i_{-1/2}|0\rangle$ with the right-moving helicity $\pm\frac{1}{2}$ pair $\bar{\psi}^\mu_{-1/2}|0\rangle$ are all physical. (Note that the same 18 free fermions were present in the original $N=8$ model as here, but 12 of the above 18 states were projected out by the sum over spin structures, leaving only an abelian gauge group.)

We conclude the discussion of the $SU(2)^6$ model with the remark that just as we have fermionized X^i, we can now bosonize the entire system of 20 left-moving fermions $\psi^i, \chi^i, \phi^i, \psi^\mu$ and 12 right-moving fermions $\bar{\chi}^i, \bar{\phi}^i$. The spin structures of all these fermions are summed over together, so the zero-mode eigenvalues for the respective bosons belong to the set

$$(V_{10}; A_6) + (A_{10}; V_6) + (C_{10}; S_6) + (S_{10}; C_6), \qquad (3.11)$$

where $A_{2l}, V_{2l}, S_{2l}, C_{2l}$ are defined in (3.5). In our convention the last left-moving boson (H^{10}) corresponds to the two light-cone Minkowski fields ψ^μ, $\mu = 1,2$. The set (3.11) is not a lattice because we are using light-cone gauge here; however, had we used the covariant formulation and incorporated the bosonized superconformal ghost zero-modes, we would have found an odd self-dual lorentzian lattice of signature (11,7) [28,29].

3.3. SU(4) ⊗ SU(2) AND SO(5) ⊗ SU(3) MODELS

In this subsection we will construct the other two $N = 4$ models of refs. [6,7] as orbifolds. The SU(4) ⊗ SU(2) model can be obtained by twisting the $N = 8$ $\Gamma^{6,6}(D_6)$ model by the group $P = P_L = Z_4$ generated by

$$\omega_L = \text{diag}(C_2, C_2, 1, 1), \qquad v_L = (0^4, \tfrac{1}{2}, 0), \qquad \omega_R = 1, \qquad v_R = 0. \quad (3.12)$$

Here ω_L and v_L are written in the orthonormal basis for $\Gamma^{6,6}(D_6)$ described above, and $C_2 = \begin{pmatrix} 0 & 1 \\ 1 & 0 \end{pmatrix}$ exchanges two bosonic coordinates. Note that ω_L appears to have order two, but the actual order of the twist (3.12) acting on the Hilbert space is four, due to cocycle factors which appear in twisted vertex operators [30]. The square of this transformation is precisely the Z_2 transformation used in the previous subsection to construct the $SU(2)^6$ model. Therefore we can alternatively describe the SU(4) ⊗ SU(2) model as a Z_2 orbifold of the $SU(2)^6$ model by re-interpreting (3.12) as a Z_2 transformation of the 2d conformal fields of the latter model. Since the $SU(2)^6$ model involves 18 real left-moving fermions ψ^i, λ^i, μ^i all on the same footing, it is convenient to write the Z_2 action directly on these fermions. Clearly ω_L just exchanges $\psi^1 \leftrightarrow \psi^2$, $\psi^3 \leftrightarrow \psi^4$, $\lambda^1 \leftrightarrow \lambda^2$, $\lambda^3 \leftrightarrow \lambda^4$, $\mu^1 \leftrightarrow \mu^2$, $\mu^3 \leftrightarrow \mu^4$; whereas v_L takes $\psi^5 \leftrightarrow \psi^5$, $\lambda^5 \leftrightarrow -\lambda^5$, $\mu^5 \to -\mu^5$.

We would like to show that this model is equivalent to the SU(4) ⊗ SU(2) model of [6,7]. The general strategy is to alternately bosonize and fermionize the model defined by (3.12), until it is rewritten in terms of fermions which are totally untwisted, i.e. have the same set of boundary conditions (spin structure) as the original $SU(2)^6$ model. Then one writes the world-sheet supersymmetry generator $T_F(z)$ in terms of the new, untwisted fermions, and one discovers that T_F^{int} takes the form (3.9) with group G = SU(4) ⊗ SU(2) – the form in which the model has been described in refs. [6,7].

The first step is to diagonalize the Z_2 action of P on the fermions by defining

$$\tilde{\psi}^1 = \sqrt{\tfrac{1}{2}}(\psi^1 - \psi^2), \qquad \tilde{\psi}^2 = \sqrt{\tfrac{1}{2}}(\psi^1 + \psi^2),$$

$$\tilde{\psi}^3 = \sqrt{\tfrac{1}{2}}(\psi^3 - \psi^4), \qquad \tilde{\psi}^4 = \sqrt{\tfrac{1}{2}}(\psi^3 + \psi^4), \tag{3.13}$$

plus the same definitions with ψ replaced by λ and by μ. Thus the 8 fermions $\lambda^5, \mu^5, \tilde{\psi}^1, \tilde{\psi}^3, \tilde{\lambda}^1, \tilde{\lambda}^3, \tilde{\mu}^1, \tilde{\mu}^3$ are odd under P, whereas the remaining fermions are even. The next step is to re-bosonize the new set of 20 real fermions (i.e. including also the two fields ψ^μ) such that the action of P on the 10 bosons is a pure translation. That is, we let

$$\sqrt{\tfrac{1}{2}}\, i(\lambda^5 \pm i\mu^5) = \mathrm{e}^{\pm iH^1},$$

$$\sqrt{\tfrac{1}{2}}\, i(\tilde{\psi}^1 \pm i\tilde{\psi}^3) = \mathrm{e}^{\pm iH^2},$$

$$\sqrt{\tfrac{1}{2}}\, i(\tilde{\lambda}^1 \pm i\tilde{\lambda}^3) = \mathrm{e}^{\pm iH^3},$$

$$\sqrt{\tfrac{1}{2}}\, i(\tilde{\mu}^1 \pm i\tilde{\mu}^3) = \mathrm{e}^{\pm iH^4},$$

$$\sqrt{\tfrac{1}{2}}\, i(\tilde{\psi}^2 \pm i\tilde{\psi}^4) = \mathrm{e}^{\pm iH^5},$$

$$\sqrt{\tfrac{1}{2}}\, i(\tilde{\lambda}^2 \pm i\tilde{\lambda}^4) = \mathrm{e}^{\pm iH^6},$$

$$\sqrt{\tfrac{1}{2}}\, i(\tilde{\mu}^2 \pm i\tilde{\mu}^4) = \mathrm{e}^{\pm iH^7},$$

$$\sqrt{\tfrac{1}{2}}\, i(\psi^5 \pm i\psi^6) = \mathrm{e}^{\pm iH^8}, \tag{3.14}$$

the remaining 2 bosons being the original coordinate $X^6 = H^9$ ($\partial X^6_{\mathrm{L}} = \lambda^6\mu^6$), plus one boson ($H^{10}$) made by combining the two transverse 4d fermions ψ^μ. Using the bosonization dictionary (3.7), one sees that the action of P on the new bosons is to shift by $\frac{1}{2}$ in the $H^{1,2,3,4}$ directions, and to leave the other 6 left-moving bosons alone. We also need to describe the set of zero-mode eigenvalues for the new bosons *before* P acts on the system, and here we must include the right-moving zero-modes as well. But the diagonalization (3.13) does not alter the coherent sum over spin structures for the 20 left-moving and 12 right-moving fermions, so clearly the zero-mode eigenvalues before shifting are given by the same set as for the $SU(2)^6$ model, namely eq. (3.11).

What effect does a shift of the form $((\frac{1}{2})^4, 0^6; 0^6)$ have on this system? The answer is that it gives rise to a new set of zero-mode eigenvalues which, like the $SU(2)^6$ set, would form an odd self-dual lorentzian lattice of signature (11, 7) had we used the covariant formulation. But all such lattices are unique up to automorphism [31], and in our case the automorphism turns out to be a matrix acting just on the first four of the ten left-moving coordinates H^i (the shifted directions): $M^{ij}=M^{ji}=(M^{ij})^{-1}=\delta^{ij}-\frac{1}{2}$. In other words, we define four new bosons

$$\tilde{H}^i(z)=M^{ij}H^j(z), \qquad i, j=1,\dots,4, \tag{3.15}$$

to replace the four shifted bosons $H^i(z)$, $i=1,\dots,4$; then the set of zero-mode eigenvalues for the newest set of bosons, $\tilde{H}^i$, $i=1,\dots,4$, H^i, $i=5,\dots,10$, plus the same six right-moving bosons as before, is just the $SU(2)^6$ model's set (3.11).

Finally we re-fermionize the bosons. The four new bosons give us 8 new real fermions

$$\eta^i=-i\sqrt{\tfrac{1}{2}}\left(e^{i\tilde{H}^i}+e^{-i\tilde{H}^i}\right), \qquad \xi^i=-\sqrt{\tfrac{1}{2}}\left(e^{i\tilde{H}^i}-e^{-i\tilde{H}^i}\right), \qquad i=1,\dots,4, \tag{3.16}$$

while from the old bosons we recover the 12 left-moving fermions $\tilde{\psi}^2$, $\tilde{\psi}^4$, ψ^5, ψ^6, $\tilde{\lambda}^2$, $\tilde{\lambda}^4$, λ^6, $\tilde{\mu}^2$, $\tilde{\mu}^4$, μ^6, and ψ^μ ($\mu=1,2$) which were left invariant by P, as well as all 12 right-moving fermions $\bar{\lambda}^i$ and $\bar{\mu}^i$. Again we have a set of 20 + 12 real fermions that are free and untwisted, and just as in the original $SU(2)^6$ model we have a coherent sum over their spin structures. The *only* difference between the two models is that after the repeated bosonization and re-fermionization the world-sheet supersymmetry generator $T_F(z)$ no longer has the same form in terms of the new set of fermions as it had in terms of the old ($SU(2)^6$) set of fermions. Clearly the change in T_F is restricted to the piece $T_F^{(15)}$ that is composed of the 15 twisted fermions ψ^i, λ^i, μ^i, $i=1,\dots,5$, in terms of which $T_F^{(15)}$ takes the form (3.9) with group $G=SU(2)^5$. We will now see that in terms of the 15 untwisted fermions $\tilde{\psi}^2$, $\tilde{\psi}^4$, ψ^5, $\tilde{\lambda}^2$, $\tilde{\lambda}^4$, $\tilde{\mu}^2$, $\tilde{\mu}^4$, η^i and ξ^i $T_F^{(15)}$ is given by (3.9) with $G=SU(4)$ instead.

First we rewrite $T_F^{(15)}$ in terms of the intermediate set of fermions introduced in eq. (3.13):

$$T_F^{(15)}=-\tfrac{1}{2}\Big\{\sqrt{\tfrac{1}{2}}\left[\tilde{\psi}^2(\tilde{\lambda}^1\tilde{\mu}^1+\tilde{\lambda}^2\tilde{\mu}^2)+\tilde{\psi}^1(\tilde{\lambda}^1\tilde{\mu}^2+\tilde{\lambda}^2\tilde{\mu}^1)\right]$$
$$+\sqrt{\tfrac{1}{2}}\left[\tilde{\psi}^4(\tilde{\lambda}^3\tilde{\mu}^3+\tilde{\lambda}^4\tilde{\mu}^4)+\tilde{\psi}^3(\tilde{\lambda}^3\tilde{\mu}^4+\tilde{\lambda}^4\tilde{\mu}^3)\right]+\psi^5\lambda^5\mu^5\Big\}. \tag{3.17}$$

Then we use the two bosonizations (3.14), (3.16) and the relation (3.15) between H^i and $\tilde{H}^i$ to rewrite bilinears of the fermions in (3.17) which are odd under P as bilinears in η^i and ξ^i:

$$\tilde{\lambda}^{\binom{1}{3}}\tilde{\mu}^{\binom{1}{3}}=\tfrac{1}{2}\left(\pm\eta^1\eta^2\mp\xi^1\xi^2+\eta^3\eta^4+\xi^3\xi^4\right),$$
$$\tilde{\psi}^{\binom{1}{3}}\tilde{\lambda}^{\binom{1}{3}}=\tfrac{1}{2}\left(\pm\eta^1\eta^4\mp\xi^1\xi^4+\eta^2\eta^3+\xi^2\xi^3\right),$$
$$\tilde{\psi}^{\binom{1}{3}}\tilde{\mu}^{\binom{1}{3}}=\tfrac{1}{2}\left(\pm\eta^1\eta^3\mp\xi^1\xi^3+\eta^2\eta^4+\xi^2\xi^4\right),$$
$$\lambda^5\mu^5=\tfrac{1}{2}\left(\eta^1\xi^1-\eta^2\xi^2-\eta^3\xi^3-\eta^4\xi^4\right). \tag{3.18}$$

Substituting (3.18) into (3.17), one writes $T_F^{(15)}$ as a trilinear product of the untwisted fermions, and it is easy to check that it indeed has the form (3.9) for the group SU(4). That the four-dimensional gauge group is SU(4) ⊗ SU(2) then follows from the SKM algebra construction [14, 42, 27] mentioned above, plus the fact that all 18 supercurrents correspond to physical states.

The remaining $N=4$ model of refs. [6,7] – the one with gauge group SO(5) ⊗ SU(3) – can also be constructed as an orbifold. The construction – and the proof of equivalence – will closely parallel those just presented for the SU(4) ⊗ SU(2) case. Once again we start from the $SU(2)^6$ model, but now we twist it by $P = P_L = Z_3$ generated by

$$\omega_L = \mathrm{diag}(C_3,1,1,1), \qquad v_L = \left(\left(\tfrac{1}{3}\right)^6\right), \qquad \omega_R = 1, \qquad v_R = 0.$$

As usual, ω_L and v_L are written in an orthonormal basis for $\Gamma^{6,6}(D_6)$, and C_3 cyclically permutes the first three bosonic coordinates. Our first step is to write the action of P on the fermions ψ^i, χ^i and ϕ^i of the $SU(2)^6$ model (we have to use complex fermions here). The rotation ω_L permutes the fermions: $\psi^1 \to \psi^2 \to \psi^3 \to \psi^1$, and similarly for χ^i and ϕ^i, $i=1,2,3$, and the shift v_L takes $\psi^i \to \psi^i$, $\chi^i \to \omega\chi^i$, $\phi^i \to \bar{\omega}\phi^i$, $i=1,\ldots,6$, where $\omega \equiv e^{2\pi i/3}$. Therefore the linear combinations

$$\tilde{\psi}^i = \sqrt{\tfrac{1}{3}} \sum_{j=1}^{3} \omega^{i\cdot j}\psi^j, \qquad \tilde{\chi}^i = \sqrt{\tfrac{1}{3}} \sum_{j=1}^{3} \omega^{(i+1)\cdot j}\chi^j, \qquad \tilde{\phi}^i = \sqrt{\tfrac{1}{3}} \sum_{j=1}^{3} \omega^{(i-1)\cdot j}\phi^j \qquad (3.19)$$

for $i=1,2,3$, together with $\psi^{4,5,6}$, $\chi^{4,5,6}$ and $\phi^{4,5,6}$, diagonalize the action of P: $\tilde{\psi}^2$, $\tilde{\chi}^2$, $\tilde{\phi}^2$ and $\chi^{4,5,6}$ acquire the phase ω under P, $\tilde{\psi}^1$, $\tilde{\chi}^1$, $\tilde{\phi}^1$ and $\phi^{4,5,6}$ acquire $\bar{\omega}$, and $\tilde{\psi}^3$, $\tilde{\chi}^3$, $\tilde{\phi}^3$ and $\psi^{4,5,6}$ are left invariant.

Next we bosonize:

$$\begin{aligned}
&\tilde{\psi}^2 = -i\,e^{iH^1}, && \tilde{\psi}^1 = -i\,e^{-iH^1}, && \chi^4 = -i\,e^{iH^4}, && \phi^4 = -i\,e^{-iH^4},\\
&\tilde{\chi}^2 = -i\,e^{iH^2}, && \tilde{\chi}^1 = -i\,e^{-iH^2}, && \chi^5 = -i\,e^{iH^5}, && \phi^5 = -i\,e^{-iH^5},\\
&\tilde{\phi}^2 = -i\,e^{iH^3}, && \tilde{\phi}^1 = -i\,e^{-iH^3}, && \chi^6 = -i\,e^{iH^6}, && \phi^6 = -i\,e^{-iH^6}, \qquad (3.20)
\end{aligned}$$

with the remaining, P-invariant, fermions combining into $H^7,\ldots,H^{10}$. With this choice of bosonization P acts like a shift on the H^i; in particular, $H^1,\ldots,H^6$ are shifted by $\frac{1}{3}$, whereas $H^7,\ldots,H^{10}$ are left invariant. Just as in the SU(4) ⊗ SU(2) case, one finds that the shift $((\frac{1}{3})^6, 0^4; 0^6)$ gives rise to a set of zero-mode eigenvalues for the H^i which becomes identical to the original $SU(2)^6$ set (3.11) after one performs an appropriate lattice automorphism. In this case the matrix for the automorphism affects only the first six H^i: $M^{ij} = M^{ji} = (M^{ij})^{-1} = \delta^{ij} - \frac{1}{3}$. So, we redefine the six shifted bosons, $\tilde{H}^i(z) = M^{ij}H^j(z)$ $(i=1,\ldots,6)$, and then re-fermionize them:

$$\eta^i = -i\sqrt{\tfrac{1}{2}}\left(e^{i\tilde{H}^i} + e^{-i\tilde{H}^i}\right), \qquad \xi^i = -\sqrt{\tfrac{1}{2}}\left(e^{i\tilde{H}^i} - e^{-i\tilde{H}^i}\right), \qquad i = 1,\dots,6. \quad (3.21)$$

(Re-fermionization also recovers the P-invariant fermions $\tilde{\psi}^3$, $\tilde{\chi}^2$, $\tilde{\phi}^1$, $\psi^{4,5,6}$ and ψ^{μ} as well as the right-moving fermions.) Once again the model is defined as a coherent sum over spin structures for all of these 20 + 12 fermions, and only the form of the supersymmetry generator T_F^{int} has changed.

In order to reconstruct the new T_F^{int} we first write it in terms of the fermions that diagonalize P (eq. (3.19)):

$$T_F^{\text{int}} = -\tfrac{1}{2}\Bigg\{\sqrt{\tfrac{1}{3}}\left[\tilde{\psi}^3(\tilde{\chi}^3\tilde{\phi}^3 + \tilde{\chi}^1\tilde{\phi}^2 + \tilde{\chi}^2\tilde{\phi}^1) + \tilde{\psi}^1(\tilde{\chi}^1\tilde{\phi}^1 + \tilde{\chi}^2\tilde{\phi}^3 + \tilde{\chi}^3\tilde{\phi}^2)\right.$$

$$\left. + \tilde{\psi}^2(\tilde{\chi}^2\tilde{\phi}^2 + \tilde{\chi}^1\tilde{\phi}^3 + \tilde{\chi}^3\tilde{\phi}^1)\right] + \sum_{i=4}^{6}\psi^i\chi^i\phi^i\Bigg\}. \quad (3.22)$$

Then we rewrite T_F^{int} in terms of the final set of fermions $\tilde{\psi}^3$, $\tilde{\chi}^3$, $\tilde{\phi}^3$, $\psi^{4,5,6}$, η^i and ξ^i, using the two bosonizations (3.20), (3.21) and the relation (3.15) between H^i and $\tilde{H}^i$ to rewrite not only the fermion bilinears

$$\begin{aligned}
\tilde{\psi}^1\tilde{\chi}^2 + \tilde{\psi}^2\tilde{\chi}^1 &= \eta^1\eta^2 + \xi^1\xi^2, \\
\tilde{\chi}^1\tilde{\phi}^2 + \tilde{\chi}^2\tilde{\phi}^1 &= \eta^2\eta^3 + \xi^2\xi^3, \\
\tilde{\phi}^1\tilde{\psi}^2 + \tilde{\phi}^2\tilde{\psi}^1 &= \eta^3\eta^1 + \xi^3\xi^1, \qquad\qquad (3.23)\\
\chi^i\psi^i = \eta^i\xi^i - \tfrac{1}{3}\sum_{j=1}^{6}\eta^j\xi^j, &\qquad i = 4,5,6 \quad (\text{no sum over } i),
\end{aligned}$$

but also one trilinear pair

$$\tilde{\psi}^1\tilde{\chi}^1\tilde{\phi}^1 + \tilde{\psi}^2\tilde{\chi}^2\tilde{\phi}^2 = \sqrt{\tfrac{1}{2}}\left(\eta^4\eta^5\eta^6 - \eta^4\xi^5\xi^6 - \xi^4\eta^5\xi^6 - \xi^4\xi^5\eta^6\right). \quad (3.24)$$

Substituting (3.23) and (3.24) into (3.22) then gives T_F^{int} in the desired form (3.9) with G = SO(5) ⊗ SU(3). The 10 SO(5) fermions are $\eta^{1,2,3}$, $\xi^{1,2,3}$, $\tilde{\psi}^3$, $\tilde{\chi}^3$, $\tilde{\phi}^3$ and $\sqrt{\tfrac{1}{3}}(\psi^4 + \psi^5 + \psi^6)$; the 8 SU(3) fermions are $\eta^{4,5,6}$, $\xi^{4,5,6}$ and the remaining two linear combinations of $\psi^{4,5,6}$.

3.4. OTHER $N = 4$ MODELS

Let us now describe a new $N = 4$ orbifold model; its gauge group is $SU(3)^2$, which is not contained in the gauge groups of the previous models. Our starting point is the $N=8$ model that has zero modes of X^i taking values on the lattice

$$\Gamma^{6,6}(\mathrm{E}_6) \equiv (\mathbf{1};\mathbf{1}) + (\mathbf{27};\mathbf{27}) + (\overline{\mathbf{27}};\overline{\mathbf{27}}), \tag{3.25}$$

where **1**, **27** and $\overline{\mathbf{27}}$ denote the three conjugacy classes of E_6 weights. Class **1** is the root lattice of E_6; it consists of weights of E_6 representations of zero triality. Weights of representations with triality ± 1 make classes **27** and $\overline{\mathbf{27}}$. The easiest way to visualize the weight lattice of E_6 is to describe it in terms of weights of the maximal subgroup $\mathrm{A}_2 \otimes \mathrm{A}_2 \otimes \mathrm{A}_2$:

$$\mathbf{1} = (\mathbf{1},\mathbf{1},\mathbf{1}) + (\mathbf{3},\mathbf{3},\mathbf{3}) + (\overline{\mathbf{3}},\overline{\mathbf{3}},\overline{\mathbf{3}}),$$

$$\mathbf{27} = (\mathbf{1},\mathbf{3},\overline{\mathbf{3}}) + (\overline{\mathbf{3}},\mathbf{1},\mathbf{3}) + (\mathbf{3},\overline{\mathbf{3}},\mathbf{1}),$$

$$\overline{\mathbf{27}} = (\mathbf{1},\overline{\mathbf{3}},\mathbf{3}) + (\mathbf{3},\mathbf{1},\overline{\mathbf{3}}) + (\overline{\mathbf{3}},\mathbf{3},\mathbf{1}).$$

Each of the A_2 lattices* is a two-dimensional hexagonal lattice; the three 2-planes spanned by these sublattice are orthogonal to each other.

Unlike $\Gamma^{6,6}(\mathrm{D}_6)$, the lattice $\Gamma^{6,6}(\mathrm{E}_6)$ has no simple interpretation in terms of free fermions, so we are forced to describe this $N=8$ model and its orbifolds in bosonic language. (The formalism for describing asymmetric orbifolds bosonically has been worked out in [8]; we will draw on the results of that paper when necessary.) The $N=4$ model with gauge group $\mathrm{SU}(3)^2$ is obtained by twisting the above $N=8$ model by $\mathrm{P}=\mathrm{P_L}=\mathrm{Z}_3$ which is generated by $(\omega_\mathrm{L}, v_\mathrm{L}; 1, 0)$, where ω_L rotates the plane of the first A_2 through an angle of $\frac{4}{3}\pi$, and v_L is a weight vector for the triplet representation of the second A_2. Note that ω_L^3 is a 4π rotation, and hence is trivial on space-time fermions as well as bosons, and $3v_\mathrm{L} \in \Gamma^{6,6}(\mathrm{E}_6)$, so that $(\omega_\mathrm{L}, v_\mathrm{L};1,0)$ indeed has order 3.

Let us follow our bosonic description of the $\mathrm{SU}(2)^6$ model and focus on the massless spectrum, beginning with the untwisted sector. Since the SO(6) rotation ω_L does not belong to SU(3), it breaks all four left-moving supersymmetries and in fact leaves no massless (R, NS) states. The surviving (NS, NS) and (NS, R) states again form an $N=4$ supergravity multiplet, and also a $\mathrm{U}(1)^4$ super-Yang-Mills multiplet. The rank is reduced from 6 to 4 because only four of the six internal directions – those tangent to the second and third A_2 root lattice planes – are left invariant by ω_L. The generators of $\mathrm{U}(1)^4$ form the Cartan subalgebra of the $\mathrm{SU}(3)^2$ gauge group; gauge bosons corresponding to root vectors come from the twisted sectors.

* We use A_2 to refer to certain sublattices of $\Gamma^{6,6}(\mathrm{E}_6)$ to avoid confusion with the SU(3) gauge group factors.

Following the formalism of ref. [8], we begin the description of the twisted sectors by defining I – the sublattice of $\Gamma^{6,6}(\mathrm{E}_6)$ which is left invariant by ω_{L} – and its dual lattice I^*. Using $(\mathrm{A}_2)^3$ notations for the left-moving bosonic zero modes and E_6 for the right-moving ones, we can write I as

$$I = (\vec{0}, \mathbf{1}, \mathbf{1}; \mathbf{1}) + (\vec{0}, \mathbf{3}, \bar{\mathbf{3}}; \mathbf{27}) + (\vec{0}, \bar{\mathbf{3}}, \mathbf{3}; \overline{\mathbf{27}}),$$

The dual lattice I^* is obtained by adjoining the vector f which is a weight vector for the 3 representation on both the second and third A_2 lattices, and has zero components in the remaining directions. Hence the index of I in I^* is $|I^*/I| = 3$. This index is important because it affects the overall degeneracy factor D for states in the twisted sector, which is the generalization to an asymmetric orbifold of the number of fixed points (tori) of a symmetric twist (ω, v). The general formula is [8]

$$D = \sqrt{\frac{\det'(1-\omega)}{|I^*/I|}}, \tag{3.26}$$

where $\det'(1-\omega)$ omits the unit eigenvalues of $\omega = (\omega_{\mathrm{L}}, \omega_{\mathrm{R}})$ and hence is equal to 3 in the present case, so that $D = 1$ here.

The formula (3.26) may be obtained by making the modular transformation $\tau \to -1/\tau$ on the trace of $(\omega_{\mathrm{L}}, v_{\mathrm{L}}; \omega_{\mathrm{R}}, v_{\mathrm{R}})$ in the untwisted sector. The same transformation also shows that in the twisted sector the bosonic zero-mode eigenvalues $(p_{\mathrm{L}}, p_{\mathrm{R}})$ lie on the shifted lattice $I^* + v$. In $N = 4$ models massless states are constrained to have $p_{\mathrm{R}} = 0$ while the value of $\frac{1}{2}p_{\mathrm{L}}^2$ is fixed by the requirement that the left-moving energy of the state also vanishes. In the case at hand we find that massless states in the singly-twisted sector have

$$p_{\mathrm{L}} \in (\vec{0}, \mathbf{3}, \mathbf{1}) + (\vec{0}, \mathbf{1}, \bar{\mathbf{3}}) + (\vec{0}, \bar{\mathbf{3}}, \mathbf{3})$$

and $p_{\mathrm{L}}^2 = \frac{2}{3}$. These are precisely six such vectors p_{L}: $(\vec{0}, 3, \vec{0}) + (\vec{0}, \vec{0}, \bar{3})$. (Here 3 ($\bar{3}$) refers to the (anti)-triplet weights only rather than to the entire class of weights for the appropriate A_2 sublattice.) The negatives of these 6 vectors lie in $I^* + 2v_{\mathrm{L}}$ and therefore give rise to the 6 massless states in the doubly-twisted sector; together the 12 states make up the roots of $\mathrm{SU}(3)^2$.

To verify that the 4d gauge group really is $\mathrm{SU}(3)^2$ we should construct the supercurrents for the 16 states and compare their operator product expansions with (2.8). The supercurrents for the 4 generators of the Cartan algebra are simply iDX^i, $i = 3, 4, 5, 6$ (coordinates of the second and third A_2 planes, which are left invariant by ω_{L}). The supercurrents for the 12 root vectors are the products of a dimension $\frac{1}{6}$ *twist superfield*★, which twists the supercoordinates $DX^{1,2}$, and dimension $\frac{1}{3}$ soliton operators $e^{ip_L \cdot X_L}$, where p_L is a triplet or an anti-triplet weight vector of the second or third A_2. Without going into the details of twist superfield operator products, it is easy to see that the supercurrent operator products have the right form. Clearly the supercurrents form two mutually anticommuting sets of eight, according to whether they involve the second or third A_2, because the soliton operator products vanish otherwise. Look at one of these two sets. The charges of the 6 twisted states under the Cartan currents ∂X^i ($i = 3, 4$, say) are proportional to p_L^i. But these vectors (the $3 + \bar{3}$ on the A_2 weight lattice) form a regular hexagon about the origin, just as the roots of SU(3) do, so the Cartan-root operator products are correct up to overall normalization. (The unusual normalization arises because we are constructing an SU(3) Kac-Moody algebra with $k = 3$ rather than $k = 1$.) Finally the root-root operator products have the correct general form just as a consequence of 'momentum' conservation for the soliton operators; the twist superfields provide the appropriate fractional powers of $z - w$ to make the right singularity. In fact the representation of the $k = 3$ non-supersymmetric SU(3) Kac-Moody algebra which is provided by the above SU(3) supercurrents is identical to the parafermionic construction of ref. [33]. The generalized commutation relations of parafermions have appeared in the mathematical literature under the name of Z algebras [34]. The soliton operators appearing in the construction for central charge k are $e^{i\alpha \cdot X/\sqrt{k}}$, with α a root; i.e., the length^2 of the root vectors which appear are effectively rescaled by a factor of $1/k$ from the usual normalization (length$^2 = 2$ for long roots). In the SU(3) case at hand, the parafermions are particular combinations of twist fields and the fermions $\psi^{3,4,5,6}$.

Another $N = 4$ orbifold model – with 4d gauge group $SO(5) \otimes SU(2) \otimes SU(2)$ – can be constructed by a different twisting of the same $N = 8$ $\Gamma^{6,6}(E_6)$ model as above. This time we have $P = P_L = Z_6$, generated by an ω_L which rotates the first A_2 plane by $\frac{2}{3}\pi$ rather than $\frac{4}{3}\pi$, so that it has order 6 acting on space-time fermions. It is accompanied by a shift v_L which is half of a weight vector for the triplet (3) representation of the second A_2 and half of a $\bar{3}$ vector for the third A_2. Without going into the details of the construction, we will simply say which generators of the gauge group come from which sector. As usual, Cartan generators reside in the untwisted sector. Long roots of SO(5) – which have the appropriate length$^2 = \frac{2}{3}$ for $k = 3$, i.e., $\hat{k} = 0$ – come from the doubly-twisted sector and its anti-sector. Two of the short roots of SO(5) (length$^2 = \frac{1}{3}$) come from the singly-twisted sector and its anti-sector, while the other two come from the triply-twisted sector. The latter sector also contains all four roots of the two SU(2) factors; these roots have length$^2 = 1$, which corresponds to $k = 2$, i.e. also $\hat{k} = 0$.

Both of these E_6-based $N = 4$ models have gauge groups of dimension 16 rather than the 18 of the first three models. Nevertheless, all simple SKM factors involved

* The dimension of the twist field equals the difference between vacuum energies of the twisted and the untwisted sectors. See ref. [32] for the theory of twist fields and superfields.

are minimal, i.e. have $\hat{k}=0$ and thus can be fully described in terms of 16 free fermions (cf. the next section). Hence the super-stress tensor associated with the SKM algebras has central charge $\hat{c}^{\mathrm{SKM}}=\frac{16}{3}$ in each case (cf. eq. (2.12)), and both models have to incorporate a left-over super-Virasoro algebra with $\hat{c}^{\mathrm{L}}=\frac{2}{3}$ – one in the discrete series of unitary algebras with $\hat{c}<1$. In addition to $\hat{c}^{\mathrm{L}}$, we can also determine which eigenvalues h^{L} of L_0^{L} occur in the spectra of the two models by using the physical state condition $L_0=\bar{L}_0$, which implies that $h^{\mathrm{L}}+h^{\mathrm{SKM}}=\bar{h}^{\mathrm{int}}$ (modulo $\frac{1}{2}$), or (modulo 1) in the (R, R) sector. Comparing the spectrum of the appropriate L_0^{SKM} with the spectrum of $\bar{L}_0$ – the latter is the same for both models and also for the $N=8$ $\Gamma^{6,6}(\mathrm{E}_6)$ model – yields the same result for both models: NS states have $h^{\mathrm{L}}=0$ or $\frac{1}{6}$ (modulo $\frac{1}{2}$) while Ramond states have $h^{\mathrm{L}}=\frac{1}{24}$ or $\frac{3}{8}$ (modulo 1). Note that while all these values of h^{L} are allowed for unitary representations of the $\hat{c}=\frac{2}{3}$ algebra [40,19], there are other allowed values of h^{L}, namely $h^{\mathrm{L}}=\frac{1}{16}$ (modulo $\frac{1}{2}$), that do not occur here. However, the values of h^{L} that we found are the only ones allowed if the left-over algebra is extended to an (untwisted) $N=2$ super-Virasoro algebra [35]. And in fact there is a good reason for the existence of a global N = 2 supersymmetry on the world sheet: the U(1) current $\Sigma_{i=1}^{3}\bar{\psi}^i\psi^i$ that generates it is preserved by any abelian twist.

The fact that both E_6 orbifolds yield the same left-over algebra may appear surprising, but it can easily be explained by comparing the respective twist groups $\mathrm{P}=\mathrm{P}_{\mathrm{L}}$: The square of the Z_6 generator we have used to make the $\mathrm{SO}(5)\otimes\mathrm{SU}(2)^2$ model is precisely the Z_3 twist that generates the $\mathrm{SU}(3)^2$ model. This means that we should be able to start directly from the $\mathrm{SU}(3)^2$ model, twist it by a Z_2 group, and get the $\mathrm{SO}(5)\otimes\mathrm{SU}(2)^2$ gauge group. If the Z_2 element twists only the fermions that make up the SKM algebra, then the left-over algebra will indeed be the same. The required twist does exist: it acts on supercurrents of both SU(3)'s as the adjoint representation of the matrix diag$(-1,-1,+1)$. Eight out of the sixteen currents are left invariant by this Z_2 twist and form the untwisted sector of the new model; the group they generate is $\mathrm{SU}(2)^2\otimes\mathrm{U}(1)^2$. The twisted sector also contains eight states; their quantum numbers with respect to the untwisted group are $(\mathbf{2},\mathbf{2},0,0)+(\mathbf{1},\mathbf{1},\pm1,\pm1)+(\mathbf{1},\mathbf{1},\pm1,\mp1)$. It is easy to see that the first four twisted states enlarge the $\mathrm{SU}(2)^2\equiv\mathrm{SO}(4)$ group to SO(5) while the last four are the non-Cartan generators of the $\mathrm{SU}(2)^2$ algebra.

Unlike the $\mathrm{SU}(3)^2$ case, an $\mathrm{SO}(5)\otimes\mathrm{SU}(2)^2$ SKM algebra with $\hat{c}\leqslant 6$ does not have to be minimal: One can have $\hat{k}=1$ for one of the SU(2) factors, while keeping $\hat{k}=0$ for the other one and for SO(5). This combination would have $\hat{c}^{\mathrm{SKM}}=6$ and thus would not need a left-over algebra. Such a model can in fact be constructed as an $N=4$ orbifold. Again without going into the details, we can outline the construction as follows: We start with the $N=8$ model whose lattice is $\Gamma^{6,6}(\mathrm{D}_5+\mathrm{A}_1)$. We twist it by $\mathrm{P}=\mathrm{P}_{\mathrm{L}}=\mathrm{Z}_6$ generated by a $\frac{2}{3}\pi$ rotation ω_{L} that permutes the first three coordinates of the D_5 lattice, accompanied by a shift v_{L} which is one third of a vector weight for D_5 (along the fourth or fifth D_5 coordinate) plus one third of a root for A_1. It is a straightforward procedure to construct the properly normalized root diagram for the non-Cartan generators of the gauge group; we find that its geometry indeed fits $SO(5) \otimes SU(2)^2$. Moreover, the root lengths squared are $\frac{2}{3}$ and $\frac{1}{3}$ for the SO(5) roots, 1 for the roots of one of the SU(2)'s and $\frac{2}{3}$ for the other; this translates into $k = (3, 2, 3)$ or $\hat{k} = (0, 0, 1)$ for the respective gauge group factors.

For our final example of a $P = P_L$ orbifold we present an $N = 4$ model with gauge group G_2. This time both bosonic and fermionic techniques will be useful. As usual, our starting point is an $N = 8$ model; its lattice is $\Gamma^{6,6}(A_2 + A_2 + D_2)$. First, we fermionize the two dimensions corresponding of the D_2 sublattice and make a Z_2 twist that unifies the spin structure of the resulting fermions with the spin structure of the ψ^i (cf. the $SU(2)^6$ model). This twist reduces the supersymmetry to $N = 4$ and enlarges the gauge group to $SU(2)^2 \otimes U(1)^4$; both SU(2) factors are minimal. Second, we make a Z_3 twist that leaves the SU(2)'s alone, rotates the first A_2 sublattice by $\frac{4}{3}\pi$ and shifts the second A_2 by a weight vector in the **3** class. This time we use the bosonic arguments we've developed for the $SU(3)^2$ model and find that instead of the A_2^2 lattice we now have a minimal SU(3) SKM algebra, plus a left-over algebra with $\hat{c} = \frac{4}{3}$ (presumably a sum of two unitary super-Virasoro algebras with $\hat{c} = \frac{2}{3}$). So far we have built a model with gauge group $SU(3) \otimes SU(2)^2$, which is interesting by itself since it contains the standard model gauge group.

We have promised a G_2 model, however, so another twist is due. The SU(3) Lie algebra has an outer Z_2 automorphism – complex conjugation in an antihermitian basis – that leaves an SO(3) subalgebra invariant and changes the signs of the other five generators. The $SU(2)^2$ algebra also has an outer Z_2 automorphism – exchange of the two SU(2)'s. Combining these two automorphisms together, we leave six of the 14 generators of $SU(3) \otimes SU(2)^2$ invariant and change the signs of the other eight. Now we can use the fact that we have a minimal SKM algebra which can be described in terms of free fermions, since we already know how to treat a Z_2 twist which changes the signs of eight real fermions and leaves everything else invariant. The untwisted sector gives rise to an $SU(2) \otimes SU(2)$ SKM algebra, which is however highly non-minimal: the first SU(2) (subalgebra of the SU(3)) has $k = 4 \cdot 3 = 12$ or $\hat{k} = 10$ while the second one has $k = 2 \cdot 2 = 4$ or $\hat{k} = 2$. The twisted sector yields another eight generators which transform with respect to the untwisted subalgebra as (4, 2); this completes the adjoint representation of G_2. If we now compute the root lengths of G_2, we find that the short roots have length$^2 = \frac{2}{12}$ while the long roots have length$^2 = \frac{2}{4}$; this corresponds to $k = 4$ or $\hat{k} = 0$, which agrees with the requirement that $\hat{c}(G_2) = \frac{14}{3}$. (The remaining $\hat{c}^L = \frac{4}{3}$ is supplied by the same left-over algebra as in the $SU(3) \otimes SU(2)^2$ model – we have not twisted the left-over piece).

Our main reason for constructing a G_2 model was to show that all allowed 4d gauge groups that we have explicitly listed in lines (A)–(F) of (2.14) do appear in $N = 4$ orbifold models. Using the same techniques, but starting from different lattices $\Gamma^{6,6}(\mathcal{G})$ and twisting by other cyclic groups $P = P_L = Z_l$, we have constructed $N=4$ models with all the other allowed gauge groups, except those having odd rank. The latter failure is due to the fact that abelian twists always preserve a Cartan subalgebra of an even dimension; presumably non-abelian twist groups $P = P_L$ can be used to make at least some of the allowed odd-rank gauge groups.

3.5. CHIRAL $N=1$ MODELS

By now we have constructed several models with gauge groups large enough to contain the standard model, but they were non-chiral because of $N=4$ supersymmetry. Our next question is whether there are chiral models in the type II superstring framework. The answer to this question is 'yes': we will now give an example of a chiral 4d model; this model has $(SU(2) \otimes U(1))^2$ gauge group and $N=1$ space-time supersymmetry.

The starting point is the previous $N=4$ $SO(5) \otimes SU(3)$ model, which is most conveniently described here in terms of untwisted free fermions, using the form (3.9) for the supersymmetry generator T_F^{SKM}. We twist this model by a group $P = Z_4$ which acts on both left- and right-movers, in order to correlate left-moving gauge quantum numbers with right-moving space-time helicities. In order to get a chiral theory, we need to break at least three of the four (right-moving) space-time supersymmetries. This is accomplished by the right-moving rotation

$$\omega_R = \mathrm{diag}(C_2^-, C_2^-, -1, -1),$$

with $C_2^- = \begin{pmatrix} 0 & -1 \\ 1 & 0 \end{pmatrix}$, and written in the usual orthonormal basis for the D_6 lattice. One can rewrite ω_R in terms of the linear combinations of the right-moving fermions $\bar\psi^i, \bar\chi^i, \bar\phi^i$ which diagonalize it, and then bosonize those fermions. One finds that ω_R is equivalent to the shift

$$\tilde v_R = \left(\tfrac14, \tfrac14, \tfrac12, \tfrac12, 0, 0; \tfrac14, \tfrac14, -\tfrac12; 0\right)$$

acting on the new bosons. The semicolons separate the bosons arising from $\bar\chi^i$ and $\bar\phi^i$; $\bar\psi^i$; $\bar\psi^\mu$, respectively. The distinction between $\bar\psi$ and $\bar\chi, \bar\phi$ is important because the spin structures for $\bar\psi$ are summed over separately from the rest of the fermions.

The action of P on the left-moving fermions must be a group automorphism in order to preserve the world-sheet supersymmetry generator (3.9). In our case, we rotate the SU(3) fermions by (the adjoint representation of) the matrix $\mathrm{diag}(i, i, -1)$, while the corresponding SO(5) matrix is $\mathrm{diag}(C_2^-, C_2^-, 1)$. This twist leaves $G = SU(2) \otimes U(1) \otimes SU(2)' \otimes U(1)'$ as the unbroken 4d gauge group. If we bosonize the left-moving fermions in a particular way we can describe the action of P on them as a shift:

$$\tilde v_L = \left(\tfrac14, \tfrac14, 0, 0; \tfrac14, \tfrac14, \tfrac12, 0, 0; 0\right),$$

acting on bosons. Here the semicolons separate the bosons H^i; $\tilde H^i$; H^{10} arising from

η^a; $\tilde{\eta}^a$; ψ^μ, respectively, where η^a ($\tilde{\eta}^a$) are the fermionic parts of the SU(3) (SO(5)) supercurrents.

Since P acts as a pure shift on the 10 left-moving and 10 right-moving bosons we have defined, it is a straightforward exercise to list all the massless states, sector by sector, characterizing them by their 10 + 10 bosonic zero-mode eigenvalues. However, in order to compute the gauge charges of the states, we need to identify the generators $(I_3, Y; I_3', Y')$ of the Cartan subalgebra of G as linear combinations of the fields ∂H^i and $\partial \tilde{H}^i$. Defining I_3^i by $I_3(z) = \Sigma_{i=1}^4 I_3^i \partial H^i(z)$, etc., and using the known charges of states in the adjoint representations, we find that

$$\begin{aligned} I_3^i &= \left(\tfrac{1}{2}, -\tfrac{1}{2}, 1, 0\right), \qquad & I_3^{i\prime} &= \left(\tfrac{1}{2}, -\tfrac{1}{2}, 0, 1, 0\right), \\ Y^i &= (1,1,0,0), \qquad & Y^{i\prime} &= (1,1,2,0,0). \end{aligned} \tag{3.27}$$

Gauge charges are given by inner products of these vectors with the vectors of H_0^i and $\tilde{H}_0^i$ eigenvalues for the massless states. Because of $N=1$ supersymmetry, we need only list the spectrum of massless positive-helicity fermions other than gauginos; the list below gives their SU(2) ⊗ U(1) ⊗ SU(2)′ ⊗ U(1)′ charges, preceded by the number of times they appear in the spectrum:

untwisted sector: $1(\mathbf{1}, 0; \mathbf{1}, \pm 2), 2(\mathbf{2}, 1; \mathbf{1}, 0), 2(\mathbf{1}, 0; \mathbf{2}, 1)$;
singly-twisted sector: $2(\mathbf{2}, -\tfrac{1}{2}, \mathbf{1}, \tfrac{3}{2}), 2(\mathbf{1}, \tfrac{1}{2}; \mathbf{2}, \tfrac{1}{2}), 2(\mathbf{2}, -\tfrac{1}{2}; \mathbf{1}, -\tfrac{1}{2}), 2(\mathbf{1}, \tfrac{1}{2}; \mathbf{2}, -\tfrac{3}{2})$;
doubly-twisted sector: $1(\mathbf{1}, \pm 1; \mathbf{1}, \pm 1), 1(\mathbf{2}, 0; \mathbf{2}, 0), 2(\mathbf{2}, 0; \mathbf{1}, -1), 2(\mathbf{1}, -1; \mathbf{2}, 0)$.

One can check that the spectrum is anomaly free.

A second chiral $N=1$ model very similar to the first one can be constructed from the $N=4$ $SU(2)^6$ model by twisting by exactly the same group $P = Z_4$ as above. To be precise, P acts as the same shift in an appropriate bosonic picture; its action as an isomorphism of the SKM algebra is of course different. Specifically, P acts on the generators of four of the six SU(2)'s as $\exp(\tfrac{1}{2} i\pi I_3)$, on the fifth SU(2) as $\exp(i\pi I_3)$, and the sixth SU(2) is left invariant by P. Thus the 4d gauge group of the second model is $SU(2) \otimes U(1)^5$, and all massless (non-gauge) fermions are SU(2) singlets. Here we just list their U(1) charges:

untwisted sector: $1(\pm 1, 0, 0, 0, 0), 2(0, \underline{1, 0, 0, 0})$;
singly-twisted sector: $2(\pm \tfrac{1}{2}, \underline{-\tfrac{3}{4}, \tfrac{1}{4}, \tfrac{1}{4}, \tfrac{1}{4}})$;
doubly-twisted sector: $1(0, \underline{\tfrac{1}{2}, \tfrac{1}{2}, -\tfrac{1}{2}, -\tfrac{1}{2}}), 1(0, \tfrac{1}{2}, \tfrac{1}{2}, \tfrac{1}{2}, \tfrac{1}{2}), 1(0, -\tfrac{1}{2}, -\tfrac{1}{2}, -\tfrac{1}{2}, -\tfrac{1}{2})$,
$2(0, \underline{\tfrac{1}{2}, -\tfrac{1}{2}, -\tfrac{1}{2}, -\tfrac{1}{2}})$.

An underline indicates that all distinct permutations of the charges appear. Again the spectrum is anomaly free. One can similarly twist the $N=4$ SU(4) ⊗ SU(2) model by the same group P; however in this case the $N=1$ model obtained turns out to be non-chiral.

The results of this section can be summarized as follows: Using asymmetric orbifolds, we have constructed several modular-invariant classical solutions of the type II superstring which lead to supersymmetric gauge theories in four dimensions.

In particular, we have reproduced the three models of Bluhm et al. [6]. The additional models of Kawai et al. [7] can be constructed as asymmetric orbifolds as well, using the three $N=4$ models as intermediate steps.* We have also constructed other $N=4$ models that are not equivalent to free fermions; one of these models has gauge group $SU(3)^2$ which contains the standard model group $SU(3) \otimes SU(2) \otimes U(1)$. Finally, we have given two *chiral* models, in which massless fermions transform as a non-self-conjugate representation of the gauge group.

4. Gauge quantum numbers of massless particles

In the previous section we have seen how to construct several models from the type II superstring which have realistic gauge groups, and also models with chiral fermions, but we have found no models with realistic group representations for the massless fermions. In this section we will see that this problem is a general feature of models based on the type II superstring, and is not due merely to our lack of imagination in constructing models via asymmetric orbifolds. In fact, *no classical vacuum of the type II superstring can be consistent with the particle spectrum of the standard model.* More specifically, we will show that in any four-dimensional model whose gauge group contains $SU(3) \otimes SU(2) \otimes U(1)$, massless triplets of SU(3) (such as quarks) cannot coexist with massless SU(2) doublets (such as left-handed leptons). To prove this general result we return to the algebraic approach used in sect. 2. In that section we classified the different super-Kac-Moody (SKM) algebras which can be incorporated into a type II model in a way consistent with unitarity and the existence of massless chiral fermions. In this section we will see that unitarity restricts the possible representations of a given SKM algebra, and therefore also limits the possible gauge quantum numbers of massless particles, such that SU(3) triplets are incompatible with SU(2) doublets.

Our starting point is the obvious remark that every state in the spectrum of a type II model belongs to some representations of the (left-moving) SKM algebra associated with the 4d gauge group. This just means that the supercurrent modes $\mathcal{J}_r^a$ and J_n^a act within the Hilbert space of states. The SKM representations of interest for string theory are those in which the spectrum of the 'left-moving energy' operator L_0 is bounded from below. Since the positive-frequency modes of $\mathcal{J}^a(z)$ and $J^a(z)$ when applied to a state lower its energy, by successively applying these operators to any state one must eventually arrive at a state $|\boldsymbol{r}\rangle$ which is annihilated by all of these operators:

$$\forall r, n > 0: \qquad \mathcal{J}_r^a|\boldsymbol{r}\rangle = J_n^a|\boldsymbol{r}\rangle = 0\,; \tag{4.1}$$

$|\boldsymbol{r}\rangle$ is called a *highest weight* or *primary* state with respect to the SKM algebra. Also

* While this paper was being typed we received [36], in which models similar to those of Kawai et al. are constructed using slightly different techniques.

$|\boldsymbol{r}\rangle$ must represent the zero-mode subalgebra of the SKM algebra. In the NS sector this subalgebra is just the ordinary Lie algebra generated by J_0^a, so

$$J_0^a|\boldsymbol{r}\rangle \equiv T_{(r)}^a|\boldsymbol{r}\rangle \tag{4.2}$$

for some representation matrices $T_{(r)}^a$; eq. (4.2) defines the gauge group representation $\boldsymbol{r}$ under which the state $|\boldsymbol{r}\rangle$ transforms. In the R sector $|\boldsymbol{r}\rangle$ must also represent the Clifford algebra of fermionic zero-modes $\mathscr{J}_0^a$. We will discuss only NS sector SKM representations in this section, however – we know from sect. 2 that all R states are massive whenever the SKM algebra is non-abelian.

All states in a model can be constructed by applying the negative-frequency modes $\mathscr{J}^a_{-r}$, J^a_{-n} $(r, n > 0)$ to the various highest weight states. The collection of states

$$\mathscr{J}^{a_1}_{-r_1} \dots \mathscr{J}^{a_k}_{-r_k} J^{a_1}_{-n_1} \dots J^{a_l}_{-n_l}|\boldsymbol{r}\rangle \tag{4.3}$$

for a particular highest weight state $|\boldsymbol{r}\rangle$ is called a *highest weight representation*. Thus each state belongs to some highest weight representation, or is a linear combination of states in different highest weight representations. Since the operators $\mathscr{J}^a_{-r}$, J^a_{-n} carry an adjoint index a, all states in a highest weight representation transform in the same way under the center of the ordinary Lie group. In particular, for the Lie group SU(3) (SU(2)), states can only have nonzero triality (half-integer isospin) if some highest weight state does also.

In fact, as long as we are interested only in massless particles, we can ignore the non-primary states in (4.3), with the exception of gauge bosons and their superpartners. The reason is as follows: left-moving energies of states in (4.3) differ from each other by integers or half-integers. Thus if the superfield $\boldsymbol{r}(z, \theta)$ that makes a highest-weight state $|\boldsymbol{r}\rangle$ has positive conformal dimension h_r – which means that $|\boldsymbol{r}\rangle$ has energy greater than $-\frac{1}{2}$ – then all states in (4.3) other than $|\boldsymbol{r}\rangle$ itself have strictly positive energies, i.e. are massive. The only exception occurs when $\boldsymbol{r}(z, \theta)$ is the identity operator of dimension 0. In this case $|\boldsymbol{r}\rangle$ is just the usual NS vacuum state $|0\rangle$ (which transforms trivially under the Lie algebra), and the adjoint states $\mathscr{J}^a_{-1/2}|0\rangle$ are massless. These states give rise to the 4d gauge bosons when tensored with the helicity ± 1 right-moving modes $\bar{\psi}^\mu_{-1/2}|\bar{0}\rangle$, but they can also yield massless scalars or fermions in the adjoint representation of the group if tensored with right-moving modes of appropriate helicity. For example, gauginos are made by tensoring $\mathscr{J}^a_{-1/2}|0\rangle$ with $|\bar{\alpha}\rangle$, which is the same right-moving Ramond state that is tensored with $\psi^\mu_{-1/2}|0\rangle$ in order to make a gravitino. Actually this gaugino $\leftrightarrow$ gravitino relation works both ways: If there are massless fermions of type $\mathscr{J}^a_{-1/2}|0\rangle \otimes |\bar{\alpha}\rangle$, then $\psi^\mu_{-1/2}|0\rangle \otimes |\bar{\alpha}\rangle$ is a gravitino, hence there is a space-time supersymmetry, and these adjoint fermions are in fact gauginos. If there is more than one $|\bar{\alpha}\rangle$, then there is $N > 1$ supersymmetry; such theories are necessarily non-chiral. On the other hand, if there is a massless scalar whose vertex contains a left-moving supercurrent $J^a(z,\theta)$, then on dimensional grounds this vertex also contains a superfield of dimension $(0,\frac{1}{2})$, i.e. a right-moving supercurrent; as we have seen in section 2, such supercurrents also destroy 4d chirality. Therefore, in chiral models all massless states other than gauge bosons and gauginos should be primary with respect to the SKM algebra.

At this point we have reduced the question 'What gauge quantum numbers can a massless particle have?' to the task of classifying unitary highest weight representation of SKM algebras and describing how the highest weight states transform with respect to the Lie algebra. This has been done by Goddard and Olive [37] for the ordinary Kac-Moody algebras and generalized to SKM algebras by Kac and Todorov [14]; the following couple of paragraphs summarize the results of [37, 14] which are relevant to this discussion.

Unitarity of a highest weight representation is the requirement that all the states (4.3) have non-negative norm. The hermitian conjugation needed to define the norm in a 2d field theory exchanges incoming and outgoing states on the 2d cylinder; therefore it exchanges $z \leftrightarrow 1/\bar{z}$ in the complex plane. If we choose a basis in which the generators $T^a = J_0^a$ of the Lie algebra G are hermitian, then the supercurrents $J^a(z,\theta)$ are also self-adjoint under hermitian conjugation, which means that their modes satisfy

$$(\mathcal{J}_r^a)^\dagger = \mathcal{J}_{-r}^a\,, \qquad (J_n^a)^\dagger = J_{-n}^a\,. \tag{4.4}$$

Now consider the state $(J_{-1}^a + iJ_{-1}^b)|\boldsymbol{r}\rangle$ in the highest weight representation generated from a highest weight state $|\boldsymbol{r}\rangle$. Its norm can be computed as*

$$\left\|(J_{-1}^a + iJ_{-1}^b)|\boldsymbol{r}\rangle\right\|^2 = \langle \boldsymbol{r}|k_{\mathrm{G}} - 2f_{\mathrm{G}}^{abc}J_0^c|\boldsymbol{r}\rangle\,, \tag{4.5}$$

and the requirement that the norm is non-negative imposes an upper limit on the eigenvalues of T_r^c. Since the latter eigenvalues are completely determined by the Lie algebra for any of its representations, the choice of allowed Lie representations for Kac-Moody primary states is severely limited for small k_{G}. For example, for SU(2) non-negativity of (4.5) implies $|I_3| \leqslant \frac{1}{2}k$, which means that only representations of isospin $I \leqslant \frac{1}{2}k$ are allowed for the highest weight states. For other groups one replaces $2I$ with the so-called representation level $\lambda_{\mathrm{G}}(\boldsymbol{r})$ which is defined in terms of weight vectors of $\boldsymbol{r}$ [37]; for $\mathrm{G}=\mathrm{SU}(N)$, $\lambda_{\mathrm{G}}(\boldsymbol{r})$ is equal to the number of columns in the Young tableau of $\boldsymbol{r}$ (which for $N=2$ indeed equals to $2I(r)$). To summarize, highest weight states of a unitary representation must have

$$\lambda_{\mathrm{G}}(r) \leqslant k_{\mathrm{G}}\,. \tag{4.6}$$

On the other hand, representations satisfying (4.6) are *integrable* – they occur in the

* It suffices to consider the case where G is simple.

spectrum of a string propagating on the group manifold (with its radius quantized in terms of k) [38] – so they are indeed unitary.

The condition (4.6) was derived using bosonic currents only, so it applies to any Kac-Moody algebra, not necessarily supersymmetric. In fact, for the SKM case a stronger condition can be derived, namely

$$\lambda_G \leqslant \hat{k} \equiv k - C_G(\text{adj}) \tag{4.7}$$

for all highest weight NS states; the condition for Ramond states is different. This stronger restriction follows from the fact that an SKM algebra contains a second non-supersymmetric Kac-Moody subalgebra. It is generated by the currents

$$\hat{J}^a(z) \equiv J^a(z) + \frac{i}{k} f^{abc} :\mathcal{J}^b(z)\mathcal{J}^c(z): \tag{4.8}$$

and has central charge equal to $\hat{k}$. Note that in order for this second Kac-Moody algebra to have any unitary representations one must have $\hat{k} > 0$. Moreover, in the minimal case $\hat{k} = 0$ this second algebra degenerates: there are no primary states other than $|0\rangle$ and all $\hat{J}^a_{-n}|0\rangle$ are null. What this means is that the currents $\hat{J}^a(z)$ actually vanish; hence the original bosonic currents $J^a(z)$ can be written as bilinears of $\mathcal{J}^a(z)$ (cf. (3.10)). But the $\mathcal{J}^a(z)$ are free fermions (up to normalization), so their algebra has a unique highest weight state in the NS sector, namely the usual Neveu-Schwarz vacuum, which is of course a gauge singlet.

In the general case one can rewrite the states (4.3) occurring in a highest weight representation of the SKM algebra using the negative-frequency modes of $\hat{J}^a(z)$ instead of $J^a(z)$:

$$\mathcal{J}^{a_1}_{-r_1} \cdots \mathcal{J}^{a_k}_{-r_k} \hat{J}^{a_1}_{-n_1} \cdots \hat{J}^{a_l}_{-n_l} |\boldsymbol{r}\rangle . \tag{4.9}$$

Because the $\hat{J}^a(a)$ commute with the fermionic currents $J^a(z)$, this shows that the Hilbert space is the tensor product of a free fermion part, $\mathcal{J}^{a_1}_{-r_1} \ldots \mathcal{J}^{a_k}_{-r_k}|\boldsymbol{r}\rangle$, and an ordinary Kac-Moody part, $\hat{J}^{a_1}_{-n_1} \ldots \hat{J}^{a_l}_{-n_l}|\boldsymbol{r}\rangle$. Again, the free fermion part has a unique highest weight state in the NS sector, so there is a one-to-one correspondence between unitary NS representations of the SKM algebra with central charge k and unitary representations of the ordinary Kac-Moody algebra with central charge $\hat{k}$, with identical gauge transformation properties for the corresponding primary states $|\boldsymbol{r}\rangle$. This is one of the main results of refs. [14, 20].

Having learned the restrictions on unitary highest weight representations of SKM algebras, we can now apply them to the problem at hand – building realistic superstring models. As we have seen in sect. 2, existence of gluons implies an SU(3) SKM algebra, but now eq. (4.7) tells us that the existence of quarks – massless color triplets – requires the SKM algebra to be non-minimal: $\hat{k}_3 \geqslant 1$, or $k_3 \geqslant 4$.* Simi-

* Note that if the SU(3) is a subalgebra of a bigger SKM algebra, the latter algebra could be minimal. For example, a minimal SU(4) algebra ($k = 4$ or $\hat{k}_4 = 0$) contains a non-minimal SU(3) subalgebra (also $k = 4$ but $\hat{k}_3 = 1$).

larly, existence of $W^\pm$ and Z^0 gauge bosons implies an SU(2) algebra while existence of weak doublets requires $\hat{k}_2 \geqslant 1$, or $k \geqslant 3$. We will now see that while each of the conditions $\hat{k}_2 \geqslant 1$ and $\hat{k}_3 \geqslant 1$ is quite innocuous by itself, together they spell a disaster. Indeed, when substituted into eq. (2.12), they yield

$$\hat{c}(\mathrm{SU}(3) \otimes \mathrm{SU}(2) \otimes \mathrm{U}(1)) \geqslant 6 \cdot \tfrac{2}{3}, \tag{4.10}$$

which exceeds the $\hat{c}(\mathrm{SKM}) \leqslant 6$ limit that any 4d models based on the type II superstring must obey. Thus we are left with the following unpleasant alternatives:

♠ Giving up weak doublets. If $\hat{k}_2 = 0$, then one can have $\hat{k}_3 = 1$ (this allows quarks) in a model with $\hat{c}(\mathrm{SKM}) = 6$.

♣ Giving up quarks. If $\hat{k}_3 = 0$, then one can have $\hat{k}_2 = 1$ or $\hat{k}_2 = 4$. The former choice allows weak doublets and requires a $\hat{c} = \frac{2}{3}$ left-over algebra to achieve $\hat{c}^{\mathrm{int}} = 6$. The second alternative has no left-over algebra and allows massless states of higher weak isospin.

♦ Giving up the U(1) gauge boson; then one can have $\hat{k}_3 = 1$ and $\hat{k}_2 = 2$ and get $\hat{c}(\mathrm{SU}(3) \otimes \mathrm{SU}(2)) = 6$. Curiously, $\hat{k}_2 = 2$ allows for massless scalars or fermions to be weak triplets as well as doublets, so perhaps a model containing, say, the Georgi-Glashow model could be constructed.

♥ Or, perhaps, giving up one of the space-time dimensions, instead of a U(1) factor (since both contribute 1 to $\hat{c}$).

Obviously none of the above sacrifices is phenomenologically acceptable, and the only real alternative is to give up on the type II superstring itself. Despite its ultimate failure, the type II superstring theory comes tantalizingly close to producing a realistic model: any of the important features of the standard model, such as chiral fermions, gauge groups containing the SU(3) ⊗ SU(2) ⊗ U(1), massless quarks or leptons, can be achieved. It is just that we can never have all these features at once, i.e. in the same model. Although limit (4.10) exceeds the critical value $\hat{c}^{\mathrm{int}} = 6$ by a rather modest amount $\delta\hat{c} = \frac{2}{3}$, this is enough to render any classical vacuum* of the type II superstring incapable of reproducing all the particles of the standard model in its massless spectrum.

5. Conclusions

In this paper we have explored the phenomenological possibilities for classical vacua of the type II superstring and have found them lacking. Previous work [6, 7] exhibited four-dimensional models with gauge groups containing the standard model's one. We have shown explicitly that those models are in fact asymmetric

* The smallness of $\delta\hat{c}$ makes one hope that perhaps quantum string effects can increase the critical dimension of the superstring so that (4.10) could be accommodated. At present this possibility is of course pure speculation, whose test will have to wait until we understand quantum string effects much better than we do now.

orbifolds. In addition we have constructed asymmetric orbifold models with other gauge groups which contain the standard model's group, but which are not subgroups of the previously realized groups, and also models with chiral fermions. The latter constructions are important because they provide an argument [8] that at least some asymmetric orbifolds cannot be interpreted as supersymmetric non-linear sigma models. The argument assumes that a Kaluza-Klein mechanism is responsible for the Cartan subalgebra of any 4d gauge group generated by sigma model compactification of the type II superstring. Then it can be shown [5] that all 4d fermions transform non-chirally under the gauge group. Hence constructions resulting in chiral models, such as those in sect. 3, must be inherently different from supersymmetric sigma models.

The main result of this paper, however, was to show, under some very mild assumptions as to what constitutes a model based on the type II superstring, that it is *not* possible to obtain a phenomenologically acceptable model. In particular, our proof did not depend on any particular compactification scheme, or even on whether the non-Minkowski degrees of freedom could be interpreted as internal dimensions at all, so long as they gave rise to a superconformal 2d field theory. In view of the physical importance of the question of whether realistic type II superstring vacua exist, and also the fact that one can construct classical vacua which are not too far from the standard model in their massless particle content, it might be worthwhile to see if any of the assumptions leading to the no-go result could be relaxed.

In any case, the general approach described here applies as well to the $(0,1)$ superconformal theory describing the compactified heterotic string, although restrictions from unitarity will be much less stringent. The most obvious application is to show that abelian dimension $(0,\frac{1}{2})$ supercurrents in a 4d heterotic string model destroy 4d chirality [20] – non-abelian supercurrents eliminate all massless fermions [19]. So all gauge symmetry arises from a left-moving non-supersymmetric Kac-Moody algebra with $c^{\mathrm{KM}} \leqslant 22$. Because so many quasi-realistic heterotic string models satisfy this constraint, one must ask subtler questions than just whether massless fermions exist with the gauge quantum numbers of quarks and leptons. One will undoubtedly need to apply to the problem more than just the string tree-level considerations of sections 2 and 4, for example by incorporating as well the constraints of modular invariance; even then one may be unable to progress without a better understanding of the dynamics of compactification.

Finally, some of the specific constructions we gave in sect. 3 may be of use in constructing heterotic string models. For example, the two orbifolds of the $N=8$ model based on the lattice $\Gamma^{6,6}(\mathrm{E}_6)$ demonstrate how to incorporate members of the $N=1$ or $N=2$ unitary discrete series into modular invariant string models in a rather non-trivial fashion (i.e. not just taking direct products of discrete systems which are individually modular invariant). They might therefore be useful prototypes for similar constructions in the heterotic string as well. In general, modular invariant systems involving members of the discrete series need not have any geometric interpretation at all. It may prove fruitful to study the exact circumstances under which such models can be interpreted in terms of orbifolds or perhaps other generalized geometries.

Two of the authors (L.D. and V.K.) would like to thank Jeff Harvey and Tom Banks for helpful discussions; C.V. would like to thank Paul Ginsparg for discussions.

References

[1] M. Green and D. Gross, eds., Unified string theories (World Scientific, 1986);
M. Green, J.H. Schwarz and E. Witten, Superstring theory, (2 volumes) (Cambridge University Press, 1986)

[2] D.J. Gross, J.A. Harvey, E. Martinec and R. Rohm, Phys. Rev. Lett. 54 (1985) 502, Nucl. Phys. B256 (1985) 253; B267 (1986) 75

[3] P. Candelas, G. Horowitz, A. Strominger and E. Witten, Nucl. Phys. B258 (1985) 46

[4] L.J. Dixon, J.A. Harvey, C. Vafa and E. Witten, Nucl. Phys. B261 (1985) 671, B274 (1986) 285

[5] E. Witten, *in* Quantum field theory and the fundamental problems of physics, Proc. of 1983 Shelter Island Conference, ed. R. Jackiw, N.N. Khuri, S. Weinberg and E. Witten (MIT Press, 1985); also Princeton preprint PUPT-1024 (1986)

[6] R. Bluhm, L. Dolan and P. Goddard, Nucl. Phys. B289 (1987) 364

[7] H. Kawai, D.C. Lewellen, and S.-H.H. Tye, Cornell preprint CLNS 87/760

[8] K.S. Narain, M.H. Sarmadi and C. Vafa, Nucl. Phys. B288 (1987) 551

[9] L. Castellani, R. D'Auria, F. Gliozzi and S. Sciuto, Phys. Lett. 168B (1986) 47

[10] L. Brink, P. di Vecchia and P. Howe, Phys. Lett. 65B (1976) 471;
S. Deser and B. Zumino, Phys. Lett. 65B (1976) 369;
A.M. Polyakov, Phys. Lett. 103B (1981) 211

[11] R. Brower, Phys. Rev. D6 (1972) 1655;
P. Goddard and C. Thorn, Phys. Lett. 40B (1972) 235;
E.F. Corrigan and P. Goddard, Nucl. Phys. B68 (1974) 189

[12] C. Lovelace, Phys. Lett. 135B (1984) 75;
D. Friedan and S.H. Shenker, unpublished talk at the Aspen Summer Institute (1984);
D. Friedan, Z. Qiu and S.H. Shenker, *in* Proc. of the 1984 Santa Fe Meeting of the APS Division of Particles and Fields, ed. T. Goldman and M. Nieto (World Scientific, 1985);
E. Fradkin and A. Tseytlin, Phys. Lett. 158B (1985) 316;
C. Callan, D. Friedan, E. Martinec and M. Perry, Nucl. Phys. B262 (1985) 593;
A. Sen, Phys. Rev. Lett. 55 (1985) 1846; Phys. Rev. D32 (1985) 2794

[13] D. Friedan, E. Martinec and S.H. Shenker, Nucl. Phys. B271 (1986) 93

[14] V.G. Kac and I.T. Todorov, Comm. Math. Phys. 102 (1985) 337

[15] E. Martinec, Phys. Rev. D30 (1984) 1198

[16] M. Mueller and E. Witten, Phys. Lett. 182B (1986) 28

[17] J. Fuchs, Nucl. Phys. B286 (1987) 455

[18] V. Knizhnik and A. Zamolodchikov, Nucl. Phys. B247 (1984) 83;
H. Sugawara, Phys. Rev. 170 (1968) 1659

[19] D. Friedan, Z. Qiu and S.H. Shenker, Phys. Lett. 151B (1985) 37

[20] D. Friedan and S.H. Shenker, unpublished

[21] V.G. Kac, Advances in Math. 35 (1980) 264

[22] D. Freed and C. Vafa, Harvard preprint HUTP-86/A090

[23] K.S. Narain, Phys. Lett. 169B (1986) 41

[24] K.S. Narain, M.H. Sarmadi and E. Witten, Nucl. Phys. B279 (1987) 369

[25] C. Vafa, Nucl. Phys. B273 (1986) 592

[26] L.J. Dixon and J.A. Harvey, Nucl. Phys. B274 (1986) 93
[27] I. Antoniadis, C. Bachas, C. Kounnas and P. Windey, Phys. Lett. 171B (1986) 51
[28] J. Cohn, D. Friedan, Z. Qiu and S.H. Shenker, Nucl. Phys. B278 (1986) 577
[29] W. Lerche and D. Lust, Phys. Lett. 187B (1987) 45
[30] J. Lepowsky, Proc. Natl. Acad. Sci. USA 82 (1985) 8295
[31] P. Goddard and D. Olive, *in* Vertex operators in mathematics and physics, ed. J. Lepowsky et al. (Springer, 1984)
[32] L. Dixon, D. Friedan, E. Martinec and S.H. Shenker, Nucl. Phys. B282 (1987) 13;
S. Hamidi and C. Vafa, Nucl. Phys. B279 (1987) 465
[33] D. Gepner, Nucl. Phys. B, to be published
[34] J. Lepowsky and R.L. Wilson, Inv. Math. 77 (1984) 199
[35] W. Boucher, D. Friedan and A. Kent, Phys. Lett. B172 (1986) 316
[36] W. Lerche, B.E.W. Nilsson and A.N. Schellekens, CERN preprint CERN-TH-4692/87
[37] P. Goddard and D. Olive, *in* Unified string theories, ed. M. Green and D. Gross (World Scientific, 1986)
[38] D. Gepner and E. Witten, Nucl. Phys. B278 (1986) 493
[39] D. Friedan, Z. Qiu and S.H. Shenker, *in* ref. [12]
[40] D. Friedan, Z. Qiu and S.H. Shenker, *in* Vertex operators in mathematics and physics, ed. J. Lepowsky et al. (Springer, 1984); Phys. Rev. Lett. 52 (1984) 1575
[41] P. di Vecchia, V.G. Knizhnik, J.L. Petersen and P. Rossi, Nucl. Phys. B253 (1985) 701
[41] P. Windey, Comm. Math. Phys. 105 (1986) 511

References

[1] A. Belavin, A. Polyakov and A. Zamolodchikov, Nucl. Phys. B 241 (1984) 333.

[2] D. Friedan, E. Martinec and S. Shenker, Nucl. Phys. B 271 (1986) 93.

[3] D. Friedan, in: Unified String Theories (Proc. Santa Barbara workshop 1985), M. Green and D. Gross, eds. (World Scientific, Singapore, 1986).

[4] S. Shenker, in: Unified String Theories (Proc. Santa Barbara workshop 1985), M. Green and D. Gross, eds. (World Scientific, Singapore, 1986).

[5] E. Martinec, String Calculus: Conformal Field Theory as a Tool in String Field Theory, Lectures given at the 1987 Spring School on Superstrings, Trieste, University of Chicago preprint EFI 87-45.

[6] M. Peskin, Introduction to String and Superstring Theory 2, SLAC-PUB 4251, Lectures presented at the 3rd Advanced Study Institute in Elementary Particle Physics, Santa Cruz (1986).

[7] H. Nicolai and A.N. Schellekens, in: Proc. 5th Adriatic Meeting on Superstrings, Anomalies and Unification, Dubrovnik (1986).

[8] J.E. Paton and H.M. Chan, Nucl. Phys. B 10 (1969) 516.

[9] L. Clavelli, P.B. Cox and B. Harms, Phys. Rev. Lett. 61 (1988) 787.
Z. Bern and D. Dunbar, Type-I superstrings in less than ten dimensions. 2. Finiteness, Niels Bohr Institute preprint NBI-HE-88-49.

[10] C. Hull and E. Witten, Phys. Lett. B 160 (1985) 398.

[11] M. Ademollo et al., Phys. Lett. B 62 (1976) 105; Nucl. Phys. B 111 (1976) 77; B 114 (1976) 297.

[12] P. Ramond and J. Schwarz, Phys. Lett. B 64 (1976) 75.

[13] M. Pernici and P. van Nieuwenhuizen, Phys. Lett. B 169 (1986) 381.

[14] A. Schwimmer and N. Seiberg, Phys. Lett. B 184 (1987) 191.

[15] A. Parkes, Phys. Lett. B 184 (1987) 19.

[16] Y. Kikuchi and C. Marzban, Phys. Rev. D 36 (1987) 2583.

[17] H. Kawai, D. Lewellen, J.A. Schwartz and S.-H.H. Tye, Nucl. Phys. B 299 (1988) 431.

[18] R. Bluhm, L. Dolan and P. Goddard, Nucl. Phys. B 309 (1988) 330.

[19] V.G. Kac, Lecture Notes in Physics, vol. 94 (Springer, Berlin, 1979), p. 441.
B.L. Feigin and D.B. Fuchs, Funct. Anal. Appl. 16 (1982) 114.

[20] D. Friedan, Z. Qiu and S. Shenker, Phys. Rev. Lett. 52 (1984) 1575; in: Vertex Operators in Mathematics and Physics, J. Lepowsky, S. Mandelstam and I. Singer, eds. (Springer, New York, 1985).

[21] P. Goddard, A. Kent and D. Olive, Phys. Lett. B 152 (1985) 88; Commun. Math. Phys. 103 (1986) 105.

[22] J.F. Gomes, Phys. Lett. B 171 (1986) 75.

[23] R. Dijkgraaf and E. Verlinde, Modular invariance and the fusion algebra, in the Proceedings of the Annecy workshop on Conformal Field Theory (March 1988) (Nucl. Phys. Proceedings Supplement).

[24] P. Di Francesco, P. Saleur and J.-B. Zuber, Nucl. Phys. B 290 (1987) 527.

[25] T. Jayaraman and K. Narain, Correlation functions for minimal models on the torus, CERN preprint CERN-TH.5166/88 (1988).

[26] J. Bagger, D. Nemeschansky and J.-B. Zuber, Phys. Lett. B 216 (1989) 320.

[27] J.H. Choi and I.G. Koh, Genus two characters of the Ising model, Korea Inst. Sci. preprint KAIST-88/12.

[28] A. Zamolodchikov, Theor. Math. Phys. 65 (1986) 1205.

[29] L. Dixon, D. Friedan, E. Martinec and S. Shenker, Nucl. Phys. B 282 (1987) 13.

[30] S. Hamidi and C. Vafa, Nucl. Phys. B 279 (1987) 465.

[31] L. Dixon, Some world-sheet properties of superstring compactifications on orbifolds and otherwise, Princeton preprint PUPT-1074.

[32] F. Bais, P. Bouwknegt, K. Schouten and M. Surridge, Nucl. Phys. B 304 (1988) 348, 371.

[33] K. Yamagishi, Phys. Lett. B 205 (1988) 466.

[34] A. Bilal and J.-L. Gervais, Phys. Lett. B 206 (1988) 412.

[35] P. Bouwknegt, Phys. Lett. B 207 (1988) 295.

[36] E. Verlinde, Nucl. Phys. B 300 (1988) 360.

[37] C. Vafa, Phys. Lett. B 206 (1988) 421.

[38] G. Anderson and G. Moore, Commun. Math. Phys. 117 (1988) 441.

[39] S. Mathur, S. Mukhi and A. Sen, Nucl. Phys. B 312 (1989) 15; Phys. Lett. B 218 (1988) 176.
S. Mathur and A. Sen, Phys. Lett. B 213 (1988) 303.
[40] N. Seiberg and G. Moore, Phys. Lett. B 212 (1988) 451; Nucl. Phys. B 313 (1988) 16.
[41] R. Brustein, S. Yankielowicz and J.-B. Zuber, Nucl. Phys. B 313 (1988) 21.
[42] K.-H. Rehren and B. Schroer, Nucl. Phys. B 312 (1989) 715.
[43] E. Witten, Quantum field theory and the Jones polynomial, Commun. Math. Phys. 121 (1989) 351.
[44] R. Dijkgraaf, C. Vafa, E. Verlinde and H. Verlinde, The operator algebra of orbifold models, Harvard preprint HUTP-88/A052 (1988).
[45] P. Goddard and D. Olive, in: Vertex Operators in Mathematics and Physics (Springer, Berlin, 1984); Int. J. Mod. Phys. A 1 (1986) 303.
[46] D. Gepner and E. Witten, Nucl. Phys. B 278 (1986) 493.
[47] V.G. Kac, Infinite Dimensional Lie Algebras, second edition (Cambridge University Press, 1985).
[48] E. Witten, Commun. Math. Phys. 92 (1984) 451.
[49] A. Cappelli, C. Itzykson and J.-B. Zuber, Nucl. Phys. B 280 (1987) 445; Commun. Math. Phys. 113 (1987) 1.
[50] Th. Ohrndorf, Nucl. Phys. B 281 (1987) 470.
[51] L. Alvarez-Gaumé, G. Moore and C. Vafa, Commun. Math. Phys. 106 (1986) 1.
[52] P. Ginsparg and C. Vafa, Nucl. Phys. B 289 (1987) 414.
[53] E. Verlinde and H. Verlinde, Nucl. Phys. B 288 (1987) 357.
[54] W. Lerche, A. Schellekens and N. Warner, Phys. Rep. 177 (1989) 1.
[55] F. Englert and A. Neveu, Phys. Lett. B 163 (1985) 349.
[56] S. Elitzur, E. Gross, E. Rabinovici and N. Seiberg, Nucl. Phys. B 283 (1987) 413.
[57] K. Narain, Sarmadi and E. Witten, Nucl. Phys. B 279 (1986) 369.
[58] D. Freed and C. Vafa, Commun. Math. Phys. 110 (1987) 349; Addendum, Commun. Math. Phys. 117 (1988) 349.
[59] K. Bardakci, E. Rabinovici and B. Säring, Nucl. Phys. B 299 (1988) 151.
[60] R. Dijkgraaf, E. Verlinde and H. Verlinde, Commun. Math. Phys. 115 (1988) 649.
[61] P. Ginsparg, Curiosities at $c=1$, Harvard preprint HUTP-87/A068 (1987).
[62] I. Antoniadis and C. Bachas, Nucl. Phys. B 298 (1988) 586.
[63] J. Bagger, D. Nemeschansky, N. Seiberg and S. Yankielowicz, Nucl. Phys. B 289 (1987) 349.
[64] P. Goddard, D. Olive and A. Schwimmer, Phys. Lett. B 157 (1985) 393.
[65] A.N. Schellekens, Phys. Lett. B 199 (1987) 427.
[66] A. Casher, F. Englert, H. Nicolai and A. Taormina, Phys. Lett. B 162 (1985) 121.
[67] F. Englert, H. Nicolai and A.N. Schellekens, Nucl. Phys. B 274 (1986) 315.
[68] W. Lerche, A. Schellekens and N. Warner, Phys. Lett. B 214 (1988) 41.
[69] W. Boucher, D. Friedan and A. Kent, Phys. Lett. B 172 (1986) 316.
A. Sen, Nucl. Phys. B 278 (1986) 289; B 284 (1987) 423.
T. Banks, L.J. Dixon, D. Friedan and E. Martinec, Nucl. Phys. B 299 (1988) 613.
T. Banks and L. Dixon, Nucl. Phys. B 307 (1988) 93.
[70] J. Lauer, D. Lüst and S. Theisen, Nucl. Phys. B 304 (1988) 236; B 309 (1988) 771.
[71] W. Lerche, Nucl. Phys. B 308 (1988) 102.
[72] D. Lüst and S. Theisen, Superstring partition functions and the characters of exceptional groups, preprint MPI-PAE/PTh 41/88 (1988).
[73] M. Cvetič, in the Proceedings of "String '88", College Park, Maryland (May 1988) (World Scientific, Singapore, 1988).
[74] S. Ferrara, Recent issues on the moduli space of 2d superconformal field theories, CERN-TH.5151/88.
[75] N. Seiberg, Nucl. Phys. B 306 (1988) 286.
[76] S. Cecotti, S. Ferrara and L. Girardello, Nucl. Phys. B 308 (1988) 436.
[77] M. Cvetič, J. Louis and B. Ovrut, Phys. Lett. B 206 (1988) 229.
[78] M. Cvetič, Phys. Rev. Lett. 59 (1987) 2829.
[79] A. Font, L. Ibañez, H. Nilles and F. Quevedo, Nucl. Phys. B 307 (1988) 109.
[80] S. Chaudhuri and J.A. Schwartz, Phys. Lett. B 219 (1989) 291.
[81] D. Gepner, String theory on Calabi–Yau manifolds: the three generation case, Princeton preprint PUPT-88-0085 (1988).
[82] A. Lutken and G. Ross, Phys. Lett. 213 (1987) 152.
[83] M. Lynker and R. Schimmrigk, Phys. Lett. 208 (1988) 216; Phys. Lett. B 215 (1988) 681.
[84] B. Greene, C. Vafa and N. Warner, Calabi–Yau manifolds and renormalization group flows, Harvard preprint HUTP-88/A047.
[85] J. Distler and B. Greene, Nucl. Phys. B 309 (1988) 295.

[86] T. Eguchi and A. Taormina, Extended superconformal algebras and string compactifications, CERN-TH.5123/88, to appear in the Proceedings of the Spring School in Superstrings, Trieste (April 1988).
[87] M. Green and J. Schwarz, Phys. Lett. B 149 (1984) 117.
[88] M.F. Atiyah and I.M. Singer, Ann. Math. 87 (1968) 485, 546; 93 (1971) 119, 139.
M.F. Atiyah and G.B. Segal, Ann. Math. 87 (1968) 531.
O. Alvarez, I. Singer and B. Zumino, Commun. Math. Phys. 96 (1984) 409.
Y.-S. Wu, A. Zee and B. Zumino, Nucl. Phys. B 239 (1984) 477.
L. Alvarez-Gaumé and P. Ginsparg, Ann. Phys. (NY) 161 (1985) 423.
[89] A.N. Schellekens and N.P. Warner, Phys. Lett. B 177 (1986) 317; Phys. Lett. B 181 (1986) 339; Nucl. Phys. B 287 (1987) 317.
[90] P. Ginsparg, Informal string lectures, to appear in the Proceedings of the U.K. Institute for Theoretical High Energy Physics, Cambridge, 16 Aug.–5 Sept. 1988, Harvard preprint HUTP-87/A077.
[91] W. Lerche, B.E.W. Nilsson, and A.N. Schellekens, Nucl. Phys. B 289 (1987) 609.
W. Lerche, B.E.W. Nilsson, A.N. Schellekens and N.P. Warner, Nucl. Phys. B 299 (1988) 91.
[92] E. Witten, Phys. Lett. B 149 (1984) 351.
[93] K. Pilch and A.N. Schellekens, Nucl. Phys. B 259 (1985) 637.
[94] M. Dine, N. Seiberg and E. Witten, Nucl. Phys. B 289 (1987) 589.
J. Atick, L. Dixon and A. Sen, Nucl. Phys. B 292 (1987) 109.
M. Dine, L. Ichinose and N. Seiberg, Nucl. Phys. B 293 (1987) 253.
[95] J.J. Atick and A. Sen, Nucl. Phys. B 296 (1988) 157.
[96] P. Ramond, Phys. Rev. D 3 (1971) 2415.
A. Neveu and J.H. Schwarz, Nucl. Phys. B 31 (1971).
[97] K. Bardakci and M. Halpern, Phys. Rev. D 3 (1971) 2493.
[98] M. Green and J. Schwarz, Nucl. Phys. B 181 (1981) 502; B 198 (1982) 252, 441.
[99] D. Gross, J. Harvey, E. Martinec and R. Rohm, Nucl. Phys. B 267 (1986) 75.
[100] J. Shapiro, Phys. Rev. D 5 (1972) 1945.
[101] W. Nahm, Nucl. Phys. B 114 (1976) 174; B 120 (1977) 125.
[102] E. D'Hoker and D. Phong, Nucl. Phys. B 278 (1986) 225.
[103] M. Halpern, Phys. Rev. D 12 (1975) 1684.
[104] I.B. Frenkel and V.G. Kac, Inv. Math. 62 (1980) 23.
G. Segal, Commun. Math. Phys. 80 (1981) 301.
[105] L. Clavelli and J. Shapiro, Nucl. Phys. B 57 (1973) 490.
[106] J. Scherk and J. Schwarz, Phys. Lett. B 57 (1975) 463.
E. Cremmer and J. Scherk, Nucl. Phys. B 103 (1976) 399.
[107] J. Schwarz, Phys. Rep. 89 (1982) 225.
[108] P. Forgacs, L. Palla, Z. Horvath and P. Vecsernyés, Nucl. Phys. B 308 (1988) 477.
[109] E. Witten, in: Proceedings of the Shelter Island II Conference, N. Khuri et al., eds. (MIT Press, Cambridge, MA, 1984).
[110] L. Palla, in: Proceedings of the 1978 Tokyo Conference on High Energy Physics.
[111] S. Randjbar-Daemi, A. Salam and S. Strathdee, Nucl. Phys. B 214 (1983) 491; Phys. Lett. B 132 (1983) 35; Nucl. Phys. B 242 (1984) 447.
[112] C. Wetterich, Nucl. Phys. B 222 (1983) 20; B 223 (1983) 109; B 234 (1984) 413.
[113] P. Candelas, G. Horowitz, A. Strominger and E. Witten, Nucl. Phys. B 258 (1985) 46.
[114] M. Green, J. Schwarz and P. West, Nucl. Phys. B 254 (1985) 327.
[115] P. Frampton, H. van Dam and K. Yamamoto, Phys. Rev. Lett. 54 (1985) 1114.
[116] J. Distler and B.R. Greene, Nucl. Phys. B 304 (1988) 1.
[117] H. Kawai, D. Lewellen and S. Tye, Phys. Rev. Lett. 57 (1986) 1832.
[118] W. Lerche and D. Lüst, Phys. Lett. B 187 (1987) 45.
[119] A. Cohn, D. Friedan, Z. Qiu and S. Shenker, Nucl. Phys. B 278 (1986) 577.
[120] M. Mueller and E. Witten, Phys. Lett. B 182 (1986) 28.
[121] P. di Vecchia, V.G. Knizhnik, J.L. Petersen and P. Rossi, Nucl. Phys. B 253 (1985) 701.
P. Goddard and D. Olive, Nucl. Phys. B 257 (1985) 245.
I. Antoniadis, C. Bachas, C. Kounnas and P. Windey, Phys. Lett. B 171 (1986) 51.
[122] H. Kawai, D. Lewellen and S. Tye, Int. J. Mod. Phys. A 3 (1988) 279.
[123] A.N. Schellekens and N.P. Warner, Nucl. Phys. B 308 (1988) 397.
[124] A. Chamseddine and J.-P. Derendinger, Nucl. Phys. B 301 (1988) 381.
[125] F. Gliozzi and P. Fre, Phys. Lett. B 208 (1988) 203.

[126] Y. Kazama and H. Suzuki, Phys. Lett. B 216 (1989) 112; Mod. Phys. Lett. A 4 (1989) 235.
[127] Da-Xi Li, Phys. Rev. D 34 (1986) 3780.
[128] C.S. Lam, Int. J. Mod. Phys. A 3 (1988) 913.
[129] D. Bailin and A. Love, Phys. Lett. B 207 (1988) 151.
D. Bailin, D. Dunbar and A. Love, Phys. Lett. B 212 (1988) 23.
[130] D. Markushevich, M. Olshanetsky and A. Perelomov, Commun. Math. Phys. 111 (1987) 247.
[131] L. Ibañez, J. Mas, H. Nilles and F. Quevedo, Nucl. Phys. B 301 (1988) 157.
L. Ibañez, J.E. Kim, H. Nilles and F. Quevedo, Phys. Lett. B 191 (1988) 282.
A. Font, L. Ibañez, H. Nilles and F. Quevedo, Phys. Lett. B 210 (1988) 101.
[132] I. Antoniadis, J. Ellis and J. Hagelin, Phys. Lett. B 208 (1988) 209.
[133] P. Frampton and M. Ubriaco, Phys. Rev. D 38 (1988) 1341.
[134] A.N. Schellekens and N.P. Warner, Nucl. Phys. B 313 (1988) 41.
[135] S. Chaudhuri, H. Kawai and S. Tye, Chiral bosonic supercurrents and four-dimensional fermionic string models, Cornell preprint CLNS 88/852.
[136] J. Casas and C. Muñoz, Phys. Lett. B 209 (1988) 343.
[137] J. Schwarz, Superconformal symmetry and superstring compactification, CALTECH preprint CALT-68-1531.
[138] P. Ginsparg, Phys. Rev. D 35 (1987) 648.
[139] R. Bluhm and L. Dolan, Phys. Lett. B 169 (1986) 347.